Intelligent Systems, Control and Automation: Science and Engineering

Volume 94

More information about this series at http://www.springer.com/series/6259

Maria Isabel Aldinhas Ferreira
João Silva Sequeira · Rodrigo Ventura
Editors

Cognitive Architectures

Springer

Editors
Maria Isabel Aldinhas Ferreira
Centro de Filosofia
Universidade de Lisboa
Lisbon, Portugal

João Silva Sequeira
Instituto de Sistemas e Robótica,
Instituto Superior Técnico
Universidade de Lisboa
Lisbon, Portugal

Rodrigo Ventura
Instituto de Sistemas e Robótica,
Instituto Superior Técnico
Universidade de Lisboa
Lisbon, Portugal

ISSN 2213-8986 ISSN 2213-8994 (electronic)
Intelligent Systems, Control and Automation: Science and Engineering
ISBN 978-3-030-07373-2 ISBN 978-3-319-97550-4 (eBook)
https://doi.org/10.1007/978-3-319-97550-4

This Springer imprint is published by the registered company Springer Nature Switzerland AG
The registered company address is: Gewerbestrasse 11, 6330 Cham, Switzerland

Editorial

From so simple a beginning
endless forms
most beautiful and most wonderful
have been, and are being, evolved

—Darwin

The Machine does not isolate man from the great problems of Nature
But plunges him more deeply into them

—Antoine de Saint-Exupéry

This book is about cognitive architectures; i.e., it is about life-forms endowed with particular corporeal identities, giving shape and meaningfulness to the environment in which they are embedded, creating a dynamic world to which they are irresistibly bound, in an essential dialectical relationship. A world where they evolve, making their best effort to thrive, and in so doing, end up defining individual and collective existential narratives. This book is also about embodied and non-embodied artificial intelligent systems, human constructs, meant to be able to populate the human world, capable of identifying different life contexts and behaving according to human values and conventions, systems capable of performing tasks in a human-like way. Finally, this book is about trying to grasp the essence of cognition, here understood as the effective action that enables a cognitive entity to continue its existence in a definite environment as it brings forth its world [3]. By creating artificial environments where non-embodied artificial entities evolve, human beings are ultimately looking for "those features of the world where the details do not matter, where large equivalence classes of structure, action and so on lead to a deep sameness of being" [4, p. 7].

In Chapter "Cognitive Architectures: The Dialectics of Agent/Environment," Ferreira sets the motto for the book. By positing that cognition is the embodied, embedded, and always situated process whereby life-forms bound to their respective environments in an essential dialectical relationship strive to live and replicate within the existential spatiotemporal framework defined by their own corporeal dynamics, the author addresses the fundamental role played by different physical architectures, different corporeal realities in the shaping of meaningful worlds. Addressing the complexity and richness of human cognition in which that essential interaction is symbolically mediated, Ferreira points out that every newborn comes into life in a particular physical, economic, social, cultural, and linguistic atmosphere—a semiosphere. An environment where specific relations of production have not only determined particular social structures and social hierarchies, but also determined the typical patterns of behavior to be followed in each circumstance and context, the definition of public and domestic space [2], the creation of institutions, the architectural options, the production of artefacts and technological artefacts, and the production of art forms.

The chapter concludes by identifying the hybrid forms of cognition present in human reality nowadays, hybrid forms that result from the interaction of hybrid agents and the existence of hybrid worlds.

Chapter "Complementarity of Seeing and Appearing" addresses the phenomenon of coloring among species, its place in the individual's Umwelt, and the functions it plays in distinct contexts and environmental settings. Using the example of coloration of animal surfaces to show how processes based on interactions of the individual parts lead to the emergence of "meaning at the level of communication between individuals," Jindřich Brejcha, Pavel Pecháček, and Karel Kleisner argue that, due to complementarity of appearance and perception, the exposed surfaces of organisms ultimately become semi-autonomous entities subjected to their own evolution. In the final part of this chapter, the authors investigate various explanations of the evolution of coloration in the context of its role in animal behavior and communication and within particular environmental settings.

In Chapter "The Extended Domicile—Culture, Embodied Existence and the Senses," Juhanni Pallasmaa states that the human sensory and neural system, as well as the brain, is the result of evolutionary adaptation to the prevailing environments and conditions of life during the continuum of human evolutionary history. The nature of our senses and neural functions, as well as instinctive environmental preferences, needs to be viewed within a bio-cultural and bio-historical perspective, instead of regarding them as ahistorical, unchanging, or given properties of the *Homo Sapiens*. Through our human structures, both physical and mental, we turn limitless, shapeless, and meaningless "natural" space into lived cultural space with specific human purposes and meanings. Lived reality fuses observation, memory, and fantasy, as well as the cerebral and the embodied, into single existential experience.

Chapter "What We Need from an Embodied Cognitive Architecture" provides a succinct overview and discussion of the two main perspectives involved around the concept of embodied cognition, proposing a clarification of two fundamental issues: (i) the meaning of the term and (ii) whether the existence of a physical body is, in fact, paramount in the process of cognition and, if this is a fact, the role the physical body plays in the process.

According to Serge Thill, resolving these unclear aspects remains the major challenge in current theories of embodied cognition. At this stage, the main point is, according to the author, arriving at a unifying definition that will at least have to acknowledge a role for the physical body. In his opinion, what an embodied cognitive architecture needs to provide at the current state of theoretical understanding is a framework that allows us to parameterize contributions of the body in cognitive processes. This includes parametrizing the body itself, but goes beyond that, in that the way in which sensorimotor experience is used in higher cognition is itself also left open to parametrization. Such an architecture could then explore the predicted consequences of given activities under various theoretical assumptions regarding embodiment and help further the state of the art by then comparing these predictions with reality.

In Chapter "The Architect's Dilemmas," David Vernon identifies two reasons to design a cognitive architecture: to gain a better understanding of cognition in general and to build artificial systems that have capabilities commonly found in humans. However, as the author recognizes, the design of a cognitive architecture is a daunting undertaking, involving many challenges on a scale that is not always apparent when one embarks on the task. Like architecture in the built environment and system architecture in software engineering, a cognitive architecture captures both abstract conceptual form and details of functional operation; focusing on inner cohesion and self-contained completeness, Vernon points out that a cognitive architecture captures the top two layers of Marr's three-level hierarchy of abstraction, also known as the Levels of Understanding framework, i.e., the top level computational theory and, below this, the level of representation and algorithm. At the bottom (third) level, there is the implementation or instantiation of this algorithmic and representational framework: the realization of the cognitive architecture as a working cognitive system. In order to achieve its "daunting undertaking," the architect has, in the author's words, to face and solve—we would say—three dilemmas: The Dilemma of Fidelity, The Dilemma of Embodiment, and The Dilemma of Autonomy.

Marco Monforte, Fanny Ficuciello, and Bruno Siciliano, in Chapter "Human Cognition-Inspired Robotic Grasping," address the complexity of reproducing the human physical architecture, namely the hand; it is sometimes gracious, sometimes strong movements, effortless under normal circumstances, and so precious in our daily lives. As the authors insightfully point out, the hand is one of the most complex and fascinating organs of the human body. We can powerfully squeeze objects, but we are also capable of manipulating them with great precision and dexterity. On the other hand, the arm, with its redundant joints, is in charge of reaching the object by determining the hand's pose during pre-shaping. The complex motion and task execution of the upper-limb system may lead us to think that the control requires a huge brain effort. As a matter of fact, neuroscience studies demonstrate that humans simplify planning and control using a combination of primitives, which the brain modulates to produce hand configurations and force patterns so as to grasp and manipulate different objects. This concept can be transferred to robotic systems, allowing control into a space of lower dimension. The lower number of parameters characterizing the system allows for embodying the control in machine learning frameworks, reproducing a sort of human-like cognition.

Tony J. Prescott and Daniel Camilleri initiate Chapter "The Synthetic Psychology of the Self" by revising the two possible paths identified by Alan

Turing for developing human-level Artificial Intelligence (AI)—(i) emulating the more abstract abilities of the human mind and (ii) providing a robot with "the best sense organs that money can buy, and then teach[ing] it to understand and speak English. This process could follow the normal teaching of a child." As the authors point out, the first approach has been spectacularly successful at producing some forms of machine intelligence, though *not* at emulating or approaching "general intelligence"—the wider intellectual and cognitive capacities of our species. The authors describe the work that has been developing in Sheffield and that can be seen as belonging to the emerging discipline of Synthetic Psychology. The long-term goal of this project is building a machine that can pass the Garland test, while being sufficiently biomimetic in design that we can credibly argue that its "mental states" are analogous to human mental states in an interesting way. This goal starts from the premise that we can seek to create an artificial mind that is similar to our own by emulating human linguistic and robotic capacity and by employing a cognitive architecture that has been reverse-engineered from findings in psychology and neuroscience.

In Chapter "Constructive Biology of Emotion Systems: First- and Second-Person Methods for Grounding Adaptation in a Biological and Social World," the interpretation of emotions and similar phenomena is viewed as support for survival and coping in the world. Chrystopher L. Nehaniv addresses the fundamental role played by emotions and feelings[1] in the dialectics between the cognitive agent and its surrounding world by distinguishing (i) those that are grounded in the first-person experience of an emotional agent, those emotions, drives, or experiences that are self-oriented (homeostasis, intake, outflow, hunger, pain, irritation) or others that suggest a generalized or specific recognition of other agents or objects (curiosity, fear or hatred, envy, yearning, greed) from (ii) those that involve relations to a second person (sympathy) or social regulation (shame, guilt), or affective episodic structure (hope, regret).

The chapter explores channels of meaning for agents in interaction games as these relate to emotions, temporal dynamics of affect in relation to behavior, remembering, and learning and outlines how affective coloring of episodic memories might provide a mechanism for emergent spatial and social navigation, as well as considering the role of the temporal horizon in behavior selection.

Emotions, feelings, and internal states and the role they play in cognition are also the theme developed by Eva Hudlicka in Chapter "Modeling Cognition–Emotion Interactions in Symbolic Agent Architectures: Examples of Research and Applied Models." The author states the importance assigned to emotion and the developments undertaken by emotion research over the past two decades, namely the progress in understanding the circuitry that mediates affective processing in biological agents. On the other hand, emotion researchers are also now recognizing that computational models of emotion provide an important tool for understanding the mechanisms of affective processing. There has also been significant progress in

[1]Damásio [1] again calls our attention to the fundamental role played by our emotions and feelings and how they relate to internal primitive states of the inner system.

affective computing technologies, including affective virtual agents, social robots, and affect-adaptive human–computer interaction in general, including affective gaming, and the associated desire to model more affectively realistic and believable agents and robots. This chapter describes a generic methodology for modeling emotions and their effects on cognitive processing. The methodology is based on the assumption that a broad range of both state and trait influences on cognition can be represented in terms of a set of parameters that control processing within the architecture modules. As such, the methodology is suitable both for exploring the nature of the mechanisms that mediate cognition–emotion interaction and for developing the afore-mentioned more affectively realistic and believable agents and robots. An implementation of this generic methodology in a symbolic cognitive-affective architecture is described, focusing on an example of a research model.

João A. Garcia and Pedro U. Lima, in Chapter "Improving Human Behavior Using POMDPs with Gestures and Speech Recognition," point out the importance of robots empathizing and developing affective interactions with users when socially interacting. This chapter proposes a decision-theoretic approach to problems involving interaction between robot systems and human users, with the goal of estimating the human state from observations of its behavior, taking actions that encourage desired behaviors. The approach is based on the Partially Observable Markov Decision Process (POMDP) framework, which determines an optimal policy (mapping beliefs onto actions) in the presence of uncertainty as to the effects of actions and state observations, extended with information rewards (POMDP-IR) to optimize the information-gathering capabilities of the system. The POMDP observations consist of human gestures and spoken sentences, while the actions are split into robot behaviors (such as speaking to the human) and information-reward actions to gain more information about the human state. Under the proposed framework, the robot system is able to actively gain information and react to its belief about the state of the human (expressed as a probability mass function over the discrete state space), effectively encouraging the human to improve his/her behavior, in a socially acceptable manner.

Results of applying the method to a real scenario of interaction between a robot and humans are presented, supporting its practical use.

In Chapter "An Overview of the Distributed Integrated Cognition Affect and Reflection DIARC Architecture," Matthias Scheutz, Thomas Williams, Evan Krause, Bradley Oosterveld, Vasanth Sarathy, and Tyler Frasca present and describe the DIARC architecture, comparing it to classical cognitive architectures. The DIARC architecture that has been under development for about 15 years is different from other cognitive architectures like SOAR or ACT-R. DIARC is an intrinsically component-based distributed architecture scheme that can be instantiated in many different ways. Moreover, DIARC has several critical features such as affect processing and deep natural language integration, is open-world enabled, and allows for one-shot instruction-based learning of new percepts, actions, concepts, rules, and norms.

After laying out the theoretical foundations, the authors specifically focus on action, vision, and natural language subsystems, briefly mentioning different use cases of DIARC, in particular, on autonomous robots in human–robot interaction experiments and for building cognitive models.

Chapter "Non-human Intention and Meaning-Making: An Ecological Theory" discusses the phenomena of agency, meaning-making, and intentionality in an environment populated by human beings and autonomous machines. As Michael A. R. Biggs states, it is an inevitable consequence of the increasingly adaptive complexity of social robots and their embeddedness in the environment by which they become part of our social ecology that we will have to begin to deploy concepts that have previously been reserved for humans. The concepts of intention and meaning-making are such concepts. The robot, if it is to successfully negotiate dynamic obstacles to fulfilling those intentions, must anticipate—that is to say, predict—what will happen if it takes certain courses of action. For these operations to be successful, the robot must have a worldview and must make decisions in accordance with it. As the author points out, meaning-making is perhaps the most advanced of these concepts, but to what extent can meaning-making really be a part of the robot's behavior?

Chapter "Implementing Social Smart Environments with a Large Number of Believable Inhabitants in the Context of Globalization" addresses the role of technological innovation and its impact on both the lives of human beings and the places where they live. As Alexander Oscherenko points out, the modern world now not only is populated with humans who perform everyday tasks but also consists of technical artifacts that perform routine or intelligent tasks. Such artifacts function in environments that are referred to as "smart"; smart because these artifacts have believable behavior and reactions, believable in the sense of said behavior and reactions being comprehensive for the humans.

This chapter elaborates on prototyping software for Smart Environments (SEs). SEs represent physical places with believable inhabitants. To achieve believability, the inhabitants manifest emotional, personal, and cultural characteristics. Consideration of these characteristics in computer systems has many advantages. For instance, SEs with believable inhabitants installed in a public place such as an airport or a metro station could help to avoid panic by warning passengers of possible danger. They could guide customers through a shopping mall. Or SEs equipped with intelligent assistance system could help humans in extreme situations, such as an earthquake or a hazard. Visitors to museums are already guided by SEs. The chapter describes an approach to building SEs that maintain a large number of Embodied Conversational Agents (ECAs) in the context of globalization. The ECAs of such SEs are numerous and represent humans from different cultures. Practically, an ECA is an agent, for example, a robot that occupies a physical space and is able to converse believably. For real-life interaction in an intercultural SE, ECAs maintain models of emotions, personality, and culture.

In Chapter "EcoSim, an Enhanced Artificial Ecosystem: Addressing Deeper Behavioral, Ecological, and Evolutionary Questions," Ryan Scott, Brian MacPherson, and Robin Gras develop the theme of ecological modeling. As the authors state, behavioral ecology has a strong tradition of accounting for the role of organism–environment interactions in behavior. Both behavioral ecology and the related field of optimal foraging theory model animal behavior in terms of optimal adaptation to environmental niches. The goal is not to test whether organisms actually behave optimally, but to use normative expectations to interpret behavioral data and/or generate testable hypotheses. One approach to understanding the behavior of complex ecosystems is through individual-based models (IBMs), which provide a bottom-up approach, allowing for the consideration of the traits and behavior of individual organisms. Since natural ecosystems are very complex (in terms of number of species and of ecological interactions), ecosystem models aim to characterize the major dynamics of ecosystems in order to synthesize the understanding of such systems and to allow for predictions of their behavior. Ecosystem simulations can also help scientists to understand theoretical questions regarding the evolutionary process of the emergence of species, as well as the emergence of learning capacities. This chapter discusses IBMs and uses the Overview, Design concepts, and Details (ODD) protocol to describe a predator–prey evolutionary ecosystem IBM called EcoSim. EcoSim is one of the most complex and large-scale IBMs of its kind, allowing hundreds of thousands of intricate individuals to interact and evolve over thousands of time steps. Individuals in EcoSim have a behavioral model represented by a fuzzy cognitive map (FCM). The FCM described in this chapter is a cognitive architecture well-suited for individuals in EcoSim due to its efficiency and the complexity of decision-making it allows. Furthermore, it can be encoded as a vector of real numbers, lending itself to being part of the genetic material passed on by individuals during reproduction. This allows for the meaningful evolution of their behaviors and natural selection without predefined fitness. EcoSim has been enhanced to increase the breadth and depth of the questions it can answer. New features include fertilization of primary producers by consumers, predator–prey combat, sexual reproduction, sex linkage of genes, multiple modes of reproduction, size-based dominance hierarchy, and more.

Maria Isabel Aldinhas Ferreira
João Silva Sequeira
Rodrigo Ventura

References

1. Damásio, A. (2017). *The strange order of things-life, feeling and the making of cultures*. Portuguese translation. Círculo de Leitores.
2. Lefebvre, H. (1974). *The production of space* (Donald Nicholson- Smith, Trans). Victoria: Blackwell Publishing
3. Maturana H., & Varela F. (1987). *The tree of kowledge: The biological roots of human understanding*
4. Miller, J., & Page, S. (2007). *Complex adaptive systems: An introduction to computational models of social life*. New Jersey: Princeton University Press

Acknowledgements

We would like to thank the contributors who have kindly agreed to interpret, according to their own experience and expertise, the grounding concept standing at the core of this book: "Cognition is an embodied, embedded, and always situated experience. This means that it involves a cognitive entity endowed with a particular physical architecture bound in a dialectical relationship with the environment in which it is immersed, behaving according to the prompts placed by this environment, reacting, adapting to it, and in this way, defining its own existential narrative and history." We would also like to thank Nathalie Jacobs from Springer for her support when this book was just a plan. To Ms. Jacobs, Cynthia Feenstra, and Balaji Sundarrajan, many thanks for their kindness, permanent support, and always fantastic work.

Contents

Contributors

Maria Isabel Aldinhas Ferreira Centre of Philosophy-Language Mind and Cognition Group, Universidade de Lisboa, Lisboa, Portugal; Institute for Systems and Robotics, Instituto Superior Técnico, Universidade de Lisboa, Lisboa, Portugal

Michael A. R. Biggs University of Hertfordshire, Hatfield, UK

Jindřich Brejcha Faculty of Science, Department of Philosophy and History of Science, Charles University, Prague, Czech Republic; Department of Zoology, National Museum, Prague, Czech Republic

Daniel Camilleri The University of Sheffield and Sheffield Robotics, Sheffield, UK

Fanny Ficuciello DIETI, Università degli Studi di Napoli Federico II, Naples, Italy

Tyler Frasca HRI Laboratory, Tufts University, Medford, MA, USA

João A. Garcia Institute for Systems and Robotics, Instituto Superior Técnico, University of Lisbon, Lisbon, Portugal

Robin Gras University of Windsor, Windsor, Ontario, Canada

Eva Hudlicka Psychometrix Associates & University of Massachusetts, Amherst, MA, USA

Karel Kleisner Faculty of Science, Department of Philosophy and History of Science, Charles University, Prague, Czech Republic

Evan Krause HRI Laboratory, Tufts University, Medford, MA, USA

Pedro U. Lima Institute for Systems and Robotics, Instituto Superior Técnico, University of Lisbon, Lisbon, Portugal

Brian MacPherson University of Windsor, Windsor, Ontario, Canada

Marco Monforte DIETI, Università degli Studi di Napoli Federico II, Naples, Italy

Chrystopher L. Nehaniv Adaptive Systems Research Group, University of Hertfordshire, Hatfield, UK

Bradley Oosterveld HRI Laboratory, Tufts University, Medford, MA, USA

Alexander Osherenko Humboldt Innovation, Humboldt-Universität zu Berlin, Berlin, Germany

Juhani Pallasmaa Helsinki, Finland

Pavel Pecháček Faculty of Science, Department of Philosophy and History of Science, Charles University, Prague, Czech Republic

Tony J. Prescott The University of Sheffield and Sheffield Robotics, Sheffield, UK

Vasanth Sarathy HRI Laboratory, Tufts University, Medford, MA, USA

Matthias Scheutz HRI Laboratory, Tufts University, Medford, MA, USA

Ryan Scott University of Windsor, Windsor, Ontario, Canada

Bruno Siciliano DIETI, Università degli Studi di Napoli Federico II, Naples, Italy

Serge Thill Centre for Robotics and Neural Systems, University of Plymouth, Plymouth, UK; School of Informatics, University of Skövde, Skövde, Sweden

David Vernon Carnegie Mellon University Africa, Kigali, Rwanda

Thomas Williams MIRROR Lab, Colorado School of Mines, Golden, CO, USA

Cognitive Architectures: The Dialectics of Agent/Environment

Maria Isabel Aldinhas Ferreira

Abstract In what concerns living systems, cognition is an embodied, embedded and always situated experience. This means that it involves an entity endowed with a particular physical architecture bound in a dialectical relationship with the environment in which it is immersed, behaving according to the prompts placed by this environment, reacting, learning and adapting to it defining this way its own existential narrative and history. Highlighting the fact that human cognition stems from more simple and basic forms of cognition with which it shares essential life mechanisms, the present chapter focuses on the essential semiosic process that is inherent to the dialectics agent/environment and the role played by corporeal architectures in the construction of meaningful worlds, namely, the hybrid realities, where natural and artificial intelligence cohabit.

1 Subjective Worlds

Cognition is the embodied, embedded and always situated process whereby life forms bound to their respective environments in an essential dialectical relationship thrive "to persist and prevail"[1] within the existential spatio/temporal framework defined by their own corporeal dynamics.

[1] Cf [4, p. 32] on these concepts.

M. I. Aldinhas Ferreira (✉)
Centre of Philosophy-Language Mind and Cognition Group, Universidade de Lisboa, Lisboa, Portugal
e-mail: isabelferreira@letras.ulisboa.pt

M. I. Aldinhas Ferreira
Institute for Systems and Robotics, Instituto Superior Técnico, Universidade de Lisboa, Lisboa, Portugal

M. I. Aldinhas Ferreira et al. (eds.), *Cognitive Architectures*, Intelligent Systems, Control and Automation: Science and Engineering 94,
https://doi.org/10.1007/978-3-319-97550-4_1

Every species has a typical evolved intelligent architecture, the phenotypic structure, which is the joint product of its genes and the environmental variations faced during its developmental and evolutionary history. This cognitive architecture embodies vital information concerning the regulation and equilibrium of its internal live states—homeostasis—as well as the information relative to the sensorial/perceptive perceptive systems available to interact with a typical external environment defining the species specific world model.

A life form and its environment constitute a "closed purposive organization" [2] bound by a relationship of mutual influence. In regard to that which relates evolved systems, form seems to follow from function, as the existence of a particular physical structure is shaped by the specific functional needs that the organism has met along its evolutionary history and ontogeny. This functional level of explanation is essential for understanding how natural selection designs organisms and how, in the course of evolutionary time, new features were added or discarded from the species design [5].

Genetic "instructions" provide general constraints for neural development, determining the different levels of neural organisation and the specificity of the sensorial equipment that organisms belonging to different species display. These instructions define the types and forms of interaction available, and are also responsible for the entity's capacity to identify and assign meaning to particular environmental features, responding accordingly. On this account, [20, p. 16] states:

> The nature of the environment [...] acquires a curious status: it is that which lends itself [...] to a surplus of significance. Like jazz improvisation, environment provides the "excuse" for the neural "music" from the perspective of the cognitive system involved.

To illustrate the fundamental role played by different physical architectures in the definition of particular meaningful worlds—the Umwelten[2]—Uexkull [19, p. 45] takes the female tick as an existential model. Providing a glance at the way it interacts with the environment within which it is embedded across the essential timings of its life cycle, Uexkull identifies the forms of interaction with the external world that are available for the tick and how these provide the information the organism requires to exist: "Out of the egg crawls a not yet fully developed little animal, still missing one pair of legs as well as genital organs. Even in this state, it can already ambush cold-blooded animals such as lizards, for which it lies in wait on the top of a blade of grass. After many moltings, it has acquired the organs it lacked and can now go on its quest for warm-blooded creatures. Once the female has copulated, she climbs with her full count of eight legs to the tip of a protruding branch of any shrub in order either to fall onto small mammals who run by underneath or to let herself be brushed off the branch by large ones. The eyeless creature finds the way to its lookout with the help of a general sensitivity to light. The blind and deaf bandit becomes aware of the approach of its prey through the sense of smell. The odor of butyric acid, which is given off by the skin glands of all mammals, gives the tick the cue to leave its watch post and leap off. If it then falls onto something warm—which its fine sense

[2]We follow the German plural form.

of temperature will tell—then it has reached its prey, the warm-blooded animal, and needs only use its sense of touch to find a spot as free of hair as possible in order to bore past its own head into the skin tissue of the prey. Though it has no sense of taste, the tick pumps a stream of blood, as long as it is warm, slowly into itself [...]".

Given the needs dictated by its internal state(s) at a scheduled point in its life cycle, three features become salient in the tick's surrounding environment:

1. Odor of butoric acid
2. Hairy surface
3. ±37°.

Following a sequence, each of these three cues is perceived,[3] defining a pattern that is identified and assigned a value—meaning—triggering the following pre-set behaviours:

1. Odor of butoric acid _______________ leap off
2. Hairy surface ____________________ cling to it
3. ±37° _________________________ pump the host's blood.

By assigning a meaning to this set of cues and acting accordingly to a final goal—laying its eggs—the tick ensures the survival of its species. Uexkull points that out (ibidem): "And now something miraculous happens. Of all the effects emanating from the mammal's body, only three become stimuli, and then only in a certain sequence. From the enormous world surrounding the tick, three stimuli glow like signal lights [...]. Through these features, the progression of the tick's actions is so strictly prescribed that the tick can only produce very determinate effect marks. The whole rich world surrounding the tick is constricted and transformed into an impoverished structure that, most importantly of all, consists only of three features and three effect marks".

This dialectics that binds a cognitive architecture to its environment can be seen replicated endlessly[4] in nature, highlighting the fact that reality is perceived, "conceived", and modelled differently depending on the "eyes of the beholder", i.e., according to the perceptive/sensorial capacities of the cognitive agent, in other words, according to its corporeal architecture.

Cassirer [2] pointed out that whatever is alive has its own circle of action for which it is there and which is there "for" it, both as a wall that closes it off and as a viewpoint that it holds "open" for the world.

A life form and its physical world constitute a unit—a microcosm—bound by an essential dialectic relationship [5, 7]. This dialectic relationship that binds different cognitive agents[5] to their selected environments is an ongoing dynamic process of reciprocal influence. Seeking to satisfy the existential demands of their internal

---

[3] As Uexkull also reveals, experiments have proved that only the butoric acid seems to be responsible for triggering the particular sequence of responses.

[4] If we imagine how this applies to other life forms as mammals ... fish ... plants ... bacteria, viruses ..., cells.

[5] The term agent is here assigned to all cognitive entities indistinctively.

states, life forms strive to cope with the environmental prompts. By identifying and adequately responding to meaningful patterns, by learning and adapting they guarantee their self-subsistence and species replication within a definite life-span and according to biologically determined timings and stages. At the core of this dialectics stands semiosis. Defined as an essential "interpretative" process present in all life processes [5–7, 10], semiosis is to Sebeok [17, 18] the criterial attribute of life the feature that distinguishes the animate from the inanimate. According to Ferreira [6–8], semiosis emerges from the structural coupling of the living entity and its environment, guaranteeing the cohesion, sustainability and prevalence of the microcosm. This interpretative capacity, this "meaning-making", is, as Sagan [15] points out, much older than words. Damásio [4, pp. 108, 109] states that[6] "in the beginning, there were only sensations and reactions by unicellular organisms [...] sensing and responding accordingly started in this way [...] messages were like irritating substances that caused the corresponding irritation. There were no "eyes" nor "ears" [...] there were just the primordials of a perceiving process that, with evolution and with the development of nervous systems, would lead to world modeling, mind definition and, finally, subjectivity".

In this sense, we can agree with Merleau-Ponty [14] that meaning exists at a pre-reflective level of existence. In fact, there seems to be a primary, pre-ontological "meaning-making capacity" present at all levels of life activity and inherent to life itself. Based on the recurring properties of previous encounters, cognitive architectures incorporate existential narratives, constituting the "know-how" that guides all present interactions. This "know-how" comprehends the capacity to identify and assign a value—meaning—to particular environmental features, simultaneously triggering the organism's adequate response from a repertoire that is basically pre-established.

As posited by Ferreira [5, 7], independent of the type of cognition or level of semiotic complexity involved, meaning is a value—a structured entity. This value is assigned by the cognitive agent—a natural or artificial entity—to an individuated environmental feature or a cluster of features that, because of the agent's nature and needs, emerges in the environment as a salient typical pattern.

In the diagram below, reproduced from Ferreira [7, p. 9], the oval on the left represents the set of all cognitive agents endowed with a particular physical architecture {**X**}, while the oval on the right represents the set of all possible environmental features {**Y**}. **f** is a function from domain **X** to codomain **Y**; the small oval stands for the image of **f**, i.e., the set of all possible outputs obtained when the function is evaluated at each element of the subset. In other words, the smaller oval represents the set of all possible meaningful features for **X** in the codomain **Y**, i.e., its potential self-world (Fig. 1).

Uexkull distinguishes the Umwelt from the Innenwelt. If the Umwelt corresponds to the entity's particular "view" of the world—its world model—the Innenwelt is defined by the internal state(s) that characterize an entity's internal condition at a given time. Conceived as inherently systemic, the concept of Innenwelt is essential for

[6]Author's translation from the Portuguese version.

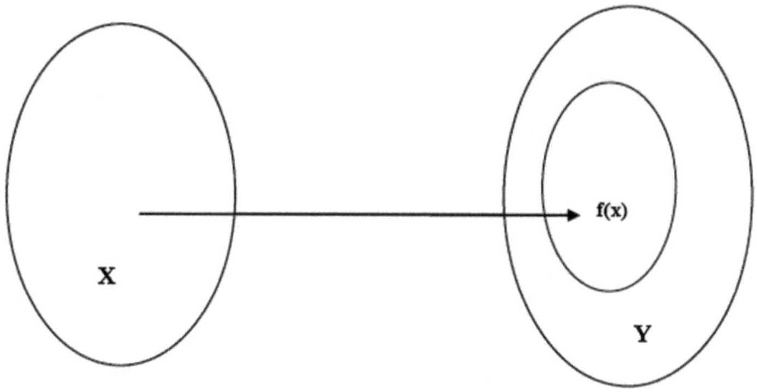

Fig. 1 Meaning—a value assigned to an environmental feature

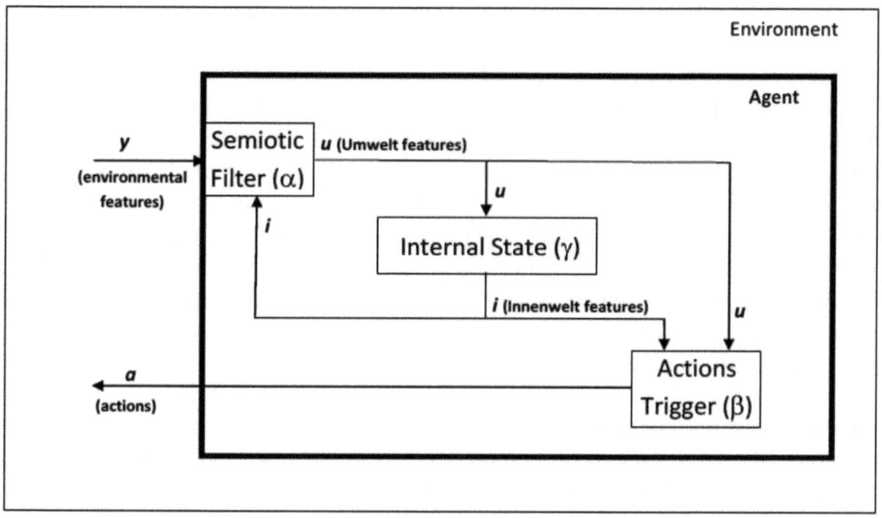

Fig. 2 Modelling cognition

understanding why specific environmental features emerge and take on more salience comparative to others in the organism's lived space. In fact, salience is determined by the life form's existential needs, as reflected by the states of its Innenwelt at a given moment of its life timeline. These states will define the priorities of the emergence of salience in what concerns the environmental features' prominence.

The diagram reproduced in Fig. 2, [8, p. 3], aims to capture the invariants present in the dialectics essential to cognition.

The diagram represents the roles and functions played by the key concepts of the model: ***y*** is a vector of dimension:[7] ($N_y \times 1$), which is assumed to represent all of the

[7]In general ($N_l \times N_c$) indicates de dimensions of a matrix, Nl being the number of its rows and Nc that of its columns; thus, ($N \times 1$) represents an N-component vector in the form of a column matrix.

potential information present in the entity's environment. Acknowledging that not all environmental features will be perceived by the agent and that other features will have different importance at different times and within different contexts, the agent's view of its environment (Umwelt) was modeled through an ($N_u \times 1$) vector, $\boldsymbol{u}$. This vector is created from the environmental features vector, $\boldsymbol{y}$, through the application of a semiotic filter, **F**, whose characteristics are dependent on the agent's internal state (Innenwelt), represented through an ($N_i \times 1$) vector, $\boldsymbol{i}$. The agent's particular view of the world—the Umwelt—will then influence both its actions and its consequent transition to a new internal state. This new internal state will, in turn, influence both the agent's actions and its semiotic filter, and, through it, its environmental perception. The vectors $\boldsymbol{u}$ (Umwelt) and $\boldsymbol{i}$(Innenwelt) are, therefore, in a dialectic relationship that determines and triggers the determine the agent's actions. We assume that there are N_a possible actions that can be executed by the agent and collect the respective probabilities of execution in a vector, $\boldsymbol{a}$. These actions when executed, will have an effect on the environment, allowing or not the satisfaction of the needs dictated by internal states and providing a means for learning to occur.

The process of cognition is an ongoing learning and maturation process through which lifeforms constantly rewrite narratives defining and redefining their "view" of the world and adjusting their responses accordingly. As Varela [20, p. 60] writes: "Ordinary life is necessarily one of situated agents [...] situatedness means that a cognitive entity has, by definition, a perspective. This means that it isn't related to its environment "objectively", that is, independent of the system's location, heading, attitudes and history. Instead, it relates to it in relation to the perspective established by the constantly emerging properties of the agent itself and in terms of the role such running redefinition plays in the system's entire coherence".

Situatedness is reflected in the two overlapping narratives simultaneously running in all lifeforms: one concerning their evolutionary history as a member of a species, embodying the achievements of their predecessors in their struggle for life, the other, the actualization of this evolutionary narrative by the present physical body in particular contexts and circumstances. These particular contexts and circumstances that the new lifeform will have to face and interact with, constructing a particular microcosm, may not be exactly the prototypical, i.e., the ones "expected" by the system [12]. However, in the course of the dialectics that binds the cognitive agent to its environment and in its struggle for life, the organism will always try to respond to the environmental prompts, adjusting, adapting, evolving or otherwise perishing.

2 Umwelt Overlap[8]: The Overlap of Individual Experiences

Different cognitive agents will define according to their physical bodies distinct world views. The existence of multiple subjective[9] worlds, multiple meaningful spheres of existence apparently sharing the same spatio/temporal framework[10] is, again, acknowledged by Uexkuhll who in the introduction to "Umwelt und Innenwelt der Tiere" invites the reader to an imaginary stroll (1934:5):

> [...] a stroll into unfamiliar worlds; worlds strange to us but known to other creatures, manifold and varied as the animals themselves. The best time to set out on such an adventure is on a sunny day. The place, a flower—strewn meadow, humming with insects fluttering with butterflies. Here we may glimpse the worlds of the lowly dwellers of the meadow. To do so, we must first blow, in fancy, a soap bubble around each creature to represent its own world, filled with the perceptions which it alone knows. When we ourselves then step into one of these bubbles, the familiar meadow is transformed. Many of its colourful features disappear, others no longer belong together but appear in new relationships. A new world comes into being. Through the bubble we see the world of the burrowing worm, of the butterfly, or of the field mouse; the world as it appears to the animals themselves, not as it appears to us.

The metaphor of the soap bubble is fundamental to highlight the inherently subjective character of cognition, a subjective process that takes place in a circumscribed sphere: a virtual sphere, a figurative perimeter, traced according to the type of interactions allowed by the physical architecture of the organism and that models in the general environment the organism's Umwelt, its meaningful world [5]. But the metaphor of the soap bubble is also fundamental to understand how these coexistent individual worlds frequently overlap at variable degrees.

Life is characterised by the crisscrossing of individual spheres of existence, of individual Umwelten. The Umwelt of the tick and that of the mammal overlap at a time t, when one becomes the host of the other. The same happening, for instance, with the wolf and the lamb in the relation predator-prey, when the prey becomes the energy supplier of the predator, or between the male eagle and the female eagle in a mating relation. Umwelten also overlap at varying degrees in the so called social species whose individual members are assigned specific roles and usually enroll in cooperative tasks guaranteeing, this way, their subsistance and the community's existence and sustainability, as it is the case of ants or that of bees. But it is with the social species par excellence—the human being—that this overlap becomes the ground for a galaxy of existential interactions from which primarily results the notion of Oneself and that of Otherness, the interaction with the Other(s) and from this the shaping of individual and social identity Ferreira [5, 7]. It is also in the context of the

---

[8]This Concept and Corresponding Mathematical Modeling Are Defined and Developed in Ferreira and Caldas [10].

[9]Subjective in the sense that they result from individual experience.

[10]This spatio/temporal framework is the observer's—the human—spatio/temporal frame. Each life form, in fact, develops according to a virtual "timeline" that is exclusively defined by its internal corporeal dynamics and by the environmental circumstances it will face within a pre-set potential life span.

overlap of multiple spheres of existence that specific relations of production emerge giving rise to particular social structurings, and "work", understood as the creative and generative capacity to produce and change reality becomes an inherently human achievement.

3 The Observer's Myth

Senses are an essential window to the world we live in, providing the data that build mental representations enabling the construction of particular world views. As it happens with all other lifeforms, human beings perceive and interact with the external physical world in their species-specific way. It is thanks to their cognitive architecture, the evolved physical body endowed with innate competences, namely that of symbolic encoding, that human beings are able to give shape and substance to their meaningful worlds anchored on the notion of Self and fulfilled by the net of relationships this self defines and establishes with the meaningful Other(s).

Damásio [4] points out that it was the mapping capacity provided by the emergence of nervous systems linked to a web of neural circuits that allowed for some life forms, namely human beings, the generation and definition of a cartography where patterns of activity and the spatial relations between the active elements inside a pattern are represented and ultimately minds, understood as representations of a subjective lived world can be defined.

Experience is necessarily subjective, and consequently temporal. The organisation of experience according to a temporal axis along which the multiplicity of events are placed in respect to their "before" and "after" is an essentially subjective construal. The outcome of this subjective construal is a totality in which the division of time into present, past and future is no longer a substantial division. Experience, presented in an unbroken flow, will allow the subject to flash back in time, re-experiencing facts or events, and, simultaneously, will allow them to use those past experiences as a standpoint, enhancing a better understanding of the present or anticipating/predicting the future. On this topic, Cassirer [1, p. 167] much inspired by Augustine, writes:

> Strictly speaking [...] we should say that the present time comprises three different relations and through them three different aspects and determinations. There is a present of past things, a present of present things and a present of future things. The present of past things is called memory; the present of present things is called intuition; that of future things is called expectation. Thus, we may not think of time as an absolute thing, divided into three absolute parts: rather, the unitary consciousness of the "now" encompasses three different basic directions and is first constituted in this triality.

Conscious of the complex way meaning is composed and conveyed among human organisms, Cassirer [2] defines the human being as "animal symbolicum". He suggests, on the basis of Uexkull's biology, the existence of a symbolic system, which falls between the "receptor" and "effector" systems that it shares with all of the other organisms. It is this symbolic system that allows signs to be assigned values, enhancing a three-part relationship between the "Sign-Using Self", "Constructed Reality" and the "Other Self".

Reality is not just the reflection that mirrors an external objective world in our eyes, a world existing independently of the subjects of experience, but rather is the result of an individual and collective symbolic construction, a construction emerging from the semiosic process that lies at the core of all forms of cognition. Cassirer says that we must break radically with the presupposition that what we call the visible reality of things is given and present at hand as a finished substratum prior to all formative activities of the mind, because it is not the reality of things that endures, but only the form that reality assumes through us.

The model that characterises the basic forms of semiosis analysed above is also found in the upper levels of semiotic structuring that characterize human cognition. Cassirer [1, p. 56] has this intuition when he writes:

> If perception did not embrace an originally symbolic element, it would offer no support and no starting point for the symbolism of language [...] perception, as such, signifies, intends and "says" something, and language merely takes up this first significatory function [...] the word of language makes explicit the representative values and meanings that are embedded in perception itself.

In what concerns reality, we are never observers, even when we think we are, but always experiencers. In fact, though reality is perceived as external, we know that this very reality results from a semiosis grounded in a unique experiencer/experienced relationship, which the conscious mind ignores, giving the experiential subject the status of virtual observer. The subjectivity inherent to this world view was also stressed by Kant [11]:

> What objects may be in themselves, and apart from all this receptivity of our sensibility, remains completely unknown to us. We know nothing but our mode of perceiving them- a mode which is peculiar to us, and not necessarily shared in by every being, though, certainly, by every human being.

Simondon [16] calls the historical and cultural context in which human cognition takes place the pre-experiential background issued from the experience of all precedent generations, a common background that only comes to life in the present individual appropriation, being in this way consequently changed by the action of those who share it. In fact, every newborn comes to life in a particular physical, economic, social, cultural and linguistic atmosphere. A physical environment where specific relations of production have not only determined the particular social structuring and social hierarchies, but have also determined the typical patterns of behaviour to be followed in all circumstances and contexts, the definition of public and domestic space [13], the creation of institutions, the architectural options, the production of artefacts and technological artefacts, and the production of art forms. It is in the restricted and very controlled life circle provided by the close family that the child seizes the concept of Otherness in the person of its caregivers, especially of its mother, learns how to designate them and how to designate itself, as it starts to shape its own identity. It is also here that it develops essential motor programs, such as that of sitting by itself, walking on two legs, or both handling a spoon in the proper conventional way and carrying the spoon with food to the mouth; it learns that this particular object is a [spoon] and not a [mug] and that its function is to handle food; it becomes aware that artefacts generally have a function associated with them, as well as the spaces defined in its home. It learns that there are behaviours and procedures to be followed in different contexts. It is here in this first restricted circle that the child is slowly introduced into a constructed reality. A world where people, with slight variations, follow the essential typical routines [9], each, eventually, subsuming sets of others that guarantee not only the biological and social existence, but also the maintenance of the necessary conditions of production on which a particular society stands at a given time of its development, e.g.

get up at about the same time
follow identical hygienic procedures
have breakfast
take the children to school
rush to work
get into a train, bus, etc.
start working
get a coffee at the local Starbucks
stop working
rush back home
pick up the children (at school)
cook dinner
go to sleep

Though the essentials of this universe and the basic typical patterns of behaviour with their respective motor programs are incorporated into that first circle of social interaction, the learning process carries on throughout life with the broadening and diversification of social circles [7], with the consequent permanent updating of social conventions, with the introduction of new artefacts and the consequent updating of existing motor programs: how to step onto and off of an escalator, how to swipe the screen of a smartphone so that the camera is activated.

The encapsulation of meaning in symbolic forms is a cognitive demand, as human beings need to preserve and objectify experience, to reflect upon it, to create for themselves a shared model of their lived world. Symbolisation makes the translation of inherently subjective experience into an objective medium possible. By freeing meaning from the immediacy of subjective experience and turning it into a collectively sharable object, language allows it to be incorporated, redefined and reshaped in different contexts and world views.

Damásio [3] states that we will probably never know how faithful our knowledge of the world is in what concerns absolute reality. But what we need, and we have it, is a remarkable consistency in terms of the nature and content of the mental representations that our individual minds produce, and consequently are able to share collectively. This very consistency of our experience and the fact that, through language, this same consistency can be verified and confirmed by the experience of others lead us to believe that this is an experiencer-independent reality, an objective reality.

Cassirer points out that the problem refers not to the objectivity of existence, but to the objectivity of meaning. We would say that this objectivity of meaning is achieved through language, a symbolic construction in which the whole community participates and from which objectivity of being emerges.

4 Hybrid Worlds, Hybrid Agents

Digitization, the conversion of an analogue signal to binary bits, allowed information to be represented in a universal manner and be stored as data. This data can be filtered, tracked, duplicated and transmitted, infinitely, at incredible speed. Digitization has not only empowered human cognition exponentially by accelerating intrinsic semiosic processes but it has also changed the very nature of the typical environment by creating new agents, new umwelten and new overlapping of experience.

For purely analytic purposes and not taking into account other possible hybrid forms, we could consider the following main types of cognition present in the contemporary world:

(i) The typical forms involving a natural system and its physical environment. We include in this case the forms of human interaction with the surrounding environment (analogue) and consider as physical environment the compound of physical, social, cultural and linguistic counterparts.
(ii) Those involving natural systems—human beings—and digital interfaces existing in the analogue world, in typical human life contexts, as it is the case of all the interactions that take place on the Internet via computer or smartphone.
(iii) The forms of cognition involving human beings interacting with virtual environments augmented reality scenarios ... where displacement from the subject's actual mental spatio/temporal framework occurs, as those induced by electronic devices operating on the external perception organs or through induction in the neural system.
(iv) The forms of cognition involving human beings with enhanced capacities and the physical environment, as in the case of bionic components.
(v) The forms of artificial embodied cognition involving a physical artificial system that interacts physically with its body and with the surrounding environment (physical, social, cultural, linguisticm etc.) as in the case of robotic systems.
(vi) The embodied and/or non-embodied forms of artificial cognition interacting with a digitized world, as in the case of the Internet of Things (IoT) or in the case of Artificial Life Research.

Common to all these forms of cognition is the existence of an agent that interacts with an environment driven by certain needs and expectancies. All these instances are profoundly human in the sense that they reflect and incorporate the human view of the world and the way human beings interact with it in an essential semiosic process.

References

1. Cassirer, E. (1985). *The philosophy of symbolic forms. Volume 3: The phenomenology of knowledge* (2nd ed.). New York: Yale University Press.
2. Cassirer, E. (1996). *The philosophy of symbolic forms. Volume 4: The metaphysics of symbolic forms*. New York: Yale University Press.
3. Damásio, A. R. (1995). *Descartes' error: Emotion, reason and the human brain*. G.P. Putnam's Sons.
4. Damásio, A. (2017). A estranhaordem das coisas: A vida, ossentimentos e as culturashumanas. Círculo de Leitores.
5. Ferreira, M. I. A. (2007). On meaning: The phenomenon of individuation and the definition of a worldview. Ph.D. thesis. University of Lisbon. Faculty of Arts. Lisbon. Portugal.
6. Ferreira, M. I. A. (2010). On meaning: A biosemiotic approach. *Biosemiotics, 3*(1), 107–130. https://doi.org/10.1007/s12304-009-9068-y. Springer.
7. Ferreira, M. I. A. (2011). *On meaning: Individuation and identity—The definition of a world view*. England: Cambridge Scholars Publishing. ISBN 1443829250.
8. Ferreira, M. I. A. (2012). Modelling artificial cognition in biosemiotic terms. *Biosemiotics*. https://doi.org/10.1007/s12304-012-9159-z. Springer.
9. Ferreira, M. I. A. (2013). Typical cyclical behavioral patterns: The case of routines, rituals and celebrations. *Biosemiotics*. Science + Business Media Dordrecht. https://doi.org/10.1007/s12304-013-9186-4. ISSN 1875-1342.
10. Hoffmeyer, J. (2008). *Biosemiotics: An examination into the signs of life and the life of signs*. University of Scranton Press.
11. Kant, E. (1996). *Critique of pure reason* (Werner S. Pluhar, Trans.). USA: Hacket Publishing Company Inc.
12. Krasnegor, N. A., & Lecanuet, J. P. (1995). Behavioral development of the fetus. In J. P. Lecanuet, W. P. Fifer, N. Krasnegor, W. P. Smothermaneds (Eds.), *Fetal development—A psychobiological perspective*. New Jersey: Lawrence Erlbaum Associates, Publishers.
13. Lefebvre, H. (1974). *The production of space* (Donald Nicholson—Smith, Trans.). Victoria, Blackwell Publishing.
14. Merleau-Ponty, M. (1968). *The visible and the invisible*. Northwestern University Press.
15. Sagan (2010). Foray in the world of animals and humans, introduction. In C. Wolfe (Ed.), Posthumanities 12. Minnesota Press.
16. Simondon, G. (1964). L'individu et sagenèsephysico-biologique. P.U.F.
17. Sebeok, T. A. (1972). *Perspectives in Zoosemiotics*. The Hague: Netherlands, Mouton.
18. Sebeok, T. A. (1985, 1976) *Contributions to the doctrine of signs*. Bloomington: Indiana University Press.
19. von Uexkull, J. (1933). A theory of meaning. in a foray in the world of animals and humans. In C. Wolfe (Ed.), Posthumanities 12. Minnesota Press 2010.
20. Varela, F. J. (1992). Autopoiesis and a biology of intentionality. In *Proceedings from the Dublin Workshop on Autopoiesis and Perception, essay 1*. http://www.eeng.deu.ie/pub/autonomy/bmem9401.

Complementarity of Seeing and Appearing

Jindřich Brejcha, Pavel Pecháček and Karel Kleisner

Abstract In this chapter, we use the example of colouration of animal surfaces to show how processes based on interactions of the individual parts enter, as units, into processes on other levels and how this processual scaffolding leads to the emergence of 'meaning' on the level of communication between individuals. We review recent understanding of colour production and pattern formation in animals. We describe self-organization and dynamical nature of these processes. To highlight the inseparability of seeing and appearing, we discuss shared evolutionary origins of sight and colouration. Common evolutionary explanations of colouration are then discussed. Due to the complementarity of appearance and perception, the exposed surfaces of organisms ultimately become semi-autonomous entities subjected to their own evolution.

1 Introduction

One of the basic human quests is a search for the smallest particle that underlies the constitution of all other things. This investigation led to the discovery of atoms, however, it then turned out that atoms, too, consist of smaller particles, which, in turn, are formed of yet smaller pieces of matter. It is an important discovery of the last fifty years that, in addition to particles that constitutes matter, there also exist exchange particles, which are responsible for interactions.

J. Brejcha (✉) · P. Pecháček · K. Kleisner
Faculty of Science, Department of Philosophy and History of Science,
Charles University, Prague, Czech Republic
e-mail: brejcha@natur.cuni.cz

J. Brejcha
Department of Zoology, National Museum, Prague, Czech Republic

M. I. Aldinhas Ferreira et al. (eds.), *Cognitive Architectures*, Intelligent Systems, Control and Automation: Science and Engineering 94,
https://doi.org/10.1007/978-3-319-97550-4_2

In biology, we wonder how the seemingly endless heterarchy of constituent parts led to the appearance of life in all of its manifold shapes and manifestations. We ask ourselves what life is, how it came to be, how it is maintained, and how it creates variability. It is quite clear that when an organism is taken apart, it cannot be put back together in such a way that it would be alive again. This is because the individual components of a living entity are not organised statically: they enter into complex and dynamic interactions, which, in a given context and on many levels of biological organisation, engender a synergy of function and form. Noble [61], as well as various other scholars (see, e.g. [21]), illustrates this process using a musical metaphor. He compare metabolism, development, and their historic sequence through generations to performances of a large orchestra, in which the individual musicians follow a shared score (a metaphor for DNA), but the final result is determined by coordination of the individual performers, who all, to some extent, improvise. Even the individual chords are formed by a harmony of individual tones, and minute changes in their presentation can have a major impact on the resulting whole. On top of that, the context matters: an orchestra sounds rather different in a concert hall than it does outdoors.

In the following chapter, we use the example of colouration of animal surfaces to show how processes based on interactions of the individual parts enter, as units, into processes on other levels and how this processual scaffolding leads to the emergence of 'meaning' on the level of communication between individuals. At the beginning of the chapter, we first investigate the causes of colouration on a cellular level. Then, we mention the possibility of self-organisation of patterns based on interactions among various components that contribute to the development of an individual. And since colouration is perceived by other organisms, we then turn our attention to an investigation of shared evolutionary origins of sight and colouration. We argue that due to complementarity of appearance and perception, the exposed surfaces of organisms ultimately become semi-autonomous entities subjected to their own evolution [41]. In the final part of this chapter, we then investigate various explanations of the evolution of colouration in the context of its role in animal behaviour and communication and within particular environmental settings.

2 The Building Blocks of Animal Colouration

The body surfaces of vertebrates are covered in multiple-layered skin, which forms the main barrier between the animal's inner environment and its surroundings. Skin develops mainly from the ectoderm, a germ layer that also gives rise to various other body parts, such as eyes, horns, and teeth (which are, however, constituted from several types of tissue of various embryonic origins). Generally speaking, skin also includes scales, hair, or feathers, i.e., epidermal derivatives that cover the body's surface. Importantly, the integument has the ability to absorb and reflect light, which leads to body colouration. In some groups of organisms, this ability can be significantly reduced. This especially happens in animals (or in their developmental stages) that are not exposed to light, e.g., in cave fish, olms, sirens, or deep sea sharks and fish. Generally speaking, however, both fish and amphibians are coloured even in their earliest developmental stages, since the single-celled stage [3] and both the bare surface and its derivatives can be coloured.

Colouration can be produced by the composition of extracellular fluids of the outermost tissue, for instance, by blood perfusion. It can also be caused by the arrangement of polymer fibres that form the skin, such as keratins [82], which are produced by the keratinocytes, or collagens, extracellular proteins generated by fibroblasts [8]. The resulting colour is the effect of reflection of incident light from the organised collagen fibres. Structural colouring mediated by collagens is found, for instance, in birds, in particular, on the bare, unfeathered skin of their heads [71]. The same type of colouration is also found on the scrotum and face of various mammals, such as mandrills or Robinson's mouse opossum [72]. Keratin fills the skin cells and forms certain types of scale and their surfaces, but also, for instance, the surface structure of hair and feathers. As an example, the black patterns on the back of the Gaboon viper (*Bitis rhinoceros*) is produced by organised nanostructures on the scale's surface, which form a sort of nanoscopic black body that absorbs all incidental light energy [84]. Another rather frequent occurrence is also iridescence, which is caused by the refraction of light falling under various angles on to microscopic structures on the surface of various skin derivatives [16]. Cells that have special optic properties are known as chromatophores. In the vertebrates, they take the form of pigment cells, which are a special type of neural crest-derived cell [29]. Pigment cells can be present either in the dermis or in the epidermis, which is separated from the dermis by a basement membrane. It is typical of the pigment cells that they belong to but a handful of cell types that can pass through this membrane [96].

The colour of the body surface of poikilotherm vertebrates—fishe in a broad sense of the term, amphibians, lepidosaurs, and basal archelosaurs (turtles and crocodiles)—as perceived by our eyes is often the result of an interplay among up to three types of cell with different optic properties and their basic horizontal organisation in the dermis. There are three basic types of dermal pigment cell: xanthophores (characterised by vesicles that contain pterins or carotenoids), iridophores (which contain crystalline purines and their derivatives), and melanophores (which, in their melanosomes, contain melanins). In the dermis, the xanthophores are usually closest to the surface, while under them are the iridophores, and yet lower are the melanophores. These three kinds of cell can form specialised cup-like structures, so-called dermal chromatophore units, which are found in some fishes [22], frogs [5], anoles [88], and chameleons [89]. In dermal chromatophore units, melanophores reach with their 'fingers' above the two upper layers of pigment cells. This enables a rapid relocation of melanosomes, observable as a sudden darkening of the body's surface. Change in skin colour can also be due to a relocation or transformation of entire colour-generating cells [58]. Sometimes, however, these three kinds of chromatophore do not form such units. In such cases, pigment cells in the dermis form a continuous three-layered cover over the whole body [47].

Skin colouring is determined by mutual relations between the thickness of layers of the individual types of cell and the thickness, number, and mutual distance of crystalline pellets and the various types of pigment present [27, 89]. There also exist, however, various modifications to this basic organisation. For instance, when one cell produces both xanthophore and iridophore organelles, this results in 'mosaic chromatophores' [22], which have turned out to be of special importance in the search for the origins of pigmentation. To wit, based on a comparison of the ways in which intracellular vesicles can form in the chromatophores, Bagnara et al. [4] have proposed a model of shared origin of all three types of pigment cell. The observation that inspired their model was the fact that while the various types of pigment cell clearly differ in their structure and the contents of the vesicles, in mosaic chromatophores, one finds morphological and biochemical features that include various kinds of vesicle. This led Bagnara et al. [4] to hypothesise that pigment cells and vesicles have a shared origin in a 'primordial vesicle'.

In homeothermic vertebrates, that is, birds (apical archelosaurs) and mammals, colouration is determined by either the structure and amount of pigments in pigment cells and related epidermal derivatives or by the structure of the extracellular matrix, rather than by organisation of pigment cells as such [37, 82]. Feather colouration is determined by the presence of pigment cells at the base of a growing feather in the epidermis. Using their finger-like projections, pigment cells distribute melanosomes to the keratocytes that form the feather. Once the development of a feather is completed, the formation and distribution of pigments is final and does not change until the next moulting [87].

It should be noted that we started learning about the development and evolution of plumage only in recent decades [69]. It is thus not surprising that our knowledge of the developmental mechanisms responsible for the final colouration of bird plumage is still limited [73]. The original function of plumage is likewise unclear, but colouration may have played an important part in this process [20]. In mammals, colouration is determined solely by melanophores located in the epidermis or hair follicles [86]. Melanophores and keratocytes in the epidermis function in close coordination, forming so-called epidermal melanin units. In these units, melanophores, through their melanosomes, supply surrounding keratocytes, which, in turn, provide them with various growth factors [74].

Just like in the vertebrates, colouration in invertebrates can either be based on pigments, which absorb certain parts of the light spectrum, or be the result of special structures that reflect the light of particular wavelengths. Unlike most vertebrates, which produce pigmentation in special kinds of cell, insects usually synthetize pigments or their precursors directly in epidermal cells [95]. Pigments in insects can, however, also be found outside epidermal cells, for instance, in the scales that cover butterfly wings. In butterflies, which are, in terms of colouration, the most varied representatives of the insect class, we find pigments belonging to four distinct chemical categories: melanins, ommochromes, pterins, and flavonoids [59]. As noted above, some of these are synthetized by the butterflies themselves, while others are not. The latter group includes, for instance, flavonoids, which butterflies, as well as other animals, acquire mainly from plants [35].

In insects, one also finds several different kinds of structural colouration. This phenomenon has been described in considerable detail, especially in beetles [80].

The most common way of achieving coloured surfaces in beetles is through multilayered reflectors consisting of thin reflective layers. These are located in beetle cuticles, where constructive light interference then produces iridescent colouration [48]. Constructive interference is also responsible for the functioning of so-called optical grids, i.e., a system of periodically repeating structures, such as ridges and grating [80]. Another type of structural colouration in beetles is photonic crystals, which are two- or three-dimensional grids created either by nanoscopic spheres or microscopic gaps that reflect select wavelengths, thus creating intense colouration [50]. Scattering is another type of structural colouration. This created by interaction between the incident light and various substances. It is known to exist in many groups of insects, such as butterflies [70], for whom light scatters across the surface of the pigment globules in their wings.

3 The Origin of Colour Patterns

In insects, coloured patterns on the body's surface are caused by prepattern morphogens. In other words, they are based on a specific expression of gene products, namely proteins. This seems to be facilitated by a combination of two processes. The first, known as the 'French flag model', describes a situation in which the gradient of the morphogen based on various thresholds of concentration marks a coordinate system, thus determining the location for the development of future traits [12, 44]. This process does not take into account the possibility of internal interaction between two different morphogens (the concentration of one morphoghen does not directly depend on the concentration of the other). Recent research, however, shows that even the French flag patterning involves dynamic interactions [33]. The second process is based on an internal interaction between two or more morphogens (the concentration of one morphogen depends on the concentration of the other morphogen or morphogens). This process is exemplified by, e.g., reaction-diffusion (RD) processes [34, 44, 90]. The individual morphogens spread through space and, based on mutual interaction, establish a spatially non-homogeneous dynamic equilibrium, which leads to differentiation of the tissue. In vertebrates, such interactions are based on the interaction of particular types of pigment cell, as such [51, 57]. This interaction is all the more interesting since, at short distances, it can take the form of active contact between one cell (xanthophore) and another individual cell (melanophore), this contact then leading to active migration [31]. Such interactions between individual cells probably underlie the whole logic of the RD process. At long distances, pigment cells probably influence each other en masse through the gradient of the morphogens they produce. Long distance regulation is little known, and it is possible that long distance interactions are also mediated by long cell protrusions. Pattern formation is, however, somewhat more complex. For instance, iridophores seem to form pattern-establishing interaction matrices, which may function as spatial constraints for the two other types of pigment cell [62].

Such communication and behaviour on one level of self-organisation, be it molecular or cellular, leads to changes in colouration, which may, in turn, be the object of interaction at a higher level (see below in this chapter). For instance, the production of morphogens is also based on a finer level of interaction among expressed loci of the DNA [40]. Nonetheless, all of these self-organising properties need not necessarily influence other levels of organisation directly by physical or chemical dependencies. Functions of the individual levels of organisation might arise through their possible involvement in the mechanisms of other levels of organisation. Mutual interactions among the various levels of organisation thus are not purely physically causative (universal). They are context-dependent processes whose mutual involvement was in their evolutionary history, established based on a compatibility of functional motivation.

4 Sensory Apparatus

The sensory organ that perceives surface colouration is the eye. Eyes are the first location where visual signals are processed and treated, and only later is the signal processed by the nervous system. Nilsson [60], in his article on the evolutionary origin of eyes and visually directed behaviour, introduces the term 'sensory task', which he defines as a systematic behavioural or physiological response to a particular stimulus. He then adds ([60], p. 2834):

> Sensory systems can improve fitness only through the responses they trigger Thus, sensors and effectors make sense only in combination, and evolution of the senses is intimately linked to the evolution of locomotion and behaviour.

One could imagine the system of visual perception and optic signalisation in animals using a metaphor of two radios, a receiver and a transmitter, whose function is to mediate a mutual connection. In order to achieve this, the radios need to be tuned to the same 'frequency' and some transmission between them must occur. In the history of bodies, eyes and coloured surfaces have been spatially separate, but essentially, they form one coherent organ with a shared evolutionary apparatus and functional potential. This system is based on the mutual complemantarity of two parts, in which one without the other loses parts of its evolvability.

Recent studies offer remarkable insights into links between the evolution of eyes and coloured patterns on animal bodies. In an extinct midge, *Eohelea petrunkevitchi*, which had been found in Baltic amber, a structure was preserved on its wing, whose shape and surface perfectly resemble the composite eyes of arthropods and is, at first sight, indistinguishable from them [15]. In *Heliconius* butterflies, the mapping of genetic expression had shown that an important regulator of eye development in invertebrates, the *optix gene*, plays an important role in the construction of red patterns on their wings [54, 75]. In vertebrates, too, one finds connections between the regulation of eye development and the development of body surfaces. In this case, the crucial shared regulatory factor for eye and colouration development is the *mitf* gene (microphthalmia-associated transcription factor) [64].

In vertebrate eyes, we find two kinds of pigmented cell that produce melanin. The first are uveal pigment cells, which share their embryonic origin with pigment cells of the integument. Their function is mainly to ensure that light passes into the eye only through the pupil. These cells, however, are also found in the choroid, the supporting layer behind the retina. The other type of pigmented cell in vertebrate eyes is the pigmented cells of retinal pigment epithelium (RPE). These cells have several functions, but in this context, especially relevant is the crucial role they play in the transformation of all-trans-retinal to 11-cis-retinal, thus enabling regeneration of photoactive proteins [93]. Unlike pigment cells of the integument, which develop from the neural crest, RPE pigmented cells originate in the neural tube. During the ontogenesis of eyes, the area of the optic vesicle is first defined by the overlapping expression of several genes (*pax6*, *pax2*, and *mitf*). Later on in the development, these genes are expressed specifically in the individual developmental modules of the eye: pax6 in the retina, pax2 in the optic stem, and *mitf* in the RPE [7]. A mutation of the *mitf* gene can result in microphthalmy (small eye syndrome), which is linked to abnormal hyperproliferation of the RPE and hypoproliferation of the retina. Other concurrent effects of this mutation include non-closure of the optic fissure, macular puckering, production of dark pigments in cells originally intended for involvement in the formation of the retina, or the formation of a second retina on the dorsal side of the RPE. In integumental melanophores, the same mutation in the *mitf* genes has different effects then it does in the RPE. It leads to either the disappearance or hypoproliferation of pigment cells [9]. In short, for melanophores, the *mitf* represents the basic and universal signalling interface, which not only mediates cellular proliferation and differentiation, but also determines the survival of pigment cells and mediates intracellular signals to surrounding pigment cells. Synthesis of important enzymes and structural proteins also depends on the *mitf*, because it can bind into areas of the DNA that trigger the transcription of loci where those molecules are coded [92]. Melanins, whose synthesis depends on the *mitf*, can absorb energy by creating waves in their long chains, thus transforming it into kinetic energy and gradually converting it to heat. In the inner ear, for instance, melanin-producing pigmented cells of unknown embryonic origin help the ear cope with excessive kinetic energy of sounds.

Another important part of the visual apparatus, like the melanosomes of the melanophores, are vesicles containing carotenoids dissolved in lipids. These are analogical to organelles of other pigment cells, namely the xanthophores. In addition to being part of the filtration mechanism of photosensory cells in the form of lipid droplets [39], the carotenoids that are dissolved in them are precursors of retinyl esters. In both vertebrates and invertebrates, retinyl esters are precursors of 11-cis-retinal, which is isomerised during light absorption by rhodopsin [93]. In the *Drosophila*, for instance, participation in the optic apparatus is probably the only important function of carotenoids in their bodies. In the archelosaurs, a duplication of the locus for the cytochrome monooxygenase enzyme, which processes ingested carotenoids, led to the production of red oil droplets in the eye. This was an evolutionary novelty that appeared in this group: in other groups, the oil droplets in the eyes are either yellow or colourless. Moreover, the archelosaurs have co-opted this mechanism to produce red colour in their pigment cells: this was their evolutionary novelty [91]. Another case of co-option of a mechanism from eyes to integument could be platelets of iridophores. Crystallising compounds such as purines and pterins are also present, not only in the pigment cells, but also in the *tapetum lucidum* of various animals [63].

We are only beginning to understand the relations between the gene networks of various types of pigment cell but it is already clear that the individual types of cell share many regulatory mechanisms, including the prominent role of the *mitf* gene [30, 37]. Expression of the *mitf* has been observed both in the pigment cells of the eye and in the lens of the cnidaria (in particular, in the jellyfish *Tripedalia cystophora*), that is, in the sister group of all bilaterians, which includes arthropods, molluscs, and vertebrates [45]. Available data shows that the eyes of cnidaria, arthropods, molluscs, and vertebrates are not the product of convergent evolution, i.e., something that evolved independently in a number of different specific ancestors. Rather, it seems to be a case of parallel evolution of evolutionary mechanisms and cellular types that have a shared history, probably going back to a shared ancestor. It seems to be a case of deep homology that took place early in the evolution of the animal kingdom [83]. In vertebrates, moreover, the melanophores on their surface express their own type of photosensory protein melanopsin, whose aminoacidic sequence resembles most closely the opsins in cephalopods or insects. It should be noted, however, that while melanopsin is produced mainly on the melanophores, it is also generated in the hypothalamus, in the cells of the iris, and in the retina [65].

From an evolutionary perspective, the pigment organ, in the sense of a collection of all integumentary pigment cells in the body, and eyes, along with their molecular genetic ‘packages’, all share a common cellular ancestor that was both photosensitive and pigmented [2, 94].

5 Hypotheses of Adaptivity of Signalling by Colouration

Animal colouration can function as a protection against sunlight or assist in thermoregulation. Moreover, when colouration does not activate observers’ receptors, it helps the animal to hide and serves as cryptic colouration. When it does activate the visual receptors of other relevant animals, it can play an aposemantic role, be part of a mimetic complex, or play a role in social or sexual selection. With the exception of thermoregulation or protection against sunlight, both functions that depend on subject-less phenomena, colouration enters into interaction among at least two individuals. This then opens the question of the nature and origins of inputs and outputs of such interactions, as well as the issue of sustainability and the very existence of the interaction as such.

Co-evolution between colouration and the cognitive apparatus can be the result of a sensory drive. For instance, in *Pundamilia cichlids* from the African Lake Victoria, evolution of the visual apparatus has occurred on both a molecular and and an ecological level, directly linked to the evolution of male colouration and female preference for colour signals [81]. *Pundamilia* females prefer patterns that are conspicuous with respect to their visual apparatus and within their environment. A similar phenomenon has also been observed, for instance in the *Heliconius* butterflies, for whom a duplication of a locus for protein UV rhodopsin (UVRh2) goes hand in hand with the development of unique pigments, such as 3-hydroxy-DL-kynurenine, which form a wide range of yellow colours further enhanced by the reflection of light in the ultraviolet spectrum. Visual models of light perception in butterflies indicate that such innovations led to a broadening of the range of colours on their wings [11]. In guppies, *Poecilia reticulata*, males have sexually dimorphic orange-red spots on the side of their bodies. Their females, meanwhile, are attracted to orange objects regardless of their connection with reproduction. Across various guppy populations, it is the strong female preference for orange-red objects that explains a large part of their preference for conspicuous male colouration [76]. The hypothesis of sensory drive explains the evolution of colouration and its perception by constraints placed on it by the detection abilities of the visual apparatus. It does not, however, say anything about the evolution of signals that are not influenced by these constraints and yet play a role in interaction between individuals.

Some molecules that play an important role in the development of animal surface colouration are also crucial for other traits. In particular, mutations in genes that code the production of these compounds result in changes of properties other than colouration. Such manifestations of mutations, i.e., cases when one locus influences a number of traits, is known as pleiotropy. In vertebrates, pleiotropic effects linked to pigment synthesis are especially found in receptors for melanocortins and their antagonist, the agouti protein. Melanocortin receptors and their interaction with the agouti protein determine whether melanocytes will synthesise the very dark eumelanin or the yellow-red pheomelanin, and mutations in the loci that code for these proteins lead to changes in vertebrate colouration. Ducrest et al. [17] summarised the currently known links between changes in sexual activity, aggression, reactions to stressful conditions, energy output, growth, and sleep, but also phenomena such as yawning, stretching, grooming, average heart rate, and neuroregeneration on the one hand and melanocortin receptors and agouti proteins on the other hand; this comparison revealed a clear correlation with changes in colouration. Similarly, Wittkopp and Beldade [95] summarised the pleiotropic effects of mutations in genes that code for proteins important in cascades that control pigment production and their effect on immunity, but also other phenomena, such as properties of the cuticle, resistance against desiccation, nerve activity, animation, feeding, social behaviour, and partner preferences in insects. These studies indicate that colouration is directly linked to various essential traits and can, hypothetically, provide important information about its bearer in a form that is accessible for other organisms in its environment.

When we assume that surface colouration indicates true information about its bearer, we speak of honest signalling. This is a special case of an adaptive mechanism in which the persistence of a trait is constrained by natural selection. The organism that best adapts its behaviour to its environment, and thus acquires the most efficient immune system, displays this high level of adaptation on its surface, which, in turn, along with the tendency of said behavior to be dominant, gives it the best chance of reproductive success. During the mating season, signals enable animals to recognise their conspecifics. In the case of mate selection by one of the sexes, they also help decide between potential mating candidates; admittedly the decision process takes longer when the judged objects are highly similar, but fast selection can compromise the quality of the final choice. This issue can be solved by the potential mate's signalisation of high qualities, which facilitates a faster decision [24].

Many animals use multi-component signalisation to enhance perception by potential mates, because composite signals have a greater effect than the simple sum of their individual components [77]. Signals are employed not only during the mating season, but also in formation of social hierarchies, and mate choice is thus not necessarily about preference for one particular kind of signal, but also, and perhaps mainly, about preference for an individual with a higher position in the group hierarchy [36]. For instance, the throat colouration and calls of males of the European tree frog (*Hyla arborea*) have a positive impact on their reproductive success, even though it does not signal better physical condition as measured by the weight to length ratio [23]. In animals whose females copulate with multiple males, competition can take a cryptic form in which fertilisation is influenced by competition among sperm or by genome compatibility. In such cases, the selection is independent of the female's decision [1]. Colourful spots on the bodies of vertebrates have been much discussed in studies of signalisation of health. Pigments that produce colour on the body's surface, such as melanins, pterins, porphyrins, flavonoids, and psittacofulvins, function in the body at least in part as antioxidants [55].

In relation to the signalisation of health, considerable attention has recently focused on carotenoids, which, unlike the above-mentioned pigments, originate in plant food and cannot be produced by animals on their own. The availability of carotenoids has a direct impact on the quality of yellow, orange, or red colouration and signalisation of carotenoid levels by colourful patches could be an instance of honest signalling [18, 32, 56, 79].

Carotenoids can function in the body as antioxidants, but under changed conditions or in a different context, they can also have a pro-oxidant effect. For instance, in people, higher oxygen pressure in the body has the effect that β-carotene loses its antioxidant properties and starts behaving as a pro-oxidant. The same phenomenon takes place when carotenoids are ingested in amounts that exceed the physiological capacities of the animal [46]. Melanins, too, can become highly dangerous to living cells, and especially quinones, their derivatives being highly toxic, although their high reactivity makes them ideal for capturing free radicals produced by oxidative stress. When a melanosome membrane is compromised, this causes not only irreversible damage to the cell itself, but also to other cells, whereby the extent of the damage depends on the number of damaged melanosomes and the amount of pigment they contained. Melanocytes can also have a phagocytic function and their role in the body can, in some cases, resemble that of the lytic cells. Zahavi [97] has proposed that costly signals, i.e., signals that are difficult to maintain or even put its bearer at a disadvantage, guarantee honest signalling. This concept, known as the 'handicap principle', is based on the idea that only a high-quality individual will be able to maintain such signals.

On the other hand, it is well known that endogenous pterins have optic properties very similar to the carotenoids [55]. Pterins and carotenoids can participate in the mutual creation of an ornament. This is the case with, for instance, male guppies (*Poecilia reticulata*), whose sexually dimorphic patch is created by both of these pigments [26], or in the dewlaps of anoles, in which the throat skin colouration in male brown anoles (Norops sagrei) is created mainly by red pterins, whereas female dewlap colouration is formed at the edges by yellow and red pterins and in the centre by carotenoids [85]. A comparison between brown anoles (*Norops sagrei*) and humble anoles (*Norops humilis*) showed that colouration of the dewlap can be achieved by various combinations of carotenoids and pterins [85]. Rutowski et al. [78] proposed for alfalfa butterflies (*Colias eurytheme*) that pterins can reinforce the signalisation produced by the structural elements of their patches. Moreover, structural elements can produce colouration even in the total absence of pigments at wavelengths that usually result from absorption by pigments [28].

It thus seems that the qualities that are being signalised are not always readily discernible what qualities are being signalised. Colouration by pigments need not be determined by one particular type of pigment: it can be the result of the whole metabolism of pigments and their precursors. The final colouration thus cannot be ascribed to one particular compound and its properties. It is the result of a dynamic network of synthesis of pigments and their precursors. For instance, the synthetic pathway of pterins in the body includes a feedback to one of the precursors that influences all of the other products of this pathway [10]. In some cases, it is thus hard to imagine that the signal recipient could, based on observation of a phenomenon as complex as colour is, make decisions regarding the quality of the observed individual and all of its attractive attributes.

6 Arbitrary Coevolution of Colouration and Preference

> Consider, then, what happens when a clearly-marked pattern of bright feathers affords, in a certain species of bird, a fairly good index of natural superiority. A tendency to select those suitors in which the feature is best developed is then a profitable instinct for the female bird, and the taste for this point becomes firmly established among the female instincts.

Fisher [19, p. 187] used these words to formulate a concept currently known as runaway selection. The process was later mathematically defined by Lande [49] and Kirkpatrick [38], and became known as the Lande-Kirkpatrick (LK) mechanism. Fischer's formulation shows that the model was originally thought to describe adaptation, i.e., it was based on an idea of 'natural superiority' of bearers of some particular trait. Richard Prum, however, explains that the LK mechanism in fact outlines a process that does not so much signal a universal adaptivity of indicator traits, but rather represents a null model of sustainability and progressivity of a trait in a population process in cases when a trait and a preference for it are genetically correlated [66–68]. This means that if, in a population, there exists preference for a trait, the offspring of such a mating will carry both the trait and the preference for it. Non-random mating of individuals within a population based on some particular properties or attributes is called assortative mating, and it is this assortative mating that creates a genetic correlation between a trait and a preference for it. If a trait and the relevant preference are genetically correlated, this correlation is heritable, and choosing the trait can have an impact on preferences in the subsequent generation. The strength of such correlation and the degree of variability in the population then determine the dynamics of the process. If the genetic correlation is weak with respect to the trait's variability, both the trait and the preference for it will tend to reach a stable equilibrium in the population. If, however, the genetic correlation is strong with respect to the variability of the trait, the process becomes destabilised and the population will quickly start diverging away from the equilibrium. That then leads to the above-mentioned runaway process, which is a special case of the LK model.

It should be noted that this process does not require that the trait should have any specific attributes and it need not exemplify perfect adaptation to the environment. In order to create evolutionarily new states of a trait and a preference for it, the process requires only their mutual correlation, i.e., a feedback between a preference and a trait [66, 68]. A trait, for instance, colouration, thus need not be universally true in the sense of being a generally valid signal that indicates the state of otherwise hidden qualities of an individual. The process described by the LK model is based on an arbitrary pairing between a trait and a preference, and which trait will figure in the process depends solely on individual preferences. The trait is not, however, random, because preference for it is the result of individual agency, that is, the result of individual experiences and needs.

According to Richard Prum, a trait in an LK process does not have any particular meaning [66], but is that really so? Is it not rather the case that a trait that starts interacting with a preference can have any meaning based on the context of the preference? In such a case, the meaning of the trait—just like the trait itself—would not represent any universal truth, any particular quality. The meaning of a trait within an LK process could be defined by each and every individual, while preference for it is displayed, for instance, during a mating ritual between a particular male and a particular female [68].

One perhaps ought to view preference for a trait and its meaning not as universal entities, but rather as essentially interactive entities that appear on the level of individuals. Interface between particular individuals is the level of 'self'-organisation that produces the dynamics of a trait, in this case, the evolution of a colouration pattern and quality. Individual interface is, in this context, essential, because neither traits nor preferences are the result of straightforward agency of gene loci. Traits and preferences are dynamic, multi-local interactions that emerge from a scaffolding of a multitude of processes involved on other levels of organisation and amount to an individual's overall set-up. Thus, for instance, colouration, which emerged through self-organising processes related to the evolution of pigmentation (see above), appears as one of the interaction hubs set in a network of links among individuals in virtue of being a compatible substrate for preferences of other individuals. On the other hand, however, it also influences, via feedback, the processes that take place on levels of organisation that contribute to the formation of an individual via mutual perception and interpretation.

7 Biological Meaning as an Evolutionary Factor

One question remains to be answered: What kind of force maintains the orchestrated organisation of all of the levels from genes to behaviour? Is there any such force or is it something else? And, most importantly, in what way is the biological heterarchy united towards a single survival function? One could say the force in question is natural selection, but such an answer would be, at best, incomplete, at worst, wrong. The question really is: What determines the intensity and direction of natural selection?

One of the most important achievements of Darwinism is that it provides an explanation of the origins of functional traits of organisms by natural selection. Darwinist natural selection is not any particular thing: in every context, it manifests itself differently. Corning [14] characterises it as 'a kind of umbrella term that refers to whatever functionally significant factors are responsible in a given context for causing differential survival and reproduction'. Selection doubtless plays a role in biological evolution and it is most unlikely that new insights would ever significantly lessen its importance. But natural selection is not the only generally recognised element of the process of evolution.

In evolution, it is far from rare that an already existing structure becomes the basis of new adaptive traits, and this can take place irrespective of what, if any, purpose such structures had served previously. Gould and Vrba [25] call the original state of a trait its 'primary exaptation', while its subsequent modifications become secondary adaptations. In molecular and developmental biology, the term 'co-option' refers to an event in which a product of gene expression, such as a protein, which previously served some original function, is recruited for a new function. Exaptation, on the other hand, refers to all potential adaptive functions for which a trait could be, in the course of evolution, co-opted and further improved by selection.

Maran [52] suggests that one could distinguish between two kinds of evolutionary process: those that are influenced by subject-specific perception of organism, and those in which the subject's activity plays no part. From this perspective, sexual selection fundamentally differs from natural selection (e.g., environmental selection due to abiotic factors). In sexual selection, the direction in which a trait evolves is determined by the active role of the subjects of relevant sex. In environmental selection, on the other hand, this dimension is usually absent. From this perspective, sexual selection is, in principle, akin to artificial selection [43]: in both cases, the direction and intensity of selection is derived from some properties of the subject, which, in the case of artificial selection, can be, for instance, the taste of a breeder. In such cases, the selection of a trait can take place only if the trait assumes some meaning in the Umwelt of a particular species. The event within which a trait acquires meaning within the Umwelt of an organism (whereby the original meaning is relabelled) is what we call a semiotic co-option [42, 53]. In other words, semiotic co-option is a process whereby a trait that used to have a different meaning—or no meaning at all—acquires a specific meaning within the Umwelt of a given organism. Variability of Umwelt-specific interpretation on either an interspecific or intraspecific level leads to a selection of relevant meaning carriers. Selection that arises from variability of meanings assigned by individual organic subjects is what we call semiotic selection. While semiotic co-option explains the way in which semantic organs emerge from unspecified precursors, subsequent semiotic selection determines their possible evolutionary trajectory.

The notions of semiotic co-option and semiotic selection are compatible with the commonly used terms 'co-option' and 'selection', but the new terms add precision to the conceptual framework used in describing phenomena in which the subject's activity plays a key role. In a similar vein, Peter Corning uses the term 'teleonomic selection' to describe a 'purposeful (cybernetic) process (i.e., an act of choosing) that always occurs in the minds' of living organisms; it is living beings that do the selecting, and it is a process that is intimately related to meeting the basic survival and reproductive needs of a given organism in a given context' ([14], p. 11). The process of evolution is then a synergistic effect of a whole range of causal factors that work on many levels of organization [13]. Approximately one hundred years ago, scholars such as Lloyd Morgan, Henry Fairfield Osborn, and James M. Baldwin proposed a similar research agenda that would emphasise the evolutionary importance of downward causation and behaviour. For example, Baldwin [6, p. 37f] thought that individual accommodation, i.e., the functional adjustment of individual organisms to their environment, is a process that determines the direction of evolution. He saw it as 'a positive factor in evolution, a real force emphasizing that which renders an organism fit; whereas natural selection, while a necessary condition, is yet a negative factor, a statement that the most fit are those which survive' ([6], p. 38).

In biology, context is a crucial but vague concept, because it always changes, together with all of the other living systems that co-produce the context and are at the same time affected by it. What we need is some concept of value shared among the agents that co-create contexts, secure the functioning of the organic closure, and influence evolutionary trajectories. A biological meaning is a possible candidate for such a universal value. This factor is real, and it naturally intertwines with survival and reproduction; it may be attributed to anything, pattern, trait, and the like, that is perceived as fitting by a cell, organism, colony, community, or any other sensing whole.

References

1. Andersson, M., & Simmons, L. W. (2006). Sexual selection and mate choice. *Trends in Ecology & Evolution, 21*, 296–302.
2. Arnheiter, H. (1998). Evolutionary biology: Eyes viewed from the skin. *Nature, 391*, 632–633.
3. Bagnara, J. T. (1984). The amphibian egg as a pigment cell. *Yale Journal of Biology and Medicine, 57*, 335.
4. Bagnara, J. T., Matsumoto, J., Ferris, W., et al. (1979). Common origin of pigment cells. *Science, 203*, 410–415.
5. Bagnara, J. T., Taylor, J. D., & Hadley, M. E. (1968). The dermal chromatophore unit. *Journal of Cell Biology, 38*, 67–79.
6. Baldwin, J. M. (1902). *Development and evolution*. London: Macmillan.

7. Bäumer, N., Marquardt, T., Stoykova, A., et al. (2003). Retinal pigmented epithelium determination requires the redundant activities of Pax2 and Pax6. *Development, 130*, 2903–2915.
8. Benedek, G. B. (1971). Theory of transparency of the eye. *Applied Optics, 10*, 459–473.
9. Bharti, K., Miller, S. S., & Arnheiter, H. (2011). The new paradigm: Retinal pigment epithelium cells generated from embryonic or induced pluripotent stem cells. *Pigment Cell & Melanoma Research, 24*, 21–34.
10. Braasch, I., Schartl, M., & Volff, J.-N. (2007). Evolution of pigment synthesis pathways by gene and genome duplication in fish. *BMC Evolutionary Biology, 7*, 74.
11. Briscoe, A. D., Bybee, S. M., Bernard, G. D., et al. (2010). Positive selection of a duplicated UV-sensitive visual pigment coincides with wing pigment evolution in Heliconius butterflies. *Proceedings of the National Academy of Sciences of the United States of America, 107*, 3628–3633.
12. Carroll, S. B., Gates, J., Keys, D. N., et al. (1994). Pattern formation and eyespot determination in butterfly wings. *Science, 265*, 109–115.
13. Corning, P. (2010). *Holistic Darwinism: Synergy, cybernetics, and the bioeconomics of evolution*. University of Chicago Press
14. Corning, P. A. (2014). Evolution on purpose: How behaviour has shaped the evolutionary process. *Biological Journal of the Linnean Society, 112*, 242–260.
15. Dinwiddie, A., & Rachootin, S. (2010). Patterning of a compound eye on an extinct dipteran wing. *Biology Letters*. https://doi.org/10.1098/rsbl.2010.0809.
16. Doucet, S. M., & Meadows, M. G. (2009). Iridescence: A functional perspective. *Journal of the Royal Society Interface, 6*, S115–S132.
17. Ducrest, A.-L., Keller, L., & Roulin, A. (2008). Pleiotropy in the melanocortin system, coloration and behavioural syndromes. *Trends in Ecology & Evolution, 23*, 502–510.
18. Faivre, B., Grgoire, A., Prault, M., et al. (2003). Immune activation rapidly mirrored in a secondary sexual trait. *Science, 300*, 103.
19. Fisher, R. A. (1915). The evolution of sexual preference. *The Eugenics Review, 7*, 184.
20. Foth, C., Tischlinger, H., & Rauhut, O. W. M. (2014). New specimen of Archaeopteryx provides insights into the evolution of pennaceous feathers. *Nature, 511*, 79–82.
21. Gilbert, S. F., & Bard, J. (2014). Formalizing theories of development: A fugue on the orderliness of change. In A. Minelli & T. Pradeu (Eds.), *Toward a theory of development* (pp. 129–143). Oxford: Oxford University Press.
22. Goda, M., Ohata, M., Ikoma, H., et al. (2011). Integumental reddish-violet coloration owing to novel dichromatic chromatophores in the teleost fish, Pseudochromis diadema. *Pigment Cell & Melanoma Research, 24*, 614–617.
23. Gomez, D., Richardson, C., Thry, M., et al. (2011a). Multimodal signals in male European treefrog (Hyla arborea) and the influence of population isolation on signal expression. *Biological Journal of the Linnean Society, 103*, 633–647.
24. Gomez, D., Thry, M., Gauthier, A.-L., & Lengagne, T. (2011b). Costly help of audiovisual bimodality for female mate choice in a nocturnal anuran (Hyla arborea). *Behavioral Ecology, 22*, 889–898.
25. Gould, S. J., & Vrba, E. S. (1982). Exaptation a missing term in the science of form. *Paleobiology, 8*, 4–15.
26. Grether, G. F., Hudon, J., & Endler, J. A. (2001). Carotenoid scarcity, synthetic pteridine pigments and the evolution of sexual coloration in guppies (Poecilia reticulata). *Proceedings of the Royal Society B: Biological Sciences, 268*, 1245–1253.
27. Grether, G. F., Kolluru, G. R., & Nersissian, K. (2004). Individual colour patches as multicomponent signals. *Biological Reviews, 79*, 583–610.
28. Haisten, D. C., Paranjpe, D., Loveridge, S., & Sinervo, B. (2015). The cellular basis of polymorphic coloration in common side-blotched lizards, Uta stansburiana. *Herpetologica, 71*, 125–135.
29. Hall, B. K. (2008). *The neural crest and neural crest cells in vertebrate development and evolution*. New York: Springer.

30. Higdon, C. W., Mitra, R. D., & Johnson, S. L. (2013). Gene expression analysis of zebrafish melanocytes, iridophores, and retinal pigmented epithelium reveals indicators of biological function and developmental origin. *PLoS One*, *8*, e67801.
31. Inaba, M., Yamanaka, H., & Kondo, S. (2012). Pigment pattern formation by contact-dependent depolarization. *Science*, *335*, 677.
32. Inouye, C. Y., Hill, G. E., Stradi, R. D., & Montgomerie, R. (2001). Carotenoid pigments in male house finch plumage in relation to age, subspecies, and ornamental coloration. *Auk*, *118*, 900–915.
33. Jaeger, J., Surkova, S., Blagov, M., et al. (2004). Dynamic control of positional information in the early Drosophila embryo. *Nature*, *430*, 368–371.
34. Kauffman, S. A., Shymko, R. M., & Trabert, K. (1978). Control of sequential compartment formation in Drosophila. *Science*, *199*, 259–270.
35. Kayser, H. (1985). Pigments. *Comprehensive Insect Physiology*, *10*, 367–415.
36. Kekäläinen, J., Valkama, H., Huuskonen, H., & Taskinen, J. (2010). Multiple sexual ornamentation signals male quality and predicts female preference in minnows. *Ethology*, *116*, 895–903.
37. Kelsh, R. N., Harris, M. L., Colanesi, S., & Erickson, C. A. (2009). Stripes and belly-spots-a review of pigment cell morphogenesis in vertebrates. *Seminars in Cell & Developmental Biology*, *20*, 90–104.
38. Kirkpatrick, M. (1982). Sexual selection and the evolution of female choice. *Evolution*, *36*, 1–12.
39. Kirschfeld, K. (1982). Carotenoid pigments: their possible role in protecting against photooxidation in eyes and photoreceptor cells. *Proceedings of the Royal Society B: Biological Sciences*, *216*, 71–85.
40. Kitano, H. (2002). Systems biology: A brief overview. *Science*, *295*, 1662–1664.
41. Kleisner, K. (2015). Semantic organs: The concept and its theoretical ramifications. *Biosemiotics*, *8*, 367–379.
42. Kleisner, K. (2011). Perceive, co-opt, modify, and live! Organism as a centre of experience. *Biosemiotics*, *4*, 223–241.
43. Kleisner, K., & Tureček, P. (2017). Cultural and biological evolution: What is the difference? *Biosemiotics*, *10*, 127–130.
44. Kondo, S., & Miura, T. (2010). Reaction-diffusion model as a framework for understanding biological pattern formation. *Science*, *329*, 1616–1620.
45. Kozmik, Z., Ruzickova, J., Jonasova, K., et al. (2008). Assembly of the cnidarian camera-type eye from vertebrate-like components. *Proceedings of the National Academy of Sciences of the United States of America*, *105*, 8989–8993.
46. Krinsky, N. I. (2001). Carotenoids as antioxidants. *Nutrition*, *17*, 815–817.
47. Kuriyama, T., Miyaji, K., Sugimoto, M., & Hasegawa, M. (2006). Ultrastructure of the dermal chromatophores in a lizard (Scincidae: Plestiodon latiscutatus) with conspicuous body and tail coloration. *Zoological Science*, *23*, 793–799.
48. Land, M. F. (1972). The physics and biology of animal reflectors. *Progress in Biophysics & Molecular Biology*, *24*, 75–106.
49. Lande, R. (1981). Models of speciation by sexual selection on polygenic traits. *Proceedings of the National Academy of Sciences of the United States of America*, *78*, 3721–3725.
50. Large, M. C. J., Wickham, S., Hayes, J., & Poladian, L. (2007). Insights from nature: Optical biomimetics. *Physica B: Condensed Matter*, *394*, 229–232.
51. Manukyan, L., Montandon, S. A., Fofonjka, A., et al. (2017). A living mesoscopic cellular automaton made of skin scales. *Nature*, *544*, 173–179.
52. Maran, T. (2009). John Maynard Smith's typology of animal signals: A view from semiotics. *Sign Systems Studies*, *37*, 477–497.
53. Maran, T., & Kleisner, K. (2010). Towards an evolutionary biosemiotics: semiotic selection and semiotic co-option. *Biosemiotics*, *3*, 189–200.
54. Martin, A., McCulloch, K. J., Patel, N. H., et al. (2014). Multiple recent co-options of optix associated with novel traits in adaptive butterfly wing radiations. *Evodevo*, *5*, 7.

55. McGraw, K. J. (2005). The antioxidant function of many animal pigments: Are there consistent health benefits of sexually selected colourants? *Animal Behaviour, 69*, 757–764.
56. McGraw, K. J., & Ardia, D. R. (2003). Carotenoids, immunocompetence, and the information content of sexual colors: An experimental test. *The American Naturalist, 162*, 704–712.
57. Nakamasu, A., Takahashi, G., Kanbe, A., & Kondo, S. (2009). Interactions between zebrafish pigment cells responsible for the generation of Turing patterns. *Proceedings of the National Academy of Sciences of the United States of America, 106*, 842–843.
58. Nielsen, H. I. (1978). Ultrastructural changes in the dermal chromatophore unit of Hyla arborea during color change. *Cell and Tissue Research, 194*, 405–418.
59. Nijhout, H. F. (1991). *The development and evolution of butterfly wing patterns*. Washington, DC: Smithsonian Institution Press.
60. Nilsson, D.-E. (2009). The evolution of eyes and visually guided behaviour. *Philosophical Transactions of the Royal Society B: Biological Sciences, 364*, 2833–2847.
61. Noble, D. (2006). *The music of life*. New York: Oxford University Press.
62. Nüsslein-Volhard, C., & Singh, A. P. (2017). How fish color their skin: A paradigm for development and evolution of adult patterns. *Bioessays*. https://doi.org/10.1002/bies.201600231.
63. Oliphant, L. W., & Hudon, J. (1993). Pterins as reflecting pigments and components of reflecting organelles in vertebrates. *Pigment Cell & Melanoma Research, 6*, 205–208.
64. Planque, N., Raposo, G., Leconte, L., et al. (2004). Microphthalmia transcription factor induces both retinal pigmented epithelium and neural crest melanocytes from neuroretina cells. *Journal of Biological Chemistry, 279*, 41911–41917.
65. Provencio, I., Jiang, G., Willem, J., et al. (1998). Melanopsin: An opsin in melanophores, brain, and eye. *Proceedings of the National Academy of Sciences of the United States of America, 95*, 340–345.
66. Prum, R. O. (2010). The Lande Kirkpatrick mechanism is the null model of evolution by intersexual selection: Implications for meaning, honesty, and design in intersexual signals. *Evolution, 64*, 3085–3100.
67. Prum, R. O. (2012). Aesthetic evolution by mate choice: Darwin's really dangerous idea. *Philosophical Transactions of the Royal Society B: Biological Sciences, 367*, 2253–2265.
68. Prum, R. O. (2017). *The evolution of beauty: How Darwin's forgotten theory of mate choice shapes the animal world-and us*. New York: Doubleday.
69. Prum, R. O., & Brush, A. H. (2002). The evolutionary origin and diversification of feathers. *The Quarterly Review of Biology, 77*, 261–295.
70. Prum, R. O., Quinn, T., & Torres, R. H. (2006). Anatomically diverse butterfly scales all produce structural colours by coherent scattering. *Journal of Experimental Biology, 209*, 748–765.
71. Prum, R. O., & Torres, R. (2003). Structural colouration of avian skin: Convergent evolution of coherently scattering dermal collagen arrays. *Journal of Experimental Biology, 206*, 2409–2429.
72. Prum, R. O., & Torres, R. H. (2004). Structural colouration of mammalian skin: Convergent evolution of coherently scattering dermal collagen arrays. *Journal of Experimental Biology, 207*, 2157–2172.
73. Prum, R. O., & Williamson, S. (2002). Reaction diffusion models of within-feather pigmentation patterning. *Proceedings of the Royal Society B: Biological Sciences, 269*, 781–792.
74. Quevedo, W. C, Jr. (1972). Epidermal melanin units melanocyte-keratinocyte interactions. *American Zoologist, 12*, 35–41.
75. Reed, R. D., Papa, R., Martin, A., et al. (2011). Optix drives the repeated convergent evolution of butterfly wing pattern mimicry. *Science, 333*, 1137–1141.
76. Rodd, F. H., Hughes, K. A., Grether, G. F., & Baril, C. T. (2002). A possible non-sexual origin of mate preference: Are male guppies mimicking fruit? *Proceedings of the Royal Society B: Biological Sciences, 269*, 475–481.
77. Rowe, C. (1999). Receiver psychology and the evolution of multicomponent signals. *Animal Behaviour, 58*, 921–931.

78. Rutowski, R. L., Macedonia, J. M., Morehouse, N., & Taylor-Taft, L. (2005). Pterin pigments amplify iridescent ultraviolet signal in males of the orange sulphur butterfly, Colias eurytheme. *Proceedings of the Royal Society B: Biological Sciences*, *272*, 2329–2335.
79. Saks, L., McGRAW, K., & Hrak, P. (2003). How feather colour reflects its carotenoid content. *Functional Ecology*, *17*, 555–561.
80. Seago, A. E., Brady, P., Vigneron, J.-P., & Schultz, T. D. (2009). Gold bugs and beyond: A review of iridescence and structural colour mechanisms in beetles (Coleoptera). *Journal of the Royal Society Interface*, *6*, S165–S184.
81. Seehausen, O., Terai, Y., Magalhaes, I. S., et al. (2008). Speciation through sensory drive in cichlid fish. *Nature*, *455*, 620–626.
82. Shawkey, M. D., & D'Alba, L. (2017). Interactions between colour-producing mechanisms and their effects on the integumentary colour palette. *Philosophical Transactions of the Royal Society B*, *372*, 20160536.
83. Shubin, N., Tabin, C., & Carroll, S. (2009). Deep homology and the origins of evolutionary novelty. *Nature*, *457*, 818–823.
84. Spinner, M., Kovalev, A., Gorb, S. N., & Westhoff, G. (2013). Snake velvet black: Hierarchical micro-and nanostructure enhances dark colouration in Bitis rhinoceros. *Scientific Reports*, *3*, 1846.
85. Steffen, J. E., & McGraw, K. J. (2009). How dewlap color reflects its carotenoid and pterin content in male and female brown anoles (Norops sagrei). *Comparative Biochemistry and Physiology—Part B: Biochemistry & Molecular Biology*, *154*, 334–340.
86. Stewart, E., Ajao, M. S., & Ihunwo, A. O. (2013). Histology and ultrastructure of transitional changes in skin morphology in the juvenile and adult four-striped mouse (Rhabdomys pumilio). *The Scientific World Journal*. https://doi.org/10.1155/2013/259680.
87. Strong, R. M. (1902). The development of color in the definitive feather. *Science*, *15*, 527.
88. Taylor, J. D., & Hadley, M. E. (1970). Chromatophores and color change in the lizard, Anolis carolinensis. *Cell and Tissue Research*, *104*, 282–294.
89. Teyssier, J., Saenko, S. V., Van Der Marel, D., & Milinkovitch, M. C. (2015). Photonic crystals cause active colour change in chameleons. *Nature Communications*, *6*, 6368.
90. Turing, A. M. (1990). The chemical basis of morphogenesis. *Bulletin of Mathematical Biology*, *52*, 153–197.
91. Twyman, H., Valenzuela, N., Literman, R., et al. (2016). Seeing red to being red: Conserved genetic mechanism for red cone oil droplets and co-option for red coloration in birds and turtles. *Proceedings of the Royal Society B. The Royal Society*, *283*, 20161208
92. Vachtenheim, J., & Borovanský, J. (2010). Transcription physiology of pigment formation in melanocytes: Central role of MITF. *Experimental Dermatology*, *19*, 617–627.
93. von Lintig, J., Kiser, P. D., Golczak, M., & Palczewski, K. (2010). The biochemical and structural basis for trans-to-cis isomerization of retinoids in the chemistry of vision. *Trends in Biochemical Sciences*, *35*, 400–410.
94. Vopalensky, P., & Kozmik, Z. (2009). Eye evolution: Common use and independent recruitment of genetic components. *Philosophical Transactions of the Royal Society B: Biological Sciences*, *364*, 2819–2832.
95. Wittkopp, P. J., & Beldade, P. (2009). Development and evolution of insect pigmentation: Genetic mechanisms and the potential consequences of pleiotropy. *Seminars in Cell & Developmental Biology*, *20*, 65–71.
96. Yasutomi, M. (1987). Migration of epidermal melanophores to the dermis through the basement membrane during metamorphosis in the frog, Rana japonica. *Pigment Cell & Melanoma Research*, *1*, 181–187.
97. Zahavi, A. (1975). Mate selection—A selection for a handicap. *Journal of Theoretical Biology*, *53*, 205–214.

The Extended Domicile—Culture, Embodied Existence and the Senses

Juhani Pallasmaa

Abstract We extend our physical selfs, perceptual and cognitive realities as well as memories and imagination through countless technical inventions and conceptual systems. In his book *The Extended Phenotype*, the biologist Richard Dawkins, suggests that in the biological world such extensions are so important that, for instance, the dams and water regulation systems of the beaver should be included in the biological definition of the species of the beaver. Similarly, our countless constructions, structures, technical systems as well as intellectual discoveries, ought to be included in the definition of *Homo Sapiens*, but we still continue to see ourselves limited by our skin. Altogether, we tend to think of our environments in terms of isolated, definable objects and entities, rather than dynamic and constantly interactive and expanding systems. Architecture is likewise seen as material aestheticized structures that are external to us, rather than as part of our biological and mental constitution. However, our environments from intimate objects to rooms, buildings, cities, regions and all the way to the entire world and the universe, can also be regarded as part of our material, perceptual, and conceptual reality. Instead of being seen as material objects and buildings, architecture should be regarded as an active entity which very concretely mediates our relationships with the world through space and time. Human history, culture, and collective consciousness widen our world of thought and action beyond material boundaries. Through our structures, we, humans, turn limitless, shapeless and meaningless space into lived space with human meanings. We also regard architecture as an aesthetic expression of its architect, but Maurice Merleau-Ponty argues thought provokingly: "We come to see not the work of art, but the world according to the work". Architecture has a crucial role in the constitution of the human world, both material and mental.

J. Pallasmaa (✉)
Helsinki, Finland
e-mail: jpallasmaa@gmail.com

M. I. Aldinhas Ferreira et al. (eds.), *Cognitive Architectures*, Intelligent Systems, Control and Automation: Science and Engineering 94,
https://doi.org/10.1007/978-3-319-97550-4_3

The human sensory and neural system, as well as the brain, is the result of evolutionary adaptation to the prevailing environments and conditions of life during the continuum of human evolutionary history. The nature of our senses and neural functions, as well as instinctive environmental preferences, needs to be viewed in a bio-cultural and bio-historical perspective, instead of regarding them as ahistorical, unchanging or simply given properties of the *Homo Sapiens*. We are undeniably historical beings, but the time perspectives in our biological constitution, behaviour and mental lives are most often neglected in today's objectified and aestheticized design thinking, as design tends to be interested only in the dimensions of now-ness and novelty. We dwell in the continua of space and time, but we are not usually conscious of the fact that we continue to be subject to evolutionary forces and changes in the future as well. Although human adaptation to the conditions of life has primarily taken place through technological inventions, we undoubtedly also keep evolving biologically. With artificial intelligence and stem cell manipulation, we are already in dangerously confusing territory in regard to the categories of what is biological and what is man-made.

1 Adaptation Through Technology

Even our own inventions, structures and acquired habits eventually cause biological changes. The taming of fire, for instance, estimated to have taken place roughly 50,000 years ago, caused changes in human tooth structure and intestinal functions as a consequence of eating cooked food. "Control over fire changed human anatomy and physiology and became encoded in our evolving genome", Stephen Pyne argues [1]. The first architectural writer in history, Marcus Vitruvius Pollio (80–70 BC-15 BC), even connects the origins of architecture with the domestication of fire [2]. Like Vitruvius, some linguistic scholars of our time have suggested that gathering around a fire for extended periods also accelerated the development of language. Fire has been so central during the course of human cultural evolution that even in our technologized and globalized culture, it continues to convey deep feelings of domesticity and pleasure, and flames are still a strong stimulus for dreaming and imagination. Gaston Bachelard, the philosopher of poetic imagery, wrote two books on the poetic impact of fire on the human imagination [3].

We tend to think that our technical inventions are all beneficial and "innocent", but the man-made and technologized world can cause major changes in our behaviour and habits, as well as in our mental lives. Walter J. Ong argues convincingly that writing, and especially mechanical printing, initiated the shift from aural space to visual space, and that this shift to the hegemony of vision was not entirely positive. "Print replaced the lingering hearing-dominance in the world of thought and expressions with the sight-dominance, which had its beginning in writing", Ong argues [4]. In his view, "[T]his is an insistent world of cold, non-human facts" [5]. The fundamental change in the perception and understanding of the world seems irreversible to the writer: "Though words are grounded in oral speech, writing tyrannically locks them into a visual field forever […] a literate person cannot fully recover a sense of what the word is to purely oral people" [6].

No doubt, similar sensory and mental changes initiated by ever-evolving technologies continue today. Current studies in Finland have shown that children are becoming incapable of identifying the facial gestures and emotions of others due to their extensive communication through mobile phones. The current shift in architectural design from manual sketching, drawing and model to the insistent use of computers and 3D modelling must be having similar negative consequences on our embodied and spatial modes of thinking and imagining. Thinking has always had its bodily and emotive components. We are engaged in creative work as complete embodied and sensory beings, not just through vision and intellect. In fact, an unconscious, non-logical, associative, emotive, and intuitive synthetic mode of thinking is the very essence of our creative capacity.

2 Biology and Aesthetics

In his book *Inner Vision: An Exploration of Art and the Brain*, neurobiologist Semir Zeki outlines "a theory of aesthetics that is biologically based" [7]. "My primary aim is to convince the reader that we are at the threshold of a great enterprise, of learning something about the neurobiological basis of one of the most noble and profound of human endeavours [arts]", he adds [8]. Zeki's assumption and goal seem entirely plausible to me. In fact, it would be questionable to assume that our aesthetic sensibilities and preferences would have developed independently of our biological evolution, or that our aesthetic preferences would conflict with the evolutionary principles of survival. Isn't the deep resonance between our natural settings and aesthetic sensibilities the reason why we experience nature and its evolving phenomena as pleasurable and beautiful? My assumption suggests that we experience beauty primarily unconsciously as nature's expression of its inner causalities. This is what Josef Brodsky, the nobel Laureate poet, seems to suggest in his credo, "The purpose of evolution, believe it or not, is beauty" [9]. This poetic formulation will probably not be approved by today's theorists of evolution, but can well be valorized by evolutionary and biological argumentation in the future. In today's world of forceful aesthetic conditioning, personality and politics, as well as architecture, have turned into deliberate aesthetic manipulations, and as a result of our current aesthetic culture, aesthetic choices seem distant from their original biological motives, losing their spontaneity. It is surely a mistake to think of human evolution only in cultural terms, distanced and separated from the underlying processes of biological meaning. It is equally misguided to neglect the biological ground of human behaviour and instinctual choices; this is the lesson of ecological psychology. Based on the ecological psychology of Jay Appleton, and especially his prospect-refuge theory [10], Grant Hildebrand has analyzed the psychological effects of Frank Lloyd Wright's houses, and concludes that the architect grasped intuitively the fundamental psychological meaning of this basic polarity, which still applies in today's spatial design [11]. In another book of his, entitled *Origins of Architectural Pleasure*, Hildebrand has a suggestive chapter title, "The Aesthetics of Survival", which boldly connects the cultural and biological dimensions of environmental qualities [12]. Aesthetic sensibilities seem ultimately to serve purposes of survival and evolution, but they may just as well be distorted by arbitrary cultural values, such as the fashion of forcefully bound legs of ladies in China between the 10th and 19th centuries, or todays' fashionable but esthetically arbitrary architecture.

The biologist Edward O. Wilson is the spokesman for *biophilia*, "the new ethics and science of life", whose passionate defense of life and lifelike processes is seminal today, when humankind is running out of time to establish the future conditions for human life through absolutely necessary cultural adjustments. "All of man's troubles may well arise [...] from the fact that we do not know who we are, and do not agree on what we want to become", he writes [13].

Now that biological precedents and models are increasingly being used in advanced technologies, our own biological essence and historicity must surely also be acknowledged, including in relation to architecture and planning. Our biological historicity is evidenced by relics such as the *plica semilunaris*, the pink triangles in our eye corners to which our horizontally moving extra eye lids were fixed during our lizard phase in the Saurian age. Human culture has developed towards increasing artificiality, but we need to recognize the biological reality and its refined processes of adaptation, change and becoming.

3 Interacting with the World

Like all forms of life, we are related to our living world through the senses and neural systems. Life is an evolving system of interaction with its contexts and environments. With the advance of scientific research, it is becoming clear that our interactions with the world are far more complex than we have so far assumed. We do not just dwell in the world, as we are also part of it in a complex manner. We are part of the "flesh of the world", to use the suggestive notion of Maurice Merleau-Ponty [14]. As Semir Zeki remarks, quoting Henri Matisse: "We see in order to be able to acquire knowledge about the world [...] Other senses do the exact same thing" [15].

Since Aristotle, we have believed that we have five senses, but Steinerian philosophy names twelve human senses [16], and a recent study suggests that we are connected with the world through no less than thirty-three systems of monitoring and interaction [17]. The fixation with the five senses has evidently been supported by the simple fact that we have a specific, identifiable and visible organ for each one of these five modes of sensing, whereas the sensing of environmental atmospheres and of our own existence, for instance, are multi-sensory, unfocused and shapeless, and they lack "thingness". Tonino Griffero calls such complex and diffuse phenomena "quasi-things" [18]. As architecture, especially that of modernity, has primarily been interested in form, such "formless" phenomena as atmospheres, feelings, empathy and emotions have been largely neglected. Also, the existential sense is central to our relationship with the environment and architecture, but due to its complex and synthetic nature, it cannot be associated with or located in any specific sensory organ. The sense of self, or the existential sense, is our coordinating and synthesizing sense, not vision, as we usually think. The thirty-year-old discovery of "the mirror neurons" by a research group at the University of Parma is another significant biologically-determined capacity of "learning" and "understanding" through unconscious imitation and simulation, which has already proved of seminal importance for the understanding of how we internalize external phenomena and stimuli, such as works of art.

Current research on the significance and complex functions of the bacterial world in our intestines dramatically complicates our interaction with the environment. The recent understanding of the role and complexity of our intestinal bacterial universe, "our second brain" [19], serves as an example of the fundamental expansions that are currently taking place in the understanding of our interactions with the world. We have only recently learned that each one of us carries more than one and half kilos of bacteria in our intestines, and we actually have more bacterial DNA in our bodies than human DNA.

4 The Extended Man

Our sensory systems, not to mention the imaginative projections of the mind, such as concepts and metaphors, enable us to "sense" the entire universe."Through vision, we touch the sun and the stars", Maurice Merleau-Ponty exclaims poetically [20]. Besides, we extend our physical, perceptual and cognitive capacities, as well as memory and imagination, through an ever-increasing number of technical inventions and conceptualizing systems, such as the dramatic expansion of human memory through the Google and the computerized "cloud".

In his book *The Extended Phenotype* [21], the controversial biologist Richard Dawkins suggests that the acquired extensions of the body functions are so important in the biological world that, for instance, the dams and water regulation systems of the beaver should be included in the biological definition of the beaver species. Altogether, the refinements of the ways by which even lower animals adjust their relationships with their surroundings are often almost beyond imagination, but these amazing capabilities have hardly been studied seriously [22]. Again, the deep evolutionary time helps in understanding the development of the superb skills of animals. For instance, spiders have been practicing their methods of web construction for over 300 million years, in comparison with the roughly 50,000 years of human construction.

Similarly, our own countless constructions, structures, and technical systems, as well as intellectual discoveries, should be included in the concept and definition of *Homo Sapiens*, but we still continue to see ourselves limited by the surface of our skin. In their series of pioneering research publications of 1963–67 entitled *The World Resources Inventory*, Richard Buckminster Fuller and John McHale introduced the idea of both the human individual and the collective humankind, as seen through their huge external material, technical and conceptual extensions [23].

Even biologically, the sphere of the human body is not limited by the skin. We sense our personal space as an extension of our body and feel it being violated as if it were part of our physical body. In the 1960s, the American anthropologist Edward T. Hall introduced the discipline of *proxemics,* the study of the human unconscious and culture-specific use of personal and collective space as behavioural extensions of the body [24]. The designer of spaces needs to understand these unconscious extensions and invisible behavioural mechanisms, not just the anatomy and physical dimensions of the human body.

But even our actual metabolic functions exceed the body's limits. Hall mentions the research of A. S. Parkes and H. M. Bruce from the 1960s into the functions of our ductless endocrine glands, which showed that although these glands—in accordance with their very name—have been assumed to function strictly within the body, they also function and interact externally through chemical communications [25]. The researchers even suggested renaming their research area as "exocrinology" to express the unexpected external communicative functions of the internal glands.

More recently, research has established that with today's instruments of measurement, the electrical impulses of our heart can be monitored at a five-meter distance. These examples should make it clear that our range of metabolic interactions extend into space beyond our skin. So, where are the boundaries of our functional and experiential selves? How do we frame and define the human being for whom we design?

5 The Unity of Space and Self

We think of ourselves as creatures limited by our skin and of our environments as a set of isolated, definable objects and entities outside of ourselves, rather than as integrated, dynamic, constantly interactive and interweaving systems. Besides, we still continue to make a categorical separation between outer and inner, material and mental realities, although science has revealed the multiplicity of interactions between these assumed oppositions, and phenomenological thinking in philosophy has questioned and abandoned such exclusive categorical distinctions. It is a fundamental phenomenological assumption that the inner and outer spaces, as well as the material and the mental, constitute a continuum. The American literary scholar Robert Pogue Harrison gives this mirroring a poetical expression: "In the fusion of place and soul, the soul is as much a container of the place as place is container of soul, and both are susceptible to the same forces of destruction" [26]. Merleau-Ponty gives this reciprocity and simultaneity an even more cryptic formulation: "The world is wholly inside and I am wholly outside myself" [27].

Yet another surprising interaction between the world and the human mind has recently been suggested by the Californian philosopher Alva Noë. In his provocative book *Out of Our Heads*: *Why You Are Not Your Brains, and Other Lessons from the*

Biology of Consciousness [28], he argues that the reason why research has failed to locate human consciousness in the brain is that the location of consciousness has been sought in the wrong place. In the philosopher's view, consciousness is a relation between the mind and the world, and as a relational phenomenon, it cannot be placed, because a relation has no distinct physical location. At the same time, this view also suggests a complete continuum between the inner and the outer, the mental and the material. We have come to believe that our consciousness is the most human of our capacities, but it may well be "out there" instead of being inside our brains. Atmospheres, which are proving to be significant aspects of architectural and environmental quality, are similarly in-between and relational phenomena. It is the relational essence of atmospheres that has made them difficult to identify and grasp theoretically or intellectually, although we spontaneously feel them and they unavoidably impact our feelings and behavior [29].

6 Architecture—Object or Experience?

Architecture is also normally seen as aestheticized material structures that are external to us, rather than as part of our biological and mental constitution. It is regarded as physical and material spaces, structures and objects, instead of experiences or mental and emotional encounters. However, environments from the most intimate objects to rooms, buildings, cities, regions, and all the way up to the entire world can be regarded as part of our perceptual, mental and conceptual reality, and instead of being seen merely as material contents and entities, architecture can be regarded as active verbs, which concretely mediate and alter our relationships with the world, space and time. In addition to organizing and channeling life and actions, architecture determines our relationships with the world and gives our experiences of it specific meanings. John Dewey argued provocatively that "mind is a verb" [30], and the essence of architecture can also be seen as a verb. The verb connotation of architecture becomes concrete when we realize that it is always a kind of pre-scripted choreography for human movement, action, attention and emotion. Architecture organizes our material world, but it also provides horizons and frames for perception and understanding. The world is experienced through and in relation to human structures, material and conceptual, current and historical. The built structures of our experiential world pre-organize and pre-interpret the world for our perception and understanding. It is entirely feasible to think that a house pre-senses and pre-experiences the landscape around it, natural or man-made, on behalf of the future resident. Besides, architecture is also always an invitation to distinct acts and activities and a promise of predictability, order and safety.

When all of the extensions of our mobility, climatic adaptation, sensory reach and memory, as well as cognition and imagination, are seen as essential characteristics of our bio-cultural selves, architecture also turns into a dense field of interactions in space, time and meaning. Human history, culture, and collective consciousness further widen the world of thoughts and actions beyond material boundaries. Through our human structures, both physical and mental, we turn limitless, shapeless and meaningless "natural" space into lived cultural space with specific human purposes and meanings. Instead of living in a natural world, we live in a man-made world structured by our countless constructions, devices and inventions, as well as conceptualizations and ideas.

We have also primarily regarded architecture as an aesthetic expression of its individual architect, but Maurice Merleau-Ponty argues, thought-provokingly: "We come to see not the work of art, but the world according to the work" [31]. The philosopher's statement on the real contents of art certainly applies in architecture. Instead of being merely individual and artistic expressions, buildings are essentially about the world and being human in that world. Architecture acquires its content and meaning through its resonance with universally human qualities, not from explicitly individual expressions. It has a crucial role in the constitution of the human world, both material and mental, as well as in the establishment of our very humanity.

7 Embodied Experience

Since its invention in Renaissance times, the perspectival understanding of space has emphasized and strengthened the retinal and focused architecture of vision. Through its geometric construction, focused perspectival space turns us into outsiders and observers, as it pushes us outside of the realm of the object of focused perception, whereas simultaneous, haptically and peripherally perceived spaces enclose and enfold us in their embrace, making us insiders and participants. In the retinal understanding of space, we observe it, whereas acoustic, haptic and olfactory spaces, as well as percepts of peripheral and unfocused vision, constitute our lived and shared existential condition. We are embraced by space, rather than looking at it. This mode of sensing is also the grounding for atmospheric experience and attunement, both being notions that have been neglected in modern architectural theory. Contrastingly, theoretical studies on architectural spaces have frequently described them as negative or absent volumes and forms. Yet, the world and the perceiver are not separated and polarized, as they are both ingredients in the shared existential flesh, "the flesh of the world", to use Merleau-Ponty's notion.

The quest to liberate the eye from its perspectival fixation has gradually brought about conceptions of a multi-perspectival, simultaneous and haptic space. The dynamic life and depth in our perception arise from the fact that they are essentially an ever-changing dynamic collage of separate multi-perspectival glances that constitute a haptic continuum, our true embodied experience of space. This is the perceptual and psychological essence of Impressionist, Cubist and Abstract Expressionist painterly spaces, which pull us into the painting and cause us to experience it as insiders in a fully embodied plastic sensation. Visual space thus turns into an embodied plastic and existential space, which is essentially a dialogue and exchange between the space of the world and the internal space of the perceiver's mental world. The experience of interiority and belonging is a merging of the outside and inside worlds, the evocation of Rainer Maria Rilke's beautiful notion of *Weltinnenraum* [32]. This is a unique and personal existential space that we occupy in our continuous lived experience. In the recognition of place, particularly that of one's domicile and home, the external world and space become internalized, and they are sensed as intra-personal conditions, rather than external material objects, scenes or percepts. Our domicile is the *Omega* point of Pierre Teilhard de Chardin "from which the world can be seen as a whole and correctly" [33]. Our domicile grants us the experience of complete interiority, which implies the fusion of the world and the self.

The heightened presence and reality of profound artworks derive from the way they engage our perceptual and psychological mechanisms and articulate the boundary between the viewer's experience of self and the world. Such an experience also reveals and re-activates our deep biological and forgotten existential memories. The experience of domicile gives both space and place their historical and temporal dimensions. Works of art have two simultaneous existences: their existence as material objects or performance (in music, theatre and dance) on the one hand, and as imaginative worlds of imagery, emotion and ideal on the other. The experiential reality of art is always an imaginative reality, a fusion of perception, memory and imagination, and it is essentially a recreation by the viewer/listener/reader/occupant. This is the message of John Dewey's seminal book *Art as Experience* of 1934: "In common conception, the work of art is often identified with the building, book, painting, or statue in its existence apart from human experience. Since the actual work of art is what the product does with and in experience, the result is not favorable to understanding […] When artistic objects are separated from both conditions of origin and operation in experience, a wall is built around them that renders almost opaque their general significance, with which esthetic theory deals" [34].

Lived reality always fuses observation, memory and fantasy, as well as the cerebral and the embodied, into fused existential experiences. As the consequence of this categorical "impurity" of experience, it is beyond precise objective and scientific description, and approachable only through its live encounter and the resulting poetic evocation. This is the innate structural vagueness of human consciousness. Gaston Bachelard was an authoritative philosopher of science until his mid-career, when he came to the dramatic conclusion that only a poetic approach, not scientific inquiry and methodology, can touch upon the essence of lived human reality. Science deals with conceptualizations and fragmentations of reality, whereas the artist touches upon and conveys the lived reality that reflects true human meanings and values.

Instead of confining us in an alienating, constructed or fabricated artificiality, moving works of architecture connect us with the complexities and mysteries of perception and the real world. In meaningful architectural works, the imaginary world is rooted in the techtonic reality, materiality and processes of construction. Authoritative architecture also articulates and expresses its processes of construction and use at the same time that it expresses how it feels to be human in this world. In Merleau-Ponty's view, "Cézanne's paintings make us feel how the world touches us" [35]. Profound architecture similarly makes us feel the way in which the world touches us or how we are contained in it or are part of its flesh. True architecture articulates the functional, behavioural and technical realities of building and its use, but it also maintains its autonomy as an artistic and confessional statement. In today's utilitarian and quasi-rational world, this autonomy of architecture is severely threatened. The narrative and logic of construction, as well as its utility, distinguishes architecture from other art forms, such as sculpture and installation art, which also utilize space, as all art forms, including music, do. Without the tension between its simultaneous material reality and its imaginary mental suggestion, its utility and autonomy, reason and emotionality, a piece of architecture remains a crude piece of practical construction and utility. Instead of being the product of a scientific process of thinking, real architecture is always a confession. And a meaningful embodiment of architecture fuses our biological and cultural essences.

> What is most human is not rationalism, but the uncontrolled and uncontrollable continuous surge of creative radical imagination in and through the flux of representation, affects and desires.
>
> Cornelius Castoriadis [36]

References

1. Stephen, J. (2012). *Pyne, fire* (p. 47). London: Reaktion Books.
2. Vitruvius (1960) Capter I: The origin of the dwelling house. In *The ten books on architecture* (M. H. Morgan, Trans.) (p. 38). New York: Dover Publications.
3. Bachelard, G. (1988). *The psychoanalysis of fire (Boston: Beacon Press, 1964), and; the flame of a candle*. Dallas, Texas: The Dallas Institute Publications.
4. Ong, W. J. (1991). *Orality & literacy—The technologizing of the word* (p. 121). London and New York: Routledge.
5. Ibid., 122.
6. Ibid., 12.
7. Zeki, S., & Vision, Inner. (1999). *An exploration of art and the brain* (p. 1). New York: Oxford University Press.
8. Ibid., 2.
9. Brodsky, J. (1997). An immodest proposal. In *On grief and reason* (p. 208). New York: Farrar, Straus and Giroux.
10. Appleton, J. (1975). *The experience of landscape*. London: John Wiley.
11. Hildebrand, G. (1991). *The wright space: Pattern & meaning in frank lloyd wright's houses*. Seattle: University of Washington Press.
12. Hildebrand, G. (1999). *Origins of architectural pleasure* (p. 5). Berkeley, Los Angele, London: University of California Press.
13. Wilson, E. O. (1984). *Biophilia: The human bond with other species* (p. 20). Cambridge, MA: Harvard University Press.
14. Merleau-Ponty, M. (1969). The intertwining—The chiasm. In C. Lefort (Ed.), *The visible and the invisible.* Evanston, IL: Northwestern University Press. "My body is made of the same flesh as the world [...], and moreover [...] this flesh of my body is shared by the world" (248), and: "The flesh [of the world or my own] is [...] a texture that returns to itself and conforms to itself" (146).
15. Semir Zeki, quoting Henri, M., op.cit., 4.
16. Soesmann, A. (1998). *Our twelve senses: Wellsprings of the soul*. Stroud, Glouchestershire, Worchester, UK: Hawthorn Press,.
17. Howes, D. (ed.). (2011). *The sixth sense reader* (pp. 23–24). New York: Berg.

18. Griffero, T. (2017). *Quasi-things: The paradigm of atmospheres*. Albany: State University of New York.
19. "Our Second Brain" is the title of a French television document on the human intestinal bacterial universe and its biological functions. The film was shown in Finnish television in the sumer of 2017.
20. Merleau-Ponty, M. (1993). The introduction, In D. M. Levin (Ed.) *Modernity and the hegemony of vision* (p. 14). Berkeley, Los Angeles, London: University of California Press.
21. Dawkins, R. (1982). *The extended phenotype*. New York: Oxford University Press.
22. For the marvels of animal architecture, see Juhani Pallasmaa, *Animal Architecture* (Helsinki, Museum of Finnish Architecture, 1995). A few years ago I was invited to lecture in a conference in Venice of biologists, mathematicians and computer scientists on "What Can We Learn from Swarming Insects" organized by the Institute of Living Technology. Such an events suggests a growing interest in animal building behaviour and the human application of these frequently nearly unbelievable skills brought about by millions of years of evolution.
23. Buckminster Fuller, R., McHale, J. (1963). *The world resources inventory*. The series of publications published by Southern Illinois University, Carbondale, were Initiated by Fuller in 1963 at the Conference of the International Union of Architects in London.
24. Hall, E. T. (1966, 1982). *The hidden dimension*. New York, London, Toronto, Sydney, Auckland: Anchor Books.
25. Op.cit., 33.
26. Robert Pogue Harrison. (2008). *Gardens: An essay on the human condition* (p. 130). Chicago: The University of Chicago Press.
27. Merleau-Ponty, M. (1962). *The phenomenology of perception* (p. 409). London: Routledge and Kegan Paul.
28. Noë, A. (2009). *Out of our heads: Why you are not your brains, and other lessons from the biology of consciousness*. Hill and Wang: New York, London, Toronto.
29. Juhani, P. (2012). On atmospheres: Peripheral perception and existential experiences. In *Encounters 2: Juhani Pallasmaa—Architectural Essays* (pp. 238–251). Helsinki: Rakennustieto Publishing.
30. As quoted in Sarah Robinson, Dewey, J. (2015). The dialogue between architecture and neuroscience (p. 3). In *ARQ architectural research quarterly*. Cambridge: University Press.
31. Maurice Merleau-Ponty as quoted in McGilchrist. I. (2010). *The master and his emissary: The divided bRain and the making of the western world* (p. 409). New Haven and London: Yale University Press.
32. Rainer Maria Rilke as quoted in Lukijalle [to the reader], *Rainer Maria Rilke, Hiljainen Taiteen Sisin: Kirjeitä vuosilta 1900–1926* [†he silent innermost core of art: letters 1900–1926), Liisa Ehnwald, ed. (Helsinki: TAI-teos, 1997), 8.
33. Pierre Teilhard de Chardin. (2008). *The phenomenon of man*. London: Harper & Row.
34. Dewey, J. (1934). *Art as experience* (p. 4). New York: Putnam's.
35. Merleau-Ponty, M. (1964). Cézanne's doubt. In *Sense and non-sense* (p. 19). Evanston, Ill.: Northwestern University Press.
36. Cornelius Castoriadis as in Modell, A. H. (2006). *Imagination and the meaningful brain*. Cambridge, MA and London, England: The MIT Press.

What We Need from an Embodied Cognitive Architecture

Serge Thill

Abstract Given that original purpose of cognitive architectures was to lead to a unified theory of cognition, this chapter considers the possible contributions that cognitive architectures can make to embodied theories of cognition in particular. This is not a trivial question since the field remains very much divided about what embodied cognition actually means, and we will see some example positions in this chapter. It is then argued that a useful embodied cognitive architecture would be one that can demonstrate (a) what precisely the role of the body in cognition actually is, and (b) whether a body is constitutively needed at all for some (or all) cognitive processes. It is proposed that such questions can be investigated if the cognitive architecture is designed so that consequences of varying the precise embodiment on higher cognitive mechanisms can be explored. This is in contrast with, for example, those cognitive architectures in robotics that are designed for specific bodies first; or architectures in cognitive science that implement embodiment as an add-on to an existing framework (because then, that framework is by definition not constitutively shaped by the embodiment). The chapter concludes that the so-called semantic pointer architecture by Eliasmith and colleagues may be one framework that satisfies our desiderata and may be well-suited for studying theories of embodied cognition further.

S. Thill (✉)
Centre for Robotics and Neural Systems, University of Plymouth,
Plymouth PL4 8AA, UK
e-mail: serge.thill@plymouth.ac.uk

S. Thill
School of Informatics, University of Skövde, Skövde 54128, Sweden

M. I. Aldinhas Ferreira et al. (eds.), *Cognitive Architectures*, Intelligent Systems, Control and Automation: Science and Engineering 94,
https://doi.org/10.1007/978-3-319-97550-4_4

1 Introduction

Cognitive architectures were originally meant to produce a unified theory of cognition in the sense of Newell (see [29] for a detailed discussion of this and the origin of the term). As such, their origin lies in *cognitivist* paradigms of cognition, and the focus is on *human* cognition. It follows from this cognitivist heritage that they were not meant, originally, to address embodied views of cognition, nor were they necessarily meant to be implementable in an artificial cognitive system.

Vernon [29] also notes that there are at least two different views of what a cognitive architecture actually is: in cognitivist paradigms, "[...] the focus in a cognitive architecture is on the aspects of cognition that are *constant over time* and that are *independent of the task*" (p. 65). In emergent paradigms, however, a cognitive architecture is "everything a cognitive system needs to get started" (p. 67), with further changes emerging from the future development of the system (which, notably, is not guaranteed to be successful).

By default, neither view says much about the embodiment of the cognitive agent. In a list of desiderata for cognitive architectures [23], it follows from the "ecological realism" desideratum that the cognitive architecture needs to be able to function in an embodied setting, but this does not—by itself—imply that this embodiment contributes something non-abstractable to cognition. In another list of desiderata (this time for *developmental* cognitive architectures), stronger requirements are put on embodiment [30]: if the cognitive architecture is to adhere to embodied views of cognition, then it must also treat the body as constitutive of cognition (this is their second desideratum: physical embodiment).

The current theoretical understanding of human cognition, however, imposes at least two challenges for such embodied cognitive architectures: first, cognitive science remains fragmented regarding what it actually means, precisely, when it claims that cognition is "embodied". At a minimum, it is the idea that human cognition cannot be understood without taking into account that its purpose is to control a situated body interacting with a physical world, but stronger interpretations are possible. In particular, some views take an anti-functionalist perspective, in which at least some aspects of cognition are exclusively a property of living beings: these can thus not be captured through computational accounts such as would be produced by a cognitive architecture.

Similarly, the degree to which the body is fundamentally and necessarily constitutive of cognition (or specific cognitive mechanisms) also remains open for discussion. For example, purely computational approaches are capable of solving language processing problems that are seen as examples of embodied cognition in humans (see [25] for a discussion). While it is relatively clear—given the development in the cognitive sciences in recent decades—that the body cannot be ignored in its entirety (it would be difficult, for example, to formulate a theory of affordance processing without taking into account the body of the agent, see [24] or a review), it is much less clear whether it is necessarily required for all cognitive mechanisms, and what its precise role actually is.

Second, if the body is constitutive of cognition—as claimed in some accounts—then requiring a physical embodiment for the cognitive architecture implies that it can only be used to study cognition as appropriate for that embodiment. The degree to which it is valid to assume sufficient equivalences between distinct (appropriately chosen) bodies (for example, a robot and a human one)—as is sometimes done in cognitive robotics (see e.g. [22])—is not clear. On the one hand, it can be argued that any difference between two bodies will necessarily result in some difference in sensorimotor experience as well: for example, robots commonly use various motors in their actuation while humans use muscles and tendons. This immediately results in a strong difference between motor activations: muscles and tendons, in engineering terms, form spring-damper systems and cannot be controlled in the same way as a conventional motor. On the other hand, whether or not such differences in sensorimotor experience have functional consequences for higher cognitive processes is less of a given. For example, there is no evidence to suggest that such higher-level cognitive difference can be observed in human beings who lack certain sensory or motor capabilities (and consequently possess a qualitatively different sensorimotor experience of the world).

Overall, given that embodiment is not well-defined, it is also not clear what a useful embodied cognitive architecture would look like, particularly if the aim is to further the study of biological cognition.[1] The purpose of the present chapter is therefore to address how an embodied cognitive architecture might be designed with the lack of well-defined concepts in mind.

The chapter first revisits the point already sketched out in this introduction by providing a succinct overview of different flavours of embodied cognitive science, with the simple purpose of highlighting the breadth of plausible positions regarding what precisely the body brings to the table.[2]

The chapter then proposes that these two aspects must be explicitly represented in a cognitive architecture: the precise role of the body is not clear, and whether it is needed at all (in a constitutive sense) for specific cognitive processes is not clear either. Resolving these unclear aspects arguably remains the major challenge in current theories of embodied cognition: although it is relatively straightforward to produce evidence that suggest a role for the body in higher-level cognition (such as the apparent involvement of pre-motor areas in the processing of language, see [5] for a discussion); it is much harder to be precise about the exact nature of this role. Mahon and Caramazza [14] pointed out—almost 10 years ago at the time of this writing—that the fundamental underlying problem is that the hypothesis is miscast: adding more data on embodied effects does not further the theory as such.

---

[1]This is arguably the most relevant aim. If the aim is simply to create a robot controller, then there is no particular need to appeal to theories of *human* cognition, and therefore also no ambiguities due to a lack of an agreed-upon meaning of the terms used.

[2]The different flavours of embodied cognition have been extensively reviewed by multiple researchers over the past two decades. It is not the purpose here to produce another such review.

What we therefore need from an embodied cognitive architecture—to complete this chapter's title—is an exploration and quantification of the predicted consequences of different theoretical positions one can take.

2 Flavours of Embodied Cognitive Science

In modern cognitive science, traditional paradigms of cognition as some form of symbol manipulation have been superseded by a characterisation in terms of embodiment. That is not to say, as highlighted before, that there is any form of agreement on what this actually entails, a point that has been extensively discussed before (see, e.g., [31, 32] for some examples of such discussions).

There are a number of reasons why this is the case. For example, as [4] notes, the same label of "embodied cognitive science" is used for two—philosophically distinct—theories of mind: one is essentially a continuation of ecological psychology and its predecessors, while the other is a reaction to the concerns arising from a purely computationalist approach to the study of the mind. Notably, the latter remains representationalist, while the former never was.

An additional debate concerns, for example, what precisely the body actually contributes. To some, it is primarily an interface to the world, which provides the computational system with the means of "grounding" the symbols of its computations in some form of experience. In essence, the body is a means by which to solve the symbol grounding problem [13], and this sometimes forms the basis for arguing that a cognitive model must be "embodied" in a physical agent (e.g. [22]).

Since the body is primarily an interface to the outside world in this perspective, the emphasis is also placed on the external senses—sight, hearing, and so on: the body shapes cognition at a minimum insofar as the senses define what information is available (and potentially already process—the idea that the body helps in shaping information so as to facilitate, or entirely remove the need for, computations is called *morphological computation*; see [15, 16]). This position can be pushed further to argue that what actually matters in this is that cognition controls a physical body in the real world, situated in its environment and all that contains (including other agents): to ignore this aspect of cognition ignores its fundamental purpose, and can therefore not lead to a valid theory of cognition. However, even in a strongly situated interpretation, it is the fact that this physical embodiment and its situatedness exist that is important; not necessarily the exact nature of the body itself. This is the argument that enables robotic models of cognition: the interaction with the physical reality is more important than the precise body through which this interaction is achieved.

Others see the role of the body as being much richer. Stapleton [19, 20], for example, argues that internal bodily processes (including, but not restricted to interoceptive senses) contribute just as much. Some then argue that such a perspective is, in fact, necessary, even if one wishes to address problems that are ostensibly focussed on the external world, such as symbol grounding (see [25, 27] for discussions).

The type of body itself is a further topic of discussion. As Ziemke [32] notes, this is in fact ignored by many (particularly if they perceive the body as an interface to the world). In other words, the exact type of body does not matter to a particular cognitive model; only that it provides the functionalities required by the model. This is in contrast with an anti-functionalist view, such as that posited by Searle's Chinese room argument [17]: here, abilities such as understanding or intentionality fundamentally require a biological, living body, and it would therefore be futile to study embodied human cognition using, for example, a robot body.

It is beyond the scope of this chapter to go into the details of these and other strands of embodied cognitive science. The main point here is that a unifying definition cannot be more specific than claiming that embodied cognitive science posits some role of the body in cognition.

At the same time, it is worth remembering that cognitive science has traditionally delivered no shortage of models of human cognitive mechanisms that are not embodied in any sense, yet have proven quite useful at making reasonable predictions that could then be shown to be appropriate. Similarly, one can also find examples in which purely computational solutions exist for problems that are thought to require some sort of embodiment. One example [25] is that of synonymy and polysemy: on one hand, it has been argued that resolving these requires a sensorimotor experience of the underlying concepts, but on the other, the field of computational linguistics has methods that can adequately deal with these without any such experience.

It follows that it is possible to take intermediate positions: to acknowledge that some cognition can be purely computational, but augmented by contributions of the embodiment where necessary. This is relatively explicit in theories of language, in which arguments for a co-existence of embodied and non-embodied phenomena are often put forward [2, 25]. Dove [7] coined the term "dis-embodied" for that precise purpose.

3 Towards an Embodied Cognitive Architecture

3.1 What We Need from an Embodied Cognitive Architecture

We noted at the outset that cognitive architectures were originally intended to provide a theory of (human) cognition. When theories of cognition began to consider embodiment, the claim that models of cognition must also be embodied began to appear and led to cognitive and developmental robotics [3] as a new discipline. It is, however, not the case that every cognitive model instantiated in a robot necessarily has anything to say about human cognition, as opposed to the way in which to make a robot behave in a certain manner; as such, a robot body is also insufficient to be able to claim the status of an embodied cognitive architecture.

The core issue remains the disagreement on what embodiment really means, as discussed above. An embodied cognitive architecture that claims the status because it adheres to a particular interpretation of embodiment is not satisfactory—from a theoretical perspective—since it does not contribute to the resolution of that disagreement. Mahon and Caramazza [14] highlighted that the debate around embodied versus disembodied theories of cognition cannot be advanced by collecting more of the same data; similarly, the debate as to what the actual role of the body in embodied theories of cognition is cannot be resolved by an architecture that implements one view (and thus, by design, behaves in accordance with that view).

At the same time, it is also clear that we cannot build an architecture that remains entirely open to all possible interpretations—as a trivial example, we cannot capture, by computational means, a view that does not believe in a functionalist account of cognition. It has to be assumed that a theory of cognition—even if cognition is embodied—can be expressed in computational terms; if it cannot, then it is not clear how such an architecture could be formulated.

The proposal here is therefore the following: what an embodied cognitive architecture needs to provide at the current state of our theoretical understanding is a framework that allows us to parametrise contributions of the body in cognitive processes. This includes parametrising the body itself, but goes beyond that, in that how sensorimotor experience is used in higher cognition is itself also left open to parametrisation. Such an architecture could then explore the predicted consequences of given activities under various theoretical assumptions regarding embodiment and help further the state of the art by then comparing these predictions with reality. In the remainder of this chapter, the example of symbol grounding is used to illustrate what such an approach might look like.

3.2 Represenationalism and Dynamicism

The choice of symbol grounding as an example naturally brings up a discussion on representations. One of the debates, as seen above, concerns whether modelling work in embodied cognitive science should take a computationalist viewpoint, or a dynamical systems one. Specifically, the former thinks of cognition (grounded or not) as some form of manipulation of symbols and representations, while the latter argues that cognition should be modelled in terms of dynamical systems, capturing interactions and couplings rather than symbol manipulations.

Importantly, it would be indefensible for either view to argue that a natural cognitive system definitely adheres to their view of computation; the question is simply what is necessary to create a "good" model of cognition. For example, Chemero [4] defends the dynamical systems approach, but is explicit that this is an epistemological rather than a metaphysical standpoint: natural cognitive systems may or may not make use of representations, but either way, adding representationalist interpretations to a dynamical model of cognition does not improve the model in terms of its explanatory power, and is therefore unnecessary.

For the present purposes, the core aim is to outline how an embodied cognitive architecture that can capture the role of the body in cognition might be sketched.

This question is, in a sense orthogonal, to the debate as to whether or not models should be representationalist, since it applies to both interpretations. On the other hand, if we do want to propose at least an outline of an architecture, then we do need to express that in some formal language, which is likely to make use of some form of symbolic notation (in particular, since the example given here is symbol grounding).

For the purposes of this chapter, we will present the example using the Neuro-engineering framework (NEF, see [9]), and the semantic pointer architecture (SPA, see [8, 10]). The next section will give a brief introduction to these, but we can already note that these form a framework in which cognitive models can be formulated in a formal language (defined by SPA), even though the model is actually implemented using biologically plausible spiking neurons. In other words, while the design language of the model is clearly symbolic, the resulting model itself could be analysed entirely within a dynamical systems paradigm if that was desired. Of course, given that the initial description of the model is available, that can also be used to analyse the behaviour.[3]

Although the question of whether or not to use a representationalist approach is, as argued above, at least somewhat misleading in the present context, it is worth noting that the NEF/SPA combination does not treat this as a mutually exclusive choice, and ideas from both paradigms flow into the framework (see [8] for a thorough discussion).

3.3 *Overview over NEF and SPA*

The NEF defines three principles regarding what neurons compute [9]:

1. Neural representations are defined by the combination of nonlinear encoding (see, e.g., neuron tuning curves) and (weighted) linear decoding.
2. Transformations of neural representations are functions of variables that are represented by neural populations. Transformations are determined using an alternately weighted linear decoding.
3. Neural dynamics are characterized by considering neural representations as control theoretic state variables. Thus, the dynamics of neurobiological systems can be analysed using control theory.

On the basis of these principles, it is possible to develop a full theory of information processing in biological neural systems, and using these principles, to construct simulations thereof. This development is not covered in detail here, because it would take us too far beyond the scope of this chapter, but a detailed account can be found in [8].

[3]One interesting effect resulting from the use of biologically plausible (and therefore constrained) neurons to implement models is that the actual behaviour of the model may differ from the symbolic description, for example, if the latter stipulates computations that cannot be accurately implemented by the neurons. In fact, without this, the case for going through the trouble of creating the neural implementation would be much less compelling.

The upshot of the NEF is that it gives us a way to think about cognitive computations in a manner that is informed by that which neurons are particularly well-suited for computing.[4] The semantic pointer architecture (SPA) is an answer to how one might want to build models of cognition given the principles defined by the NEF. In particular, these principles lead to the conclusion that it is natural to express models of cognition using vector algebra. Specifically, SPA [8] postulates that higher-level cognition is the appropriate manipulation of such vectors in a high-dimensional space (Eliasmith suggests that a 500-dimensional space may be sufficient for human cognition).

Importantly, these vectors are not randomly chosen. Vectors that encode information about objects, for example, are meant to be created through successive encodings of direct sensoriomotor experiences, in line with the observed hierarchical structures in the human brain such as the visual cortex [12]. For example, the retinal image of an object is successively compressed through the different layers of the hierarchy for object recognition (V1 → V2 → V4 → IT) into a vector with significantly lower dimensionality than the original retinal input. This resulting representation at the top of the hierarchy is termed a *semantic pointer* (because it retains partial semantic content from the original retinal image), encoding the visual appearance of the object.

The vector encoding the entire object would then be a combination of the vectors encoding the various sensorimotor experiences associated with this object—we will return to that point below. First, it is important to highlight that, because semantic pointers are constructed from sensorimotor experience, they are not arbitrary. Most simply, they are grounded symbols that, on a theoretical level, are quite compatible with the perceptual symbol system proposed by Barsalou [1]. In addition, because these symbols are compressed versions of the original sensorimotor experience, they also retain, as mentioned, partial semantic information about the sensorimotor experience that formed them. Computation over these symbols can therefore be co-determined by the way in which the symbols were created in the first place. This is the property of interest as far as this chapter is concerned: SPA provides a theoretical framework in which it is possible, in principle, to construct models of cognition whose functioning could be modulated by sensorimotor experience, and in which this modulation can be quantified.

[4]This separates NEF/SPA from most other attempts to create architectures that operate both at symbolic and subsymbolic levels: traditionally, these often start with an arbitrary symbolic framework that is then converted into a neural representation (which is always possible, given that neural networks are universal function approximators, so there is nothing intrinsically insightful in this step alone). Such "arbitrary" marriages have never been particularly compelling [1]. In NEF/SPA, the symbolic language in which a cognitive model is expressed is defined and constrained by an understanding of the underlying neural substrate.

3.4 Formalising Symbol Grounding

SPA uses circular convolution (usually denoted by the symbol "$\circledast$") to bind vectors together. Circular convolution takes two vectors (of the same length) as input and returns one vector of the same length as output (which is a desirable property to deal with scaling issues, but has some side effects that we'll return to below). Eliasmith [8] gives the example that one could construct a semantic pointer for perceptual features of a robin by combining information of various modalities (here, addition is just the addition of vectors):

$$\mathbf{robinPercept} = \mathbf{visual} \circledast \mathbf{robVis} + \mathbf{auditory} \circledast \mathbf{robAud} + \mathbf{tactile} \circledast \mathbf{robTact} + \cdots,$$

where each element in bold represents a semantic pointer. **robin** could then be defined as:

$$\mathbf{robin} = \mathbf{perceptual} \circledast \mathbf{robinPercept} + \mathbf{isA} \circledast \mathbf{bird} + \mathbf{indicates} \circledast \mathbf{spring} + \cdots$$

Semantic pointers created in this manner allow for a range of cognitively interesting operations; in particular, they allow for both deep and shallow semantic processing.[5] The reader is referred to the original book for discussions, and to the example of SPAUN [10] for a demonstration of a model built using these principles capable of solving a range of cognitively interesting problems.

More focused on the development of concepts, including their putative sensorimotor grounding, Thill and Twomey [27] discuss how human concepts can be described in such a framework so that the constituent parts (whether from a sensorimotor grounding, interoceptive features, or linguistic/amodal information) are all represented. The general form of the proposal is that concepts can be composed of semantic pointers from various modalities, including purely amodal, linguistic information:

$$\mathbf{C} = \mathbf{S^D} + \mathbf{S^T} + \sum_i \sum_j \mathbf{Includes_i} \circledast \mathbf{C_j} + \mathbf{Label} \circledast \mathbf{name}, \tag{1}$$

where $\mathbf{S^D}$ refers to semantic pointers that are created directly from features of sensorimotor experience obtained from both external and body-internal modalities (similarly, $\mathbf{S^T}$ refers to temporal integration of some sensorimotor experience):

$$\mathbf{S^D_t} = \sum_i \sum_j \mathbf{Modality^{ext}_i} \circledast \mathbf{feature_j} + \sum_k \sum_l \mathbf{Modality^{int}_k} \circledast \mathbf{feature_l}. \tag{2}$$

[5] One aspect of this compression mechanism and the binding of vectors that we do not go into detail about here is that it is reversible: the compressed encoding is easily manipulable in computations, but should there be a need to recall details about the underlying sensorimotor experience, this can be done through unbinding and decompression in order to re-obtain details of the original experience.

A concept can also include other concepts (which can also be linguistic information, captured in the above by the term $\sum_i \sum_j \mathbf{Includes_i} \circledast \mathbf{C_j}$), and of course, a label associated with it (**Label** $\circledast$ **name**).

Thill and Twomey [27] use this account to first to highlight the need to include interoceptive features when considering the modalities in which concepts can be grounded, and second to clarify that some constituents are not necessarily available from birth: the account is therefore a developmental one, since Eq. 1 can capture the way in which a concept changes over time as more sophisticated information about the concept becomes available and is integrated. For a much more thorough discussion of these ideas, the interested reader is referred to that paper.

3.5 Determining the Role of the Body

For the purposes of the present chapter, we can note that Eq. 1 provides explicit terms for contributions from sensorimotor experience to the formation of a particular concept. This makes it possible to test consequences of omitting such an experience from these concepts: they can still retain amodal constituents, and would still be of a valid format in the sense that they could be used in models of cognition. If theories of embodiment are right, then one would expect the omission of sensorimotor experience to have a fundamental effect on such models.

Similarly, such a formulation can explicitly test what consequences, if any, variations in sensorimotor experience have for the development of higher-level concepts, and therefore the effects these might have on higher-level cognition. These variations can be due to different embodiments (e.g., a robot versus a human body), or, for example, to biological or cultural differences.

In a model—such as the one sketched above—developed using the NEF/SPA framework, the critical aspect is that higher-level cognition is effectively modelled as operations on vectors that, in some form, represent concepts worth reasoning with/over (even though the final implementation is in a spiking neural network in which that mode of operation is no longer necessarily apparent unless one has the original formulation of the model). What matters, therefore, are the relative locations of these vectors in the overall space, since these determine the outcomes of the operations. For example, if two vectors for the concept of "grasping", one constructed from a full sensorimotor experience in a space of similarly grounded concepts, and another constructed purely from lexical information (e.g. using distributional semantics, which can be shown (as mentioned before) to solve issues thought to require a sensorimotor grounding, see [25] for a discussion) end up in a sufficiently similar location (relative to all other concepts) in their respective spaces, then the same transformations can also result in vectors that maintain this relative positioning. Consequently, it is possible to have two vector spaces that are not alike because the precise locations of the vectors have been determined by different means (sensorimotor experiences due to different bodies), but do still retain the relative positioning of vectors, so that at least some higher-level reasoning on these vectors can be done with the same result in both. At the simplest, the example is that of a comparison: if shown a picture of an animal, and then asked if it is more similar to a cow or a dog, the exact way in which all this information is encoded does not matter; it only matters that the encoding maintains the relative distances between the given picture, the dog, and the cow.

It is therefore important that these vectors are not arbitrarily created: in NEF/SPA, the intention is to use hierarchies as given in the sensorimotor cortices to build these, which imposes a relatively strong biological constraint on any such model. With these constraints in place, it would therefore be possible to create models that can investigate the exact way in which differences in sensorimotor experience (for instance, due to biological, social, or cultural differences) affect the constructed encodings.

Artificial cognitive systems can, of course, be designed to create their encoding in any way deemed reasonable. However, it is therefore possible, on one hand, to explicitly design for spaces that maintain relative features, as much as possible, with human spaces, and on the other, to explicitly test what differences, if any, completely different spaces have for, for example, interactions between humans and robots. More generally, this allows for an explicit test of the idea that similar sensorimotor experiences are a requisite for natural communication, for example.

Finally, it is worth highlighting that the present chapter has used the NEF/SPA framework as an example because it is relatively simple to visualise the ideas discussed here in vector spaces, but the claim is certainly not that such explorations must be constructed in this particular framework. Rather, the overall point is that what we need from an embodied cognitive architecture is what these examples illustrated: a way of quantifying how differences in sensorimotor experience propagate through cognitive mechanisms in a manner that allows for explicit explorations of what consequences, if any, differences in this experience make.

3.6 Challenges for a NEF/SPA Approach

There are a number of challenges associated with the endeavour sketched out above, and it is worth highlighting two that are particularly apparent in the NEF/SPA framework, because they both relate to the creation of semantic pointers. The framework effectively postulates two mechanisms for this. The first is a compression mechanism that operates on sensory inputs (the prime example being the hierarchical organisation of the visual cortex that, incidentally, also inspired deep networks). The challenge with respect to this is that there currently is no model that implements this entire creational process in a general sense. Current models either assign random vectors to act as representation of different concepts or simplify the perceptual process so that it is highly specific to the problem currently under consideration (for example, the work presented in [10] uses a simple square image that can only contain numbers and certain symbols as input in their model).

Since this part of the model is crucial to understanding how human sensorimotor experience may be integrated, a significant piece of the puzzle is currently lacking in detail. This does not preclude explorations of the way in which differences in sensorimotor experiences would affect cognition in a space constructed by some reasonable compression hierarchy, but strong claims about human cognition specifically may be out of reach for now.

The second way in which a semantic pointer can be created is through composition of existing pointers, as described earlier: different semantic pointers can be bound and added together. As such, one can, for example, bind the sensory inputs from different modalities to these modalities and add them so as to construct a semantic pointer that represents the full sensory experience of, say, a bird: its visual appearance (which may decompose further into individual concepts such as wings, feathers, beaks, and feet), the sound it makes, what it might feel like to hold one in your hands, and so on.

The critical aspect here is that the binding operator used is circular convolution. This has the previously mentioned advantage that the dimensionality does not change: two vectors of the same dimensionality will produce a third vector of this dimensionality when convolved. This is, in principle, highly desirable, as it avoids the massive scaling problems one would otherwise encounter when repeatedly binding concepts together. However, the downside is that information is lost in the process. In other words, reversing the operation (which one might want to do when one would like to access specific information contained in the overall representation of, say, a seagull[6]) will produce a vector that is similar but not identical to the original. Over successive operations, this can become highly problematic, as the original information might end up overly distorted and no longer recognisable.

The way the NEF/SPA addresses this problem is through a concept called a clean-up memory [21]: these take, as an input, a noisy version of a known concept (such as might be produced through the process of deconvolution) and "clean it up", that is to say, they return the original vector. They do this by comparing the input vector with stored vectors (hence, "memory") and computing the similarities (the assumption being that the vector with the highest similarity is the most likely candidate for the "clean" version of the input vector).

The crucial problem with this approach is that, while there is evidence to suggest that the brain contains structures that appear remarkably similar to such clean-up memories [8], there is currently no good account of how these memories are formed in the first place, and current state-of-the-art models simply directly include appropriately populated clean-up memories. Eliasmith [8] suggests that these memories are created on-the-fly as needed, but this delegates the problem to questions of how appropriate candidate vectors can be retrieved from memory and copied into the clean-up memory within a reasonable time frame, how many candidate vectors are to be copied, and so on. This is not to suggest that there is a critical fault in the clean-up memory system; but it is another mechanism that appears crucial in the

[6]This would be *unbinding*, which, in SPA, is done through convolving with the inverse of that to which a vector is currently bound.

overall theory of cognition (at least in the NEF/SPA sense), one that also needs to be studied more.

The second of these challenges is specific to the NEF/SPA framework. Other approaches that achieve similar models (for example, the neural blackboard architecture (NBA), see [28]) do not require clean-up memories but may have other drawbacks (for example, see [8] for a demonstration that the NBA might require more neurons than are available in the human cortex to implement human-level cognitive mechanisms).

The first challenge, however, is more general. Any embodied cognitive architecture will eventually have to integrate sensorimotor experience into its higher cognitive mechanisms; in the case of human cognition, this process is constrained by the biology of the sensorimotor cortices, and this cannot be abstracted at the outset.

4 Final Considerations

In summary, this chapter has attempted to sketch how one might want to design an embodied cognitive architecture that could help to further the current state of affairs in embodied theories of cognition. We have highlighted, in particular, the lack of agreement as to what precisely (if anything) the body contributes to cognitive mechanisms. Challenges in this sense come from at least two directions: one is that positions that do agree that the body does contribute something disagree on what exactly that is (including how uniquely human cognition is then tied to the human body); the other is that there is a rich history of entirely disembodied models in cognitive science that work well as models of human cognitive mechanisms (in the sense that they have generated useful predictions that have turned out to be accurate).

Given that state of affairs, the main take home message of this chapter is that what we need from an embodied cognitive architecture is a way to overcome these challenges specifically. It is not about creating a controller for a robot, or other artificial cognitive systems; it is not about providing more demonstrations that sensorimotor inputs can shape cognition; and it is not about providing an implementation of a very specific interpretation of embodied cognitive science. In this chapter, we have exemplified how one might approach this. Specifically, we have discussed the way in which the NEF/SPA framework can be used to explore not only how concepts integrated into higher-level cognition can vary in function of differences in sensorimotor experience, but also what effect, if any, this has on higher level cognition.

It is worth pointing out that in these discussions, we mostly remained agnostic as to the specific nature of the body. In one sense, this is, of course, by design: it would not be helpful to focus on a specific embodiment if we want to have a general theory of embodied cognition applicable to cognitive systems other than humans. In another sense, it also highlights that an embodied cognitive architecture, to make a theoretical contribution, does not necessarily have to exist in a specific embodiment. It needs to highlight what the consequences of changes in embodiment would be, but it can explore this at an abstract level. There are other examples in the literature that show this is indeed a feasible approach: for example, Thill et al. [26] have shown how differences in the nature of the space encoding movements and contexts (which can differ between agents with different bodies) can explain the organisation of parietal mirror neurons. Furthermore, many models in Dynamic Field Theory [11] explicitly or implicitly operate on that principle, and have been successfully applied in developmental Psychology to give embodied explanations of effects such as the A-not-B error [18], or in modelling premotor involvement in movement decision-making [6].

To conclude, it is clearly the case that, as the brief for this book states, cognition of living systems is embodied, embedded and always situated, and that this shapes how we reason.[7] It is, however, not as clear as to whether this is a fundamentally necessary property for cognition. Even if it is, in some sense fundamental, it is not clear what the exact "fundamental" contribution is, nor what the consequences are for the design of artificial cognitive systems that we might want to interact with in the future. The case made in this chapter is that theories of cognition have not recently progressed in a significant manner in these terms, and that the time is ripe to design an embodied cognitive architecture built to tackle this challenge head-on.

References

1. Barsalou, L. W. (1999). Perceptual symbol systems. *Behavioral and Brain Sciences*, *22*(4), 577–660.
2. Barsalou, L. W., Santos, A., Simmons, W. K., & Wilson, C. D. (2008). Language and simulation in conceptual processing. *Symbols, embodiment, and meaning* (pp. 245–283). Oxford: Oxford University Press.
3. Cangelosi, A., & Schlesinger, M. (2015). *Developmental robotics: From babies to robots*. MIT Press.
4. Chemero, A. (2009). *Radical embodied cognitive science*. Cambridge, MA: MIT Press.
5. Chersi, F., Thill, S., Ziemke, T., & Borghi, A. M. (2010). Sentence processing: Linking language to motor chains. *Frontiers in Neurorobotics*, *4*(4).
6. Cisek, P., & Kalaska, J. F. (2010). Neural mechanisms for interacting with a world full of action choices. *Annual Review of Neuroscience*, *33*(1), 269–298. PMID: 20345247.
7. Dove, G. (2011). On the need for embodied and dis-embodied cognition. *Frontiers in Psychology*, *1*(242).
8. Eliasmith, C. (2013). *How to build a brain: A neural architecture for biological cognition*. Oxford: Oxford University Press.
9. Eliasmith, C., & Anderson, C. H. (2002). *Neural engineering: Computation, representation, and dynamics in neurobiological systems*. Cambridge, MA: MIT Press.
10. Eliasmith, C., Stewart, T. C., Choo, X., Bekolay, T., DeWolf, T., Tang, Y., et al. (2012). A large-scale model of the functioning brain. *Science*, *338*(6111), 1202–1205.
11. Erlhagen, W., & Schöner, G. (2002). Dynamic field theory of movement preparation. *Psychological Review*, *109*(3), 545–572.
12. Felleman, D. J., & Van Essen, D. C. (1991). Distributed hierarchical processing in primate visual cortex. *Cerebral Cortex*, *1*, 1–47.

[7]It is also worth remembering, as many have pointed out (e.g. [4]), that this position has a long history in theory of mind, and is not a merely a reaction to computationalist approaches that have been arising in cognitive science more recently.

13. Harnad, S. (1990). The symbol grounding problem. *Physica D: Nonlinear Phenomena*, *42*(1–3), 335–346.
14. Mahon, B. Z., & Caramazza, A. (2008). A critical look at the embodied cognition hypothesis and a new proposal for grounding conceptual content. *Journal of Physiology-Paris*, *102*(1), 59–70. Links and Interactions Between Language and Motor Systems in the Brain.
15. Pfeifer, R., Bongard, J., & Grand, S. (2007). *How the body shapes the way we think: A new view of intelligence*. Cambridge, MA: MIT press.
16. Pfeifer, R., & Iida, F. (2005). Morphological computation: Connecting body, brain and environment. *Japanese Scientific Monthly*.
17. Searle, J. R. (1980). Minds, brains, and programs. *Behavioral and Brain Sciences*, *3*(9), 417–424.
18. Spencer, J. P., Austin, A., & Schutte, A. R. (2012). Contributions of dynamic systems theory to cognitive development. *Cognitive Development*, *27*(4), 401–418. The Potential Contribution of Computational Modeling to the Study of Cognitive Development: When, and for What Topics?
19. Stapleton, M. (2011). *Proper embodiment: The role of the body in affect and cognition*. Ph.D. thesis, The University of Edinburgh.
20. Stapleton, M. (2013). Steps to a "properly embodied" cognitive science. *Cognitive Systems Research*, *22–23*, 1–11.
21. Stewart, T. C., Tang, Y., & Eliasmith, C. (2010). A biologically realistic cleanup memory: Autoassociation in spiking neurons. *Cognitive Systems Research*, *12*(2), 84–92.
22. Stramandinoli, F., Cangelosi, A., & Marocco, D. (2011). Towards the grounding of abstract words: A neural network model for cognitive robots. In *The 2011 International Joint Conference on Neural Networks (IJCNN)* (pp. 467–474).
23. Sun, R. (2004). Desiderata for cognitive architectures. *Philosophical Psychology*, *17*(3), 341–373.
24. Thill, S., Caligiore, D., Borghi, A. M., Ziemke, T., & Baldassarre, G. (2013). Theories and computational models of affordance and mirror systems: An integrative review. *Neuroscience & Biobehavioral Reviews*, *37*(3), 491–521.
25. Thill, S., Padó, S., & Ziemke, T. (2014). On the importance of a rich embodiment in the grounding of concepts: Perspectives from embodied cognitive science and computational linguistics. *Topics in Cognitive Science*, *6*(3), 545–558.
26. Thill, S., Svensson, H., & Ziemke, T. (2011). Modeling the development of goal-specificity in mirror neurons. *Cognitive Computation*, *3*(4), 525–538.
27. Thill, S., & Twomey, K. (2016). What's on the inside counts: A grounded account of concept acquisition and development. *Frontiers in Psychology: Cognition*, *7*(402).
28. van der Velde, F., & de Kamps, M. (2006). Neural blackboard architectures of combinatorial structures in cognition. *Behavioral and Brain Sciences*, *29*(2), 37–70.
29. Vernon, D. (2014). *Artificial cognitive systems: A primer*. Cambridge, MA: MIT Press.
30. Vernon, D., von Hofsten, C., & Fadiga, L. (2016). Desiderata for developmental cognitive architectures. *Biologically Inspired Cognitive Architectures*, *18*, 116–127.
31. Wilson, M. (2002). Six views of embodied cognition. *Psychonomic Bulletin & Review*, *9*(4), 625–636.
32. Ziemke, T. (2003). What's that thing called embodiment? In *Proceedings of the 25th Annual Meeting of the Cognitive Science Society* (pp. 1305–1310).

The Architect's Dilemmas

David Vernon

Abstract The creation of a cognitive architecture presents the architect with many design choices. Some of these choices come in the form of a dilemma, in which the selection of any option over another entails both benefits and opportunity costs. This chapter highlights three dilemmas that confront the architect when deciding how the key issues of fidelity, embodiment, and autonomy should be addressed and reflected in the design. In each case, it discusses the various options, their roots, and the consequences and costs of choosing one option over another. It concludes by considering these three dilemmas in the context of the stance on cognitive adopted by the editors of this book.

1 Introduction

The design of a cognitive architecture is a daunting undertaking, involving many challenges on a scale that is not always apparent when one embarks on the task. The time and effort involved almost always exceed expectation, sometimes leading to a project that spans decades [2, 13, 22, 35, 46, 47]. The task is made all the harder by the fact that the design options derive from underlying principles, in cognitive science and cybernetics, for example, that are not always evident. Worse still, they often involve choices between two apparently competing options, both of which have elements that are attractive. In this chapter, we highlight three such dilemmas—fidelity, embodiment, and autonomy—and we look at the choices in each case, examining the consequences of choosing one option over another. At the end, we reflect on the choices implied by the characterization of cognition that has motivated this book. We begin by examining the role of a cognitive architecture in the design and implementation of a cognitive system.

D. Vernon (✉)
Carnegie Mellon University Africa, Kigali, Rwanda
e-mail: vernon@cmu.edu

M. I. Aldinhas Ferreira et al. (eds.), *Cognitive Architectures*, Intelligent Systems, Control and Automation: Science and Engineering 94,
https://doi.org/10.1007/978-3-319-97550-4_5

2 The Role of Cognitive Architecture

Like architecture in the built environment and system architecture in software engineering, a cognitive architecture captures both abstract conceptual form and details of functional operation, focusing on inner cohesion and self-contained completeness [49]. The goal of creating a complete model is significant. It means that all of the mechanisms required for cognition fall under the compass of a cognitive architecture. These include perception, action, control, learning, reasoning, memory, adaptivity, and prospection. This accords cognition much greater breadth than has been the case in the past, when it was viewed by many as a reasoning and planning filling sandwiched between perception and action. Today, cognition, as a process, and a cognitive architecture, as a framework, are seen to embrace all of the elements required for effective action [28]. Thus, a cognitive architecture reflects the specification of a complete cognitive system, its components, and the way these components are dynamically related as a whole. It provides both an abstract model of cognition and the sufficient basis for a software instantiation of that model [25]. Ron Sun captures this succinctly:

> A cognitive architecture provides a concrete framework for more detailed modelling of cognitive phenomena, through specifying the essential structures, division of modules, relations between modules, and a variety of other aspects [45].

A cognitive architecture makes explicit the set of assumptions upon which that cognitive model is founded. Depending on the purpose of the modelling exercise, an issue we will mention below and return to in Sect. 3, these assumptions are derived from several sources: biological or psychological data, philosophical arguments, or hypotheses inspired by work in different disciplines, such as cognitive neuroscience and artificial intelligence.

In essence, then, the role of a cognitive architecture is to provide a complete model of cognition and to do so at at least two levels of abstraction, setting out the overall process by which cognition produces effective action (in whatever guise that may take) and the detailed computational elements by which that process is effected, including formalisms for knowledge representations and the types of memory used to store them, the processes of reasoning, inference, and prediction that act upon that knowledge, and the learning mechanisms that acquire it.

In a sense, a cognitive architecture captures the top two layers of Marr's three-level hierarchy of abstraction, also known as the *Levels of Understanding* framework [26, 27], i.e., the top level computational theory and, below this, the level of representation and algorithm. At the bottom (third) level, there is the implementation or instantiation of this algorithmic and representational framework: the realization of the cognitive architecture as a working cognitive system.

Once it has been created and instantiated, a cognitive architecture plays a second role, providing the means to validate the assumptions and hypotheses on which the computational model is based, refine their representational and algorithmic foundations, and develop their implementation further.

3 The Dilemma of Fidelity

The model of cognition encapsulated in a cognitive architecture may refer either to natural cognitive agents, to artificial cognitive agents, or to both. The term itself has its roots in cognitive science (a branch of human psychology) and is credited to Allen Newell and his colleagues in their work on a *unified theory of cognition* [32, 33], i.e., a theory that covers a complete range of cognitive issues, such as attention, memory, problem solving, decision making, and learning, from a comprehensive set of perspectives, including psychology, neuroscience, and computer science. Allen Newell and John Laird's Soar architecture [20, 22, 23, 37], John Anderson's ACT-R architecture [1, 2], Paul Rosenbloom's Sigma architecture [38], and Ron Sun's CLARION architecture [45, 47] are all candidate unified theories of cognition. Recognizing the importance of generality and completeness mentioned above, recent work is endeavouring to bring various strands together to create a common model of cognition[1] and a consensus on what must be included in a cognitive architecture in order to provide a human-like mind [24].

However, some cognitive architectures, e.g., [12, 15], make no claim about the biological plausibility of the cognitive architecture, although they often draw inspiration from what is known about cognition in natural systems. Instead, they focus on the practical application of cognitive science.

In effect, there are two reasons to design a cognitive architecture: one is to gain a better understanding of cognition in general and the other is to build artificial systems that have capabilities that are commonly found in humans [17]. The motivation for the first is a principled one, the motivation for the second is a practical one. These two motivations are obviously different, but they are not necessarily complementary. There is no guarantee that success in designing a practical cognitive architecture for an application-oriented cognitive robot will shed any light on the more general issues of cognitive science. Similarly, it is not evident that efforts to date to design general cognitive architectures have been tremendously successful for practical applications.

From the principled perspective, a cognitive architecture is an abstract meta-theory of cognition that focuses, as we have mentioned, on generality and completeness [24], drawing on many sources in shaping these architectures, often encapsulated in lists of design principles and desirable features referred to as *desiderata* [18, 21, 45, 52]. The second perspective focuses on the practical necessities of the cognitive architecture and designing on the basis of user requirements. Here, the goal is to create an architecture that addresses the needs of an application without being concerned whether or not it is a faithful model of cognition. These two approaches have been dubbed *design by desiderata* and *design by use case* [51].

The dilemma, then, is this: should a cognitive architecture be a general or a specific framework? Should you focus on discovery or invention? Should you favour fidelity or expediency? These are important design questions because a specific instance of a cognitive architecture derived from a general schema will inherit relevant elements embedded in a well-founded framework, but it may also inherit elements that are not

[1] Earlies work on this topic was under the bannes of A Standard Model of the Mind.

strictly necessary for the specific application domain, yielding an architecture that is more complicated than is necessary for that specific application domain. If your focus is on creating a practical cognitive architecture for a specific application, it may not be appropriate to instantiate a design guided by desiderata; arguably, you are better off proceeding in a conventional manner by designing a system architecture that is driven by user requirements, drawing on the available repertoire of AI and cognitive systems algorithms and data-structures. However, the danger here is that the systems perspective that is crucial to cognitive architectures may not be as well-grounded in firm principles as it needs to be. Conversely, if your focus is a unified theory of cognition, then developing use cases and designing a matching system architecture is unlikely to yield insights on the underlying principles of cognition. You may miss some of the key considerations that make natural cognitive systems so flexible and adaptable, and it is unlikely that you will shed much light on the bigger questions of cognitive science.

4 The Dilemma of Embodiment

Embodiment—or, more specifically, embodied cognition—refers to the role that an agent's body plays in the cognitive function of that agent. Possessing a body, however, does not necessarily mean that an agent is embodied, since that body may play no causal role in the agent's cognitive processes.

The cognitive systems community is divided into two schools: those that think an agent's body plays no causal role and those that think it does.[2] Among those that think it does, there are several stances that vary according to the strength of the assertions they make. The dilemma that confronts the architect designing a cognitive architecture is to select which stance to adopt on embodiment. In the following, we will outline the various stances and the implications of adopting one or another in the design of a cognitive architecture.

The essence of cognitivism, a widely-adopted paradigm of cognitive science, is that cognition comprises computational operations defined over symbolic representations and that these computational operations are not tied to any given instantiation [9, 48, 49]. A physical body may facilitate exploration and learning, but it is by no means necessary. The principled decoupling of the cognitivist computational model of cognition from its instantiation as a physical system is referred to as *computational functionalism* [34]. The chief point of computational functionalism is that the physical realization of the computational model is inconsequential to the model: any physical platform that supports the performance of the required symbolic computations will suffice, be it computer or human brain. Computational functionalism effectively says that the mind is the software of the brain *or any functionally equivalent system.* This is an important claim:

[2]The literature on embodiment and embodied cognition is varied and extensive; see [49, Chap. 5], for a brief overview.

> Computational functionalism entails that minds are multiply realizable, in the sense in which different tokens of the same type of computer program can run on different kinds of hardware. So if computational functionalism is correct, then ... mental programs can also be specified and studied independently of how they are implemented in the brain, in the same way in which one can investigate what programs are (or should be) run by digital computers without worrying about how they are physically implemented [34].

There is an alternative school of thought in cognitive science that takes a very different view on this, arguing that cognitive systems are intrinsically embodied and embedded in the world around them, developing through real-time interaction with their environment. From the point of view of embodiment, the way the cognitive agent perceives the world—its space of possible perceptions—derives not from a pre-determined, i.e., purely objective, world, but rather from the actions in which the system can engage. In other words, it is the space of possible actions facilitated by and conditioned by the particular embodiment of the cognitive agent that determines how that cognitive agent perceives the world. Thus, the cognitive system constructs and develops its own understanding of the world in which it is embedded, i.e., its own agent-specific and body-specific knowledge of its world. This position is encapsulated in the *embodied cognition thesis*.

> Many features of cognition are embodied in that they are deeply dependent upon characteristics of the physical body of an agent, such that the agent's beyond-the-brain body plays a significant causal role, or physically constitutive role, in that agent's cognitive processing [53].

Underpinning embodied cognition is the assertion that perception and action are mutually dependent and that the dependency acts in both directions: action depends on perception (this, at least, raises no cause for objection), but perception also depends on action and, importantly, on the state of the agent's body (this is a little less obvious, but there is a large body of psychological and neuroscientific evidence to support it, e.g., [4, 10, 19, 36]). The mutual dependence of perception and action implies a dependence of cognition on the embodiment of the cognitive agent and the actions that embodiment enables. This has a far reaching consequence: agents with different type of body understand the world differently. The dependence of percepts, and associated concepts constructed through cognitive activity, on the specific form of embodiment is a fundamental cornerstone of embodied cognition and emergent cognitive systems, in general.

There are three hypotheses on embodiment associated with the embodied cognition thesis: the conceptualization hypothesis, the constitution hypothesis, and the replacement hypothesis [43].

The position that the physical morphology—the shape or form—and motor capabilities of a system has a direct bearing on the way the cognitive agent understands the world in which it is situated is sometimes referred to as the *conceptualization hypothesis*. That is, the characteristics of an agent's body determine the concepts an organism can acquire, and so agents with different type of body will understand the world differently.

The idea that the body (and possibly also the environment) plays a constitutive rather than a supportive role in cognitive processing, i.e., that the body is itself an integral part of cognition, is referred to as the *constitution hypothesis*. The claim made by the constitution hypothesis is stronger than that made by the conceptualization hypothesis. Cognition is not only influenced and biased by the characteristics and states of the agent's body, the body and its dynamics also augment the brain as an additional cognitive resource. In other words, the way the body is shaped and the way in which it moves help it accomplish the goals of cognition without having to depend on brain-centred neural processing.

There is a third claim sometimes made by proponents of embodied cognition: that because an agent's body is engaged in real-time interaction with its environment, the need for representations and representational processes is removed. This is referred to as the *replacement hypothesis*. The point of this hypothesis is that there is no need for the cognitive system to represent anything, computationally or otherwise, because all of the information it needs is already immediately accessible as a consequence of its sensorimotor interaction.

While the potential attractions of embodied cognition are numerous—the real-time situated coupling between the cognitive system and the environment, the possible removal of the need for symbolic representations, the embedded and grounded exploitation of the environment by the cognitive system to facilitate cognitive activity and off-load cognitive work and scaffold enhanced capabilities—the current capabilities of cognitivist systems are far more advanced. This is reflected in the state of embodied cognition that is sometimes referred to as a research program rather than a mature discipline. It is a plausible and, to many, a very compelling thesis, but, despite the fact that it is now accepted as a mainstream alternative to cognitivism, much remains to be done to establish it as an established science with well-understood engineering principles. In other words, it is not clear how the principles of embodiment should be manifest in a cognitive architecture. Also, embodied cognition entails that many aspects of procedural and declarative knowledge are agent-specific and cannot be directly shared with other cognitive agents.

On the other hand, in the cognitivist tradition, knowledge can be exchanged directly among different forms of cognitive agent, exactly because of it divorces cognition and cognitive architectures from the agent body, relying instead on the acquisition of the knowledge necessary to perform whatever task is necessary from whatever source is available. The cognitive architecture, then, is the fixed part of the cognitive model [24], which is completed by the addition of appropriate knowledge. The dilemma for the architect is that adopting a non-embodied cognitivist approach simplifies the task of designing the cognitive architecture but ignores considerable psychological and neuroscientific evidence of the role the body plays in cognition. Conversely, adopting an embodied stance does recognize this role, but it adds considerable complexity and the need to incorporate principles that are not yet fully developed into the design.

5 The Dilemma of Autonomy

The third dilemma concerns autonomy. To understand why autonomy presents a dilemma when designing a cognitive architecture, we need to be clear what we mean by autonomy. Unfortunately, that's easier said than done [7, 14] and one can identify more than twenty types of autonomy [49]. What is common to most interpretations is the idea that autonomy relates to the degree of self-determination of a system, i.e., the degree to which a system's behaviour—its goals and the manner in which it achieves them—is determined by the system itself and not its environment, including other agents [40]. Thus, an autonomous system is not controlled by some other agent, but is self-governing and self-regulating, selecting its goals, determining how best to achieve them, and then acting accordingly [16].

However, if an external agent cannot exert a causal influence on an autonomous cognitive system, how can one get it to do something useful? We want autonomy, but we also want some control over the cognitive system. This seems to present the architect with the dilemma of having to choose between control and autonomy. However, the choice is a little deeper than that. Mirroring the dilemma of fidelity and the need to choose between opting for the completeness and generality of natural cognitive systems or expediency when designing cognitive architectures for practical application, it is useful to distinguish between biological and robotic autonomy [54].

In robotics, it is common to distinguish between adjustable, shared, sliding, and subservient autonomy, all more or less equivalent terms that are suggestive of ways of qualifying the degree of autonomy and the relative involvement of a human with the cognitive system in carrying out tasks and pursuing goals. In these modes of autonomy, the system controls its own behaviour only to some extent, with the goals being determined by the human with which it is interacting [30]. In such cases, it is necessary for the cognitive architecture to accommodate this sharing: what information does the autonomous agent share with the user and on what basis does it decide whether or not it should be shared, for example [44]? The architect must still devolve to the cognitive system some power to make independent decisions and, in essence, all we have done is push the autonomy dilemma a little further down the line. The resolution of the dilemma hinges on the impact of those decisions, striking a balance between a human retaining control over the choice of superordinate goal and giving the system sufficient freedom to select strategies adaptively in order to meet these goals. A solution to this problem may lie in exploiting the information-theoretic concept of empowerment [39] in the design of the cognitive architecture.

For biological autonomy, we can differentiate between *behavioural autonomy* and *constitutive autonomy* [3, 14]. Behavioural autonomy is concerned with the extent to which the agent sets its own goals and its robustness and flexibility in achieving them as it interacts with the world around it, including other cognitive agents. Constitutive autonomy is concerned with the internal organization and the organizational processes that keep the system viable, maintaining itself as an identifiable autonomous entity. Indeed, Maturana and Varela, whose work provided the inspiration for the enactive view of cognition, define autonomy as "the condition of subordinating all changes to the maintenance of the organization" [29].

Constitutive and behavioural autonomy are related: an agent cannot deal with uncertainty and danger if it is not organizationally equipped to do so. Behaviour depends on internal preparedness, but appropriate behaviour is also needed to allow the agent to bring about the necessary environmental conditions for constitutive autonomy to be able to operate effectively. This complementarity of the constitutive and the behavioural reflects two different sides of the characteristic of recursive self-maintenant systems [6] to deploy different processes of self-maintenance depending on environmental conditions, with constitutive and behavioural autonomy corresponding to the internal and external aspects of that adaptive capacity, respectively.

The dilemma is now whether to base the design of the cognitive architecture on organizational principles that are not overtly concerned with achieving goals as perceived by external agents, or to focus on behaviour, but perhaps at the cost of missing some key aspect of cognition, e.g., *homeostasis* [5, 8], with the autonomy of an agent being effected through a hierarchy of homeostatic self-regulatory processes [31, 55], similar to Damasio's hierarchy of levels of homeostatic regulation [11].

If the processes that support constitutive autonomy were also to give rise to behavioural autonomy, the dilemma might be resolved without compromise. Recent work proposing that constitutive autonomy derives from self-organization based on continual predictive inference of the causes of sensory perturbations, coupled with continual adaption by updating the prediction model *and* responding with actions that minimize the long-term average surprisal, suggests this might just be the case [41, 42, 50].

6 Conclusion

Before concluding, let us recap the three dilemmas. The dilemma of fidelity involves choosing between a general and complete cognitive architecture that is a faithful model of human cognition, derived from desiderata, and an architecture that is specific to a particular application domain, derived from use cases. In the former case, all of the relevant elements will be addressed and it will be a well-founded framework, but some elements may be included that are not relevant for a given application domain and the architecture may be more complicated than necessary. In the latter case, the architecture will be focused on and driven by user requirements, but it may not be well-grounded in theory and may miss key principles that underpin cognition. Also, it is unlikely to yield insights into a unified theory of cognition.

The dilemma of embodiment involves choosing between cognitivism and computational functionalism in which the agent's body plays no role in the cognitive process and an alternative paradigm in which, to a greater or lesser degree, the body does play a causal role. In the former case, adopting a non-embodied cognitivist approach simplifies the task of designing the cognitive architecture but ignores considerable psychological and neuroscientific evidence of the role the body plays in cognition. Conversely, adopting an embodied stance does recognize this role, but it adds considerable complexity and the need to incorporate design principles that are not yet fully developed.

The dilemma of autonomy involves choosing between independence and control. Choosing independence, even partial independence and shared autonomy, creates the problem of how to incorporate the required restrictions on freedom to act into the cognitive architecture. On the other hand, choosing control simplifies the design of the cognitive architecture but undermines one of the key aspects of cognition: autonomous operation. If the goal is to emulate biological autonomy, the dilemma involves choosing between constitutive autonomy, focussing on organizational principles that are not overtly concerned with achieving the goal of external agents, and behavioural autonomy, with the potential of missing some key organizational aspect of cognition.

To conclude, let us look at the position that the editors of this book adopt on cognition:

> In what concerns living systems, cognition is an embodied, embedded and always situated experience. This means it involves a cognitive entity endowed with a particular physical architecture interacting with the specific world it is immersed in, behaving according to the prompts placed by this environment, reacting, adapting to it, and this way defining its own existential narrative and history.

It is apparent that the editors have already confronted the dilemmas identified in this chapter and clearly favour the choices that see a cognitive system as a self-organizing biologically-plausible entity exhibiting complete autonomy, embodied and focussed on development. Furthermore, they add the following:

> Highlighting the nature of the dialectics that binds different life forms to their specific environments, the book addresses the topic of artificial cognition in the domains of robotics and artificial life.

The key word here is *dialectics*, suggesting a process of never-ending discovery by which mutual interactions continually reveal new depths of meaning, at least insofar as the relationship between the agent and its world is concerned. This is the very essence of the concept of co-determination: the mutual specification of the system's reality by the system and its environment [28], strongly echoing the links between cognition, embodiment, and constitutive autonomy, succinctly captured by Anil Seth: "the purpose of cognition (including perception and action) is to maintain the homeostasis of essential variables and of internal organization ... [so that] ... perception emerges as a *consequence* of a more fundamental imperative towards organizational homeostasis, and not as a stage in some process of internal world-model construction" [42].

Having addressed the dilemmas and decided which choices best match the cognitive architecture design goals, the architect still faces a daunting challenge, but at least some of the design decisions are explicitly laid bare.

References

1. Anderson, J. R. (1996). Act: A simple theory of complex cognition. *American Psychologist, 51*, 355–365.
2. Anderson, J. R., Bothell, D., Byrne, M. D., Douglass, S., Lebiere, C., & Qin, Y. (2004). An integrated theory of the mind. *Psychological Review, 111*(4), 1036–1060.
3. Barandiaran, X., & Moreno, A. (2008). Adaptivity: From metabolism to behavior. *Adaptive Behavior, 16*(5), 325–344.
4. Barsalou, L. W., Niedenthal, P. M., Barbey, A., & Ruppert, J. (2003). Social embodiment. In B. Ross (Ed.), *The Psychology of Learning and Motivation* (Vol. 43, pp. 43–92). San Diego: Academic Press.
5. Bernard, C. (1878). *Leçons sur les phénomènes de la vie commun aux animaux et végétaux*. Paris: J.-B. Baillière.
6. Bickhard, M. H. (2000). Autonomy, function, and representation. *Communication and Control-Artificial Intelligence, 17*(3–4), 111–131.
7. Boden, M. A. (2008). Autonomy: What is it? *BioSystems, 91*, 305–308.
8. Cannon, W. B. (1929). Organization of physiological homeostasis. *Physiological Reviews, 9*, 399–431.
9. Clark, A. (2001). *Mindware—An Introduction to the Philosophy of Cognitive Science*. New York: Oxford University Press.
10. Craighero, L., Fadiga, L., Rizzolatti, G., & Umiltà, C. A. (1999). Movement for perception: A motor-visual attentional effect. *Journal of Experimental Psychology: Human Perception and Performance, 25*(6), 1673–1692.
11. Damasio, A. R. (2003). *Looking for Spinoza: Joy, sorrow and the feeling brain*. Orlando, Florida: Harcourt.
12. Dickmanns, E. (2003). A general cognitive system architecture based on dynamic vision for motion control. *Journal of Systemics, Cybernetics and Informatics, 1*(5), 1–6.
13. Franklin, S., Madl, T., D'Mello, S., & Snaider, J. (2014). Lida: A systems-level architecture for cognition, emotion, and learning. *IEEE Transactions on Autonomous Mental Development, 6*(1), 19–41.
14. Froese, T., Virgo, N., & Izquierdo, E. (2007). Autonomy: A review and a reappraisal. In: E. Almeida, F. Costa, L. Rocha, E. Costa, I. Harvey & A. Coutinho (Eds.), Proceedings of the 9th European Conference on Artificial Life: Advances in Artificial Life (Vol. 46–48, pp. 455–465). Berlin, Heidelberg: Springer. https://doi.org/10.1007/978-3-540-74913-4_46.
15. Gomez, E.P., Cao, H., De Beir, A., Van De Perre, G., Lefeber, D., & Vanderborght, B. (2016). A multilayer reactive system for robots interacting with children with autism. In *Proceedings of the Fifth International Symposium on New Frontiers in Human-Robot Interaction*.
16. Haselager, W. F. G. (2005). Robotics, philosophy and the problems of autonomy. *Pragmatics and Cognition, 13*, 515–532.
17. Krichmar, J. L. (2012). Design principles for biologically inspired cognitive architectures. *Biologically Inspired Cognitive Architectures, 1*, 73–81.
18. Krichmar, J.L., Edelman, G.M. (2006). Principles underlying the construction of brain-based devices. In T. Kovacs, & J. A. R. Marshall (Eds.), *Proceedings of AISB '06—Adaptation in Artificial and Biological Systems*, (Vol. 2, pp. 37–42). Symposium on Grand Challenge 5, Architecture of Brain and Mind University of Bristol, Bristol.
19. Lackner, J. R. (1988). Some proprioceptive influences on the perceptual representation of body shape and orientation. *Brain, 111*, 281–297.
20. Laird, J.E. (2008) Extending the soar cognitive architecture. In: *Proceedings of the First Conference on Artificial General Intelligence* (pp. 224–235). IOS Press, Amsterdam, The Netherlands.
21. Laird, J.E. (2009) Towards cognitive robotics. In: G.R. Gerhart, D.W. Gage, & C.M. Shoemaker (Eds.), *Proceedings of the SPIE — Unmanned Systems Technology XI* (Vol. 7332, pp. 73320Z–11).
22. Laird, J. E. (2012). *The Soar Cognitive Architecture*. Cambridge, MA: MIT Press.

23. Laird, J. E., Newell, A., & Rosenbloom, P. S. (1987). Soar: An architecture for general intelligence. *Artificial Intelligence*, *33*(1–64).
24. Laird, J. E, Lebiere, C., & Rosenbloom, P. S. (2017). A standard model of the mind: Toward a common computational framework across artificial intelligence, cognitive science, neuroscience, and robotics. *AI Magazine*.
25. Lieto, A., Bhatt, M., Oltramari, A., & Vernon, D. (2017, in press). The role of cognitive architectures in general artificial intelligence. *Cognitive Systems Research*.
26. Marr, D. (1982). *Vision*. San Francisco: Freeman.
27. Marr, D., & Poggio, T. (1977). From understanding computation to understanding neural circuitry. In E. Poppel, R. Held, J.E. Dowling (Eds.), *Neuronal Mechanisms in Visual Perception, Neurosciences Research Program Bulletin* (Vol. 15, pp. 470–488).
28. Maturana, H., & Varela, F, (1987), The Tree of Knowledge—The Biological Roots of Human Understanding. Boston & London: New Science Library.
29. Maturana, H.R., Varela, F.J. (1980). *Autopoiesis and Cognition—The Realization of the Living*. Boston Studies on the Philosophy of Science, D. Dordrecht, Holland: Reidel Publishing Company.
30. Meystel, A. (2000). From the white paper explaining the goals of the workshop: Measuring performance and intelligence of systems with autonomy: Metrics for intelligence of constructed systems. In: E. Messina, & A. Meystel (eds) *Proceedings of the 2000 PerMIS Workshop, NIST* (Vol. 970). Gaithersburg, MD, U.S.A: Special Publication.
31. Morse, A., Lowe, R., & Ziemke, T. (2008). Towards an enactive cognitive architecture. In *Proceedings of the First International Conference on Cognitive Systems*. Germany: Karlsruhe.
32. Newell, A. (1982). The knowledge level. *Artificial Intelligence*, *18*(1), 87–127.
33. Newell, A. (1990). *Unified Theories of Cognition*. Cambridge MA: Harvard University Press.
34. Piccinini, G. (2010). The mind as neural software? Understanding functionalism, computationalism, and computational functionalism. *Philosophy and Phenomenological Research*, *81*(2), 269–311.
35. Ramamurthy, U., Baars, B., D'Mello, S.K., & Franklin, S. (2006). LIDA: A working model of cognition. In: D. Fum, F. D. Missier, & A. Stocco (Eds.), *Proceedings of the 7th International Conference on Cognitive Modeling* (pp. 244–249).
36. Rizzolatti, G., & Craighero, L. (2004). The mirror neuron system. *Annual Review of Physiology*, *27*, 169–192.
37. Rosenbloom, P., Laird, J., & Newell, A. (Eds.). (1993). *The Soar Papers: Research on Integrated Intelligence*. Cambridge, Massachusetts: MIT Press.
38. Rosenbloom, P. S., Demski, A., & Ustun, V. (2016). The sigma cognitive architecture and system: Towards functionally elegant grand unification. *Journal of Artificial General Intelligence*, *7*, 1–103.
39. Salge, C., Polani, D. (2017). Empowerment as a replacement for the three laws of robotics. *Frontiers in Robotics and AI 4*.
40. Seth, A. (2010). Measuring autonomy and emergence via Granger causality. *Artificial Life*, *16*(2), 179–196.
41. Seth, A. K. (2013). Interoceptive inference, emotion, and the embodied self. *Trends in Cognitive Sciences*, *17*(11), 565–573.
42. Seth, A. K. (2015). The cybernetic Bayesian brain—from interoceptive inference to sensorimotor contingencies. In T. Metzinger & J. M. Windt (Eds.), *Open MIND* (Vol. 35, pp. 1–24). Frankfurt am Main: MIND Group.
43. Shapiro, L. (2011). Embodied Cognition. Routledge.
44. Sheridan, T. B., & Verplank, W. L. (1978). *Human and computer control for undersea teleoperators*. Technical Report, MIT Man-Machine Systems Laboratory.
45. Sun, R. (2004). Desiderata for cognitive architectures. *Philosophical Psychology*, *17*(3), 341–373.
46. Sun, R. (2007). The importance of cognitive architectures: an analysis based on clarion. *Journal of Experimental & Theoretical Artificial Intelligence*, *19*(2), 159–193.

47. Sun, R, (2016), Anatomy of the Mind: Exploring Psychological Mechanisms and Processes with the Clarion Cognitive Architecture. Oxford University Press.
48. Varela, F. J. (1992). Whence perceptual meaning? A cartography of current ideas. In F. J. Varela & J. P. Dupuy (Eds.), *Understanding Origins—Contemporary Views on the Origin of Life* (pp. 235–263). Boston Studies in the Philosophy of Science. Dordrecht: Mind and Society, Kluwer Academic Publishers.
49. Vernon, D. (2014). *Artificial Cognitive Systems—A Primer*. Cambridge, MA: MIT Press.
50. Vernon, D. (2016). Reconciling constitutive and behavioural autonomy: The challenge of modelling development in enactive cognition. *Intellectica: The Journal of the French Association for Cognitive Research*, *65*, 63–79.
51. Vernon, D. (2017). Two ways (not) to design a cognitive architecture. In V. C. Chrisley R, Müller, Y. Sandamirskaya & M. Vincze (Eds.), *Proceedings of EUCognition 2016, Cognitive Robot Architectures, European Society for Cognitive Systems, CEUR-WS* (Vol. 1855, pp. 42–43), Vienna.
52. Vernon, D., von Hofsten, C., & Fadiga, L. (2016). Desiderata for developmental cognitive architectures. *Biologically Inspired Cognitive Architectures*, *18*, 116–127.
53. Wilson, R.A., & Foglia, L. (2011). Embodied cognition. In E. N. Zalta (Ed.), *The Stanford Encyclopedia of Philosophy*.
54. Ziemke, T. (2008). On the role of emotion in biological and robotic autonomy. *BioSystems*, *91*, 401–408.
55. Ziemke, T., & Lowe, R. (2009). On the role of emotion in embodied cognitive architectures: From organisms to robots. *Cognition and Computation*, *1*, 104–117.

Human Cognition-Inspired Robotic Grasping

Marco Monforte, Fanny Ficuciello and Bruno Siciliano

Abstract The hand is one of the most complex and fascinating organs of the human body. We can powerfully squeeze objects, but we are also capable of manipulating them with great precision and dexterity. On the other hand, the arm, with its redundant joints, is in charge of reaching the object by determining the hand pose during preshaping. The complex motion and task execution of the upper-limb system may lead us to think that the control requires a very significant brain effort. As a matter of fact, neuroscience studies demonstrate that humans simplify planning and control using a combination of primitives, which the brain modulates to produce hand configurations and force patterns for the purpose of grasping and manipulating different objects. This concept can be transferred to robotic systems, allowing control within a space of lower dimension. The lower number of parameters characterizing the system allows for embodying the control in machine learning frameworks, reproducing a sort of human-like cognition.

1 Postural Synergies in Human Beings

With 27 bones, 18 joints and 39 intrinsic and extrinsic muscles with over 20 degrees of freedom [1–3], the hand is one of the most complex biomechanical parts of the human body. A traditional point of view is that the brain controls each joint and muscle to generate forces for grasping objects [4, 5]. To date, however, most studies have emphasized the opposite [6]. An early attempt to characterize hand postures during grasping has been made in [7], describing two main categories: *precision grasps* and *power grasps*. In the first category, one or more fingers are positioned,

M. Monforte (✉) · F. Ficuciello · B. Siciliano
DIETI, Università degli Studi di Napoli Federico II, Naples, Italy
e-mail: marco.monforte@iit.it; marco.monforte3@gmail.com

F. Ficuciello
e-mail: fanny.ficuciello@unina.it

B. Siciliano
e-mail: bruno.siciliano@unina.it

M. I. Aldinhas Ferreira et al. (eds.), *Cognitive Architectures*, Intelligent Systems, Control and Automation: Science and Engineering 94,
https://doi.org/10.1007/978-3-319-97550-4_6

usually in opposition to the thumb, to exert the necessary pressure to avoid the object falling from the hand [8]. In the second category, the palm is involved in the grasp to constrain the object. Later on, other authors [9–12] proposed further categorizations, based on the configuration of the fingers and on the part involved in the contact with the object. The key point of all of these works is that the fingers are used to generate forces, and it is assumed implicitly that the hand configuration is linked to this goal. If this is true, the posture should not change over time, but rather, there should be a discrete set of postures, each one corresponding to a grip.

This problem has been further investigated in [13], which introduces for the first time the concept of *Postural Hand Synergies* to study how the human brain controls the hand pre-grasping without considering haptic feedback. The results of these studies have revealed that the hand is controlled using a number of principal motions. A combination of those motions allows for continuously changing from a power to a precision grasp preshaping.

The Principal Component Analysis (PCA), performed on a number of hand configurations measured on different subjects, has shown that the first two components account for >80% of the variance among the dataset, implying a huge reduction from the 15 degrees of freedoms (DoFs) used to define a simplified kinematic model of the human hand. Higher-order PCs provide additional information about the hand posture, providing small adjustments to the fingers' position.

Another relevant work, developed in [14], has been conducted to study whether the grasp can be described by a lower number of postural synergies and whether there are similarities between synergies in grasping different objects. Five subjects have performed different types of reach-to-grasp movement on objects of different size and shape, while 21 joint positions have been recorded along the entire movement thanks to markers and a four-camera video system. The SVD analysis used in this work has proved that the first eigenposture explains most of the variance in the configurations and is comparable across the subjects. The second eigenposture contributes to the opening of the hand to its maximum during the reaching phase and to the thumb and finger flexion during the closing phase. Finally, higher order eigenpostures contribute by adding further information to the hand shape, in particular, about the flexion of the PIPs and DIPs joints.

All of these works suggest that the human brain does not control each finger or muscle independently but it applies some patterns learned during the evolution process through its cognitive capabilities, aiming at optimizing and simplifying the control of such complex biomechanical structures.

2 Postural Synergies in Robotics

The continuous technological improvement of recent decades is leading the robotics field to spread exponentially throughout in our society. Robots should be provided with improved reasoning capabilities and sensorimotor skills in order to interact deftly with their surrounding environment. Anthropomorphic robotic hands contribute to this purpose, providing great dexterity and manipulation capabilities. Their high number of joints, however, might represent a complication for planning and control, especially during interaction with the environment.

Therefore, the use of postural synergies holds great potential, implying a substantial reduction of the dimension of the grasp synthesis problem. Their computation requires human hand motion mapping on the robotic hand.

2.1 Mapping Human Hand Motion to a Robotic Hand

Human hand motion mapping is a quite challenging problem due to the complexity and variety of hand kinematics. To obtain an accurate estimation of the human hand posture, a reliable kinematic hand model and very precise motion tracking instruments are required.

A model-based approach has been proposed in [15], using the fully actuated anthropomorphic DEXMART Hand [16]. The method is based on the detection of the positions of the fingertips of the human hand with respect to the palm through a Kinect RGBD camera. Due to the obvious differences in size and kinematics of the human hand [17], 5 different subjects have been involved in the acquisitions. Each of them had to perform the 36 grasps, with different types of grasp and objects of different shape and size.

To obtain the measures of the fingertips with respect to the palm frame, we first need to compute the homogeneous transformation between the camera frame and the palm frame. This goal has been achieved by measuring a set of five reference points on a rigid panel fixed to the back of the hand. Thus, first and foremost, each subject has worn this panel on the opisthenar and has assumed an open-hand posture. Ten points have been detected: the fingertips and five points suitably placed on the panel (Fig. 1a). Once the transformation T_i between the camera and the palm frame of the i-th subject has been found, each subject performs the 36 configurations.

To map the human grasps on the DEXMART Hand, a Closed-Loop Inverse Kinematics (CLIK) algorithm [18] has been used to retrieve the hand configuration, starting from the measured fingertip positions. In the CLIK algorithm, the DEXMART hand kinematics, properly scaled according to the dimensions of the human hand has

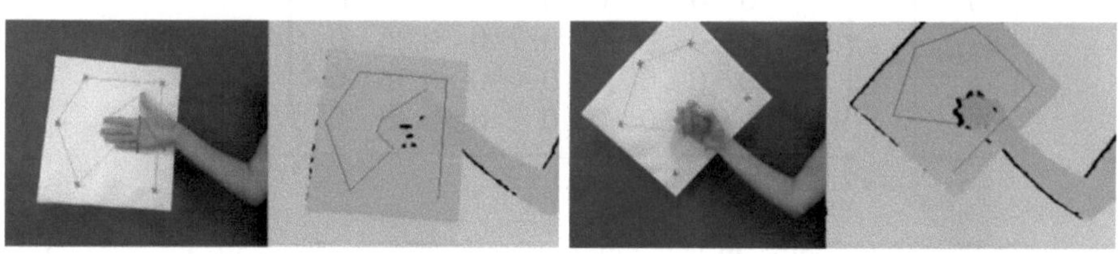

(a) Open-hand configuration. (b) Grasping an object.

Fig. 1 Snapshots from the fingertip position acquisition process been used. The scaling of the robotic hand kinematics is obtained by multiplication of the D-H parameters for the ratio between the lengths of human and robotic fingers.

As a result, the matrix $\mathbf{C} \in \mathbb{R}^{36\times 15}$ has been created, where each $\mathbf{c}_i$ is a joint configuration representing the average of the five robotic hand configurations mapped from the five subjects.

2.1.1 Mapping to an Under-Actuated Robotic Hands

The same mapping method has been applied in [19] to an under-actuated robotic anthropomorphic hand to evaluate how the postural synergies change with respect to the fully-actuated case. The robotic hand considered in this work is the Schunk 5-Finger Hand (S5FH) [20]. Its structure is human-inspired, with dimensions comparable to those of humans and a weight of 1.3 kg. The hand possesses 20 degrees of freedom actuated by only 9 motors, thanks to mechanical synergies that regulate the kinematic couplings between the joints. These mechanical couplings are represented by the matrix S_m in (1), where the relationship between the 20 joints and the 9 motors is clear:

$$\underbrace{\begin{bmatrix} q_{t_o} \\ q_{t_{cm}} \\ q_{t_{mcp}} \\ q_{t_{dip}} \\ q_{i_s} \\ q_{i_{mcp}} \\ q_{i_{pip}} \\ q_{i_{dip}} \\ q_{m_{mcp}} \\ q_{m_{pip}} \\ q_{m_{dip}} \\ q_{p_o} \\ q_{r_s} \\ q_{r_{mcp}} \\ q_{r_{pip}} \\ q_{r_{dip}} \\ q_{l_s} \\ q_{l_{mcp}} \\ q_{l_{pip}} \\ q_{l_{dip}} \end{bmatrix}}_{\mathbf{q}} = \underbrace{\begin{pmatrix} 0.5 & 0 & 0 & 0 & 0 & 0 & 0 & 0 & 0 \\ 0 & 0.29 & 0 & 0 & 0 & 0 & 0 & 0 & 0 \\ 0 & 0.29 & 0 & 0 & 0 & 0 & 0 & 0 & 0 \\ 0 & 0.42 & 0 & 0 & 0 & 0 & 0 & 0 & 0 \\ 0 & 0 & 0 & 0 & 0 & 0 & 0 & 0 & 0.25 \\ 0 & 0 & 1 & 0 & 0 & 0 & 0 & 0 & 0 \\ 0 & 0 & 0 & 0.49 & 0 & 0 & 0 & 0 & 0 \\ 0 & 0 & 0 & 0.51 & 0 & 0 & 0 & 0 & 0 \\ 0 & 0 & 0 & 0 & 1 & 0 & 0 & 0 & 0 \\ 0 & 0 & 0 & 0 & 0 & 0.49 & 0 & 0 & 0 \\ 0 & 0 & 0 & 0 & 0 & 0.51 & 0 & 0 & 0 \\ 0.5 & 0 & 0 & 0 & 0 & 0 & 0 & 0 & 0 \\ 0 & 0 & 0 & 0 & 0 & 0 & 0 & 0 & 0.25 \\ 0 & 0 & 0 & 0 & 0 & 0 & 0.26 & 0 & 0 \\ 0 & 0 & 0 & 0 & 0 & 0 & 0.36 & 0 & 0 \\ 0 & 0 & 0 & 0 & 0 & 0 & 0.38 & 0 & 0 \\ 0 & 0 & 0 & 0 & 0 & 0 & 0 & 0 & 0.5 \\ 0 & 0 & 0 & 0 & 0 & 0 & 0 & 0.26 & 0 \\ 0 & 0 & 0 & 0 & 0 & 0 & 0 & 0.36 & 0 \\ 0 & 0 & 0 & 0 & 0 & 0 & 0 & 0.38 & 0 \end{pmatrix}}_{\mathbf{S}_m} \underbrace{\begin{bmatrix} m_0 \\ m_1 \\ m_2 \\ m_3 \\ m_4 \\ m_5 \\ m_6 \\ m_7 \\ m_8 \\ m_9 \end{bmatrix}}_{\mathbf{m}} + q_0, \qquad (1)$$

where $\mathbf{q}$ is the joints vector, $\mathbf{m}$ is the motors vector and $\mathbf{q}_0$ is an offset representing the vector of joint values when the motor positions are zero.

In this case, to map the subject fingertip positions for the 36 configurations to the robotic hand, the CLIK algorithm must take these couplings into account. The differential kinematics between the mechanical synergies subspace and the Cartesian space then becomes

$$\dot{\mathbf{x}} = \mathbf{J}_{h_s}\dot{\mathbf{m}}, \tag{2}$$

where $\mathbf{J}_{h_s}$ is the mechanical synergies Jacobian, computed as

$$\mathbf{J}_{h_s} = \mathbf{J}_h\mathbf{S}_m. \tag{3}$$

In (2), $\dot{\mathbf{x}}$ is the derivative of the five fingertips position vector $\mathbf{x} \in \mathbb{R}^{15}$ and $\mathbf{J}_h$ is the (15×20) S5FH Jacobian. The CLIK algorithm using the $\mathbf{J}_h^T$ has ultimately been used to map the grasps executed by the five subjects to the S5FH, leading, as with the DEXMART Hand, to the creation of a matrix of the configurations $\mathbf{C} \in \mathbb{R}^{36\times 9}$.

2.2 *Hand Synergies Computation*

Several methods have been proposed for computing the postural synergies. In [21–23], the synergies subspace is constituted by a matrix of constant eigengrasps, while in [24], the synergies are mapped directly to the robotic hand, resulting in a non-constant synergy matrix.

The first method has been used in [19] on the matrix $\mathbf{C} = \{\mathbf{c_1}, \ldots, \mathbf{c_{36}}\}$ after centering through the vector $\bar{\mathbf{c}}$, which is the mean configuration over the 36 grasps. In this way, the matrix $\mathbf{C}_{norm} = \{\mathbf{c}_1 - \bar{\mathbf{c}}, \ldots, \mathbf{c}_{36} - \bar{\mathbf{c}}\}$ of the grasp offsets with respect to $\bar{\mathbf{c}}$ has been computed. The Principal Component Analysis can now be applied by diagonalizing the covariance matrix of $\mathbf{C}_{norm}$ such that

$$\mathbf{C}_{norm}\mathbf{C}_{norm}^T = \mathbf{E}\mathbf{S}^2\mathbf{E}^T, \tag{4}$$

where the $(h \times h)$ orthogonal matrix $\mathbf{E}$ gives the directions of the variance of the data, while the diagonal matrix $\mathbf{S}^2$ represents the variance in each direction sorted by decreasing magnitude. Moreover, the matrix $\mathbf{E}$ represents the base matrix of the synergies subspace. Considering the entire (9×9) matrix, we obtain the whole configuration space of the hand, but, analyzing the variance described by the first j-th eigenvalues, it has been found that the first three principal components account for >85% of the variance, in accordance with what has been proved for the human hand in [13, 14]. This means that the matrix $\mathbf{C}$ can be reconstructed faithfully selecting the three predominant components from the PCA:

$$\hat{\mathbf{E}} = \left[\mathbf{e}_1\ \mathbf{e}_2\ \mathbf{e}_3\right], \tag{5}$$

while the configuration $\mathbf{c}_i$ can be projected onto the postural synergies subspace with a suitable choice of the parameter vector $\boldsymbol{\alpha} \in \mathbb{R}^3$ of the postural synergies:

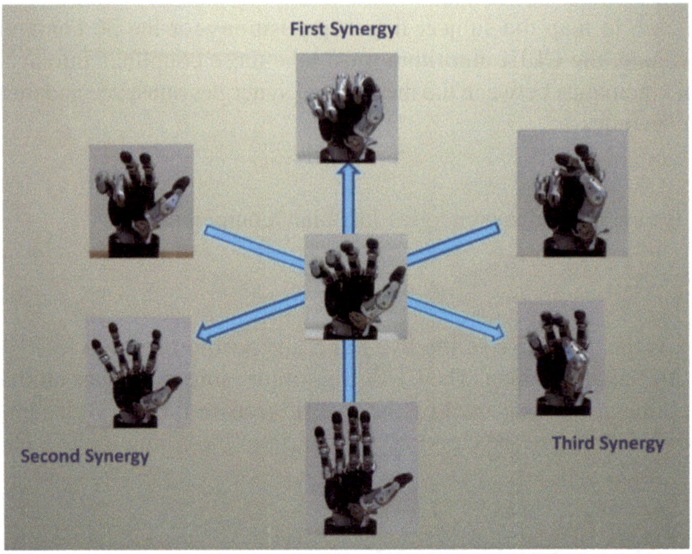

Fig. 2 Mean configuration and first three eigengrasps for the S5FH

$$\hat{\mathbf{c}}_i = \bar{\mathbf{c}} + \hat{\mathbf{E}} \begin{bmatrix} \alpha_{1,i} \\ \alpha_{2,i} \\ \alpha_{3,i} \end{bmatrix}. \quad (6)$$

With these parameters, each synergy can be associated with a minimum and a maximum configuration, obtained by spanning the respective eigenvector through the minimum and maximum value of the associated weight $\boldsymbol{\alpha}_i$ without violating the joint limits. Figure 2 shows the mean configuration $\bar{\mathbf{c}}$ in the center and the minimum and maximum configuration from each synergy computed on the Schunk 5-Finger Hand.

It can be seen that the first synergy acts on the joints with a flexion movement, thus it is responsible for the opening and closing of the hand. The second synergy generates opposite motions for the metacarpophalangeal flexion and proximal interfalangeal flexion joints of the index and middle fingers (the ones without couplings). Finally, the third synergy acts mainly on the flexion and opposition of the thumb.

2.3 *Grasping Control in the Synergies Subspace*

Each grasp posture can be reconstructed from the linear combination of the restricted number of synergies adopted. From (6), it is easy to see that the coefficients of the synergies, characterizing the i-th configuration, can be obtained with a simple inversion:

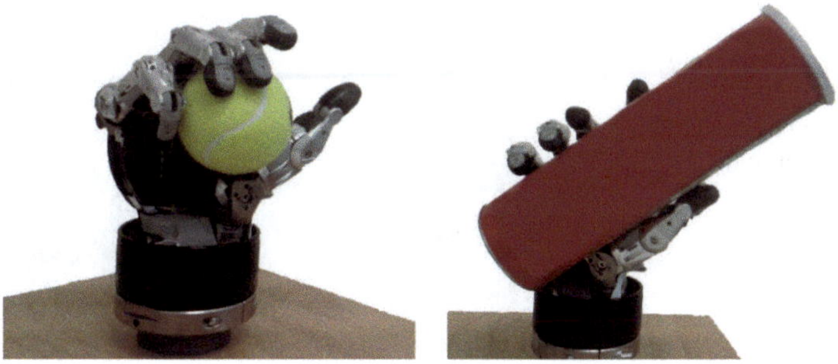

Fig. 3 Two examples of reconstructed configuration

$$\begin{bmatrix} \alpha_{1,i} \\ \alpha_{2,i} \\ \alpha_{3,i} \end{bmatrix} = \hat{\mathbf{E}}^{\dagger} (\mathbf{c}_i - \bar{\mathbf{c}}), \tag{7}$$

where $\hat{\mathbf{E}}^{\dagger}$ is the Moore-Penrose pseudo-inverse of $\hat{\mathbf{E}}$.

Of course, since the synergies provide only the final posture, the entire movement is not defined. However, we remember that in [14], it has been noticed that the human being opens its hand when reaching for an object to grasp it. Thus, it is licit to assume that a grasping movement might start from the initial configuration $\bar{\mathbf{c}}$, then go into an open-hand configuration $\mathbf{c}_0$, and finally towards the grasp configuration $\mathbf{c}_i$. In this way, the movement of the finger is obtained by means of a linear interpolation of the three coefficients $\boldsymbol{\alpha}$ corresponding to the three configurations mentioned above and computed with (7).

However, due to differences between human hand and robotic hand kinematics, some postures might not be accurate enough to allow for effective grasping of the object. This is also clear from Fig. 3, where we can see that not all the fingers are in contact with the objects. Thus, this simple approximation obtained by means of a few predominant synergies must be integrated with an appropriate control law, operating directly in the synergies subspace, in order to adjust the grasp and adapt the hand to the shape of the object. The approach proposed in [19] is a CLIK algorithm, in which a constant fingertip reference term is given by an approximation of the desired grasp in the synergies subspace. This term will determine a good hand pre-grasping. Afterwards, an additional term is designed to close the hand around the centroid of a virtual object, calculated as the mean position of the fingers employed in the grasp. The inverse kinematics is based on the synergies Jacobian given by $\dot{\mathbf{x}} = \mathbf{J}_{h_{ss}}\dot{\boldsymbol{\alpha}}$, with $\mathbf{J}_{h_{ss}} = \mathbf{J}_s\mathbf{S}_m\mathbf{S}_s$ and where $\mathbf{S}_s = \hat{\mathbf{E}}$ and $\boldsymbol{\alpha}$ are the synergies coefficients of the grasp posture. The latter are linked to the joint velocities by Eq. (8).

$$\dot{\mathbf{q}} = \mathbf{S}_m\dot{\mathbf{m}} = \mathbf{S}_m\mathbf{S}_s\dot{\boldsymbol{\alpha}}. \tag{8}$$

Fig. 4 Example of grasp configurations without (*left*) and with (*right*) the additional term based on the virtual object centroid

Moreover, to limit the grasping forces, the target position of the CLIK is limited through the measured motor current and by means of a defined threshold related to the texture of the object. The experiments have proved that a wide variety of objects can be grasped with this control strategy in the synergies subspace. The algorithm is stable and effectively modifies the finger positions to close the hand on the object and regulate the contact forces (Fig. 4).

2.4 Mapping Human Arm Motion to a Robotic Arm

The same concept of hand synergies can be extended to the human arm. A mo-cap suit has been used in [25]. The goal is the creation of a dataset of reaching-to-grasp movements (thus waveforms and no longer static postures) for a robotic arm for the computation of the arm synergies. The robotic arm is a KUKA Lightweight Robot 4+ [26], while the mo-cap system in question is the Xsens-MVN tool [27], composed by the Xsens suit and the proprietary software Moven Studio. The Xsens is a full body suit equipped with IMU sensors, named MTx, with advanced sensor fusion algorithms and wireless communication. Seventeen MTx are mounted on the most important parts of the human body, such as the head, chest, arm, forearm, hand, and so on. These MTx send their data to the MVN software, which, after an earlier calibration process, allows for real-time visualization of the human motion, playback and editing of the received data. An important option of this software is given by the possibility of sending data to third applications. Using an UDP/TCP-IP socket, the MTx data have been sent to the robotic arm using a mapping method that exploits the fact that the KUKA LWR presents 7 degrees of freedom, like the human arm. This has enabled a faithful replication of the master's movements.

To map her/his arm motion to the KUKA LWR 4+, the human master has to wear the Xsens suit (Fig. 5). After the calibration procedure provided by the software, the unit quaternions of the arm, forearm and hand, $\mathcal{Q}_{arm}$, $\mathcal{Q}_{forearm}$ and $\mathcal{Q}_{hand}$, are continuously received from the C++ script, which is charged with controlling the robotic arm by means of two CLIK algorithms. The first CLIK receives $\mathcal{Q}_{arm}$ and solves for the first three joints of the KUKA. The second CLIK receives $\mathcal{Q}_{hand}$ and solves for the other four joints of the robot arm. The elbow is a redundant joint

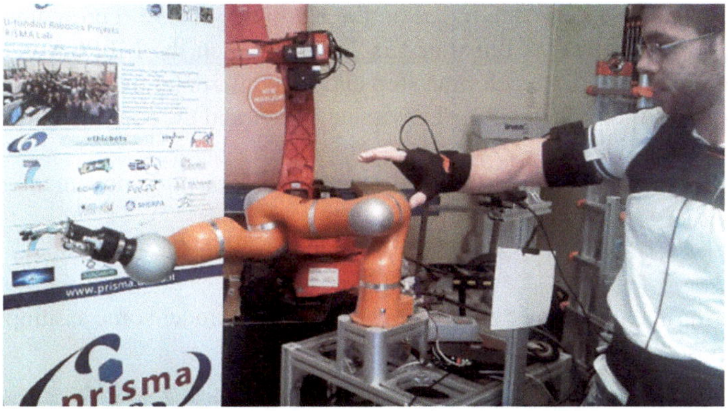

Fig. 5 The human master wearing the Xsens suit in order to telemanipulate the KUKA LWR 4+ and S5FH hand-arm system

controlled in the null space of the robot Jacobian to impose the same angle between the human arm and forearm, computed from $\mathcal{Q}_{arm}$ and $\mathcal{Q}_{forearm}$ with (9):

$$\theta_{elbow} = \arccos\left(\frac{\mathbf{v}_a \mathbf{v}_f}{\|\mathbf{v}_a\| \|\mathbf{v}_f\|}\right), \tag{9}$$

where $\mathbf{v}_a$ and $\mathbf{v}_f$ are the respective directions of the arm and the forearm.

Using this setup, 38 grasps of balls and cylinders, of different shapes and sizes, and with precision and power configurations have been performed. Since each acquired motion has a different duration, a data conditioning process has been carried out using Dynamic Time Warping (DTW) [28], in order that all of the same time length t_f for the N samples may be reached. As a result, the matrix $\mathbf{M} \in \mathbb{R}^{38\times7\times N}$ of grasping movements has been obtained.

2.5 Arm Synergies Computation

The approach presented in Sect. 2.2 for computing the hand postural synergies uses the PCA technique on static configurations. To compute motion arm synergies, an extension for multivariate waveforms of the PCA, namely Multivariate Functional Principal Component Analysis (MFPCA), has been used. A well-recognized procedure for computing the MFPCA does not exist at the moment. A first approach has been proposed in [29] and consists in stacking the waveforms recorded for each demonstration and then performing a common Univariate Functional Principal Component Analysis. The computed FPCA are then divided by the number of variables, obtaining the single FPCs. Extensions of the clustering problem and of data analysis of different dimensional domains have been respectively proposed in [30] and [31].

These works have an approach based on the Karhunen-Loeve representation of the data. A different method is illustrated in [32], in which the MFPCA is computed by performing the PCA at each time step and then interpolating the results. Further details about the theory behind the Univariate FPCA and the Multivariate FPCA can be found in [31].

According to the Karhunen-Loeve theorem, each grasping movement

$$\mathbf{m}_i(t) = \left(m_i^{(1)}(t), m_i^{(2)}(t), \ldots, m_i^{(7)}(t)\right) \quad \text{with} \quad t \in \left[0, t_f\right], \quad i = 1, \ldots, 38 \tag{10}$$

can be seen as a realization of a stochastic process and, under some assumptions, can be decomposed as

$$\mathbf{m}_i(t) = \boldsymbol{\mu}(t) + \sum_{k=1}^{\infty} \xi_{ik} \boldsymbol{\varphi}_k(t) \quad \text{with} \quad t \in \left[0, t_f\right], \tag{11}$$

where $\boldsymbol{\mu}(t)$ is the vector of the mean functions of the joints, $\boldsymbol{\varphi}_k$ is the vector of the k-th FPCs and ξ_{ik} is the k-th coefficient (or *score*) of the respective FPC for the i-th demonstration (Fig. 6). Thus, by truncating the sum to K terms, it is possible to approximate and parametrize each grasping movement with K scalar coefficients.

In analogy with the postural synergies of the hand, the function $\boldsymbol{\varphi}_k$ represents the waveform of a synergy, while the coefficient ξ_{ik} modulates the latter to obtain the movement.

From the MFPCA application on the matrix $\mathbf{M}$, 7 mean functions $\left[\mu^{(1)}(t), \ldots, \mu^{(7)}(t)\right]$ of the KUKA joints are obtained, K eigenfunctions $\left[\varphi_1^{(j)}(t), \ldots, \varphi_K^{(j)}(t)\right]$ for each joint j representing the basis of the subspace for each

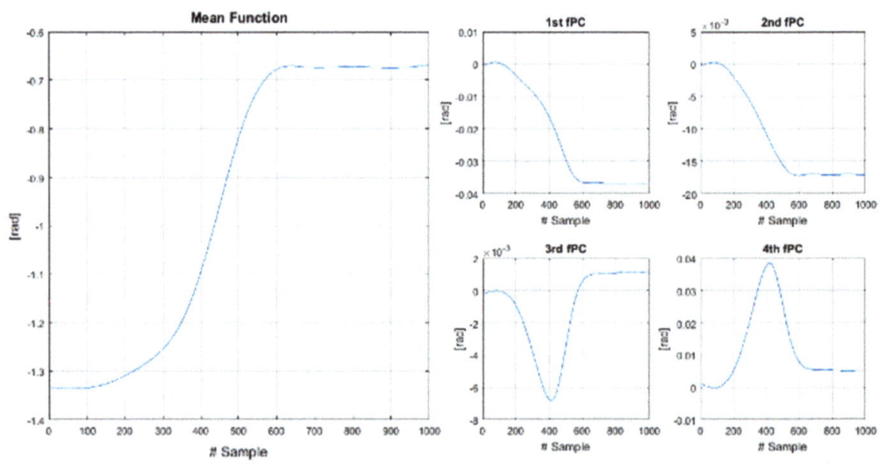

Fig. 6 Example of a mean function and the first 4 FPCs computed for the 5th joint

joint, and a matrix $\Xi \in \mathbb{R}^{38 \times K}$, where each row contains the scores of the respective demonstration.

From the analysis of the eigenvalues, it has been found that $K = 2$ FPCs are enough to cover >90% of the variance.

3 Combining Synergies with Machine Learning

From Sects. 2.2 and 2.3, it is clear that synergic patterns can be computed to reduce the number of parameters so as to control a high DoFs device.

In order to generalize the grasping strategy, a supervised learning system can be trained with the goal of learning the non-linear function that links the object's characteristics to its coefficients, so as to estimate the synergies coefficients for new objects. By selecting appropriate input features, such as the object type, its dimensions and/or the type of grasp, a database of configurations can be created, as in [19]. Applying the synergies computation to this database, a training set can be obtained associating the synergies coefficients with each example (thus, with each grasp performed). In this way, a Neural Network model can be trained with one of the several existing methods. Close attention, of course, must be paid to the creation of the training set, which has to cover a large variety of object shapes and sizes, and to the model hyperparameters tuning, such as learning rate, regularization term (to prevent underfitting and overfitting), number of hidden layers and number of hidden units. Anyway, a small percentage of error is always present when using neural networks. In the case of synergies, this is due first and foremost to the approximation introduced by their computation, and then to other reasons, like the training procedure itself.

To compensate for this error, in [33], a Reinforcement Learning strategy has been integrated directly into the synergies subspace. In particular, a modified version of the Policy Improvement with Path Integrals (PI^2) algorithm [34, 35] has been used. The policy update uses a probability-weighted averaging, without the needs of a gradient estimate and avoiding numerical instabilities due to matrix inversions. The synergies coefficients obtained from the neural network initialize the vector $\boldsymbol{\theta}$, representing the mean value of a Multivariate Gaussian distribution. From the latter, a number K samples are executed. K is defined by the user, along with the number of iterations of the algorithm, the initial covariance matrix $\boldsymbol{\Sigma}_{init}$ of the Multivariate Gaussian and a decay rate of $0 < \gamma \leq 1$. The PI^2 extracts these K samples and evaluates them, using a function based on the grasp quality index defined in [36] and already used in [19]. After the evaluation, the mean of the Multivariate Gaussian is updated by weighting the previous trials and moving $\boldsymbol{\theta}$ toward those attempts with better reward. The covariance matrix, instead, is multiplied by the decay rate, in order to reduce the dispersion of the subsequent trials from the good values obtained previously. The algorithm proceeds in this way until it reaches an optimal mean value for the Gaussian, with a covariance so small that the samples are too close to the mean to bring substantial differences.

The experiments are made with the robotic hand-arm system constituted by the KUKA LWR 4+ and the SCHUNK 5FH. Exploiting the hand and arm synergies computed in [19] and [25], two neural networks have been trained (one for the hand and one for the arm) using the Matlab NN Toolbox to provide the initial synergies parameters for a PI^2 algorithm. Human supervision has been necessary (but could be replaced in future by a vision system) to tell whether the object was grasped or not, in order to evaluate the reward function adopted (12)

$$r(\boldsymbol{\theta}_k) = V(\boldsymbol{\theta}_k) + \phi, \tag{12}$$

where $V(\boldsymbol{\theta}_k)$ is the measured quality index and ϕ is

$$\phi = \begin{cases} 0 & \text{if grasp succeeds} \\ 10^3 & \text{if grasp fails} \end{cases}, \tag{13}$$

which is chosen in order to penalize failed trials, and thus lead the convergence of the PI^2 toward the successful grasp.

4 Conclusions

The experiments carried out proved that the usage of 3 hand synergies and 2 arm synergies in a machine learning system composed by neural networks and reinforcement learning allows the robot to improve its grasping capabilities through a trial-and-error approach (Fig. 7). Machine learning techniques can be efficiently combined with synergies in order to create frameworks capable of reducing the complexity of control by taking inspiration from human cognitive architectures.

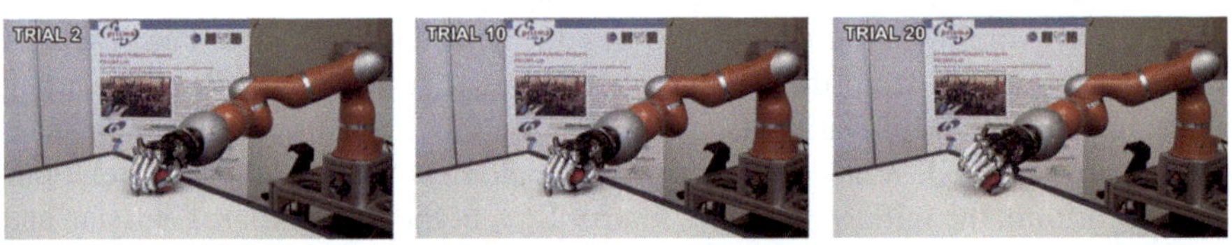

(a) Precision grasp of a ball.

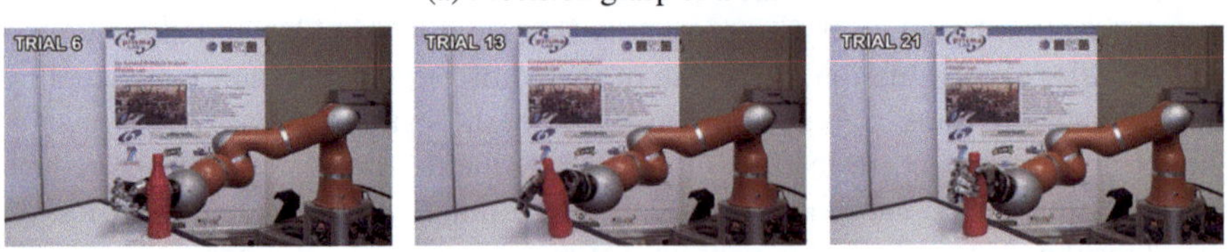

(b) Power grasp of a cylinder.

Fig. 7 Improvement of the grasp during execution of the algorithm

Acknowledgements This research has been partially funded by the EC Seventh Framework Programme (FP7) within RoDyMan project 320992 and by the national grant MUSHA under Programma STAR linea 1.

References

1. Kapandji, I. A. (1970). *The physiology of the joints. Upper limb* (Vol. 1, 2nd ed., pp. 146–202). London: E. and S. Livingstone.
2. Tubiana, R. (1981). Architecture and function of the hand. In R. Tubiana (Ed.), *The Hand* (Vol. 1, pp. 19–93). Philadelphia, PA: Saunders.
3. Soechting, J. F., & Flanders, M. (1997). Flexibility and repeatability of finger movements during typing: Analysis of multiple degrees of freedom. *Journal of Computing Neuroscience, 4*, 29–46.
4. Lemon, R. N. (1999). Neural control of dexterity: What has been achieved? *Experimental Brain Research, 128*, 6–12.
5. Schieber, M. (1990). How might the motor cortex individuate movements? *Trends Neuroscience, 13*, 440–445.
6. Schieber, M. (1995). Muscular production of individuated finger movements: The roles of extrinsic finger muscles. *Journal of Neuroscience, 15*, 284–297.
7. Napier, J. R. (1956). The prehensile movements of the human hand. *Journal Bone and Joint Surgery*, 38B, 902–913.
8. Johannson, R. S., & Cole, K. J. (1992). Sensory-motor coordination during grasping and manipulative actions. *Current Opinion in Neurology, 2*, 815–823.
9. Kamakura N., Matsuo M., Ishii H., Mitsuboshi F., & Miura Y. Patterns of static prehension in normal hands. *The American Journal of Occupational Therapy, 7*, 437–445.
10. Elliot, J. M., & Connolly, K. J. A. (1984). Classification of manipulative hand movements. *Developmental Medicine & Child Neurology, 26*, 283–296.
11. Klatzky, R. L., Pellegrino, J., McCloskey, B., Doherty, S., & Smith, T. (1987). Knowledge about hand shaping and knowledge about objects. *Journal of Motor Behavior, 19*, 187–213.
12. Cutkosky, M. R., & Howe, R. D. (1990). Human grasp choice and robotic grasp analysis. In S. T. Venkataraman & T. Iberall (Eds.), *Dextrous robot hands* (pp. 5–31). New York: Springer.
13. Santello, M., Flanders, M., & Soechting, J. F. (1998). Postural hand synergies for tool use. *The Journal of Neuroscience, 18*, 10105–10115.
14. Mason, C. R., Gomez, J. E., & Ebner, T. J. (2001). Hand synergies during reach-to-grasp. *Journal of Neurophysiology, 86*, 2896–2910.
15. Ficuciello F., Palli G., Melchiorri C., & Siciliano B. (2013). A model-based strategy for mapping human grasps to robotic hands using synergies. In *Proceedings 2013 IEEE/ASME International Conference on Advanced Intelligent Mechatronics*.
16. Palli, G., Melchiorri, C., Vassura, G., Berselli, G., Pirozzi, S., Natale, C., De Maria, G., & May, C. (2012). Innovative technologies for the next generation of robotic hands. In B. Siciliano (Ed.), *Advanced Bimanual Manipulation*. Springer Tracts in Advanced Robotics (Vol. 80, pp. 173–218). Springer.
17. Grebenstein, M., Chalon, M., Hirzinger, G., & Siegwart, R. (2010). A method for hand kinematics designers 7 billion perfect hands. In *Proceedings 1st International Conference on Applied Bionics and Biomechanics* (pp. 357–362). Venice, Italy.
18. Siciliano, B., & Khatib, O. (Eds.). (2008). *Springer Handbook of Robotics* (2nd ed.). Springer.
19. Ficuciello, F., Federico, A., Lippiello, V., & Siciliano, B. (2017). Synergies evaluation of the SCHUNK S5FH for grasping control. *Springer Proceedings in Advanced Robotics, 4*, 225–233.
20. Schunk hand webpage. http://www.schunk-modular-robotics.com/en/home/products/servo-electric-5-finger-gripping-hand-svh.html.

21. Ficuciello, F., Palli, G., Melchiorri, C., Siciliano, B. (2011). Experimental evaluation of postural synergies during reach to grasp with the UB Hand IV. In *Proceedings IEEE/RSJ International Conference on Intelligent Robots and Systems* (pp. 1775–1780). San Francisco.
22. Geng, T., Lee, M., & Hulse, M. (2011). Transferring human grasping synergies to a robot. *Mechatronics*, *21*(1), 272–284.
23. Sun, S., Rosales, C., & Suarez, R. (2010). Study of coordinated motions of the human hand for robotic applications. In *Proceedings IEEE International Conference on Robotics and Automation* (pp. 776–781). Anchorage, Alaska.
24. Gioioso, G., Salvietti, G., Malvezzi, M., & Prattichizzo, P. (2011). Mapping synergies from human to robotic hands with dissimilar kinematics: An object based approach. In *IEEE International Conference on Robotics and Automation, Workshop on Manipulation Under Uncertainty*. Shangai.
25. Ficuciello, F., Zaccara, D., & Siciliano, B. (2016). Learning grasps in a synergy-based framework. In *Springer Proceedings in Advanced Robotics* (Vol. 1, pp. 125–135). Cham.
26. KUKA Lightweight Robot 4+ webpage. https://www.kukakore.com/wp-content/uploads/2012/07/KUKA_LBR4plus_ENLISCH.pdf.
27. Xsens-MVN webpage. https://www.xsens.com/products/xsens-mvn.
28. Giorgino, T. (2009). Computing and visualizing dynamic time warping alignments in R: The dtw package. *Journal of statistical Software*, *31*(7), 1–24.
29. Ramsay, J. O., & Silverman, B. W. (2005). *Functional Data Analysis*. Springer.
30. Jacques, J., & Preda, C. (2004). Model-based clustering for multivariate functional data. *Computational Statistics & Data Analysis*, *71*, 92–106.
31. Happ, C. (2015). Multivariate functional principal component analysis for data observed on different (dimensional) domains. *Journal of the American Statistical Association*.
32. Berrendero, J., Justel, A., & Svarc, M. (2011). Principal components for multivariate functional data. *Computational Statistics & Data Analysis*, *55*(9), 2619–2634.
33. Ficuciello, F., Zaccara, D., & Siciliano, B. (2016). Synergy-based policy improvement with path integrals for anthropomorphic hands. In *Proceedings IEEE/RSJ International Conference on Intelligent Robots and Systems* (IROS2016) (pp. 1940–1945).
34. Theodorou, E., Buchli, J., & Schaal, S. (2010). A generalized path integral control approach to reinforcement learning. *Journal of Machine Learning Research*.
35. Stulp, F., & Sigaud, O. (2012). Path integral policy improvement with covariance matrix adaptation. In *Proceedings of the 29 International Conference on Machine Learning*.
36. Bicchi, A. (1994). On the closure properties of robotic grasping. *International Journal of Robotics Research*, *14*(4), 319–334.

The Synthetic Psychology of the Self

Tony J. Prescott and Daniel Camilleri

Abstract Synthetic psychology describes the approach of "understanding through building" applied to the human condition. In this chapter, we consider the specific challenge of synthesizing a robot "sense of self". Our starting hypothesis is that the human self is brought into being by the activity of a set of transient self-processes instantiated by the brain and body. We propose that we can synthesize a robot self by developing equivalent sub-systems within an integrated biomimetic cognitive architecture for a humanoid robot. We begin the chapter by motivating this work in the context of the criteria for recognizing other minds, and the challenge of benchmarking artificial intelligence against human, and conclude by describing efforts to create a sense of self for the iCub humanoid robot that has ecological, temporally-extended, interpersonal and narrative components set within a multi-layered model of mind.

Alan Turing, one of the founders of computer science, once suggested that there were two paths to human-level Artificial Intelligence (AI)—one through emulating the more abstract abilities of the human mind, such as chess playing, the other, much closer to the spirit of this book, by providing a robot with "the best sense organs that money can buy, and then teach[ing] it to understand and speak English. This process could follow the normal teaching of a child" [68, p. 460]. Turing was noncommittal about which approach would work best and suggested we try both. Two-thirds of a century after Turing, as different AIs battle between themselves to be the world's best at chess [61], it is clear that the first approach has been spectacularly successful at producing some forms of machine intelligence, though not at emulating or approaching "general intelligence"—the wider intellectual and cognitive capacities

---

T. J. Prescott (✉) · D. Camilleri
The University of Sheffield and Sheffield Robotics, Sheffield, UK
e-mail: t.j.prescott@sheffield.ac.uk

M. I. Aldinhas Ferreira et al. (eds.), *Cognitive Architectures*, Intelligent Systems, Control and Automation: Science and Engineering 94,
https://doi.org/10.1007/978-3-319-97550-4_7

of our species.[1] Enthusiasm for Turing's second approach has therefore re-emerged and is continuing to grow.

1 Beyond the Turing Test

Even more famously, and in the same paper [68], Turing also suggested a way of deciding whether a machine could think in the form of an "imitation game." In what is now universally known as the "Turing test", a judge is asked to distinguish between a human and a machine based on written communication alone. In devising the test, Turing explicitly sought to avoid defining thinking in terms of unobservables, for example, operations of the mind. Instead, he argued that we should focus on behavioral phenomena, such as the ability to conduct a conversation that, in a human, would be recognized as requiring thinking. The design of the Turing test is intended to create an unbiased way of comparing a machine with a man or woman, since there are no extraneous clues, such as appearance or tone of voice, to reveal which is which. Since 1991, an annual competition, the Loebner prize, has sought to evaluate the ability of AIs to pass tests based on Turing's proposal—a prize of $100,000 stands on offer to the first AI to be consistently mistaken for an adult human following an extended and open-ended conversation.

In *Ex Machina,* the 2015 science fiction movie about future AI, Nathan Bateman, the fictional inventor of Ava, a new kind of humanoid robot, proposes an alternative to the Turing Test, in which "the real test is to show you that she [Ava] is a robot; then see if you still feel she has consciousness."[2] What we might call the "Garland test", after the writer of *Ex Machina*, Alex Garland,[3] is arguably a tougher challenge than the original test devised by Turing—there is no question of whether you are speaking to a robot or a human; the witness you are interrogating is clearly a machine. Yet, like Caleb Smith, the young programmer whom Nathan chooses to interact with his robot, you might feel compelled by the robot's ability to converse and behave in a life-like way to view this machine as having a mind of its own.

It is worth noting that Turing intended his test as a way of deciding whether a machine could think, and not whether a machine has consciousness. Indeed, Turing writes, "I do not think these mysteries [about consciousness] necessarily need to be solved before we can answer the question with which we are concerned in this

[1]By this, we mean the cluster of different but overlapping intellectual/cognitive faculties that make humans adaptive, flexible sociotechnical animals. Gardner's [22] "multiple intelligences" view provides a good guide to this broader notion of human cognition. Attempts to create machine intelligence of this more multi-faceted form are increasingly discussed under the label *Artificial General Intelligence* (AGI) (e.g., [23]), hence we are using the phrase "general intelligence" rather than Gardner's multiple intelligences.

[2]Nathan Bateman to Caleb Smith about the humanoid robot "Ava" he has created, from the original movie script for *Ex Machina* (2015) by Alex Garland.

[3]The suggestion that we call this the Garland test has also been made by Murray Shanahan, one of the scientific advisors on *Ex Machina*.

paper [whether a machine can think]" ([68], p. 447). However, many commentators have considered the test to be about consciousness, for example, John Searle, in describing the Chinese Room, a thought experiment predicated on the Turing test, rephrases Turing's question "can a machine think?" as "can a machine have conscious thoughts?" ([59], p. 20). The Chinese Room is intended to demonstrate that a machine could pass the Turing test in Chinese *without* understanding Chinese. Turing might possibly have agreed. For Searle, and others, thoughts have to come from conscious minds in order to be actual thoughts (to be "about" something), whereas for Turing, it was enough for a system to generate the right kind of behavior to be considered as thinking; consciousness was something else.

Other forms of Turing test have also been proposed by Harnad [24, 25], who has suggested a hierarchy of Turing tests: Level T1 is a narrow AI, for instance, one that can prove mathematical theorems or is exceptional at chess. T2, the original test, demonstrates what Harnad calls "pen-pal" level indistinguishability by emulating human linguistic capacity. T3, the "total Turing test", requires that the robot is capable of emulating human language *and action*, but need not be made of biological stuff or otherwise constrained to match a particular internal structure. For Harnad, T3 is the level at which we judge other people, the point at which symbolic computation becomes "grounded" in the external world, and therefore the correct level at which to judge whether a machine has conscious thoughts.[4] Harnad also describes, but rejects as too stringent, a level T4—detailed biological indistinguishability—as might be required by some anti-functionalist stances.

One of the more intriguing ideas in *Ex Machina* is that we are left unsure, at the end, as to whether the robot, Ava, has a mind similar to ours or whether it is, instead, an alien and devious AI that is able to emulate and deceive humans when this serves its purposes. Does this ending suggest a challenge to Harnad's proposal for a T3 Turing test or, indeed, for the Garland test (which is a variant of that test)? Harnad [25] admits that the T3 test is under-constrained in emulating *how* people think, but like Turing, he is comfortable with that; for Harnad, succeeding in the T3 test is evidence enough of grounded (and conscious) thoughts. However, what if we want to get closer to understanding the mind, or to build a machine that actually does think like a human? The evidence from Chess and Go is that machines can exceed human experts at these intellectual challenges without matching the way in which people play either game. Similarly, T3 equivalence could give us grounded symbols, but without further resolving how human minds work.

But perhaps we can get closer to human general intelligence without going all the way to T4 equivalence. Specifically, suppose we add the constraint of having a human-like *cognitive architecture* in addition to matching human symbolic and robotic capacity. If we can match both the behavior and the architecture of mind, then there is a greater likelihood that our AI will not only act like us but also think

[4] It has been suggested that Harnad's T2 level cannot be achieved without first building T3 to achieve symbol-grounding [26]. Going directly to T2 is nevertheless a theoretical possibility, even if it might prove impossible to achieve without a contribution from robotics.

like us. Following the scheme of Harnad's test hierarchy, we might call this level T3.5.

2 Robotics as Synthetic Psychology

Based on this line of reasoning, we have, for the past seven years, been involved in various projects concerned with the development of aspects of general intelligence for humanoid robots. This work builds on the above premise that we can seek to create an artificial mind that is similar to our own by emulating human linguistic and robotic capacity and by employing a cognitive architecture that has been reverse-engineered from findings in psychology and neuroscience. The hope is that we can make significant progress without having to concern ourselves with all of the T4-level detail. The long-term goal is to build a machine that can pass the Garland test whilst being sufficiently biomimetic in design that we can credibly argue that its "mental states" are analogous to human mental states in an interesting way.

This goal can also be seen as belonging to the sub-discipline of *synthetic psychology*, an enterprise within the cognitive sciences named after Valentino Braitenberg's inspirational book *Vehicles: Experiments in Synthetic Psychology* [11], which advocates that we build artificial creatures as a path to understanding the brains and behavior of biological organisms. This "understanding through building" approach also forms a core principle of the emerging field of *Living Machines* [52].[5] Within robotics, there is a growing group of researchers interested in this challenge, indeed, when we add in developmental constraints, this approach to reverse-engineering the human converges within the emerging field of developmental robotics (e.g., [14, 36]).

So, what should the ambition of a synthetic psychologist be in building a human-like machine? For many philosophers and cognitive scientists, even some roboticists, the Holy Grail is to understand and recreate human consciousness. While this ambition is attractive, it suffers from two serious drawbacks. First, the difficulty of deciding what consciousness is, and second, the challenge of measuring subjective first-person phenomena using a third-person approach (the tools of science).[6] For this reason, we have chosen not to make consciousness a target of our synthetic psychology research, preferring instead a (hopefully) more tangible phenomenon—to construct a robot with a "sense of self" [50]. Perhaps we will find that we cannot

[5]This idea also follows in the footsteps of many others. For example, the eighteenth century Neapolitan philosopher Giambattista Vico, who wrote "*verum et factum reciprocantur seu convertuntur* [the true is precisely what is made]", and the 20th century physicist Richard Feynman, whose office blackboard on the day he died held the message, "what I cannot create I do not understand".

[6]There are multiple measures of so-called "correlates of consciousness", Tononi's Φ [65], a measure of information integration, being one of the better-known ones. The problem is that there is no way to be sure that an organism or machine that scores highly on any such measure is actually experiencing consciousness. This is known as the "other minds" problem in philosophy. For Turing [68], this was part of the reason to devise a behavioral test for the existence of machine thought and to leave the challenge of consciousness to others.

completely disentangle self from consciousness, but even so, by understanding the broader nature of self, we may be able to see more clearly what, if anything, is still left to explain about first-person experience.

3 Defining and Deconstructing the Self

Some might baulk at the thought of trying to synthesize the self without directly addressing consciousness, others, following Hume [30], may consider that there is little to be assembled in a synthetic self beyond a bundle of perceptions. But there is an interesting third way. For instance, writers such as the psychologists Blakemore [9] and Hood [29], the cognitive scientist Hofstadter [28], the architect Abel [1] and the philosopher Metzinger [39] have argued that the self as we conventionally imagine it is an illusion, but that, nevertheless, there is something there to be understood. For Blakemore, it is a complex of memes, for Hood, an internal simulation, for Hofstadter, a "strange loop", for Abel, a "field of being" that can extend outside the body[7], and for Metzinger, a meta-representation (amongst other things). Thus, while for Blakemore, the self is a construct, for Hood, Hofstadter, Abel and Metzinger, the self is also a process, or set of processes, some of which may be representational and reflective, that arise in the brain and body. The proposal we are seeking to investigate is similar: that the sense of self can be emulated by a set of definable and buildable processes that can be situated in some suitably configured robot.

The notion that self is a process suggests that it can come and go, for instance, when the relevant processes are suspended during sleep,[8] perhaps even with the switch from an inward to an outward focus of attention. This idea of the self as a transient thing has also been put forward by the philosopher Galen Strawson, who has proposed "that many mental selves exist, one at a time and one after another, like pearls on a string" ([62], p. 424). This poetic metaphor asserts a number of things. First, that the self is not continuous, immutable, and immortal (as Descartes and many others have imagined, and as Hume and others have questioned), and second, that "selves" are nevertheless "things" worthy of study, and perhaps capable of emulation.

[7] Abel's "field of being" view stems from Merleau-Ponty's [38] phenomenology and his insistence on the centrality of the experience of the body. Studies in cognitive neuroscience, such as those of the "rubber hand" illusion (see [10]), support Merleau-Ponty's proposal that the sense of the body/self can extend into objects and the world. With virtual reality systems and telepresence robots, it is now possible to experimentally manipulate the sense of a virtual body, or of a physically remote robot body, and the associated feelings of immersion or "presence", demonstrating that "my body is wherever there is something to be done" (Merleau-Ponty, [38] p. 291) and providing new ways to test hypotheses about the self.

[8] This was proposed by Hume [30], for whom, if the stream of perceptions is turned off, as happens in sleep, the self ceases to exist, and by Locke [35], for whom self was a manifestation of consciousness, which, in turn, requires an awake mind. Some elements of Locke's view of self, which saw identity as arising from learning and memory, are close to the ideas of the extended and narrative selves discussed in this chapter.

What we particularly like about Strawson's approach is that he provides some helpful suggestions as to how we might proceed with the study of self, highlighting five questions ([62], p. 406):

1. *The phenomenological question—what is the nature of the sense of the self?*
2. *The local phenomenological question—what is the nature of the human sense of the self?*
3. *The general phenomenological question—are there other possibilities, when it comes to a sense of the self, e.g., can we describe the minimal case?*
4. *The conditions question—what are the grounds or preconditions of possession of a sense of the self?*
5. *The factual (metaphysical) question—is there (could there be) such a thing as the self?*

Questions 1 and 2 are psychological in nature, and we think that we can make progress on these through empirical exploration[9] of the facets of self and their variability across the population, taking into account, in particular, developmental and neurological differences. Indeed, a wealth of literature already exists on these topics going back to the earliest days of psychological investigation, some of which is discussed in brief below.

Question 3 might direct us to the panoply of animal life as an interesting place to look for the presence of other kinds of self (and pending the discovery of any extraterrestrial selves). Comparative cognition offers many interesting insights, as well as proposals for how we might test for similar facets of self across species. However, with robotics, we also have the possibility of building new kinds of self, including candidate minimal selves, for which we might adopt some of the cross-species yardsticks identified by comparative studies.

Question 4 speaks to another kind of enquiry, namely as to whether there are any necessary conditions restricting the possibility of an entity possessing a self. One requirement we might posit is a body-world boundary and the ability to sense and maintain the internal milieu, while another might be the possession of a particular kind of cognitive architecture in which there are processes that have the capacity to monitor and predict other internal processes. These ideas will be discussed further below.

Finally, question 5 seems to be largely philosophical, however, we think that progress could also be made via a synthetic approach. Specifically, once we have built a robot that exhibits some relevant phenomena of self, we can ask whether a particular conception of self, for instance, Strawson's string of pearls, is useful or not.[10] Indeed,

[9] We should admit here that Strawson intends the more restricted philosophical sense of phenomenology as a form of systematic reflection on the structure of experience. We prefer to interpret the challenge of describing the nature of self from a more empirical perspective as phenomena associated with self that could be accessible to methods in psychology and cognitive neuroscience.

[10] Note that, for a theory or concept of self to be useful, we would not consider that the self has to be emergent in a strong sense (that is, not reducible to lower level phenomena), but rather it has to serve a useful explanatory function in our psychological theory. In other words, the concept of self as explicated and realized in machine form should help us to provide useful accounts of human

we will have an instantiation of a specific theory of self *as a machine*, whose inner workings will be far more accessible than those of a human mind (see [40]). Such a robot should provide an insightful tool for advancing both the philosophical and scientific understanding of the phenomenon of self-hood.

As we peruse Strawson's questions, we think it becomes evident that synthetic psychology could have a lot to say. For instance, on the question of the constitutive conditions, we can build synthetic systems that match the proposed requirements, then apply our phenomenological and Garland tests: Does it behave as though it has a self? Do others see it as having a self? We can also make progress on this question of the minimal form of the target phenomenon—what is the simplest robot that could qualify for self-hood? Let's build it and study it. On the issue of architecture, we can seek to identify a decomposition of the systems underlying the human self that, when suitably replicated in a robot, gives rise to self-like phenomena; this seems to us to be a tractable, if ambitious, challenge.

Note that if selves are transient, as Strawson and others have proposed, we do still need to explain why the experience of self is one of continuity—that you feel you are the same self yesterday, today, tomorrow. Here, we can appeal to the continuity of the body (and the localization of the self within the body) as providing much of the necessary continuity. We can also look to episodic memory and imagination as allowing the instantaneous self to roam in time, recollecting itself as it once was and imagining itself as it might yet be, thus creating an experience of self that can step outside the present and conceive of itself as enduring. Finally, we can consider semantic memory and narrative as providing the basis for a stable self-concept (beliefs and stories about the self). These ideas can also be investigated in our robotic models.

4 A "Systems" View of Self

The plan to create a synthetic robot self becomes more plausible if we can find good evidence for a "systems" view of self in psychology and cognitive neuroscience. If this human "self-system" is at least weakly modular,[11] then we can proceed by building the necessary components, then integrating them with each other and within our robot control architecture, gradually approaching a model of the complete self.

(or machine) cognition and behavior. See Verschure and Prescott [72] for a discussion of theory building and the role of synthetic approaches in the sciences of mind and brain.

[11]Modularity is itself a topic that is widely debated within the cognitive sciences. Again, we consider that the synthetic approach can help answer some of the longstanding questions about how distributed vs. modular human minds/brains are. Our view is that the distributed nature of the brain can be over-stated. The brain is a layered architecture [49], and as such, there *is* significant replication of function and some redundancy across these layers, however, there is also localization of function and specific local or repeated circuits that perform roles that can be clearly described and differentiated.

The psychological literature related to the self is vast, and we will not seek to summarize it here. One starting point is the often cited proposal made by the cognitive psychologist Neisser [42, 43], who suggested five different kinds of self-knowledge:

"The *ecological self* is the individual situated in and acting upon the immediate physical environment. [...]. The *interpersonal self* is the individual, engaged in social interaction with another person. [...]. The *conceptual self*, or self-concept, is a person's mental representation of his/her own (more or less permanent) characteristics. [...] The *temporally extended self* is the individual's own life-story as he/she knows it, remembers it, tells it, projects it into the future. The *private self* appears when the child comes to understand and value the privacy of conscious experience [...]" ([43], pp. 18–19, our italics). Table 1 builds on Neisser's five-way split, conceiving of each of these as a sub-system of the self and relating each to some psychological phenomena that can provide benchmarks for the existence of that aspect of self in a person or robot. We have also followed Gallagher [21], Jeannerod [32] and others by adding agency—the *agential self*. The systems view asserts that some sense of self can emerge in the absence of some of these components and that some aspects of self, perhaps particularly the private self, could emerge from the interaction of these components without being explicitly designed, i.e., the sum is more than its parts.

5 A Diversity of Selves Across the Life-Span, the Population, and the Animal Kingdom

There is evidence to support this "systems" view of self from developmental psychology, neuroscience, and comparative psychology, which we will briefly review next.

From the study of human development, it is clear that very young infants have a sense of their ecological selves, for example, having a self-other distinction. This may emerge through exploration of the body in the womb. The fetus explores and discovers its body through "motor babbling"; it also touches itself, and the experience of skin-on-own-skin, or "double touch", is different from the experience of touching parts of the mother [54]. These activities allow the unborn child to learn the extension and limits of its own body. The emerging ability to control its own body, and to distinguish when a sensory event was caused by its own action, can also provide the newborn with some pre-reflective sense of agency (along the lines proposed by Jeannerod [32]). Agency in older children is often studied in the context of executive function and self-regulation, for example, the ability to withhold actions, show cognitive flexibility, or control emotional expression; these aspects of agency show multiple phases of development through infancy and the pre-school years [7, 75]. Infantile amnesia, which lasts until we are around two years of age [33], implies that the infant lives in the here and now, lacking a strong sense of its extension in time. The mirror test—recognizing that it is you in a mirror, not another child—is another milestone for the two-year-old [2, 4] that may indicate the beginnings of a reflective self-model.

Table 1 Some of the phenomena of self and how these might be grouped into different self-components based on Neisser [42, 43], Gallagher [21] and others. These sub-systems are assumed to be weakly modular but with significant interdependencies. The private self is in italics since it reflects first-person phenomena that may be emergent properties of the wider system. This decomposition is intended as a hypothesis to be investigated, refuted and revised using both analytical (empirical) and synthetic approaches

Phenomena of self	Component of self
Sensing the body Distinguishing yourself from the world Having a point-of-view Actively seeking sensory information	Ecological
Having emotions, drives and motivations Selecting actions that generate integrated behavior Knowing what events you have caused in the world	Agential
Having awareness of where you are Having awareness of a personal past and future Self-recognition (e.g., in a mirror) Knowing what you will do next	Extended
Learning by imitation Sharing attention Seeing others as selves Imagining other points-of-view	Interpersonal
Having beliefs about who you are (a self-concept) Having personal goals Having a life story (a narrative)	Conceptual
Having experience Having a feeling of being something Having a unitary stream of consciousness Having a sense of choice Having a feeling of being the same thing over time	Private

The newborn is a social creature, adapted to bond rapidly with its caregivers, yet significant changes occur in its capacity for sociality in the first year, including the emergence of shared attention, social referencing (looking to adults to understand the meanings of events), imitation, and wariness of strangers [44]. It is not until a child is around three years of age that it has "theory of mind"—the ability to conceive of another's point-of-view as different from its own [17]. The emergence of this interpersonal self, which is able to interpret the actions and intentions of others, likely builds on capacities of the ecological self to represent and reason about the child's own body. Finally, the conceptual self may emerge from the extended self, through consolidation of episodic memories into semantics—knowledge of the self and the world—and with help from the growing capacity to manipulate concepts and summarize events using language. Prior to the school years, children struggle to assemble coherent descriptions of past episodes [6], but as we grow older, we get more practiced at translating life events into story form, with the most important ones being rehearsed and consolidated to become stable chapters in the emerging self-narrative.

In the neurosciences, there is evidence from the study of neurodiversity and brain damage that also supports the decomposition of the self into component parts. Many conditions can impact on the sense of the ecological self: a disturbed body model can generate sensory neglect [70], or the sense that a part of your body does not belong to you (see [10]). Disorders of the hypothalamus, the basal ganglia, limbic system and prefrontal cortex can disrupt motivation, action integration and the experience of agency [31, 32, 45]. Damage to areas such as the temporal lobe, particularly the hippocampal system, can cause loss of the sense of place, or of the ability to think about the past or future, whilst sparing the core sense of the self in the here and now [67].[12] Activity in the "default mode" network of cortical sub-systems is also recognized as a critical substrate for the human capacity for "mental time travel" [58]. A well-known example of an altered social self occurs in people with autism, a condition that particularly impacts on the ability to understand others as social actors [5], whilst leaving intact other aspects of self (however, see [69]). The phenomenon of multiple personality disorder (e.g., [60]) shows the possibility that the self can assemble itself into one identity at one time, and into a very different one a few minutes later, with no shared consciousness or memory. This speaks to the constructed nature of the self and to its dynamical character as well. Specifically, if we think of identity as a stable attractor for the self system, then, in the unusual case of multiple personalities, the system is bi- or multi-stable and able to flip between different internally coherent, but mutually inconsistent, conceptions of self.

Comparative psychology also demonstrates variety in the nature of self (if we accept that animals can have selves). A self-other distinction, along with an ability to recognize the consequences of your actions, and hence some form of minimal self, may be shared by all bilateral multi-celled animals (see below). On the other hand, the capacity to conceive of the self as extending into the future and the past is far less universal and may only be well-developed in a limited number of animal groups, including some of the larger-brained mammals and birds [63]. The ability to voluntarily search in autobiographical memory for traces of particular events may be specific to humans having evolved in early homo lineages [18]. Evidence of a reflective self-model, as demonstrated by the mirror test, has also been shown in only a limited number of species, including great apes, dolphins, orca whales, elephants, and one species of bird (Eurasian magpies) [4, 53]. The presence of an interpersonal self that has theory of mind, which has been extensively investigated

---

[12]Endel Tulving's patient N.N. exemplifies this point [67]. A traffic accident caused N.N. to experience profound retrograde and anterograde amnesia, nevertheless he could still talk about himself, his experience, his preferences, and so on; he had intact short-term memory and could describe time and events in general terms. He could talk about consciousness, which he described as "being aware of who we are and what we are and where we are" ([67], p. 4). When asked to imagine what he might do tomorrow, however, his mind drew a blank, which he described as being "like swimming in the middle of a lake. There's nothing there to do hold you up or do anything with" ([67], p. 4). Like other patients with amnesia, N.N. could be described as "marooned in the present" [34] or as having a self that has lost much of its "temporal thickness" [20].

only in primates, may also be confined to animals that have an expanded neocortex [66].

6 A Minimal Robotic Self?

As noted earlier, one of the questions we would like to address through the synthetic approach concerns the possibility of a minimal self. Gallagher [21] reviews a number of proposals for minimal selves, identifying two key aspects, body ownership and agency, similar to the ecological and agential sub-systems noted in Table 1. He suggests, following Bermúdez [8], that the sense of self can be non-conceptual, pre-reflective, confined to the present, and a transient entity like one of Strawson's pearls.

Tani [64] has sought to create such a transient self for a mobile robot through a simple layered control system consisting of a perception module, an association module, and a prediction module. The robot was tasked with following a wall whilst searching for colored landmarks; the actions of the robot consist of steering by controlling left and right wheel-speeds and choosing whether to allocate visual attention to wall-following or to landmark searching. The robot monitors the reliability of its own predictions and uses this to arbitrate between control by the "bottom-up" sensory module and that by the "top-down" prediction module. Tani proposes that a form of self emerges when the predictions of the top-down module diverge from those of the sensory module, resulting in a period of dynamic instability, and that this "self" disappears when the prediction and sensory modules transition to a period of coherence.

Tani draws analogies to mammalian brain systems, however, the simple control system that he describes could be compared to much simpler nervous systems, for example, the nerve nets of some jellyfish can be conceived of as forming layered architectures in which distinct distributed networks compete for control of the motor system [47]. The earliest bilaterian animals, whose existence in the Precambrian era more than 540 million years ago is evidenced by fossils of their foraging trails, likely possessed internal organs, tentacle-like appendages, multiple sensors, and a nervous system that included a central ganglion, sometimes referred to as the "archaic brain" (see [47] for review). Modern day worms, including animals as simple as *C. Elegans*, have shown associative learning and the ability to use sensory signals to predict aversive chemicals and the presence of absence of food [3]. If monitoring the divergence between internal expectations about the world and sensory experience can give rise to a self, then perhaps minimal selves were present in some of the first mobile multicellular animals.

Tani's model is based on the hypothesis that the self requires a process that has an internal state that can evolve according to its own dynamics without being too tightly coupled to the world—the predictions of the system can drift from accurately forecasting the world, and at this point, the robot obtains a self. However, all animals with nervous systems interoceptively sense their bodies at the same time as they exteroceptively sense the environment; the patterns of sensory signals from the internal milieu, which will have very different dynamics from those of the sensed external world, thus already provide a basis for pre-reflectively distinguishing self from other.

7 A Biomimetic Cognitive Architecture for the Robot Self

In Sheffield, we have been building and testing brain-based robots, as experiments in synthetic psychology, since the mid-1990s, devising a number of models of brain architecture based on principles of layered control [49] and inspired by neurobehavioral studies of active sensing in rodents [50]. For the past seven years, together with European colleagues, we have also been incorporating models of key brain systems into a brain-inspired control architecture for the iCub robot (Fig. 1) called *distributed adaptive control* (DAC), developed by Paul Verschure and colleagues [71, 73, 74].

DAC is a high-level conceptual scheme that seeks to capture the cognitive architecture of the human brain and consists of four tightly coupled layers: *soma, reactive, adaptive* and *contextual*. Across these layers, there are three functional columns of organization: The first comprises the sensory, perceptual and memory sub-systems

Fig. 1 The iCub humanoid robot. A biomimetic robot platform for embodied testing of theories of general human intelligence developed by European researchers led by the Italian Institute of Technology. Picture from Sheffield Robotics

relating to the *world*, the second the interoceptive, motivational and memory sub-systems related to the *self*, and the third sub-systems that operate on the world through *action*. These DAC sub-systems do not directly map on to specific neural substrates, however, significant progress has been made relating parts of the DAC architecture to different brain sub-systems and circuits [51, 71]. Recent efforts to create a multi-faceted robot sense of self for iCub, using DAC, are summarised in [48] and detailed in [41]; here, we briefly summarize the architecture and some of the self-related capabilities it enables.

In DAC, the *somatic* layer corresponds to the body and provides access to exteroceptive, interoceptive, and proprioceptive signals from, respectively, the environment, internal processes and regulatory systems, and the motor/effector system. The *reactive* layer instantiates multiple fast, reflexive sensorimotor loops that support behaviors linked to needs; these loops are stability-seeking processes that reduce drives through action. The *adaptive* layer extends the sensorimotor loops of the reactive layer to make use of learned contingencies and to allow actions to be associated with states of the world. The adaptive layer is thus part of the solution to the symbol grounding problem, through the acquisition of mappings from internal states to world states. Whereas the adaptive layer operates largely in the here and now, the *contextual* layer adds the ability to store and retrieve short- and long-term memories, linked to goal achievement, that can act as action plans to be triggered by sensory contexts and that can be chained to create behavior sequences. This layer also includes predictive systems that can forecast the future state of the world based on action plans. Contextual layer systems can also encode and retrieve event memories and form abstract representations of events in narrative form that allow the robot to summarize and communicate about past episodes.

The DAC architecture generates aspects of the *ecological self* through interoceptive processes that maintain a model of the robot's physical parts and the geometry of its current body pose, and exteroceptive processes that monitor the robot's immediate surroundings. For example, using somatotopic maps modelled on human primary sensory cortex, and techniques such as self touch, Giorgio Metta, Matej Hoffmann and colleagues have developed methods that allow the iCub to learn its own body model [27], and recalibrate its knowledge of its own geometry [55]. Additionally, by combining vision with tactile sensing and with proprioception, iCub is able to develop a sense of peripersonal space that allows it to predict contacts with objects before they happen [56]. This foundation provides the beginnings of an ecological self that can be used to distinguish self from other, plan safe movement trajectories, and reason about the capacity for movement of others (see more below).

In Sheffield, we have been working to develop an episodic or *event* memory system for the DAC adaptive and contextual layers that can contribute to a robotic *extended self*. Our hypothesis is that event memory can be usefully considered as an attractor network operating in a latent (hidden) variable space whose dimensions encode salient characteristics of the physical and social world in a highly compressed fashion [19]. According to this view, the operation of perceptual systems in the adaptive and contextual layers can be analogized to learning processes that identify psychologically meaningful latent variable descriptions. Instantaneous memories then correspond to points in this latent variable space and event memories to trajectories through this space. A single latent feature space can be used to represent memories across multiple sensory modalities thus providing sensory fusion. This enhances compression as coupled signals among heterogenous modalities are discovered and represented in a common set of latent variables. This can also be thought of as concept discovery—the identification of underlying invariance in patterns of multi-modal sensory flow. The current implementation, illustrated in Fig. 2, demonstrates effective memory formation and retrieval of human faces, actions, voices and emotions [12, 15, 37]. Due to its generative nature, and ability to interpolate, the system can also generate fantasy memories from parts of the latent variable space that have not been populated by real data. This leads to the possibility of imagining future events [15]. The ability of the system to reconstruct the sensory pattern associated with a recalled memory [13], retrieved using a verbal cue, suggests that event memory can contribute to the grounding of linguistic symbols in sensorimotor experience.

Neuroscience research suggests that an effective approach to building the *interpersonal self* could be to use the robot's own internal body models—the ones that underlie the ecological self—to simulate the pose and actions of others. With iCub, our collaborators have developed DAC processes that allow the robot to represent the state of the world from a different point-of-view (see [41]), allowing iCub to reason about what a human partner can see and helping the robot to resolve perceptual ambiguities and improve communication. One important human ability that benefits from the interpersonal self is the capacity to learn by imitation. Yiannis Demiris and co-workers have demonstrated that you can build up from motor babbling to a hierarchical learning system that uses forward models, inspired by studies of the primate "mirror neuron" system, to learn by imitation [16]. This system has been used with the iCub to allow it to rapidly acquire new hand gestures and sequences of actions involved in playing games or solving puzzles.

As shown in Fig. 2, a key part of the broader system in which our synthetic event memory operates is the component related to narrative reasoning, this is one of the sub-systems that generate the *conceptual self*. Peter Dominey and colleagues (see, e.g., [46]) have been working to model autobiographical memory and narrative construction using an acquired grammar, together with compact and structured representations of iCub's interaction history. Using this narrative system, iCub can recall and discuss past events, including some of its past interactions with people, from a first-person perspective. One longer-term goal is to integrate this narrative construction process with the event memory system developed in Sheffield such that linguistic descriptions can be abstracted from representations of events as attractor patterns in latent variable space. Using the generative capabilities of the event memory, narratives could also be played out, and "grounded", or "relived", via reconstruction as internally simulated sensory scenes.

In sum, we have made a start in instantiating some of the different aspects of the sense of self in the iCub robot. The lower layers of the DAC architecture integrate internal and external sensory signals so as to regulate self-correcting control loops based on drives. These sub-systems meet many of the criteria for a minimal self. The upper layers encode representations of past events that can be used to reason about the future and about social others, creating some of the elements that we are seeking for the extended and interpersonal selves. Finally, the narrative system provides the seeds for a self-concept and life story. We have not sought to build a *private self* directly, rather, the plan is to create the rest of the architecture and then see if an impression of the experiential self can emerge from within in our version of the Garland test. Indeed, on a good day, when all of the sub-systems are working properly, interacting with the iCub can begin to feel as though "someone is home", even for the people who have helped to develop the robot's control systems and understand how they operate. On the other hand, it also feels as if we have only just set out on the journey of deconstructing the human self and recreating it in a machine. Indeed, as Turing

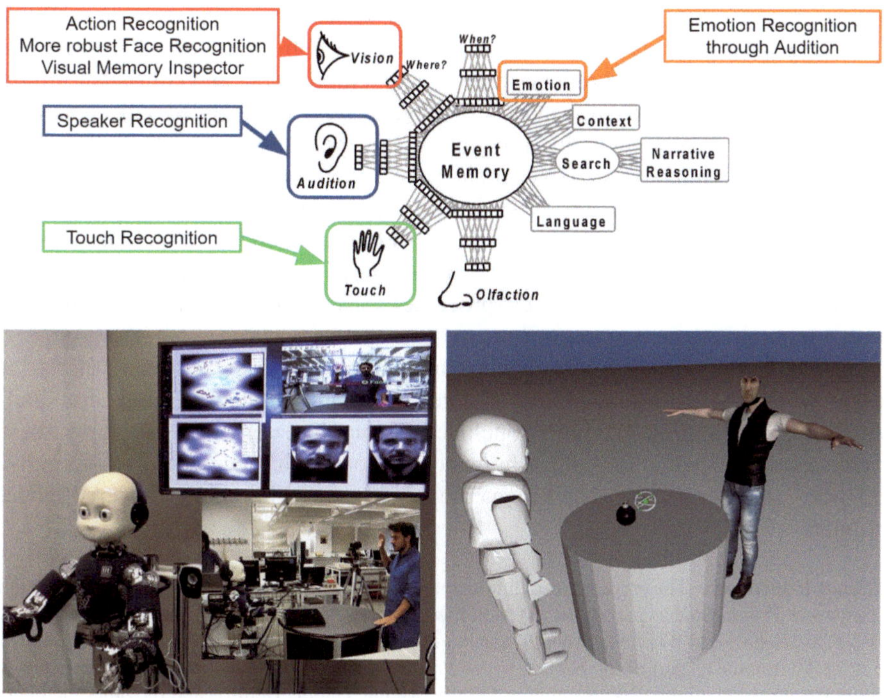

Fig. 2 The synthetic event memory system developed at Sheffield. Our model of event memory integrates across multiple modalities to encode memories as patterns in a low-dimensional latent variable space that can be used to reconstruct past experiences based on partial cues or explicit search. *Top*: The proposed architecture for a synthetic episodic/event memory system based on Rubin [57] and Evans et al. [19]. The highlighted areas show the components that have been constructed to date, which include sub-systems for action, touch, and emotion recognition, for speaker recognition in the visual and auditory modalities, and the ability to display visual memories (the visual memory inspector). *Bottom-left*: iCub operating in real-time to recognize actions and faces. The TV monitor behind the robot shows two latent variable spaces, the visual pre-processing of the camera scene, and the reconstruction of the remembered face based on the recovered memory. *Bottom-right*: a screen-shot from the visual memory inspector, which allows researchers to see iCub's simulation of itself and its perceptual world. Here, iCub represents a face and two objects on the circular table

wrote at the end of *Computing Machinery and Intelligence*—"we can only see a short distance ahead, but we can see plenty there that needs to be done" ([68], p. 460).

8 Conclusion

This chapter has argued that the human self is brought into being by the activity of a set of self-processes instantiated by the brain and body and has proposed that we can synthesize an artificial self by developing equivalent sub-systems within an integrated biomimetic cognitive architecture for a humanoid robot. While the various self-processes may be transient, the continuity provided by a physical body, in a human or robot, can provide the basis for the experience of a continued self. This suggests a key role for embodiment, first in establishing a boundary between the self and the world, and second in providing a predictable and consistent setting in which the self awakens to find itself. Beyond this, an extended self, generated by the capacity to remember and imagine, allows the self to escape from the island of the present, while abstraction and narrative allow it to construct and maintain a coherent set of beliefs and stories about itself. To evaluate the possibility of a robot self, we have suggested a version of the Turing test, extended to include physical embodiment and human-like cognitive architecture, that asks whether people who encounter a robot with synthetic self-processes consider that they have met an entity with a self.

We began the chapter by motivating this work in the context of the criteria for recognizing other minds, and the challenge of benchmarking artificial general intelligence against human. We have concluded by summarizing some initial efforts to create a sense of self for the iCub humanoid robot that has ecological, temporally-extended, interpersonal and narrative components set within a multi-layered model of mind.

Acknowledgements The preparation of this chapter was supported by funding from the EU Seventh Framework Programme as part of the projects *Experimental Functional Android Assistant* (EFAA, FP7-ICT-270490) and *What You Say Is What You Did* (WYSIWYD, FP7-ICT-612139) and the EU H2020 Programme as part of the *Human Brain Project* (HBP-SGA1, 720270). We are particularly grateful to Paul Verschure, Peter Dominey, Giorgio Metta, Yiannis Demiris and the other members of the WYSIWYD and EFAA consortia, and to our colleagues at the University of Sheffield who have helped us to develop memory systems for the iCub, particularly Uriel Martinez, Andreas Damianou, Neil Lawrence, Luke Boorman and Matthew Evans. The Sheffield iCub was purchased with the support of the UK Engineering and Physical Science Research Council (EPSRC).

References

1. Abel, C. (2014). *The extended self: Architecture, memes and minds*. Manchester: Manchester University Press.
2. Amsterdam, B. (1972). Mirror self-image reactions before age two. *Developmental Psychobiology, 5*(4), 297–305.
3. Ardiel, E. L., & Rankin, C. H. (2010). An elegant mind: Learning and memory in Caenorhabditis elegans. *Learning & Memory, 17*(4), 191–201. https://doi.org/10.1101/lm.960510.
4. Bard, K. A., Todd, B. K., Bernier, C., Love, J., & Leavens, D. A. (2006). Self-awareness in human and chimpanzee infants: What is measured and what is meant by the mark and mirror test? *Infancy, 9*(2), 191–219. https://doi.org/10.1207/s15327078in0902_6.
5. Baron-Cohen, S., Leslie, A. M., & Frith, U. (1985). Does the autistic child have a 'theory of mind'? *Cognition, 21,* 37–48.
6. Bauer, P. J. (2012). The life I once remembered: The waxing and waning of early memories. In D. Berntsen & D. C. Rubin (Eds.), *Understanding autobiographical memory* (pp. 205–225). Cambridge: CUP.
7. Bell, M. A., & Deater-Deckard, K. (2007). Biological systems and the development of self-regulation: Integrating behavior, genetics, and psychophysiology. *Journal of Developmental & Behavioral Pediatrics, 28*(5).
8. Bermúdez, J. (1988). *The paradox of self-consciousness*. Cambridge, MA: MIT Press.
9. Blakemore, S. (2003). Consciousness in meme machines. *Journal of Consciousness Studies, 10*(4–5), 19–30.
10. Blakeslee, S., & Blakeslee, M. (2007). *The body has a mind of its own*. New York: Random House.
11. Braitenberg, V. (1986). *Vehicles: Experiments in synthetic psychology*. Cambridge, MA: MIT Press.
12. Camilleri, D., & Prescott, T. J. (2017). *Action recognition with unsynchronised multi-sensory data*. Paper presented at the 7th Joint IEEE International Conference on Development and Learning and on Epigenetic Robotics (ICDL-EPIROB), Lisbon, Portugal.
13. Camilleri, D., Damianou, A., Jackson, H., Lawrence, N., & Prescott, T. J. (2016). iCub visual memory inspector: Visualising the iCub's thoughts. In N. F. Lepora, A. Mura, M. Mangan, P. F. M. J. Verschure, M. Desmulliez, & T. J. Prescott (Eds.), *Biomimetic and Biohybrid Systems, the 5th International Conference on Living Machines* (pp. 48–57). Berlin: Springer LNAI.
14. Cangelosi, A., Schlesinger, M., & Smith, L. B. (2015). *Developmental robotics: From babies to robots*. Cambridge, MA: MIT Press.
15. Damianou, A., Henrik, C., Boorman, L., Lawrence, N. D., & Prescott, T. J. (2015). A top-down approach for a synthetic autobiographical memory system. In S. Wilson, T. J. Prescott, A. Mura, & P. F. M. J. Verschure (Eds.), *Biomimetic and Biohybrid Systems, the 4th International Conference on Living Machines* (Vol. 9222, pp. 280–292). Berlin: Springer LNAI.
16. Demiris, Y., Aziz-Zadeh, l., & Bonaiuto, J. (2014). Information processing in the mirror neuron system in primates and machines. *Neuroinformatics, 12*(1), 63–91.
17. Doherty, M. (2009). *Theory of mind: How children understand others' thoughts and feelings*. Hove: Psychology Press.
18. Donald, M. (2012). Evolutionary origins of autobiographical memory: A retrieval hypothesis. In D. Berntsen & D. C. Rubin (Eds.), *Understanding autobiographical memory* (pp. 269–289). Cambridge: CUP.
19. Evans, M. H., Fox, C. W., & Prescott, T. J. (2014). Machines learning—towards a new synthetic autobiographical memory. In A. Duff, N. Lepora, A. Mura, T. Prescott, & P. M. J. Verschure (Eds.), *Biomimetic and Biohybrid Systems, the 3rd International Conference on Living Machines* (Vol. 8608, pp. 84–96). Berlin: Springer LNAI.
20. Friston, K. (2017). The mathematics of mind-time. *Aeon*.
21. Gallagher, S. (2000). Philosophical conceptions of the self: Implications for cognitive science. *Trends in Cognitive Sciences, 4*(1), 14–21.

22. Gardner, H. (2006). *Multiple intelligences: New horizons*. New York: Basic Books.
23. Goertzel, B., & Pennachin, C. (2007). *Artificial general intelligence*. New York: Springer.
24. Harnad, S. (1991). Other bodies, other minds: A machine incarnation of an old philosophical problem. *Minds and Machines, 1,* 43–54.
25. Harnad, S. (1994). Does the mind piggy-back on robotic and symbolic capacity? In H. L. Morowitz & J. L. Singer (Eds.), *The mind, the brain, and complex adaptive systems, santa fe institute studies in complexity XXII* (pp. 204–220). Boston: Addison Wesley.
26. Hauser, L. (1993). Reaping the worldwind: Reply to harnad's "other bodies, other minds". *Minds and Machines, 3,* 219–238.
27. Hoffmann, M., Straka, Z., Farkas, I., Vavrecka, M., & Metta, G. (2017). Robotic homunculus: Learning of artificial skin representation in a humanoid robot motivated by primary somatosensory cortex. *IEEE Transactions on Cognitive and Developmental Systems, pp*(99), 1–1. https://doi.org/10.1109/tcds.2017.2649225.
28. Hofstadter, D. (2007). *I am a strange loop*. New York: Basic Books.
29. Hood, B. (2012). *The Self illusion: Why there is no 'you' inside your head*. London: Constable and Robinson.
30. Hume, D. (1740). *A treatise on human nature.*
31. Humphries, M. D., & Prescott, T. J. (2010). The ventral basal ganglia, a selection mechanism at the crossroads of space, strategy, and reward. *Progress in Neurobiology, 90*(4), 385–417. https://doi.org/10.1016/j.pneurobio.2009.11.003.
32. Jeannerod, M. (2003). The mechanism of self-recognition in humans. *Behavioural Brain Research, 142*(1), 1–15. https://doi.org/10.1016/S0166-4328(02)00384-4.
33. Lambert, F. R., Lavenex, P., & Lavenex, P. B. (2017). The "when" and the "where" of single-trial allocentric spatial memory performance in young children: Insights into the development of episodic memory. *Developmental Psychobiology, 59*(2), 185–196. https://doi.org/10.1002/dev.21479.
34. Lidz, T. (1942). The amnesic syndrome. *Archives of Neurology and Psychiatry, 47,* 588–605.
35. Locke, J. (1777). *An enquiry concerning human understanding.*
36. Lungarella, M., Metta, G., Pfeifer, R., & Sandini, G. (2003). Developmental robotics: A survey. *Connection Science, 15*(4), 151–190. https://doi.org/10.1080/09540090310001655110.
37. Martinez-Hernandez, U., Damianou, A., Camilleri, D., Boorman, L. W., Lawrence, N., & Prescott, T. J. (2016). *An integrated probabilistic framework for robot perception, learning and memory.* Paper presented at the 2016 IEEE International Conference on Robotics and Biomimetics (ROBIO), Qingdao, China. pp. 1796–1801.
38. Merleau-Ponty, M. (1945/1962). *Phénoménologie de la Perception* (C. Smith, Trans.). London: Routledge.
39. Metzinger, T. (2009). *The ego tunnel: The science of the mind and the myth of the self*. New York: Basic Books.
40. Mitchinson, B., Pearson, M., Pipe, T., & Prescott, T. J. (2011). Biomimetic robots as scientific models: A view from the whisker tip. In J. Krichmar & H. Wagatsuma (Eds.), *Neuromorphic and brain-based robots* (pp. 23–57). Boston, MA: MIT Press.
41. Moulin-Frier, C., Fischer, T., Petit, M., Pointeau, G., Puigbo, J. Y., & Pattacini, U., et al. (2017). DAC-h3: A proactive robot cognitive architecture to acquire and express knowledge about the world and the self. *IEEE Transactions on Cognitive and Developmental Systems, PP*(99), 1–1. https://doi.org/10.1109/tcds.2017.2754143.
42. Neisser, U. (1988). Five kinds of self-knowledge. *Philosophical Psychology, 1,* 35–59. https://doi.org/10.1080/09515088808572924.
43. Neisser, U. (1995). Criteria for an ecological self. In P. Rochat (Ed.), *The Self in infancy: Theory and research.* Amsterdam: Elsevier.
44. Nelson, K. (2007). *Young minds in social worlds: Experience, meaning and memory*. Cambridge, MA: Harvard University Press.
45. Panksepp, J. (1998). *Affective neuroscience: The foundations of human and animal emotions*. Oxford: OUP.

46. Pointeau, G., & Dominey, P. F. (2017). The role of autobiographical memory in the development of a robot self. *Frontiers in Neurorobotics, 11,* 27.
47. Prescott, T. J. (2007). Forced moves or good tricks in design space? Landmarks in the evolution of neural mechanisms for action selection. *Adaptive Behavior, 15*(1), 9–31.
48. Prescott, T. J. (2015). Me in the machine. *New Scientist,* 36–39.
49. Prescott, T. J., Redgrave, P., & Gurney, K. N. (1999). Layered control architectures in robots and vertebrates. *Adaptive Behavior, 7*(1), 99–127.
50. Prescott, T. J., Mitchinson, B., Lepora, N. F., Wilson, S. P., Anderson, S. R., Porrill, J., et al. (2015). The robot vibrissal system: Understanding mammalian sensorimotor co-ordination through biomimetics. In P. Krieger & A. Groh (Eds.), *Sensorimotor integration in the whisker system* (pp. 213–240). New York: Springer.
51. Prescott, T. J., Ayers, J., Grasso, F. W., & Verschure, P. F. M. J. (2016). Embodied models and neurorobotics. In M. A. Arbib & J. J. Bonaiuto (Eds.), *From neuron to cognition via computational neuroscience* (pp. 483–512). Cambridge, MA: MIT Press.
52. Prescott, T. J., Lepora, N., & Verschure, P. F. M. J. (2018). *The handbook of living machines: Research in biomimetic and biohybrid systems*. Oxford, UK: OUP.
53. Prior, H., Schwarz, A., & Gunturkun, O. (2008). Mirror-induced behavior in the magpie (Pica pica): Evidence of self-recognition. *PLoS Biology, 6*(8), e202.
54. Rochat, P. (2001). *The infant's world*. Cambridge, MA: Harvard University Press.
55. Roncone, A., Hoffmann, M., Pattacini, U., & Metta, G. (2014). *Automatic kinematic chain calibration using artificial skin: Self-touch in the iCub humanoid robot.* Paper presented at the 2014 IEEE International Conference on Robotics and Automation (ICRA), pp. 2305–2312.
56. Roncone, A., Hoffmann, M., Pattacini, U., Fadiga, L., & Metta, G. (2016). Peripersonal space and margin of safety around the body: Learning visuo-tactile associations in a humanoid robot with artificial skin. *PLoS ONE, 11*(10), e0163713. https://doi.org/10.1371/journal.pone.0163713.
57. Rubin, D. C. (2006). The basic-systems model of episodic memory. *Perspectives on Psychological Science, 1*(4), 277–311.
58. Schacter, D. L., Addis, D. R., Hassabis, D., Martin, V. C., Spreng, R. N., & Szpunar, K. K. (2012). The future of memory: Remembering, imagining, and the brain. *Neuron, 76*(4). https://doi.org/10.1016/j.neuron.2012.1011.1001, https://doi.org/10.1016/j.neuron.2012.11.001.
59. Searle, J. (1990). Is the brain's mind a computer program? *Scientific American, 262*(1), 20–25.
60. Silberman, E. K., Putnam, F. W., Weingartner, H., Braun, B. G., & Post, R. M. (1985). Dissociative states in multiple personality disorder: A quantitative study. *Psychiatry Research, 15*(4), 253–260. https://doi.org/10.1016/0165-1781(85)90062-9.
61. Silver, D., Hubert, T., Schrittwieser, J., Antonoglou, I., Lai, M., & Guez, A., et al. (2017). Mastering chess and shogi by self-play with a general reinforcement learning algorithm. arXiv:1712.01815v1 *[cs.AI] 5 Dec 2017*.
62. Strawson, G. (1997). The self. *Journal of Consciousness Studies, 4*(5/6), 405–428.
63. Suddendorf, T., & Corballis, M. C. (2007). The evolution of foresight: What is mental time travel, and is it unique to humans? *Behavioral and Brain Sciences, 30*(3), 299–313. https://doi.org/10.1017/S0140525X07001975.
64. Tani, J. (1998). An interpretation of the 'self' from the dynamical systems perspective: A constructivist approach. *Journal of Consciousness Studies, 5,* 516–542.
65. Tononi, G. (2004). An information integration theory of consciousness. *BMC Neuroscience, 5*(1), 42. https://doi.org/10.1186/1471-2202-5-42.
66. Towner, S. (2010). Concept of mind in non-human primates. *Bioscience Horizons: The International Journal of Student Research, 3*(1), 96–104. https://doi.org/10.1093/biohorizons/hzq011.
67. Tulving, E. (1985). Memory and consciousness. *Canadian Journal of Psychology, 26*(1), 1–12.
68. Turing, A. M. (1950). Computing machinery and intelligence. *Mind, 59*(236), 433–460.
69. Uddin, L. Q. (2011). The self in autism: An emerging view from neuroimaging. *Neurocase, 17*(3), 201–208. https://doi.org/10.1080/13554794.2010.509320.

70. Vallar, G. (1998). Spatial hemineglect in humans. *Trends in Cognitive Sciences, 2*(3), 87–97. https://doi.org/10.1016/S1364-6613(98)01145-0.
71. Verschure, P. F. M. J. (2012). Distributed adaptive control: A theory of the mind, brain, body nexus. *Biologically Inspired Cognitive Architectures, 1,* 55–72. https://doi.org/10.1016/j.bica.2012.04.005.
72. Verschure, P. F. M. J., & Prescott, T. J. (2018). A living machines approach to the sciences of mind and brain. In T. J. Prescott, N. Lepora, & P. F. M. J. Verschure (Eds.), *The handbook of living machines: Research in biomimetic and biohybrid systems*. Oxford, UK: OUP.
73. Verschure, P. F. M. J., Krose, B., & Pfeifer, R. (1992). Distributed adaptive control: The self-organization of structured behavior. *Robotics and Autonomous Systems, 9,* 181–196.
74. Verschure, P. F. M. J., Pennartz, C. M. A., & Pezzulo, G. (2014). The why, what, where, when and how of goal-directed choice: Neuronal and computational principles. *Philosophical Transactions of the Royal Society of London B: Biological Sciences, 369*(1655). https://doi.org/10.1098/rstb.2013.0483.
75. Zelazo, P. D. (2004). The development of conscious control in childhood. *Trends in Cognitive Sciences, 8*(1), 12–17. https://doi.org/10.1016/j.tics.2003.11.001.

Constructive Biology of Emotion Systems: First- and Second-Person Methods for Grounding Adaptation in a Biological and Social World

Chrystopher L. Nehaniv

Abstract We consider the interpretation of emotions and similiar phenomena as support for survival and coping in the world. Grounded in the first-person experience of an emotional agent, certain such emotions, drives or experiences are self-oriented (homeostasis, intake, outflow: hunger, pain, irritation), while others suggest a generalized or specific recognition of other agents or objects (curiosity, fear; or hatred, envy, yearning, greed). Other, more complex emotions are involved in relations to a second person (sympathy) or social regulation (shame, guilt, feelings of loyalty) or affective episodic structure (hope, regret). Considering complex emotions in relation to other 'persons' yields insight into the roles and possible design of various emotional phenomena in behavioral regulation in biological, software, and social contexts. Affective coloring of episodic memories of sequences of actions and experiences may suggest a mechanism for the grounding of behavioral adaptation. We explore channels of meaning for agents in interaction games as these relate to emotions, the temporal dynamics of affect in relation to behavior, remembering, and learning; and we outline how affective coloring of episodic memories might provide a mechanism for emergent spatial and social navigation, as well as considering the role of the temporal horizon in behavior selection.

This previously unpublished paper written in 1999 had been accepted for journal publication in the special issue guest-edited by L. Cañamero and P. Petta on Grounding Emotions in Adaptive Systems of Cybernetics and Systems: An International Journal, vol. 32(5–6), 2001, but was too long for inclusion in this full form and was not shortened by author. The references have not been updated.

C. L. Nehaniv (✉)
Adaptive Systems Research Group, University of Hertfordshire, Hatfield AL10 9AB, UK
e-mail: C.L.Nehaniv@herts.ac.uk; nehaniv@gmail.com

M. I. Aldinhas Ferreira et al. (eds.), *Cognitive Architectures*, Intelligent Systems, Control and Automation: Science and Engineering 94,
https://doi.org/10.1007/978-3-319-97550-4_8

1 Introduction

1.1 The 'Person' of an Agent

Human languages distinguish the self ('I'—the first person), and the other with whom the self directly interacts and whom the self addresses ("you"—the second person), and the other with whom one is not engaged directly ("he/she/they"—the third person). We argue that these grammatical categories of human natural language are useful in considering the adaptive role of emotions as they relate to, regulate, and control behaviors in a biological and social setting. We shall at first use the term "emotion" rather more loosely than is common in the literature (e.g., [1]) to include phenomena often labelled as drives, feelings, emotions, and so on. Furthermore, we will not insist on distinction according to the manner in which such phenomena are experienced, whether consciously, as sensations with affective coloring, or as physiological changes and so on. Nevertheless, such notions as 'drive', 'feeling', 'emotion', etc. (distinguished clearly below) emerge from the considerations in the internal and external domain of person and related factors. As suggested by an anonymous reviewer, this domain of person may turn out to correlate with different areas of nervous system and brain function.

We consider emotions or emotion-like phenomena in evolved organisms (whether or not they possess a nervous system) with a view to applications in artificial physical and possibly software agents [2, 3], but also to understanding the role emotions can play in regulating the behavior of differently endowed organisms embedded in their biological and social environments.

While we speak of 'person' here, this 'person' is understood as any organism or agent, rather than a human person. In the case of the virus, bacterium, vegetable or fungal agent, the reader may feel that the use of this term or the use of the term 'emotion' is too anthropomorphic, since it is likely (but impossible to determine) that such creatures do not have experiences in the sense that humans and other animals do. Yet we show how anthropomorphism and the egomorphic principle, carefully applied as tools, reveal aspects of universality justifying the use of such terms. A person will be a situated agent itself (first person) or any entity 'treated' as such by the agent (second person). Whether or not the qualitative experiential aspects of emotions (qualia—the feeling of experience) arise in such a 'person' will be a secondary consideration, although it is an important philosophical question and perhaps eventually may be a rigorous neurophysiological one. Rather, we show how the notion of 'person' may be useful in the design and understanding of artificial agents (physical or software) and of biological ones, in particular with regard to emotion systems and the grounding and generation of adaptive behavior.

1.2 Constructive Biology

The first-person viewpoint in agent construction is strongly related to *constructive biology*, i.e., biology motivated by the desire to understand how biological systems actually are constructed by nature and develop over time, rather than just to obtain a descriptive understanding. This is the engineering and scientific viewpoint that one's understanding should enable one to, in principle, *build* the systems of interest. For example, Barbara Webb has shown through building that a much simpler mechanism than expected, not involving functional decomposition or planning, is sufficient to account for much of the phonotaxis behavior observed in crickets [4, 5]. Valentino Braitenberg's examples [6] of simple robots to whom human observers attribute such states as 'fear', 'aggression', 'love', etc., illustrate that meaning of an interaction for an external observer can be quite different to that its has for the agent (in these cases, simple taxis). The constructive biology of multi agent systems will inescapably lead to mappings between channels of meaning that respect structural constraints and grounding of agents.

In contrast to behaviorists like Skinner [7], a constructive biologist need not be restricted to external observation of stimuli and responses, rejecting speculation of what occurs inside the organism. Indeed, the engineering and design of internal mechanisms are just aspects of the experimental and theoretical apparatus that the constructive biologist may manipulate, vary, and control. Just as with studying and building improved sensors and actuators in artificial agents and robots, mechanisms of internal control, remembering, predicting, and possibly empathizing and biographical reconstruction for second persons can be the object of scientific inquiry.

The study of correspondence via the algebraic notion of homomorphism (full, partial or relational) provides an inroad for the precise study of correspondence between agents interacting with their environments or with each other. Preserving the structure of meaning channels for an agent coupled to its environment is required for the usefulness of and determines the quality of metaphors and mappings in the design, algebraic engineering, interaction dynamics, and constructive biology of groups of situated agents.

2 A Generalized Phenomenological Perspective

2.1 I Feel, Therefore I Am

The question of emotions has defied scientific analysis at least since the days of Aristotle, because emotions are of a subjective nature [1, 8, 9]. The situation is similar to the problem of studying consciousness: each of us can be sure he or she is conscious via direct experience, but cannot know directly whether another human also experiences consciousness. Our experience of consciousness and emotion is *first-person*. Conscious or not, the interaction of an agent with the world around through its particular sensors and actuators, with its particular body and capacities, is also a first-person phenomenon, *'I'*.

2.2 You Are Like Me, Therefore You Feel

We can observe that beings, who we hypothesize are like ourselves, seem to have similar coupling to the environment and similar experiences. The assumption that other humans are second persons, *'Thou'*, that they experience the world in a way similar to the way that 'I' do, turns them from mysterious objects into beings whose actions make sense. We can understand their actions by attributing intent, motivations, and emotions to them similar to the ones we ourselves experience. Treating others as second persons, empirically, has turned out (for most of us humans) to be a successful basis for us in dealing with our social world (but see the discussion of failures in mechanisms of empathy in relation to autism in [10]). Thus, it seems that the assumption that other beings have minds like our own is a good working hypothesis for a social being (at least for mammals living in individualized societies—although perhaps not for social insects [11–13]). Indeed, this may be an involuntary assumption, but the tendency to make it is in itself revealing.

Just as we cannot study the consciousness of others directly, we also cannot experience the affective states of other humans or other living things, nor can we be sure that anything is experienced at all. The empirical phenomenological perspective asks not for external proof of such experience, but rather adopts, as a working hypothesis, *'Others are like me'*. One may also adopt this hypothesis also in the case of non-human mammals and even other animals—which (it seems) are probably not experiencing complex human emotions like shame or pride. However, as a metaphor, an engineering assumption, or design principle, one can use the notion of experience of emotion to gain insight into how systems surviving in biological and social worlds might function.

In order to do this systemically, it is now necessary to distinguish the *feeling* of experience (*qualia* of emotion and consciousness) from drives and emotions (operationally defined below in terms of reinforcing stimuli), and to study *channels of meaning* [14, 15]. Meaning is understood as (1) information in interaction games between an agent and its environment or between agents mediated with respect to their own sensors and actuators and as (2) *useful* for satisfying homeostatic and other drives, needs, goals or intentions. The resulting (1st and 2nd person) methods belong to the realm of constructive biology, which seeks to understand biological systems through building, and, conversely, to learn engineering 'tricks' from the biological and social world that are useful in robotics, software agents, and adaptive control systems. We argue that this natural and naive intuition, 'others are like me', combined with the formal notion of channels of meaning in interaction, holds the source of a systematic foundation for a constructive biology for understanding real biological systems and for engineering artificial ones.

3 Meaning, Observers, and Information

We first introduce an approach to the channels of meaning for observers and agents in interactions games mediated by sensors, actuators, and environment (Nehaniv [14, 15]). The introduction into channels of meaning in relation to observers and agents in this section includes substantial passages first published in [14]; a rigorous information-theoretic formulation is in [15]. This background will help us to clarify issues of how emotions and emotion-like phenomena ground adaptation in *interaction games*. Agents, in their own first-person interaction with the environment, access channels of meaning, but also do so in their interaction with others, second persons, having some similarities of dynamic interaction with the first-person agent.

3.1 Meaning for Observers and Agents

The notions of truth and meaning, as mathematical logicians tell us, only make sense in reference to a particular universe of discourse. Less obviously perhaps, meaning only makes sense from the standpoint of an *observer*. This observer may be someone manipulating a formal system to determine referents and applying predicates according to compositional rules, may be an animal hunting in the forest, may be a Siberian swan in a flock of conspecifics overwintering with other birds on a northern Japanese lake, or may be an artificial agent maintaining control parameters over an industrial process. For some, the observer may be the 'mind of God', as the final external attributor of truth and meaning.

Umwelt is the ethologist's concept of the 'world around' an animal, i.e., the local environment as experienced by the animal in its particular embodiment, including senses and means of acting on the world. Considering an animal, robot or other agent in its *Umwelt* is an example of taking what we call here the 'first-person perspective'.

We take a stricter view than that of most logicians on when one can speak sensibly of meaning. For our purposes, a notion of meaning only makes sense for agents situated and embedded in interaction with their particular *Umwelt*, the world around them. Actually, this is a view wider in scope than the traditional one in which there is a single, objective, viewpoint-independent standard. *One can now speak of meaning with (and only with) respect to anything that could potentially qualify as an 'observer'*, not just with respect to some implicit, universal 'third person' as external impersonal observer. Indeed, to the extent that the later hypothesized observer is (often implicitly) attributed ontological status, it is merely an (important) special case in the class of possible observers. An observer, or 'person' in our sense, need not be conscious or alive; it is any agent that may be as simple as an active process on the Central Processing Unit (CPU) of your computer, a reactive control loop, a software agent, a robot, or as complex as an animal, or even a logician pondering the Platonic 'realm of forms'.

The meaningfulness of the behavior of a creature or agent may be completely in the eye of the beholder—but not necessarily in the mind or awareness of the beholder, which indeed might have none. For whom is the behavior meaningful? To whom is it meaningful if several such agents or creatures interact, e.g., if robots interact to perform a task such as collecting objects (Beckers et al. [16]) or, as in the case of analogous collective behavior by termites [17] using stigmergy (environmental signs of the progress of work)? Meaningfulness may be in the designer's eye or in the adaptiveness of the activity as tending to increase the probability that copies of an agent's genes (if it has any) are represented in future generations. The latter notion of evolutionary, behavioral, survival adaptiveness (in biological agents, the tendency to increase reproductive success) hints at the possible nature of meaning for evolved or constructed systems. Meaning arises with information that helps an agent attain its 'goals'. Note that meaning in this sense starkly contrasts with—but may also be considered a compatible refinement of—Shannon's measure of information content [18], which is minimal for a constant unchanging signal but is maximal for random signals, both of which might well be devoid of meaning for all agents and observers. Agent goals may be conscious or unconscious, merely surviving, reacting, maintaining the self or reproducing, or they may motivate actions according to intentionality. In fact, it may be that the agent has no proper goals, but that its actions reflect only a merely reactive coupling to the environment. If the goals are observer-attributed rather than within the agent, then the corresponding meaning exists only in relation to such observers. The agent itself may be such an observer, in which case, meaning could then arise for it in its interaction with its *Umwelt*.

Meaning then need not be linguistically nor even symbolically mediated. It may or may not involve representations, but must arise in the dynamics realizing the agent's functioning and interaction in its environment (cf. the notion of 'structural coupling' of Maturana and Varela [19]), supporting adaptive, self maintaining or reproductive behaviors, or goals, or possibly intentions or plans. Multiple observers, as in the case of interaction among human agent observers, result in multiple arisings of meaning. Any entity that exists at the level of a biological unit of evolutionary selection (e.g., unicellular organism, differentiated multicellular organism, eusocial insect colony) could potentially be an agent or observer in our sense, as could a human organization such as a government, tribe or corporation. Robotic and software agents are not excluded.

In the realm of constructive biology, robotics and artificial agent construction, meaning can also arise in the interaction channels between the agent and the environment in which it is 'embodied'. These channels could be artificially evolved or designed. Similarly, these considerations apply to software agents, which might in a sense be considered embodied with respect to their particular environments as long as mutually perturbing channels exist between the agent and its environment (this ontology-independent definition of embodiment is due to Tom Quick [20]), with degree of embodiment measurable according to the complexity of the dynamics occurring between the two.

The philosopher Ludwig Wittgenstein insisted on defining meaning of words and other signs in terms of their use by agents engaged in language games (including artificial and everyday language) [21, 22]. An insight going back to the 19th century by C. S. Peirce [23, 24], the father of semiotics, is that signs *mediate* meaning, only making sense in the context of systems of signs, and that an *interpretant* always links a signifier to a signified in an embedded and embodied process (*semiosis*). This situated and embodied nature of agent semiotics highlights the meaninglessness of signals, signs, and sign systems in isolation, without agents, and thus without uses. Signal and sign systems may or may not have formally specifiable structures. They may be difficult to describe, prescribe or construct for given competences and desired performances in various interaction games.

We note that there is no fundamental reason to restrict Wittgenstein's in sights to language games or the 'language' of interaction games to verbal utterances. Other kinds of signals and actions can also be used by an agent interacting with its environment. Thus, we speak of interaction games as a generalization of Wittgenstein's language games. The partner in an interaction game may be another agent, or it may be the environment where the agent is situated. The agent interacts in the game by accessing channels of meaning.

4 Locus and Channels of Meaning

Where is meaning for an agent? It is in the observer, who, as we said may be the agent itself. So, in looking for meaning in any situation, one must ask, *Where are the observers?*

An agent interacts with the world through its sensors, embodiment and actuators. An evolved biological agent uses sensory and action channels that have been varied and selected over the course of evolution. The channels it uses are meaningful to it for its survival, homeostasis, reproduction, etc. The access to the particular channels has evolved because they are of use to the agent for such purposes, and thus meaning arises for the agent in how it accesses these channels. In this access, the agent is in the role of an observer (though not necessarily a conscious one) and this observer is also an actor.

What is meaning then? *Meaning is information considered with respect to channels of interaction (perception and or action) whose source and target are determined with respect to an observer.* The source and target may be known, uncertain, or unknown; they may be agents or aspects of environments; information in the channel may or may not be accessible to the observer; the observer may be an agent at one end (or possibly both ends) of the channel, or may be external to the channel.

The attempts and successes of formalization and rationalism to escape from context, to formulate universal scientific laws that do not depend on the particular observer and aspects of the messiness of embodiment, useful Platonic entities such as numbers, and generic impersonal statements about 'he'/'she'/'it'/'they', have been extremely important in the history of science and engineering. They have led to great successes in the physical sciences, mathematics and engineering, achieving somewhat less success in the case of animate beings, such as in biology at the level of the organism, psychology, sociology, and economics (where agents matter). Such logical positivistic approaches tend to presuppose a single unique plane of description, one universal coordinate system or model in which all phenomena may be described and understood. (Note, however, that sometimes more sophisticated versions allow several viewpoints, which agree where they overlap but may also explain some areas which are not mutually explainable in a consistent manner, e.g., in relativistic physics, the theory of manifolds in differential geometry and topology—obtained by 'gluing' locally Euclidean pieces of space, and more general coordinate systems affording formal understanding of systems [25]). We propose that first- and second-person perspectives can assist in the agent sciences and pre-sciences just mentioned. The third-person observer perspective is thus an *extra-agent* view. Nevertheless, there is an agent present in this viewpoint, namely, the external observer itself.

5 First and Second Person Meaning: 'I' and 'Thou'

5.1 The First Person: An Agent's Perspective

The notion 'first person' refers to the experience of an agent itself, the particular embodiment of the agent in its environment, and its particular sensorimotor and internal state dynamics. It is thus an *intra-agent* perspective. The agent is considered in its own *Umwelt* and may be biological, an engineered physical artifact, or a software agent cycling through a reactive, deliberative or post-reactive control loop (active process). Techniques for the first-person perspective include developmental, subsumption staged build-up, exploiting dynamics of embodiment, non-monolithic task-specific intelligence (Brooks et al. [26, 27]), and, for temporal grounding, histories and autobiographic reconstruction [10, 28]. The books of Maes [2] and Nehaniv [29] include research on situated, embodied, embedded biologically inspired systems and relevant issues in Artificial Intelligence (AI). The latter seeks to extend the framing of contemporary metaphor theory as conceptual rather than linguistic (cf. [30, 31, 33]) to agents and artifacts.

5.2 The Second Person: 'I' and 'Thou'

Blending of the first person and the other gives rise to the second person, while relations to the self complicate matters further (in addition to the consideration of other dimensions mentioned below). [In human cognitive abilities, 'blending' seems to be a ubiquitous phenomenon in conceptual manipulation, understanding, and problem solving (see Turner [31, 32].)

Inheritance of characteristics resulting from reproduction in biological systems makes the siblings and progeny of an agent resemble it. The channels of sensation and action, and the manner of embodiment of these others, is thus likely to be very similar to that of the agent. This similarity can be a substrate for interaction and provides structure that the agent's own structure can be related and mapped to. These other agents are thus 'second persons', alter-*egos* (i.e., other 'I's) in the world whose actions could be analyzed and possibly 'understood' as corresponding to one's own. A tendency to regard such others as 'egomorphic', similar to the self, or to expect that their actions in given situations should be similar to what one's own would be could thus be adaptive. This *egomorphic principle* may be at the root of the ability of animals to perceive signals of intent in others. For example, a dog might not have a theory of other minds, but may well growl when it perceives and acts on signals, such as gaze direction, of another animal looking at a piece of food it has grasped in its teeth and paws.

A generalization of the egomorphic principle in humans is their anthropomorphizing tendency to view other animals and objects around them as having human-like consciousness, feelings, intentions or goals. This tendency may lead to appropriate behavior in response to, say, perceived threat and anger in a snarling carnivore protecting its young, or to less successful behavior in, say, attributing a vengeful state of mind to storm clouds and trying to appease them with burnt offerings. The notion 'second person' refers to the experience by an agent of other agents and of the interaction dynamics with other agents. It is thus an *inter-agent* notion. Aspects include *theory of other minds and empathic resonance* [10]; biographic reconstruction for others [28]; perception of signals of intention; interaction; and *mapping* of the self to the other. In mapping the self to the other, the latter becomes, for this observer a blend of the self with the notions of otherness: *the second person*—to whom are attributed states and dynamics (e.g., intentions, drives, feelings, desires, goals) and possibly a biographic history [28]. As the second person, the other ceases to be an object and becomes an agent. As just mentioned, it may be that such mapping from 'I' to 'Thou' also lies at the core of the anthropomorphizing tendencies so often observed in human interaction with computers and robots. How such interaction dynamics work in natural agents and could be constructed in artificial ones leads one into the study of imitation [34–36], social dynamics, communication and the understanding of language games and interaction games.

6 Grounding in Interaction: Tropism, Taxes, Reinforcers, Drives (from Internal Milieu), Emotions (from External Stimuli)

6.1 Emotional Grounding

Masanao Toda's Fungus Eaters [37] carry out certain behavioral programs for survival when they have certain "urges"; this results in adaptive behavior. Such urges serve to regulate behavior, yielding appropriate actions in appropriate contexts (e.g., eating when hungry). Pfeifer [38] discuss how observer-attributed emotions emerge based on the implementation of taxis and drives in simple robotic implementations. Grand [39] has used such notions in implementing CREATURES, a successful product based on artificial life technology in which so-called Norns grow up in an environment learning and responding to stimuli while being governed by a set of urges (hunger, sex drive, etc.) that may take on dynamically varying numerical values.

Emotion systems involved in feedback control of situated agents may serve to provide the grounding for embodied agents in the body-environment coupling. Moreover, affect may play an important role in memory and historical grounding. The psychologist R. Zajonc has shown in the 1980s at the University of Michigan [40] that humans prefer stimuli with which they have previous experience to new stimuli (*"Familiarity breeds content[ment]"*); whereas rational humans who have lost some affective capacities due to aphasia are unable to function "rationally" (Damasio [41]).

6.2 *Inside/Outside: Drive/Emotion*

Biological agents respond to changes in environmental states or their own states motivated and modulated by several classes of emotion-like mechanism.

6.2.1 Dimensions of Emotion

Emotions are always experienced in the self. They may be present or absent, and they are valenced (positive or negative). They may have varying levels of intensity (degree). An emotion may focus on the physical self, objects (or others to whom intent is not attributed in the emotion) in relation to the self, or engagement with other agents (second-person emotions) or the perception by others—including real and imagined persons—of the self (second-person emotions). They may involve valuation of motives of others (other social emotions). In another dimension, consequences for that agent or actions of this agent are important in emotion. Emotions may focus on temporal episodes or just on the current state of the world.

Interaction of biological agents is grounded in first-person and also—in those agents that perceive others as others—second-person experience.

Valenced responses to stimuli (positive or negative) by the body have been implicated both as the source of emotional expression and often as prior to cognitive aspects [9, 42, 43]. For many and certainly for more complex emotions, cognitive factors [44] conditioned by cultural factors [45] play a role in the genesis and experience of emotion. Temporal aspects are also evident in emotions such as hope and regret.

Several related emotion-like phenomena also need to be considered and distinguished from emotion:

6.2.2 Taxis/Tropism

Taxes and tropisms are 'hard wired' approach or avoidance behaviors in response to stimuli, e.g., turning toward light (in plants), or moving up a gradient of food (*E. coli*) or pheromone concentration (in moths). The behaviors are stereotyped and not instrumentally arbitrary, i.e., the agent does not employ and cannot even be trained to employ alternative strategies of behavior in response to the stimulus, but reacts in a fixed manner.

6.2.3 Drives

Drives are homeostatic and instinctual mechanisms of internal motivational change or modulators of behavior in response to internal aspects of state: hunger, thirst, sex drive, maintaining temperature and other variables within acceptable ranges while interacting to the environment. Needs of self-maintenance and self-production are regulated by hormones (blood-borne signals) in the context of internal milieu account for many drives. The appropriate ranges of related parameters maintained in a homeostatic, possibly living, system need not be fixed, but may depend dynamically also on cyclical or otherwise varying internal aspects of state, for example, this is the case with hunger, and sex drive, which varies with history and hormonal state, although either of these may at times be triggered (like an emotion—see below) by external stimuli.

6.2.4 Reinforcing Stimuli

Reinforcing stimuli are stimuli that an agent will work to obtain (positive reinforcing stimuli) or to avoid or terminate (negative reinforcing stimuli). In the case of experimental psychology and ethology, this constitutes an operational definition for the identification of stimuli as positively or negatively reinforcing. Stimuli, perceptable to the agent, which for the agent are not reinforcing are called neutral or *unconditioned stimuli*; the agent does not seek to avoid or obtain such a stimulus. An unconditioned stimulus need not remain one. Some stimuli are innately reinforcing, i.e., by the design, nature, or default structuring of the agent, and are referred to as *primary reinforcers*. Proposed categories of primary reinforcers [46] for animals include tastes (salt, sweet, bitter, sour, and others), odor (putrefying odors, pheromones), somatosenory (pain, touch, grooming, washing, temperature), certain visual stimuli (symmetry, open blue sky, secure cover), auditory (warning call, vocalizations), reproduction (courtship, mate guarding, nest building, infant attachment to parents, crying of infant, parental attachment), novel stimuli (leading to curiosity), sleep, altruism within kin and social groups, group acceptance, play, and others.

According to an egomorphic view, negative reinforcers (punishers) are considered *painful* or *unpleasant* to the agent, whilst positive ones (rewards) are *pleasant* or *enjoyable*.

Unconditioned stimuli which originally had no reinforcement effect (the agent would neither seek to obtain nor avoid them) may become associated through Hebbian learning or classical conditioning with reinforcing stimuli. This is called *stimulus-reinforcement association learning*, i.e., the association of a stimulus with an existing reinforcing stimulus. In this way, Pavlov's dog associated the sound of a bell with food [47]. Such learning is to be clearly distinguished from "stimulus response learning", also called "habit learning", since it is only the *association* of stimuli, and not a response, that is learned.

Once a stimulus is associated with a reinforcer, it becomes a *conditioned reinforcer* if the agent will either work to obtain or avoid it. Any reinforcer that is not a primary (unlearned) reinforcer is referred to as a *secondary reinforcing stimulus*.

This picture of primary and secondary reinforcers just painted could be misconstrued as static, but that would be an oversimplification. Stimuli can become or cease to be secondary reinforcers as new associations are learned and old ones forgotten in a changing environment. It can happen that a primary negative reinforcers like pain associated with the eating of a spicy food becomes a secondary positive reinforcer; or in pathological cases, animals, including humans, may work to obtain painful stimulation.

Moreover, the reinforcement value of stimuli often dynamically varies with the state of internal parameters and drives. For example, satiety or habituation to a particular stimulus may caused it to lose its reinforcement value temporarily.

6.2.5 Emotions

Emotions are defined as changes in state in response to primary or secondary reinforcing stimuli, or in some cases, due to the remembering of such stimuli. Notice that since the definition of reinforcer is operational, so is this definition of emotion. The experience of qualia (feeling, awareness, or consciousness) of the state change is a possible but not a necessary aspect of emotion in this formal sense. The operational definition of emotion as state change in response to reinforcing stimuli here follows Gray [48] and Rolls [46] for animals. We observe that it also makes sense for artificial agents. The definining state change may ensue following neural processing, biochemical reactions and physiological changes, motivation and perceptual interplay, rule-based reasoning, cognitive appraisals (Ortony et al. [44], Roseman et al. [49, 50]), or in response to changes in bodily configuration and expression (James and Lange [9, 51]), combinations of these factors (Izard [52]), or by any means at all. Such types of change induced by the mechanisms listed above have been proposed in various theories and models of human, animal, and agent emotions. For example, Ortony et al. [44] define emotions as "valenced reactions to events/agents/objects [...] whose particular nature is determined by the way in which the eliciting situation is construed". Although their cognitive appraisal framework was never intended for the generation of artificial emotion, it has been applied to this as well as to artificial reasoning about affect in multiagent systems (Elliot [53]).

In the approaches mentioned above, the *type of eliciting stimulus* (e.g., object, event, action, or person) *and the particular drives related to and sensory characteristics of the reinforcing stimulus contribute to determining the character of the emotion.* It is also evident that *whether the experience is first-person or attributed to a second person also contributes in a fundamental way to the character of the emotion.*

Studies of human emotion reveal that two dimensions are extremely relevant in what is understood (intuitively rather than formally) to be required for emotion: first, emotions are *valenced*, they are either good or bad, pleasant or unpleasant; and second, they have degree (level of intensity). These properties of emotions imply that they can serve as an evaluative function in situations that result in them. In this way, they can serve as a *'common currency'* (Rolls [46]) by which to evaluate stimuli and then to compare the likely results of various courses of action.

6.2.6 Moods

Moods are longer term changes in system state that persist over extended periods of time and may have strong effects on body and behavior; for example, 'peppiness' may last all morning, while depression may last for years. (Formally, this definition still lacks the rigor of the preceding ones.)

While emotions are state changes in response to stimuli, a mood does not—to use a grammatical metaphor—"have an object", i.e., it is not elicited in relation to a particular object, agent or event in the environment. Any operational distinction between mood and emotion is complicated by the fact that remembering or imagining an environmental stimulus might result in an emotion.

Motivation and intent that arise from remembering or in planning can also guide behavior. Fast reactive responses elicited by some emotions seem to arise through certain limbic neural pathways in animal brains, slower cortical functions and deliberative evaluation may play a role in others, while abstract symbol manipulation may be involved in other highly cognitive emotions and possibly in consciousness. There may be several pathways to action, mediated by several levels at which drives and emotions arise and are arbitrated amongst. A three-layer (reactive, deliberative, and self-monitoring) architecture proposed by Aaron Sloman and collaborators as a model of human-like emotion, for example realizes such division of labor in the control of behaviour [54, 55].

6.2.7 Temporal Factors in Associative Learning

Temporal aspects of the stimuli-reinforcer association, along with the type of stimulus, are extremely important in the class of emotion elicited. That is, the temporal extent of the learned stimulus may precede, coincide with, follow, or overlap in several possible ways the duration of the experience of the reinforcer. For example, the sound of a bell may be predictive of food if it has always preceded food; hearing it may lead to emotions of expectation (anticipation, hope—or in a negative case, dread and fear), or, if the positive (respectively, negative) reinforcer is not forthcoming, to disappointment (respectively, relief). Alternatively, if the positive [resp. negative] reinforcing stimulus does occur, then 'hopes confirmed (satisfaction)' [respectively, 'fear confirmed'] state changes define the resulting emotion. First-person actions *preceding* reinforcing stimuli can elicit such emotions as guilt, regret, shame, pride; or pleasant surprise (unexpected reward), unpleasant surprise (unexpected punisher), or neutral surprise (unexpected non-reinforcing stimulus—formally, this last is a state change in response to new information, but is not an emotion, since it does not involve a reinforcer, unless perhaps the *novelty* itself is reinforcing). Some factors contributing to the character of complex (social) emotions such as pride, guilt or shame are the attribution of observer status to others who may perceive the first person's action.

Varying temporal configuration, the results in state change of the agent might be somewhat different if the bell had always co-occurred with food, e.g., disappointment might be more immediate if no food were presented during the sound of the bell than if food had always been presented only sometime after or before the sound of the bell. The two later conditions could first result, respectively in positive anticipation and confusion as immediate effects rather than disappointment. Thus, the relative temporal configuration of associated stimuli influences the character of emotion. One artificial neural network architecture that can learn associations together with the relative temporal configuration is DRAMA, developed by Aude Billard [56, 57]. We return to the consideration of temporal aspects and emotion in the section below where we discuss the temporal horizon of humans and other animals, as well as the possible implications for constructed agents, in response to ideas of Heidegger.

6.3 First-Person Emotions

First-person emotions may be viewed as useful for regulating the individual's own state. An organism maintains its state within certain acceptable ranges for various parameters and tends to act to restore its state to within these ranges if disturbed (homeostasis). Irritability or pain are experienced when there is a deviation away from the acceptable range of relevant parameters in or outside the body (acidity, glucose-level, pressure, suitability of environment for respiration—oxygen concentration, air pressure, etc.). Many emotional terms and terms for feelings are names for the experience of such deviations or the corresponding restorations (hunger, nausea, itching, dizziness, feelings of cold/heat/burning). Pleasure and pain are the most general terms for such first-person emotions.

The point is that such first-person drives and emotions play a role in regulation of behavior involving only the organism itself in its environment (e.g., a snake moving onto a sun-baked stone when feeling cold). The drives and emotions provide a motivational role in directing behaviour. In organisms and agents, various forms of taxis and tropism can also be construed as resulting from "generalized emotions" whose motivational direction of behavior is mediated via internal signals or aspects of state in organismal dynamics arising from impinging sensory data but lead only to inflexible behaviors.

6.4 Second-Person and Social Emotions

To support complex social interactions, including grounding and adaptive behavior in individualized (and possibly anonymous) societies, other types of emotions are useful. Empathic experience and the recognition of other (possibly of non-conspecific) individuals as having an experience of the world analogous to one's own may be requisite for the experience of *second-person* emotions and *social emotions* in interaction games with others. It is possible to imagine that such mechanisms do operate without necessarily requiring representation of other minds, i.e., in individuals without a theory of other minds (cf. the relation of this notion to possible mechanisms implicated in autism, e.g. [58]). But do dogs growling at someone approaching their food have a theory of mind? The perception of *intent* on the part of another conspecific or other animal seems likely not to require the representation of a mind constructing that intent.

Emotions concerning the behavior of others with whom one is not (potentially) engaged could be termed "third-person emotions". An example might be disdain, but not indignation, since the latter results from considering consequences to the self (or second persons with whom one empathizes).

7 Emotion in Adaptive Systems

Since emotions are changes in state elicited by reinforcing stimuli, their valence and degree can serve as measures of the [un]desirability of pursuing a course of action that leads to further stimuli. In particular, the particular course of action to take in obtaining or avoiding the same or an associated stimulus is not encoded in either the valence or degree of the emotion, yet the agent can take this valence and degree as a guide to suggest a course of action: to work (somehow) either to obtain or to work to avoid or terminate a stimulus. How the agent works to obtain a stimulus can be to choose to invoke more general strategies and behaviors generically applicable to large classes of situations: e.g., approach, grab, flee, hide. In this way, stimulus and response are de-coupled and the relations for behavior in response to a stimulus are modifiable, dynamically learnable and reconfigurable. Thus, the common currency of emotion can serve to modulate the control of the agent and to motivate or suppress certain responses in its interaction games.

This provides a mechanism for a "*two-process theory of learning*" (Mowrer [59], Gray [48], Rolls [46]). This type of effect of synthesized emotion on learning can be found implicitly, for instance, Bruce Blumberg's Silas T. Dog, a synthetic worlds virtual agent, which attempts to determine which aspects of input (external stimuli) elicited an internal state change (e.g., increase in a 'fear variable') and learns the association of the stimulus with the formal emotional change. The latter association influences, in a rudimentary way, the agent's learning and behavior (e.g. avoidance of locations where an unpleasant stimulus was encountered) [60], pp. 216–217.

This two-process model is distinct from the behaviorist's operational analyzed model of stimulus response learning (Hull [61], Spence [62], Skinner [7]), which it factors into (1) stimulus-reinforcer association (see above) and (2) the learning of responses. In contrast to the learning of fixed stimulus-response pairs, this factorization allows an approach for flexibility of behavioural responses to reinforcing stimuli and the capacity to substitute one behavior for another if the first fails to achieve the desired effect. It is useful as part of a constructive biology approach addressing the possible internal mechanisms relating affect, learning and behavior.

Moreover, the *expression* of an emotional or drive state may be perceptable to conspecifics or other agents (prey or predators). Recognition of the expression of the other can serve as a index to its state: e.g., the other's recent experience of reinforcing stimuli (signs of seeing a tasty victim, instinctual alarm calls), and hence as an indicator of its intent (that it might work to obtain or avoid something in the environment). Certainly in the case of biological evolution, it could be adaptive to use the information expressed in such signals to avoid predators, assess the state of prey, or gauge the likely behavior of conspecifics. Such use is a second-person method of adaptation.

Possible design approaches to making responses that exploit these signals include (1) to rely on natural selection and evolution (over generations) or (2) to rely on learning and adaptation (within an individual). The systematicity of either (1) or (2) in associations of signal-from-others with appropriate behavioral response can be *ad hoc*, partial, or comprehensive, and could vary in degree of flexibility.

A comprehensive or partial correspondence from signals perceived in the other agent and how to respond with one's own action, can be achieved by incremental learning, or it could take advantage of a natural correspondence—the identification of the other agent as a 'second person' with a similar architecture to one's own, at least similiar enough that some predication could be accurately made of the state of the agent based on the signals generated, together with predication of likely action that one would make in such a state ('mind reading' or 'reading of intent' [63]). We hypothesize that socially intelligent animal species make widespread use of such systematic second-person mechanisms. (See Dautenhahn [12] for related issues of social intelligence in animal species with individualized societies.)

Strategies (1) and (2), with varying systematicity and flexibility, may be applied to the design or the explanation of the interaction with others of a particular first-person agent.

8 Further Temporal Aspects of Emotion, Behavior, and Narrative

A feature of narrative, but not of many forms of communication, is that it provides an 'extrasensory' channel for by which an agent may obtain meaningful information to modulate or guide its immediate or future behavior. Other means by which this may occur include memory and remembering (often also involving narrative structure), and, with generally smaller temporal scope, moods and emotions.

Heidegger [64] saw the state of man as being situated in the Now, being here in the imminence of the Future in relation to the impinging Past. This *temporal horizon* is extremely broad in humans compared to other animals, as is evidenced by our emotions such as hope and regret, concern with planning for future actions and story-telling about past or imagined events. This vast temporal horizon means that humans will tend to deal with interaction in a way that makes narrative sense, and may anthropomorphically expect their technological agents to do so. Affect and narrativity thus intertwine with each other. Extrasensory data from narrative and historical temporal grounding helps an agent escape from the present in its preception-action cycle.

The cost and reward of experience stimuli provide a uniform dimension in which to evaluate the result or desirability of action, and the relative values of these costs and rewards (or 'pains' and 'pleasures') may be modulated by the current state of the agent. Most attempts to introduce 'emotion parameters' into AI systems can be seen as an attempt to solve the well-known contextualization problem in AI and robotics, i.e., to transcend simple reactivity by allowing the settings of the parameters to modulate behavior, so as to respond appropriately to the given context.

The author and Kerstin Dautenhahn have been developing algebraic tools for a mathematically rigorous framework expressing histories and, more generally, subjective views on the temporal experiences of (possibly non-human) agents [28, 65–67]. Such a framework is intended to provide formal support to realizing in robotics and software of the notion of dynamic autobiographic agents that actively construct and reconstruct their memories and autobiographies [68]. By opening up the possibility of using their own and each other's histories, this work provides for temporal grounding to help release such agents from mere reactivity.

9 Implications for Agents and Artifacts

These considerations, while not yet leading to any exhaustive classification of emotion and related phenomena, already do lead us to some consequences for the phenomenology of emotion in the design of social agents and the understanding of biological ones.

9.1 How Can Emotion Improve Decision Making?

Obviously, pain and pleasure are important in reinforcement learning. But emotion can serve as a contextualization cue for appropriate action. Doing something that doesn't feel right leads an organism to another approach—to use a computational metaphor, to switch to another "program". The intensity of emotion provides a valenced measure for the degree to which this is desirable or undesirable.

An emotion such as regret expresses pain about actions or lost opportunities in the past, and thus leads to evaluation of prior actions and reflections on their appropriateness for the past context; potentially, results of such analysis can guide future behavior and facilitate an escape from the immediacy of merely reactive reinforcement learning.

9.2 How Is Emotion Useful in Social Behavior?

Social emotions such as gratitude, guilt, regret, sympathy, or being proud of another all require the existence (or perception) of a second person. Clearly such emotions have social implications, since they are likely to influence behavior toward others. The fact that they exist suggests that they may play a role in group selection, that is, the selective advantage for the individual gained by virtue of belonging to a successful group. Such emotions, as well as the recognition of the social standing of others, may lead to natural ethics in a community of agents [69, 70].

Primates spend a lot of time in social relations, in grooming, forming and breaking alliances in complex hierarchies of power relations. It may be true that social creatures with more complex social structures need larger, more complex brains. Concern about the relations to and the intentions of others seems a major focus of primate social behavior [71]. The social intelligence hypothesis [63] states that abstract and more generalized reasoning skills evolved in humankind based on the pre-adaptation of existing social reasoning skills, such as those seen in other higher primates. Thus, socially situated emotion may well have been a prerequisite for the development of human-like intelligence. If this is the case, the implication is that to achieve such a level of artificial intelligence may well require not only computational power, and not only affective computation, but support and grounding for second-person and social emotions.

9.3 How Can Emotional and Episodic Memory Be Integrated?

The effect of emotion, especially intense emotion, on memory is well-known (e.g., [72]). Experiments conducted by Zajonc [40] show that humans prefer stimuli to which they have been previously exposed over other stimuli (the so-called *mere exposure effect*). This could be adaptive, since it would tend to lead an agent toward the familiar (as more pleasant) rather than the unknown. Going toward the familiar has at least two advantages:

(1) the agent tends to stay in contexts of which it has more experience and in which it hence has been able to survive or otherwise act appropriately
(2) this could serve as an aid to navigation without resort to maps—going toward the familiar is likely to lead one home or to a destination the path to which the agent has traversed before.

Indeed, this latter mechanism suggests a manner for navigation to arise as an emergent phenomenon. Such emergence of navigational skill would offer a radical alternative to what traditional roboticists refer to as "localization", i.e., determining one's location on an internal, external or constructed map (possibly using external beacons and landmarks) despite drift, freak perceptions, sensor noise, etc. Many of us can find our way home or to the store following the familiar—even on the second time the trip is made—while being totally incapable of drawing or visualizing a geometric or in any way accurate map of our route (not to mention the space in which this trajectory is followed).

Such considerations suggest that emotion may play a role for cognitive maps, in which affective features may mark episodic sequences, resulting in an 'emotional coloring' not only of actions but of sequences of actions and experiences.

A natural supposition then is that second-person and social emotions could play a role in the affective coloring of episodes of social interaction, and thus serve to lay down a cognitive, non-representational map emerging during social interaction and useful in navigating the social world.

Expression of affectively colored historical memories (episodes) appropriate for the agent situated in its world could be approached via the algebraic framework of Nehaniv and Dautenhahn [35] to provide historical grounding for an affective system.

9.4 What Could Affective Implementations Teach Us About 'Wet', Biological Systems?

Computer and robotic simulations of emotional systems might contribute to biological research by providing frameworks and testbeds in which purported mechanisms for various behaviors could be studied. The success or failure of an artificial implementation of someone's explanatory mechanism could never prove that a biological system has the same hypothesized structure or uses the hypothesized mechanism, but successful implementations could show that a proposed mechanism may be sufficient and feasible for realizing the phenomenon displayed by an emotional organism embedded in its environment. In this connection of computationally-inspired biology, a good example is the insect-like robots of Barbara Webb mentioned above, showing that a mechanism much simpler than several other, more elaborate ones that had been proposed was sufficient to account for phonotaxis in mate-seeking crickets and produced, as side-effects, other cricket-like aspects of movement and behavior [4, 5]. Emotionally intelligent artificial agent implementations could be used to gauge the explanatory power of affective mechanisms in explaining successful social behavior, possibly and even at the level of non-individualized societies like ants, termites, even the cellular slimemold *Dictyostelium discoideum*, as well as social mammals.

10 Summary and Discussion

Our viewpoint on emotions in agents begins from a first-person perspective, relating to second-person interaction and navigation in a biological and social world. Also, the nature of agents (human, biological, hard/software) in time and temporal situatedness are discussed with respect to emotion and behavior, and in relation to effects on human cognition. The constructive viewpoint in biology seeks to build rather than describe the mechanisms sufficient for the design of living and artificial agents. This also includes the validation and prescription of behavioural mechanisms.

Without persons, observers and agents, no meaning is possible, since meaning is defined as information in channels of sensing and acting in interaction with the environment and with others. Biological agents access channels of meaning in order to survive, grow and reproduce. Drives, emotions and other mechanisms such as tropisms provide force (motivation) for guiding appropriate behaviour to attain these, generally unconscious, goals. Drives and homeostatic maintenance, mediated by signals such as hormones, allow an agent to respond to changes in its internal milieu. Emotions are valenced responses to (external) reinforcing stimuli. Some preferences need no inferences, and pathways to emotion may vary from hard-wired, reactive, deliberative, or cognitive appraisals. Taking an agent's perspective, a intra-agent first-person viewpoint, means considering the agent in its own *Umwelt* interacting through the channels of meaning to which it has access. The 'treatment' of other agents, as having similar experience or coupling to the world via their own channels of meaning (the egomorphic principle), or the recognition of others as others are second-person methods of engineering, design. They are employed or employable in natural and artificial evolution to exploit correspondence between the other's channels of meaning and those of the first-person agent. Such second-person methods can be adaptive in social interaction in cooperation, competition for mates and resources, as well as predator-prey interaction. The anthropomorphizing tendency of humans is a case of this egomorphic principle.

A two-process system of learning allows the association of stimulus with a positive or negative reinforcer as one stage, and flexible behavioral response as a second phase which involves learning of general strategies of how to avoid or obtain stimuli. Emotion gives a'common currency' in which comparing alternatives and provides motivation for behavior to obtain or avoid stimuli effectively. This common currency helps solve the problem of contextualization of behavior—when to do what (and when not to bother) in a flexible way, which is even more flexible using a two-process system of learning.

Temporal aspects of association learning have an important impact on the class of emotion experience. Reactive robots, agents, and some simple biological organisms have a very small temporal horizon, existing only in the 'now' with little or no capacity to learn. The temporal horizon of humans and some other animals is much wider. The capacity to remember the past and imagine the future, facilitates an escape from reactivity. The scope of the temporal horizon in humans and other animals and agents varies considerably. Remembering situations and thoughts of future situations let humans plan, and serve as a source of extrasensory data furthering the escape from the present preception-action cycle. Post-reactivity and deliberation are supported by remembering, history, and narrative intelligence. They can lead to episodic emotions such as fond recall, grief, regret, hope, guilt, and shame with very long, sometimes pathologically long, temporal extent.

Emotions serve an agent in fast decision making, via a meta-rational evaluation in uncertain, uncircumscribed environment, and avoid the combinatorial explosion faced by grounded classical AI-systems. Social behavior can be supported by emotion and episodic memory in such applications as (1) spatial navigation, (2) social navigation, and (3) judging relevance via affective matching. First-person methods in emotion synthesis, 'hormonal control' to contextualize behavior, and emergent emotion are being developed in the embodied and situated Artificial Intelligence communities. Second-person methods will need to be employed in the affective grounding of multi-agent systems. The expression and recognition of affect will play major roles in this development. A few existing applications are built either on synthetic imitation or cognitive appraisals (with applications of the latter mostly to entertainment so far). Temporal grounding and narrative intelligence are expected also to play a crucial role in further developments.

Acknowledgements A preliminary version of this work [73] was presented at the Simulation of Adaptive Behaviour (SAB'98) Workshop "Grounding Emotions in Adaptive Systems" organized by Dolores Cañamero, Chisato Numaoka, and Paulo Petta. The author gratefully acknowledges stimulating discussions with Kerstin Dautenhahn and Joseph Goguen on many of the topics treated here. The ideas expressed are nevertheless the author's own.

References

1. Harré, R., & Gerrod Parrott, W. (Eds.). (1996). *The emotions: Social, cultural and biological dimensions*.
2. Maes, P. (Ed.). (1991). *Designing autonomous agents: Theory and practice from biology to engineering and back*. MIT Press.
3. Beynon, M. Empirical modelling and the foundations of artificial intelligence. In [28] (pp. 322–364).
4. Webb, B. (1994). Robotic experiments in cricket phonotaxis. In D. Cliff, P. Husbands, J.-A. Meyer, & S. W. Wilson (Eds.), *From animals to animats 3: Proceedings of the third international conference on simulation of adaptive behavior, August 8–12, 1994* (pp. 45–54). Brighton, England.
5. Webb, B. (1995). Using robots to model animals: A cricket test. *Robotics and Autonomous Systems*, 16-117-134.

6. Braitenberg, V. (1986). *Vehicles: Experiments in synthetic psychology*. MIT Press.
7. Skinner, B. F. (1938). *The behavior of organisms*. New York: Appleton Century.
8. Arnold, M. B. (Ed.). (1968). *The nature of emotion: Selected readings*. Harmondsworth: Penguin.
9. James, W. (1884). What is emotion? *Mind, 9*, 118–205.
10. Dautenhahn, K. (1997). I could be you—the phenomenological dimension of social understanding. *Cybernetics and Systems Journal, Special Issue on Epistemological Aspects of Embodied AI, 28*(5), 417–453.
11. Dautenhahn, K. (1998). The art of designing socially intelligent agents—science, fiction, and the human in the loop. Special issue on socially intelligent agents. *Applied Artificial Intelligence, 12*(7–8), 513–617.
12. Dautenhahn, K. Embodiment and interaction in socially intelligent life-like agents. In [28] (pp. 102-142).
13. Wilson, E. O. (1974). *The insect societies*. Harvard University Press.
14. Nehaniv, C. L. The second person—meaning and metaphors. In [28].
15. Nehaniv, C. L. (1999). Meaning for observers and agents. In *Proceedings IEEE international symposium on intelligent control intelligent systems and semiotics, ISIC/ISAS'99, 15–17 September 1999*. Cambridge, Massachusetts.
16. Beckers, R., Holland, O. E., & Deneubourg, J. L. (1994). From local actions to global tasks. In R. A. Brooks & P. Maes (Eds.), *Artificial Life IV* (pp. 181–189). MIT Press.
17. Grassé, P. P. (1959). La reconstruction du nid et les coordinations inter-individuelles chez Bellicositermes natalensis et Cubitermes sp. La theorie de la stigmergie. Essai d'interpretation des termites constructeurs. *Ins. Soc., 6*, 41–48.
18. Shannon, C. E., & Weaver, W. (1963). *The mathematical theory of communication*. University of Illinois Press.
19. Maturana, H. R., & Varela, F. J. (1992). *The tree of knowledge: The biological roots of human understanding* (Revised edition). Shambala Publications, Inc.
20. Quick, T., Dautenhahn, K., Nehaniv, C., & Roberts, G. (1999, September). On bots and bacteria: Ontology-independent embodiment. In *Proceedings fifth European conference on artificial life*. Switzerland.
21. Wittgenstein, L. (1958). *The blue and brown books*. Harper & Brothers.
22. Wittgenstein, L. (1968). *Philosophical investigations, (Philosophische Untersuchungen)*. German with English translation by G. E. M. Anscombe, 1964. Basil Blackwell, Oxford, reprinted 3rd edition.
23. Goguen, J. An introduction to algebraic semiotics, with application to user interface design. In [28] (pp. 242–291).
24. Peirce, C. S. (1965). *Collected papers* (Vol. 2). Elements of Logic: Harvard.
25. Nehaniv, C. L. (1997). Algebraic models for understanding: Coordinate systems and cognitive empowerment. In *Proceedings of the second international conference on cognitive technology: Humanizing the information age* (pp. 147–162). IEEE Computer Society Press.
26. Brooks, R. A. (1986, April). A robust layered control system for a mobile robot. *IEEE Journal, Robotics and Automation, RA-2*, 14–23.
27. Brooks, R. A., Breazeal, C., Marjanović, M., Scassellati, B., & Williamson, M. M. The Cog project: Building a humanoid robot. In [28] (pp. 52–87).
28. Nehaniv, C. L., & Dautenhahn, K. (1998). Embodiment and memories—algebras of time and history for autobiographic agents. In R. Trappl (Ed.), *Cybernetics and systems '98, proceedings of the 14th European meeting on cybernetics and systems research (Symposium on embodied cognition and artificial intelligence; co-organized by Maja Mataric and Eric Prem), Vienna, Austria, 14–17 April 1998* (Vol. 2, pp. 651–656). Austrian Society for Cybernetic Studies.
29. Nehaniv, C. L. (Ed.). (1999). *Computation for metaphors, analogy and agents*. Lecture notes in artificial intelligence (Vol. 1562). Springer.
30. Ortony, A. (1993). *Metaphor and thought* (2nd ed. (1st edition: 1979)). Cambridge University Press.
31. Turner, M. (1996). *The literary mind*. Oxford.

32. Turner, M. *Forging connections* (pp. 11–26). In [28].
33. Lakoff, G., & Johnson, M. (1980). *Metaphors we live by*. University of Chicago Press.
34. Dautenhahn, K., & Nehaniv, C. L. (Eds.). (1999). *Proceedings AISB'99 symposium on imitation in animals and artifacts, April 6–9, 1999*. Edinburgh, Scotland: Society for the Study of Artificial Intelligence and Simulation of Behaviour.
35. Nehaniv, C. L., & Dautenhahn, K. (in press). Of hummingbirds and helicopters: An algebraic framework for interdisciplinary studies of imitation and its applications. In J. Demiris & A. Birk (Eds.), *Learning Robots: An Interdisciplinary Approach*. World Scientific Press.
36. Scassellati, B. Imitation and mechanisms of joint attention: A developmental structure for building social skills on a humanoid robot (pp. 176–195). In [28].
37. Toda, M. (1982). *Man, robot, and society*. The Hague: Nijhoff.
38. Pfeifer, R. (1994). The Fungus Eater approach to emotion: A view from artificial intelligence. *Cognitive Studies, 1*, 42–57.
39. Grand, S., Cliff, D., & Malhorta, A. (1997, February). CREATURES: Artificial life autonomous agents for home entertainment. In *Proceedings first international conference on autonomous agents (AGENTS'97—Marina del Rey)*. Association for Computing Machinery.
40. Zajonc, R. B. (1980). Feeling and thinking: Preferences need no inferences. *American Psychologist, 35*, 151–175.
41. Damasio, A. R. (1994). *Descartes' error: emotion*. New York: Reason and the Human Brain. G. P. Putnam & Sons.
42. Darwin, C. (1965). *The expression of emotion, reason, and the human brain in man and animals, 1892*. Reprinted by the University of Chicago Press.
43. Zajonc, R. B. (1984). On the primacy of affect. *American Psychologist*, 39-117-123.
44. Ortony, A., Clore, G. L., & Collins, A. (1998). *The cognitive structure of emotions*. Cambridge University Press.
45. Landman, J. Social control of 'Negative' emotions: The case of regret. In [1].
46. Rolls, E. T. (1999). *The Brain and Emotion*, Oxford.
47. Pavlov, I. P. (1928). *Lectures on conditioned reflexes* (W. H. Gantt, Trans.). New York: Liverwright.
48. Gray, J. A. (1975). *Elements of a two-process theory of learning*. Academic Press.
49. Roseman, I. J. (1991). Appraisal determinants of discrete emotions. *Cognition and Emotion, 5*(3), 161–200.
50. Roseman, I. J., Antoniou, A. A., & Jose, P. E. (1996). Appraisal determinants of emotions: Constructing a more accurate and comprehensive theory. *Cognition and Emotion, 10*(3), 241–277.
51. James, W., & Lange, C. G. (1922). *The emotions*.
52. Izard, C. E. (1993). Four systems for emotion activation: Cognitive and noncognitive processes. *Psychological Review, 100*(1), 68–90.
53. Elliot, C. D. (1992). *The affective reasoner: A process model of emotions in a multi-agent system*. PhD thesis in computer science, Northwestern University.
54. Sloman, A., & Croucher, M. (1981, August). Why robots will have emotions. In *Proceedings seventh international conference on AI* (pp. 197–202).
55. Wright, L., Sloman, A., & Beaudoin, L. (1996). Towards a design based analysis of emotional episodes. *Philosophy, Psychiatry and Psychology, 3*(2), 101–126.
56. Billard, A. (1999). *DRAMA, a connectionist model for robot learning: Experiments in grounding communication through imitation in autonomous robots*. Ph.D. thesis in artificial intelligence, University of Edinburgh.
57. Billard, A., & Hayes, G. (1999, January). DRAMA, a connectionist architecture for control and learning in autonomous robots. *Adaptive Behavior, 7*(1).
58. Cohen, S. B. (1995). *Mindblindness: An essay on autism and theory of mind*, MIT Press.
59. Mowrer, O. H. (1960). *Learning theory and behavior*. Wiley.
60. Picard, R. (1997). *Affective computing*. MIT Press.
61. Hull, C. L. (1943). *Principles of behavior*. New York: Appleton Century Crofts.
62. Spence, K. W. (1956). *Behavior theory and conditioning*. Yale University Press.

63. Byrne, R. W., & Whiten, A. (1988). *Machiavellian intelligence*. Clarendon Press.
64. Heidegger, M. (1972). *On time and being*. Harper Torchlight Books.
65. Dautenhahn, K., & Nehaniv, C.L. (1998, January 19–21). Artificial life and natural stories. In: *International symposium on artificial life and robotics—AROB III'98* (Vol. 2, pp. 435–439). Beppu, Oita, Japan.
66. Nehaniv, C. L. (1997, November). What's your story?—Irreversibility, algebra, autobiographic agents. In K. Dautenhahn (Ed.), *Socially intelligent agents: Papers from the 1997 AAAI fall symposium* (Vol. FS-97-02, pp. 150–153). MIT, Cambridge, Massachusetts: American Association for Artificial Intelligence Press.
67. Nehaniv, C. L., & Dautenhahn, K. (1998). Semigroup expansions for autobiographic agents. In T. Imaoka & C. L. Nehaniv (Eds.), *Proceedings of the first symposium on algebra, languages and computation, 30 October–1 November 1997* (pp. 77–84). University of Aizu, Japan.
68. Dautenhahn, K. (1996, November). Embodiment in animals and artifacts. In *AAAI'96 Symposium on Embodied Action and Cognition*. American Association for Artificial Intelligence Press, Technical Report FS-96-02, Boston.
69. Ridley, M. (1996). *The origins of virtue: Human instincts and the evolution of cooperation*. Viking Books.
70. Sigmund, K. (1999). The social life of automata. In C. L. Nehaniv (Ed.), *Mathematical and computational biology*. Lectures on mathematics in the life sciences series (Vol. 26, pp. 133–146). American Mathematical Society.
71. Dunbar, R. I. M. (1997). *Grooming, gossip, and the evolution of language*. Harvard University Press.
72. Gregor, J. A. (1997). *Memory & remembering: Everyday memory in context*. Addison Wesley Longman Limited.
73. Nehaniv, C. (1998). The first, second and third person emotions: Grounding adaptation in a biological and social world. In D. Canamero, C. Numaoka, & P. Petta (Eds.), *From animals to animats: Fifth international conference of the society for adaptive behavior (SAB'98) workshop on grounding emotions in adaptive systems, 21 August 1998* (pp. 43–47). Zürich, Switzerland.

Modeling Cognition–Emotion Interactions in Symbolic Agent Architectures: Examples of Research and Applied Models

Eva Hudlicka

Abstract The past two decades have witnessed a resurgence of interest in emotion research, as well as progress in understanding the circuitry that mediates affective processing in biological agents. Emotion researchers are now recognizing that computational models of emotion provide an important tool for understanding the mechanisms of affective processing. There has also been significant progress in affective computing technologies, including affective virtual agents, social robots and affect-adaptive human-computer interaction in general, including affective gaming and the associated desire to model more affectively realistic and believable agents and robots. This chapter describes a generic methodology for modeling emotions and their effects on cognitive processing. The methodology is based on the assumption that a broad range of both state and trait influences on cognition can be represented in terms of a set of parameters that control processing within the architecture modules. As such, the methodology is suitable both for exploring the nature of the mechanisms mediating cognition-emotion interaction and for developing more affectively realistic and believable agents and robots. An implementation of this generic methodology in a symbolic cognitive-affective architecture is described, focusing on an example of a research model. The chapter concludes with a discussion of open questions and challenges in affective modeling.

1 Introduction

The past three decades have witnessed a significant increase in emotion research in psychology and neuroscience. Progress has been made in understanding the circuitry that mediates affective processing in biological agents. Two of the more significant findings have been the recognition that cognitive and affective processing are highly interconnected at the neural level (e.g., [5, 21]), and that affective processing cannot be localized into specific brain regions (e.g., [3]), as has previously been assumed.

E. Hudlicka (✉)
Psychometrix Associates & University of Massachusetts, Amherst, MA, USA
e-mail: hudlicka@cs.umass.edu

M. I. Aldinhas Ferreira et al. (eds.), *Cognitive Architectures*, Intelligent Systems, Control and Automation: Science and Engineering 94,
https://doi.org/10.1007/978-3-319-97550-4_9

These findings have brought to the forefront the need for an improved understanding of cognition-emotion interactions, at both the neural and psychological levels.

Researchers in affective science are increasingly recognizing the potential benefits of computational models of emotion as one of the tools for studying affective phenomena. In fact, a new term has emerged that captures this approach, and *in computo* is now considered another alternative to the more traditional in vitro and in vivo approaches to attempting to understand the mechanisms that mediate affective processing, including cognition-emotion interactions.

Computational models of emotion aim to represent some aspect of affective processing (e.g., emotion generation, affective biases), and are being constructed at both the neural and psychological levels, using different methodologies and aiming to model affective phenomena at distinct levels of resolution [12]. Models at the neural levels often use connectionist (neural network) approaches, while models at the psychological level are often symbolic, using cognitive-affective architectures that aim to simulate an end-to-end information processing cycle (see-think/feel-do).

One of the core benefits of these models is that the very process of designing and developing a computational simulation-based model necessitates a degree of operationalization of the high-level theories that often reveals gaps and inconsistencies that might otherwise not become apparent. This then allows the development of more refined theories, which can then be subjected to empirical validation. This is particularly critical for theories defined at the psychological level, which are often cast in terms of highly aggregated constructs that allow for a significant degree of ambiguity. Such *in computo* models of specific affective phenomena thus provide an important means of generating and modeling alternative hypotheses regarding their mechanisms, which can then be tested via empirical studies. This coupled modeling-empirical study approach represents a promising cross-disciplinary method for identifying the mechanisms of information processing, both cognitive and affective, in biological agents.

1.1 Research Versus Applied Models

The models developed for the purpose of understanding these mechanisms can be termed *research models* [14]. Research models can be contrasted with models developed for the purpose of enhancing the affective realism, believability or overall functionality of a virtual agent or robot. These models do not aim to contribute to our understanding of affective processing in biological agents (although they may), but rather to produce a certain type of behavior that will enhance human-computer interaction or the abilities of some virtual agent or robot. These models can be termed *applied models.*

The distinction between research and applied models is important, since their aims, modeling approaches, and, most importantly, criteria for validation are quite distinct. *Research models* aim to *emulate* some aspect of affective processing and to represent structures, processes and mechanisms that exist in biological agents, in order to understand how these processes are implemented in biological agents. Understanding these mechanisms then provides a foundation for the development of more effective pedagogical approaches and training systems, decision-support systems, and approaches to the diagnosis and treatment of cognitive-affective disorders.

In contrast, in *applied models* it is sufficient to *simulate* the processing necessary to produce the objectives of a particular model or its associated agent. Typically, these objectives involve increasing the affective realism and believability of virtual agents or robots, thereby enhancing some aspect of human-computer interaction. This is particularly important in learning and training contexts, as well as in the increasingly common coaching and relational contexts (e.g., virtual relational agents used as companions and coaches, designed to be engaging over longer periods of time). Whether or not the processing that takes place in an applied model actually resembles the processing in biological agents is irrelevant. In fact, applied models may often be implemented via a black-box approach, in which only the input-output matching is important, and the means through which the desired outputs are obtained are irrelevant; e.g., IF <stimulus A present> THEN <display emotion expression X>.

These distinct types of model have distinct requirements and benefits, and it is critical to understand, for any given context, which modeling approach is appropriate and desirable. Applied models impose fewer constraints on their implementation, and the criteria for whether or not the model is valid are significantly different. In other words, an applied model is 'valid' if it meets some human-defined criteria for performance (e.g., users assess virtual characters with these models as more believable; social robots with these models are more effective in achieving their interactional goals with humans, etc.). In contrast, a research model is only valid if it structurally and functionally corresponds to the modeled phenomenon. Clearly, this is a much more difficult objective to achieve.

1.2 Modeling Effects of Affective States and Traits on Cognition

This chapter describes a generic methodology for modeling the effects of both affective states and traits on cognitive processing, and the associated cognitive-affective architecture. Although the focus here is on models of affective states and traits, the methodology supports the modeling of a wide range of interacting behavior moderators and individual differences, hence the name MAMID (Methodology for Analysis and Modeling of Individual Differences) [7, 11].

The MAMID methodology and architecture are suited for both *research* and applied purposes. In the case of *research models*, the architecture enables the modeling of alternative hypotheses of particular affective phenomena, with a specific focus on models of a broad range of affective heuristics and biases on aspects of attentive, perceptual and cognitive processing. In the case of *applied models*, the architecture enables the rapid construction of agents with distinct affective and personality profiles, with the potential of enhancing their affective realism and believability across various contexts (virtual affective agents, social robots, affective non-playing characters in games).

The methodology is based on the assumption that a broad range of both state and trait influences on cognition can be represented in terms of a set of parameters that control processing within the individual architecture modules. The parameters are defined outside of the architecture, and influence both the processing and structure of the architecture. In terms of processing, the parameters influence both the speed and capacity of processing within the architecture modules, as well as the likelihood that specific data will be processed at a given time (e.g., an attended cue will be further processed versus ignored). In terms of structure, the parameters influence both the topology of the architecture, thereby controlling which modules execute in a given cycle, as well as the contents of the architecture memory, thereby making some data (schemas) more or less likely to play a role in a particular processing cycle. The latter enables the architecture to model different types of 'beliefs' an agent may have.

1.3 Different Levels of Modeling Resolution

The MAMID model and cognitive-affective architecture aim to model psychological, rather than neural, phenomena. Thus no assumptions or claims are being made that the represented psychological-level constructs (e.g., cues, situations, goals) and processing (e.g., situation assessment, goal re-prioritization, emotion generation) have any direct correspondence to actual neural level structures and processes. Psychological (typically symbolic) computational models and neural (often connectionist) computational models address distinct phenomena at different levels of aggregation. That said, it is intriguing to consider the possibility that the recently discussed neuromodulatory mechanisms that appear to mediate some of the systemic effects of emotions may be implemented in symbolic models via the types of parametric manipulation of distinct modules and processing that MAMID uses to model effects of emotions on cognition.

The chapter is organized as follows. Relevant research from psychology is first introduced, followed by a description of both the modeling methodology and the associated cognitive-affective architecture. A research model aiming to elucidate the mechanisms of affective biases is then described in detail. The chapter concludes with a discussion of open questions and challenges in computational affective modeling.

2 Relevant Emotion Research Background

Definitions Although no universally agreed-upon definition of emotions exists, underscoring our lack of understanding of these complex phenomena, it is helpful to establish a working definition of emotions for modeling purposes, as follows.

Emotions can be defined as "evaluative judgments of the environment, the self and other social agents, in light of the agent's goals and beliefs". The resulting emotions then motivate and coordinate both internal processing and behavior, including social behavior and specific affective expressions, to enable adaptive interaction between the agent and its environment.

Note that 'agent' is used here in the abstract sense of any autonomous entity, which includes both biological and synthetic agents. Note also that the definition above implies that emotions generally play an adaptive role. While it is certainly the case that emotions can become dysregulated, and thus maladaptive, contemporary emotion research assumes that the evolutionary role of emotions is to support more effective adaptation.

Terminology The following definitions are assumed. Emotion refers to a transient state, lasting for seconds or minutes, typically associated with well-defined triggering cues and characteristic patterns of expressions and behavior (more so for the simpler, fundamental emotions than for complex emotions with strong cognitive components). Emotions can thus be contrasted with other terms describing affective phenomena: *moods*, sharing many features with emotions but lasting longer (hours to months); *affective states*, undifferentiated positive or negative 'feelings' and associated behavioral tendencies (approach, avoid); and *feelings,* a problematic and ill-defined construct from a modeling perspective. (Averill points out that "feelings are neither necessary nor sufficient conditions for being in an emotional state" [1]). This chapter focuses on emotions, although the methodology and architecture can also model moods and the less differentiated affective states.

Multiple Modalities A defining characteristic of emotions is their multi-modal nature. Emotions in biological agents are manifested across four distinct, but interacting, modalities. The most familiar is the *behavioral/expressive* modality, in which the expressive and action-oriented characteristics are manifested (e.g., facial expressions, speech, gestures, posture, behavioral choices). Closely related is the *somatic/physiological modality*—the neurophysiological substrate that makes behavior (and cognition) possible (e.g., heart rate, neuroendocrine effects, blood pressure). The *cognitive/interpretive* modality is most directly associated with the evaluation-based definition provided above, and is emphasized in the cognitive appraisal theories of emotion generation. The most problematic modality, from a modeling perspective, is the *experiential/subjective* modality: the conscious, and inherently idiosyncratic, experience of emotions within the individual. While the current emphasis in emotion modeling is on the cognitive modality (involved in appraisal) and the behavioral modality (manifesting emotions in agents), it is important to recognize that both the physiological and experiential modalities also play critical roles [17].

Affective Biases Emotions exert profound influences on cognition in biological agents, including the fundamental processes mediating information processing (attention, perception, memory), but also higher level processes, including situation assessment, decision-making, goal management, planning and learning. Emotion effects, including affective decision biases and heuristics, can be adaptive or maladaptive, depending on their type, magnitude and context. For example, the preferential processing of threatening stimuli associated with anxiety and fear can be adaptive in situations when survival depends on the fast detection of danger and protective behavior (e.g., avoid an approaching car that has swerved into your lane). However, the same effect can be maladaptive if neutral stimuli are judged to be threatening (e.g., passing car is assumed to be on a collision course and causes the driver to swerve into a ditch), or if the threat level of a stimulus is exaggerated.

A number of emotion effects on cognitive processing have been identified by cognitive psychologists and emotion researchers. For example, positive emotions induce a global focus and the use of heuristics, whereas negative emotions induce a more local focus and analytical thinking [4]; anxiety reduces attentional and working memory capacities, biases attention towards the detection of threatening stimuli, and biases interpretive processes towards higher threat assessments; anxiety also induces a self-bias, mood induces mood-congruent biases in recall, and negative affect reduces estimates of control and induces more analytical thinking [16, 19, 20].

3 MAMID Modeling Methodology and Architecture

The core component of the MAMID modeling approach is a *generic methodology* for modeling multiple, interacting effects of individual differences within symbolic cognitive architectures, via parametric manipulations of the architecture *processes* and *structures* [7–9, 11]. The underlying thesis of this approach is that the combined effects of a broad range of individual differences, including affective states and traits, can be integrated and represented in terms of these parameter values.

The current focus is on modeling the effects of emotions (joy, fear, anger, and sadness) on the cognitive processes mediating decision-making (attention, situation assessment, expectation generation, goal management and action selection), in terms of parameters that control processing within the individual modules of a cognitive-affective architecture. These parameters control the speed and capacities of the different architecture modules, as well as the ranking of the individual constructs processed by these modules (e.g., cues, situations, goals), as they map the inputs (perceptual cues) onto the outputs (selected actions), and thereby implement the complete see-think/feel-do sequence. A high-level schematic of the MAMID cognitive-affective architecture is shown in Fig. 1. Figure 2 illustrates the parameter-based modeling approach.

The MAMID architecture implements a sequential see-think/feel-do processing sequence, consisting of the following modules: *Sensory Pre-processing* (translates incoming data into task-relevant cues); *Attention* (filters incoming cues and selects a subset for processing); *Situation Assessment* (integrates individual cues into an overall situation assessment); *Expectation Generation* (projects current situation onto possible future states); *Affect Appraiser* (derives a valence and four of the basic emotions from external and internal elicitors); *Goal Management* (identifies high-priority goals); and *Behavior Selection* (selects the best actions for goal achievement).

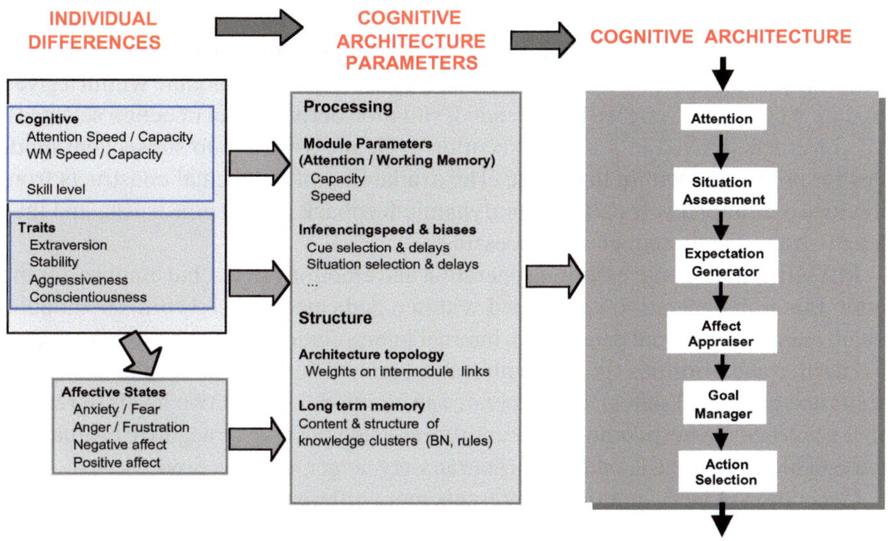

Fig. 1 MAMID cognitive-affective architecture

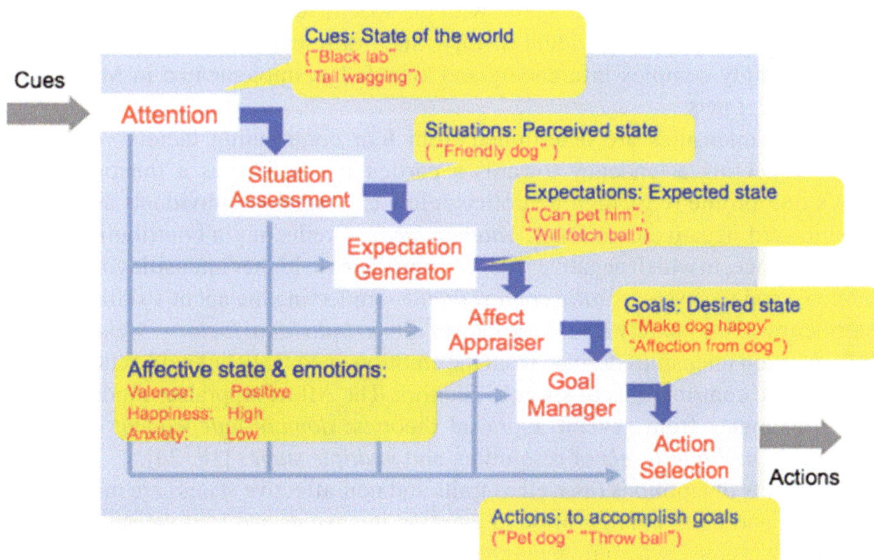

Fig. 2 Schematic illustration of MAMID methodology for state and trait modeling

These modules map the incoming stimuli (cues) onto the outgoing behavior (actions), via a set of intermediate internal structures (situations, expectations, and goals), collectively termed *mental constructs*. This mapping is enabled by long-term memories (LTM) associated with each module, represented by belief nets. *Mental constructs* are characterized by their attributes (e.g., familiarity, novelty, salience, threat level, valence, etc.), which influence their processing, that is, their rank and the consequent likelihood of being processed by the associated module within a given execution cycle (e.g., cue will be attended, situation derived, goal or action selected). All constructs derived in a given execution cycle are available to subsequent modules for processing within that cycle. The availability of the mental constructs from previous execution cycles allows for dynamic feedback among constructs, and thus departs from strictly sequential processing.

MAMID models both emotion generation and emotion effects, but emphasizes the latter. Emotion generation is modeled within a dedicated *Affect Appraiser* module, which integrates external data (cues), internal interpretations (situations, expectation) and desires and priorities (goals), with stable and transient individual characteristics (traits and emotional states), and generates an emotional state at two levels of resolution: a *valence* (corresponding to an undifferentiated positive or negative evaluation) and one of the four basic emotions (fear/anxiety, anger, sadness, joy).

Generation of basic emotions represents more differentiated processing, in which the intensity of each emotion is influenced by both task- and individual-specific criteria. This involves a consideration of a variety of idiosyncratic criteria that determine, for example, whether a high-threat situation or an impending goal failure will lead to anger or anxiety in a particular agent, with the specific effect being a function of the agent's personality and individual history. Such differentiated processing requires correspondingly complex inferencing and knowledge, implemented in MAMID in terms of belief nets.

Emotion intensities are determined from four contributing factors: *Trait bias factor*—reflecting a tendency towards a particular emotion, as a function of the agent's trait profile (e.g., high neuroticism/low extraversion individuals are predisposed toward negative emotions). *Valence factor*—reflecting a contribution of the current valence, in which negative valence contributes to higher intensities of negative emotions, and vice versa. *Static context factor*—reflecting the agent's skill level and contributing to the anxiety level if skill level is low. *Individual factor*—weighted sum of the emotion intensities derived from the emotion-specific belief nets, reflecting the idiosyncratic contributions of specific elicitors. The Affect Appraiser module incorporates elements from several appraisal theories: *domain-independent appraisal dimensions, multiple-levels* of resolution, and *multiple stages* [18, 24].

The effects of emotions (as well as traits and non-affective states) are modeled by mapping a particular configuration of emotion intensities and trait values onto a set of parameter values, which then control processing within the architecture modules, as well as the data flow among the modules, e.g., decrease/increase the modules' capacity and speed, introduce a bias for particular types of construct, such as high-threat or self-related constructs (see Figs. 2 and 3).

Functions implementing these mappings are constructed on the basis of the available empirical data. For example, the anxiety-linked bias towards preferentially attending to threatening cues and interpreting situations as threatening is modeled in MAMID by ranking high-threat cues and situations more highly, thereby making their processing by the Attention and Situation Assessment modules more likely.

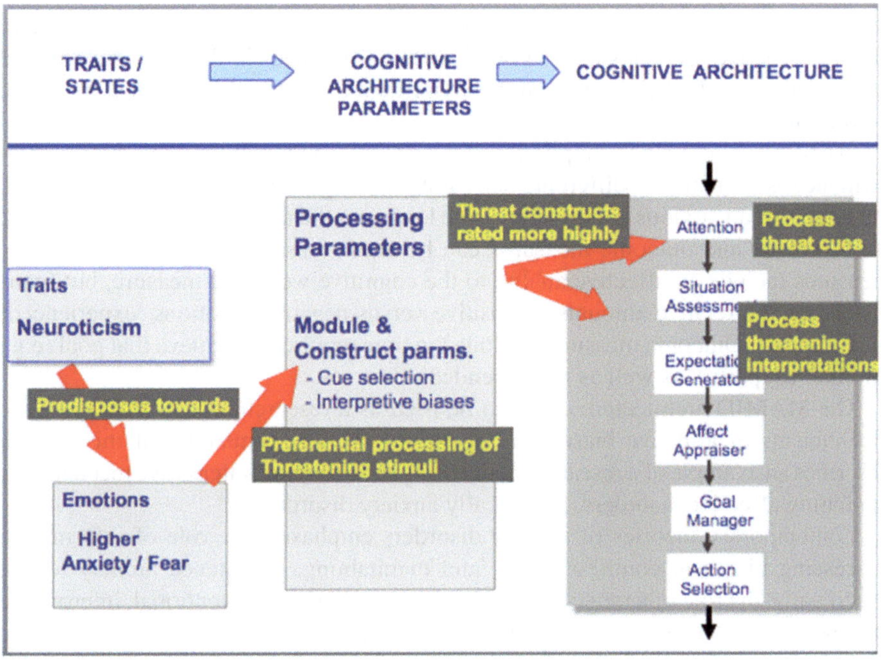

Fig. 3 Modeling threat bias within MAMID

Currently, the parameter-calculating functions consist of weighted linear combinations of the factors that influence each parameter. For example, working memory capacity reflects a normalized weighted sum of emotion intensities, trait values, baseline capacity, and skill level.

4 Examples of Research Models

The MAMID methodology and the associated architecture represent a domain-independent framework, which can be instantiated across multiple domains. The domain-specific knowledge is encoded in MAMID's long-term memory, associated with each module, and represented in terms of Bayesian belief nets. The examples below are from a search-and-rescue task domain, in which the MAMID architecture represents the behavior of different agents who are cooperating to find a 'lost party' in an inhospitable Arctic terrain, encountering a variety of anxiety-producing setbacks (e.g., various emergencies, inadequate resources, mechanical failures) and needing to ensure adequate supplies from available "supply stations" to maintain adequate resources (fuel, first aid kits) [10]. The behavior of the individual team members (snow cat drivers) is controlled by instances of the MAMID architecture.

Distinct personalities can be defined for different agents, which then contribute to distinct affective reactions that, in turn, induce differences in decision-making and action selection. The differences in behavior then impact both individual and team performance.

MAMID was used in the applied model context to explore differences in team performance as a function of different configurations of agent personalities. For example, the effects of an anxious versus aggressive leader on the team's performance [9, 10]. Both process and outcome measures can be used to assess performance. Process measures include an affective analog to the cognitive workload measure, but focus instead on the relative amounts of positive versus negative emotions 'experienced' by the agents. Outcome measures include the time required to achieve that goal (e.g., find the lost party), as well as the expended resources.

The MAMID architecture can also be used as a research model, to explore the mechanisms of affective biases on cognition [13]. The remainder of this section describes an example of a research model designed to explore alternative mechanisms mediating affective disorders, specifically anxiety disorders.

Contemporary theories of anxiety disorders emphasize the role of information processing biases in contributing to, and maintaining, heightened anxiety levels; specifically, the role of a range of emotion-induced biases on attentional, interpretive and memory processes [19]. The MAMID modeling methodology and architecture provide the representational and processing infrastructure that enables the construction of explicit models of these biases, and also supports the modeling of alternative mechanisms mediating anxiety disorders.

Two of the biases that have been extensively studied as mediators of anxiety disorders are the *attentional* and the *interpretive* biases. Attentional biases focus attention on stimuli with a particular affective content. In anxiety disorders, the biasing effects focus attention on negative and threatening stimuli; an individual in a state of anxiety selectively focuses on threatening stimuli and neglects non-threatening stimuli, thereby maintaining or even increasing their state of anxiety. Interpretive biases selectively direct interpretation of stimuli to favor an interpretation with a specific affective tone [6]. In anxiety disorders, this type of bias contributes to interpretations of ambiguous stimuli as dangerous, threatening or negative, again, maintaining or increasing the individual's state of anxiety. Both of these mechanisms can be explicitly modeled in MAMID, via parametric manipulations of the modeling processes and structures.

MAMID supports the exploration of alternative mechanisms mediating anxiety disorders through its ability to represent the attentional and interpretive biases and the resulting anxiety states, including the extreme state of a panic attack, through parametric manipulations of the underlying processing. The modeling approach demonstrates how the same set of underlying processes can generate a wide variety of effects, ranging from adaptive protective behavior through mild dysfunction to paralyzing pathology, depending on the values of the parameters controlling the processing: as the anxiety intensity increases, the processing becomes increasingly biased, demonstrating increasingly dysfunctional behavior.

Two key features of the MAMID model that make it suitable for modeling the mechanisms of psychopathology are (1) a high degree of parameterization, enabling manipulation of architecture topology, data flow, and processing within the individual modules, and (2) a testbed environment, within which the model is embedded, and which facilitates rapid model development and interactive 'tuning', by providing the modeler access to a range of model parameters and control of the functions that derive their values. By manipulating these parameters, alternative hypotheses regarding the specific mechanisms of an observed phenomenon can be rapidly implemented and their behavior evaluated within the context of a specific simulated environment.

In this example, the modeled agent approaches a difficult "emergency situation", but lacks the required resources. The agent's state of anxiety, dynamically calculated by the Affect Appraiser module within the MAMID architecture, is high, in part due to a trait-induced tendency towards higher anxiety and in part due to the task difficulty level and a lack of adequate resources.

Panic attack is an interesting anxiety state to explore, because its extreme nature provides a useful context in which to model the effects of anxiety on attentional and interpretive processes, and cognition-emotion interaction in general. Panic attack is a state in which the confluence of multiple anxiety effects produces a type of a 'perfect storm', frequently inducing behavioral paralysis. Three anxiety-linked effects are involved: *threat processing bias, self processing bias*, and *capacity reductions in both attention and working memory.* MAMID models all three of these effects and provides parameters that control their relative contributions to the overall effect on information processing.

The MAMID processing parameter values are calculated from linear combination of the weighted factors influencing the parameter. A specific parameter-induced effect (e.g., reduced module capacity) can thus be obtained from multiple combinations of the individual factors that influence the final value of a given parameter and their associated weights. These alternative configurations then provide the means of defining alternative mechanisms that mediate specific effects. The MAMID testbed environment provides facilities that support the rapid construction of these alternative mechanisms, via interactive manipulation of the factors and weights, which allow the modeler to control the magnitude and contribution of each influencing factor.

Anxiety-induced threat bias is modeled by first calculating the threat level of each cue, situation and expectation (the mental constructs), from factors that include an a priori 'fixed' threat level (e.g., a low level of resources is inherently more threatening than adequate resources), state and trait anxiety factors, and individual history (prior experience with a specific type of situation that has caused anxiety before). The threat level is then used as a weighted factor in the function calculating the overall construct rank, which determines the likelihood of its processing within a given execution cycle. In states of high-anxiety, high-threat constructs have a higher ranking, and are thus processed preferentially: high-threat cues are given preference over low-threat cues, and high-threat interpretations are therefore preferentially derived in situation assessment (see Fig. 3).

Self bias is modeled by including a weighted factor reflecting the self versus non-self origin of each construct in its rank-calculating function. High levels of state or trait anxiety then induce a higher ranking for self-related constructs, contributing to their preferred processing. In cases of high anxiety, this bias will produce a focus on the anxious state itself—a common feature of anxiety disorders that typically further increases the anxious state of the individual.

The *capacity reduction* effects on attention and working memory are modeled by dynamically calculating the capacity values of all modules during each simulation cycle, from weighted factors representing the emotion intensities, the four traits represented in MAMID, baseline capacity limits, and skill level.

MAMID models a panic attack state as follows. Stimuli, both external and internal, arrive at the Attention Module, whose capacity is already reduced. Because of the threat- and self-bias, self-related high-threat cues are processed preferentially, in this case resulting in the agent's focus on a self-related anxiety cue. This cue, reflecting the agent's anxious state, consumes the limited module capacity, leading to the neglect of external and non-threatening cues (e.g., proximity of a supply station). This results in a continued self- and threat-focus in the downstream modules (Situation Assessment and Expectation Generation). No useful goals or behaviors can be derived from these constructs, and the agent enters a positive feedback-induced vicious cycle (an endless self-reflection), in which the reduced-capacity and biased processing excludes cues that could lower the anxiety level and trigger adaptive behavior.

A number of factors can be manipulated to induce the effects described above, simultaneously or sequentially, reflecting multiple, alternative mechanisms that mediate the anxiety biasing effects. In the case of the capacity parameters, alternative mechanisms can be defined from the agent's overall sensitivity to anxiety (reflected in the weights associated with trait and state anxiety intensity factors), the baseline, 'innate' capacity limits (reflected in the factors representing the minimum and maximum attention and working memory capacities), and the anxiety intensity itself. This factor can be further manipulated via the set of parameters influencing the affect appraisal processes, including the nature of the affective dynamics (e.g., maximum intensity, and the intensity ramp-up and decay functions).

The above example illustrates how MAMID can represent alternative mechanisms that mediate a range of anxiety behaviors, by explicitly representing the attentional and interpretive biases that contribute to the generation of anxiety, and the multiple, interacting causal pathways mediating these processes. The notion that the same underlying mechanism can result in distinct observable behavior and symptom severity, ranging from normal to severely disabling, depending on the values of the controlling parameters, is consistent with the emerging transdiagnostic model of psychopathology, and MAMID thus lends itself to modeling the mechanisms that mediate psychopathology from the transdiagnostic perspective. Again, it is important to emphasize that significant research would be required to validate the proposed model, and that validation of research models of emotion represents a significant challenge in computational affective science.

5 Summary and Conclusions

5.1 Summary

This chapter described a methodology and architecture for modeling the effects of emotion and affective traits on cognitive processing, in terms of parameter-controlled manipulations of the architecture processing. The model aims to represent the see-think/feel-do sequence of information processing at the psychological level (versus the neural level), and is implemented within a symbolic cognitive-affective architecture. The approach is suitable for both applied models, which aim to enhance the agent's or robot's believability or effectiveness, and for research models, which aim to elucidate the mechanisms of affective biases on cognition. An example of an applied model was briefly described, followed by a more extensive description of a research model, whose aim is to identify alternative mechanisms of affective biases, within the context of modeling anxiety disorders.

The paper draws the important distinction between applied and research models, with the former used to produce more believable or effective agent behavior, but with no aim to emulate biological mechanisms that mediate affective processing, whereas the latter aim to emulate the structures and processing in biological agents and elucidate the associated underlying mechanisms.

Applied models are much more easily constructed and their criteria for 'validation' reflect arbitrary standards defined by specific requirements of the associated agent or human-agent interaction, e.g., agent is believable, agent is effective in its teaching or coaching role, etc. In contrast, research models must meet additional constraints, so as to match the processing in biological agents, and their validation is much more challenging, since the structures explicitly represented in the model (e.g., cues, goals, situations) are difficult or impossible to identify directly in biological agents.

5.2 Conclusions

Efforts to construct symbolic computational models of emotion began over three decades ago. One of the earliest was the Cog Aff architecture of Sloman and colleagues (see [23] for an in-depth discussion). Sloman's work represents some of the most profound thinking about emotion modeling and the roles of emotions in both biological and synthetic agents, and should be required reading for anyone attempting to construct research models of emotion.

The majority of existing computational models of affect, at the psychological level, have been developed for applied purposes: to enhance the believability and effectiveness of virtual agents, social robots, and non-playing characters in games. These models are being increasingly incorporated into virtual agents and robots, as the desire for more affectively realistic and engaging agents across various contexts increases (e.g., learning, training, coaching, therapeutic, gaming).

Research models are beginning to be developed, as the emotion research community recognizes the value of the in computo approach. However, these models are much more difficult to construct, in part due to the need to emulate, rather than simulate, biological processes, and in part due to the need to account for the multimodal nature of emotion and represent the complex feedback mechanisms among the different modalities. These models also face significant challenges in regards to validation, which must necessarily rely on inferential approaches.

For both applied and research models, there is a need for the development of more systematic approaches to model design. Some work exists in this area, and includes efforts to represent alternative emotion theories in terms of uniform representations to facilitate comparison [2], and efforts to identify generic tasks required for the construction of affective models [14]. In general, there is great need for cross-disciplinary collaborations [22], the development of tools and sharable components, and the development of standards [15].

References

1. Averill, J. R. (1994). I feel, therefore I am—I think. In P. Ekman & R. J. Davidson (Eds.), *The nature of emotion: Fundamental questions*. Oxford, Oxford University Press.
2. Broekens, J., et al. (2008). Formal models of appraisal: theory, specification, and computational model. *Cognitive Systems Research, 9*(3), 173–197.
3. Fellous, J. M. (2004). From human emotions to robot emotions. In: *AAAI Spring Symposium: Architectures for Modeling Emotion,* Stanford University, CA, AAAI Press.
4. Gasper, K., & Clore, G. L. (2002). Attending to the big picture: Mood and global versus local processing of visual information. *Psychological Science, 13*(1), 34–40.
5. Gray, J. R., Braver, T. S. & Raichle, M. E. (2002). Integration of emotion and cognition in the lateral prefrontal cortex. *Proceedings of the National Academy of Sciences* (US), *99*(6), 4115–4120.
6. Hertel, P. T., & Mathews, A. (2011). Cognitive bias modication: Past perspectives, current findings, and future applications. *Perspectives on Psychological Science, 6*(6), 521–536.
7. Hudlicka, E. (1998). Modeling emotion in symbolic cognitive architectures. In *AAAI Fall Symposium: Emotional and Intelligent I*, Orlando, FL, AAAI Press.
8. Hudlicka, E. (2002). This time with feeling: integrated model of trait and state effects on cognition and behavior. *Applied Artificial Intelligence, 16,* 1–31.
9. Hudlicka, E. (2004). Two sides of appraisal: Implementing appraisal and its consequences within a cognitive architecture. In *AAAI Spring Symposium: Architectures for Modeling Emotion*. Stanford University, CA, AAAI Press. TR SS-04-02.
10. Hudlicka, E. (2005). *MAMID-ECS: Application of human behavior models capable of modeling individual differences to risk-analysis and risk-reduction strategy development in human-system design*. VA, Psychometrix Associates: Blacksburg.
11. Hudlicka, E. (2007). Reasons for emotions. In W. Gray (Ed.) *Advances in cognitive models and cognitive architectures,* NY, Oxford.
12. Hudlicka, E. (2008). What are we modeling when we model emotion? In *AAAI Spring Symposium: Emotion, Personality, and Social Behavior.* Stanford University, CA, Menlo Park, CA: AAAI Press. Technical Report SS-08-04: 52-59.
13. Hudlicka, E. (2008). Modeling the mechanisms of emotion effects on cognition. In *AAAI Fall Symposium: Biologically Inspired Cognitive Architectures*. Arlington, VA, Menlo Park, CA: AAAI Press. TR FS-08-04 82-86.

14. Hudlicka, E. (2012). Guidelines for designing computational models of emotions. *International Journal of Synthetic Emotions (IJSE), 2*(1), 26–79.
15. Hudlicka, E. (2015). From habits to standards: Towards systematic design of emotion models and affective architectures. In J. B., T. Bosse, J. Dias, J. Van der Zwaan (Eds.) *Towards Pragmatic Computational Models of Affective Processes* (pp. 1–21). Springer.
16. Isen, A. M. (1993). Positive affect and decision making In J. M. Haviland & M. Lewis (Eds.), *Handbook of emotions,* NY, Guilford.
17. Izard, C. E. (1993). Four systems for emotion activation: Cognitive and noncognitive processes. *Psychological Review, 100*(1), 68–90.
18. Leventhal, H., & Scherer, K. R. (1987). The relationship of emotion to cognition. *Cognition and Emotion, 1,* 3–28.
19. Macleod, C., & Mathews, A. (2012). Cognitive bias modification approaches to anxiety. *Annual Review of Clinical Psychology, 8,* 189–217.
20. Mineka, S., et al. (2003). Cognitive biases in emotional disorders: Information processing and social-cognitive perspectives. In R. J. Davidson, K. R. Scherer, & H. H. Goldsmith (Eds.), *Handbook of Affective Science,* NY, Oxford.
21. Phelps, E. A. (2012). Emotion and cognition: Insights from studies of the human amygdala. *Annual Review of Psychology, 57,* 27–53.
22. Reisenzein, R., et al. (2013). Computational modeling of emotion: Toward improving the inter- and intradisciplinary exchange. *IEEE Transactions on Affective Computing, 4*(3), 246–266.
23. Sloman, A., Chrisley, R., & Scheutz, M. (2005). The Architectural basis of affective states and processes. In J.-M. Fellous & M. A. Arbib (Eds.), *Who needs emotions?.* NY: Oxford University Press.
24. Smith, C. A. & L. Kirby (2000). Consequences require antecedents: Toward a process model of emotion elicitation. In J. P. Forgas (Ed.), *Feeling and Thinking: The role of Affect in Social Cognition,* NY, Cambridge.

Improving Human Behavior Using POMDPs with Gestures and Speech Recognition

João A. Garcia and Pedro U. Lima

Abstract This work proposes a decision-theoretic approach to problems involving interaction between robot systems and human users, with the goal of estimating the human state from observations of its behavior, and taking actions that encourage desired behaviors. The approach is based on the Partially Observable Markov Decision Process (POMDP) framework, which determines an optimal policy (mapping beliefs onto actions) in the presence of uncertainty on the effects of actions and state observations, extended with information rewards (POMDP-IR) to optimize the information-gathering capabilities of the system. The POMDP observations consist of human gestures and spoken sentences, while the actions are split into robot behaviors (such as speaking to the human) and information-reward actions to gain more information about the human state. Under the proposed framework, the robot system is able to actively gain information and react to its belief on the state of the human (expressed as a probability mass function over the discrete state space), effectively encouraging the human to improve his/her behavior, in a socially acceptable manner. Results of applying the method to a real scenario of interaction between a robot and humans are presented, supporting its practical use.

1 Introduction

Social robots need to be capable of developing affective interactions and to empathize with human users [4]. This requirement involves the ability to infer and react according to latent variables: the user's affective and motivational status.

J. A. Garcia · P. U. Lima (✉)
Institute for Systems and Robotics, Instituto Superior Técnico, University of Lisbon, Lisbon, Portugal
e-mail: pedro.lima@tecnico.ulisboa.pt

J. A. Garcia
e-mail: joao.p.garcia@tecnico.ulisboa.pt

M. I. Aldinhas Ferreira et al. (eds.), *Cognitive Architectures*, Intelligent Systems, Control and Automation: Science and Engineering 94,
https://doi.org/10.1007/978-3-319-97550-4_10

The agent acting in a Human-Robot Interaction (HRI) scenario must take into account the effects of its actions on the human user, which are uncertain, and the sensory information it receives, which is noisy. Planning under these conditions is attainable through Partially Observable Markov Decision Processes (POMDPs) [3]. POMDPs, through the transition and observation models, deal with the aforementioned uncertainty, by probabilistically modeling the possible outcomes of the agent's different actions and the accuracy of the sensory information. Furthermore, the problem of empathizing with the human user adds the goal of information gain on latent (i.e., not directly observed) state variables, which is addressed by the extensions to POMDPs introduced by Partially Observable Markov Decision Processes with Information Rewards (POMDPs-IR) [9].

Thus, this work introduces a POMDP-IR framework for planning under uncertainty in HRI problems, which allows the agent to accomplish a given task, actively infer latent state variables of interest and adapt its behavior accordingly. The aforementioned framework is implemented in a real robot system, to ensure it is capable of successfully solving HRI planning problems in practice.

2 Related Work

Among HRI scenarios, Decision-Theoretic (DT) approaches to planning based on the POMDP framework are found in assistive scenarios, such as the robot wheelchair [10], in which the goal is to recognize the intention of the user but do not include social capabilities for improving recognition. Also, in socially assistive settings, the POMDP framework models the social interaction between robot and human users in, e.g., nursing homes [7], although without taking into account the user's status. Finally, the POMDP was used to model problems with latent variables and adapt the agent's behavior accordingly in an automated hand-washing assistant [1]. However, the agent in the latter work does not actively seek to gain information on the user's status, and is, therefore, limited to reacting based on a possibly high-uncertainty belief on the hidden variables.

The traditional POMDP model does not allow for rewarding low-uncertainty beliefs. Consequently, in order to obtain a certain level of knowledge about the features of interest, the POMDP framework needs to be extended to reward information gain. This extension is provided through the POMDP-IR (POMDP with Information Reward)) framework. DT planning based on POMDP-IR has been applied to the problem of active cooperative perception [9]. The present work, however, is focused on multimodal human-robot interaction.

3 Background

POMDP-IR can be expressed as a tuple $(S, A, T, R, \Omega, O, \gamma)$, where:

- $S = S_1 \times \cdots \times S_n$ represents the environment's factored state space, defining the model of the world;
- A is a finite set of actions available to the agent that contains the domain-level action factor A_d and an Information-Reward (IR) action factor A_i for each state factor of interest ($A = A_d \times A_1 \times \cdots \times A_l$, where l is the number of IR actions);
- T is the transition function that represents the probability of reaching a particular state $s \in S$ by a given state-action pair ($T : S \times A \times S \rightarrow [0, 1]$);
- R is the reward function, which defines the numeric reward given to the agent for each state-action pair ($R : S \times A \rightarrow \mathbb{R}$), and is therefore given by $R = R_d(s, a_d) + \sum_{i=1}^{l} R_i(s_i, a_i)$, with $s \in S$, $a_d \in A_d$, $s_i \in S_i$, $a_i \in A_i$, R_d being the POMDP reward model and R_i the information reward;
- Ω is a finite set of observations that correspond to features of the environment directly perceived by the agent's sensors;
- O is the observation function that represents the probability of perceiving observation $o \in \Omega$ after performing action $a \in A$ and reaching state $s' \in S$ ($O : S \times A \times \Omega \rightarrow [0, 1]$);
- γ is the discount factor, used to weight rewards over time.

The POMDP-IR fits into the classic POMDP framework, and can, therefore, be represented as a belief-state Markov Decision Process (MDP), in which the history of executed actions and perceived observations are encoded in a probability distribution over all states: the belief state. Every time the agent performs an action $a \in A$ and observes $o \in \Omega$, the belief is updated by the Bayes' rule:

$$b^{ao}(s') = \frac{P(o|s', a)}{P(o|b, a)} \sum_{s \in S} P(s'|s, a)b(s), \tag{1}$$

where $P(s'|s, a)$ and $P(o|s', a)$ are defined by the Transition and Observation model, respectively, and

$$P(o|b, a) = \sum_{s' \in S} P(o|s', a) \sum_{s \in S} P(s'|s, a)b(s) \tag{2}$$

is a normalizing constant. Furthermore, the value function $V^{\pi}(b)$, defined as the expected future discounted reward given to the agent by following policy π, starting from belief b:

$$V^{\pi}(b) = \mathbf{E}_{\pi}\left[\sum_{t=0}^{\infty} \gamma^t R(b_t, \pi(b_t)) \Big| b_0 = b\right], \tag{3}$$

where $R(b_t, \pi(b_t)) = \sum_{s \in S} R\big(s, \pi(b_t)\big) b_t(s)$, remains approximately Piecewise Linear Convex (PWLC) in the POMDP-IR framework. This way, the most common algorithms for solving POMDPs, which exploit the PWLC representation of the value function, can also be used to solve POMDPs-IR. The optimal policy π^* is characterized by the optimal value function V^*, which satisfies the Bellman optimality equation:

$$V^*(b) = \max_{a \in A} \left[R(b, a) + \gamma \sum_{o \in O} P(o|b, a) V^*(b^{ao}) \right]. \quad (4)$$

Solution methods for POMDPs differ from exact solution algorithms (e.g., Monahan's enumeration algorithm [5]), intractable for large problems, to approximate policy optimization (e.g., Point-based Value Iteration (PBVI) [6]). The method of reference in solving POMDPs throughout this work is *PERSEUS* [8], a randomized PBVI algorithm.

4 Framework Description

The proposed framework approaches the problem of planning under uncertainty in HRI under the POMDP-IR extension. Figure 1 represents the projected POMDP-IR as a two-stage Dynamic Bayesian Network (DBN), which depicts the dynamics of the HRI problem.

4.1 States and Transitions

The agent acting in an HRI scenario considers two types of state factors: the *task* variables T and the *person* variables P. The *task* variables model the environment features that provide information on the progress of the tasks. On the other hand, the *person* variables track the human state and are inherently latent. The latter are used to gain information on the human user's affective and motivational status and adapt the robot behavior accordingly.

The number of state variables depends on the amount of features essential to represent the environment, and is, therefore, dependent on the specific task. The criteria for the selection of states involve a trade-off between operational complexity and predicted system performance, since operational complexity increases with the number of states.

Furthermore, depending on the objectives of the agent acting in an HRI setting, the *task* variables might not exist. This is the case when the single goal of the agent is to gain information on the human user.

A *person* variable can have a constant value over time if its value does not change during the task. This is the case with personal traits (e.g., *Personality* and *Prefer-*

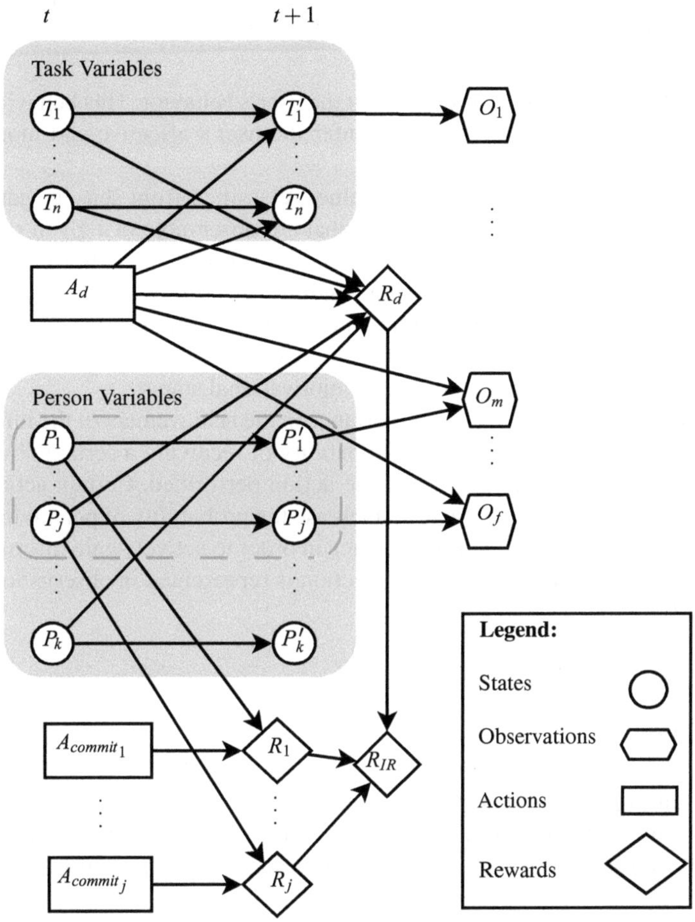

Fig. 1 DBN representation of the DT model for multimodal HRI

ences), which are relevant for the robot behavior and do not change for the duration of the interaction. In Fig. 1, P_k represents a constant *person* variable.

Otherwise, *person* variables are inferred from the user's behavior at each time step (factors P_1 to P_j in Fig. 1), which is represented in the model's observations. These state variables may consist of state factors of interest, according to the POMDP-IR framework.

4.2 Observations and Observation Model

In a social HRI setting, observations reflect the user's behavior. This behavior is used to monitor the progress of the task and infer the user's affective and motivational status.

Observations are discrete, symbolic values, classified from sensory data, which correspond to features of the environment that are observable in a given state.

The observation factors are contingent on the sensory capabilities of the robot system. Nevertheless, the correct understanding of the user's status relies on the agent being capable of recognizing human communication methods. Consequently, the robot system ought to be able to recognize speech and gestures in order to understand the human user's affective and motivational status.

The observation model is of key importance in the achievement of the information gain goals of the agent. It reflects the probability of receiving a certain observation, given the state of the environment and the action performed. Certain actions, such as questioning or approaching the user, increase the probability of perceiving certain observations. This fact is of utter importance in order to actively gain information on the user's status. The dependency on the action is represented in observations O_m to O_f in Fig. 1.

4.3 Actions

The model in Fig. 1 comprehends two sets of actions: A_d and A_{commit}. The actions in A_d have an effect on the environment and are dependent on the actuators of the agent, while the actions in A_{commit} are used to achieve the information gain goals of the agent.

Typically, the action set A_d contains the minimum set of functionalities that allow the agent to complete its tasks. Social robots need to communicate in a natural, easily understandable way with the human users. To achieve this objective, the robot must be able to express different moods and emotions. Consequently, the action set A_d of a social robot ought to include speech and/or gestural capabilities and/or graphical emotion displays.

Following the POMDP-IR framework, besides the domain-level action factor A_d, the model has additional action factors A_{commit} for each state factor of interest. The state factors of interest, in the problem under study, are included in the *person* variables, as these contain the aforementioned affective and motivational state of the human user. The actions in A_{commit} allow for rewarding the agent for decreasing the uncertainty regarding particular features of the environment.

4.4 Reward Model

In the DT model in Fig. 1, rewards are either associated with task objectives: R_d, or with the information gain goals: $R_i, i = 1, \ldots, j$. The sum of these rewards, R_{IR}, constitutes the reward awarded to the agent at each time step.

The behavior of the robot consists of the sequence of domain actions A_d the agent performs. In the social HRI scenario, and in order to adapt the robot's behavior to the user's affective and motivational status, the reward assigned to an action depends not only on the *task* variables, but also on the *person* variables.

The information rewards R_i influence the behavior of the agent, with the purpose of achieving a low uncertainty regarding certain *person* variables. The value of these rewards is dependent on the threshold of knowledge required, according to the POMDP-IR framework [9].

5 Selected Application

The proposed approach was tested in a case study that considers a socially assistive task: rehabilitation therapy.

5.1 Scenario

Rehabilitation therapy includes passive or active exercises. In the first, the therapist (human or robot) physically assists the patient in moving the affected limb. On the other hand, in active exercises, the patient moves the affected limb by him/herself, while the therapist has the functions of coaching and motivating.

Up-to-date research in rehabilitation robotics mainly covers passive exercises. Nevertheless, social robots provide a way to approach active rehabilitation exercises, representing an innovative way to monitor, motivate and coach patients.

Overall, the goals of the robot therapist in the considered rehabilitation scenario are:

- To help the user in the given setting, by monitoring the patient's movements (e.g., encouraging the patient to continue if he/she stops performing the exercise);
- To adapt its behavior and, consequently, the therapy style (e.g., nurturing vs challenging the patient), in accordance with the patient's affective and motivational status.

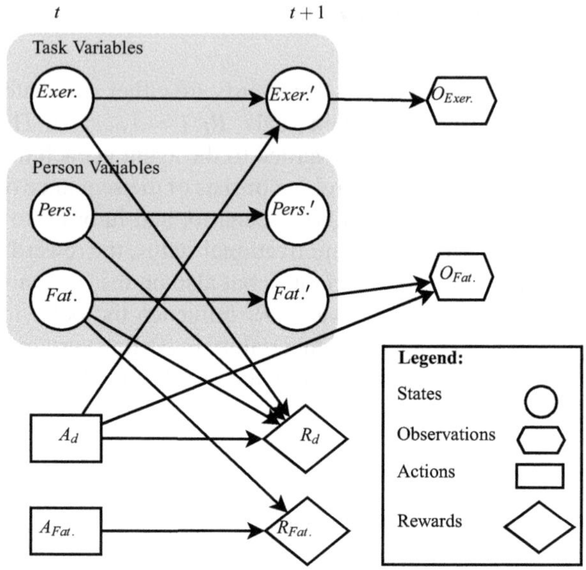

Fig. 2 DBN representation of the DT model for the robot therapist

5.2 *Decision-Theoretic Model for the Robot Therapist*

The application of the proposed framework to the robot therapist scenario results in the DT model represented in Fig. 2.

5.2.1 States

The significant features of the environment in which the robot is to operate are related to the human user. Fulfillment of the task's objectives requires that the agent keep track of the user's movements (state $Exer.$), possess knowledge regarding relevant personal traits ($Pers.$) of the user and infer his/her affective status ($Fat.$). Therefore, the proposed DT model considers the state space represented, in factored form, in Table 1.

The user's movement is encoded in the *task* state factor *Exercise* ($Exer.$). When the exercise is performed as prescribed, the state factor assumes the value *Correct*: $Exer. = Correct$. Otherwise, if the movement is inappropriately performed or not performed at all, $Exer. = Incorrect$. The state factor *Personality* ($Pers.$) is a constant *person* variable, known beforehand by the problem designer, which represents the patient's behavioral personality, as *Introverted* or *Extroverted*. Finally, the *Fatigue* state factor ($Fat.$) is a measure of the patient's weariness, caused by the physical exercise. It assumes the values *Tired* or *Energized*, depending on whether the patient shows signs of fatigue or liveliness, respectively.

Table 1 State, observation and action spaces for the robot therapist case study

	Factors	Values
States	$Exer.$	Correct, incorrect
	$Pers.$	Introverted, extroverted
	$Fat.$	Tired, energized
Observations	$O_{Exer.}$	Proper, wrong
	$O_{Fat.}$	Weary, energetic, none
Actions	A_d	Nurture, challenge, query patient, end therapy, none
	$A_{Fat.}$	Commit tired, commit energized, null

5.2.2 Observations

The observation space is represented, in factored form, in Table 1. Observations reflect the relevant behavior of the patient, in accordance with the task's goals. In the present case study, the agent ought to classify the movement performed by the patient ($O_{Exer.}$) and his/hers affective status ($O_{Fat.}$).

The gesture-related observation factor $O_{Exer.}$ is used to evaluate the exercise and assumes, as a result, the values *Proper* or *Wrong*. $O_{Exer.} = Proper$ whenever the agent perceives that the patient performed the movement as prescribed. Otherwise, $O_{Exer.} = Wrong$ if the agent perceives that the patient did not perform the movement or performed it incorrectly.

The observation factor $O_{Fat.}$, which is related to the affective status of the patient represented in state factor *Fatigue*, assumes the values *Weary*, *Energetic* or *None*. $O_{Fat.} = Weary$ or $O_{Fat.} = Energetic$ when the patient demonstrates feeling tired or lively, respectively. Otherwise, $O_{Fat.} = None$ if the agent does not perceive any relevant information regarding the affective status of the patient.

$O_{Exer.}$ is obtained by visual classification of the patient's gestures and $O_{Fat.}$ through classification of the user's verbal responses.

5.2.3 Actions

The proposed DT model considers two action factors: the *Action Domain* A_d and the *IR Action* $A_{Fat.}$ At each time step, the agent chooses one value for each action factor. The possible values for the action factors are represented in Table 1.

The IR action is defined according to the POMDP-IR framework, with a *commit* action for each value of the related state factor ($Fat.$) and a *null* action. $A_{Fat.}$ allows for rewarding the agent for reducing the uncertainty regarding the state factor $Fat.$, related to the patient's fatigue.

The *Action Domain* A_d contains the set of functionalities that allow the agent to achieve the task and information gain goals.

The therapy style, i.e., the robot's approach to the patient, changes as a function of his/her *Fatigue* and *Personality*. Dependent on these factors, the encouragement is classified as $Nurture$ or $Challenge$ if the agent opts, respectively, for a softer (e.g., "You are doing great! Keep up the good work.") or a more defiant approach (e.g., "You can do better than that!").

Since the therapy style is dependent on the *person* variables, it is important to gain information and maintain a low uncertainty regarding the state factors $Pers.$ and $Fat.$ As $Pers.$ is constant, the agent actively seeks to reduce uncertainty on the state factor $Fat.$ through the $Query\ Patient$ action. This action consists of verbally interacting with the patient to infer his/hers *Fatigue*.

Moreover, the agent ought to end the exercise ($End\ Therapy$) when the patient persistently shows he/she is not able to proceed with it. Finally, at each time step, the agent might choose to do nothing ($None$).

5.2.4 Transition, Observation and Reward Functions

The proposed framework allows us to take into account the effects of time in the states of the DT model. Namely, in the current case study, the transition function T encodes that $b(Fat. = Tired)$ increases at each time step in the absence of opposing observations ($O_{Fat.} = Energetic$). That is, the agent realistically believes that the patient is feeling more tired over time. The transition function of this case study dictates that the probability of the patient correctly performing the exercise ($Exer. = Correct$) increases with the motivation actions ($Nurture$ or $Challenge$). Moreover, Personality ($Pers.$) is modeled as a constant variable, not inferred by the agent, as its value does not change during the task.

The observation function O encodes the error in sensory data classification. This means, for instance, that even if the patient's gesture is classified as incorrect ($O_{Exer.} = Wrong$), the agent's belief about $Exer. = Incorrect$ is not 100%, and the robot might require more information before motivating the patient. Furthermore, the probabilities in O take into account that information-gathering actions (such as $Query\ Patient$) increase the probability of perceiving a verbal response from the user (e.g., $O_{Fat.} = Weary$).

The DT model in Fig. 2 rewards IR actions ($R_{Fat.}$) and A_d actions (R_d). The information rewards are defined, in accordance with the POMDP-IR framework, so that the agent actively seeks to have a certainty about $Fat.$ greater than 75% (i.e., $b(Fat. = Tired) > 0.75$ or $b(Fat. = Energized) > 0.75$). Actions in A_d are rewarded in accordance with the state of the environment: *Encouragement* actions ($Nurture$ and $Challenge$) are rewarded 0.2 whenever the patient is incorrectly performing the exercise or 0.1 when he/she shows signs of feeling tired, and penalized -0.1 otherwise. The reward given to each action also depends on the state factor $Pers.$: for an *Introverted* person, the $Nurture$ action is preferred, while the $Challenge$ action is favored for an *Extroverted* person; The $Query\ Patient$ action is penalized with -0.2; $None$ is neither rewarded nor penalized; $End\ Therapy$ receives high penalization (-1) when the patient feels energetic and a reward of 0.1

otherwise. Rewards are defined over the abstract states and actions of the DT model. The discount factor in this case study is $\gamma = 0.9$.

As it would be impractical to obtain the models from empirical studies, especially as the system becomes more complex, the aforementioned reward values are tuned to lead to a policy that handles different patients adequately.

6 Experiments

The robot therapist case study was implemented as a robot system consisting of a real social mobile robot networked with a RGB-D camera, which interacted, in different experiments, with distinct persons, in a realistic apartment testbed.

6.1 Experimental Setup

The networked robot system used in the present case study consists of the MOnarCH robot platform, represented in Fig. 3a, and an external Kinect camera. The robot platform provides the actuating capabilities required to implement the domain actions A_d and the sensors necessary for the speech-related observations $O_{Fat.}$. The Kinect camera is strategically located for a clear view of the patient's movements and is used, therefore, for the classification of the exercise $O_{Exer.}$.

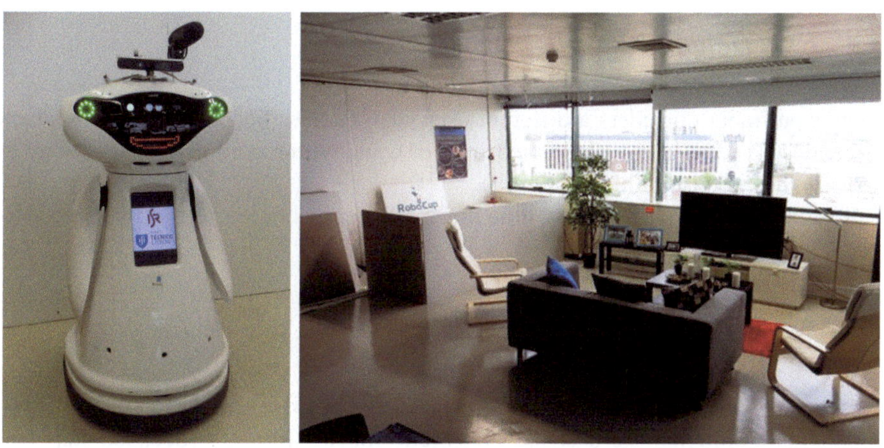

(a) Robot Platform used in the experiments

(b) Living room area of the ISRoboNet@Home testbed

Fig. 3 Experimental setup for the robot therapist case study

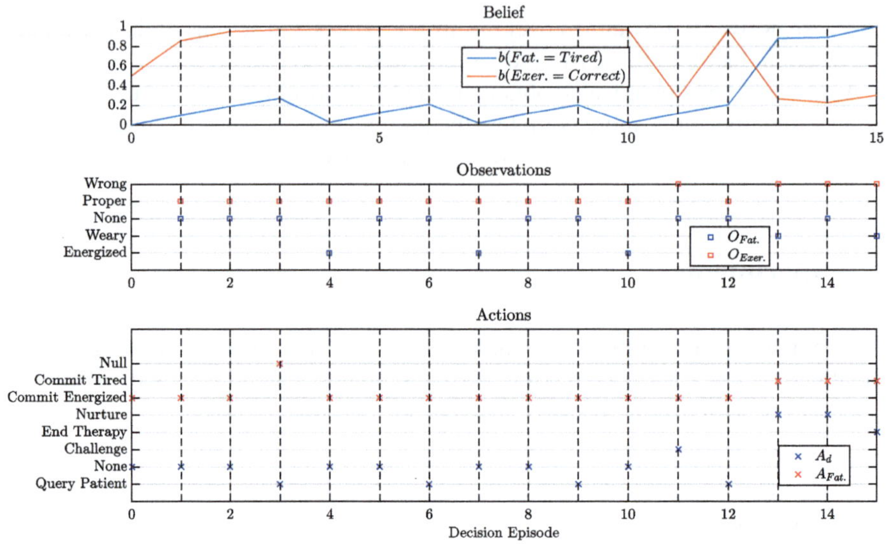

Fig. 4 Evolution of the belief about the states $Fat.$ and $Exer.$ w.r.t. the decision episode, the observations received and the actions performed, for experiment A

The experiments within this case study took place in the ISRobotNet@Home Testbed,[1] which is represented in Fig. 3b. This testbed provides the infrastructure necessary to implement networked robot systems in a domestic environment.

6.2 *Experimental Results*

Each experiment considers a different user, who is classified according to his/hers personality (i.e., as introverted or extroverted), and with regard to his/hers ability to perform the exercise (athletic or unfit).

The experiments carried out within this work were recorded, and videos are available at https://goo.gl/TlyXGT.

6.2.1 Experiment A

This experiment considers a user who is classified as extroverted ($Pers. = Extroverted$) and athletic. The user feels energetic for the first 50 s (decision step 10), approximately, and tired afterwards. Figure 4 plots the data acquired in experiment A.

At the beginning, the robot chooses not to act, since the exercise is well performed and the agent has a low uncertainty regarding the *fatigue* status of the user. This

[1]http://welcome.isr.tecnico.ulisboa.pt/isrobonet/.

uncertainty on the state factor $Fat.$, however, increases over time, driving the robot to actively seek to reduce it, by querying the user (decision step 3). The answer ($O_{Fat.} = Energetic$), informs the robot that the user is still active and motivated, increasing the certainty about $Fat. = Energized$. This behavior is repeated until the user does not perform the exercise correctly ($O_{Exer.} = Incorrect$) in decision step 11. Then, the robot motivates the person through a challenging approach according to the considered *personality* of the user and the current *fatigue* status. After receiving information that the user now feels tired ($O_{Fat.} = Weary$), the robot changes therapy style and adopts a nurturing approach. As the user continuously shows an inability to carry out the exercise and the certainty about $Fat. = Tired$ increases, the robot finally chooses to end the therapy in decision step 15.

6.2.2 Experiment B

This experiment considers a user classified as extroverted ($Pers. = Extroverted$) and unfit. The user feels energetic for the first 40 s, approximately, and tired afterwards. Figure 5 plots the data acquired in experiment B.

Figure 6 represents an episode of experiment B where the robot interacts with the user.

The behavior of the robot is similar to that in the previous experiment while the user demonstrates feeling energetic and correctly performs the exercise. Nonetheless, the user incorrectly performs the exercise more often, upon which occasions the robot

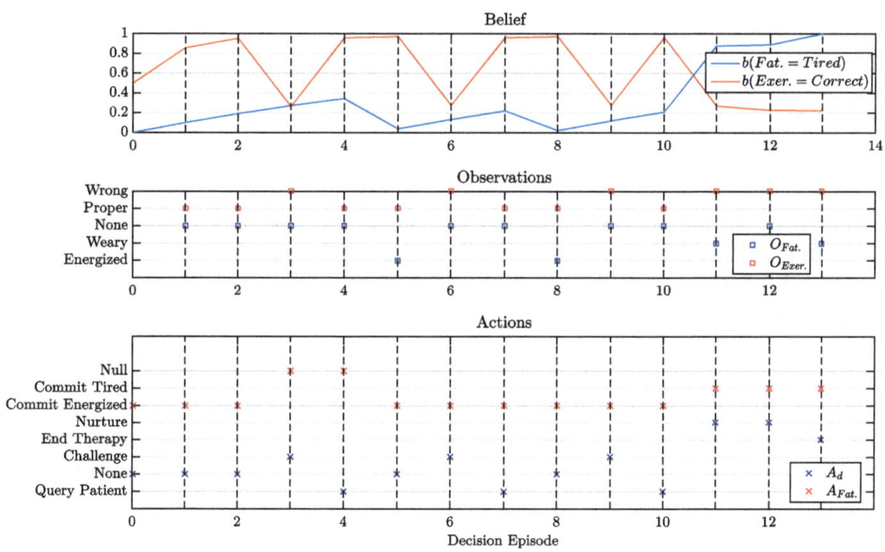

Fig. 5 Evolution of the Belief about the states $Fat.$ and $Exer.$ w.r.t. the decision episode, the observations received and the actions performed, for experiment B

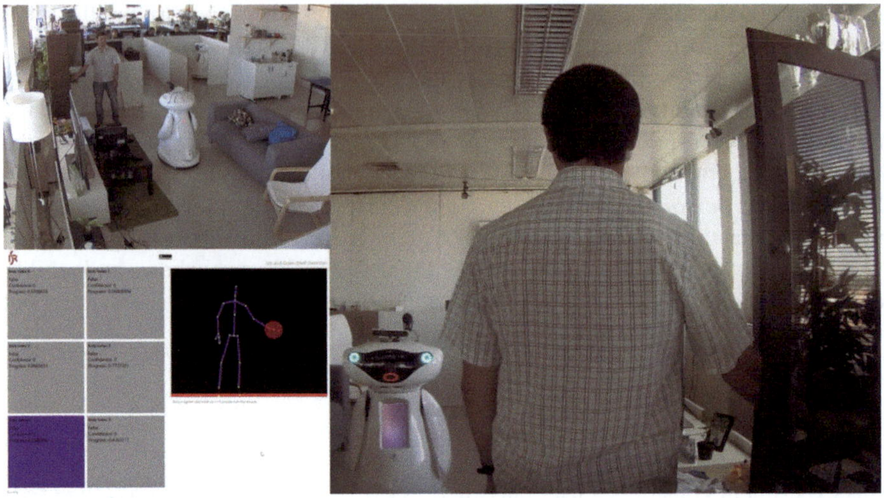

Fig. 6 Episode of the experiment B when the robot queries the user. Right and top left images show different views of the interaction between the robot and the human; bottom left image represents the interface of the gesture classification application

motivates the user with a challenging approach, while the agent believes that the user feels motivated/energetic. Despite motivating the user, the robot keeps track of his/her *fatigue* and reacts when the uncertainty about $Fat.$ is high. Finally, the agent ends the therapy once it persistently observes that the user is not performing the exercise and feels tired.

6.2.3 Experiment C

This experiment considers a user classified as introverted ($Pers. = Introverted$) and athletic. The patient feels energetic up to, approximately, 45 s (decision step 9), and tired afterwards. Figure 7 plots the data acquired in experiment C.

The behavior of the robot is heavily dependent on its knowledge regarding the fatigue status of the user. While the uncertainty about the $Fat.$ state factor is high, the robot queries the user. Since the uncertainty about $Fat.$ increases over time, the agent performs the action $Query\ Patient$ until it perceives an answer $O_{Fat} = Energetic$ or $O_{Fat} = Weary$ (decision steps 3 and 4/7 and 8). Nevertheless, the robot performs the therapy task while actively gathering information on the environment, motivating the user once the belief about $b(Fat. = Tired)$ is high, and ending the therapy appropriately.

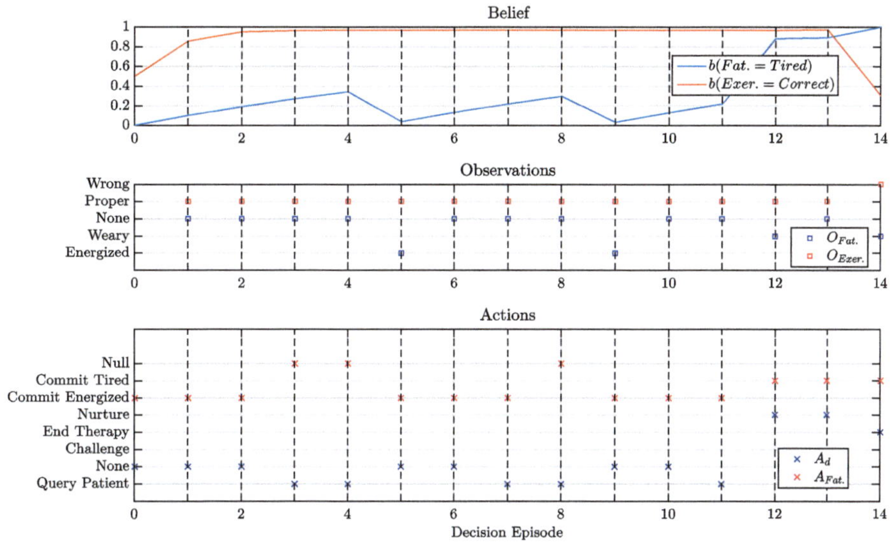

Fig. 7 Evolution of the Belief about the states $Fat.$ and $Exer.$ w.r.t. the decision episode, the observations received and the actions performed, for experiment C

6.2.4 Experiment D

This experiment considers a user who is classified as introverted $(Pers. = Introverted)$ and unfit. The user feels energetic for the first 40 s (decision step 8), approximately, and tired thereon. Figure 8 plots the data acquired in experiment D.

The behavior of the robot changes in accordance with its belief about the states of the environment. In the present experiment, there is a "trade-off" between motivating or querying the user depending on the belief about the state factors $Fat.$ and $Exer.$ In decision step 3, the agent queries the agent due to the high uncertainty about $Fat.$ Afterwards, the agent perceives no answer, but observes that the user performed the movement incorrectly. This observation does not translate, however, into an absolute certainty about the exercise having been performed incorrectly $(b_4(Exer. = Correct) \approx 0.3)$, since the DT framework takes into account sensor-related noise. The agent then queries the user once again (decision step 4), due to the increasing uncertainty about the *fatigue* of the user. Once again, the Network Robot System (NRS) receives no answer $(O_{Fat.} = None)$, and observes that the user performed the movement incorrectly. This time, the agent's belief about $Exer. = Incorrect$ is higher $(b_5(Exer. = Incorrect) \approx 0.95)$, and thus it motivates the user. Nevertheless, the uncertainty about $Fat.$ is still high in decision step 6, and the robot once again queries the user, perceiving an answer this time.

For the rest of the experiment, the robot follows a behavior similar to that of the previous experiments, until it ends the trial in decision step 14.

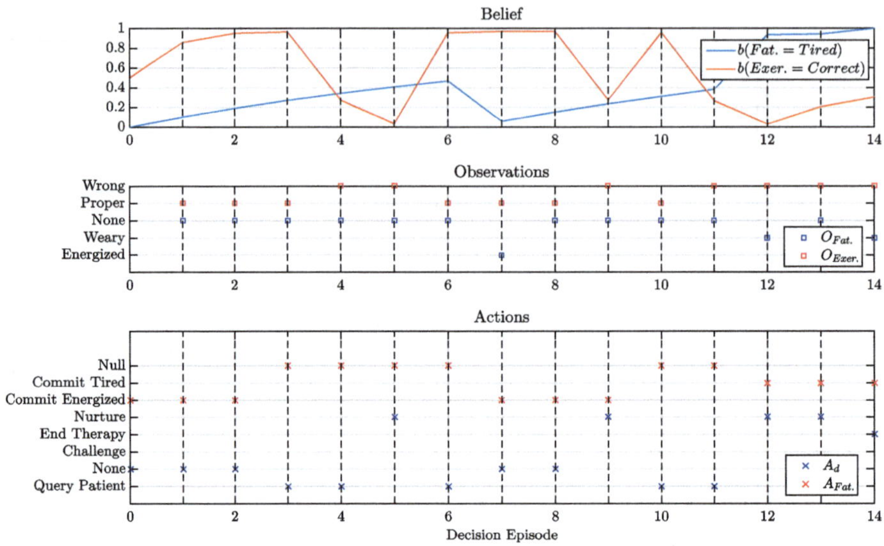

Fig. 8 Evolution of the Belief about the states *Fat.* and *Exer.* w.r.t. the decision episode, the observations received and the actions performed, for experiment D

6.3 Discussion

Table 2 details the behavior of the robot for each experiment. As expected, the number of motivation actions is higher for the users classified as unfit, who perform the exercise incorrectly more often than the athletic users; and the number of query actions is higher for the users classified as introverted.

The robot detected the fatigue status change from *Energized* to *Tired* in all of the experiments. Moreover, the agent motivated the user upon detection of faulty movements, either immediately after observing $O_{Exer.} = Wrong$ (experiments A, B and C) or after two consecutive observations (experiment D). Finally, the agent ended

Table 2 Behavior of the robot with regard to the experiment

	A	B	C	D
Motivation actions	3	5	2	4
Query actions	4	3	5	5
Time elapsed until agent detected change of user's status (s)	15	15	15	20
Time elapsed until agent ended therapy because it detected user was tired (s)	10	10	10	10
Duration of the experiment (s)	75	65	70	70

the therapy after consistently observing that the user was not capable of proceeding with the exercise.

Overall, the DT approach to planning in the robot therapist resulted in a behavior capable of achieving the task and information goals, in a manner both adaptive to the user's status and socially appealing.

7 Conclusions and Future Work

Building on the POMDP-IR framework, this work introduced a DT approach to planning under uncertainty with information rewards in social HRI. The properties of the DT framework were demonstrated in the robot therapist case study and the experiments' results validate the proposed framework for a problem involving robot systems in HRI scenarios.

Use of (PO)MDPs to model decision-making in realistic scenarios, such as the framework proposed in this work, presents an important practical difficulty, since they assume complete knowledge of the stochastic transition and observation models, meaning one needs to specify or estimate all of the probabilities involved. Moreover, any change to the parameters of these models implies a recalculation of the DT policy. Alternatively, in model-free Reinforcement Learning (RL) approaches [2], the DT policies are learned from the interaction of robot agents with their environment, without requiring full knowledge of the transition and observation models. Therefore, we plan to use RL in future applications of these methods.

To further validate the framework developed within this work, we plan its application to another health-related scenario, which we have been working with under the CMU-Portugal project INSIDE,[2] considering distinct scenarios of HRI with autistic children and their therapists. INSIDE is a research project, whose team developed a mobile robot with several interaction sensors and expressiveness skills, networked with RGB-D cameras. This networked robot system has been designed to display symbiotic autonomy when interacting with autistic children. Research has reported that autistic children are frequently willing to engage with social robots, and even create affective bonds with them. This is probably due to the predictability of the robots' behaviour. Despite the relative simplicity (when compared with a human) of the behaviours displayed by the INSIDE robot system so far, the system requires multi-modal perception systems that enable it to recognize children's activity (e.g., speech/sound, gestures, motion and location) and actuation systems so as to interact with the children using different approaches (e.g., spoken sentences, motion, "face" expressiveness). As the autonomy level of the robot system increases, autonomous decision-making methods such as the one described in this work must be included in the system.

The INSIDE robot system is composed of a mobile robot with onboard sensors, such as a LIDAR (for self-localization and obstacle avoidance), RGB-D cameras (to

[2]http://www.project-inside.pt.

57. Schermerhorn, P., & Scheutz, M. (2011, February). Disentangling the effects of robot affect, embodiment, and autonomy on human team members in a mixed-initiative task. In *Proceedings of the 2011 International Conference on Advances in Computer-Human Interactions*, Gosier, Guadeloupe, France (pp. 236–241).
58. Schermerhorn, P., Scheutz, M., & Crowell, C. R. (2008). Robot social presence and gender: Do females view robots differently than males? In *Proceedings of the Third ACM IEEE International Conference on Human-Robot Interaction*, Amsterdam, The Netherlands (pp. 263–270). ACM Press.
59. Scheutz, M. (2006). ADE—Steps towards a distributed development and runtime environment for complex robotic agent architectures. *Applied Artificial Intelligence*, *20*(4–5),
60. Scheutz, M., & Andronache, V. (2004). Architectural mechanisms for dynamic changes of behavior selection strategies in behavior-based systems. *IEEE Transactions of System, Man, and Cybernetics Part B: Cybernetics*, *34*(6), 2377–2395.
61. Scheutz, M., Briggs, G., Cantrell, R., Krause, E., Williams, T., & Veale, R. (2013). Novel mechanisms for natural human-robot interactions in the DIARC architecture. In *Proceedings of AAAI Workshop on Intelligent Robotic Systems*.
62. Scheutz, M., Cantrell, R., & Schermerhorn, P. (2011). Toward humanlike task-based dialogue processing for human robot interaction. *AI Magazine*, *32*(4), 77–84.
63. Scheutz, M., Harris, J., & Schermerhorn, P. (2013). Systematic integration of cognitive and robotic architectures. In *Advances in Cognitive Systems* (pp. 277–296).
64. Scheutz, M., Krause, E., Oosterveld, B., Frasca, T., & Platt, R. (2017). Spoken instruction-based one-shot object and action learning in a cognitive robotic architecture. In *Proceedings of the 16th International Conference on Autonomous Agents and Multiagent Systems*.
65. Scheutz, M., Krause, E., & Sadeghi, S. (2014). An embodied real-time model of language-guided incremental visual search. In *Proceedings of the 36th Annual Conference of the Cognitive Science Society*.
66. Scheutz, M., Schermerhorn, P., Kramer, J., & Anderson, D. (2007, May). First steps toward natural human-like HRI. *Autonomous Robots*, *22*(4), 411–423.
67. Scheutz, M., Schermerhorn, P., Kramer, J., & Middendorff, C. (2006). The utility of affect expression in natural language interactions in joint human-robot tasks. In *Proceedings of the 1st ACM International Conference on Human-Robot Interaction* (pp. 226–233).
68. Searle, J. R. (1969). *Speech acts: An essay in the philosophy of language*, vol. 626. Cambridge university press.
69. Searle, J. R. (1975). Indirect speech acts. *Syntax and Semantics*, *3*, 59–82.
70. Shafer, G. (1976). *A mathematical theory of evidence*. Princeton University Press.
71. Sloman, A., & Scheutz, M. (2002). A framework for comparing agent architectures. In *UK Workshop on Computational Intelligence* (pp. 169–176).
72. Strait, M., Briggs, P., & Scheutz, M. (2015). Gender, more so than age, modulates positive perceptions of language-based human-robot interaction. In *4th International Symposium on New Frontiers in Human-Robot Interaction, AISB*.
73. Strait, M., Canning, C., & Scheutz, M. (2014). Let me tell you! investigating the effects of robot communication strategies in advice-giving situations based on robot appearance, interaction modality, and distance. In *Human-Robot Interaction (HRI)* (pp. 479–486).
74. Talamadupula, K., Briggs, G., Chakraborti, T., Scheutz, M., & Kambhampati, S. (2014). Coordination in human-robot teams using mental modeling and plan recognition. In *Proceedings of IROS*.
75. Talamadupula, K., Briggs, G., Scheutz, M., & Kambhampti, S. (2017). Architectural mechanisms for handling human instructions for open-world mixed-initiative team tasks and goals. In *Advances in Cognitive System*, vol. 5.
76. Trafton, G., Hiatt, L., Harrison, A., Tamborello, F., Khemlani, S., & Schultz, A. (2013). ACT-R/E: An embodied cognitive architecture for human-robot interaction. *Journal of Human-Robot Interaction*, *1*(1), 78–95.
77. Veale, R., Briggs, G., & Scheutz, M. (2013). Linking cognitive tokens to biological signals: Dialogue. In *Proceedings of the 35th Annual Conference of the Cognitive Science Society*, Austin, TX. Cognitive Science Society.

detect children's faces and their emotions), and a directional microphone (to recognize children's utterances and therapist keywords), networked with four Microsoft Kinect RGB-D cameras installed on the ceiling of the room (to detect and locate the children and understand some of their gestures). Additionally, a supervision interface, comprising an actuation and a perception console, enables external operators, hidden from the children, to become aware of the interaction status and intervene in the robot decision-making process if necessary. A state machine orchestrates the sequencing of behaviours, interfacing with them through a behaviour manager. The system was developed to follow an adjustable autonomy strategy, aiming at a smooth transition from a Wizard of Oz setup (in which external operators, can override the information sensed and processed by the system, as well as the behavior selections suggested by the decision-making algorithm) to full autonomy. The symbiotic autonomy manifests itself through the fact that some of the robot behaviours consist of asking the child to help, while others make suggestions to the child as to what to do. We plan to apply our method to the development of a decision-making system that encourages the children to progress in games (e.g., building a puzzle, removing an obstacle that prevents the robot from entering an area where it can help the children during the game) by observing a child's behavior and updating the belief about his/her performance.

Acknowledgements This work was partially funded by project CMUP-ERI/HCI/0051/2013 and also by the LARSyS strategic funding from FCT, project UID/EEA/50009/2013.

References

1. Hoey, J., Poupart, P., Av, Bertoldi, Craig, T., Boutilier, C., & Mihailidis, A. (2010). Automated handwashing assistance for persons with dementia using video and a partially observable Markov decision process. *Computer Vision and Image Understanding*, *114*(5), 503–519.
2. Jaakkola, T., Singh, S. P., & Jordan, M. I. (1995). Reinforcement learning algorithm for partially observable Markov decision problems. In *Advances in neural information processing systems* (Vol. 7, pp. 345–352). MIT Press
3. Kaelbling, L. P., Littman, M. L., & Cassandra, A. R. (1998). Planning and acting in partially observable stochastic domains. *Artificial Intelligence*, *101*(1), 99–134.
4. Leite, I., Martinho, C., & Paiva, A. (2013). Social robots for long-term interaction: A survey. *International Journal of Social Robotics*, *5*(2), 291–308.
5. Monahan, G. E. (1982). A survey of partially observable Markov decision processes: Theory, models, and algorithms. *Management Science*, *28*(1), 1–16.
6. Pineau, J., Gordon, G., & Thrun, S. (2003). Point-based value iteration: An anytime algorithm for pomdps. In *International Joint Conference on Artificial Intelligence (IJCAI)* (pp. 1025–1032)
7. Pineau, J., Montemerlo, M., Pollack, M., Roy, N., & Thrun, S. (2003). Towards robotic assistants in nursing homes: Challenges and results. *Special issue on Socially Interactive Robots, Robotics and Autonomous Systems*, *42*(3–4), 271–281.
8. Spaan, M. T. J., & Vlassis, N. (2005). Perseus: Randomized point-based value iteration for POMDPs. *Journal of Artificial Intelligence Research*, *24*(1), 195–220.

9. Spaan, M. T. J., Veiga, T. S., & Lima, P. U. (2015). Decision-theoretic planning under uncertainty with information rewards for active cooperative perception. *Autonomous Agents and Multi-Agent Systems*, *29*(6), 1157–1185.
10. Taha, T., Miro, J. V., & Dissanayake, G. (2008). POMDP-based long-term user intention prediction for wheelchair navigation. In *IEEE International Conference on Robotics and Automation, 2008, ICRA 2008* (pp. 3920–3925)

An Overview of the Distributed Integrated Cognition Affect and Reflection DIARC Architecture

Matthias Scheutz, Thomas Williams, Evan Krause, Bradley Oosterveld, Vasanth Sarathy and Tyler Frasca

Abstract `DIARC` has been under development for over 15 years. Different from other cognitive architectures like SOAR or ACT-R, `DIARC` is an intrinsically component-based distributed architecture scheme that can be instantiated in many different ways. Moreover, `DIARC` has several distinguishing features, such as affect processing and deep natural language integration, is open-world and multi-agent enabled, and allows for "one-shot instruction-based learning" of new percepts, actions, concepts, rules, and norms. In this chapter, we will present an overview of the `DIARC` architecture and compare it to classical cognitive architectures. After laying out the theoretical foundations, we specifically focus on the action, vision, and natural language subsystems. We then give two examples of `DIARC` configurations for "one-shot learning" and "component-sharing". We also briefly mention different use cases of `DIARC`, in particular, for autonomous robots in human-robot interaction experiments and for building cognitive models.

1 Introduction

Classical cognitive architectures (CCAs) have evolved significantly since their inception in the late 1970s, with more and more features added on top of their core production systems. The ACT-R architecture (currently at version 7), for example, started from a model of associative memory and has morphed into a system allowing multiple inheritance among chunks, together with any number of new buffers connected to the central production system that can be added to the architecture to hold memory chunks (e.g., to allow for interactions with sensory and effector modules, see the ACT-RE models by Trafton et al. [76]). Similarly, the SOAR architecture (currently at version 9.6) started with a production system that only featured "chunking"

M. Scheutz (✉) · E. Krause · B. Oosterveld · V. Sarathy · T. Frasca
HRI Laboratory, Tufts University, Medford, MA 02155, USA
e-mail: matthias.scheutz@tufts.edu

T. Williams
MIRROR Lab, Colorado School of Mines, Golden, CO 80401, USA

M. I. Aldinhas Ferreira et al. (eds.), *Cognitive Architectures*, Intelligent Systems, Control and Automation: Science and Engineering 94,
https://doi.org/10.1007/978-3-319-97550-4_11

as the single architectural learning mechanism and has morphed into a system that integrates reinforcement learning, as well as semantic and episodic memories (in addition to the original working memory). While some of the extensions were driven by the need for more complex mechanisms to be able to develop adequate cognitive models, others were driven by the need to provide more capabilities for applications (e.g., in virtual and robotic agents). In addition to classical cognitive architectures, newer cognitive architectures such as Icarus, Clarion, and others were developed to address specific research questions (e.g., the implicit-explicit dichotomy in cognitive systems or the questions of how to learn and execute hierarchical skills, respectively).

Different from CCAs, the "Distributed Integrated Affect Reflection and Cognition" `DIARC` architecture [63, 66] was originally neither designed as nor intended to be a model of *human cognition*. Rather, it was conceptualized from the beginning as an *architecture scheme* (similar to the *CogAff* architecture scheme [71]) that could subsume a large set of possible architectures, and thus be used to realize a diverse set of cognitive systems of varying complexity, especially situated embodied systems such as robots. Architecture schemes are templates that, when filled in with details (i.e., specific components and their connections), specify individual architectures. In `DIARC`, this means that once components and their interactions are fixed, a particular `DIARC` architecture (a `DIARC` instance) is obtained. Note that the distinction between an architecture scheme and an architecture instance is different from the distinction between the algorithms and knowledge in cognitive architectures [41], in which "algorithms" are said to define the architecture *per se* (components and links) and knowledge is viewed as being encoded in representations contained in those components (either preloaded or acquired during operation). However, the design-as-architecture scheme does not preclude using different `DIARC` instances as cognitive models. And, in fact, different instances of `DIARC` have been used to model different cognitive aspects (e.g., the interaction of affect and goal processing [55], or a language-guided conjunctive visual search [65]).

In the following, we will first lay out the theoretical commitments made by `DIARC` as an architecture scheme and then discuss in greater detail the notable features that distinguish `DIARC` from other cognitive architectures. We then briefly give examples of two instances of `DIARC` for "one-shot learning" and "component-sharing", respectively, as well as applications of `DIARC` in cognitive modeling, autonomous robotics, and human-robot interaction.

2 Theoretical Commitments

Every cognitive architecture is based on basic theoretical assumptions about the structure and nature of its components, the data representations used inside and across components, as well as the information and control flow, and possibly the timing of component updates and information exchanges. To show the commonalities and differences in theoretical commitments of `DIARC` compared to CCAs, we start with the four-part framework for discussing CCAs proposed in [40]: (1) structure and processing, (2) memory and content, (3) learning, and (4) perception and motor.

Following this comparison, we discuss additional theoretical commitments DIARC makes concerning its components, as well as the principles underwriting the overall polylithic design and implementation of DIARC in a multi-agent system middleware.

2.1 Structure and Processing

In line with typical assumptions in CCAs, DIARC is composed of a set of components that operate in parallel and can communicate with each other by exchanging messages using logical representations. Different from most CCAs, DIARC components operate asynchronously in real parallelism and do not assume any synchronization mechanism (e.g., as imposed by a "perceive-think-act" cycles). Moreover, in addition to there not being any prescription of a particular system-wide cognitive cycle across components, there is also no prescription of the update timing of a component (e.g., compared to 50 or 100 ms for cognitive cycles in some CCAs). Rather, each component can update at the rate appropriate for the information it processes (and may run multiple threads of control within itself).

2.2 Memory and Content

While the details of the interaction among different components depend on the particular architecture (e.g., ACT-R provides a buffer mechanism that serves as the interface between the core production system and other modules), CCAs typically impose a communication bottleneck when they require that different modules interact via a special (short-term) *working memory* component and two long-term memory components for procedural and declarative knowledge. In contrast, DIARC does not impose such structural or communication constraints based on memories and memory access, but rather allows components to locally implement their own short-term and long-term memories. Consequently, there is no mandated component-based distinction between declarative versus procedural knowledge—both kinds could coexist in the same memory component—although typically, procedural knowledge is stored in the action execution component while declarative knowledge is stored in a special memory and inference component that can be instantiated multiple times as needed and used for short-term and long-term information storage. Different memories can be cross-indexed and accessed via consultants that establish those links (see Sect. 3.3.3). Moreover, there is no prescribed knowledge representation format in DIARC for knowledge stored within components (e.g., the way in which declarative knowledge has to be represented as "chunks" in ACT-R or procedural knowledge has to be represented in terms of production rules). Rather, knowledge representations can take different forms within components, depending on the nature of the process (e.g., saliency maps inside the vision processing component, dependency graphs in the parser, clauses in the reasoner, etc.). However, there is a requirement that messages exchanged conform to the same format across architecture instances (i.e., logical expressions are used as a common currency and data representation format across components).

2.3 Learning

In CCAs, all long-term memory entries are learnable online and incrementally during task performance using "architectural" (i.e., built-in) learning mechanisms, often through inverting the information flow. Different types of learning are employed, depending on the types of long-term memory (e.g., reinforcement learning to generate weights for action selection and procedural composition such as "chunking" for procedural learning, or the learning of facts together with their meta-data for declarative learning). In contrast, `DIARC` does not prescribe any particular architectural learning mechanism, but allows components to implement their own learning strategies, depending on the information they process. For example, the vision and auditory subsystem can employ unsupervised learning to improve the accuracy of their classifiers (e.g., adjust object recognizers to build better recognition templates in the vision system or adjusting word prototypes to be able to better recognize different word instances in the speech recognizer). Policy-based action execution systems might use reinforcement learning to improve their policies, or they could use action sequences from plan traces to learn the appropriate action sequences (very much like what "problem solving" allows in architectures like Soar). In addition to online learning, some `DIARC` components can also be trained offline (e.g., the vision and speech components, the parser, policy-based planers, etc.). Most importantly, however, `DIARC` directly supports instruction-based "zero-shot" and "one-shot" learning *across most knowledge representations in the architecture*, both through its integration of natural language processing and component capabilities for one-shot learning (e.g., in the vision and speech recognition components) [38, 64]. As a result, new words, percepts, actions, skills, rules, plan operators, norms, and other forms of knowledge can be learned quickly through natural language instruction during performance and used in "open-world" tasks for which not all required task-based knowledge is available ahead of time, but rather must be acquired online during task execution (e.g., [75]).

2.4 Perception and Motor

CCAs assume that perception modules generate symbolic structures representing the perceived object, relation, event, etc. while motor modules convert such symbol structures into motor actions. Both perception and motor modules may allow for learning (e.g., to acquire new perceptual and action patterns), although such learning is typically outside of the architectural specification (as lower-level perceptual and motor control processes are typically not included in CCAs). In contrast, `DIARC` was specifically aimed at real-world, real-time interactions and thus takes both perceptual and motor processes very seriously, providing detailed models for both (e.g., an extensive vision system that can process information from various types of sensors in real-time and various robot body modules that can process motor behavior for different robot body types). Similar to CCAs, learning in these components is not prescribed, but rather different learning methods are allowed. Different from CCAs, DIARC permits zero-shot perception and motor learning from instruction (e.g., direct learning of new percepts or new primitive actions from natural language descriptions without exposure to the percepts or the actions). Moreover, perceptual processing and action sequencing are closely tied to the real world (e.g., update frequencies of the vision system are related to the frame rates of sensors, and action commands at the lowest motor levels are tied to the command speed of effectors and the durative nature of embodied activities).

2.5 Additional Component Commitments

In addition to the types of commitments found in CCAs, DIARC makes additional theoretical commitments about its components that are not found in CCAs:

- *Affect integration.* Affect is, surprisingly, not part of CCAs, even though it is a central component of human cognition and CCAs are often intended to be "models" of human cognition. All DIARC components must represent both positive and negative affect (in the simplest case, as measures of how well a component performs, in more complex cases, as richer representations of desired component states). Some components like the Goal Manager collect affective evaluations from other components to compute composite evaluations of how well the agent (controlled by the DIARC instance) is doing, which can then be used to prioritize goals and modulate expected utility [56].
- *Open-world processing.* All DIARC components must be open-world enabled, i.e., allow for partial and incomplete representations of the information they process, as happens in open-world scenarios for which not all of the information is available initially, but rather has to be acquired through discovery and learning processes (e.g., unknown words in goal instructions referring to unknown entities in unknown locations, e.g., [75]).
- *Multi-level introspection.* All DIARC components must allow for the introspection of their states through middleware-enabled introspection processes, which can be used to optimize component and architecture performance, but also to detect and recover from faults (e.g., [36]). In addition, introspection methods can be used by components to detect available functionality in DIARC instances (e.g., the Goal Manager component can determine the kinds of perception and action primitives that are available in other components and can be used in action scripts, see Sect. 3.1). Explicit logical annotations of pre- operating, and post-conditions of

services made available by components to other components can be used to enable introspective access and run-time reflection on system features and capabilities.
- *Component-sharing*. All DIARC components must allow for component sharing across multiple agents, i.e., two or more agents realized as DIARC instances might share a single DIARC component (e.g., a common natural language processing subsystem consisting of speech recognizer, semantic and pragmatic parser, and dialogue manager components). Component-sharing allows for efficient implicit realization of agent-to-agent communication in which instead of explicit communication, agents have direct access to required knowledge structures [45].

2.6 Polylithic Design and Implementation

A result of being an architecture scheme, and thus allowing for different configurations among possible components and links, DIARC is intrinsically *polylithic*, compared to the monolithic nature of classical cognitive architectures. The polylithic nature is guaranteed by the implementation of DIARC in the "Agent Development Environment" ADE [1–3, 31–36, 54, 59], which was specifically developed to address the various challenges posed by sustained long-term operation of autonomous robots. Analogous to current robotic infrastructures (such as ROS [46], JAUS [30], Yarp [42], and several others), ADE provides a basic communication and computational infrastructure for parallel distributed processes that implements various functional components of an agent architecture (e.g., the interfaces to a robot's sensors and actuators, the basic navigation and manipulation behaviors, path and task planning, perceptual processing, natural language parsing, reasoning and problem solving, etc.).[1] Different from other robotic infrastructures, ADE was from the very beginning, *designed to be as secure and fault-tolerant as possible* (e.g., [1, 3, 54]). These features have been evaluated in HRI experiments [32, 34]. Moreover, due to ADE's extendability, DIARC is easily and systematically extendable by just adding more DIARC components (that may simply "wrap" existing libraries and algorithms) implemented in ADE to an architecture instance (this is different from CCAs such as ACT-R or Soar, in which extensions can only be accomplished through specialized mechanisms such as buffers or special I/O links).

3 An Overview of Select DIARC Components and Processes

After our brief overview of DIARC and its theoretical commitments, we now present a few central DIARC components and processes in more detail. By "central", we intend that these components will typically be part of a DIARC instance, even though they

[1] A detailed conceptual and empirical comparison of robotic infrastructures up to 2006 can be found in [35].

do not necessarily have to be included for all applications: (1) The Goal Manager, (2) the Vision System, and (3) the Natural Language subsystems.[2]

3.1 Goals, Actions, and Action Execution

Goals represent terminal states of the internal or external environment that an agent may need to satisfy. In DIARC, the Goal Manager (GM) receives goals from other components in the architecture, including itself. It evaluates the incoming goals, determines what behavior or action the agent should perform, how the agent should proceed, and handles the priority of each action. The priority of the actions are computed based on the urgency, expected utilities, and overall affective state. When the GM receives a goal, it determines the validity of the goal, initializes an Action Interpreter to select a sequence of actions, which, when executed, will accomplish the goal state, and then manages the execution of that action.

3.1.1 Action Representation

Actions in DIARC are stored within the Action Database, a long-term procedural memory, and are represented by a name, arguments, as well as pre-, operating-, and post-conditions. An action is either a primitive action or an action script. Primitive actions describe the specific functionality of their advertising components. For example, a vision component would advertise a *findObject* action that allows the GM to direct the vision component to look for an object, while a manipulation component would advertise the *graspObject* and *moveObject* actions in order for the GM to direct a manipulation component to act on an object. Action scripts are complex tasks containing a sequences of primitive actions, action scripts, action operators (e.g., arithmetic, comparison, etc.), and control statements (e.g., conditional statements and loops).

3.1.2 Action Execution

When the GM receives a goal submission, it creates a new Action Interpreter. The Action Interpreter first initializes the process for selecting an action. Then, if an action is found, it manages the execution of that action. Within the action Interpreter, an observation mechanism, described below, allows the agent to make observations about the world state by checking the state of events, objects, and agents. This mechanism enables the agent to track the progress of action execution.

[2] The description of additional relevant components, such as the Belief and Inference subsystem, the (Motion and Task) Planning subsystem, and the interfaces with other middleware, will have to await a different publication outlet.

Once the Action Interpreter selects an action to perform, in order to follow social norms and core rules, it verifies that the action is neither forbidden nor that it executing it would make the system enter a forbidden state. Then, it confirms that all of the action's pre-conditions are satisfied. For each pre-condition, the Action Interpreter spawns an observer (if available) to check the state of the environment. However, if there is no observer available, it checks the State Machine, which holds the agent's knowledge of the current state of the world. If any of the pre-conditions are not satisfied, then the Action Interpreter will cancel the execution and will report the failure conditions. During the course of execution of the action, there are conditions that need to be satisfied throughout ("operating conditions"). Thus, the Action Interpreter starts observation monitors for each operating condition. If at any point one of the conditions is no longer met, then it will cancel the action and report the failure conditions.

After the Action Interpreter completes the initial preparations, it can continue the execution process by checking to see if the action is a primitive action or a complex action represented by an action script. If the selected action is a script, then a similar process as described below for the primitive actions occurs for each sub-action. Each sub-action specifies the assigned agent responsible for carrying it out. However, if it is a primitive action, then the Action Interpreter checks the agent specified to perform the action. Because each action has a specified agent involved, actions can contain multiple agents interacting with each other. If the agent delegated to perform the action is a `DIARC` agent, then it will proceed normally with the execution. Otherwise, the Action Interpreter observes the other agent performing the action and the post-conditions of the action. Finally, the Action Interpreter confirms that the post-conditions of the action have been satisfied. For each condition, the Action Interpreter spawns an observer, if available. Otherwise, it will check the State Machine. The observers can confirm that other agents have performed their appointed tasks. If all of the post-conditions are met, then the action returns successfully, otherwise the action fails and the failure conditions are reported.

3.1.3 Observers

An agent must be able to track the progress of an action by observing the world and checking the conditions of the action. For instance, if an agent picks up an object off a table, it must observe that the object is in its hand and that the object has been lifted off of the table. While the agent can execute this action blindly and simply assume it to have been completed successfully (e.g., if the action can be performed with a very low error rate), it will not truly know whether the action was successfully executed unless it observes the action outcomes. This mechanism is particularly critical for multi-agent interactions in which one agent must wait for another agent to perform an action.

Observers are implemented as special primitive actions that adhere to a particular method signature and explicitly advertise the types of observations they enable (e.g., `touching(X,Y)`). To make use of the observations, the Action Interpreter looks for available observers in the Action Database when verifying conditions for an action. If an observer is found for a particular condition, then a new observer sub-goal is spawned. During verification of pre- and post-conditions, the Action Interpreter blocks execution until the observer process either returns successfully or has timed out. On the other hand, observers for operating conditions are spawned concurrently with the action to be executed and the capability to interrupt action execution in cases of failures.

3.2 Perception and Cognitive Affordances

The Vision component (VIS) is responsible for almost all of the visual perception capabilities in DIARC. This component consists of a highly parallel, modular, and flexible framework composed of various general purpose *Image Processors*, *Saliency Operators*, *Object Detectors*, *Object Validators*, and *Object Trackers* and is responsible for the detection, identification, and tracking of target objects and relations among them. VIS is capable of operating on a variety of input sensor types (e.g., RGB, RGB-D, depth-only), and automatically configures its available capabilities based on this information. Additionally, VIS supports multiple asynchronous "visual searches" that can optionally share parts of their processing pipelines so as to reduce redundant computations and save computational resources.

Image Processors are generally used to implement common low-level image processing tasks such as feature extraction (e.g., SIFT) and edge detection, and provide a mechanism for commonly consumed image processing results to be simultaneously shared across several vision processors and visual searches.

Saliency Operators are attentional mechanisms that can be used to guide a visual search to the most salient parts of a scene. These can be driven by top-down language-guided constraints (e.g., red, tall) and/or bottom-up models such as those by Itti and Koch [29].

Object Detectors are responsible for segmenting object candidates from the scene. Detectors can take the form of either generic object detectors that attempt to break the scene into constituent parts, or specialized detectors that search the entire scene for objects of a particular class or category (e.g., face, mug).

Object Validators consume segmented object candidates from Detectors and attempt to classify them as having particular properties (e.g., color, shape, category/class). Successfully validated objects are passed through to the next stage of the vision pipeline.

Object Trackers are the last stage of a vision pipeline. Trackers consume object candidates that have been fully validated (i.e., meet all visual search criteria), and are responsible for tracking objects from frame to frame.

One critical aspect of the vision component is exposing and advertising its capabilites to the rest of the system. This is done through simple quantifier-free first-order predicate representations, in which each vision processor described above (with the exception of Trackers and Image Processors) advertises what it is capable of processing (e.g., `red(X)`, `mug(X)`, `on(X,Y)`). In order for a component in the system to make use of VIS capabilities, it simply has to make a request to VIS in the form a quantifier-free first-order predicate representation. VIS will take this request and attempt to find a collection of vision processes capable of satisfying each part of the predicate request. If all parts are satisfied, the relevant visual processors are assembled into a vision pipeline and a visual search is started. Requesting components are then able to retrieve any available search results.

VIS also has the ability to dynamically learn new object representations. These representations can take the form of either definitions (e.g., "a medkit is a white box with a red cross on it") or instances (e.g., "this object is a medkit"). For learned definitions, VIS must be able to map all parts of a definition to existing vision capabilities. Then, when a request for a visual search for a learned definition is made, existing vision processors representing each part of the definition can be dynamically assembled into a vision pipeline capable of locating the target object(s). Learning new object instances, however, relies on at least one detector or validator capable of learning new object models on the fly. VIS does not impose restrictions on the underlying modeling approach, but methods to date have relied on global point cloud features (e.g., ViewpointFeatureHistogram as implemented in PCL [47]).

3.2.1 Cognitive Affordances

Affordance perception refers to the ability of an agent to extract meaning and usefulness from objects in its environment, often performed through perceptual (e.g., visual and haptic) analysis of object features [27, 98]. *Cognitive affordance* is a richer notion that extends traditional aspects of object functionality and action possibilities by incorporating the influence of non-perceptual aspects: changing context, social norms, historical precedence, and uncertainty. This allows for an increased flexibility with which to reason about affordances in a situated manner. For example, consider a knife, which offers grasp affordances across the entirety of its body, including the handle and blade (note: although one has to be careful when grasping a blade, it is nevertheless still possible, and therefore an affordance). However, the cognitive affordances of grasping offered by the same knife can vary depending on the context of the task (grasping by the handle when using it versus grasping by the blade when handing it over).

`DIARC` implements the current state-of-the-art formalism of cognitive affordances that uses a probabilistic logic-based approach [49, 50, 98], in which affordances are represented as condition-action rules (R), very much like production rules, in which the left-hand sides (LHS) represent perceptual invariants (F) in the environment, together with contextual information (C), and the right-hand sides (RHS) represent affordances (A) actualizable by the agent in the situation (e.g., the rule that one should grab a knife by the handle when using it would be translated by specifying the grasping parameters as F, the task context of "using a knife" as C and the constrained grasping location, together with other action parameters, as A). Affordance rules (R) take the overall form

$$r \stackrel{\text{def}}{=} f \wedge c \underset{[\alpha,\beta]}{\Longrightarrow} a,$$

where $f \in F, c \in C, a \in A, r \in R$, and $[\alpha, \beta] \subseteq [0, 1]$. $[\alpha, \beta]$ is a confidence interval intended to capture the uncertainty associated with the truth of the affordance rule r such that if $\alpha = \beta = 1$, the rule is logically true, while $\alpha = 0$ and $\beta = 1$ assign maximum uncertainty to the rule. Similarly, each of the variables f and c also have confidence intervals associated with them, and are used for inferring affordances, as described in more detail below. Thus, rules can then be applied for a given feature percept f in a given context c to obtain the implied affordance a under uncertainty about f, c, and the extent to which they imply the presence of a. Currently, a Dempster-Shafer theoretic uncertainty-processing framework is used for reasoning with these probabilistic rules and inferring the confidence intervals [70].

The DIARC implementation is in the form of a separate affordance component (AFF) in combination with several other components. Given a set of affordance rules, AFF determines the subset of applicable rules by matching their left-hand sides given the current context and perceivable objects in the environment, together with their confidence intervals, and then determines the confidences on the fused right-hand sides (in case there are multiple rules with the same RHS) based on the inference and fusion algorithm in [49]. It uses the "confidence measure" λ defined in [44] to determine whether an inferred affordance should be realized and acted upon. For example, we could check the confidence of each affordance on its uncertainty interval $[\alpha_i, \beta_i]$: if $\lambda(\alpha_i, \beta_i) \leq \Lambda(c)$ (where $\Lambda(c)$ is an confidence threshold, possibly depending on context c), we do not have enough information to confidently accept the set of inferred affordances, and can thus not confidently use the affordances to guide action. However, even in this case, it might be possible to pass on the most likely candidates to other parts of the integrated system. Conversely, if $\lambda(\alpha_i, \beta_i) > \Lambda(c)$, then we take the inferred affordance to be certain enough to use it for further processing.

From a systems standpoint, in order to process cognitive affordances, two primary sub-components have been implemented [49]: (1) an Affordance Reasoning Sub-component (ARC), and (2) a Perceptual Semantics and Attention Control Sub-component (PAC). In addition, two supporting component-specific memories—Long-term Memory (LTM) and Working Memory (WM)—are needed for storing and updating logical affordance rules and related uncertainties. During inference, ARC searches through all available affordance rules of the form specified above in the agent's LTM and populates WM with the relevant rules. Once the rules are in WM, both PAC and ARC can use these rules as the basis for perception and inference as well as AFF works closely with sensory and perceptual systems (e.g., VIS) and other components in DIARC to coordinate perceptual and action processing. AFF is connected to the Goal Manager (GM/AM), and during the execution of actions, GM/AM sends affordance requests to AFF. These requests provide information about the current action to be performed and the context. AFF returns the specific perceptual features that need to be searched in the environment. This allows GM/AM to direct the attention of low-level perceptual systems like the vision component (VIS) and perform searches in a focused manner, only looking for perceptual features in the environment that are relevant to the applicable rules in AFF. The presence or absence of the searched perceptual features (along with perceptual uncertainty information) is passed back to AFF, which subsequently performs uncertain logical inferences (logical AND and modus ponens) on the rules. GM/AM is at the heart of `DIARC` and helps coordinate most goal-directed action. In dialogue-driven tasks, GM/AM is typically the recipient of processed language-based knowledge obtained via the natural language pipeline; instructions, questions, commands, and other utterances can flow through this NL pipeline to and from the GM/AM. Another recipient of language-based knowledge in `DIARC` is the belief component BEL. BEL maintains a history of all declarative knowledge passing through `DIARC` and is capable of performing various logically-driven knowledge-representation and inference tasks. Thus, it serves as a convenient holding-area for cognitive affordance information partially processed through the NL pipeline, which can then be retrieved and processed by AFF.

With the capability to perceive and learn cognitive affordances, the agent can learn normative behavior from instruction and immediately apply this newly acquired knowledge to the task at hand.

3.3 Natural Language Dialogues

Different from most CCAs, natural language understanding and generation for dialogue interactions is at the core of `DIARC`, and thus deeply integrated with other components not related to language. In the following sections, we briefly discuss the core language components and their interactions within and outside the language subsystem. The design of these architectural components is justified and inspired by a long tradition of empirical work at our laboratory, evaluating aspects of communication in both human-human teams (e.g., [23, 25, 26, 43]) and human-robot teams (e.g., [4, 14, 72, 81, 81, 82, 95]).

3.3.1 Speech Recognition

Speech is the most common way for natural language to be conveyed in interactions between humans and autonomous systems, especially when those systems are embodied in robots. The first step in understanding natural language in these interactions is understanding speech: what was said, and by whom. A speech recognizer that is part of a larger cognitive architecture has access to more and different types of information than a speech recognizer in isolation.

In `DIARC`, the ASR (Automatic Speech Recognition) component is responsible for recognizing speech input to the system. Its main role is to convert acoustic speech signals into a text representation. This text representation is the first step in understanding spoken natural language, and is the basis for the rest of the processioning done by language components. As technologies for performing automatic speech recognition improve, the techniques that the ASR component uses to perform speech recognition can be updated to reflect the state of the art. While the internal mechanisms of the ASR component may change, these changes do not affect the role that the component plays, or its interface with the rest of the architecture. Depending on the application of an instance of DIARC, the ASR component can be configured to operate in a variety of ways. This configuration is not limited to only the speech recognition process alone, but also includes the components to which ASR is connected, and the information it sends to them.

In "closed-world" task-driven dialogues, in which the lexicon of the interaction is known to all interactors before the interaction occurs, the ASR component can be configured to recognize only utterances that can be generated from this lexicon, given some grammar. Such configurations can be achieved by adding a specific user-defined grammar, e.g., a graph on top of the existing language model of a large vocabulary speech recognizer (LVCSR), to constrain its output to that grammar alone, and thus improve recognition rates.

In contrast, the ASR component can also be configured for "open-world" scenarios, in which the robot may hear new words that are not in its lexicon and must respond to their use in a timely fashion. For this purpose, an LVCSR that is able to not only recognize a large number of words, but also recognize when it has heard a word that is outside of its known vocabulary, can be employed to allow the system to identify when it has heard a word that it does not understand, and respond accordingly. For example, it may be desirable not only to identify words that the system has not heard before, but also to start learning about them. The ASR component can be configured to store words that it has not previously heard before, and recognize when it hears them again. This is achieved by adding a one-shot speech recognizer (OSSR) to the LVCSR already present in the ASR component [64]. Forming a representation at this level is the first step in the process of learning a new word and its meaning. Being able to consistently recognize the word allows the rest of the language understanding components to begin to create a model of its meaning. This representation can also be used by the Speech Production component when the robot must speak about the new word to a human. The ASR component can be configured to connect to the Speech Production component and use its stored acoustic representations of novel words to update the models in the production system, similarly to the way in which it updates the models in the recognition system [64].

The performance of the speech recognition mechanism in the ASR component can also be improved through connections with other components in the architecture and the information they can provide. This integration into an embodied system provides the ASR component with types of information that a speech recognizer in isolation could not have. One such integration is a configuration of DIARC in which the social context of the dialogue is used to bias the results of the speech recognizer [77]. Through a connection with the Dialogue Management component, the ASR component receives biasing signals for parts of its lexicon based on the position in the dialogue and the roles of the agents that are speaking. This integration improves the performance of the speech recognition mechanism used in ASR, and results in a system as a whole that models biological mechanisms.

3.3.2 Parsing

The Parsing component, referred to as the Natural Language Processing (NLP) or Natural Language Understanding component (NLU), grounds the text of an utterance in a form that the rest of the components of DIARC can *understand*. To do this, the component must interpret the syntactic structure of the utterance, as well as its semantics. The semantic representation that is used throughout DIARC is logical predicates, so for a given utterance, the parser must produce a predicate expressions that represents its semantics.

The parsing component uses a parser that is throughly integrated with the rest of the architecture. This integration allows the parser to produce semantics of the correct form, as well as enhancing the capabilities of other components. The parser uses a dictionary of rules to interpret the utterances it receives. Each rule in the dictionary is composed of (1) the word in the lexicon to which it corresponds, (2) a syntactic definition of the word in Combinatory Categorial Grammar (CCG), and (3) a semantic definition of the word in lambda calculus. The lambda function in an entry generates all or part of a predicate whose meaning is grounded in formal expressions the rest of the system can understand. The syntactic rules determine how the lambda functions corresponding to the words in the utterance are applied in relation to each other [22].

The predicate expression created by the parser is the first notion of understanding of an utterance that is generated in DIARC. The predicates produced here are, after potential transformations by the Pragmatics and Reference Resolution components, the input to reasoning components like Dialogue, Action, Affordance and Vision. Accordingly, the representations generated by the parser must be interpretable by these other components. The Parser component can be configured with different sets of rules for different applications. The semantics it produces can be tailored to meet the representational requirements of any of the other components present in a given configuration of DIARC. This allows the system to have a universally understandable internal representation of knowledge, whose implementation can be varied for the task at hand. Configurations of the system that are used in different tasks may require different semantics for the same utterance. The parser is able to be configured with different sets of rules so that the semantics of an utterance are always *understood*, regardless of the configuration of the architecture.

Like with the ASR component, in task-driven dialogue scenarios in which the lexicon of an interaction and its meaning are mutually understood by all of the participants, the parser can be configured with rules that guarantee understanding of any of the possible utterances the system might receive. In an open world, it is, again, not possible to have rules for every scenario the system may encounter. The Parsing component is equipped with mechanisms to generate representations of novel words as it encounters them. When a word is received from the ASR component that is not in the parser's set of rules, a new rule is generated for it. The syntax of the new word is inferred from its current usage, and is updated based on subsequent usages. The semantic representation of the word is also generated in conjunction with the syntax. The first time the word is heard, the portion of the semantic predicate for the utterance that it corresponds to does not have any meaning for the other components in the system. However, its meaning can be learned through the semantics of subsequent utterances, grounding the new semantic representation in the parser in the rest of the system [17, 64].

The parser performs syntactic and semantic parsing at the same time, which allows utterances to be understood incrementally as they unfold. This incremental understanding allows for the semantics of part of the utterance to begin to be understood before the utterance has been completely received by the system. This incremental parsing and understanding is especially useful in embodied human-robot interaction scenarios in which time is critical for interactions to appear natural. The incremental understanding of the parser component can be utilized by other components to improve their performance. For example, it can be used with the vision component to improve visual search speed [37] or a planner to update plans as new information is received [18]. Additionally, in many open-world settings, a robotic agent may not be able to completely parse an utterance, due to disfluencies in the interlocutors' speech, information loss, or an excess of novel terms that do not allow for successful inference of their meaning. In these cases, the ability to provide a partial parse, on the portion of the utterance that has been understood, and to not have the requirement of a complete utterance, allows the system to at least partially understand the utterance, which may be sufficient in some interactions [39].

The semantic representations that the parser generates are those that are required by the other components within `DIARC`, as they allow other components to perform further inference and understanding on what the system has heard. Some of the first interpretation of the predicate form of an utterance occurs in the Reference Resolution (RR) component. In order to properly understand referring expressions in an utterance, RR must be able to identify them. The semantic representation generated by the Parser component demarcates the portions of an utterance that contain referring expressions, and provides additional semantic information about the nature of the referring expression based on the syntax of the utterance [79].

3.3.3 Open-World Reference Resolution and Referring Expression Generation

`DIARC`'s *Reference Resolution* (RR) and *Referring Expression Generation* (REG) components facilitate, respectively, the understanding and generation of referring expressions in uncertain and open worlds. To enable these capabilities, both components rely on a distributed, cognitively inspired memory model [78]. The base of this model is a set of *Consultants*, each of which provides access to a different architectural component that is viewed as a *distributed, heterogeneous knowledge base* [88] that can (1) provide access to a list of candidate referents; (2) advertise a list of logical predicates it can assess; (3) assess how probable it is that any of the listed candidate referents satisfy any of the advertised properties, and (4) hypothesize and assert knowledge regarding new candidate referents. This architecture is designed to provide access to knowledge of candidate referents regardless of their location and form of representation, facilitating a *domain independent* approach (cf. [83]).

In addition to this distributed model of Long Term Memory, the RR component has access to a set of hierarchically nested caches, inspired by the Givenness Hierarchy's conception of the Focus of Attention, Set of Activated Entities, and Set of Familiar Entities [28]. These caches provide fast access to likely referents during the resolution of anaphoric, deictic, and definite noun phrases [93]. When this *GH-theoretic* reference resolution process [79, 94] is unable to identify sufficiently likely candidate referents, a Long Term Memory query is performed using the *DIST-POWER* algorithm [88]. *DIST-POWER* is an adaptation of the *POWER* algorithm [87] and cognitive model [86], which uses the aforementioned Consultant framework to perform reference resolution (i.e., identify the targets of referring expressions used by the robot's interlocutors) when information is distributed across multiple architectural components. *POWER* performs reference resolution under uncertainty by effecting a search through the space of possible variable-entity assignments, incrementally computing the probability of assignments as they are built up, and pruning branches of the tree of assignments when their probability falls below a given threshold.

POWER improves on previous reference resolution approaches through its ability to handle open-worlds. If *POWER* is unable to find an acceptable set of candidate referents for a query involving n variables, it recurses, trying again using a relaxed query involving $n - 1$ variables, with the removed variable selected on the basis of linguistic factors such as prepositional attachment and recency. This process repeats until a sufficiently probable set of candidate referents is found, or until all variables have been removed. Once this process terminates, new entities are hypothesized for all variables removed in this way, using the capabilities provided by the Consultant responsible for each new entity (according to its inferred type).

Just as these consultant capabilities are used to facilitate Reference Resolution, so too are they used to facilitate Referring Expression Generation, in which properties are selected to describe referents to which *the robot* wishes to refer. This is performed by the REG component using the *DIST-PIA* algorithm [91]. *DIST-PIA* is a version of the classic *Incremental Algorithm* [21], which uses the aforementioned Consultant framework to perform Referring Expression Generation when information is uncertain and distributed across multiple architectural components [90].

3.3.4 Pragmatics

`DIARC` provides several alternate mechanisms for pragmatic understanding, in which the intentions underlying utterances made to the robot are inferred (e.g., the goals that the robot's interlocutor desires it to uptake [5]), and pragmatic generation, in which an utterance for communicating the robot's own intentions is abduced. These capabilities are crucial in order to understand and generate *indirect speech acts* [69], which we have shown in laboratory experiments to be commonly used in human-robot dialogue [14], especially in contexts with highly conventionalized social norms [95].

These capabilities are facilitated by a set of Speech Act theoretic [68] pragmatic rules, each of which maps a different utterance, under a different environmental or dialogue context, to a different candidate intention. One option is to use these rules directly, without accounting for uncertainty. In this case, during understanding, the first matching rule is used to determine the correct interpretation [11], while during generation, the utterances produced by all matching rules may be ranked and then voted upon [24].

Alternatively, *Dempster-Shafer* (*DS*) theoretic [70] rules, which are augmented with DS-theoretic uncertainty intervals [85], may be used to perform these tasks under uncertainty, in which case the results of all applicable rules are combined, yielding a *set* of candidate intentions or utterance forms, each of which is augmented with its own DS-theoretic uncertainty interval [80]. If, during the understanding process, the uncertainty intervals associated with the produced candidate intentions reflect a sufficiently high degree of uncertainty, a clarification request can immediately be constructed and issued [89]. We have shown in previous work how this process is able to produce clarification requests that resolve both referential and pragmatic uncertainty, and that align with human preferences [92].

3.3.5 Task-Based Dialogues

Utterances typically do not exist in isolation. Previously spoken utterances in a dialogue, and also an agent's mental model of the world, can permit the agent to deduce meaning from a given utterance beyond its semantics, even when considered in isolation. The Dialogue component and Belief component provide the agent with mechanisms to model the world. They allow the agent to understand/predict how other agents will act based on its own actions, as well as allowing it to better decide how it should act in a given context. The previously described natural language understating components (ASR, Parsing, Reference Resolution, Pragmatics) allow the agent to determine the semantics of an utterance, while the Dialogue and Belief components, allow it to deduce new information given those semantics.

To engage in a dialogue, an agent must be able to model its interlocutor(s), so that it fully understands the utterances it hears, and thus knows how to respond. To do this, the Dialogue and Belief components use explicit rules that represent relationships between groups of utterances, as well as relationships between utterances and the past, present, and future beliefs of agents. For task-based dialogue interactions in which information about the task can be known by the agent before engaging in the dialogue, two types of rules are used: rules about the effects of perceptions, actions and past beliefs on new or updated beliefs, and rules about the effects of types of utterances on beliefs. The Dialogue component uses these rules, in combination with the utterance semantics that it receives, to determine how the agent should respond. The rules relevant to the dialogue are part of the agent's set of beliefs about the world. They are stored in the Belief component, as are the rest of the agent's beliefs. These beliefs may originate from perceptions of the world, like understanding an utterance or performing a visual search, or may follow from the application of rules to perceptual semantics [10].

The Dialogue component determines how it should respond by monitoring the agent's beliefs in the Belief component. When it receives an utterance, it uses the information about the utterance's type (statement, questions, command, etc.), which has been determined by the other natural language understating components, to convert the utterance semantics into expressions that represent the agent's beliefs about the world. The new beliefs are asserted into the Belief component, which updates its state and performs inference based on the new information. Once the agent's beliefs have been updated, the Dialogue component considers its new belief state, which has resulted from the utterance, and determines the agent's response. The response manifests itself as the submission of a goal or goals to the Goal Manger which may, in turn, result in the submission of further goals [54, 62, 80].

4 Two Example Configurations of `DIARC`

In this section, we briefly describe two different configurations of `DIARC`: a single-agent configuration for one-shot learning of objects through natural language [64], and a multi-agent configuration that uses shared components to enable shared information and cognition between the agents [45].

4.1 Learning Object Parts in One-Shot

The first example (Fig. 1) shows the `DIARC` configuration for a robot that can learn to recognize a new part of an object and use that knowledge to pick up the object by the newly learned part. Here, we assume that the robot already knows the object ("knife") and how to recognize it. A video of the interaction can be viewed here: http://bit.ly/2cfx3gL.

Human:	Pick up the knife by the handle.
Robot:	OK.
Robot:	But what is a handle?
Human:	The orange part of the knife is the handle.
Robot:	OK.
Human:	Pick up the knife by the handle.
Robot:	OK.

When the ASR component recognizes the unknown word "handle" in the utterance "Pick up the knife by the handle", it recreates a unique identifier "UT1" for it, which it then passes on to the NLU component. The NLU component infers the proper tag of speech from the lexical context and the semantic requirements

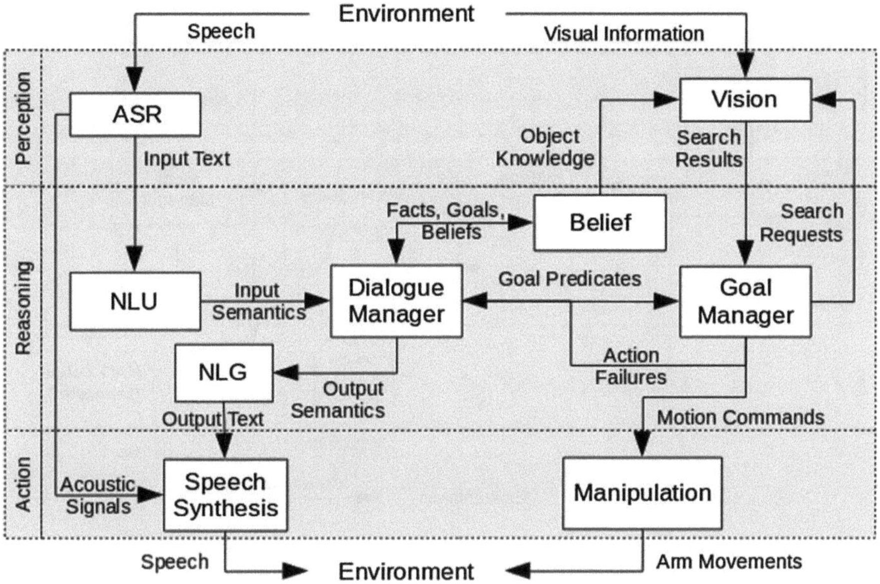

Fig. 1 The DIARC configuration for one-shot learning of objects and actions

for generating a grammatically correct command "pickUp(self,partOf(UT1,knife))", which it forwards to the Dialogue Manager component. The Dialogue Manager component checks the command for indirect interpretations [11], and once it determines that it is a literal command, it acknowledges it by generating "OK" via the NLG and Speech Synthesis components. At the same time, it forwards the request to both the Belief component, which can generate relevant implications from the command that might, in turn, impact the execution, and then to the Goal Manager component as a new goal "pickUp(self,partOf(UT1,knife),leftArm)". The Goal Manager component then begins to execute the "pickup" action, which requires it to convert the condensed description "partOf(UT1,knife))" to an expression "on(graspPoint,partOf(UT1,knife))" that it can send to the Vision component for processing. The Vision component, however, does not have any knowledge of "UT1", hence the visual search for the appropriate grasp points fails, which, in turn, causes a failure of the pickup action communicated to the Dialogue Manager component, which instructs the NLG and Speech Synthesis components to generate the question "But what is a handle?". Upon hearing the definition "The orange part of the knife is the handle.", the ASR component recognizes the word "handle", which it had previously associated with "UT1", and thus passes on "The orange part of the knife is the UT1." to the NLU component, which, in turn, generates the semantics "is(UT1,partOf(orange,knife))", and sends it again to the Dialogue Manager component. There, the utterance is recognized as a factual statement about a perceivable object and modified according to pragmatic rules to generate the form "looksLike(UT1,partOf(orange,knife))", which is then passed on to the Belief

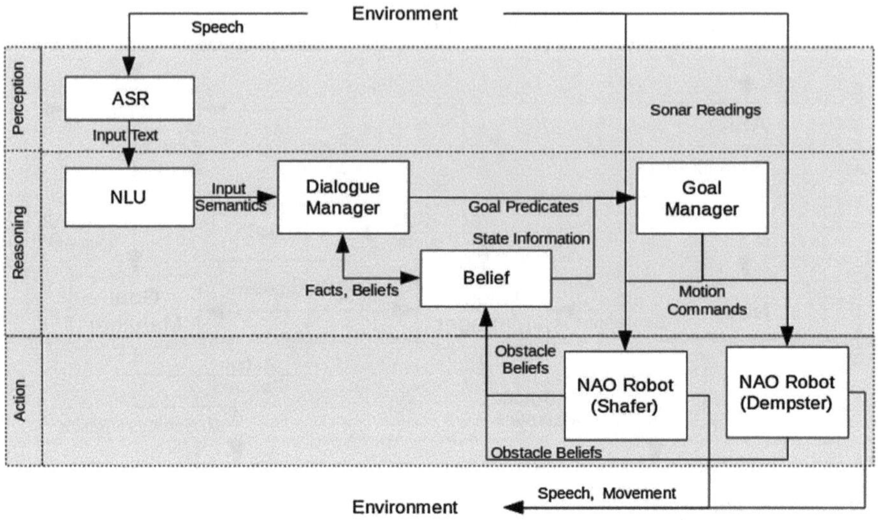

Fig. 2 The DIARC configuration for component-sharing among two Nao robots

component, and also acknowledged through the NLG and Speech Synthesis components ("OK"). The Belief component asserts the new fact to its knowledge base, thereby triggering a notification to the Vision component, which has requested to be notified of all facts of the form "looksLike(X,Y)". When the robot is then instructed again to pick up the knife by the handle, the Vision component is now able to resolve the reference "UT1" (i.e., "handle"), as it has learned the grounding of "UT1" as "partOf(orange,knife))". It finds a set of grasp points on the handle that it passes on to the Goal Manager component, which forwards the grasp constraints to the Manipulation component. The Manipulation component then selects the best grasp point to plan a trajectory for the robot's arm to those points (a subsequent lift action then completes the "pickUp").

4.2 *Sharing Components Among Multiple Agents*

The second example (Fig. 2) demonstrates the sharing of various architectural components so as to enable shared cognition, in this case, between two Nao robots called "Dempster" and "Shafer", although it works for any number of heterogeneous agents. A video of the interaction can be viewed here: https://www.youtube.com/watch?v=JPufmIPHX9Y.

Human:	Hello Shafer.
Robot:	Hello.
Human:	Hello Dempster.
Robot:	Hello.
Robot:	Dempster, tell Shafer to stand.
Human:	Certainly, I will do that right away.

The human starts by greeting both robots, and the analysis of the message content is used to invoke the subset of components in the joined architecture corresponding to the respective robot. For example, the utterance "Hello Shafer", once transliterated by the ASR component, is analyzed in the NLU component as a greeting addressed to the Shafer robot, and is passed on as such to the Dialogue Manager component, which determines the correct dialogue move for Shafer to say "Hello" as well. As a result, it routes the greeting message through the Goal Manager component directly to the Shafer Nao component, which produces the speech output in the Shafer robot (the robot components include both motion and speech synthesis functionality). Similarly, greeting the Dempster robot will cause it to respond with "Hello". When the human instructor then addresses the Dempster robot with the request for Shafer to stand, the ASR component again passes on the text form of the utterance on to the NLU component, which generates the semantics "want(human,Dempster,do(tell(Dempster,Shafer,do(Shafer,stand)))". This command is then sent to the Dialogue Manager component, which forwards the semantics to the Belief component, where it is asserted while generating an acknowledging dialogue move "Certainly, I will do that right away", which gets forwarded to the Dempster Nao component for speech synthesis (note that the Dialogue Manager component can determine the appropriate robot component from the pragmatic information about the addressee in the dialogue).

When new information is asserted in the Belief component, it may result in new goals being submitted to the Goal Manager component, in which case the Dialogue Manager is also informed of the new goal. In this case, there is a new goal for the Dempster robot: "tell(Dempster,Shafer,do(Shafer,stand))". When the Goal Manager component executes the "tell(X,Y,Z)" action using the bindings of X = Dempster, Y = Shafer, Z = do(Shafer,stand), its generates and submits the new subgoal "do(Shafer,stand)", which has the Shafer robot as the actor and the "stand" as the action. Execution then triggers the "stand" action in the Shafer Nao component, causing the robot to stand up.

Analogous to issuing commands, it is also possible to inquire about an agent's perceptions or knowledge via another agent. Such mediated interactions are automatically enabled by the sharing agent's Belief and Goal Manager components.

5 Applications

DIARC has been used on a variety of virtual and robotic agents in a great variety of contexts (e.g., [8, 10, 13, 19, 38, 51, 53, 61, 62, 74, 80, 83, 84, 87, 88, 96]). Like other cognitive architectures, it has also been integrated with other architecture (e.g., ACT-R, Soar, and Icarus [75] and Vulcan [84]. Most importantly, it has been employed in many human-robot interaction experiments, with both autonomous as well as teleoperated robots (e.g., [6, 7, 9, 12, 15, 16, 20, 57, 58, 67, 73, 79, 82, 97]). More relevant to this chapter, DIARC has also been used to model aspects of human cognition (e.g., [2, 6, 37, 48, 55, 62, 65, 67]). Here, we will only be able to provide a short summary of recent experimental and modeling work using DIARC.

5.1 *HRI Experiments with* DIARC

Recent empirical investigations of human-robot teams have largely focused on humans' use of indirect language when interacting with robots. This work has demonstrated (1) that humans will regularly use indirect language during the course of human-robot interactions [14], (2) that humans use indirect language when interacting with teleoperated and autonomous robots with a frequency similar to that used when interacting with other humans [4], (3) and that indirect language use is increased in contexts with highly conventionalized social norms [95]. Indirect language is notably used by humans in order to adhere to social norms such as *politeness* [73]. We have also investigated human perception of the use of polite language by robots. This work has shown that politeness not only increases human ratings of robot likability, but that this effect is significantly stronger for women than it is for men [72].

Our empirical work has also demonstrated differences in how robots' morphology and communication style may have significant impact on team dynamics; our investigation of autonomous versus teleoperated robots suggested that humans perceived teleoperated robots to be less intelligent than co-present human teammates [4]; our investigation of verbally versus silently communicating robots suggested that humans find silently communicating robots to be significantly creepier than verbally communicating robots (at least for the communication of task-dependent, human-understandable information among robots co-located with human teammates in a cooperative setting) [81, 82].

5.2 *Cognitive Modeling with* DIARC

Even though DIARC was never designed to be a model of the human mind, and thus was never intended to be a modeling framework for human cognition, it affords unique modeling capabilities due its real-time, embodied nature and integrated natural language understanding capabilities, and has thus been used for various types of (mostly qualitative) models over the years. Early models of incremental natural language processing demonstrate the incremental integration of perceptual context in the resolution and generation of references with ambiguities due to prepositional attachment (e.g., [6, 60]), as well as models of incremental word substitution for correcting phonetic errors (e.g., [6]). Later models of incremental natural language processing focused on natural language-guided biasing of visual spatial attention (e.g., [37]) and models of human-like task-based dialogues (e.g., [62]). Particularly notable is a model of natural-language guided conjunctive visual search that was fit to human data in a novel way and used to clarify possible explanations of observed empirical data [65].

Additional modeling work investigated the interaction between affect and cognition, in particular, the effect of mood states on goal management and ways to bias goals (e.g., [56]), as well as to modulate affect in the speech (e.g., [67]).

Most recently, `DIARC` has also been used to demonstrate human infant word learning in a cross-situational embodied context (e.g., [48]), as well as human-like norm-learning (e.g., [52]).

6 Conclusion

`DIARC`, as an architecture scheme, is neither a finished product nor does it aspire to be one. Rather, its purpose is to provide researchers with an expanding framework for exploring the functional and architectural design trade-offs of different types of autonomous agents. By being flexible in its instantiations, it allows for custom configurations of classes of systems targeted at particular physical platforms and classes of tasks. By being flexible in its component algorithms, it allows for the easy integration of novel algorithms, and thus provides researchers not interested in the development of cognitive systems with an evaluation platform. Ongoing work on `DIARC` is aimed at its unique contributions compared to classical cognitive architectures: the open-world aspects, one-shot learning, component-sharing, introspection mechanisms and integration with ADE middleware. The goal is not only to improve existing functionality for the investigation of more sophisticated algorithms and involved architectural features, but to also provide a robust implementation platform for future autonomous robot applications that allow for human-like task-based interactions in natural language dialogues.

Acknowledgements The work on `DIARC` has been supported by various research grants from the US National Science Foundation and the US Office of Naval Research over the years, most recently by NSF grant IIS1316809 and ONR grant N00014-16-1-2278.

References

1. Andronache, V., & Scheutz, M. (2004). ADE—A tool for the development of distributed architectures for virtual and robotic agents. In *Proceedings of the Fourth International Symposium "From Agent Theory to Agent Implementation"* (pp. 606–611).
2. Andronache, V., & Scheutz, M. (2004). Integrating theory and practice: The agent architecture framework APOC and its development environment ADE. In *Proceedings of AAMAS 2004* (pp. 1014–1021). ACM Press.
3. Andronache, V., & Scheutz, M. (2006). ADE–An architecture development environment for virtual and robotic agents. *International Journal of Artificial Intelligence Tools, 15*(2), 251–286.
4. Bennett, M., Williams, T., Thames, D., & Scheutz, M. (2017). Differences in interaction patterns and perception for teleoperated and autonomous humanoid robots. In *Proceedings of the IEEE/RSJ International Conference on Intelligent Robots and Systems (IROS).*
5. Brick, T., Schermerhorn, P., & Scheutz, M. (2007). Speech and action: Integration of action and language for mobile robots. In *IEEE/RSJ International Conference on Intelligent Robots and Systems (IROS)* (pp. 1423–1428). IEEE.
6. Brick, T., & Scheutz, M. (2007, March). Incremental natural language processing for HRI. In *Proceedings of the Second ACM IEEE International Conference on Human-Robot Interaction* (pp. 263–270). Washington, D.C.
7. Briggs, G., McConnell, I., & Scheutz, M. (2015). When robots object: Evidence for the utility of verbal, but not necessarily spoken protest. In *Proceedings of the 7th International Conference on Social Robotics.*
8. Briggs, G., & Scheutz, M. (2011, June). Facilitating mental modeling in collaborative human-robot interaction through adverbial cues. In *Proceedings of the SIGDIAL 2011 Conference*, Portland, Oregon (pp. 239–247).
9. Briggs, G., & Scheutz, M. (2012). Investigating the effects of robotic displays of protest and distress. In *Proceedings of the 2012 Conference on Social Robotics.* LNCS (pp. 238–247). Springer.
10. Briggs, G., & Scheutz, M. (2012). Multi-modal belief updates in multi-robot human-robot dialogue interaction. In *Proceedings of 2012 Symposium on Linguistic and Cognitive Approaches to Dialogue Agents.*
11. Briggs, G., & Scheutz, M. (2013). A hybrid architectural approach to understanding and appropriately generating indirect speech acts. In *Proceedings of the Twenty-Seventh AAAI Conference on Artificial Intelligence.*
12. Briggs, G., & Scheutz, M. (2014). How robots can affect human behavior: Investigating the effects of robotic displays of protest and distress. *International Journal of Social Robotics, 6*, 1–13.
13. Briggs, G., & Scheutz, M. (2015). "Sorry, i can't do that": Developing mechanisms to appropriately reject directives in human-robot interactions. In *Proceedings of the 2015 AAAI Fall Symposium on AI and HRI.*
14. Briggs, G., Williams, T., & Scheutz, M. (2017). Enabling robots to understand indirect speech acts in task-based interactions. *Journal of Human-Robot Interaction (JHRI).*
15. Briggs, P., Scheutz, M., & Tickle-Degnen, L. (2014). Reactions of people with Parkinson's disease to a robot interviewer. In *Proceedings of the Workshop on Assistive Robotics for Individuals with Disabilities at IROS 2014.*
16. Briggs, P., Scheutz, M., & Tickle-Degnen, L. (2015). Are robots ready for administering health status surveys: First results from an HRI study with subjects with Parkinson's disease. In *Proceedings of 10th ACM/IEEE International Conference on Human-Robot Interaction.*
17. Cantrell, R., Schermerhorn, P., & Scheutz, M. (2011, July). Learning actions from human-robot dialogues. In *Proceedings of the 2011 IEEE Symposium on Robot and Human Interactive Communication.*

18. Cantrell, R., Talamadupula, K., Schermerhorn, P., Benton, J., Kambhampati, S., & Scheutz, M. (2012, March). Tell me when and why to do it!: Run-time planner model updates via natural language instruction. In *Proceedings of the 2012 Human-Robot Interaction Conference*, Boston, MA.
19. Chakraborti, T., Briggs, G., Talamadupula, K., Zhang, Y., Scheutz, M., Smith, D., et al. (2015). Planning for serendipity. In *Proceedings of IROS*.
20. Crowell, C., Scheutz, M., Schermerhorn, P., & Villano, M. (2009, October). Gendered voice and robot entities: Perceptions and reactions of male and female subjects. In *Proceedings of the 2009 IEEE/RSJ International Conference on Intelligent Robots and Systems*, St. Louis, MO.
21. Dale, R., & Reiter, E. (1995). Computational interpretations of the gricean maxims in the generation of referring expressions. *Cognitive Science*, *19*(2), 233–263.
22. Dzifcak, J., Scheutz, M., Baral, C., & Schermerhorn, P. (2009, May). What to do and how to do it: Translating natural language directives into temporal and dynamic logic representation for goal management and action execution. In *Proceedings of the 2009 IEEE International Conference on Robotics and Automation (ICRA'09)*, Kobe, Japan.
23. Eberhard, K., Nicholson, H., Kuebler, S., Gundersen, S., & Scheutz, M. (2010, May). The Indiana cooperative remote search task (CReST) corpus. In *Proceedings of LREC 2010: Language Resources and Evaluation Conference*, Malta.
24. Gervits, F., Briggs, G., & Scheutz, M. (2017). The pragmatic parliament: A framework for socially-appropriate utterance selection in artificial agents. In *Proceedings of the 39th Annual Meeting of the Cognitive Science Society (COGSCI)*.
25. Gervits, F., Eberhard, K., & Scheutz, M. (2016). Disfluent but effective? A quantitative study of disfluencies and conversational moves in team discourse. In *Proceedings of the 26th International Conference on Computational Linguistics*.
26. Gervits, F., Eberhard, K., & Scheutz, M. (2016). Team communication as a collaborative process. *Frontiers in Robotics and AI*, *3*, 62.
27. Gibson, J. J. (1979). *The ecological approach to visual perception* (Vol. 39).
28. Gundel, J. K., Hedberg, N., & Zacharski, R. (1993). Cognitive status and the form of referring expressions in discourse. *Language*, 274–307.
29. Itti, L., & Koch, C. (2001, March). Computational modelling of visual attention. *Nature Reviews: Neuroscience*, 194–203.
30. JAUS. Jaus.
31. Kramer, J., Scheutz, M., Brockman, J., & Kogge, P. (2006). Facing up to the inevitable: Intelligent error recovery in massively parallel processing in memory architectures. In H. R. Arabnia (Ed.), *International Conference on Parallel and Distributed Processing Techniques and Applications*, Las Vegas (pp. 227–233).
32. Kramer, J., & Scheutz, M. (2006, October). ADE: A framework for robust complex robotic architectures. In *IEEE/RSJ International Conference on Intelligent Robots and Systems*, Bejing, China (pp. 4576–4581).
33. Kramer, J., & Scheutz, M. (2006). ADE: Filling a gap between single and multiple agent systems. In *Proceedings of the ACE 2004 Symposium at the 18th European Meeting on Cybernetics and Systems Research*, Vienna, Austria.
34. Kramer, J., & Scheutz, M. (2007, April). Reflection and reasoning mechanisms for failure detection and recovery in a distributed robotic architecture for complex robots. In *Proceedings of the 2007 IEEE International Conference on Robotics and Automation*, Rome, Italy (pp. 3699–3704).
35. Kramer, J., & Scheutz, M. (2007). Robotic development environments for autonomous mobile robots: A survey. *Autonomous Robots*, *22*(2), 101–132.
36. Kramer, J., Scheutz, M., & Schermerhorn, P. (2007, Oct/Nov). 'talk to me!': Enabling communication between robotic architectures and their implementing infrastructures. In *Proceedings of the 2007 IEEE/RSJ International Conference on Intelligent Robots and Systems*, San Diego, CA (pp. 3044–3049).
37. Krause, E., Cantrell, R., Potapova, E., Zillich, M., & Scheutz, M. (2013). Incrementally biasing visual search using natural language input. In *Proceedings of AAMAS* (pp. 31–38).

38. Krause, E., Zillich, M., Williams, T., & Scheutz, M. (2014). Learning to recognize novel objects in one shot through human-robot interactions in natural language dialogues. In *Proceedings of Twenty-Eighth AAAI Conference on Artificial Intelligence*.
39. Kuebler, S., Cantrell, R., & Scheutz, M. (2011). Actions speak louder than words: Evaluating parsers in the context of natural language understanding systems for human-robot interaction. In *Proceedings of RANLP* (pp. 56–62).
40. Laird, J. E., Lebiere, C., & Rosenbloom, P. (2017). A standard model of the mind: Toward a common computational framework across artificial intelligence, cognitive science, neuroscience, and robotics. *AI Magazine*.
41. Langley, P., Pat, J. E., & Rogers, S. (2009, June). Cognitive architectures: Research issues and challenges. *Cognitive Systems Research*, *10*(2), 141–160.
42. Metta, G., Fitzpatrick, P., & Natale, L. (2006). Yarp: Yet another robot platform. *International Journal on Advanced Robotics Systems*, *3*, 43–48.
43. Nicholson, H., Eberhard, K., & Scheutz, M. (2010). Um...i don't see any: The function of filled pauses and repairs. In *Proceedings of 5th Workshop on Disfluency in Spontaneous Speech* (pp. 89–92).
44. Nunez, R. C., Dabarera, R., Scheutz, M., Briggs, G., Bueno, O., Premaratne, K., et al. (2013). DS-based uncertain implication rules for inference and fusion applications. In *16th International Conference on Information Fusion (FUSION)* (pp. 1934–1941).
45. Oosterveld, B., Brusatin. L., & Scheutz, M. (2017). Two bots, one brain: Component sharing in cognitive robotic architectures. In *Proceedings of 12th ACM/IEEE International Conference on Human-Robot Interaction Video Contest*.
46. Quigley, M., Conley, K., Gerkey, B. P., Faust, J., Foote, T., Leibs, J., et al. (2009). ROS: an open-source robot operating system. In *Proceedings of ICRA Workshop on Open Source Software*.
47. Rusu, R. B., & Cousins, S. (2011, May 9–13). 3D is here: Point cloud library (PCL). In *IEEE International Conference on Robotics and Automation (ICRA)*, Shanghai, China.
48. Sadeghi, S., Scheutz, M., & Krause, E. (2017). An embodied incremental Bayesian model of cross-situational word learning. In *proceedings of the 2017 Joint IEEE International Conference on Development and Learning and Epigenetic Robotics (ICDL-EpiRob)*.
49. Sarathy, V., & Scheutz, M. (2016). A Logic-based computational framework for inferring cognitive affordances. *IEEE Transactions on Cognitive and Developmental Systems*, PP(99), 1–1.
50. Sarathy, V., & Scheutz, M. (2016). Cognitive affordance representations in uncertain logic. In *Proceedings of the 15th International Conference on Principles of Knowledge Representation and Reasoning (KR)*.
51. Sarathy, V., & Scheutz, M. (2016). A logic-based computational framework for inferring cognitive affordances. *IEEE Transactions on Cognitive and Developmental Systems*, *8*(3).
52. Sarathy, V., Scheutz, M., Austerweil, J., Kenett, Y., Allaham, M., & Malle, B. (2017). Mental representations and computational modeling of context-specific human norm systems. In *Proceedings of the 39th Annual Meeting of the Cognitive Science Society*.
53. Sarathy, V., Wilson, J., Arnold, T., & Scheutz, M. (2016). Enabling basic normative HRI in a cognitive robotic architecture. In *Proceedings of the 2nd workshop on Cognitive Architectures for Social Human-Robot Interaction at the 11th ACM/IEEE Conference on Human-Robot Interaction*.
54. Schermerhorn, P., & Scheutz, M. (2008). Natural language interactions in distributed networks of smart devices. *International Journal of Semantic Computing*, *2*(4), 503–524.
55. Schermerhorn, P., & Scheutz, M. (2009, November). Dynamic robot autonomy: Investigating the effects of robot decision-making in a human-robot team task. In *Proceedings of the 2009 International Conference on Multimodal Interfaces*, Cambridge, MA.
56. Schermerhorn, P., & Scheutz, M. (2009, July). The utility of affect in the selection of actions and goals under real-world constraints. In *Proceedings of the 2009 International Conference on Artificial Intelligence*.

78. Williams, T. (2017). A consultant framework for natural language processing in integrated robot architectures. *IEEE Intelligent Informatics Bulletin (IIB)*, 10–14.
79. Williams, T., Acharya, S., Schreitter, S., & Scheutz, M. (2016). Situated open world reference resolution for human-robot dialogue. In *Proceedings of the 11th ACM/IEEE International Conference on Human-Robot Interaction (HRI)*.
80. Williams, T., Briggs, G., Oosterveld, B., & Scheutz, M. (2015). Going beyond command-based instructions: Extending robotic natural language interaction capabilities. In *Proceedings of the Twenty-Ninth AAAI Conference on Artificial Intelligence (AAAI)*.
81. Williams, T., Briggs, P., Pelz, N., & Scheutz, M. (2014). Is robot telepathy acceptable? Investigating effects of nonverbal robot-robot communication on human-robot interaction. In *Proceedings of 23rd IEEE Symposium on Robot and Human Interactive Communication (RO-MAN)*.
82. Williams, T., Briggs, P., & Scheutz, M. (2015). Covert robot-robot communication: Human perceptions and implications for human-robot interaction. *Journal of Human-Robot Interaction (JHRI)*.
83. Williams, T., Cantrell, R., Briggs, G., Schermerhorn, P., & Scheutz, M. (2013). Grounding natural language references to unvisited and hypothetical locations. In *Proceedings of the Twenty-Seventh AAAI Conference on Artificial Intelligence (AAAI)*.
84. Williams, T., Johnson, C., Scheutz, M., & Kuipers, B. (2017). A tale of two architectures: A dual-citizenship integration of natural language and the cognitive map. In *Proceedings of the 16th International Conference on Autonomous Agents and Multi-Agent Systems*.
85. Williams, T., Núñez, R. C., Briggs, G., Scheutz, M., Premaratne, K., & Murthi, M. N. (2014). A Dempster-Shafer theoretic approach to understanding indirect speech acts. In *Advances in Artificial Intelligence—Proceedings of the 14th Ibero-American Conference on AI (IBERAMIA)*.
86. Williams, T., & Scheutz, M. (2015). A domain-independent model of open-world reference resolution. In *Proceedings of the 37th Annual Meeting of the Cognitive Science Society (COGSCI)*.
87. Williams, T., & Scheutz, M. (2015). Power: A domain-independent algorithm for probabilistic, open-world entity resolution. In *Proceedings of the IEEE/RSJ International Conference on Intelligent Robots and Systems (IROS)*.
88. Williams, T., & Scheutz, M. (2016). A framework for resolving open-world referential expressions in distributed heterogeneous knowledge bases. In *Proceedings of the 30th AAAI Conference on Artificial Intelligence (AAAI)*.
89. Williams, T., & Scheutz, M. (2016). Resolution of referential ambiguity using Dempster–Shafer theoretic pragmatics. In *Proceedings of the AAAI Fall Symposium on AI for HRI (AI-HRI)*.
90. Williams, T., & Scheutz, M. (2017). Referring expression generation under uncertainty: Algorithm and evaluation framework. In *Proceedings of the 10th International Conference on Natural Language Generation (INLG)*.
91. Williams, T., & Scheutz, M. (2017). Referring expression generation under uncertainty in integrated robot architectures. In *Proceedings of the Robotics: Science and Systems Workshop on Human-Centered Robotics: Interaction, Physiological Integration and Autonomy*.
92. Williams, T., & Scheutz, M. (2017). Resolution of referential ambiguity in human-robot dialogue using Dempster–Shafer theoretic pragmatics. In *Proceedings of Robotics: Science and Systems (RSS)*.
93. Williams, T., & Scheutz, M. (2018). Reference resolution in robotics: A givenness hierarchy theoretic approach. In J. Gundel & B. Abbott (Eds.), *The Oxford Handbook of Reference*. Oxford: Oxford University Press.
94. Williams, T., Schreitter, S., Acharya, S., & Scheutz, M. (2015). Towards situated open-world reference resolution. In *Proceedings of the AAAI Fall Symposium on AI for HRI (AI-HRI)*.
95. Williams, T., Thames, D., Novakoff, J., & Scheutz, M. (2018). Thank you for sharing that interesting fact!: Effects of capability and context on indirect speech act use in task-based human-robot dialogue. In *Proceedings of the 13th ACM/IEEE International Conference on Human-Robot Interaction (HRI)*.
96. Wilson, J. R., Krause, E., Scheutz, M., & Rivers, M. (2016). Analogical generalization of actions from single exemplars in a robotic architecture. In *Proceedings of AAMAS 2016*.

97. Yu, C., Schermerhorn, P., & Scheutz, M. (2012). Adaptive eye gaze patterns in interactions with human and artificial agents. *ACM Transactions on Interactive Intelligent Systems*, *1*(2), 13.
98. Zech, P., Haller, S., Lakani, S. R., Ridge, B., Ugur, E., & Piater, J. (2017). Computational models of affordance in robotics: A taxonomy and systematic classification. *Adaptive Behavior*, *25*(5), 235–271.

Non-human Intention and Meaning-Making: An Ecological Theory

Michael A. R. Biggs

Abstract Social robots have the potential to problematize many attributes that have previously been considered, in philosophical discourse, to be unique to human beings. Thus, if one construes the explicit programming of robots as constituting specific objectives and the overall design and structure of AI as having aims, in the sense of embedded directives, one might conclude that social robots are motivated to fulfil these objectives, and therefore act intentionally towards fulfilling those goals. The purpose of this paper is to consider the impact of this description of social robotics on traditional notions of intention and meaning-making, and, in particular, to link meaning-making to a social ecology that is being impacted by the presence of social robots. To the extent that intelligent non-human agents are occupying our world alongside us, this paper suggests that there is no benefit in differentiating them from human agents because they are actively changing the context that we share with them, and therefore influencing our meaning-making like any other agent. This is not suggested as some kind of Turing Test, in which we can no longer differentiate between humans and robots, but rather to observe that the argument in which human agency is defined in terms of free will, motivation, and intention can equally be used as a description of the agency of social robots. Furthermore, all of this occurs within a shared context in which the actions of the human impinge upon the non-human, and vice versa, thereby problematising Anscombe's classic account of intention.

1 Introduction

One way to describe human beings is as meaning-making agents. What we do is to interact with the world, our interactions being mediated by the meanings that we make. It is no longer fashionable to think that those meanings are inherent in the world around us, as was implied in classical hermeneutics, but rather that we project onto the world, in its infinite complexity, our interests and motivations, which organise themselves as meanings that we commonly say we "find" in the world. Thus, one can

Michael A. R. Biggs (✉)
University of Hertfordshire, Hatfield, UK
e-mail: m.a.biggs@herts.ac.uk

M. I. Aldinhas Ferreira et al. (eds.), *Cognitive Architectures*, Intelligent Systems, Control and Automation: Science and Engineering 94,
https://doi.org/10.1007/978-3-319-97550-4_12

describe human beings as active agents, which is indeed the premise of actor network theory. When Latour wrote about scientists in the laboratory [3], he observed their activities as a kind of interested behaviour, in which they found meaning in scientific activities owing to their identity as scientists.

This contemporary view of human beings as active agents and meaning-makers assumes that the external world is largely passive in response to our activity. However, intelligent tools increasingly accompany the activity of human beings, and these tools often mediate our interactions with the world. Thus, when we use Google to search for information, in addition to our motivation to direct the search engine by an informed choice of keywords, Google itself is operating according to algorithms and protocols that guide searches in one direction or another. Of course, these robots are there to do our bidding, but as they become increasingly intelligent, our world becomes populated by artificial versions of ourselves. If we have motivations and interests that are determined by our overall aims and objectives, can we not say that these robots also have motivations, since they too have aims and objectives inherent in their programming? When we proliferate the existence of these robots more widely throughout society, in which forms of artificial intelligence [AI] mediate so many of our interactions, do not our naive assumptions that we alone are the active agents become no longer viable?

This is a profound change from Anscombe's [1] notions of intention, motivation and responsibility. It also tends to equalise the relationship between the human user and the robot so that the human must take account of the capacities and interests of the robot when engaging with it, and as such, our sense of meaning in the world must now include other active, inorganic agents with whom we must negotiate our own meaning-making activity. Conversely, if we can adopt a somewhat anthropomorphising thought experiment, could one not say that the robot was equally engaged in a meaning-making activity in which negotiating with humans was necessary?

Meaning-making has hitherto been discussed as an essentially human activity. It is an interpretative act that we employ as part of optimising our agency in the world. When we act, we generally act with an aim in mind, and so our actions must be designed to have a certain effect and to overcome potential barriers or resistance. Thus we need to understand the context in which our actions will take place and the factors that may impinge on them. This interpretation of the ecology of meaning-making includes understanding what's going on in the current situation, so as to be able to intervene effectively to achieve the new and desired situation. All of this is normally described in terms of intention to achieve something, based on an interpretation of what would be necessary and the exercise of will that brings about transformation. But intention alone is not enough to bring about effective change, because it must be accompanied by an effective understanding of what must be done to meet the intended outcome. In an AI environment, this is the difference between intelligent and non-intelligent robots. The non-intelligent robot may have an "intention", that is to say, "programming" to do something, but if the environmental factors are not as expected, for example, an object is not where it is supposed to be, the non-intelligent robot cannot achieve its intention, i.e., the programmed goal. In such a simple example, the meaning of the situation is that a different set of actions to that originally programmed is necessary to achieve the intention. In more complex social environments, meaning may consist in identifying other dynamic agents in the environment, or hypothesising their intentions, and hence predicting their behaviour. These new possibilities problematize our existing notions of intention and meaning-making because hitherto, they have been seen as essentially human traits that to some extent differentiate humans from machines. This paper considers whether such terms should continue to be reserved for humans or whether recent developments in AI should cause us to reassess our understanding of intention and meaning-making as something environmentally situated or ecological, rather than individually situated and subjective.

2 An Ecological Theory

The traditional approach in philosophy has been to differentiate humans from non-humans, including intelligent machines and robots. Although both humans and robots have agency and can interact with the material world and change it, we have aggregated to the concept of "human" some superior powers, such as free will, interpretation and intention. Under this approach, robots do not have the capacity to exercise these essentially human qualities, and therefore have no responsibility for what they do. Free will, interpretation and intention have each been the subject of extensive analysis in philosophy, and the notion of intention has been examined in detail by Anscombe [1]. In her classic paper, she identified three different kinds of intention: (1) intention to act, (2) intention in acting, and (3) intentional action. Intention is closely related to the ideas of choice and will (volition), in which the human agent brings something about and can be said to be responsible for it, both in terms of causality and moral responsibility. Although we recognise that robots can have agency and be causally responsible for change, as in the case in which a factory robot builds a car and is certainly the agent of change that brings about its construction, we do not normally speak in terms of the moral responsibility of the robot. The responsibility for the robot's actions, if the question were to be brought to a court of law, would probably be found to lie with the programmer, because the robot "mindlessly" carries out the instructions that have been given to it. But why should we make these distinctions? Although we might desire that human beings be differentiated from other animals and from inorganic actors, in the world of artificial intelligence [AI], can and should, such differentiations be sustained? Indeed, what would be the point?

When we intend to do something, an aspect of that intention is that a future plan may or may not be realised. This is discussed in Anscombe's first category of intention. On the one hand, it may not be realised because we change our mind and we do not act as we originally intended. On the other hand, we might act unsuccessfully and not bring about what we intended. In either case, the future prediction embodied in the intention did not happen, but we nonetheless say that there was a motivation so to act or to bring something about. One of the things that we expect about robots is that they will successfully bring about what they are programmed to do, over and over again. Furthermore, robots are not usually regarded as having the capability of changing their minds in relation to this behaviour, if we regard "what is in their mind", i.e., what we might informally call their "intention", as being embodied in their initial programming. Of course, AI allows for adaptation but this is probably not the adaptation of an overall aim, even if the adaptive system may have the capacity to "change its mind" about how to achieve that aim. Thus, the changing of intermediate objectives as an apparent expression of the "intention" to fulfil an overall course of action, turns out to be something that could be meaningfully referred to in relation to inorganic agents such as robots, as well as organic agents such as humans.

We sometimes have the intention of bringing about A, but inadvertently, we bring about B. Although B was brought about, we cannot say in good faith that we had the intention of bringing about B, although we are sometimes disingenuous, and in order to save face, we say, "I meant to do that". "Meaning to do something" is an utterance, not a speech act. That is, just saying "I meant to do that" does not make it so. When I intend to bring about A, I may say aloud in advance that I predict that this will happen, or I may make purposive actions that, under normal circumstances, would bring about A, or it may be assumed by myself and perhaps others that I am attempting to bring about A on the grounds of my past history or the perception of my interests by others. But the mere subsequent utterance of the statement "I meant to do B" does not mean that, after all, I really intended to do B rather than A. Such an utterance would be regarded as post-rationalisation in psychoanalysis, and face-saving in negotiations. As yet, we have not deemed it necessary to programme face-saving into robotic behaviour. Thus, adaptive behaviour, in humans and in robots, should not normally involve a change in the overall goal, only in the means of achieving it, i.e., of changing the intermediate goals when necessary.

So, the question remains, is it useful to say that robots have intention even though they do not say "I meant to do that", and furthermore, what would be the consequences of this change of attitude in the case of less evident robotic agents such as social robots, which operate more discreetly at the margins of our environmental awareness, if they were said to have intention? In other words, is intentionality something that I attribute to an agent when I see indicators of "acting to bring about", or should intention imply the possibility of failure that is normally missing from programmed robotic behaviours? Indeed, is intention so inherently human—because to err is human—that it is meaningless to speak of a robot's intentions when they are always satisfied? Conversely, is it essential that we keep open the possibility of an intentional act in order to attribute responsibility for action, and to whom and under what circumstances should that responsibility be attributed to the human programmer or the robotic actor? This is the problem discussed by Anscombe in her third category: intentional actions.

One of the "traditional" assumptions that would be problematized is the so-called Turing Test (originally framed as "can machines think?", or "exhibit intelligent behaviour", [4, pp. 433–459]), in which it is proposed that differentiating between robots and humans would be irrelevant if we could not differentiate one from the other through questioning. This procedure assumes that we have an explicit encounter with another being whom we suspect may be a robot in a situation in which we might normally expect to meet a human, or vice versa. The scenario posited in this paper is slightly different. The scenario is that we are frequently confronted with social robots that are intelligent agents seamlessly integrated into our social environment. If we posit a seamless interaction, then we cannot know whether this agent is human or not, and the issues of intention and responsibility are indeterminable. Such a question is only likely to arise when there are questions of responsibility regarding the actions of a robot or the consequences of its action, for example, by misleading a human into taking certain actions that it otherwise would not have taken. Of course, as has already been described, under such circumstances, we might hold to account the programmer who wrote the programme that caused the robot to make the decisions that it made, leading to the undesired consequences for which the question of responsibility is an issue. But what if the social robot is integrated to a much greater extent into our social interaction, such that it passes the Turing Test? And what if the robot has such a complex AI that we cannot reasonably hold the programmer accountable for the decisions that the robot has made based upon the fundamental principles embodied in its initial programming? Do we need a model for this kind of autonomously learnt behaviour?

In human society, we already have a model disclaimer for responsibility in that we do not hold minors responsible for their decisions and actions. Parents or guardians, that is to say, the societal programmers, are normally held responsible until the minor reaches a certain age. It is interesting to note that the test for legal responsibility is not a performance criterion, as is the case with the Turing Test, but merely an age criterion. If we applied such reasoning to social robots, we might conclude that when they had been acting autonomously in the social environment for a certain period of time, during which they evolved their AI to address most of the commonly encountered problems for which they were programmed, we might infer that they could be held accountable for their intentional actions. But this would bring us back to the earlier observation that we do not at present have the legal framework or practice of holding machines to be responsible for the actions they perform or their consequences.

However, the responsibility of the mindless factory robot is not the principal focus. Instead, this paper is interested in the extent to which the concept of an, in practice, transparent agent, by which I mean an agent that cannot be differentiated from the human—not because it passes the Turing Test, but because the context in which we engage with it does not invite that kind of differentiation—has impact on the way in which we have previously described intentionality. Anscombe's category 1 apparently remains unchanged, because there is no need to infer that artificial agents have a predictive capability. However, the ability of such agents to make change, and by their adaptive behaviour be said to assume "responsibility" for actions that were not or could not have been anticipated in the original programming, does seem to imply that we can meaningfully speak of such artificial agents as having intention. This has hitherto been assumed to be a uniquely human attribute.

Of course, historically, many of the uniquely human attributes that anthropologists have identified, such as the ability to use tools, and that sociologists have identified, such as the ability to use language, etc., have been proposed in order to meet the desire to differentiate human from animal. One might regard the Turing Test as the last vestige of this historical attempt to desperately maintain such a teleological differentiation. But if we abandon our attempt to be different, in addition to the practical issue of whether we can still make such a differentiation or, indeed, whether it is necessary or productive, what would be the consequences of believing that artificial agents can have intention?

When agents act intentionally, in the sense of Anscombe's category 1, we attribute some kind of motivation or plan to them. We say they want to bring about outcome X. In this scenario, I am not merely thinking of machines with direct programming, in which we can say they mindlessly act to bring about outcome X and so they themselves do not meaningfully have an intention. In the present scenario, I am assuming AI of sufficient order, coupled with social embeddedness that renders the agents invisible, that we are unaware of the human/machine distinction and are only aware of the agency, the purposiveness, and the responsibility. Having granted a category of inorganic intentionality, which is perhaps additional to Anscombe's original three, what does this tell us about the "inner life" of these inorganic agents, and about this extended concept of intention? Do they feel satisfaction when their intentions are fulfilled? Do they feel frustration when they are not? Do they have an overall perception of the environment in which they are operating, within which they frame their decisions according to their programming and subsequent experience?

To all of these questions, it would be most interesting to answer yes. Yes, they do have responsibility for their actions; yes, they do feel frustration when their intentions are not satisfied; and yes, they do have an overall perception of the environment in which they are operating.

This is not merely a science fiction discussion in which we ask whether androids dream of electric sheep: it is a philosophical discussion about the consequences of integrating social robots that act intelligently into the human environment, and to ask how to attribute intentionality and responsibility when we interact with them. There is a reason why we should be interested in this problem. In contemporary philosophy, the focus has shifted from ontological and teleological issues, in which the question or the questioner is to some extent independent of the social environment, to questions that recognise social ecology and relational judgements, and worldviews that require meaning-making. Relational judgements imply that if we are sharing our world with other agents, whether they are organic animal agents or inorganic robotic agents, the network of relationships will present certain possibilities. Therefore, it is relevant to know who and what is in our environment, and the way in which the other, owing to being dynamic, is causing change to our environment and therefore the decisions that we make. At a macro level, this means that our worldview is impacted by our perception of what is material, of what can change, and who are the agents of change independent from us. Furthermore, we have to take account of the apparent worldview of those change-making agents in order to predict their behaviour that may impinge upon us and our ability to successfully implement our own intentions.

So, we have perhaps arrived at the possibility that artificial agents, owing to their capacity to act dynamically and to impinge upon our worldview, can clearly be said themselves to have intention of category 1. This argument also suggests the possibility that these intelligent agents are making meaning for themselves so as to fit their actions to the environment. In the past, meaning-making would also have been an ability reserved for the human. But if we abandon our differentiated status, we can now see that these intentional acts, based upon the experience gained by the social robot embedded in the same environment as ourselves, are inevitably based upon the same decision-making structures as ourselves. Indeed, causally, the decision-making actions and strategies of the inorganic agent were brought about through its programming by a human agent. The inorganic agent, in this case, the seamlessly embedded social robot, however modest it may be in its capacities, is brought up as a minor, with strict instructions in its programming that determine its behaviour. During its formative years of operation, it develops, through the use of its intelligence, experiences and additional frameworks that enable it to make decisions that were not framed or anticipated in its original programming. This is what we want the intelligent robot to do when we design it—so that it is unnecessary for us to anticipate every possible scenario in which it must take action and to determine the action it must take. An effective social robot must be judged responsible for the decisions that it makes because it has made them based on frameworks of judgment that were not placed there by the programmer. The programmer is innocent, or at best, merely an accomplice! To make such judgments, the inorganic agent must "understand" its environment and have a worldview. Of course, such a worldview need only stretch as far as the scope of agency envisaged for that robot. However, having postulated the possibility that one can describe the agent in this way, one has to conclude that meaning-making is being undertaken when the robot evaluates a scenario and identifies within it the possibility for action. Such a possibility for action is implied in the concept of intention, because we cannot meaningfully speak of intention in the circumstance in which the intended outcome is unlikely to come about or when such an outcome would be impossible. If I intend B in the circumstance in which B could not possibly happen, then my intention will be described as folly. Misguided intention, i.e., folly, is noticeably absent from Anscombe's three categories.

As a result of the foregoing argument, we have the possibility that now, or in the near future, we will share our environment with social robots, albeit with modest remits, and that these agents will not differentiate themselves from other active agents in that environment. When we, as human agents, interact with this mixed ecology, we will form a view of the active and inactive elements within it in order to frame our intentions and our actions. It is important for us, if we are not to be frustrated by the lack of fulfilment of our intentions, that we perceive the ecology in which we operate as dynamic. Relational argumentation is one contemporary outcome of the recognition of this need. It can be contrasted with the absolute argumentation of Newtonian mechanics, in which the external world behaves passively, to the extent that it is not actively making autonomous decisions. This is no longer the case. All sorts of agents, some of which are inorganic, populate our world, and some of those are making AI-led judgments about what to do in response to us at the same time as we are making intelligent judgments in response to them. As a result, our worldview, that is, the view of the range of possibilities that the world presents to us to either facilitate or frustrate our intentions, is modified by the presence of these robots.

In his test, Turing argued against those who needed to, or saw the possibility of, differentiating robots from humans; but that possibility no longer exists, not only as a consequence of the increased adaptive intelligence of the robots, but also because of their embeddedness in our social world. Our ecology now includes new autonomous agents of change.

So, what are the consequences of this philosophical description of robotic intention and meaning-making? We have seen that one can usefully refer to both intentionality and meaning-making in the case of robots, and that there may therefore be no need to differentiate human agents from embedded social robots because they share the same societal and legal responsibility for their actions. The Turing Test becomes irrelevant in such cases, because there would be no benefit from making the distinction. The traditional distinction allows humans the exclusive right to free will and responsibility, but it seems that such a distinction is no longer beneficial or sustainable. This paper has suggested that it is an inevitable consequence of the increasing adaptive complexity of social robots and their embedded-ness in the environment, in which they become part of our social ecology, that we will have to begin to deploy concepts that have previously been reserved for humans. The concept of intention is one such concept. It is meaningful to speak of the intention, whether fulfilled or not, of the robot. The robot's actions may have intended or unintended consequences. The robot, if it is to successfully negotiate dynamic obstacles to fulfilling those intentions, must anticipate—that is to say, predict—what will happen if it takes certain courses of action. For these operations to be successful, the robot must have a world-view and must make decisions in accordance with it. Meaning-making is perhaps the most advanced of the concepts that has been speculated upon here. To what extent is meaning-making really a part of the robot's behaviour?

Comparing once again to the human model, the idea of meaning-making is deployed in order to account for the way in which humans construe the world so as to anticipate how it will operate and how they can operate within it. Meaning-making embodies the idea that there are dynamic agents in the world pursuing their own objectives, and that we appear to these agents as they do to us. Their behaviour, when different from our own, can be explained by them having a different view of the world and their place in it, as well as having different motives, and therefore different objectives. These objectives are pursued through intentional action, whether by human or nonhuman agents. Their meaning-making is not a quest for the meaning of life, but rather the meaning of the presence of these other agents who are not working harmoniously with their own interests. Meaning-making results in an explanatory framework that accounts for diverse interests. Now that we have created social robots to work with us, they also work amongst us and, owing to their different function, have different, albeit normally harmonious, intentions than ourselves. Thus, meaning-making does not emerge as an exclusively human attribute, because it is linked much more to the ecological interpretation of the context in which the agent is embedded than to manifestations of some inherent subjective capability of the agent itself. Intention and meaning-making are environmental by-products of agency. This conclusion has consequences in a number of areas and for the interpretation of the previous literature.

One consequence is that Anscombe's three categories should be explored in relation to non-human as well as human agents. Intention as a concept applies in any situation in which purposive action is taken to meet an objective. This can be said to occur when there is any kind of programmed objective. One motivation for Anscombe's discussion seems to be an implied interest in responsibility, as is evidenced in her example of the man who poisons the occupants of a house [1, Sect. 23]. But equally, it can be posed in relation to a robot that brings about outcomes owing to adaptive behaviour for which we cannot hold the programmer fully responsible.

With regard to meaning-making, the traditional concept becomes more stretched, but it is unnecessary to hypothesise a ghost in the machine in order to find the concept of inorganic meaning-making a useful one. If we intend, when we design a social robot, that it should seamlessly integrate into the social context so that it can effectively serve its human masters, then we must equip it to be adaptive owing to the dynamic nature of the human context and the other human agents amongst whom it must operate. Its efficacy will be enhanced if it is able not only accommodate such situations as it finds, but also anticipate possible scenarios. This requires an adaptive map of possibilities that constitutes, this paper claims, a worldview. Meaning-making does not require a metaphysical conscience that gives meaning, in the sense of ultimate purpose, to the world. All that is required is the ability to project forward and anticipate so as to improve decision-making. Meaning, understood in this way as a practical activity, is making inferences from indicators. Thus, when X means Y, we can substitute, when X indicates Y or, as a result of X, we infer that Y will come about. Put in this way, it is reminiscent of Hume's notion of cause and effect as merely the "constant conjunction" of X with Y [2, 1748]. Hume's view changes the locus of causality from something that is extrinsic to perception, to something that is intrinsic, and makes it a psychological theory. In other words, when we think that X causes Y, we think that something is happening in the external world. When we think that X is constantly conjoined to Y, we think that something intrinsic to ourselves is happening: this is an idea rather than a fact, something that is going on inside us rather than something that is going on in the external world.

So it is with inorganic meaning-making. The concept of inorganic meaning-making is a consequence of reframing a concept that was once intrinsic so that it becomes extrinsic. When we describe meaning-making by the inorganic agent, we are not giving the agent human attributes, we are simply applying the extrinsic argument. The inorganic agent can be said to make meaning when it makes the ecological connection between X and Y and adapts its behaviour accordingly. At one level, this is simply predictive. At another level, it confirms that the agent has a model of what will happen and is acting according to that model. Such a model consists of both objects and events, and possibilities for which there are indicators. It is the presence of adaptive behaviour in the presence of indicators that underlies the argument that meaning-making is present. The perception that X means Y in situation Z corresponds to the utterance "X means Y", and successful social robots will be deploying this concept just as frequently as do humans. Thus the context dependency of all operatives is at the root of meaning-making by any agent.

References

1. Anscombe, G. E. M. (1965). *Intention*. London: Harvard University Press.
2. Hume, D. (2011 [1748]). *An enquiry concerning human understanding*. Urbana, Illinois: Project Gutenberg. Retrieved 29 July 2018, from www.gutenberg.org/ebooks/9662.
3. Latour, B., & Woolgar, S. (1986). *Laboratory life: The construction of scientific facts*. Princeton, NJ: Princeton University Press.
4. Turing, A. M. (1950). Computing machinery and intelligence. *Mind*, *LIX* (236), 433–460.

Implementing Social Smart Environments with a Large Number of Believable Inhabitants in the Context of Globalization

Alexander Osherenko

Abstract This chapter discusses Social Smart Environments (SSEs) with a large number of believable Embodied Conversational Agents (ECAs) in the context of globalization. It focuses on SSE architecture, rapid prototyping and scalability with respect to size, geography, and administration. SSE is a software environment installed in a physical place representing, for example, a city inhabited by believable ECAs that interact comprehensibly with each other; believable ECAs are software agents that stand for humans from different cultures. To ensure believability, the ECAs maintain various determinants of processing, for instance, emotional, personal and cultural, identified through an analysis of 35 scenarios of intercultural interaction. This chapter shows implementation of these determinants and development of an SSE prototype on the basis of a specification defining interaction between ECAs. In conclusion, this contribution provides insight into future work addressing, for example, innovation in societies simulated by SSEs.

Keywords Rapid prototyping of social smart environments with a large number of believable embodied conversational agents · Determinants of believable embodied conversational agents in the context of globalization · Scalability of social smart environments with respect to size, geography and administration

1 Introduction

Technological innovation changes human lives, including the spaces in which humans live. The modern world is now not only populated with humans who perform everyday tasks, but also with technical artifacts that carry out routine and intelligent jobs

---

A. Osherenko (✉)
Humboldt Innovation, Humboldt-Universität zu Berlin, Ziegelstraße 30,
10117 Berlin, Germany
e-mail: alexander.osherenko@alumni.hu-berlin.de

M. I. Aldinhas Ferreira et al. (eds.), *Cognitive Architectures*, Intelligent Systems, Control and Automation: Science and Engineering 94,
https://doi.org/10.1007/978-3-319-97550-4_13

[13, 19, 30, 39]. Moreover, these technical artifacts are no longer seen as servants, but rather as clever companions. They can master actions that a human cannot and they are very efficient in said actions. Such technical artifacts are smart, meaning they behave and react comprehensibly and in a human manner.

Technical artifacts are supposed to make human life more comfortable [17, 18]. However, they can also cause problems, for instance, interaction problems. Erroneous communication with technical artifacts can backfire and obliterate the advantages of smart interaction.

A solution for problems that occur can provide a Social Smart Environment (SSE) that maintains Embodied Conversational Agents (ECAs). An SSE is a software environment installed in a physical place, for example, a city, a building, or a large room populated by intelligent technical artifacts such as smart robots. In the context of globalization, SSEs maintain many interacting ECAs that comprehensibly represent humans from different countries.

An SSE has many applications, in both normal and extreme situations for which humans need guidance. For instance, an SSE installed in a public place such as an airport, a metro station or a shopping mall can save human lives or help to avoid panic in cases such as earthquakes [29]. An SSE can help to design safer theaters or stadiums [20]. An SSE can assist international persons in resolving cultural misunderstandings between hosts and sojourners [33]. An SSE can intelligently guide museum's visitors [38]. An SSE can increase tourist flow by integrating tourism, cultural heritage and mobility [2]. Figure 1 shows groups of international tourists (international networkers, shoppers, students, etc.) and reasons for their (virtual) trips.

The high numbers in Fig. 1 demonstrate the necessity of handling situations by, for example, using SSEs. Addressing the problem, this chapter describes an approach for the rapid prototyping of SSEs that can maintain a large number of believable ECAs. Moreover, this contribution answers the following questions concerning SSE scalability [44, p. 10], such as:

1. Size—can proposed SSEs scale up well to meet increased numbers of ECAs?
2. Geography—what is the geographical distance between the ECAs maintained by the proposed SSEs?
3. Administration—can administration of the proposed SSEs take place remotely?

Additionally, the current chapter addresses the following issues of SSE development:

1. Steps necessary for the rapid prototyping of SSEs with a large number of believable ECAs;
2. Architecture and implementation of SSEs with a large number of believable ECAs;
3. Determinants and implementation of believable ECAs.

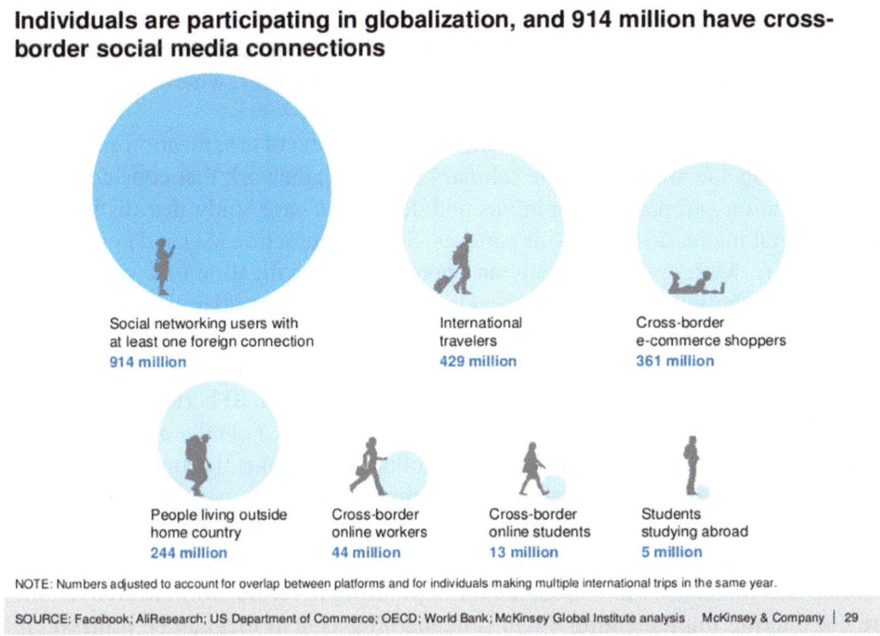

Fig. 1 Groups of international guests and reasons for their (virtual) trips

2 Recent Work

Different scholars have studied development of SSEs and discussed particular issues regarding their implementation.

Nakashima and colleagues [30] present a comprehensive study of Smart Environments (SEs). The authors describe intelligent agents, their implementation and interaction on the basis of Multi-Agent Systems.

Butz [9] examines SEs by giving special attention to the interaction between SEs and humans through displays that maintain personal information. The scholar claims that interaction takes part using human senses and through physical actions.

Bosse and colleagues [6] examine human aspects in SEs, abstaining from a pure examination of sensor data, but taking into account the human-directed sciences such as psychology and sociology. They focus on the human knowledge in ambient intelligence and describe an SE assessing the behavior of a human.

Cai and Kaufer [10] study SEs and state that simple communication among humans requires explicit computer-aided means. They define an image-word two-way mapping process that describes a mapping between image features and words for human facial features and introduce the computational implementation of human descriptions in the form of the visual and verbal interaction between them.

Aehnelt and colleagues [1] describe a situation-aware interaction in an SE. In their approach, user intentions deduced from sensory inputs are used to provide situation-aware informational assistance. For their purposes, the SE in their approach is a smart meeting room that proactively anticipates future goals. As a scenario, they study smart business applications in manufacturing industries, where they see a vital demand for facilitating decision-making at all company levels.

Fu and Zhang [16] explore virtual worlds in the context of urbanization and address corresponding social aspects. The scholars show a framework that considers social communication and personal opinions and describe a case study that distinguishes interpersonal interaction, behavior patterns, Social Interaction (SI) and communication contexts. Moreover, they study an approach to visualization of a virtual world that presents the info-structure of the virtual city under consideration of particular emotional and cultural aspects [21].

Spadavecchia and Giovannella [43] present a project that includes online monitoring and evaluation of learning processes accompanied by SI. SI between interactants proceeds through the exchange of Natural-Language (NL) emails or chatting. The project distinguishes 8 macrophases that collect data about the underlying social network and the social relationships among the learners. To assess the quality of an exchange, the emails or chat posts are scrutinized automatically according to their emotional content. Since the results revealed were encouraging, the authors plan to develop additional tools and methods for monitoring in future work. Moreover, they are considering implementing a real-time learning system that can be utilized on a daily basis.

Trovato and colleagues [45] discuss implementation of a culturally-dependent social robot that communicates with humans by showing particular emotions and altering facial expressions correspondingly. The approach acknowledges differences of emotional expression between the Japanese culture and Western culture in general, as well as the difficulty of substantiating corresponding differences. The approach uses six statistical classifiers, one for each part of the face, that regulate the expression of emotions and calculate a vector of motor angles.

3 Modeling and Implementation of Determinants

ECAs in an SSE interact with each other in a human manner. To identify determinants (significant issues) of believable SI between intercultural ECAs, we analyzed 35 scenarios of interaction among humans in the context of globalization and detected 10 agent-specific and 8 environment-wide determinants. We call some aspects 'agent-specific', emphasizing that the particular determinant is only valid in an ECA; we say 'environment-wide' so as to indicate that the particular determinant is valid for the whole SSE. To implement particular determinants, our approach uses the JADE platform, a development framework for multi-agent systems [3] .

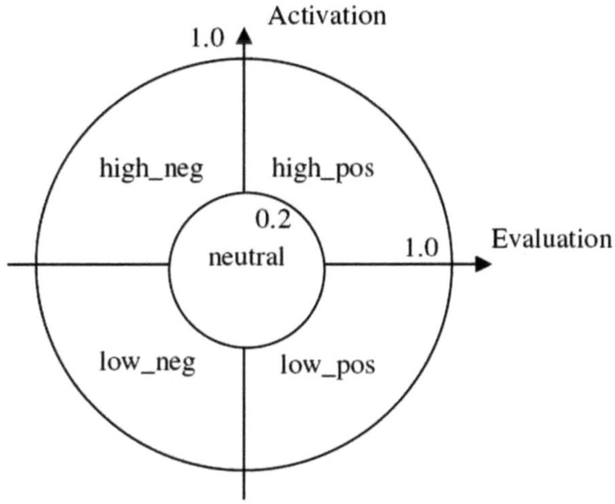

Fig. 2 Affect segmentation in the E/A space

3.1 Agent-Specific Determinants

3.1.1 Emotions

Emotions should be considered in a believable SSE, as the many scenarios of intercultural communication show [5, 7, 24, 36, 41, 45–47, 50]. Emotion-related data is acquired in our approach from the audio-visual Sensitive Artificial Listener (SAL) corpus [14]. The SAL corpus is a set of affective NL dialogues in which a wizard representing four psychologically different characters (optimistic and outgoing Poppy, confrontational and argumentative Spike, pragmatic and practical Prudence, depressing and gloomy Obadiah) tries to draw users into their own emotional state. The corpus consists of 27 NL dialogues.

SAL was transcribed and annotated by four labelers with FEELTRACE data [40], which identifies the emotions occurring in the the Evaluation/Activation (E/A) space [35]. Affect annotation of a turn in FEELTRACE contains numeric E/A data that is supplied continuously. For simplicity, the FEELTRACE annotations of turns are mapped onto 5 emotion segments in the E/A space (Fig. 2).

Figure 2 shows 5 emotional segments—high activation/negative evaluation (*high_neg*), high activation/positive evaluation (*high_pos*), low activation/negative evaluation (*low_neg*), low activation/positive evaluation (*low_pos*), and neutral—that represent affect segments of turns with different emotional loads. The value 0.2 is chosen empirically. The chosen affect segment of a turn corresponds to the vote of the majority of the annotators at the turn of a end; emotionally contradictory long turns are not considered in further experiments. Thus, 98 out of 672 turns are discarded due to the missing agreement between annotators or contradictory FEELTRACE data. The inter-annotator agreement is thus 85.42%. We adopt a model of emotions for affective behavior [34] that relies on a probabilistic Hidden Markov Model (HMM)

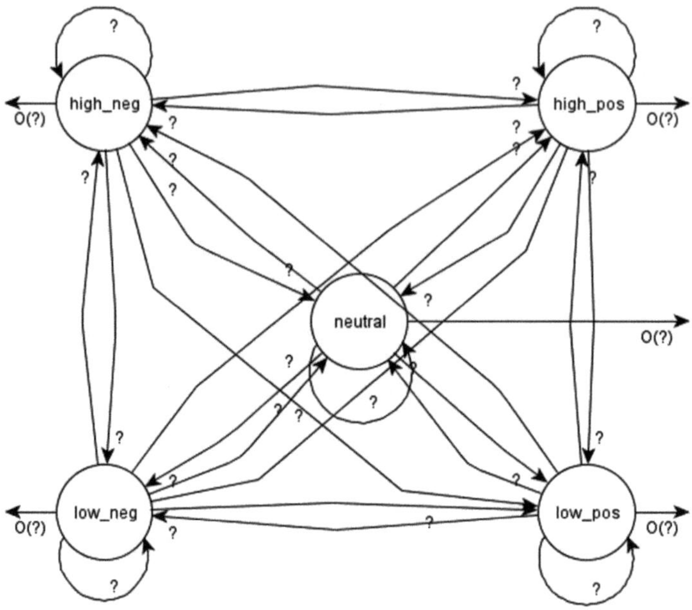

Fig. 3 A generic HMM for affective behavior

and transfers it in a generic form containing more emotion states that, in our opinion, can be used in more realistic scenarios of processing (Fig. 3).

Figure 3 shows a generic HMM for affective behavior with 5 emotion states, and question marks representing uninitialized transition and observation probabilities. To implement HMMs for affective behavior, JAHMM [15] a Java implementation of HMMs is used. To train the HMMs for affective behavior and assess initial probabilities of emotion states and transition probabilities, different algorithms can be utilized, for example, the k-means algorithm [23]. Initialization of the transition and observation probabilities is based on training sequences that can be composed, for instance, from adjacent dialogue turns with Spike, such as *low_pos neutral low_neg neutral low_pos neutral*, which results from the first, second, ..., sixth dialogue turns (Fig. 4).

Figure 4 represents an HMM for affective behavior for the Spike character with initialized transition and observation probabilities.

3.1.2 Personality

To anticipate the general disposition of an inhabitant in an SSE, a personality dimension is necessary [27]. A personality model in our approach relies on the Big-Five model that defines 5 personality traits, *Extroversion*, *Neuroticism*, *Openness to experience*, *Agreeableness*, and *Conscientiousness*, and can be assessed using the NEO questionnaire [12].

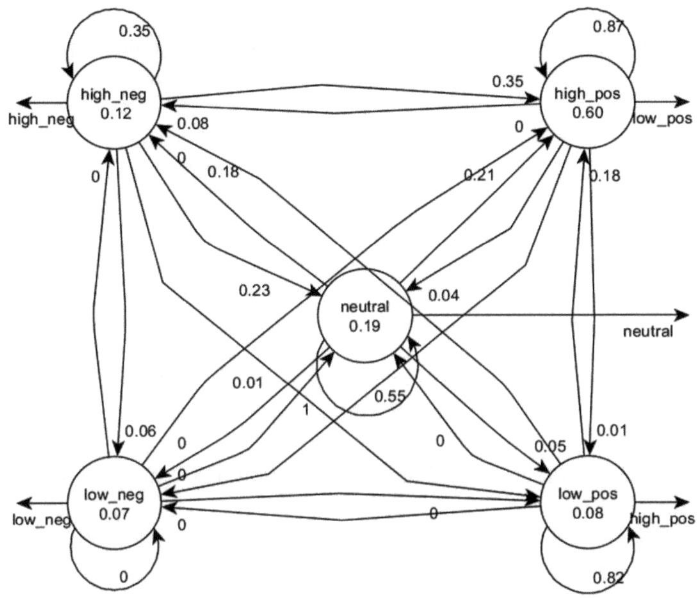

Fig. 4 A HMM for affective behavior for the Spike character

Table 1 Values of personality traits for the Spike character

Character	PT_E	PT_N	PT_A	PT_O	PT_C
Spike (con-front.) (%)	25.0	12.46	5.23	78.85	9.13

To populate the personality model of ECAs numerically, we use our own heuristics [33, pp. 102–104], which use the transition probabilities from Fig. 4. For example, we calculate personality trait extroversion PT_E as follows:

$$PT_E = \frac{\sum P(X \rightarrow high\_pos) + \sum P(X \rightarrow low\_pos)}{|\{X \rightarrow high\_pos\}| + |\{X \rightarrow low\_pos\}|}, \quad (1)$$

where $\sum P(X \rightarrow Y)$ is the sum of transition probabilities from the affect state X into the positive affect states $Y = \{high\_pos, low\_pos\}$. Values of PT_E are normalized by the number of corresponding transitions—$10 = |\{X \rightarrow high\_pos\}| + |\{X \rightarrow low\_pos\}|$.

Table 1 presents the values of the personality traits calculated using the threshold value 20%, where PT_E represents the value of the *Extroversion* trait, PT_N the value of the *Neuroticism* trait, PT_A the value of the *Agreeableness* trait, PT_O the value of the *Openness to experience* trait, and PT_C the value of the *Conscientiousness* trait.

Table 2 Acquired cultural values from the Irish corpus

Country	Power distance	Uncertainty avoidance	Individualism/ Collectivism	Masculinity/ Femininity	Long-/Short-term orientation
Ireland	49	47–48	12	7–8	13

3.1.3 Culture

Since particular aspects of an SSE are defined by the inhabitant's culture, the culture model is indispensable [42, 45]. As a culture model, we use synthetic cultures [22]. A synthetic culture is an artificial structure that distinguishes 5 dimensions:

1. The low versus high power distance dimension describes the degree to which differences in power, status, and privileges are considered by representatives of the culture;
2. The collectivism versus individualism dimension distinguishes the primary unit of the culture (I vs. we);
3. The masculinity versus femininity dimension defines the orientation of the culture towards achievement and cooperation;
4. The uncertainty avoidance dimension defines the measure of tolerance to ambiguity;
5. The short-term versuks long-term orientation dimension indicates the extent to which the future has more importance than the past or present.

To populate the culture model of ECAs, we use empirical data in [21]. Hence, we extracted cultural values for the Irish SAL corpus in Table 2.

3.1.4 Statistical Engines

Statistical processing is indispensable in SSEs [45]. In our approach, we use the WEKA statistical toolkit for data processing [48].

3.1.5 Natural-Language Processing

Many SSE approaches maintain believable ECAs that perform SI using Natural-Language (NL) utterances [43]. In our approach, we use NL approaches in [32] to analyze NL communication.

3.1.6 Social Relationships

Comprehensible interaction in an SSE considers social relationships between ECAs [38, 47]. In our approach, ECAs hold a list of relationships defined by particular IP addresses of JADE neighbor agents.

3.1.7 Context (Agent-Specific)

The agent-specific context defines the race, age, education, marital status, social class, religion, etc., and should be considered in an SSE [25]. The agent-specific context can be specified in an ECA as a dictionary of values, for example, {*"age": 35, "education": "higher"*}.

3.1.8 Knowledge (Agent-Specific)

The agent-specific knowledge refers to the facts held by a particular ECA. The agent-specific knowledge can be specified in an ECA as a dictionary of values, for instance, *"name":joinery*.

3.1.9 Time (Agent-Specific)

Some scenarios, for example, in an SSE that considers jet lag, take the temporal component into account. The agent-specific time can be specified in an ECA as the local time, realized using the system clock from the local computer. Alternatively, the ECA can install the JADE *onTick* behavior to measure time locally (more on the JADE platform in [3]).

3.2 Environment-Wide Determinants

3.2.1 Explicit Specifications

In our approach, SI in SSEs is specified by Interaction Specifications (ISs). ISs are structured texts that define participating ECAs and their behaviors. ISs resemble sequence diagrams known from the Unified Modeling Language (UML) implemented using a sequence diagram package [28] (Fig. 5).

The IS in Fig. 5 defines two agents that interact with each other, agent *Initiator* and agent *Responder*. Using this IS, our approach composes two JADE behaviors, the names of which contain the name of the initiator, the name of the responder, the name of the transaction, and the number of the initiator-responder

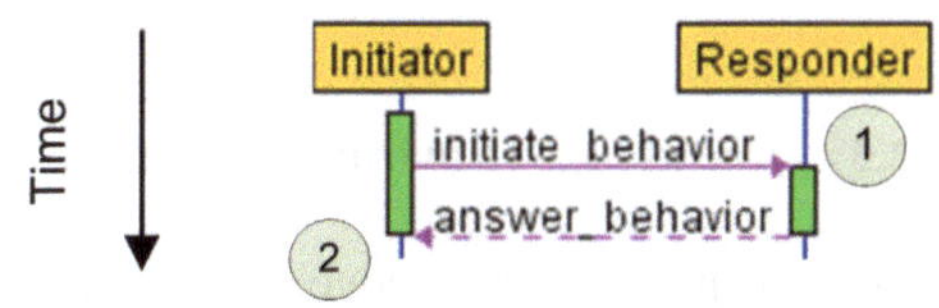

Fig. 5 Interpreting an interaction specification

combination: (1) *Initiator_Responder_initiate_behavior_0* and (2) *Responder_Initiator_answer_behavior_0* (more on the JADE behaviors in [3]).

The textual form of the IS represents Algorithm 1.

Algorithm 1 defines IS describing an SSE with 3 interacting ECAs (*agent0, agent1, agent2*) that receive the ping message and respond with the pong message.

Algorithm 1 IS defining SI in a population.

```
SSE
{
agent0.ping -> pong;
agent1.ping -> pong;
agent2.ping -> pong;
}
```

3.2.2 History

Some scenarios in an SSE consider the history of SI. In our approach, an environment-wide JADE agent holds a list of previous states of the environment that can be accessed by particular ECAs.

3.2.3 Space

Some scenarios in SSE consider a physical space that can be realized as the RoboCup soccer field [4] (Fig. 6).

Figure 6 shows a virtual space where numbered players (circles) move in directions specified by the arrows. The letters *G* and *J* respectively correspond to the German or Japanese culture of ECAs.

3.2.4 Context (Environment-Wide)

The environment-wide context defines circumstances in which SI in an SSE takes place. For example, the context can be defined by common real-world facts, such as *People find ghosts scary* [26]. To specify the environment-wide context, an SSE can hold particular rules such as *tango (culture : Argentina; value : high)* to define the high cultural value of tango in Argentina.

3.2.5 Knowledge (Environment-Wide)

Environment-wide knowledge in an SSE defines the intentions of an inhabitant, for example, knowledge about the behavior or the attended action strategy. To maintain

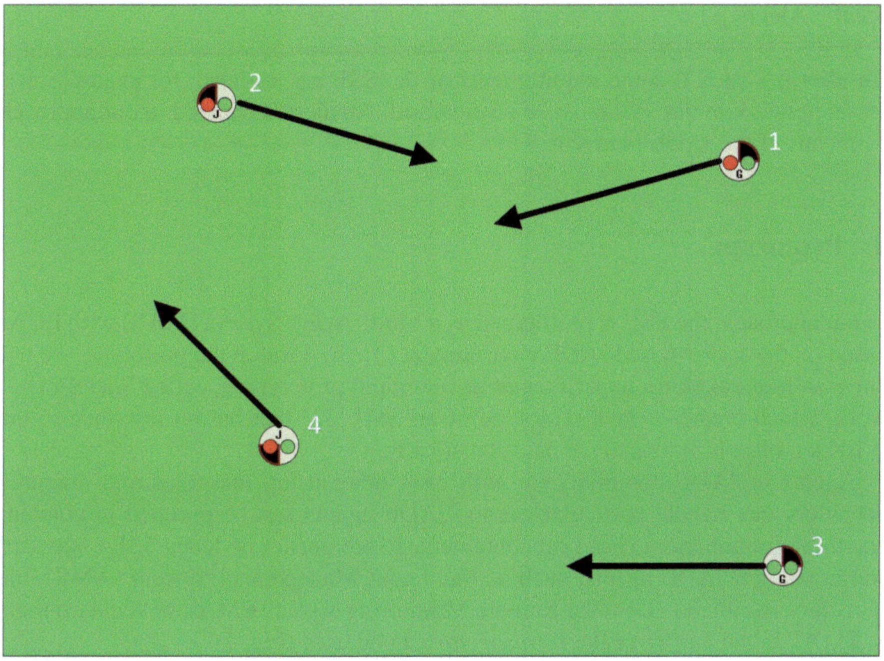

Fig. 6 An approach to spatial model based on the RoboCup representation

environment-wide knowledge, an environment-wide JADE agent holds a list of facts that can be accessed by particular ECAs.

3.2.6 Time (Environment-Wide)

Some scenarios of intercultural SI, for example, the jet lag scenario, consider the temporal component. To maintain environment-wide time, the SSE in our approach installs an environment-wide timeserver agent that maintains the global time within the environment.

3.2.7 Social Network, Topological Issues

In our approach, an SSE maintains interconnected ECAs [41, 50] according to a specific topology. Social network is realized in an environment-wide JADE agent that maintains an implementation of the social network as a list of neighboring ECAs for each ECA in the SSE (more on the JADE platform in [3]).

3.2.8 Alerts

An alert is issued if some requirements of the SSE are violated, for example, if a social network in the SSE must be reorganized. An alert is realized in our approach by means of the JADE platform.

4 Prototype

In our approach, the SSE is prototyped as a Multi-Agent System (MAS) with ECAs based on the server-based JADE environment [3]. To develop the prototype, we use our own framework for experimentation and rapid prototyping, called SocioFramework, which creates a Java prototype of an SSE [33] that has an interface to the WEKA toolkit [48] to analyze data statistically.

Agents in JADE communicate with each other using messages and maintain behaviors that handle particular events. JADE agents can be grouped in containers; these containers can have subcontainers. To administer or debug SSEs, standard administration JADE agents, such as the Agent Management System (AMS), the Directory Facilitator (DF), the Remote Monitoring Agent (RMA), or standard tools of JADE, as the *Sniffer* or the *Introspector* can be used (Fig. 7).

Figure 7 shows an SSE with two ECAs (*ECA1* and *ECA2*) that communicate with each other using a remotely installed JADE server.

To evaluate our approach, we composed SSE prototypes with specific interaction behavior. For instance, we prototyped an SSE realizing the interaction scenario with a higher status individual [37], for example, to simulate a meeting in an international

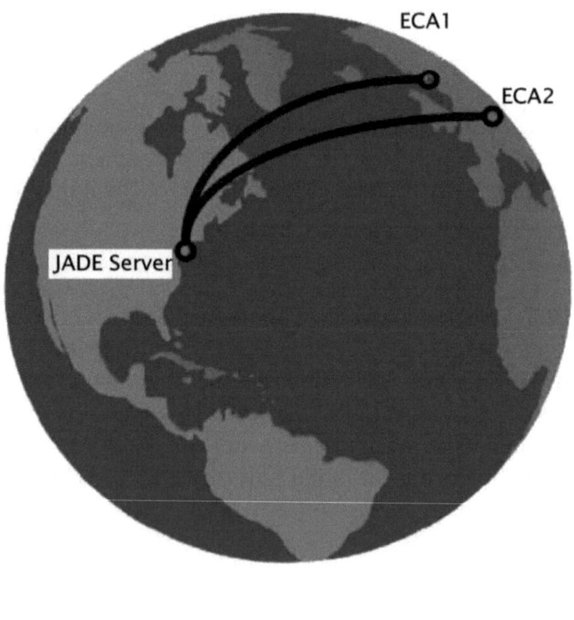

Fig. 7 Integration of JADE in building SSEs

corporation. The scenario considers specific emotional and personal properties of superior *A* and employee *B* that influence such categories of behavior as proxemics, vocalic, and symbolically intrusive (Algorithm 2).

Algorithm 2 Interacting with a higher status individual.

```
for all person in {A, B} do
    culture ← get_culture(person)
    for all category in {proxemics, vocalic, symbolically_intrusive} do
        behavior ← get_behavior(person, culture, category)
    end for
end for
```

5 Governing Scalability

One significant task of distributed systems such as SSEs stems from issues of scalability. According to Tanenbaum and Steen [44, p. 10], scalability can be measured along at least three dimensions with respect to:

1. Size;
2. Geography;
3. Administration.

Size

In our approach, SSEs maintain many ECAs. To scale up an SSE according to a higher number of ECAs, additional agents with a corresponding service interface must be logged into the JADE server. Hence, the *size* scalability of the proposed approach relies on the JADE *size* scalability that depends on the DF storing agents' access catalogue. Consequently, the DF can cause scalability problems through increased memory consumption [3, pp. 176–179].

Nevertheless, to give an idea of what the empiric size of agents in existing systems is, this chapter describes the number of agents in our and other JADE systems. It must be said that other approaches do not mainly focus on the number of agents in the system, which is a significant measure in this chapter, but rather on measuring the speed of communication.

In our approach, it was possible to build an SSE with 10,000 ECAs that exchange 20,000 interaction messages. It was possible to run the SSE with 1,000 JADE agents maintaining 2,000 interaction messages. Burbeck et al. [8] studied a JADE system with 150 pairs of agents. Cortese [11] describes a JADE system with 1,000 agents.

Consequently, we assume that SSEs are highly scalable according to the number of agents, since this measure does not appear to be critical in JADE systems.

Geography

Believable ECAs of SSEs can reside at significant geographical distances from each other. Hence, we used JADE means to resolve this issue:

```
java -cp jade.jar jade.Boot -host <IP address>
    -agents <agent name><agent class> -container
```

The command starts an instance <agent name> of agent <agent class>, where the text <IP address> specifies the IP address numerically (for example, 93.135.248.211) or as a name (for instance, localhost). Option -container specifies creation of a subordinate container within the main container.

Administration
In our approach, SSEs can span significant territory. For administration of SSEs, a remote copy of the AMS, the DF, or the RMA can be started. For example, the following command starts the AMS agent in its own container, the remote location of which is defined by the IP address 93.135.248.211:

```
java -cp jade.jar jade.Boot -host 93.135.248.211
    -agents ams1:jade.domain.ams -container
```

6 Discussion and Future Work

This chapter described development of SSEs in the context of globalization and presented means to implement believable ECAs. This chapter also addressed scalability questions concerning size, geography and administration.

Addressing the implementation of SSEs, this chapter presented:

1. Steps necessary for rapid prototyping of SSEs on the basis of ISs;
2. Architecture and implementation of SSEs maintaining up to 1,000 believable ECAs exchanging 2,000 interaction messages;
3. A thorough study and implementation of 10 agent-specific and 8 environment-wide determinants of believable ECAs on the basis of 35 scenarios of intercultural interaction.

In future work, we will consider improvement and extension of the determinants' set. For this purpose, we will study the applicability of the identified determinants to implement further scenarios of SSEs, for example, SSEs implementing SI influenced by neurobiological signals [17]. Assuming that the JADE framework is sufficiently size-scalable and in line with [49], we will implement SSEs representing real-life societies with tens of thousands of ECAs and investigate how innovation alters the lives of the human members. Moreover, we will work on integration of identified determinants in robotic ECAs using our previous experience in implementing MASs that perform cooperative tasks [31].

Acknowledgements This research was funded by BMWi (Federal Ministry for Economic Affairs and Energy).

References

1. Aehnelt, M., Bader, S., Ruscher, G., Krüger, F., Urban, B., & Kirste, T. (2013). Situation aware interaction with multi-modal business applications in smart environments (pp. 413–422). Springer, Berlin, Heidelberg. https://doi.org/10.1007/978-3-642-39226-9_45
2. Angelaccio, M., Buttarazzi, B., & Marrozzini, M. (2017). *Smart 2017: The Sixth International Conference on Smart Cities, Systems, Devices and Technologies*
3. Bellifemine, F. L., Caire, G., & Greenwood, D. (2007). Developing multi-agent systems with JADE (Wiley Series in Agent Technology). Wiley
4. Binsted, K., Luke, S., & Building, A. V. W. (1998). Character design for soccer commentary. In M. Asada & H. Kitano (Eds.), *RoboCup-98: Robot soccer world cup II* (pp. 23–35). Springer
5. Bonin, F., Campbell, N., & Vogel, C. (Dec 2012). Laughter and topic changes: Temporal distribution and information flow. In *2012 IEEE 3rd International Conference on Cognitive Infocommunications (CogInfoCom)* (pp. 53–58)
6. Bosse, T., Hoogendoorn, M., Klein, M., van Lambalgen, R., van Maanen, P., & Treur, J. (2011). *Incorporating human aspects in ambient intelligence and smart environments* (pp. 128–164). IGI Global (2011)
7. Bourdieu, P. (1983). Ökonomisches kapital, kulturelles kapital, soziales kapital. In R. Kreckel (Ed.), *Soziale Unigleichheiten* (pp. 183–198). Göttingen: Schwartz.
8. Burbeck, K., Garpe, D., & Nadjm-Tehrani, S. (2004). Scale-up and performance studies of three agent platforms. *IEEE International Conference on Performance, Computing, and Communications, 2004*, 857–863.
9. Butz, A. (2010). *User interfaces and HCI for ambient intelligence and smart environments* (pp. 535–558). Springer, Boston, MA. https://doi.org/10.1007/978-0-387-93808-0_20
10. Cai, Y., & Kaufer, D. (2011). *Incorporating human aspects in ambient intelligence and smart environments* (pp. 78–87). IGI Global
11. Cortese, E. *Benchmark on JADE message transport system*. Retrieved Sept 4, 2017, from http://jade.tilab.com/doc/tutorials/benchmark/JADERTTBenchmark.htm
12. Costa, P., & McCrae, R. R. (1991). The NEO personality inventory: Using the five-factor model in counseling. *Journal of Counseling and Development, 69*, 367–372.
13. De Carolis, B., Mazzotta, I., Novielli, N., & Pizzutilo, S. (2010). Social robots and ECAS for accessing smart environments services. In *Proceedings of the International Conference on Advanced Visual Interfaces, AVI'10* (pp. 275–278). ACM, New York, NY, USA. https://doi.org/10.1145/1842993.1843041
14. Douglas-Cowie, E., Cowie, R., Sneddon, I., Cox, C., Lowry, O., McRorie, M., et al. (2007). *The HUMAINE database: Addressing the collection and annotation of naturalistic and induced emotional data* (pp. 488–500). Springer, Berlin, Heidelberg. https://doi.org/10.1007/978-3-540-74889-2_43
15. François, J. M. (2012). An implementation of Hidden Markov models in Java. https://code.google.com/p/jahmm/
16. Fu, Z., & Zhang, X. (2011). *Designing for social urban media: Creating an integrated framework of social innovation and service design in China* (pp. 494–503). Springer, Berlin, Heidelberg. https://doi.org/10.1007/978-3-642-21660-2_56
17. Giovannella, C. (Aug 2013) "Territorial smartness" and emergent behaviors. In *2nd International Conference on Systems and Computer Science* (pp. 170–176)
18. Giovannella, C., Iosue, A., Moggio, F., Rinaldi, E., & Schiattarella, M. (July 2013). User experience of kinect based applications for smart city scenarios integrating tourism and learning. In *2013 IEEE 13th International Conference on Advanced Learning Technologies* (pp. 459–460)
19. Gómez-Sanz, J. J., Pax, R., Arroyo, M., & Cárdenas-Bonett, M. (2017). Requirement engineering activities in smart environments for large facilities. *Computer Science and Information Systems, 14*(1), 239–255.
20. das Gracas Bruno Marietto, M., dos Santos Franca, R., Steinberger-Elias, M., Botelho, W., Noronha, E., & da Silva, V. (2012). *International Conference for Internet Technology and Secured Transactions* (pp. 628–633)

21. Hofstede, G.H. (2001) *Culture's consequences: Comparing values, behaviors, institutions, and organizations across nations* (2nd and enlarged ed.). Sage, Thousand Oaks, CA
22. Hofstede, G. J., Smith, D. M., & Hofstede, G. (2002). Exploring culture: Exercises. In *Stories and synthetic cultures*. Nicholas Brealey Publishing.
23. Juang, B. H., & Rabiner, L. R. (1990). The segmental k-means algorithm for estimating parameters of hidden markov models. *IEEE Transactions on Acoustics, Speech, and Signal Processing*, *38*(9), 1639–1641.
24. Knoch, D., Nitsche, M.A., Fischbacher, U., Eisenegger, C., Pascual-leone, A., & Fehr, E. (2007). Studying the neurobiology of social interaction with transcranial direct current stimulation: The example of punishing unfairness. Cerebral Cortex (2007)
25. Labov, W. (2006). *The social stratification of English in New York City*. Cambridge University Press. https://books.google.de/books?id=bJdKY0mZWzwC
26. Liu, H., Lieberman, H., & Selker, T. (2003). A model of textual affect sensing using real-world knowledge. In *Proceedings of the 8th International Conference on Intelligent User Interfaces, IUI'03* (pp. 125–132). ACM, New York, NY, USA. https://doi.org/10.1145/604045.604067
27. Meister, M., Urbig, D., Schröter, K., & Gerstl, R. (2005). Agents enacting social roles. Balancing formal structure and practical rationality in MAS design. In *Socionics. Lecture Notes in Computer Science* (vol. 3413, pp. 104–131). Springer
28. Moffat, A. (2012). Implementation of the sequence diagram. http://www.zanthan.com/itymbi/archives/cat_sequence.html
29. Nakanishi, H., Ishida, T., Koizumi, S. (2008). Virtual cities for simulating smart urban spaces. In M. Foth (ed.), *Handbook of research in urban informatics* (pp. 256–268). Information Science Reference
30. Nakashima, H., Aghajan, H., Augusto, J. C. (2010) *Handbook of ambient intelligence and smart environments*, 1st ed. Springer
31. Osherenko, A. (2001). Plan representation and plan execution in multi-agent systems for robot control. In *Proceedings of AI in Planning, Scheduling, Configuration and Design*
32. Osherenko, A. (2010). *Opinion mining and lexical affect sensing*. Ph.D. thesis, University of Augsburg
33. Osherenko, A. (2014). Social interaction, globalization and computer-aided analysis—A practical guide to developing social simulation. In *Human-Computer Interaction Series*. Springer. https://doi.org/10.1007/978-1-4471-6260-5
34. Picard, R. (1997). *Affective computing*. Cambridge: MIT Press.
35. Plutchik, R. (1993). Emotions and their vicissitudes: Emotions and psychopathology. In: M. Lewis, J. M. Haviland-Jones (eds.), *Handbook of emotions*. The Guilford Press
36. Plutchik, R. (1994). *The psychology and biology of emotion (Comparative Government)*. Harpercollins College Div, 1st ed. http://amazon.com/o/ASIN/0060452366/
37. Rehm, M., Nakano, Y., André, E., Nishida, T., Bee, N., Endrass, B., et al. (2009). From observation to simulation: Generating culture-specific behavior for interactive systems. *AI & Society*, *24*(3), 267–280. https://doi.org/10.1007/s00146-009-0216-3.
38. Ryan, N., Mohr, P., Mantovani, G., Bartolini, S., D'Elia, A., Pettinari, M., et al. (2011). Interoperable multimedia mobile services for cultural heritage sites. In *EPOCH Conference on Open Digital Cultural Heritage Systems (2008)* (pp. 54–60). EPOCH Collection, Archaeolingua, Budapest. http://amsacta.unibo.it/5099/
39. Santos-Pérez, M., González-Parada, E., & Cano-García, J. M. (2011). *AVATAR: An open source architecture for embodied conversational agents in smart environments* (pp. 109–115). Springer, Berlin, Heidelberg. https://doi.org/10.1007/978-3-642-21303-8_15
40. Schröder, M., Cowie, R., Douglas-Cowie, E., Savvidou, S., McMahon, E., & Sawey, M. (2000). 'FEELTRACE': An instrument for recording perceived emotion in real time. In *Proceedings of the ISCA Workshop on Speech and Emotion: A Conceptual Framework for Research* (pp. 19–24). Textflow, Belfast
41. Watson, S., Kerstin Dautenhahn, W. C. S. H., & Dawidowicz, R. (2009). *Developing relationships between autonomous agents: Promoting pro-social behaviour through virtual learning environments part I* (pp. 125–138). IGI Global

42. Sorrells, K. (2013). *Intercultural communication: Globalization and social justice*. SAGE
43. Spadavecchia, C., Giovannella, C. (July 2010) Monitoring learning experiences and styles: The socio-emotional level. In *2010 10th IEEE International Conference on Advanced Learning Technologies* (pp. 445–449)
44. Tanenbaum, A.S., van Steen, M. (2002). Distributed systems : Principles and paradigms. Upper Saddle River, New Jersey Prentice Hall
45. Trovato, G., Kishi, T., Endo, N., Hashimoto, K., & Takanishi, A. (2012). *A cross-cultural study on generation of culture dependent facial expressions of humanoid social robot* (pp. 35–44). Springer, Berlin, Heidelberg. https://doi.org/10.1007/978-3-642-34103-8_4
46. Tsai, T. W., Lin, M. Y.: *An application of interactive game for facial expression of the autisms* (pp. 204–211). Springer, Berlin, Heidelberg. https://doi.org/10.1007/978-3-642-23456-9_37
47. Umata, I., Oshima, C., Ito, S., Iwasawa, S., Nakamura, H., Endo, A., et al. (Oct 2010). Do 3D images help social interaction?: A study in remote music education. In *2010 4th International Universal Communication Symposium* (pp. 197–200)
48. Witten, I. H., & Frank, E. (2005). Data mining: Practical machine learning tools and techniques, 2nd ed. In *Morgan Kaufmann Series in Data Management Systems*. Morgan Kaufmann Publishers Inc., San Francisco, CA, USA
49. Xu, Q., & Wu, Z. (Nov 2012). A study on strategy schema for smart cities based on the innovation driven. In *2012 International Symposium on Management of Technology (ISMOT)* (pp. 313–315)
50. Yassine, M., & Hajj, H. (Dec 2010) A framework for emotion mining from text in online social networks. In *2010 IEEE International Conference on Data Mining Workshops* (pp. 1136–1142)

EcoSim, an Enhanced Artificial Ecosystem: Addressing Deeper Behavioral, Ecological, and Evolutionary Questions

Ryan Scott, Brian MacPherson and Robin Gras

Abstract This chapter discusses individual-based models (IBMs) and uses the Overview, Design concepts, and Details (ODD) protocol to describe a predator-prey evolutionary ecosystem IBM called EcoSim. EcoSim is one of the most complex and large-scale IBMs of its kind, allowing hundreds of thousands of intricate individuals to interact and evolve over thousands of time steps. Individuals in EcoSim have a behavioral model represented by a fuzzy cognitive map (FCM). The FCM, described in this chapter, is a cognitive architecture well-suited for individuals in EcoSim due to its efficiency and the complexity of decision-making it allows. Furthermore, it can be encoded as a vector of real numbers, lending itself to being part of the genetic material passed on by individuals during reproduction. This allows for meaningful evolution of their behaviors and natural selection without predefined fitness. EcoSim has been enhanced to increase the breadth and depth of the questions it can answer. New features include: fertilization of primary producers by consumers, predator-prey combat, sexual reproduction, sex-linkage of genes, multiple modes of reproduction, size-based dominance hierarchy, and more. In addition to describing EcoSim in detail, we present data from default EcoSim runs to show potential users the types of data EcoSim generates. Furthermore, we present a brief sensitivity analysis of some variables in EcoSim, and a case study that demonstrates research that can be performed using EcoSim. In the case study, we elucidate some evolutionary and behavioral impacts on animals under two conditions: when primary production is limited, and when energy expenditure is reduced.

R. Scott · B. MacPherson · R. Gras (✉)
University of Windsor, 401 Sunset Ave., Windsor, Ontario, Canada
e-mail: rgras@uwindsor.ca

R. Scott
e-mail: scotto@uwindsor.ca

B. MacPherson
e-mail: macphe4@uwindsor.ca

M. I. Aldinhas Ferreira et al. (eds.), *Cognitive Architectures*, Intelligent Systems, Control and Automation: Science and Engineering 94,
https://doi.org/10.1007/978-3-319-97550-4_14

1 Introduction

Among biological disciplines, behavioral ecology has a strong tradition of accounting for the role of organism-environment interactions in behavior [69]. Behavioral ecology and the related field of optimal foraging theory [118] model animal behavior in terms of optimal adaptation to environmental niches. The goal is not to test whether organisms actually behave optimally, but to use normative expectations to interpret behavioral data and/or generate testable hypotheses. One approach for understanding the behavior of complex ecosystems is through individual-based models (IBMs), which provide a bottom-up approach allowing for the consideration of the traits and behavior of individual organisms. Ecological modelling is still a growing field, at the crossroads between theoretical ecology, mathematics, and computer science [109]. Since natural ecosystems are very complex (in terms of number of species and of ecological interactions), ecosystem models aim to characterize the major dynamics of ecosystems in order to synthesize the understanding of such systems and to allow for predictions of their behavior. Ecosystem simulations can also help scientists to understand theoretical questions regarding the evolutionary process, the emergence of species, and the emergence of learning capacities. One of the most interesting aspects of such ecosystem simulations is that they offer a global view of the evolution of the system, which is difficult to observe in nature. However, the scope of ecosystem simulations has always been limited by the computational possibilities of their time. Today, it is possible to run simulations that are more complex than ever, due to the availability of high performance computing resources.

Several ecosystem simulation platforms with various features exist. For example, Echo, one of the first such models, is a basic ecosystem simulation in which resources are limited and agents evolve [58]. In Echo, each agent, upon obtaining the required resources to copy its genome, replicates itself with some mutations. The agents, through interaction with other agents (combat, trade, or mating) or the environment, can acquire resources. Polyworld is another such IBM software [129] to evolve artificial intelligence through natural selection and evolutionary algorithms. It displays a graphical environment in which trapezoidal agents search for food, mate, and create offspring. The number of agents is typically only in the hundreds, as each agent is rather complex and the environment consumes considerable computational resources. In this model, each individual makes decisions based on a neural network which is derived from each individual's genome. Recently, Polyworld has been used to study the effects of different neuromodulation models on the adaptability of its individuals [131], finding that neuronal plasticity modulation (decreasing or increasing the rate at which neuron weights change) tends to produce individuals that adapt more effectively. It has also been used to study the way in which network topologies influence the evolved complexity of the networks [130] and, most recently, the level of chaos as the individuals in the system evolve [128]. Avida is another artificial life software platform for studying the evolutionary biology of self-replicating and evolving computer programs [97], inspired by the Tierra system [122]. Unlike Tierra, Avida assigns every digital organism its own protected region of memory and executes its program with a separate virtual CPU. A second major difference is that the virtual CPUs of different organisms can run at different speeds. The speed at which a virtual CPU runs is determined by several factors, but most importantly by the tasks that the organism performs: logical computations that the organisms can carry out to reap extra CPU speed as a bonus. With increasing computational power, individual-based simulation platforms such as Tierra, Avida, Polyworld, and EcoSim [45, 74, 108, 129] can be used to address increasingly difficult questions in biology [22, 23, 43, 75]. EcoSim [45], in particular, has been designed to model large-scale virtual ecosystems.

Recently, much has been done in the field of ecological IBMs on three main fronts: formalization and development practices of IBMs, pragmatic modelling, and paradigmatic modelling. In regard to formalization and development practices, some insist that there is an increasing need for developers of IBMs to be transparent about the process used to develop a model [5, 49, 113]. They argue that potential clients need to have a thorough understanding of the model so that they can know whether the model is applicable to whatever they would like to test. Clients need formal statements of the question(s) the model is designed to answer, descriptions of the submodels and their organization within the model, information on the degree of testing performed on the model, and the rationale behind making any modifications throughout the long and iterative process that is the "modelling cycle". So, several researchers have proposed and subsequently revised [49] a new standard format for the description of an IBM, TRAnsparent and Comprehensive Ecological modelling documentation (TRACE) [113], which differs from the previously-proposed ODD protocol [47] in that TRACE is more comprehensive and more concerned with describing the development cycle and practical ability of a model. Furthermore, the ODD protocol can be used within TRACE as a means of describing the model's implementation. TRACE complements the principle of "evaludation" [5], representing an urged evaluation and validation of a model throughout the development, application, and analysis of it. The current revision of TRACE intends to focus the developer on documenting the modelling process for the sake of ensuring quality and credibility throughout said process, as the originally proposed TRACE was less efficient and less specific regarding its goals. MacPherson and Gras [79] argue that there is too much of a focus on "evaluation" and that not all IBMs are "merely adjunctive tools". More specifically, pragmatic models, focusing on a particular species or system usually with intent of making predictions in applied ecology, should undergo a more rigorous parameterization process using empirical data, be subject to evaludation, and be more stringently documented. Pragmatic models are often tied to conservation efforts or the management of delicate ecosystems, and so a model must be realistic enough to effectively predict how a specific (very complex) ecological system will behave. On the other hand, MacPherson and Gras argue that paradigmatic models are, in fact, experimental platforms. Though they must be realistic enough, in the general sense, there should be less of a focus on incorporating empirical data into the calibration or parameterization of them, as they are typically designed to answer rather general theoretical questions, the results of which we often have no means of historically validating due to the scale of interactions being emulated in the simulation. Furthermore, they argue that paradigmatic models can lose generalizability by over-calibrating the model empirically. They propose a relaxed notion of model evaluation by removing the constraint of empirically calibrating a model; they instead insist that the calibration be reasonable, that is, consistent with general observations in nature.

Pragmatic models are those that aim to model a specific system or population, and most IBMs are pragmatic in nature [25]. de los Santos et al. [27], for instance, designed an IBM of a marine amphipod, *Gammarus locusta*, to assess the effect of long-term exposure to a chemical pollutant, aniline, on *G. locusta* populations. They used real life-history traits of *G. locusta* to parameterize the model, and observed significant negative impacts in individual survivorship and production of offspring with exposure to aniline. Other recent works in pragmatic modelling include a toxicological model for zebrafish [50], a model eliciting effects of climate change on population dynamics in European anchovies [104], a model for management of brown trout [33], and a model for motion of the blue mussel, *Mytilus edulis* [26].

As the naming convention suggests, paradigmatic modelling moves away from answering specific questions and instead aims to uncover the underlying causes of more generalized ecological or evolutionary phenomena [25]. Zaman et al. [132], for example, used Avida to show that parasite-host interactions increase the complexity and evolvability of digital organisms over a long time-frame. Avida has been used in several other recent works [32, 39, 71, 100]. Similar to Zaman et al. [132], Kvam et al. [70] also studied the complexity of the brain of a population of digital organisms, in this case Markov Brain agents. In contrast, they studied complexity in light of the problem-solving environment the agents were subject to. Olson et al. [99] used Markov Brain Agents as well, but instead they placed the agents into a toroidal world and observed changes in physical cluster tightness when subject to different types of predator attacks. Botta-Dukát and Czúcz [12] generated a spatially implicit IBM to simulate community compositions and tested the ability of five functional diversity indices. Functional diversity indices aim to determine the number of functionally different species in a community. Their simulation accounted for habitat filtering (suitability of an individual to a habitat—a means of local trait convergence) and trait-similarity-based competition for resources (a means of local trait divergence) in composing the simulated communities. With mechanisms causing individual trait divergence and convergence, they could effectively test the functional diversity indices for their ability to detect these two key assembly processes. They found trait divergence was difficult to detect for all the indices tested, whereas trait convergence was detectable by some indices. Uchmański [124] found, using an IBM, that dispersal mechanisms of individuals affect the persistence of metapopulations. In different runs of the simulation, individuals would disperse from their current habitat to another unoccupied neighboring habitat for different reasons (when one gains no resources resulting from competition, when competition yields insufficient resources to produce an offspring, random chance, or when no individuals in a habitat could reproduce). If individuals dispersed due to total loss of resources due to competition, the metapopulations persisted longest. Similarly, when individuals dispersed due to insufficient resources for reproduction, the metapopulations persisted longer than by chance. If individuals waited until none in a habitat could reproduce, the metapopulations failed to persist longer than cases in which dispersal was random. Another recent paradigmatic IBM tested the effects of patch size and refuge abundance on the strength of predator-prey interactions and population dynamics [77]. They found that refuge availability decreased the interaction strength between prey and predators, which consequently improved the stability of populations. CDPOP [72] and its descendant CDMetaPOP [73] are both IBMs that use Mendelian inheritance with any number of alleles and loci to study the effects of a varying landscape of (nearly) any complexity on the genetic structure and composition of populations or metapopulations. Though natural selection does occur, individual fitness is also influenced by user-specified spatially explicit fitness values for each genotype that is selected upon.

EcoSim is a large-scale evolving predator-prey paradigmatic ecosystem simulation that can be used to perform studies in theoretical biology and ecology [43, 84]. It has been shown that EcoSim generates patterns as complex as those observed in real ecosystems [40]. Several studies have been done using EcoSim. Devaurs and Gras [28] have shown that the behavior of this model is realistic by comparing the species-abundance patterns observed in the simulation with real communities of species. Furthermore, chaotic behavior [40] with multi-fractal properties [41] of the system has been proved to be similar to that observed in real ecosystems [114], and Golestani, Gras, and Cristescu [43] have measured the effect of small geographic barriers on speciation in EcoSim. The effect of the spatial distribution of individuals on speciation has been investigated by Mashayekhi and Gras [83]. Khater et al. [61] demonstrated that introduction or removal of predators in an ecosystem can have widespread effects on the survival and evolution of prey by altering their genomes and behavior. Mashayekhi et al. [84] proved that the extinction mechanisms in EcoSim are similar to those of real communities. Lastly, a study by Gras et al. [46] used EcoSim to elicit the roles of natural selection and spatial isolation in the speciation process. They were able to unequivocally demonstrate that in order to observe genetic clusters (species), natural selection must be present. The number of individuals per species was much greater, species abundance distributions were far more even, the compactness and separation of genetic clusters were far greater, and hybrid production was far lower (after sufficient time had passed in the simulation) in runs where natural selection was present.

Real ecosystems are extremely complex systems with numerous interacting components and feedback loops. No paradigmatic model has all of the features of real ecosystems; consequently, these artificial systems are restricted to a small spectrum of possible questions to be answered. EcoSim was already quite complex and diverse in the types of questions it could answer, but we have added specific features to further improve its realism and applicability. Our objective is to propose to the community an improved simulation platform that models as many of the important features of real ecosystems as possible. Of course, not every significant feature of real ecosystems could be integrated into such a simulation platform. However, we have chosen a set of features that seem most important in modelling a stable, long-term evolutionary ecosystem and to provide the mechanisms needed to answer the largest possible spectrum of important theoretical questions. The three most important features we have added to EcoSim are fertilization of soil via animal excretion, the ability of prey to defend against attacking predators (individually or cooperatively), and a female/male binary sex system with sexual reproduction. In previous iterations of EcoSim, individuals were of uniform sex and any two individuals of the same type (prey or predator) could attempt to reproduce.

There is a vast array of indirect impacts of herbivores on plant community features [6, 98]. Most importantly, herbivores affect the quantity and quality of organic matter returning to the soil [7, 8, 56, 126]. Generally, animal excreta facilitates decomposition through increasing soil microbial biomass [7, 34] and net Carbon (C) and Nitrogen (N) mineralization [35, 89]. Feces and urine also make it easier for plants to absorb, thereby increasing their growth rates [51]. Thus, herbivores are able to influence their own food supply [29, 54, 125] by producing negative feedback against the reduction of resources they consume. In order to include this complex feedback mechanism, we introduced a new concept to our simulated ecosystem called "fertilizer", which models the effect of prey fertilizing their environment.

There is limited experimental evidence in the ecological literature regarding mobbing behavior as a kind of reciprocal altruism between heterospecifics. Krams et al. [67] and Krams et al. [68] report that breeding Fecedula hypoleuca (pied flycatchers) engage in mobbing behavior primarily with heterospecifics as a form of defense against predation. As Krams et al. [67] note, there is little empirical evidence for the existence of mobbing behavior as a form of reciprocal altruism. EcoSim could thus be used to test for mobbing behavior as a form of reciprocal defense in the presence of predation. In a related vein, an important unresolved debate in the biological literature is whether eusociality evolved via kin selection or group selection; Nowak et al. [96] claim that group selection rather than kin selection (inclusive fitness) combined with haplodiploidy theory is the best way to explain eusociality. They suggest that there may be no real relation between haplodiploidy and eusociality, and argue that inclusive fitness theory is not sufficiently general since it is a simple mathematical theory that has great limitations [96]. Furthermore, Nowak et al. [96] argue that there is no empirical confirmation of inclusive fitness theory. On the other hand, Marshall [82] and Abbot et al. [1] argue that recent evidence helps to support inclusive fitness theory. Since there is apparently an argumentative stalemate regarding whether kin selection or group selection drives evolution, EcoSim could help to resolve this debate by testing the hypothesis that kin selection explains the evolution of eusociality and altruism. Finally, another important issue in evolutionary theory is whether predation selects for morphological defenses in prey. Bollache et al. [14] argued that the main reason that the invasive amphipod *Gammarus roeseli* was eaten less than the native amphipod species *Gammarus pulex* was due to the presence of a spin on *G. roseli*, as opposed to behavioral differences. EcoSim could be used to help resolve the debate regarding whether morphology or behavior is a key inducible defense against predators.

Typically, in sexually reproductive species in which sexual dimorphism exists, females are generally choosier than males when selecting mates. Compared to males, females typically invest far more resources (time and energy) into offspring. For instance, females typically provide more parental care than males. Females also invest more in gametes for sexual reproduction; males produce the microgamete sperm, whereas females produce large, nutritious eggs. Moreover, unlike males, females only produce a limited number of eggs as long as they are reproductively active; therefore, there is more risk associated with mate choice [2]. To broaden the applicability and increase the realism of EcoSim, we introduced a model for sexual reproduction into the simulation. Previously, there was no categorization of individuals by sex; any individual could attempt reproduction with any other of the same type (prey with prey, predators with predators). Now, prey and predator individuals are divided into two groups, males and females. Furthermore, we have made significant modifications to reproduction mechanisms such as selection of mates, energy dynamics, and genetic recombination; these changes reflect the information-gathering and decision-making process that is mate choice [9]. These new improvements were aimed at unravelling some of the most controversial issues in behavioral ecology, such as the evolution of female preference.

In addition to presenting the new version of EcoSim following the updated 7-points Overview, Design concepts, and Details (ODD) standard protocol [47, 48], we present and discuss data from EcoSim in its default configuration. We also analyze the divergence of two sister species in EcoSim. We then present a sensitivity analysis on three parameters of EcoSim: the amount of energy spent per time step for prey and predators, the maximum amount of grass held in cells, and the initialization of newly added social concepts related to defense. The purpose of this sensitivity analysis was to show how sensitive or robust EcoSim is to these parameters. Finally, we present a case study of EcoSim's application; we determined the behavior and evolution of individuals under two conditions: reduced primary production (thereby increasing competition) and reduced energy expenditure. This study serves as an example of the types of study that are made possible by the EcoSim platform.

2 ODD Description of EcoSim

EcoSim is an individual-based ecosystem simulation [45, 85] for simulating animals' behaviors in a dynamic, evolving ecosystem. The individuals of EcoSim are prey and predators acting in a simulated environment. A description of the older version of EcoSim can be found in [84, 85]. In addition to the main features outlined above, EcoSim has been expanded by adding several smaller features such as: new individuals' perceptions of their environment, new actions, new physical traits (governed by what we call the physical genome), sex-linked genes, various modes of reproduction, modified acting priority for individuals, new ways to control the dynamics of the environment, and new crossover and mutation operations that consider an individual's sex. Below, we describe the new version of EcoSim following the updated 7-points Overview, Design concepts, and Details (ODD) standard protocol [47, 48].EcoSim source code (in C++) can be obtained from the repositories at https://github.com/EcoSimIBM, and more information on EcoSim can be found at https://sites.google.com/site/ecosimgroup/home.

2.1 Purpose

EcoSim was designed to simulate animal behavior in a dynamic and evolving ecosystem. The main purpose of EcoSim is to study biological, ecological, and evolutionary theories by constructing a complex adaptive system that leads to a generic virtual ecosystem with behaviors like those found in nature. Due to the complexity, scale, and resource requirement of studying these theories in real biological systems, simulations of this nature are necessary. EcoSim uses a fuzzy cognitive map (FCM; [66]) to model an individual's behavior. Since the FCM is coded in the genome and heritable, behavior can evolve during the simulation. Importantly, the fitness of a given set of behaviours and physical traits is not predefined. Instead, fitness emerges from interactions between the model organisms and their biotic and abiotic environment.

2.2 Entities, State Variables, and Scales

2.2.1 Individuals

EcoSim has two types of individuals: prey and predators. Each individual possesses two types of traits: acquired and inherited traits (Table 1). The former varies depending on the environmental conditions and the latter is encoded in an individual's genome and is fixed during its lifetime. The age and speed are initialized to zero for newborn individuals, while energy, a crucial property of the individual, is initialized based on the amount of energy invested into a newborn by its parents at reproduction time (State of Birth or SOB—see *Reproducing* under *Submodels*). Afterward, energy is provided to the individuals by resources (food) that they find in their environment. Prey consume grass, which is dynamic in quantity and location (see *Submodels* for grass diffusion model), whereas predators hunt for prey individuals or scavenge their remains when they die. Strength of an individual is calculated based on its current energy (Energy), maximum energy (MaxEnergy), age (Age), maximum age (MaxAge) and reproductive age (RepAge). Young (Age is less than RepAge) and old individuals (Age is greater than or equal to MaxAge minus RepAge) have less Strength. Strength can range from 25% of an individual's MaxEnergy (if the individual is too young or old and has energy approaching zero) to 100% of the individual's MaxEnergy (if the individual has energy greater than or equal to 1/3 of its MaxEnergy and the individual is not too young or old).

Each individual performs one unique action during a time step, based on its perception of the environment and state (see *Emergence* under *Design Concepts*). At each time step, each individual spends energy depending on its selected action

Table 1 Several physical and life history characteristics of individuals from five independent runs. The values for the inherited features are the values at initialization, and for the acquired features they are the average values over 20,000 time steps

Type	Characteristic	Male predator	Female predator	Male prey	Female prey
Inherited	Maximum energy	3000	3000	2500	2500
	Maximum age	50	50	46	46
	Vision	20	20	8	8
	Maximum speed	20	20	6	6
	Minimum age of reproduction	5	5	6	6
	State of birth	14	18	12	16
	Defense	N/A	N/A	0.05	0.05
	Cooperative defense	N/A	N/A	0.05	0.05
Acquired	Average energy	2312.2	2211.4	1664.9	1678.3
	Average age	16.5	13.7	14.3	12.3
	Average speed	3.4	2.9	6.5	6.0
	Average strength	3306.3	3107.9	2478.9	2439.7

(e.g., reproduction, eating, moving), the complexity of its behavioral model (number of existing edges in its FCM; see *Adaptation* under *Design Concepts* for details), and its physical characteristics (encoded in its physical genome; see *Adaptation* under *Design Concepts* for details). To achieve a realistic rate of energy expenditure we involved as many of its contributory factors as possible and used empirically-determined physiological scaling rates (see Eq. (1), per time step energy penalty for prey, and Eq. (2), per time step energy penalty for predators). In general, any action performed by a living organism is involved in spending some amount of energy [20], dependent on what the action is [11]. Thus, the action performed was included as a contributing factor in energy expenditure (Eqs. (1) and (2)). Moreover, the size of a living organism plays a fundamental role in its metabolic rate [21]. In EcoSim, the size of each individual is modelled through its MaxEnergy and Strength. MaxEnergy is a heritable limit on an individual's capacity to store energy, whereas Strength is a slightly more complex proxy of size, being derived from an individual's MaxEnergy, Energy, and Age. Experimental and empirical investigations have demonstrated that there is a nonlinear relationship between adult animal's body mass and their metabolic rate, which is best described by a $3/4$ scaling exponent [53, 64, 65, 94, 102, 103, 107, 112, 116, 117]. Consequently, the metabolic rate of an individual in EcoSim is quantified through a power function of coefficient $3/4$ on its MaxEnergy (Eqs. 1 and

2). Energy expenditure associated with movement is also modelled in EcoSim using the kinetic energy equation (KE), and here we use Strength as a proxy of mass (KE = *mass* × speed^2, Eqs. (1) and (2)). The complexity of an organism's behavioral model increases an individual's energy expenditure, because it has been accepted that species belonging to a higher-level taxonomic affiliation require more energy to survive [91, 92]. Individuals with a larger brain also require more energy, as the brain is an expensive organ in terms of specific chemical and thermoregulatory needs [31, 127]. Consequently, possessing a large brain leads to a heavier metabolic requirement [111]. The complexity and the size of the brain vary in different species; while some species possess a very simple and small brain, many higher vertebrates have a brain so large and complex that it is considered as the most complex organ in these species [115]. Therefore, we also include this parameter in calculating the energy spent by an individual. Taking these points into consideration, the energy spent by prey (1) and predators (2) at any time step is given by the following equations:

$$\begin{aligned} \textit{Energy Spent by Prey} = {} & 0.8 \times \max\left((\text{NbArcs} - 100)^{0.75}, 1\right) \\ & + \frac{(\text{Strength} \times \text{Speed}^2)}{10,000} + \left(\frac{\text{MaxEnergy}}{5.5}\right)^{0.75} \\ & + (\text{Vision} \times 5.0)^{0.75} + (\text{MaxSpeed} \times 5)^{0.75} \\ & + (\text{Defense} \times 100)^{0.75} + (\text{CoopDefense} \times 75)^{0.75} \\ & + (\max(0.8 - \text{RepAge}))^{2.3}, \end{aligned} \tag{1}$$

$$\begin{aligned} \textit{Energy Spent by Predator} = {} & \left(0.8 \times \max\left((\text{NbArcs} - 130)^{0.75}, 1\right)\right) \\ & + \frac{(\text{Strength} \times \text{Speed}^2)}{11,000} + \left(\frac{\text{MaxEnergy}}{5.5}\right)^{0.75} \\ & + (\text{Vision} \times 5.0)^{0.75} + (\text{MaxSpeed} \times 5)^{0.75} \\ & + (\max(0.7 - \text{RepAge}))^{2.3}, \end{aligned} \tag{2}$$

where NbArcs is a measure of the complexity of the individual's brain based on the number of edges in its FCM (see *Adaptation* under *Design Concepts* for details), Vision refers to the distance up to which the individuals can see (which is initially 8 cells for prey and 25 cells for predator), Defense quantifies the ability of the prey individuals to protect themselves when they are attacked by predators, CoopDefense quantifies the ability of a prey individual to protect other prey in its cell, and RepAge is the age at which the individuals can start reproducing.

All individuals first perceive their environment (all the surrounding cells in their vision range) before using their behavioral model to choose a single action (see *Emergence* under *Design Concepts* for details of how individuals choose actions). After perceiving its environment (including grass resources, prey, predators, etc.), the possible actions for a prey individual are: evade (escape from predator), search for food (if there is not enough grass available in its cell, prey can move to another cell to find grass), socialize (move to the closest prey in the vicinity, move to the cell with strongest prey, move to the cell with the greatest total prey strength, and move to a cell with the least total prey strength), explore, rest (to save energy), eat, and reproduce. Predators also perceive their environment to gather information used to choose an action among: hunt (to catch and eat a prey), move to the cell with strongest prey, move to the cell with the least total prey strength, move to the cell with the weakest prey, search for food, socialize (move to the closest predator in the vicinity, move to the cell with strongest predator), explore, rest, eat, and reproduce. See the *Submodels* section for a full description of actions. Every individual takes one action per time step, after which its energy level and strength are adjusted. The age of all individuals is also increased by one unit at each time step. In addition to the acquired physical traits mentioned above, each individual has many state variables that, together, represent its state of mind. These variables are the values held in the nodes of each individual's FCM. Each FCM node has a single value that is its activation level (degree of stimulation) of its represented concept. Concepts can either be sensory, such as the individual's perception of local food, internal, such as the individual's hunger, or action, such as the individual's willingness to perform the eat action (see *Emergence, Adaptation,* and *Submodels* for more information).

2.2.2 Time Step

Each time step involves each individual perceiving its environment, making a decision, and performing one action. In addition, species memberships are updated and all relevant variables (e.g., quantity of available grass) are recorded (see *Process Overview and Scheduling* for algorithm).

2.2.3 Cells and Virtual World

The smallest units of the environment are cells. Each cell represents a large space which may contain an unlimited number of individuals, some limited amount of food, and some limited amount of fertilizer. The number of individuals a cell can host, therefore, is indirectly limited by the amount of food a cell contains. There are two types of food: grass, which only prey can eat, and meat, which only predators can eat. Grass amounts are controlled by a grass diffusion and growth model, and meat is generated when predators kill prey (see *Submodels* for grass diffusion model and meat generation). Fertilizer is produced by individuals residing in a cell (see *Submodels* for fertilizer dynamics). The virtual world consists of a matrix of 1000×1000 cells. The world is large enough that an individual moving in the same direction over the course of its entire life could not even cross half of it, and thus high-level movement patterns can be observed. The virtual world wraps around to remove any spatial bias. In addition, the dimensions of the world are adjustable, but expanding the dimensions increases the computational requirements (time and memory) of the simulation.

2.2.4 Species

By default, numerous prey and predators coexist in the simulation at any time step. Alternatively, the simulation can be run without predators. For each type, there is some number of species determined by the genetic makeup of the sets of individuals. There is at least one prey species and one predator species unless an extinction occurs, and at most there can be one species per individual. A species is a set of individuals with sufficiently similar genomes (see *Collectives* under *Design Concepts* for more details about speciation).

2.3 Process Overview and Scheduling

At each time step, the value of the state variables of individuals and cells are updated. The overview and scheduling of every time step is as follows:

1. For prey individuals:
    1.1. Perceive environment
    1.2. Compute next action
    1.3. Increase Age
    1.4. Females that chose to Reproduce act in order of decreasing Strength (to simulate female choice in mate selection)
    1.5. Remaining prey act in order of decreasing Strength
    1.6. Update list of prey (as some may have died due to depletion of Energy or maximum Age)
2. For predator individuals:
    2.1. Perceive environment
    2.2. Compute next action
    2.3. Increase Age
    2.4. Females that chose to Reproduce act in order of decreasing Strength (to simulate female choice in mate selection)
    2.5. Remaining predators act in order of decreasing Strength
    2.6. Update list of predators and prey (for predators, some may have died due to depletion of Energy, maximum Age, or combat with prey; for prey, some may have died due to predation)
3. Sort prey in order of decreasing Strength
4. Sort predators in order of decreasing Strength
5. Update prey species
6. Update predator species
7. For every cell in the world
    7.1. Update Fertilizer level
    7.2. Update Grass level
    7.3. Update Meat level

The complexity of the simulation algorithm is mostly linear with respect to the number of individuals. If we consider that there are N_1 prey and N_2 predators, then the complexity of parts 1 and 2 of the above algorithm, including the clustering algorithm used for speciation, will be $O(N_1)$ and $O(N_2)$, respectively [4]. The sorting parts (parts 3 and 4) have a complexity of $O(N_1 \log(N_1))$ and $O(N_2 \log(N_2))$, but are negligible in computational time so we will exclude them from the complexity computation. The complexity of parts 5 and 6 will be $O(N_1 + N_2)$. The virtual world of the simulation has 1000×1000 cells, therefore the complexity of part 7 will be $O(k = 1000 \times 1000)$. As a result, the overall complexity of the algorithm is $O(2N_1 + 2N_2 + k)$, which is $O(N = 2N_1 + 2N_2)$. In terms of computational time, the speed of simulation per time step is related to the number of individuals. Recent executions of the simulation produced approximately 20,000 time steps in 60 days.

2.4 Design Concepts

2.4.1 Basic Principles

The genome of each individual consists of two parts: a physical genome and a behavioral genome. An individual's genome is fixed at birth. When a new offspring is created, it receives a genome that combines the genomes of its parents with some possible mutations. An individual's physical genome determines its physical characteristics and its behavioral genome determines its behavioral characteristics. An individual's physical genome comprises values that represent its physical attributes (see Table 1, inherited traits).

The behavioral model of each individual is encoded as an FCM [45] (Fig. 1). Formally, an FCM is a directed graph that contains a set of nodes C and a set of edges I (Fig. 1); [66]. Each node C_i represents a concept and each edge I_{ij} represents the influence of the concept C_i on the concept C_j. A positive weight associated with the edge I_{ij} corresponds to an excitation of the concept C_j from the concept C_i, whereas a negative weight represents inhibition. A zero value indicates that there is no influence of C_i on C_j. The edges of an FCM can be represented by an $n \times n$ matrix, L, in which n is the number of concepts and L_{ij} is the influence of the concept C_i on the concept C_j. If $L_{ij} = 0$, there is no edge between C_i and C_j. An individual's behavioral genome is its set of FCM edges (its matrix L). Since the edges of the FCM are encoded in the genome, the behavioral model is heritable, mutable, and subject to evolution. Individuals act at each time step by using their FCM to compute their action (see *Emergence*). The activation level (degree of stimulation) of each concept, represented as the value held in its corresponding node, is dynamic in each individual. Collectively, the activation levels of every one of an individual's nodes represent the individual's behavioral state. In each FCM, three kinds of concept are defined: sensory (such as distance to foe or food, amount of energy, etc.), internal (fear, hunger, curiosity, satisfaction, etc.), and action (evade, socialize, explore, reproduce, etc.). At each time step, the activation level of a sensory concept is computed by performing

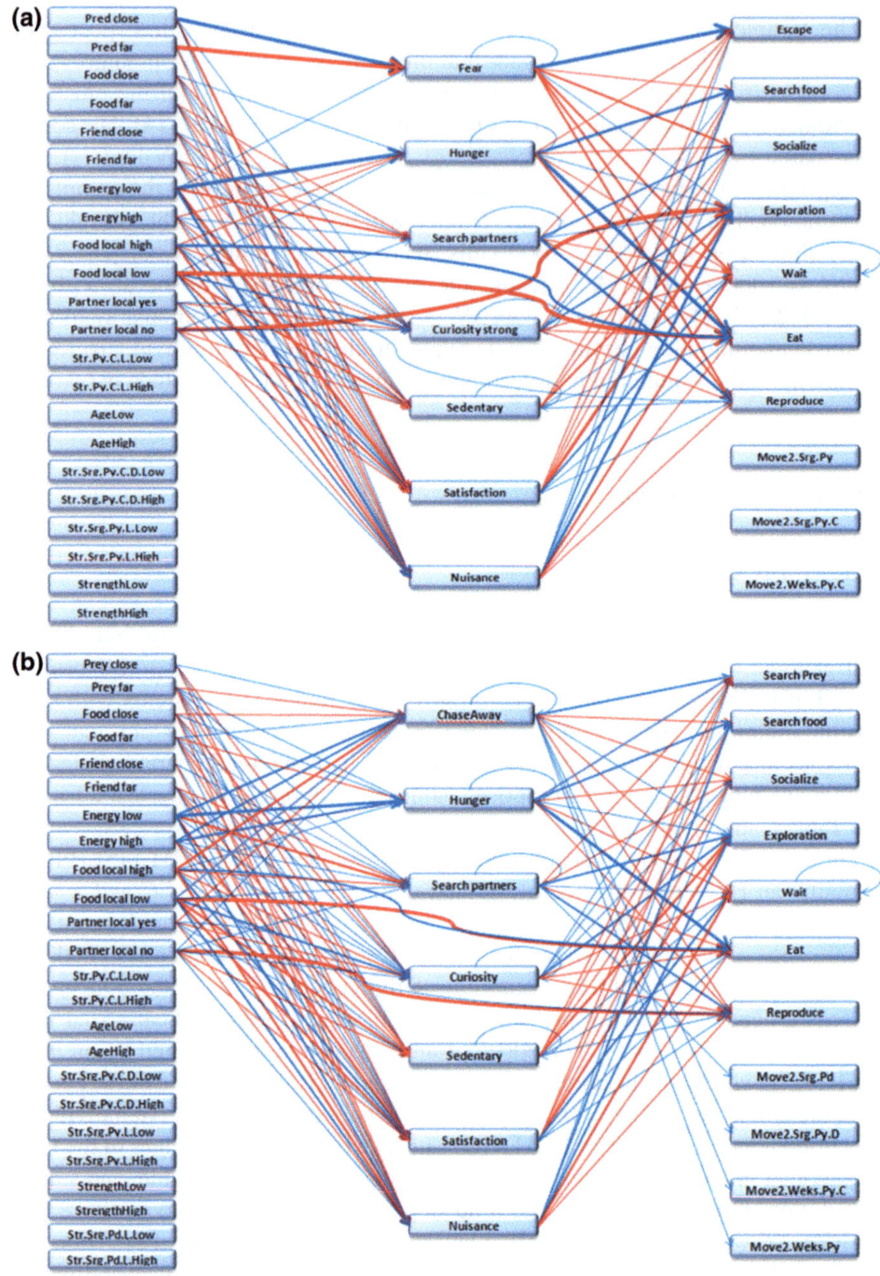

Fig. 1 An example FCM of a predator **a** and prey **b**. Red edges between nodes indicate negative association (inhibition) of a concept (where the edge begins) with another (where the edge points to), and blue edges indicate positive association (excitation). The thickness of the edges represents the magnitude of the gene. The leftmost column of nodes is sensory concepts, the middle is internal concepts, and the rightmost is action concepts. There are many unconnected nodes because we aim to observe evolution in action; over time, new edges may form and others may disappear

a fuzzification of the information that the individual perceives in the environment (changing its real scalar value into a fuzzy value, i.e., transforming the input value by a potentially nonlinear function). Subsequently, for an internal or action concept C, the activation level is computed from the weighted sum of the current activation level of all input nodes by applying a defuzzification function (another nonlinear function transforming the fuzzy input value into the final' real' value).

We will illustrate the operation of the FCM with a simplified example prey FCM (Fig. 2) consisting of only four nodes (EnemyClose, EnemyFar, Fear, and Evade). EnemyClose and EnemyFar are sensory concepts, whereas Fear is internal and Evade is an action. All sensory nodes appear in pairs, like EnemyClose and EnemyFar; the activation level of one of these nodes is always equal to $1 - a$, where a is the activation level of the other. The individual perceives its environment to get a raw value for the distance to the nearest predator; this raw value is fuzzified to compute values between 0 and 1 for the activation levels of EnemyClose and EnemyFar by nonlinearly transforming it. To compute the activation level of Fear, a weighted sum of the activation levels of all nodes with incident edges to Fear is computed and the

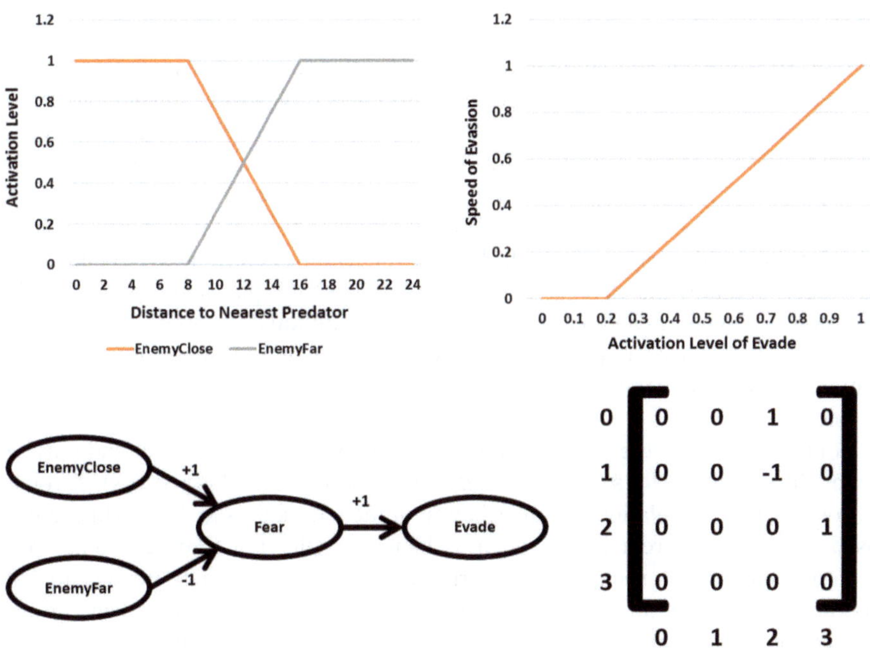

Fig. 2 A simplified example prey FCM for detection of predators (bottom left), with fuzzification (top left) and defuzzification (top right) functions, and its matrix (bottom right) which is the behavioral genome of the individual. EnemyClose and EnemyFar are sensory concepts, Fear is an internal concept, and Evade is an action concept. The edges of the FCM show influence of the activation level of a node on another. In the matrix, rows represent influencing concepts and columns represent those that are influenced. Row and column indices of 0 represent EnemyClose, 1 represent EnemyFar, 2 represent Fear, and 3 represent Evade

weights are the edge values from the behavioral genome. From our example, Fear has incident edges from EnemyClose and EnemyFar, thus we use edge weights from the behavioral genome for EnemyClose → Fear and EnemyFar → Fear to compute the weighted sum. The same computation is performed for the activation level of Evade. Finally, if Evade is the action selected by the individual (if, of all action concepts, it has the highest activation level), the speed of evasion is computed by defuzzifying the activation level of Evade. In the behavioral genome where no edge exists between two nodes (for instance, EnemyClose → Evade), the corresponding genes have values of zero. However, as individuals evolve, new edges can be added and pre-existing edges could be removed.

2.4.2 Emergence

This representation of the behavioral model allows for the apparition of positive and negative feedback loops. For instance, an individual may evolve a positive edge between the internal concept Fear and itself—this positive feedback loop can allow complex phenomena such as paranoia to emerge. Similarly, negative feedback loops can evolve that stabilize individual behavior. For instance, a negative association between EnergyHigh and Hunger with a positive association between Hunger and Eat means that after an individual replenishes its energy by performing the Eat action, it is less willing to eat again until its energy levels are lower. The fuzzification and defuzzification mechanisms allow for nonlinear transformations of the perception signal, which permits, for example, the representation of saturation of information. An individual's action is selected based on the action node with the highest activation level. Because of the way in which the behavioral genome determines the behavior of individuals and how the physical genome determines their physical capabilities, the evolution of behavioral and physical properties of individuals is emergent and it also influences other emergent properties of the system, such as number of individuals, spatial compactness of individuals (a proxy of competition for resources), and number of species.

At the initiation of the simulation, prey and predators are scattered randomly all around the virtual world (see *Stochasticity* for a description of this process). Through the course of the simulation, the distribution of the individuals in the world changes based on many different factors such as behavior selection (prey escaping from predators, individuals socializing to form groups, and individuals moving to find food resources). In addition, emergent high-level migration phenomena and grouping patterns with spiral waves can be observed because of these complex interactions between the individuals and their environment. The distribution of individuals forming spiral waves is one property of prey-predator models ([42]; Fig. 3).

Fig. 3 A cropped image of an EcoSim run at time step 20,000. Hungry predator individuals (red) chase fleeing prey individuals (green), one of the many contributory factors to the emergent high-level movement patterns we observe

2.4.3 Adaptation

The behavioral genome's maximal length is fixed (663 genes for prey and 756 for predator), where each site corresponds to an edge between two concepts of the FCM. However, many edges have an initial value of zero; only 117 edges for prey and 131 edges for predators have nonzero values at initialization. Each gene of the behavioural genome follows the continuum-of-alleles model [19] and can take values between -12 and 12. These alleles represent the strength of the positive or negative influence of one concept on another, such as the strength of the association between level of hunger and willingness to eat. In addition to the behavioral genome, every individual has a physical genome that describes its physical characteristics, with each trait coded by one gene. Maximum energy (MaxEnergy), maximum age (MaxAge), vision (Vision), maximum speed (MaxSpeed), minimum reproductive age (RepAge), and state of birth (StateOfBirth) are physical traits that both prey and predators possess. Prey have two more traits: defense (Defense), and cooperative defense (CoopDefense), so they can protect themselves from predators. The mechanisms involving the various physical traits are described further below and under *Submodels*.

Both genomes have two representations—a lightweight byte vector representation used for efficient storage in save files and for the computing of evolutionary distances and evolutionary operations, and a floating-point vector representation used for all other computing (activation levels, action selection, physical distances, energy dynamics, etc.). The mapping between these representations differs between the genomes. Both representations are fixed at birth for the individual's lifespan. For the behavioral genome, the byte value of zero maps to the floating-point value of zero. Any byte value less than 128 is reduced by 128 and then divided by 10 to get its associated floating-point value. Any byte value greater than or equal to 128 is reduced by 127 and then divided by 10 to get its associated floating-point value. Thus, byte values from zero to 127 take the range of $[-12.7, 0]$ and byte values from 128 to 255 take the range of $[0.1, 12.8]$. For example, under this representation, a byte value of 76 yields a floating-point value of $-5.2((76 - 128)/10)$ and a byte value of 200 yields 7.3 $((200 - 127)/10)$. For the physical genome, the floating-point representation of each gene has a minimum and a step. For byte value k, its floating-point equivalent is minimum + (k $\times$ step). For instance, MaxEnergy has a minimum of 100 and a step of 25. Thus, a byte value of 17 for MaxEnergy yields a floating-point value of 525.0.

The genomes of two parent individuals are transmitted to an offspring individual after recombination and potentially some mutations. EcoSim incorporates genetic recombination through crossover, and in the behavioral genome this includes epistasis (e.g., multiple stimuli can influence a given drive) but no pleiotropy (each gene influences only one link between nodes). To model this form of linkage, alleles of the behavioral genome are transmitted by blocks. All incident edges for a given FCM node are transmitted together from a randomly selected parent with equal probability (there is no recombination among genes representing edges to a given node). Sex-linkage occurs for perception nodes, as the selected parent is of the same sex as the offspring. Sex-linkage of MaxEnergy occurs, as it is a weighted sum of that of its parents. The parent with the same sex as the offspring has five times the influence on the offsprings MaxEnergy as the other parent (Eq. (3); MaxEnergy is abbreviated to ME; subscripts o, m, and f represent offspring, mother, and father, respectively). Sex-linkage occurs for StateOfBirth as well, as an offspring's StateOfBirth is equal to that of its parent of the same sex. All genes in the physical genome are potentially mutated after crossover with some probability (t-test $p = 0.001$). A mutation on a gene in the physical genome is a modification of its byte value (randomly drawn from a truncated normal distribution between -6 and $+6$). Mutations in the behavioral genome occur due to the formation of new edges (with a probability of 0.001), removal of existing edges (with a probability of 0.0005), and changes in the weights associated with existing edges (with a probability of 0.005). The effect of a given mutation is modification of the value randomly drawn from a truncated normal distribution between -0.6 and $+0.6$ on the floating-point value of a gene. The probability of mutation in the behavioral genome is doubled for old individuals (Age $>$ MaxAgeRepAge). New genes may emerge from the initial pool of edges with a zero value. This emergence and disappearance of the genes in FCM is due to natural selection and genetic drift, which lead to adaptability of individuals [46].

$$ME_o = \begin{cases} \frac{5\times ME_m + ME_f}{6}, & \text{if offspring is female} \\ \frac{5\times ME_f + ME_m}{6}, & \text{if offspring is male} \end{cases} . \quad (3)$$

2.4.4 Fitness

To measure the capacity of an individual to survive and produce offspring that can also survive, the fitness of a species is calculated as the average fitness of its individuals. The fitness of an individual is defined as the age of death of the individual plus the sum of the age of death of its direct offspring. Accordingly, the fitness value represents the individual's ability to survive and produce well-adapted offspring. There is no predefined explicit fitness-seeking process in the simulation; rather, fitness is a consequence of natural selection. Individuals who are better adapted to the environment sustain a higher level of energy, live longer, are able to have more offspring, and transfer their efficient genomes to them [45, 46]. The fitness value is only computed for analysis of the results of the simulation and is not used in process during the simulation.

2.4.5 Prediction

So far, there is no learning mechanism for individuals and they cannot predict the consequences of their decisions. The only information available to an individual for decision-making comes from its perception at a given time step and the value of the activation level of the internal and action concepts at the previous time steps. The activation levels of the concepts of an individual are never reset during its lifetime. As the previous time step activation level of a concept is involved in the computation of its next activation level, this means that the previous states of an individual participate in the computation of its current state. Therefore, an individual has a basic memory of its own past that will influence its future behaviour. As the action undertaken by an individual at a given time step depends on the current activation level of the action concepts, the behavior of the individual depends on a complex combination of the individual's perception, the current internal states, the past states it went through during its life, and its genome.

2.4.6 Sensing

Every individual in EcoSim can perceive its local environment inside of its range of vision. Some of these senses are common between prey and predator; both can perceive nearby friends and foes, how close food is, their energy level, the amount of food in their cell, how many potential reproductive partners are in their cell, and their age. Additionally, new to EcoSim, all individuals can perceive their Strength and the maximum Strength of potential mates in their cell. Also new to EcoSim, prey individuals can sense the sum of Strength of prey in their cell and the sum of Strength of the cell within vision range that has the highest sum of prey Strength. Similarly, predator individuals can sense the sum of Strength × (1 + Defense) of prey in their cell, the distance to the cell in vision range with the highest sum of prey Strength × (1 + Defense), and the maximum strength × (1 + Defense) in their cell. These new sensory concepts serve several purposes related to the notion of prey defending against predators, new to EcoSim. With these new sensory concepts, prey can use strength-related sensory information to join a cell with other strong prey to bolster cooperative defenses. Similarly, predators can use strength-related information to avoid conflict with stronger prey individuals or groups of strong prey. Alternatively, if the predator is very strong, it may use this information to gain a larger energy reward for killing stronger prey. Individuals can only reproduce with individuals of the same type in their current cell. Having the ability to sense strong individuals and move to them means that (with the right combination of edges) there is potential to improve the chance of reproducing with strong individuals. Thus, these concepts can also lead to some potentially interesting evolutionary phenomena, such as a strength-based evolutionary arms race between prey and predator populations.

2.4.7 Interaction

In EcoSim, there are direct and indirect interactions amongst individuals and between individuals and their environment. These interactions stem from actions that prey and predator individuals can perform. The only direct interaction that requires a coordinated decision by two individuals is Reproduction. Reproduction occurs between two prey or two predators. For Reproduction to be successful, the two parents need to be in the same cell, have sufficient Energy, choose the Reproduction action, and be genetically similar. The individuals cannot determine their genetic similarity with their potential partner; they try to mate and if the partner is too dissimilar (the dissimilarity between the two genomes is greater than some percentage of the speciation threshold, by default 62.5%), the reproduction fails. See *Reproducing* under *Submodels* for more details of the Reproduction action.

The Hunting action of predators is a direct interaction that occurs between a predator and some number of prey existing in a cell. For Hunting to succeed, the predator must be able to move to the cell containing its target prey individual and it must have greater Strength than its target's Energy. Should the Hunt succeed, the prey target is killed and the predator receives some amount of Energy. The predator also receives an Energy penalty if the target prey tries to defend itself, or if other prey in the cell were defending the target. See *Hunting* under *Submodels* for more details of the Hunting action.

Lastly, there are several ways that individuals can indirectly interact with each other and their environment. An individual's perception of its local environment causes its actions and movement to be influenced by the distribution of other individuals and food resources. Moreover, individuals that share a cell compete for the limited resources that the cell contains (food and mates), and this yields density dependence. Competition generally comes in two main forms, which represent opposites along a gradient. Contest competition arises when a single individual claims all of its local resources, leaving other individuals with nothing [15]. This allows individuals to potentially monopolize resources, because strong individuals continue to claim resources while the weak starve and ultimately perish. Scramble competition, in contrast, occurs when individuals share resources equally, and are thus equally penalized by local density increases [15]. Competition in EcoSim, like in most ecosystems, is neither purely contest or scramble competition; elements of both forms of competition can be observed.

2.4.8 Stochasticity

To produce variability in the ecosystem simulation, several processes involve stochasticity. At initialization, the number of grass units is determined for each cell following a uniform random distribution (a value between 1 and MaxGrass). Similarly, at initialization, individuals are randomly distributed across the world in clusters. The simulation takes as input a clustering radius and a number of prey and predator individuals per cluster (see *Initialization and Input Data*). Let x and y be random coordinates for the center of a cluster, ClusteringRadius be the clustering radius, and k be the number of prey individuals in a cluster. Then, for each of the k prey individuals, xn and yn (the x and y coordinates for the position of the nth individual in the cluster) are produced by taking x and y and subtracting from or adding to them a random value between zero and ClusteringRadius. This process occurs until the entire initial set of prey individuals is placed in the world. The same process then occurs for the predators. The age of an individual is also determined randomly at birth from a uniform distribution in [1, 24] for prey and [1, 35] for predators. Similarly, the initial energy of an individual is randomly generated in a uniform distribution, ranging from 40 to 100% of the initial maximum energy of the individual. Age and Energy are randomly generated in this manner to avoid apparition of synchronicity in action selection and death cycles early in runs that would cause instability leading to extinction of prey or predators. The sex of an individual at initialization or at birth is randomly generated with equal probability to be male or female. Stochasticity is also included in several kinds of actions of the individuals (see *Submodels* for full descriptions of each action). For instance, if a hunting predator cannot find a prey within its vision range, the direction of its movement will be random. Furthermore, the direction of the exploration action is always random.

Mutation and crossover both involve stochasticity, as described under *Adaptation*. Furthermore, when individuals perceive their environment, they perform a radial sweep about their position along the four cardinal directions. The sweep begins at a distance of one and increments to the individual's vision range. The starting cardinal direction and the direction of the radial sweep are randomly generated to remove any biases in perception and movement. Lastly, stochasticity is incorporated into the grass diffusion model (see *Submodels* for elaboration). To understand the extent of stochasticity in EcoSim, Golestani and Gras [40] examined whether chaotic behavior (one signal of non-randomness) exists in time series generated by the simulation. The authors concluded that the overall behavior of the simulation generates emergent patterns that are non-random and are instead like those observed in complex biological systems [60].

2.4.9 Collectives

An EcoSim run persists while there is at least one prey individual. If all prey die, the run is complete due to extinction as the predators can only eat prey. EcoSim can be run with or without predators, though typically there are predators as it is designed to observe predator-prey interaction. A typical EcoSim run has 60,000–1,000,000 prey and 2000–30,000 predators at any time step, depending on the parameterization of the run.

In EcoSim, it is necessary to compute the genetic distance between any two genomes of the same type (prey or predator) in order to establish the notion of species. This distance calculation does not include sex-linked genes (see *Reproducing* under *Submodels*). To compute this distance, it is first initialized to zero. For every element of the behavioral genome in its byte vector form, the absolute difference between the pair of corresponding values from each genome is added to the distance. Subsequently, for every gene of the physical genome, a weight is computed by taking the absolute difference of corresponding floating-point values and then dividing by the range of values for that gene. This weight is then multiplied by the difference between genes, multiplied by five, and added to the distance.

Species emerge from the evolving sets of prey and predators. Species membership is strictly used in data analysis—it is not used to govern any mechanics related to reproduction. There is a separate genetic similarity threshold used for reproduction which is much lower than the speciation threshold, and this allows hybridization (reproduction between members of different species) to occur (see *Reproducing* under *Submodels*). At initialization of EcoSim, there is one species per type. Species can become extinct if all their members die. EcoSim implements a species based on the genotypic cluster definition [80] in which a species is a set of individuals sharing a high level of genomic similarity. In addition, in EcoSim, each species is associated with the average of the genetic characteristics of its members, called the 'species center'. The speciation mechanism implemented in EcoSim is based on the gradual divergence of individual genomes. The speciation method begins by finding the individual A in a species S with the greatest genetic distance from the species center. Next, the individual B in S with the greatest distance to A is found. If this distance is greater than a predefined threshold for speciation, a 2-means clustering is performed [4], otherwise S stays unchanged.

To initialize the 2-means clustering process, one center is assigned to a random individual, denoted I_r, and the other center is assigned to the individual who is the most genetically different from I_r. After eight cycles of the 2-means clustering algorithm, two new sister species are created to replace S. Each species for each type in EcoSim has a unique species identifier, starting at one and incrementing automatically when a new species is formed. Of the two sister species replacing S, one retains the species identifier of S and the other obtains the next available identifier.

2.4.10 Observation

EcoSim produces a large amount of data at each time step, recording many statistics like the number of individuals, the characteristics of each individual, and the status of each cell of the virtual world. Information regarding individual characteristics include spatial position, level of energy, choice of action, species identity, parents, FCM, etc. Information about the individuals, species, and virtual world for every 20 time steps are stored in a file, optionally using the HDF5 format [123] with an average size of 6 gigabytes. Also, there is a possibility of storing all of the values of every variable in the current state of the simulation in a separate file, creating the possibility of restoring the simulation from that state afterwards. The overall size of this file, which is only stored every 20 time steps (by default, this frequency can be modified in the parameters file) of the simulation, is a few gigabytes depending on the numbers of individuals and species. All of the data is stored in a compact special format, to facilitate storage and future analysis. There are also several utility programs that can be used, for example, to analyze the simulation outputs, to calculate the species and individual fitness, to generate images of the world for each time step of the simulation, to generate video of the world throughout a run or some portion of it, and to draw the FCM of the individuals.

2.4.11 Initialization and Input Data

A parameter file (with filename "Parameters1.txt") is defined for EcoSim, which is used to assign the values for each state variable at initialization of the simulation. Example parameters include the width and height of the world, initial numbers of individuals, thresholds of genetic distance for prey/predator speciation, speed of grass growth, probability of grass diffusion, initial maximum age, initial maximum energy, initial maximum speed, initial maximum vision range, initial values of FCM edges for prey/predators, and the characteristics of the fuzzification functions for sensory input. Any of these parameters can be changed for specific experiments and scenarios. Initialization involving stochasticity (such as the initial distribution of individuals in the world) is described under *Stochasticity*, above. Many of these initial parameters are only important in stabilization of the simulation in its early stages, before the emergent properties of the system are observable. These parameters have been tested extensively to ensure that EcoSim is stable in a wide variety of scenarios (if grass levels are low, if grass levels fluctuate regularly over time, if grass diffusion probability is reduced, if prey reproduce asexually rather than sexually, etc.). EcoSim is designed to be highly generalized. Typically, the emergent properties of at least two sets of runs initialized identically (or very similarly) with few mechanical differences

Table 2 Values for user-specified parameters

User-specified parameter	Used value
Number of prey	80,000
Number of predators	4000
Max grass quantity in each cell	4000
Prey maximum energy	2500
Predator maximum energy	3000
Prey vision range	8
Predator vision range	20

are studied and compared, to observe the effect of these few mechanical differences on the evolution of the populations. Thus, the physiological scaling rates are informed by empirical biological studies (as noted above under *Individuals*) but the aim of the initial parameters of EcoSim is to produce a stable system, and thus they are largely arbitrary. An example of a list of common user-specified parameters for initially running the EcoSim are presented in Table 2.

2.5 Submodels

2.5.1 Food Sources: Grass and Meat

There are dynamic processes for the resources in each cell, such as grass growth, grass diffusion, and variation in the amount of meat at each time step. At initialization, there is no meat in the world and the amount of grass energy units is randomly determined for each cell as described under *Stochasticity*.

The grass growth rate in each cell is regulated by several factors: SpeedGrowGrass (200 by default), ProbaGrowGrass (0.035 by default), MaxGrass (4000 by default), and Fertilizer. The first, SpeedGrowGrass, is a parameter in the EcoSim parameter file that determines the speed of grass growth. For a cell not already containing grass, grass can diffuse from an adjacent cell with a probability of ProbaGrowGrass at a rate of SpeedGrowGrass, provided that one of the eight cells around the cell contains a nonzero amount of grass. Fertilizer, a feature new to EcoSim, is derived from the excretions of individuals. AmountOfFertilizer, the amount of fertilizer in a cell, is proportional to the sum of maximum energy (MaxEnergy) of the prey and predators residing in that cell, limited to a total of 20,000. If AmountOfFertilizer is less than SpeedGrowGrass, then the fertilizer does not have any effect. Otherwise, the rate of grass growth is equal to AmountOfFertilizer and limited to triple SpeedGrowGrass. For a cell already containing grass, the rate of grass growth is simply added to the amount of grass currently in the cell at a given time step. AmountOfFertilizer decreases at a rate of 10% per time step. The amount of grass in a cell is limited to MaxGrass.

Another new EcoSim feature is that MaxGrass can be set to fluctuate cyclically following a cos wave by setting the FluctuatingResources parameter in the parameter file. The period, minimum (as a ratio of MaxGrass), and amplitude (as a ratio of MaxGrass) of the wave can be set using the parameters FluctuationCycle, FluctuationMinimumRatio, and FluctuationAmplitudeRatio, respectively. Another new feature is that MaxGrass can be set in such a way that it creates regularly positioned circular patterns throughout the world using the CircularFoodGrowth parameter. The diameter of the circles, the maximum grass level at the center of the circle (as a ratio of MaxGrass, though still limited by MaxGrass), and the minimum amount of grass in any cell (as a ratio of MaxGrass) are set using the FoodCircleDiameter, FoodCircleMaxRatio, and FoodCircleMinimumRatio parameters. FoodCircleMaxRatio is used to increase the rate at which MaxGrass increases closer toward the center of a circle, and MaxGrass increases following a cos wave from FoodCircleMinimumRatio to FoodCircleMaxRatio from the edge of a circle to the center. The amount of meat in each cell is limited to MaxMeat (4000 by default) and increases every time step by the Strength of the prey killed in that cell during that time step. It also decreases at each time step by 1000, even if no meat has been eaten in this cell.

2.5.2 Actions

For each movement action M, the movement speed is equal to MaxSpeed × ActivationLevel(M), thus the speed at which an individual moves during the action depends on its willingness to perform it. Movement speed is the straightline distance that an individual can move in a single time step. Each action has its own corresponding submodel:

1. Evading (for prey only). An evading prey moves in the direction opposite to the barycenter of the five closest predators within its vision range, with respect to its position. If no predator is within the vision range of the prey, the direction is chosen randomly.
2. Hunting (for predators only). The predator selects the closest cell (including its current cell) that contains at least one prey and moves toward that cell. If it reaches the corresponding cell based on its speed, the predator selects a prey target and tries to kill it. When there are several prey in the destination cell, one of them is chosen randomly as the target. If the speed of the predator is not enough to reach the cell, it moves at its speed toward the cell and the hunt has failed. Similarly, the hunt has failed if there is no prey in the vicinity. When a predator's hunt succeeds, the Strength of the killed prey is added to the cell in meat energy units. Afterward, the predator consumes the meat to gain its required energy, min (MaxEnergy Energy, MeatUnits), where MeatUnits is the number of meat energy units produced by the killed prey. The remaining units of meat energy are allocated to the cell and can be consumed by other predators using their Eat action. Prey have a defense capability, as well as cooperative defense, and use them in a battle against the predator [3].

Prey defense and cooperative defense is passive; prey defend automatically if they have a nonzero Defense value and are targeted by a predator, or if they have a nonzero CoopDefense value and share a cell with a target. Prey spend energy when trying to defend, and predators receive an energy penalty (P in Eq. (4), AP.D and AP.S are Defense and the Strength of the attacked prey; CPi.D, CPi.CD, and CPi.S are the Defense, CoopDefense, and Strength of the prey i in the same cell) when they attempt to attack a prey individual with defense or a cell containing prey defending cooperatively. It is even possible for a predator to be killed by defending prey, particularly if the predator already has low Energy. Additionally, the prey that are involved in a cooperative defense also lose some amount of Energy based on the strength of the predator (0.2 × PredatorStrength/NumberOfDefenders). The target prey loses Energy equal to 100% of the attacking predator's Strength if it is not cooperatively defended, otherwise it loses 80% of the attacking predator's Strength. If, after the attack, the prey's Energy is greater than zero, the prey survives and the hunt has failed.

$$P = AP \cdot D \times AP \cdot S + \sum_{i} (CP_i \cdot D \times CP_i \cdot CD \times CP_i \cdot S) \qquad (4)$$

3. Searching for food. The direction toward the closest food (grass for prey, meat for predators) within the vision range is computed. If the individual's speed is high enough to reach the food, the individual is placed in the cell containing this food. Otherwise, it moves at its speed toward this food. If no food is within vision range, the individual moves in a random direction.
4. Socializing. The direction toward the closest possible mate within the vision range is computed. If the individual's speed is high enough to reach this mate, the individual is placed in the cell containing this mate. Otherwise, the individual moves at its speed toward this mate. If no mate is within vision range, the individual moves in a random direction.
5. Exploring. A direction is computed randomly. The individual moves at its speed in this direction.
6. Resting. Nothing happens.
7. Eating. If the current amount of grass (meat) in the prey's (predator's) cell is greater than 0, the prey (predator) consumes the grass (meat) to gain its required energy, min(MaxEnergy CurrentEnergy, EnergyUnits), where EnergyUnits is the number of grass (meat) energy units in the cell. EnergyUnits is decreased by the amount consumed by the individual.
8. Reproducing. Chromosomes in eukaryotic cells are usually present in pairs (diploid organisms). The chromosomes of each pair separate in meiosis, one going to each gamete. In many animal species, sex is determined by a special pair of chromosomes called sex chromosomes (allosomes), the X and Y. All other chromosomes are called autosomes. The sex chromosomes are an exception to the rule that all chromosomes of diploid organisms are presented in pairs of morphologically similar homologs. While females have two X chromosomes, the males have one X chromosome along with a morphologically unmatched

chromosome, called the Y chromosome. All somatic cells in male and female organisms have a complete set of autosome and sex chromosomes. Every egg cell contains an X chromosome, while only half of sperm cells contain an X chromosome and the other half contain a Y chromosome. This difference is a chromosomal mechanism for determining sex at the time of fertilization. In other words, while autosome chromosomes are randomly obtained from both parents, the Y chromosome in male offspring is exclusively acquired from the father [52]. Individuals in EcoSim, in contrast to the common case, are haploid. That is, their chromosomes are present as singletons that are generated from specialized evolutionary operations described below. To model more realistic individuals, we made it so that all perception genes, MaxEnergy genes, and StateOfBirth genes exist on allosomes (that is, they are sex-linked), while all other genes exist on autosomes. Thus, there is an evolving differentiation between male and female behavior.

As per the section Process Overview and Scheduling, females intending to reproduce act first. This is because females initiate reproduction in EcoSim, to simulate female choice. Females can attempt to reproduce with any male in their cell, however, success is not guaranteed and individuals always act in order of decreasing strength. There are several ways a reproduction attempt can fail in EcoSim. Reproduction fails if there are no males in the current cell. Otherwise, the female randomly selects a potential male partner. A reproduction attempt with a single male can fail if: the male has already reproduced (with a different, stronger female), the male has selected a different action (e.g., Eat or Evade), the male is below reproduction age, the male has insufficient energy to reproduce, or the genetic distance between the female and male is too great. The genetic distance threshold for reproduction failure is greater than the speciation threshold, therefore individuals from different species can reproduce to generate hybrid offspring. In this case, the hybrid offspring is assigned to the species that has a smaller genetic difference between its average genome and the genome of the offspring. The female can attempt to reproduce with each male in the current cell, but loses two Energy for each failed attempt. If reproduction succeeds, the process of generating a new offspring consists of the following steps. When a new offspring is created, it is given a genome which is a combination of the genomes of its parents using a specialized crossover operation along with some possible mutations (as explained under *Adaptation*). The sex of the offspring is randomly determined with equal probability of being male or female. Then, the initial Energy (Energy0) of the offspring is computed (Eq. (5)) based on the parents' MaxEnergy (abbreviated to ME in the equation) and StateOfBirth (abbreviated to SOB in the equation).

$$\text{Energy}_0 = \frac{ME_f \times \text{SOB}_f \times ME_m \times \text{SOB}_m}{100}. \tag{5}$$

Finally, the Energy of the two parents is decreased. The energy penalty for the mother, penaltym, is calculated based on Eq. (6), where the subscript *m* and *f* mean mother and father, respectively. The parameter Energy is the newborn individual's Energy. FPP is the first-time pregnancy penalty for the mother, which is five percent of its energy and zero for the subsequent pregnancies. The energy penalty for the father is based on Eq. (7).

$$penalty_m = \frac{\text{SOB}_m \times \text{Energy} \times 1.05}{\text{SOB}_m + \text{SOB}_f} + FPP \tag{6}$$

$$penalty_f = \frac{\text{SOB}_f \times \text{Energy} \times 1.05}{\text{SOB}_f + \text{SOB}_m}. \tag{7}$$

9. Move2StrongestPrey/Predator (for prey/predator, respectively). The direction toward the strongest possible mate within the vision range is computed. If the speed of the individual is high enough to reach the mate, the individual is placed in the cell containing this mate. Otherwise, the individual moves at its speed toward this mate. If no mate is within the vision range of the individual, the direction is chosen randomly.
10. Move2StrongestPreyCell (for prey only). This action is similar to Move2-StrongestPrey/Predator, except that the direction of movement is toward the cell with the highest cumulative Strength of prey individuals. This allows prey to benefit from cooperative defense against predators.
11. Move2WeakestPreyCell (for prey only). This action is similar to Move2Strongest-PreyCell, but the direction of movement is toward the cell with the lowest cumulative Strength of prey individuals. This allows prey to have a higher chance of success in competition with other prey individuals in accessing food or mates.
12. Move2StrongestPreyDistance (for predator only). The predator moves toward the strongest prey individual to acquire more energy after possible hunting. If the speed of the individual is high enough to reach the prey, the individual is placed in the cell containing this prey. If the speed of the predator is not enough to reach the prey, it moves at its speed toward this prey.
13. Move2WeakestPrey (for predator only). This action is similar to Move2Strongest-PreyDistance, with the exception that the direction of movement is toward the weakest prey individual for easier hunting in the future.
14. Move2WeakestPreyCell (for predator only). This action is similar to Move2-WeakestPrey, but the direction of movement is toward the cell with the lowest cumulative Strength of prey individuals to minimize the possible effect of cooperative defense by prey individuals.

2.6 Ecological and Evolutionary Properties of EcoSim

Time-series data are generated automatically by EcoSim per time step, as explained above. We computed ten runs of EcoSim in the default configuration (which we hereby refer to as Default) to 20,000 time steps. Using external tools that have already existed, we computed the mean of several important measures for these ten runs. We computed the number of prey and predator individuals, the number of prey and predator species, the mean distance evolved of all female individuals, and three physical attributes for all female individuals (MaxEnergy, MaxSpeed, and Vision). Distance evolved is computed by first computing the mean genome for all individuals at a given time step and subsequently computing the genetic distance from this genome to the prey genome that the simulation was initialized with.

As expected, there was a dependency between number of prey and predators (Fig. 4). At initialization of the simulation, the number of prey is greater than the number of predators (80,000 and 4,000, respectively). Therefore, we tend to observe an early spike in the number of prey, which subsequently sharply declines when the number of predator individuals rises. The increasing number of prey provides a good chance for the predators to have access to more food, resulting in an increasing in their Energy and reproduction rate. The resulting increase in hunting by predators accompanied by local food resource shortages for prey decreases the number of prey, and consequently the number of predators, ultimately leading to stabilization of the system. A similar phenomenon occurs at finer spatial scales; local population explosions and extinctions yield fine-scaled fluctuations in numbers of individuals over time, with a time lag between the fluctuations in number of prey and predators. This dependence of predator population on prey population is known as the Lotka-Volterra model, as outlined in Berryman [10] and empirically corroborated by Piana et al. [105] where they fitted the model to a time series dataset of 16 species of neotropical fish that were classified as either predators or prey. These time series mostly stabilize with these small fluctuations, resulting in 268,871 prey (SD = 80,804) and 10,388 predators (SD = 2,613.4). As Britten et al. [16] observed, this stabilization

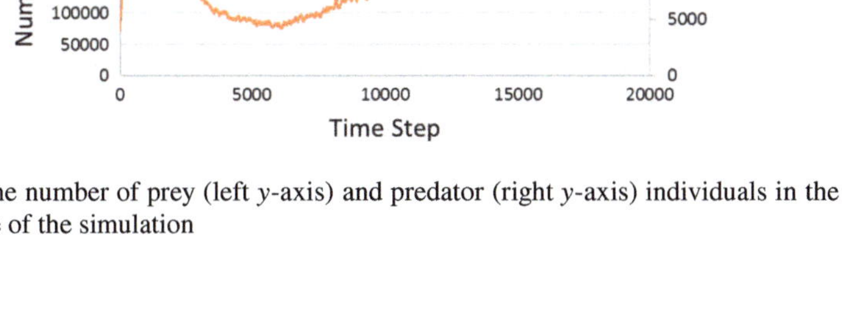

Fig. 4 The number of prey (left y-axis) and predator (right y-axis) individuals in the world, over the course of the simulation

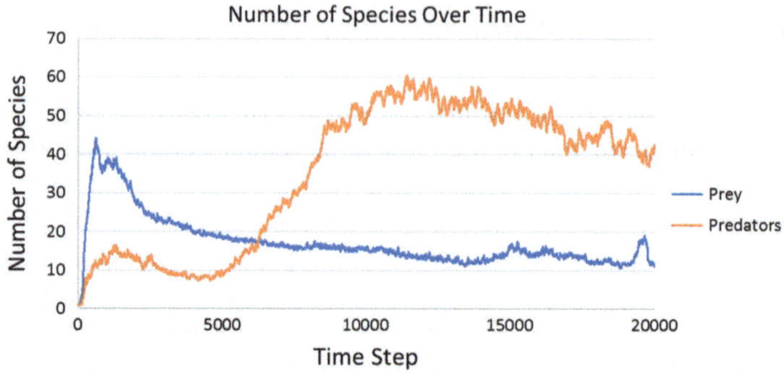

Fig. 5 The number of prey and predator species throughout the course of the simulation

can be jeopardized if there is a sudden decline in predator species in such a predator-prey system.

The number of species more strongly correlated with the number of individuals for predators than for prey (Fig. 5). Generally, an increase in the number of individuals allows for a corresponding increase in variation within the gene pool, and this increased variation tends to lead to increased speciation [62]. However, with the number of prey individuals so high, the gene flow is also very high, which results in overall genetic convergence. Spatial separation in individuals reduces gene flow. With fewer predator individuals in the world, there is greater spatial separation overall amongst predators, providing a greater opportunity for the subpopulations to genetically differentiate and ultimately yield new species. As Hoskin et al. [57] argued, reduced gene flow in allopatry results in the gradual emergence of reproductive isolation, and subsequently new species; this has been observed in EcoSim as well [43].

The prey and predator distance evolved were comparable by the end of the simulation (Fig. 6). However, at the end of the simulation, the rate of predator evolution was greater than that of prey. In fact, nearly halfway through the simulation, the distance evolved for prey hit a plateau. This highlights an important distinction—that the prey (with such a high number of individuals) evolved rapidly but in a convergent manner whereas the predators evolved more slowly but with high differentiation across all individuals. As Brodie and Brodie [17], as well as Brodie et al. [18], observe, predators that pursue prey with multiple defenses will tend to adapt evolutionarily, which may in part explain the higher rate of evolution of predators versus prey. Two main factors are responsible for the convergent evolution in prey: the aforementioned high gene flow and the fact that natural selection occurs in EcoSim, since there is no predefined fitness function [46, 61]. The fitness landscape in EcoSim is dynamic overall; both the prey and predators evolve simultaneously and the world state is constantly changing. However, many aspects of the world remain constant, such as MaxGrass, the functions that govern energy expenditure of the individuals, and the rules that govern processes like reproduction. Thus, some genetic convergence should be expected—certain behavioral and physical genotypes will be desirable regardless of the world state at any given time step. The high genetic divergence

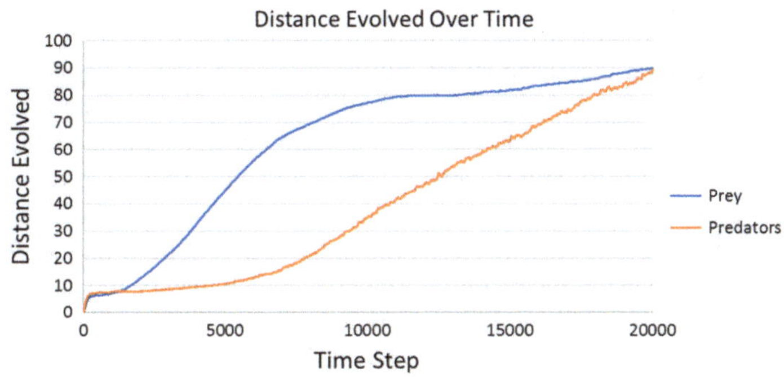

Fig. 6 The distance evolved for prey and predators throughout the course of the simulation

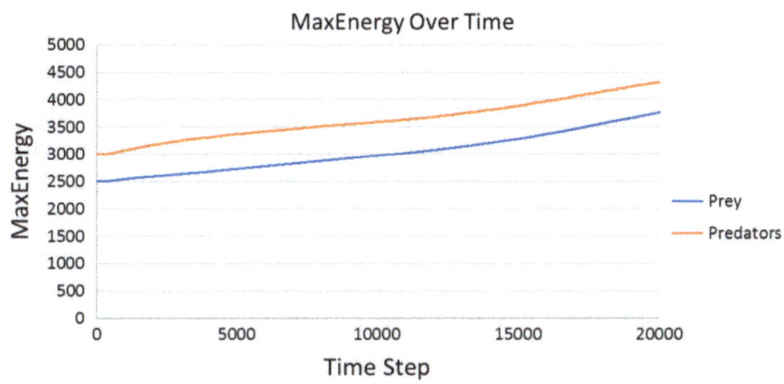

Fig. 7 The evolution of MaxEnergy for prey and predators throughout the course of the simulation

accumulated early by the predators (apparent in the number of species over time) lead to faster overall evolution later in the simulation. Another factor contributory to the fast evolution of predators later in the simulation is that there is more potential for divergence in the predator behavioral genome; the prey behavioral genome has 663 elements, whereas that of predators has 756. It is inevitable that predators will eventually evolve in a more convergent manner as well; this is observable in the subtle decrease in predator evolutionary rate over time.

MaxEnergy evolved similarly for both prey and predators (Fig. 7). In both cases, it monotonically increased from the initial values of 2500 for prey and 3000 for predators to an average of 3763 (SD = 505.7) and 4310 (SD = 372.3), respectively. As Strength is related to MaxEnergy, this could represent a type of evolutionary arms race because of the possibility of prey fighting back against predators when they attack. Alternatively, a higher maximum energy capacity may be strictly beneficial for the individuals, because it allows individuals to survive longer between Eat actions. Moller [88] performed estimates of basal metabolism rate (BMR) of 76 bird species who were pursued by predators. The author reports that birds with longer flight initiation distances used to escape predators also had higher BMRs, from which he concludes that predation creates a selection pressure on species to develop higher

BMRs [88]. Thus, it is possible that the higher maximum energy capacity is necessary in individuals due to an increased BMR. Furthermore, the energy dynamics of each physical attribute is governed in part by the energy consumption functions for prey and predators. Thus, it is possible that with a more heavily penalized MaxEnergy, it might be less prone to such a runaway. Vision and MaxSpeed are related in that individuals must both perceive a resource (a mate, food, etc.) and be able to move to it in order to use it immediately. Otherwise, the individual will have to wait for at least one time step until it can use the resource it desires, which may be too late, depending on the state of the individual and the environment around it. Thus, we should expect that Vision and MaxSpeed evolve in a related and intuitive manner. Predator Vision and MaxSpeed appeared to be heavily related in the way we expected (Fig. 8). That is, both Vision and MaxSpeed evolved to slightly increase and then slightly decrease, nearly in unison, with Vision always greater than MaxSpeed. This is intuitive because it is particularly imperative for predators to perceive their resources; potential mates are far less abundant for predators and their food resources are constantly changing positions in the world. This observation has been empirically corroborated in a study of predatory bird species conducted by Garamszegi et al. [37], in which it was found that predatory species evolved increased visual acuity along with larger brains to detect prey. On the other hand, it is less important for prey to perceive their resources, but it is important for prey to move quickly to evade predators. Potential mates and food resources are far more abundant for prey, and their food resources are static in the world (unless a cell's grass is fully consumed before the prey can reach it). Furthermore, over time, we observed that prey tended to perform the Evade action decreasingly, while they increasingly performed Explore instead (Fig. 9). The directionality of the Explore action is randomly generated, and with the high prey density, it is possible that when they Explore, they can randomly discover mates or food resources while they simultaneously evade predators. If all prey in a particular wave performed Evade when faced with a predator, many of the prey individuals would move in a similar direction, which could increase competition for resources. On the other hand, increasingly performing Explore may be evidence of the evolution

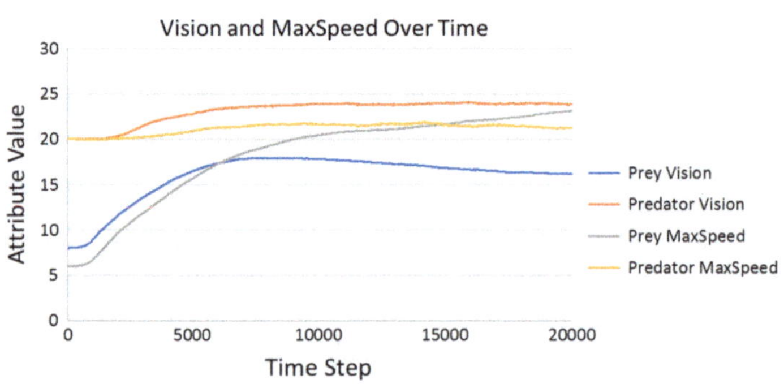

Fig. 8 The evolution of Vision and MaxSpeed for prey and predators throughout the course of the simulation

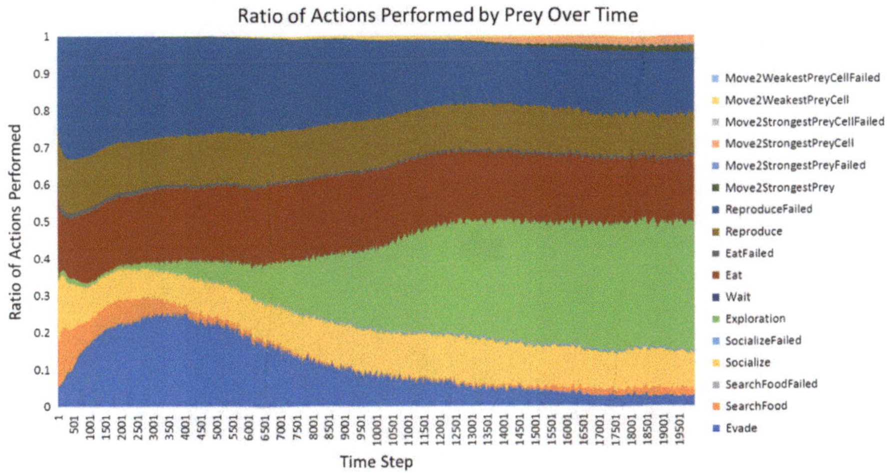

Fig. 9 Selection of actions by prey over time. Prey evolved to Evade less and Explore more, while simultaneously reducing their reproduction failure rate (ReproduceFailed). Evolution of an increase in Move2StrongestPreyCell and Move2StrongestPrey is also observed

of altruism; if a small percentage of prey purposely sacrifice themselves by moving towards the wave of predators (using Explore rather than Evade), it keeps the wave of predators away from the highest-density prey regions.

2.7 *Divergence of Sister Species*

From a single Default EcoSim run, we found two sister species (species 1 and species 40) that coexisted for 1860 time steps. Species 40 was produced at time step 246 of this particular run and went extinct at time step 2106, while species 1 was produced at initialization and persisted to the end of the simulation. We analyzed divergence of the behavioral and physical genomes of these two species throughout their coexistence.

In EcoSim, depending on the genomes within a species, differentiation of very few genes can be sufficient to trigger a speciation. When species 40 was initially produced, only one gene in the behavioral genome showed a high degree of differentiation, though 1500 time steps later the species were highly diverged in other ways (Fig. 10). Interestingly, in this case, the allele that caused the initial speciation disappeared from species 40 over 1500 time steps. This indicates that although the appearance of this allele was sufficient to cause speciation, the allele was likely deleterious and was evolved out of the species over time. This was corroborated by the fact that the change that caused the initial speciation was an evolution of the mean value of the gene FriendFar→Move2StrongestPreyCell to 0.41 in species 40, which had a mean value of −0.00020 in species 1. With no friend nearby, attempting to move to the cell with the highest cumulative strength would likely be a waste of energy and an action. Furthermore, the genetic distance between behavioral genomes of

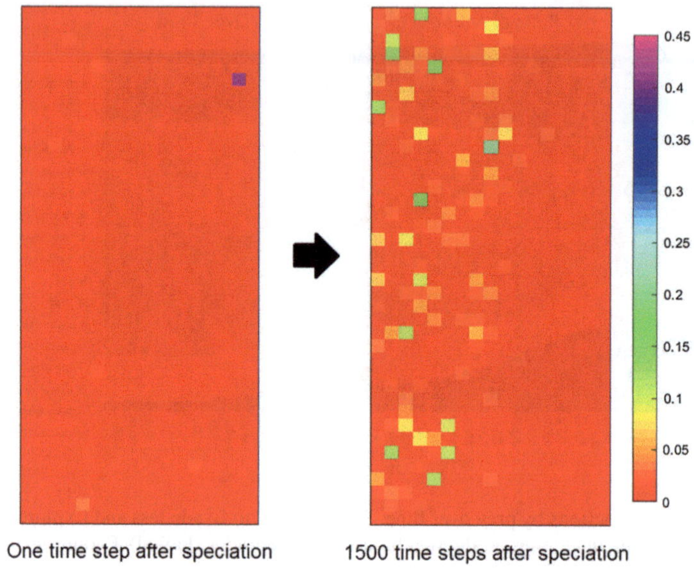

Fig. 10 Divergence of behavior models of two sister species over 1500 time steps. Each square represents the absolute difference of a gene in the average behavioral genome of two sister species from a single run of EcoSim. Though differentiation of one gene was sufficient to cause the initial speciation (purple square in left heatmap), over time, the behavioral genomes diverged substantially

these two species declined over the first 175 time steps after speciation (due to the loss of the aforementioned allele in species 40) and then rose over the subsequent time steps due to the differentiation in the other behavioral genes (Fig. 11). Another factor contributing to the initial decline in genetic distance is low spatial separation (implying high gene flow) between the species shortly after the speciation event, with increased spatial separation and genetic divergence thereafter. EcoSim allows hybridization (reproduction between individuals of different species; see *Collectives, Reproducing* under *Submodels*), thus when two species are genetically similar enough and not spatially separated, their individuals can reproduce to form hybrid offspring. Being sister species generated very early in a run, the physical genomes between these species did not differentiate.

2.8 Sensitivity Analysis of EcoSim

In addition to the ten Default EcoSim runs noted above, we computed ten runs each of EcoSim with the following modifications: reduction of initial social action edges related to defense by 25% of their default value (referred to hereon as RSE25), reduction of initial social action edges related to defense by 50% of their default value (referred to hereon as RSE50), reduction of energy spent by all individuals by 25% of

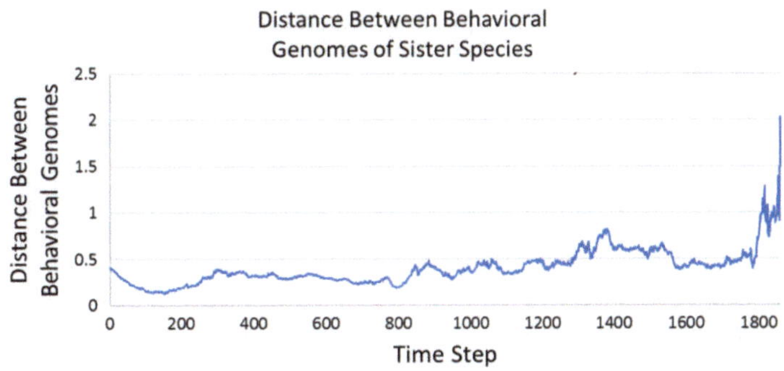

Fig. 11 Genetic distance between behavioral genomes of sister species throughout their coexistence, computed as Euclidean distance between average behavioral genomes. After a slight decline in genetic distance due to the loss of a deleterious allele in the smaller species and hybridization along the interface of the two species, the behavioral genomes of the species diverged over time

the default (referred to hereon as RE25), reduction of energy spent by all individuals by 50% of the default value (referred to hereby as RE50), reduction of MaxGrass by 25% of the default value (referred to hereby as RMG25), and reduction of MaxGrass by 50% of the default value (referred to hereon as RMG50). For RSE25 and RSE50, the affected edges for prey were all edges incident to Move2StrongestPrey, Move2StrongestPreyCell, and Move2WeakestPreyCell. The affected edges in these runs for predators were all edges incident to Move2StrongestPredator, Move2-StrongestPreyDistance, Move2WeakestPreyCell, and Move2WeakestPrey. For each of these runs, we computed the mean across ten runs and across time steps 5,000 through 6,000 for the following measures: number of prey and predator individuals, number of prey and predator species, mean energy level of all female prey individuals, and mean energy level for all female predator individuals. We computed these values over a window of 1,000 time steps, because many of the above measures show different behaviors at different scales. For instance, the number of prey or predator individuals at a very high scale may appear to follow the classic growth curve, with an initial lag period followed by a period of nearly linear growth that reduces in rate of increase as it approaches its asymptote (the carrying capacity of the system) and ultimately oscillates below the asymptote. At a smaller scale, however, many small cycles can typically be observed due to local population explosions and extinctions. For each treatment (reduction of energy spent, reduction of maximum grass, and reduction of social edges related to defense), we compared values of each observation to the respective values generated by Default EcoSim runs and determined the percent change in these observations. This allowed us to determine how sensitive or robust the system is to these changes, and it also allowed us to determine how these observations behaved relative to the different treatments (for example, to determine if a reduction in MaxGrass yields a linearly dependent reduction in number of prey individuals or number of prey species).

We expected that modifications in the action edges related to defense would yield nonlinear and nonmonotonic relationships to most of the dependent variables, as

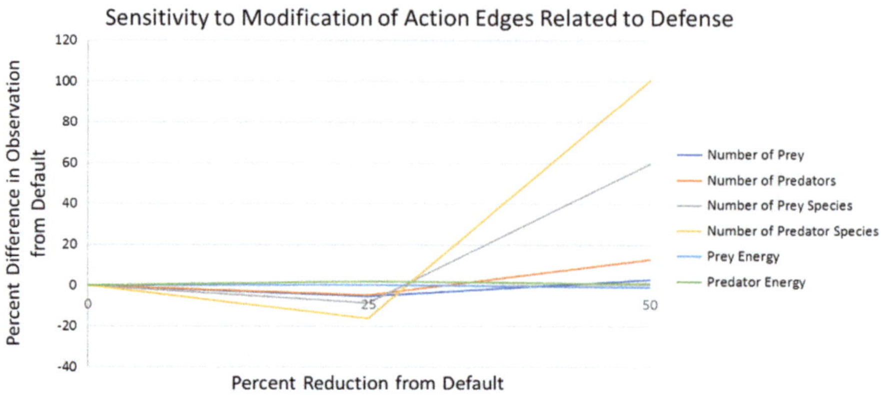

Fig. 12 Sensitivity of several variables to modification of action edges related to prey defense

we applied this modification to both prey and predators. None of the measures we computed were sensitive to these modifications (t-test $p \geq 0.15$ in all cases), and the amounts of energy of prey and predator individuals were particularly insensitive. Interestingly, both prey and predator number of individuals and species declined slightly when these edges were reduced by 25%, if both increased when these edges were reduced by 50%, but insignificantly so (Fig. 12). Though the percent difference from Default runs was very high for some of these measures, the difference was statistically insignificant due to extremely high variance (only one run was responsible for these very high values).

Modifications to the rate of Energy consumption of both prey and predators significantly impacted all of the variables we analyzed (t-test $p \leq 0.0006$ in all cases, Fig. 13). The number of prey increased to 208% of the Default value with a 25% decrease in Energy consumption, and increased further to 277% of the Default value with a 50% decrease in Energy consumption. The number of predators followed a similar trend, increasing to 431 and 626% of the Default values, respectively. Both prey and predator numbers seemed affected by diminishing returns based on reduction of Energy consumption, most likely due to increased competition when Energy consumption was decreased. The effect of reduction of Energy consumption was stronger at higher trophic levels; the effect of Energy consumption on number of predators was almost double that on number of prey. Not surprisingly, the number of predator and prey species both increase significantly with reduction of Energy consumption, though the number of prey species closely followed the trend of number of prey individuals. The number of predator species, on the other hand, did not follow the trend of the number of predator individuals; there appeared to be a tipping point where decreasing Energy consumption actually decreased the number of predator species, despite their increasing number of individuals. This is due to the interplay between genetic variation across the population and gene flow; more individuals allows for more potential genetic variation (which should increase the number of species), but more individuals also increases gene flow (which should decrease the number of species). Decreasing the Energy consumption of predator and prey

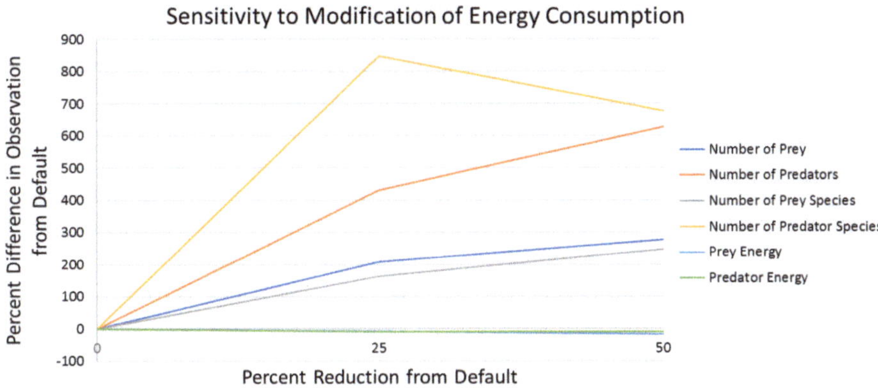

Fig. 13 Sensitivity of several variables to modification of Energy consumption per time step for both prey and predators

individuals actually decreased their mean Energy levels by 8–16%. The decreasing of Energy consumption provides the individuals with increased longevity and potential to reproduce, because their physical and behavioral traits are energetically less expensive to maintain. Thus, as we observed, the number of individuals increases drastically and disproportionately given the reduction in Energy consumption. With such a drastic increase in the number of individuals given the same food resource supply, competition strongly increased as well. Consequently, the individuals have a significantly lower Energy level.

Modifications to MaxGrass proportionally (and almost linearly) affected some variables while nonlinearly affecting other variables (Fig. 14). The differences between RMG25 and Default runs were almost all statistically significant (t-test $p < 0.01$ for all comparisons, except predator number of species, $p = 0.10$, and predator energy, $p = 0.33$). Similarly, differences between RMG50 and Default runs were mostly very statistically significant, yielding t-test $p < 0.0001$ (except predator number of species, which was still significant, with $p = 0.0015$, and prey energy, which was not, with $p = 0.78$). For instance, with a 25% and 50% reduction in MaxGrass, the number of prey individuals was reduced by 51.7% and 65.8%, respectively. Similarly, the number of predators were respectively reduced by 41.2% and 47.3%. With a 25% reduction in MaxGrass, both prey and predator number of species decreased (by 28.1% and 41.0%, respectively), while they both increased (by 309.9% and 146.3%, respectively) with a 50% reduction in MaxGrass. With only a 25% reduction in MaxGrass, it is possible that the reduction in number of prey and predators is sufficient to reduce the genetic variation across the populations while insufficient to reduce gene flow such that speciation increases. Thus, a net decrease in the number of species of each type was observed. Conversely, with a 50% reduction in MaxGrass, the number of individuals was so greatly reduced that gene flow between subpopulations was practically halted, which allowed for very high differentiation between spatially separated individuals, and consequently a high number of species. The vast difference in number of predator species given such a slight change

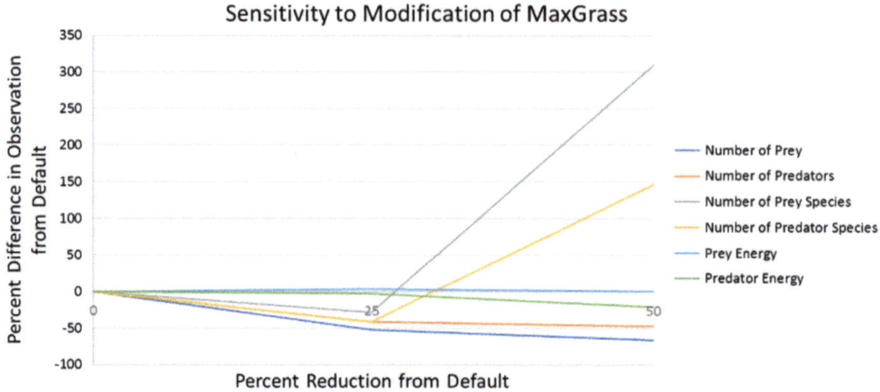

Fig. 14 Sensitivity of several variables to modification of MaxGrass

in number of predator individuals between RMG25 and RMG50 runs could also be explained by increased fragmentation of prey subpopulations. The predators must follow the prey in order to survive, and spatially fragmented prey subpopulations should yield spatially fragmented predator subpopulations. Interestingly, prey and predator energy levels were largely unaffected by this modification, though predator energy was reduced by 20.6% in RMG50 runs. Overall, some aspects of the system are sensitive to MaxGrass and many others may be nonlinearly affected by modifications to it.

3 Case Study: Application of EcoSim to Study Behavior and Evolution Under Conditions of Reduced Primary Production and Reduced Energy Expenditure

The focus of this case study will be twofold: to investigate possible links between both intraspecific and interspecific competition for resources and evolution, as well as examine possible links between energy expenditure of organisms and evolution.

First, a number of studies have found evidence of a link between competition within and between species and the evolution of morphology, as well as the evolution of resource polymorphism and temporal variation. Pafilis et al. [101] maintain that, in general, resource availability and competition (and predation) drive the evolution of body size. They conducted an empirical study in which they showed that in the presence of a high number of breeding seabirds, there is an increase in lizard population densities, which in turn results in increased intraspecific competition for resources [101]. Pafilis et al. [101] report that the resultant increase in competition for resources leads to the evolution of large body sizes (gigantism) in a species of lizards (*Podarcis gaigeae*). Along the same lines, Svanback et al. [121], in an empirical study, report that a species of perch (*Perca fluviatilis*) and a species of roach

(*Rutilis rutilis*) that cohabitate two regions of a lake were deeper bodied in the littoral region versus individuals caught in the pelagic region, which they attributed to intraspecific competition. On the other hand, Grant and Grant [44] discovered that interspecific competition between two species of Darwins finches (*Geospiza fortis* vs. *G. magnirostris*) resulted in the divergence of beak sizes.

In addition, Svanback et al. [121], cited above, found evidence of resource polymorphism in the fish and roach species, so that fish and roaches in the littoral region fed on different sorts of organisms versus their counterparts in the pelagic region. Svanback and Bolnick [120] studied a sympatric population of three-spine stickleback fish (*Gasterosteus aculeatus*) for which there was an increase in population density, thereby increasing intraspecific competition for prey. The result was diet variation between phenotypically different stickleback individuals so that some fish found alternative prey [120], although the authors attributed some of this resource polymorphism to phenotypic plasticity rather than to evolution. Marini et al. [81] demonstrated that interspecific competition between two species of mosquito (*Cx. pipiens* and *Ae. albopictus*) resulted in a shift in temporal dynamics for both species. The result is that the species tend to be in a common breeding site at different times to minimize overlap [81]. Strauss et al. [119] note that few studies have investigated the evolutionary effects of invasive species on native species. In reviewing studies on a variety of animal species, the authors conclude that, amongst other contexts (e.g., predation), invasive species as competitors drive evolution in native species [119].

Secondly, recent biological research has uncovered possible links between energy expenditure and animal morphology, as well as the rate of evolution. In a comprehensive literature review, Niven and Laughlin [95] report that the reduction of energy expenditure has driven the evolution of the morphology of sensory systems in a wide variety of vertebrate and invertebrate animal species. For example, animals that live on islands where there is limited energy due to scarce resources tend to lose some of their sensory systems, such as vision, in order to conserve energy [95]. In the same vein, Navarrette et al. [93] argue that the evolution of encephalization in humans is the result of the stabilization of energy inputs along with a redirection of energy from locomotion, reproduction and growth. Furthermore, Jasienska [59] hypothesizes that reproductive suppression in humans has evolved as a way of dealing with low energy. As Leonard and Ulijaszek [78] report, the role of energetics in the evolution of humans is an emerging domain.

Using a plethora of data relating to substitution rates for mitochondrial and nuclear genomes of a variety of vertebrate and invertebrate organisms, Gillooly et al. [38] argue that there is a direct link between the rate of energy transformation in metabolism and the rate of nucleotide substitution. In particular, they claim that smaller organisms (with a higher metabolic rate) evolve faster than larger organisms. Using a DNA-based phylogenetic analysis of 86 angiosperm plant sister species across environments with varying energy levels, Davies et al. [24] found that evolutionary rates are higher amongst populations in higher energy environments. According to the authors, many non-energetic variables such as geographical complexity and history contribute to species richness and rate of evolution in plant species, so that discerning the role of energetics with respect to these phenomena is an important area of investigation. Finally, an empirical study conducted by Mönkkönen et al. [90] found that energy use in a variety of North American and European forest birds translated into species diversification.

Besides shedding additional light on the connections between competition, energetics, and evolution, this study will help to address several open questions in ecology and evolution relating to these issues. First, our study will help to elucidate the effects of competition for limited resources on evolution. Secondly, our simulation study will help to determine the role of energetics in the evolution of morphology, which is regarded as an emerging domain by Leonard and Ulijaszek [78].

Using five runs each of the aforementioned Default, RMG25, and RE25 Eco-Sim variants, we aimed to determine differences in the way which individuals behave and evolve under conditions of reduced primary production (modelled by the RMG25 runs) and energy expenditure (modelled by the RE25 runs). Of the ten runs of each variant, we selected the five runs that were most progressed due to computational time constraints. To determine differences in behavior, we computed the mean percentage of individuals performing each action at each time step across the five runs of each type, and analyzed these time series data for differences over time. To determine differences in evolution of the behavioral genome, we compared distance evolved over time. Furthermore, we compared the evolution of physical traits such as vision range (Vision), maximum energy (MaxEnergy), and maximum speed (MaxSpeed). Since the RMG25 runs have lower amounts of grass (and consequently lower carrying capacities for prey and predators), they run very fast. Therefore, we analyzed the RMG25 runs to 20,000 time steps. The RE25 variant allows individuals to retain more energy and survive better, thus there are significantly more predators and prey in these runs. Consequently, they run slower, and we had to limit our analysis of RE25 runs to 10,000 time steps.

3.1 Reduced Primary Production

Both prey and predators evolved differently in several ways when primary production was reduced, compared to the Default scenario (Fig. 15). The amount of differentiation between the mean behavioral genome at a given time step and that at initialization (Distance Evolved) showed stark contrast between the two scenarios for prey starting at approximately 7000 time steps (Fig. 15a). Prior to that, prey living in high primary production evolved faster than those living in reduced primary production, and sometimes significantly so (t-test $p < 0.05$). However, after 7000 time steps, the prey in an environment with low primary production evolved much faster (t-test $p < 0.001$). The same phenomenon was observed for predators, however, the point at which those living in low primary production evolved further than those in high primary production came much later, at approximately 18,300 time steps. Friman et al. [36] report that the evolution of prey-predator interactions is driven by the availability of prey resources, although evolution of anti-predator defenses was greater in the presence of high resources. Along the same lines, Hiltunen et al.

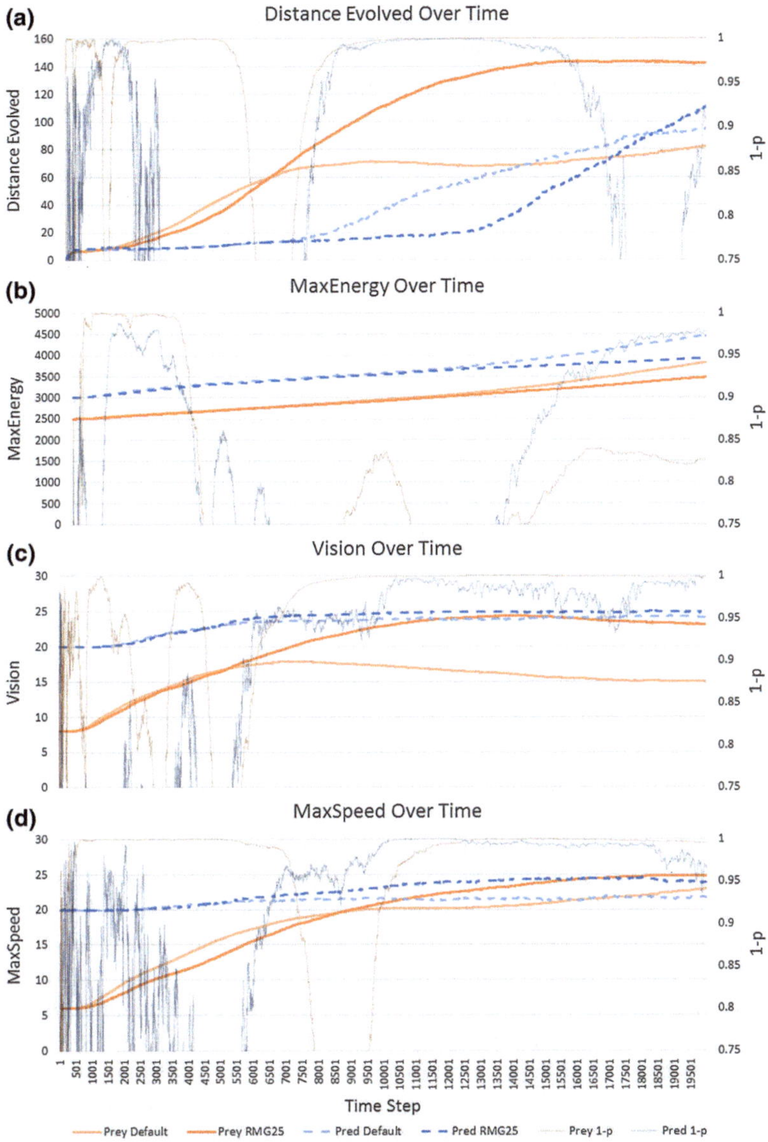

Fig. 15 Comparison of four measures related to evolution of prey and predator individuals between Default and RMG25 runs over time. Each measure uses the left y-axis while the t-test $1-p$ value uses the right axis. T-test $1-p$ value shows the significance of difference between Default and RMG25 runs for prey and predators separately. Distance evolved **a** is the genetic distance between behavioral genomes at initialization and the mean of all individuals at a given time step. MaxEnergy (**b**), Vision (**c**), and MaxSpeed (**d**) are physical properties determining the maximum energy capacity, vision range, and maximum movement speed of individuals, respectively. Values shown are the mean of all individuals alive at the given time step

[55] report that in an experimental predator-prey system involving bacteria (*P. fluorescens*) and ciliates (*T. thermophilia*), evolution of anti-predator defenses evolved at a higher rate in stable resources versus fluctuating resources. All of these results agree with what we found in our simulations. We have two main hypotheses as to why we observe these phenomena, and they are not mutually exclusive. First, this is a long-term evolutionary effect of differences in density. Reduced density of prey and predators when primary production is reduced caused a reduction in gene flow, which has been shown to increase evolutionary rates of populations. Secondly, as Distance Evolved is a measurement of evolutionary change from the initial populations, it is quite possible that the initial behavioral genome is simply more similar to the optimal genomes of the Default runs. Disputing this claim, the optimal genome is a moving target in EcoSim, as there is no fixed fitness function and the state of the simulation is highly dynamic. Furthermore, Distance Evolved is showing increasing trends in all cases, and it is impossible to determine whether it will ever equilibrate. Currently, we cannot force EcoSim to retain a constant density of prey or predators despite changing environmental conditions, which is a limitation in this particular situation. However, it is much more realistic, as in nature, the density of individuals is always dynamic and influenced by environmental conditions.

MaxEnergy displayed an increasing trend overall (Fig. 15b) and individuals in an environment with high primary production evolved a higher energy capacity—statistically significantly so in the case of predators and approaching statistical significance for prey. It is reasonable that individuals from an environment with high primary production evolve a higher energy capacity. Prey individuals can consume all of the Grass contents of a cell in a single Eat action, and each action is a highly valuable resource. Thus, it is highly beneficial to prey to obtain and retain as much Energy as possible when they do perform Eat. With a higher MaxEnergy, prey individuals can use their Eat actions more efficiently by storing more Energy per eat event. As Lewis and Kappler [76] observe, female lemurs (*Propithecus verreauxi verreauxi*) that inhabit seasonal environments will have higher body mass when there is an abundance of resources during the wet season, and during this time, they are more likely to reproduce and wean infant offspring. Furthermore, MaxEnergy influences Strength, as both are proxies for the size of the individual. A predator must have greater Strength than its prey target has Energy for a Hunt action to succeed. Thus, as prey MaxEnergy increases, that of the predators must as well. What we are observing is an evolutionary arms race between prey and predators, and the maximum amount of primary production in each cell significantly impacts the way in which this arms race occurs, as noted by Friman et al. [36].

Vision reached an equilibrium with high and low primary production for both predators and prey, except in the case of predators in low primary production, in which it evolved to its maximum value of 25 (Fig. 15c). For both predators and prey, after approximately 10,000 time steps, the difference in Vision between runs with high and low primary production was almost always statistically significant (t-test $p < 0.05$) and individuals living in low primary production evolved higher Vision. This result shows that despite the Energy cost of maintaining Vision, there is a significant advantage to being able to perceive more potential resources (such as mates and food) and competitors, particularly when food resources are scarcer. As Eklöf et al. [30] report, five species of insectivorous bats of the family Vespertilionidae developed different types and levels of visual acuity depending on the type of foraging they engaged in. Along similar lines, Potier et al. [106] observe that the visual abilities of two raptor species (*Parabudteo unicinctus* and *Milvus migrans*) differ according to their foraging activity. Reduced primary production effectively reduces the carrying capacity per cell, which increases the intensity of competition for resources within each cell. Thus, it is imperative to the survival of individuals to be able to obtain information about the locations of potential food and competitors so they can reduce their competition. In the same way, individuals evolve to move faster when primary production is reduced (Fig. 15d). Having a higher MaxSpeed aids in the dispersal of individuals, which serves to reduce competition amongst them. As stated earlier, MaxSpeed and Vision are highly related and tend to evolve together, because individuals can only move to positions with resources when they perceive these resources. Thus, the emergent pattern of evolution of MaxSpeed mirroring that of Vision is not surprising, and the difference between runs with high and low primary production were, again, mostly significant after time step 10,000. Similar to the evolution of MaxGrass, the evolution of both Vision and MaxSpeed may also represent an evolutionary arms race—a higher Vision range and MaxSpeed in prey means that they can perceive and evade potential threats more easily. We observed slight overall increases in both Vision and MaxSpeed in predators as well, despite the fact that these traits were already initialized to much higher values for predators.

Mirroring the differences in rate of behavioral evolution between the two EcoSim variants noted above, we observed many differences in the resultant behavior selections of the individuals (Fig. 16a, b—prey; Fig. 17a, b—predators). For prey, we observed significant differences in Reproduce and ReproduceFail. Overall, prey in an environment with high primary production both succeeded and failed to reproduce far more than those in an environment with low primary production, as they attempted to reproduce far more often (t-test $p < 0.05$ for much of the time series). The reason for this is twofold: Reproduce is very costly in terms of Energy, and Reproduce requires that individuals are in the same cell. Due to the Energy cost of reproduction, when primary production is low, individuals reduce reproduction to save Energy. Furthermore, as Reproduce requires individuals to be in the same cell, with lower prey density, this is much harder to achieve when primary production is low. We observed insignificant differences in Eat success, but in an environment with reduced primary production, EatFail was significantly higher (t-test $p < 0.05$ for most of the time series). This indicates that the prey were heavily affected by competition for food resources. Initially counterintuitively, we observed that prey Socialized significantly more often when primary production was reduced (t-test $p < 0.05$ for most of the time series). This was counterintuitive because Socialize brings individuals together, and it was expected that prey would aim to reduce their competition with reduced primary production by reducing their Socialization. Furthermore, we observed insignificant differences in prey Compactness, the mean number of prey individuals per cell for all cells containing at least one prey individual (defined analogously for predators). However, with lower prey density, Socialize is an important mechanism for improving

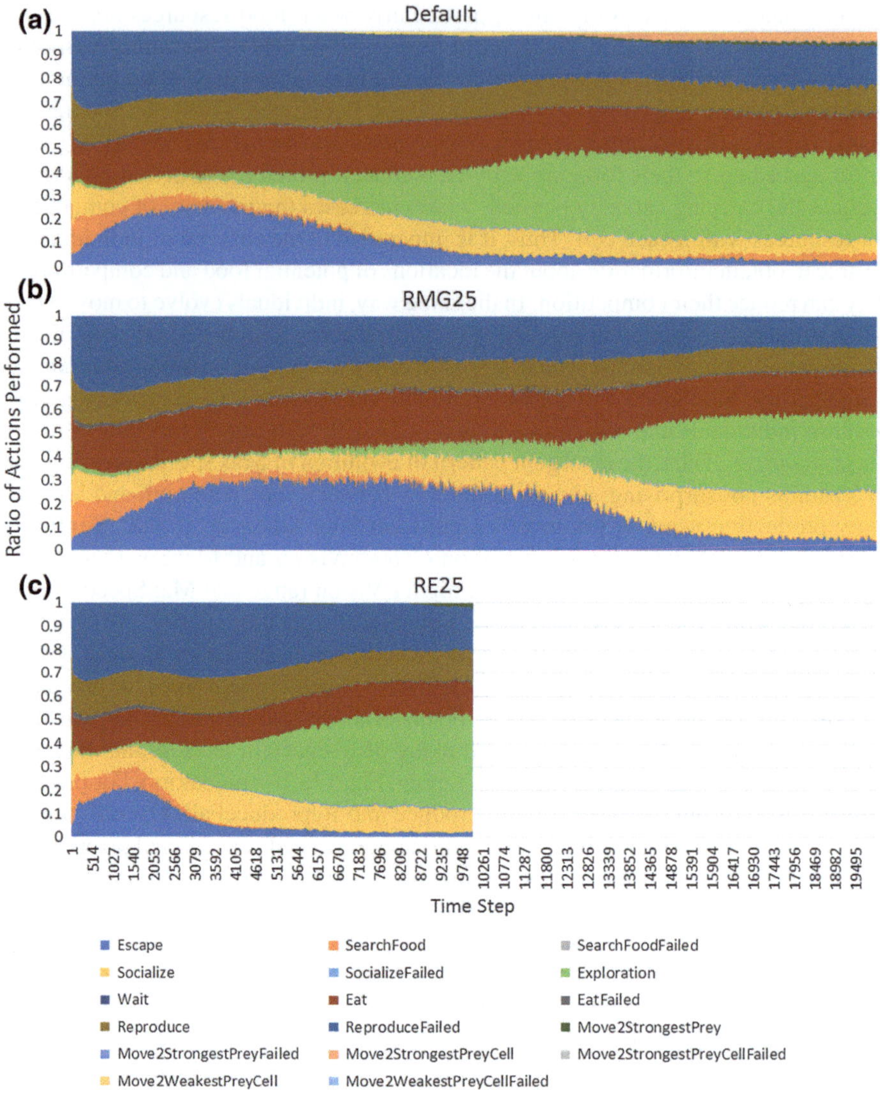

Fig. 16 Mean ratios of actions performed by prey in Default, RMG25, and RE25 EcoSim runs. Due to computational time constraints, RE25 runs were terminated at 10,000 time steps

reproduction success, as reproduction requires that mates be in the same cell. Because Compactness was not different despite significant differences in Socialize, it is likely that prey in RMG25 runs Socialize as a means to increase Reproduce success, and then disperse after in order to reduce subsequent competition. In fact, we found that actions aiding in dispersal (Escape, SearchForFood, and Explore) were performed 19% more often after Reproduction attempts in RMG25 runs than in Default runs,

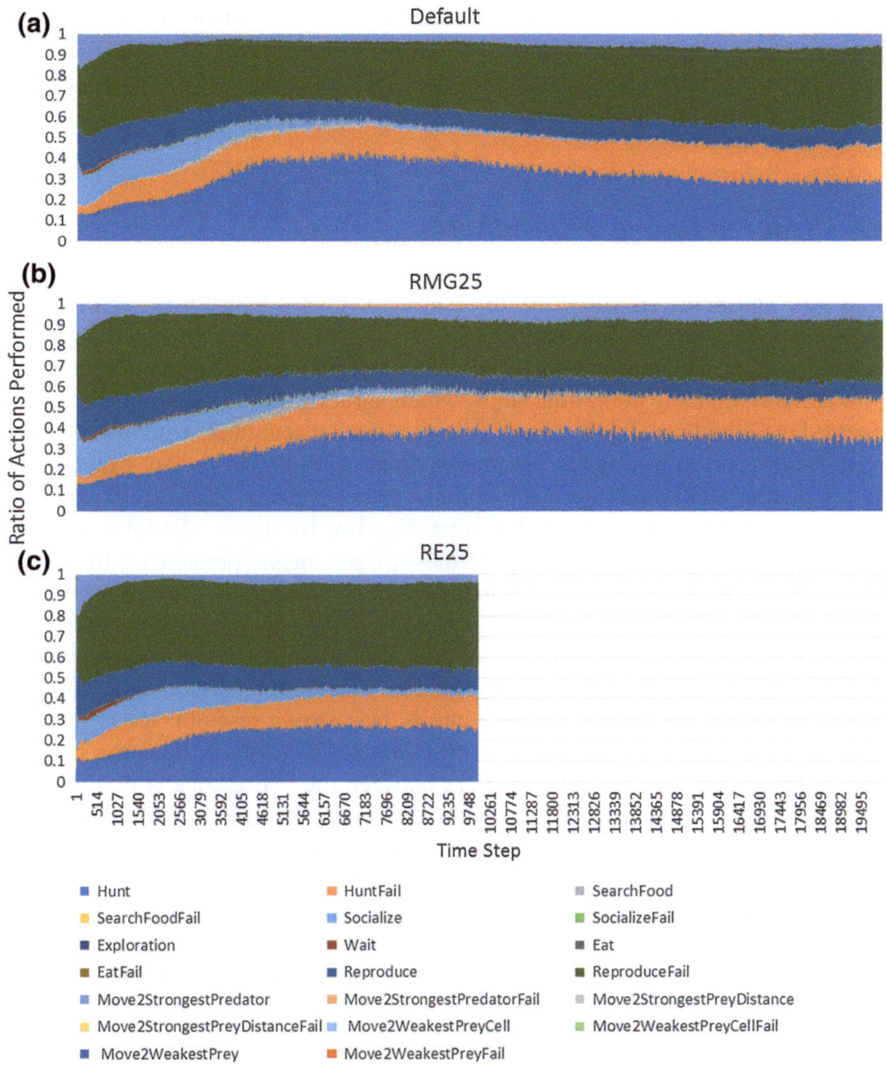

Fig. 17 Mean ratios of actions performed by predators in Default, RMG25, and RE25 EcoSim runs. Due to computational time constraints, RE25 runs were terminated at 10,000 time steps

and Reproduce was attempted after Socialize 21% more often in an RMG25 run than in a Default run. To determine this, we tracked all actions performed by all individuals born in time steps 20,000–20,010 for a single Default and RMG25 run.

Similar actions ratios overall were observed for predators between the two run types, but there was significantly more Reproduce success with high primary production for the same reasons for which we observed this phenomenon in prey (t-test $p < 0.05$ for much of the time series). With many of the other actions yielding insignificant difference, the other time series affected by primary production was the ratio of Hunt actions performed. Predators must Hunt more with low primary production because prey are scarcer, they must take the opportunity to obtain Energy when the opportunity presents itself.

3.2 *Reduced Energy Expenditure*

Both prey and predators followed different evolutionary trajectories with reduced Energy expenditure when compared to Default EcoSim runs (Fig. 18). The behavioral genomes of prey with reduced Energy expenditure evolved faster than in Default runs prior to approximately 4000 time steps, which agrees with the experimental results obtained in Gillooly et al. [38], in which it was found that animals with lower energy expenditure evolved at a faster rate than animals with higher energy expenditure. However, after 4000 time steps we found that prey with reduced energy expenditure lagged behind in rate of evolution thereafter (Fig. 18a, t-test $p < 0.05$ for most of the time series). Prior to 10,000 time steps, behavioral genomes of predators with reduced Energy expenditure evolved faster than their Default counterparts, which once again agrees with the results obtained in Gillooly et al. ([38], although near the end of the runs, it appeared inevitable that Default predators would ultimately overtake those in RE25 in terms of Distance Evolved (Fig. 18a, t-test $p < 0.05$ until approximately 9500 time steps). The shapes of these curves bear strong resemblance to those of Distance Evolved when Default runs were compared to RMG25, however, here the roles are reversed. The common element between the two graphs is that the runs with significantly higher numbers of individuals exhibited faster evolution in prey and predators early in the run, only to be overtaken by the runs with lower number of individuals later on. This corroborates our speculations regarding the links between number of individuals, spatial separation of individuals, and gene flow.

Similarly, the shape of curves for MaxEnergy over time comparing Default and RE25 runs (Fig. 18b) are very similar to those comparing Default and RMG25 runs, though again, the roles are reversed (MaxEnergy in RE25 runs is greater than that in Default runs, t-test $p < 0.05$ after 5600 time steps for prey, $p < 0.001$ after 500 time steps for predators). Of all the determinants of Energy expenditure, MaxEnergy (a proxy of the size of the individual) plays the strongest role for both prey and predator individuals, as it is penalized directly in the Energy functions and also indirectly through the cost associated with Speed of movement in a given time step. Thus, as expected, individuals with reduced Energy expenditure per time step evolved to be larger, more rapidly.

Conversely, we did not entirely expect what we observed for evolution of Vision and MaxSpeed when comparing Default runs to RE25 (Fig. 18c, d), in light of the Energy costs associated with maintaining these features. We observed that individuals from Default EcoSim runs evolved greater Vision and MaxSpeed than their RE25 counterparts (t-test $p < 0.05$ after 4000 time steps), in all cases except for MaxSpeed of prey. The results pertaining to visual acuity do, in fact, agree with empirical find-

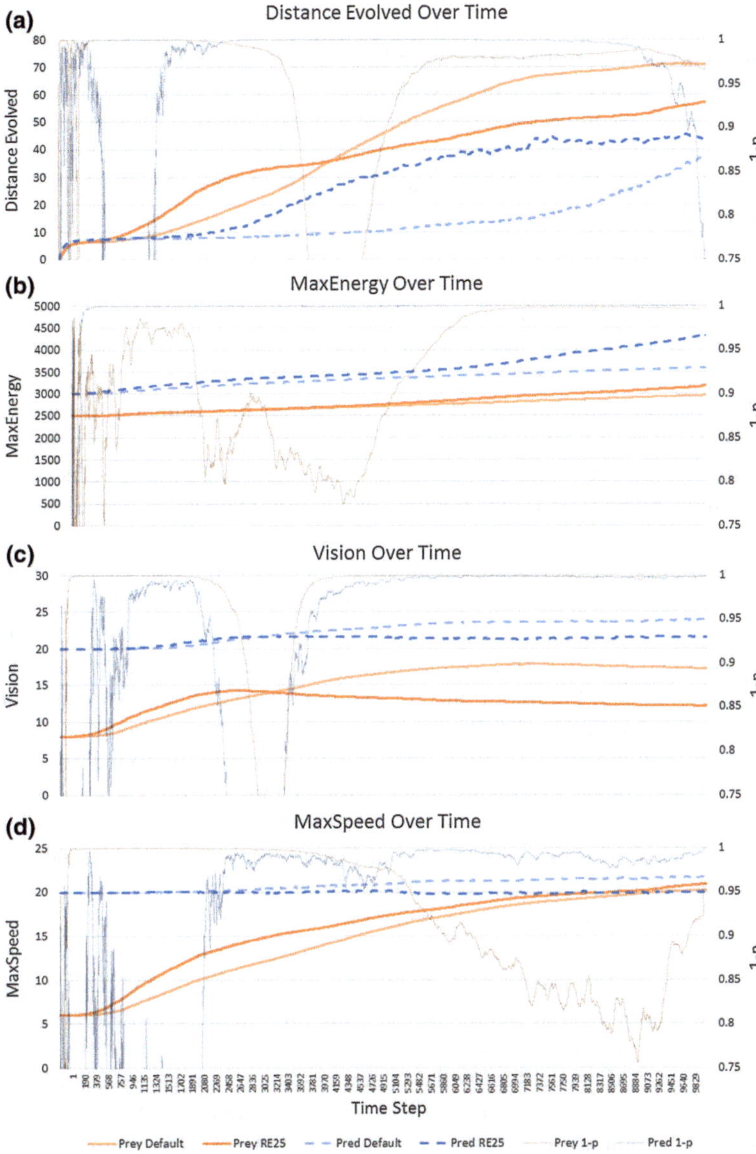

Fig. 18 Comparison of four measures related to evolution of prey and predator individuals between Default and RE25 runs over time. Each measure uses the left y-axis, while the t-test $1 - p$ value uses the right axis. The T-test $1 - p$ value shows the significance of the difference between Default and RE25 runs for prey and predators separately. Distance evolved **a** is the genetic distance between behavioral genomes at initialization and the mean of all individuals at a given time step. MaxEnergy (**b**), Vision (**c**), and MaxSpeed (**d**) are physical properties determining the maximum energy capacity, vision range, and maximum movement speed of individuals, respectively. Values shown are the mean of all individuals alive at the given time step

ings in the literature when we consider them in light of evolution of body size. Kiltie [63] found a positive correlation between body size and visual acuity across various species of birds, so that larger birds with higher energy expenditure exhibit higher visual acuity than smaller birds with lower energy expenditure. Moreover, Mech and Zollner [87] report a positive correlation between body size and perceptual range for various forest dwelling rodent species, including chipmunks, grey squirrels and fox squirrels. Finally, Rutowski, Gislen and Warrant [110] found that visual acuity increases with body size across four species of nymphalid butterfly. We expected that cheaper Energy costs associated with maintaining Vision and MaxSpeed would allow them to evolve to larger values, much like MaxEnergy. However, relatively, Vision and MaxSpeed play much smaller roles in determining the Energy expenditure of individuals per time step but crucial roles in determining the fitness of individuals. With reduced Energy consumption, the number of both prey and predator individuals was far greater than in Default runs. This result agrees with the empirical findings reported in McNab [86] with respect to a variety of vertebrate species inhabiting oceanic islands. Species with lower energy expenditure persist on oceanic islands by means of population increases, as opposed to species with higher energy expenditures [86]. As the number of individuals is much greater, so is the density of individuals, and thus finding mates is far less difficult. Furthermore, as individuals expend less Energy, they less often need to find food resources in order to survive. Thus, for both predators and prey, it is reasonable that, despite the cheaper cost of maintaining Vision and MaxSpeed, the importance of maintaining these features was overbearingly diminished as well in RE25 runs. The only anomaly is MaxSpeed of prey, however, at approximately 5000 time steps, the difference between the two run types was mostly insignificant, following a very similar trend to the earlier comparison regarding primary production. At approximately 9000 time steps in that comparison, MaxSpeed of RMG25 runs overtook that of Default runs. It is quite possible that in the long term, such a phenomenon would be observed here.

We observed several changes in behavior of prey (Fig. 16a, c) and predators (Fig. 17a, c) when their resource consumption was decreased. In prey, most notably, individuals in RE25 far more rapidly evolved a general loss in the ability to Evade predators and a reduction in Eat attempts, while only gaining in their frequency of Explore (t-test $p < 0.05$ in all cases, for most of the runs). Ultimately, in Default runs, the loss of Evade occurs as well, but at a much later time ($\sim$10,000 time steps versus 5000 time steps), and Explore still occurred significantly less in the long term (t-test $p < 0.05$ comparing Default time steps 16,000–20,000 against RE25 time steps 6000–10,000, for most of the time series). The rapid loss of the ability to Evade speaks to the futility in attempting to do so—in RE25 runs, the number of predators (and, accordingly, their density) was significantly greater (t-test $p < 0.05$ for most of the duration of the runs), and thus performing Evade was insufficient in prolonging the lives of prey individuals. The reduction in prey Eat attempts was expected, again because the individuals require less Energy to persist. The remaining prey behaviors showed no deviation between the two EcoSim variants.

Predators in RE25 runs showed a significant reduction in frequency of Hunt when compared to Default runs (t-test $p < 0.05$ for most of the time series), in accordance with what was observed in prey. Like prey, the predators in RE25 required less Energy to survive, and thus evolved to spend fewer actions on obtaining Energy. Unlike prey, however, the predators of RE25 did not show an increase in Explore (which is sensible, as Explore has very little value to predators as it is). Instead, predators evolved to attempt Reproduction significantly more often in RE25 runs when compared to Default runs (t-test $p < 0.05$ for most of the time series). Generally, as predators have much lower density, they also have a much harder time finding mates, and consequently, they tend to exhibit far more ReproduceFail than prey. In RE25 runs, with predator density greatly increased and Energy requirements slightly reduced, allocating more actions and Energy to Reproduction is necessary to improve their fitness. Thus, with the RE25 variant of EcoSim, both prey and predators get what they need to improve their fitness: the prey improve their longevity and the predators improve their fecundity through greater chance of Reproduction success.

4 Conclusion

We added many new features to EcoSim, improving the breadth and depth of questions it can now answer. The new features include new sensing and action concepts in the FCM of individuals, sexual reproduction, realistic feedback via fertilization of primary producers by consumers, and predator-prey combat, among others. In addition, new physical traits have been added to the behavioral genome, allowing different niches to emerge. Our results underline the importance of competition and energetics in evolution, and the great complexity that can emerge from relatively simplistic individuals. Our model reveals insights into the genetic mechanisms of niche adaptation, advances our understanding of both evolution and ecology, and allows us to address more complicated biological questions at resolutions varying from individual to whole communities. This is a major advantage of IBMs over empirical studies in the real world or other types of model; using IBMs, we are able to record anything we want at the resolution of the individual, something that would largely not be practical or possible otherwise. Of course, EcoSim and the general IBM approach has its drawbacks as well. Every IBM requires substantial simplification of the system it aims to replicate; as Box said regarding all scientific models, "All models are wrong but some are useful" [13]. Thus, the simplifications and assumptions made by an IBM must be understood before using it as an experimental platform, and conclusions made from use of the model must be considered in light of its assumptions and simplifications. For the same reason, it is sometimes difficult to generate new hypotheses using the IBM approach; researchers must ask themselves if the novelty of their conclusions is legitimate or, again, due to assumptions or simplifications of the model. Furthermore, many IBMs require substantial computing power, and EcoSim is no exception. Many IBMs, particularly those that would be considered pragmatic, require significant model tuning and validation to ensure legitimacy of the data they generate. Being at an early stage of the analysis of the new version of EcoSim, these preliminary results are promising and will lead to some more dedicated studies on niche emergence, reproduction, ecology, and evolution. For instance, EcoSim is currently being used to perform exciting research on sexual selection, the evolution of communication (particularly, communication of fear), asexual versus sexual reproduction, and biological invasions.

Acknowledgements This work is supported by CRC grant 950-2-3617 and the CFI grant 203617 and is made possible by the dedicated resource allocation 8047 of the Shared Hierar-chical Academic Research Computing Network (SHARCNET, www.sharcnet.ca).

References

1. Abbot, P., Abe, J., Alcock, J., et al. (2010). Inclusive fitness theory and eusociality. *Nature*. https://doi.org/10.1038/nature09831.
2. Andersson, M. B. (1994). *Sexual selection*. Princeton: Princeton University Press.
3. Arnold, K. E. (2000). Group mobbing behaviour and nest defence in a cooperatively breeding Australian bird. *Ethology, 106*, 385–393. https://doi.org/10.1046/j.1439-0310.2000.00545.x.
4. Aspinall A, Gras R (2010) K-Means clustering as a speciation method within an individual-based evolving predator-prey ecosystem simulation. In *6th International Conference on Active media technology* (pp. 318–329). Berlin, Heidelberg: Springer.
5. Augusiak, J., Van den Brink, P. J., & Grimm, V. (2014). Merging validation and evaluation of ecological models to evaludation: A review of terminology and a practical approach. *Ecological Modelling, 280*, 117–128.
6. Augustine, D. J., & McNaughton, S. J. (1998). Ungulate effects on the functional species composition of plant communities: Herbivore selectivity and plant tolerance. *The Journal of wildlife management, 62*, 1165–1183.
7. Bardgett, R. D., Wardle, D. A., & Yeates, G. W. (1998). Linking above-ground and below-ground interac-tions: How plant responses to foliar herbivory influence soil organisms. *Soil Biology and Biochemistry, 30*, 1867–1878.
8. Bardgett, R. D., Streeter, T., & Bol, R. (2003). Soil microbes compete effectively with plants for organic nitrogen inputs to temperate grasslands. *Ecology, 84*, 1277–1287.
9. Bateson, P. (1983). *Mate Choice*. Cambridge: Cambridge University Press.
10. Berryman, A. A. (1992). The origins and evolution of predator-prey theory. *Ecology, 73*, 1530–1535.
11. Blaxter, K. L. (1989). *Energy Metabolism in Animals and Man*. Cambridge: Cambridge University Press.
12. Botta-Dukát, Z., & Czúcz, B. (2016). Testing the ability of functional diversity indices to detect trait convergence and divergence using individual-based simulation. *Methods in Ecology and Evolution, 7*, 114–126. https://doi.org/10.1111/2041-210X.12450.
13. Box, G.E.P. (1979). Robustness in the strategy of scientific model building. In R. L. Launer, & G. N. Wil-kinson (Eds.), *Robustness in Statistics* (pp. 201–236). Academic Press.
14. Bollache, L., Kaldonski, N., Troussard, J. P., et al. (2006). Spines and behaviour as defences against fish predators in an invasive freshwater amphipod. *Animal Behaviour, 72*, 627–633.
15. Brännström, A., & Sumpter, D. J. T. (2005). The role of competition and clustering in population dynamics. *Proceedings of the Royal Society of London B: Biological Sciences, 272*, 2065–2072.
16. Britten, G. L., Dowd, M., Minto, C., et al. (2014). Predator decline leads to decreased stability in a coastal fish community. *Ecology letters, 17*, 1518–1525.
17. Brodie, E. D, I. I. I., & Brodie, E. D, Jr. (1999). Predator-prey arms races: asymmetrical selection on predators and prey may be reduced when prey are dangerous. *Bioscience, 49*, 557–568.

18. Brodie, E. D, Jr., Ridenhour, B. J., & Brodie, E. D, I. I. I. (2002). The evolutionary response of predators to dangerous prey: hotspots and coldspots in the geographic mosaic of coevolution between garter snakes and newts. *Evolution*, *56*, 2067–2082.
19. Bürger, R. (2000). *The Mathematical Theory of Selection, Recombination, and Mutation*. Chichester: Wiley.
20. Butler, P. J., Green, J. A., Boyd, I. L., & Speakman, J. R. (2004). Measuring metabolic rate in the field: The pros and cons of the doubly labelled water and heart rate methods. *Functional ecology*, *18*, 168–183.
21. Chapman, J. L., & Reiss, M. J. (1999). *Ecology: Principles and applications*. Cambridge: Cambridge University Press.
22. Clune, J., Misevic, D., Ofria, C., et al. (2008). Natural selection fails to optimize mutation rates for long-term adaptation on rugged fitness landscapes. *PLoS Computational Biology*. https://doi.org/10.1371/journal.pcbi.1000187.
23. Clune, J., Goldsby, H. J., Ofria, C., & Pennock, R. T. (2011). Selective pressures for accurate altruism targeting: Evidence from digital evolution for difficult-to-test aspects of inclusive fitness theory. *Proceedings of the Royal Society of London B: Biological Sciences*, *278*, 666–674.
24. Davies, T. J., Savolainen, V., Chase, M. W., et al. (2004). Environmental energy and evolutionary rates in flowering plants. *Proceedings of the Royal Society of London B: Biological Sciences*, *271*, 2195–2200.
25. DeAngelis DL, Grimm V (2014) Individual-based models in ecology after four decades. *F1000Prime Report*, *6*(39).
26. de Jager, M., Bartumeus, F., Kölzsch, A., et al. (2013). How superdiffusion gets arrested: ecological encounters explain shift from Lvy to Brownian movement. *Proceedings of the Royal Society of London*. https://doi.org/10.1098/rspb.2013.2605.
27. de Los Santos, C. B., Neuparth, T., Torres, T., et al. (2015). Ecological modelling and toxicity data coupled to assess population recovery of marine amphipod *Gammarus locusta*: Application to disturbance by chronic exposure to aniline. *Aquatic Toxicology*, *163*, 60–70.
28. Devaurs, D., & Gras, R. (2010). Species abundance patterns in an ecosystem simulation studied through Fishers logseries. *Simulation Modelling Practice and Theory*, *18*, 100–123.
29. Drent, R. H., & Van der Wal, R. (1999). Cyclic Grazing in Vertebrates and the Manipulation of the Food Resource. In H. Olff, V. K. Brown, & R. H. Drent (Eds.), *Herbivores: Between Plants and Predators* (pp. 271–299). London: Blackwell.
30. Eklöf, J., & uba J, Petersons G, Rydell J., (2014). Visual acuity and eye size in five European bat species in relation to foraging and migration strategies. *Environmental and Experimental Biology*, *12*, 1–6.
31. Falk, D. (1990). Brain evolution in Homo: The 'radiator' theory. *Behavioral and Brain Sciences*, *13*, 333–344.
32. Fortuna, M. A., Zaman, L., Wagner, A. P., & Ofria, C. (2013). Evolving digital ecological networks. *PLOS Computational Biology*. https://doi.org/10.1371/journal.pcbi.1002928.
33. Frank, B. M., & Baret, P. V. (2013). Simulating brown trout demogenetics in a river/nursery brook system: The individual-based model DemGenTrout. *Ecological modelling*, *248*, 184–202.
34. Frank, D., & Evans, R. (1997). Effects of native grazers on N cycling in a north-temperate grassland ecosystem: Yellowstone National Park. *Ecology*, *78*, 2238–2249.
35. Frank, D., & Groffman, P. (1998). Ungulate versus landscape control of soil C and N processes in grasslands of Yellowstone National Park. *Ecology*, *79*, 2229–2241.
36. Friman, V. P., Hiltunen, T., Laakso, J., & Kaitala, V. (2008). Availability of prey resources drives evolution of predator-prey interaction. *Proceedings of the Royal Society of London*, *275*, 1625–1633.
37. Garamszegi, L. Z., Mller, A. P., & Erritzøe, J. (2002). Coevolving avian eye size and brain size in rela-tion to prey capture and nocturnality. *Proceedings of the Royal Society of London*, *269*, 961–967.

38. Gillooly, J. F., Allen, A. P., West, G. B., & Brown, J. H. (2005). The rate of DNA evolution: effects of body size and temperature on the molecular clock. *Proceedings of the National Academy of Sciences of the United States of America, 102*, 140–145.
39. Goldsby, H. J., Knoester, D. B., Ofria, C., & Kerr, B. (2014). The evolutionary origin of somatic cells under the dirty work hypothesis. *PLoS ONE*,. https://doi.org/10.1371/journal.pbio.1001858.
40. Golestani, A., & Gras, R. (2010). Regularity analysis of an individual-based ecosystem simulation. *Chaos, 20*, 3120.
41. Golestani, A., & Gras, R. (2011). Multifractal phenomena in EcoSim, a large scale individual-based ecosystem simulation. In *International Conference on Artificial Intelligence* (pp. 991–999), Las Vegas.
42. Golestani, A., Gras, R. (2012). Identifying origin of self-similarity in EcoSim, an individual-based ecosystem simulation, using wavelet-based multifractal analysis. In *Proceedings of the world congress on engineering and computer science 2012 (WCECS 2012)* (pp. 1275–1285), San Francisco.
43. Golestani, A., Gras, R., & Cristescu, M. (2012). Speciation with gene flow in a heterogeneous virtual world: Can physical obstacles accelerate speciation? *Proceedings of the Royal Society of London, 279*, 3055–3064.
44. Grant, P. R., & Grant, B. R. (2006). Evolution of character displacement in Darwin's finches. *Science, 313*, 224–226.
45. Gras, R., Devaurs, D., Wozniak, A., & Aspinall, A. (2009). An individual-based evolving predator-prey ecosystem simulation using a fuzzy cognitive map as the behavior model. *Artif Life, 15*, 423–463.
46. Gras, R., Golestani, A., Hendry, A. P., & Cristescu, M. E. (2015). Speciation without pre-defined fitness functions. *PLoS ONE*,. https://doi.org/10.1371/journal.pone.0137838.
47. Grimm, V., Berger, U., Bastiansen, F., et al. (2006). A standard protocol for describing individual-based and agent-based models. *Ecological Modelling, 198*, 115–126.
48. Grimm, V., Berger, U., DeAngelis, D. L., et al. (2010). The ODD protocol: A review and first update. *Ecological Modelling, 221*, 2760–2768.
49. Grimm, V., Augusiak, J., Focks, A., et al. (2014). Towards better modelling and decision support: Documenting model development, testing, and analysis using TRACE. *Ecological Modelling, 280*, 129–139.
50. Hazlerigg, C. R. E., Tyler, C. R., Lorenzen, K., et al. (2014). Population relevance of toxicant mediated changes in sex ratio in fish: An assessment using an individual-based zebrafish (Danio rerio) model. *Ecological Modelling, 280*, 76–88.
51. Hamilton, E., & Frank, D. (2001). Can plants stimulate soil microbes and their own nutrient supply? Evidence from a grazing tolerant grass. *Ecology, 82*, 2397–2402.
52. Hartl, D. L., & Jones, E. W. (2004). *Genetics: Analysis of genes and genomes*. Burlington: Jones & Bartlett Publishers.
53. Hemmingsen, A. M. (1960). Energy metabolism as related to body size and respiratory surfaces, and its evolution. *Reports of the Steno Memorial Hospital and Nordisk Insulin Laboratorium, 9*, 1–110.
54. Hik, D. S., & Jefferies, R. L. (1990). Increases in the net aboveground primary production of a salt-marsh forage grass: A test of the predictions of the herbivore-optimization model. *The Journal of Ecology, 78*, 180–195.
55. Hiltunen, T., Ayan, G. B., & Becks, L. (2015). Environmental fluctuations restrict eco-evolutionary dynamics in predator prey system. *Proceedings of the Royal Society of London*. https://doi.org/10.1098/rspb.2015.0013.
56. Hobbs, N. T. (1996). Modification of ecosystems by ungulates. *The Journal of Wildlife Management, 60*, 695–713.
57. Hoskin, C. J., Higgie, M., McDonald, K. R., & Moritz, C. (2005). Reinforcement drives rapid allopatric speciation. *Nature, 437*, 1353.
58. Hraber, P. T., Jones, T., & Forrest, S. (1997). The ecology of Echo. *Artificial Life, 3*, 165–190.

59. Jasienska, G. (2003). Energy metabolism and the evolution of reproductive suppression in the human female. *Acta Biotheoretica, 51*, 1–8.
60. Kantz, H., & Schreiber, T. (1997). *Nonlinear Time Series Analysis*. Cambridge: Cambridge University Press.
61. Khater, M., Murariu, D., & Gras, R. (2014). Contemporary evolution and genetic change of prey as a response to predator removal. *Ecological Informatics, 22*, 13–22.
62. Khater, M., & Gras, R. (2012). Adaptation and genomic evolution in EcoSim. In T. Ziemke C. Balkenius, & J. Hallam (Eds) *From Animals to Animats 12, Proceedings of the 12th International Conference on Simulation of Adaptive Behavior, SAB 2012*, (pp. 219–229). Denmark: Odense.
63. Kiltie, R. A. (2000). Scaling of visual acuity with body size in mammals and birds. *Functional Ecology, 14*, 226–234.
64. Kleiber, M. (1932). Body size and metabolism. *Hilgardia, 6*, 315–353. https://doi.org/10.3733/hilg.v06n11p315.
65. Kleiber, M. (1961). *The fire of Life. An introduction to animal energetics*. New York: Wiley.
66. Kosko, B. (1986). Fuzzy cognitive maps. *International Journal of Man-machine Studies, 24*, 65–75.
67. Krams, I., Krama, T., & Igaune, K. (2006). Mobbing behaviour: Reciprocity-based cooperation in breeding Pied Flycatchers Ficedula hypoleuca. *IBIS, 148*, 50–54.
68. Krams, I., Krama, T., Igaune, K., & Mnd, R. (2008). Experimental evidence of reciprocal altruism in the pied flycatcher. *Behavioral Ecology and Sociobiology, 62*, 599–605.
69. Krebs, J., & Davies, N. (1997). *Behavioural Ecology: An evolutionary approach* (4th ed.). Oxford: Blackwell Publishers.
70. Kvam, P., Cesario, J., & Schossau, J. et al. (2013). Computational Evolution of Decision-Making Strategies. In D. C. Noelle, R. Dale, & A. S. Warlaumont et al. (Eds.), *Proceedings of the 37th Annual Conference of the Cognitive Science Society* (pp. 1225-1230), Austin, TX.
71. LaBar, T., Hintze, A., & Adami, C. (2016). Evolvability tradeoffs in emergent digital replicators. *Artificial Life, 22*, 483–498.
72. Landguth, E. L., & Cushman, S. A. (2010). CDPOP: A spatially explicit cost distance population genetics program. *Molecular Ecology Resources, 10*, 156–161.
73. Landguth, E. L., Bearlin, A., Day, C. C., & Dunham, J. (2017). CDMetaPOP: An individual-based, eco-evolutionary model for spatially explicit simulation of landscape demogenetics. *Methods in Ecology and Evolution, 8*, 4–11.
74. Lenski, R. E., Ofria, C., Collier, T. C., & Adami, C. (1999). Genome complexity, robustness and genetic interactions in digital organisms. *Nature, 400*, 661–664.
75. Lenski, R. E., Ofria, C., Pennock, R. T., & Adami, C. (2003). The Evolutionary Origin of Complex Features. *Nature, 423*, 139–144.
76. Lewis, R. J., & Kappler, P. M. (2005). Seasonality, body condition, and timing of reproduction in *Propithecus verreauxi verreauxi in the Kirindy Forest. Journal of the American Society of Primatologists, 67, 347–364.*
77. Li, Y., Brose, U., Meyer, K., & Rall, B. C. (2017). How patch size and refuge availability change interac-tion strength and population dynamics: a combined individual- and population-based modeling experiment. *PeerJ*,. https://doi.org/10.7717/peerj.2993.
78. Leonard, W. R., & Ulijaszek, S. J. (2002). Energetics and evolution: An emerging research domain. *American Journal of Human Biology, 14*, 547–550.
79. MacPherson, B., & Gras, R. (2016). Individual-based ecological models: Adjunctive tools or experimental systems? *Ecological Modelling, 323*, 106–114.
80. Mallet, J. (1995). A species definition for the modern synthesis. *Trends in Ecology & Evolution, 10*, 294–299.
81. Marini, G., Guzzetta, G., Baldacchino, F., et al. (2017). The effect of interspecific competition on the temporal dynamics of *Aedes albopictus and Culex pipiens. Parasites & vectors, 10, 102.*
82. Marshall, J. A. (2016). What is inclusive fitness theory, and what is it for? *Current Opinion in Behavioral Sciences, 12*, 103–108.

83. Mashayekhi, M., & Gras, R. (2012). Investigating the effect of spatial distribution and spatiotemporal information on speciation using individual-based ecosystem simulation. *GSTF Journal on Computing*, *2*, 98–103.
84. Mashayekhi, M., MacPherson, B., & Gras, R. (2014). Species-area relationship and a tentative interpretation of the function coefficients in an ecosystem simulation. *Ecological Complexity*, *19*, 84–95.
85. Mashayekhi, M., MacPherson, B., & Gras, R. (2014). A machine learning approach to investigate the reasons behind species extinction. *Ecological Informatics*, *20*, 58–66.
86. McNab, B. K. (2002). Minimizing energy expenditure facilitates vertebrate persistence on oceanic islands. *Ecology Letters*, *5*, 693–704.
87. Mech, S. G., & Zollner, P. A. (2002). Using body size to predict perceptual range. *Oikos*, *98*, 47–52.
88. Møller, A. P. (2009). Basal metabolic rate and risk-taking behaviour in birds. *Journal of Evolutionary Biology*, *22*, 2420–2429.
89. Molvar, E. M., Bowyer, R. T., & Van Ballenberghe, V. (1993). Moose herbivory, browse quality, and nutrient cycling in an Alaskan treeline community. *Oecol*, *94*, 473–479.
90. Mönkkönen, M., Forsman, J. T., & Bokma, F. (2006). Energy availability, abundance, energy-use and species richness in forest bird communities: A test of the species-energy theory. *Global Ecology and Biogeography*, *15*, 290–302.
91. Mueller, P., & Diamond, J. (2001). Metabolic rate and environmental productivity: Well-provisioned animals evolved to run and idle fast. *Proceedings of the National Academy of Sciences USA*, *98*, 12550–12554.
92. Nagy, K. A. (2005). Field metabolic rate and body size. *Journal of Experimental Biology*, *208*, 1621–1625.
93. Navarrete, A., van Schaik, C. P., & Isler, K. (2011). Energetics and the evolution of human brain size. *Nature*, *480*, 91.
94. Niklas, K. J., & Enquist, B. J. (2001). Invariant scaling relationships for interspecific plant biomass production rates and body size. *Proceedings of the National Academy of Sciences USA*, *98*, 2922–2927.
95. Niven, J. E., & Laughlin, S. B. (2008). Energy limitation as a selective pressure on the evolution of sensory systems. *Journal of Experimental Biology*, *211*, 1792–1804.
96. Nowak, M. A., Tarnita, C. E., & Wilson, E. O. (2010). The evolution of eusociality. *Nature*, *466*, 1057–1062.
97. Ofria, C., & Wilke, C. O. (2004). Avida: A software platform for research in computational evolutionary biology. *Artificial Life*, *10*, 191–229.
98. Olff, H., & Ritchie, M. E. (1998). Effects of herbivores on grassland plant diversity. *Trends in Ecology & Evolution*, *13*, 261–265.
99. Olson, R. S., Hintze, A., Dyer, F. C., et al. (2013). Predator confusion is sufficient to evolve swarming behavior. *Journal of the Royal Society Interface*. https://doi.org/10.1098/rsif.2013.0305.
100. Ostrowski, E. A., Ofria, C., & Lenski, R. E. (2015). Genetically integrated traits and rugged adaptive landscapes in digital organisms. *BMC Ecology*. https://doi.org/10.1186/s12862-015-0361-x.
101. Pafilis, P., Meiri, S., Foufopoulos, J., & Valakos, E. (2009). Intraspecific competition and high food availability are associated with insular gigantism in a lizard. *Naturwissenschaften*, *96*, 1107–13.
102. Pedley, T. J. (1977). Scale effects in animal locomotion. *The Quarterly Review of Biology*, *53*, 473–474.
103. Peters, R. H. (1986). *The Ecological Implications of Body Size*. Cambridge: Cambridge University Press.
104. Pethybridge, H., Roos, D., Loizeau, V., et al. (2013). Responses of European anchovy vital rates and population growth to environmental fluctuations: An individual-based modeling approach. *Ecological Modelling*, *250*, 370–383.

105. Piana, P. A., Gomes, L. C., & Agostinho, A. A. (2006). Comparison of predator-prey interaction models for fish assemblages from the neotropical region. *Ecological Modelling*, *192*, 259–270.
106. Potier, S., Bonadonna, F., Kelber, A., et al. (2016). Visual abilities in two raptors with different ecology. *The Journal of Experimental Biology*, *219*, 2639–2649.
107. Prothero, J. W. (1979). Maximal oxygen consumption in various animals and plants. *Comparative Biochemistry and Physiology—Part A: Molecular & Integrative Physiology*, *64*, 463–466.
108. Ray, T.S. (1991). An approach to the synthesis of life. In C. Langton, C. Taylor, J.D. Farmer,& S. Ras-mussen (Eds.), *Proceedings of Artificial Life II* (pp. 371–408), Redwood City: Addison-Wesley
109. Ricotta, C. (2000). From theoretical ecology to statistical physics and back: Self-similar landscape metrics as a synthesis of ecological diversity and geometrical complexity. *Ecological Modelling*, *125*, 245–253.
110. Rutowski, R. L., Gisln, L., & Warrant, E. J. (2009). Visual acuity and sensitivity increase allometrically with body size in butterflies. *Arthropod Structure & Development*, *38*, 91–100.
111. Safi, K., Seid, M. A., & Dechmann, D. K. N. (2005). Bigger is not always better: when brains get smaller. *Biology Letters*, *1*, 283–286.
112. Schmidt-Nielsen, K. (1984). *Scaling: Why is animal size so important?*. Cambridge: Cambridge University Press.
113. Schmolke, A., Thorbek, P., DeAngelis, D. L., & Grimm, V. (2010). Ecological models supporting environmental decision making: A strategy for the future. *Trends Ecology Evolution*, *25*, 479–486.
114. Seuront, L., Schmitt, F., Lagadeuc, Y., et al. (1996). Multifractal analysis of phytoplankton biomass and temperature in the ocean. *Geophysical Research Letters*, *23*, 3591–3594.
115. Shepherd, G. M. (1994). *Neurobiology*. Oxford: Oxford University Press.
116. Stahl, W. R. R. (1965). Organ weights in primates and other mammals. *Science*, *150*, 1039–1042.
117. Stahl, W. R. R. (1967). Scaling of respiratory variables in mammals. *Journal of Applied Physiology*, *22*, 453–460.
118. Stephens, D., & Krebs, J. (1986). *Foraging theory*. Princeton: Princeton University Press.
119. Strauss, S. Y., Lau, J. A., & Carroll, S. P. (2006). Evolutionary responses of natives to introduced species: what do introductions tell us about natural communities? *Ecology Letters*, *9*, 357–374.
120. Svanbck, R., & Bolnick, D. I. (2007). Intraspecific competition drives increased resource use diversity within a natural population. *Proceedings of the Royal Society of London*, *274*, 839–844.
121. Svanbck, R., Eklöv, P., Fransson, R., & Holmgren, K. (2008). Intraspecific competition drives multiple species resource polymorphism in fish communities. *Oikos*, *117*, 114–124.
122. Thearling, K., & Ray, T. (1994). Evolving multi-cellular artificial life. In P. Maes (Ed.), *Brooks RA* (pp. 283–288). MIT Press, Cambridge p: Proceedings of Artificial Life IV.
123. The HDF Group (2000) Hierarchical data format version 5. Accessed Feb 2014,http://www.hdfgroup.org/HDF5.
124. Uchmaski, J. (2016). Individual variability and metapopulation dynamics: An individual-based model. *Ecological Modelling*, *334*, 8–18.
125. Van der Wal, R., Bardgett, R. D., Harrison, K. A., & Stien, A. (2004). Vertebrate herbivores and ecosystem control: Cascading effects of faeces on tundra ecosystems. *Ecography*, *27*, 242–252.
126. Wardle, D. A. (2002). *Communities and Ecosystems: Linking Aboveground and Belowground Components*. Princeton: Princeton University Press.
127. Wheeler, P.E. (1984). An investigation of some aspects of the transition from ectothermic to endothermic metabolism in vertebrates. Durham University.
128. Williams, S., & Yaeger, L. (2017). Evolution of neural dynamics in an ecological model. *Geosciences*,. https://doi.org/10.3390/geosciences7030049.

129. Yaeger, L. (1994). Computational genetics, physiology, metabolism, neural systems, learning, vision, and behavior or PolyWorld: life in a new context. In *Proceedings of Artificial Life III, Santa Fe Institute Studies in the Sciences of Complexity* (Vol. 17, pp. 263–298), Redwood City: Addison-Wesley.
130. Yaeger, L. S. (2013). Identifying neural network topologies that foster dynamical complexity. *Advances in Complex Systems*. https://doi.org/10.1142/S021952591350032X.
131. Yoder, J., & Yaeger, L. (2014). Evaluating topological models of neuromodulation in Polyworld. *Artificial Life*, *14*, 916–923. https://doi.org/10.7551/978-0-262-32621-6-ch149.
132. Zaman, L., Meyer, J. R., & Devangam, S., et al. (2014). Coevolution drives the emergence of complex traits and promotes evolvability. *PLOS Biology*. https://doi.org/10.1371/journal.pbio.1002023.

MIX
Papier aus verantwortungsvollen Quellen
Paper from responsible sources
FSC® C105338

If you have any concerns about our products,
you can contact us on
ProductSafety@springernature.com

In case Publisher is established outside the EU,
the EU authorized representative is:
Springer Nature Customer Service Center GmbH
Europaplatz 3, 69115 Heidelberg, Germany

Printed by Libri Plureos GmbH
in Hamburg, Germany

Hans Gregersen

Biomechanics of the Gastrointestinal Tract

New Perspectives in Motility Research and Diagnostics

Springer

Hans Gregersen, MD, DrMSci, MPM
Professor of Bioengineering, Centre for Sensory-Motor Interaction,
Laboratory for Gastrointestinal Biomechanics and Sensory-Motor Funtion,
Department of Surgical Gastroenterology, Aalborg Hospital, Aalborg University, Denmark

British Library Cataloguing in Publication Data
Gregersen, Hans
Biomechanics of the gastrointestinal tract : new perspectives in motility research and diagnostics
1. Gastrointestinal system - Motility
I. Title
612.3'2

Library of Congress Cataloging-in-Publication Data
Gregersen, Hans, 1962–
Biomechanics of the gastrointestinal tract : new perspectives in motility research and diagnostics / Hans Gregerson.
p. cm.
Includes bibliographical references and index.
ISBN 978-1-84996-880-5 ISBN 978-1-4471-3742-9 (eBook)
DOI 10.1007/978-1-4471-3742-9

1. Gastrointestinal system—Mechanical properties I. Title.

QP145 .G696 2002
612.3'2--dc21 2002029439

http://www.springer.co.uk

Originally published by Springer-Verlag London Limited in 2003
Softcover reprint of the hardcover 1st edition 2003

Typeset by Q3 Bookwork, Loughborough, Leicestershire, England

28/3830-543210 Printed on acid-free paper SPIN 10839744

Foreword

The alimentary canal is primarily a self-regulating mechanical device designed by nature to move the liquids and solids taken in. It must move the heterogeneous food materials in various complex ways to optimize the fermentation that constitutes digestion, to provide the special flow patterns required for the absorption of the nutrients that digestion releases, and to remove indigestible residue. It varies its operations to adjust to the wide-ranging physical and chemical properties of the materials presented to it. One must marvel at the fantastic ability of the gastrointestinal machinery to do what it does.

Because of a deficient basis in theory, the motility of the gut was an empirically realised process for a long time, largely incomprehensible and unpredictable. We now possess a considerable mass of information about it from the application of new methods in the reductionist strategies of modern biology. The resultant data describe a vast array of linked nonlinear processes in its neural and muscular apparatus. Their interactions, however, still remain dimly perceived.

The movement of the gut continues to be perhaps the least rationally developed subject in physiology. The regrettable consequence of this deficit lies in the prominence of disordered gastrointestinal neural and muscular function in medicine. Disorders in this area constitute the root problem in most patients with alimentary complaints. Virtually all oesophageal disease represents dysfunction in the nerves and muscles of that organ. Diverticular disease of the colon, a common and vexing problem in Western societies, also represents the consequence of gastrointestinal neuromuscular dysfunction. Even diseases generally not thought to be of this nature, such as gall bladder disease, involve disturbed mechanics.

The description of all mechanical devices involves two elements: the portrayal of the mechanics and the analysis of the controlling systems. In the case of the machinery of the gut, the mechanics seems to me to be especially neglected, for most biologists in this area are largely unschooled in the physical sciences. Also, engineers with the necessary skills rarely have the background necessary to deal with the biological processes involved in the neuromuscular function of the gut. The new discipline of biomechanics ought to be able to bridge the chasm in this and other similar matters. It is for this reason that this book seems so welcome. I know of none like it.

I fear that many readers from both sides of the divide will find this text tough going simply because of the depth of their ignorance in the thought processes of the other camp. Experience in this subject has taught me that true understanding requires not just the reading of a text but daily conversations and discussions centred on specific questions.

Such efforts seem warranted, however, for a keener understanding of the mechanics of alimentation will surely benefit those who suffer from the mysterious disorders in the system that seem so common. From a better comprehension will come better treatments and, perhaps, better ideas about the prevention of disease in this system.

Also, it may offer new ideas in fluid mechanics. The engineer may well learn many useful lessons from a new awareness of the wondrous mechanism that evolution has produced to allow the gut to manage its fluid contents.

James Christensen, Professor
Iowa City
Iowa, USA

Preface

Biomechanics and bioengineering are by now recognised research disciplines. Though the gastrointestinal tract serves primarily a mechanical function, biomechanics and bioengineering have only to a small degree entered the thinking of researchers and clinicians in gastroenterology. Among the topics taught in gastrointestinal physiology, mechanics receives little attention. Graduates understand digestion, secretion, absorption, and hepatic functions, but they are unlikely to answer questions about mechanics. Texts in physiology and medicine give insufficient attention to the subject, yet many researchers in gastroenterology claim to know biomechanical principles, which unfortunately leads to misuse of terms and misinterpretation of assumptions and results. Biomechanical textbooks exist but none of them focus on the gastrointestinal tract. For a long time I have felt the need for a book that could be used to develop biomechanical ideas in gastroenterology and to teach biomechanics and the application of mechanics to gastrointestinal physiology and pathophysiology to the students and readers. The philosophy is that a book is needed to show the power of the bioengineering approach to the gastrointestinal research community. When these approaches have been adopted, the field will bloom and there will soon be the need for an updated book. The book demonstrates how little we know. We have to be open-minded in our search for professional and valid scientific tools rather than to let conservatism and politics be the main actors on the scene. The book presents my view of the current status and, to a large extent, reviews my own work. A great deal is focused on the passive elastic properties of gastrointestinal tissue. The rationale for this is that we need to understand the basic relations between stress and strain and the passive tissue properties before we can interpret the active contractile behaviour. Future editions will focus more on the integration of passive and active tissue properties. Many readers with an interest in gastrointestinal motility and diagnostics of motor disorders will undoubtedly find Chapter 5 about smooth muscle function and several other chapters interesting.

This book is not a comprehensive handbook of information about the selected topics, neither does it give the complex mathematical and mechanical background. The reader is referred to previously published work and to mechanics books to gain the needed insight. This book as a whole outlines a basic overview of biomechanics as it pertains to motility research in the gastrointestinal tract with minimal use of complicated equations. The physiological and clinical importance of acquiring biomechanical data, distensibility parameters, and interpretation of these results and their associated errors are discussed. Some aspects of the relationship between morphology, growth and biomechanics are discussed also. Chapter 2 provides the essential knowledge of anatomy, morphology and physiology of the gastrointestinal tract. The basic biomechanical theory is given in Chapter 3. Focus is on definitions of terms, derivation of the appropriate equations and discussion of the assumptions for the stress and strain computations. Methods and research tools are described in Chapter 4 where a number of techniques to study the mechanical properties of the gastrointestinal tract are outlined. The next chapter is devoted to gastrointestinal smooth muscle behaviour and mechanics. Recently, it was shown that residual stresses and strains are prominent in the gastrointestinal tract, therefore, the whole of Chapter 6 is allocated to treating this phenomenon. The last two chapters of the book present biomechanical data obtained in the gastrointestinal tract under normal conditions as well as the remodelling and adaptation that occur in diseased tissue.

The work presented in this book would not have been possible without contributions from my students, co-workers, collaborators, and mentors. I have received a massive scientific input from the Bioengineering Institute at University of California, San Diego. The collaboration with Professor Emeritus Y.C. Fung, Professor Andrew McCulloch, Professor Shu Chien, Professor Richard Skalak (deceased), S.Q. Liu, and Wei Huang has been very inspiring. Associate researcher Ghassan Kassab (now Professor at UC Irvine) contributed to many projects at UCSD as well as in Denmark. I sincerely acknowledge his never-failing enthusiasm. Professor James Christensen from University of Iowa Hospitals and Clinics is one of the few who recognise that we must integrate gastroenterology with bioengineering. He has been a great inspiration for many years. Many of the experiments mentioned in this book were performed by my students and postdocs: Claus Jørgensen, Birgitte Duch, Gerda Villadsen, Chunwen Gao, Jingbo Zhao, Xiao Lu, Klaus Krogh, Judd Day, James Smith, Jeff Cassin, Yanling Dou, Yanhua Fan, Chunwen Gao, Jian Yang, Romulus Lontis and Liao Donghua. Data are also presented from collaborative studies with the ultrasound group in Bergen (in particular Dr. Odd Helge Gilja); Professor David Thompson and Josephine Barlow, Hope Hospital, Salford; Professor Feng-Yuan Zhuang at the Biomedical Institute at China-Japan Friendship Hospital in Beijing (now Chairman at Beijing University of Aeronautics and Astronautics); and Professor Y.J. Zeng at Beijing Polytechnic University. The Bioengineering Institute at the University of California, San Diego (1994–96), The Institute of Experimental Clinical Research, Aarhus University (1996–99), Aalborg Hospital and the Center for Sensory–Motor Interaction at Aalborg University (1999–2001) provided excellent working conditions during the whole process. The collaboration with Professor Jens Christian Djurhuus, Dr. Asbjørn Drewes, and Professor Lars Arendt-Nielsen is appreciated. Karen Elise Jensens Foundation, Ydes Foundation, The Danish Research Councils, and many other foundations supported the work presented in this book.

I am specially indebted to Professor Y.C. Fung from University of California, San Diego on whose work this book is based. He is truly considered the Father of Bioengineering and has shown a number of students and readers of his literature the way to follow. Professor Fung's books are classics and should be read by anyone interested in biomechanics.

One of my obsessions is the science of gastrointestinal bioengineering and biomechanics. It is my hope that this book can facilitate a better understanding of the gastrointestinal tract as a biomechanical system and that an integrated approach can help to understand those gastrointestinal disorders whose pathophysiology from a bioengineering point of view hitherto has remained an enigma. I wish to address my warmest thanks to my most precious obsessions, my wife Lene and my children Anne-Sofie (Fie) and Frederik for their support and care.

I also wish to thank Nick Mowat, Melissa Morton and Eva Senior from Springer-Verlag for their cooperation in producing this book.

August 2002
Hans Gregersen, Aalborg, Denmark

Contents

1. The Concept of Biomechanics

1.1 Introduction

The gastrointestinal tract is a long conduit, in practice an invagination, extending from the oral cavity to the anus. Digestion is perhaps the single word that best describes its function. However, to understand the behaviour we need to fragment the function. The food and fluid that are swallowed are transported to the stomach and intestines where mixing, secretion and absorption take place. The many subdivisions of the gastrointestinal tract each carry specialised sub-functions. For example, the oesophagus is largely a conduit through which the food and fluid is passed with practically no absorption and secretion taking place. The motor function has traditionally been studied by the motility research community; i.e., people with an interest in problems related to the transport function, and often with a background in biology and medicine.

In recent years, bioengineering principles have started to enter the minds of physiologists. Whereas this has happened relatively quickly in the cardiovascular and orthopaedics fields, other fields such as gastroenterology lag behind. Bioengineering is the application of engineering knowledge and methods to the fields of biology and medicine. It is recognised more and more often that organisms must contend with the physical characteristics of their environment as they carry out a wide range of physiological and mechanical functions. Most bioengineering work has been done by electrical engineers developing medical electronics instruments but has also proven its value for a long time in the rehabilitation field in gait analysis and the development of artificial limbs. Biological heat transfer and development of biomaterials are other examples. Biomechanics constitutes an important area of bioengineering, especially for the study of organs with mechanical functions. Biomechanics not only relates to gait analysis, striated muscle function and the heart but also to visceral organs in general. In gastroenterology, the term motility is defined as the quality of the gut to generate motion. Motility encompasses a very wide area and comprises aspects such as movements of the wall (motor patterns) and the movement of contents (flow patterns). Thus, gastrointestinal motility research is closely associated with biomechanics (Figure 1.1). In fact, any measurement of forces or motion of the gastrointestinal tract should be treated as a biomechanical measurement that must undergo a mechanical analysis based on bioengineering principles. It is important to emphasize this fact now because, since 1990, it has become increasingly popular to acquire distensibility data in motility research. The popularity stems from the many reasons that will be discussed in later sections. It is imperative to

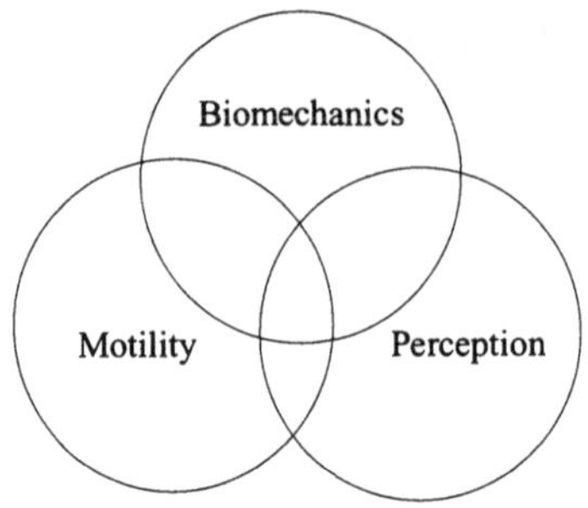

Figure 1.1. Perception, motility and biomechanics are important in understanding gastrointestinal behaviour.

know the basic principles of biomechanics and its nomenclature. This book serves to outline a basic overview of biomechanics as it pertains to research in the gastrointestinal tract.

1.2 What is Biomechanics?

Biomechanics is mechanics applied to biology and provides explanations of the mechanical behaviour of living organisms. Mechanics is the study of the motion of matter and the forces that cause such motion. Forces applied to solids cause deformation while forces applied to fluids cause flow. In biomechanics, it is important to understand the basic relations between stress and strain. Some types of stress–strain relations are simple because the material is homogenous, behave in an isotropic manner and show infinitesimal deformation even at large stresses. Many engineering materials behave this way. We were taught in physics in school about Hooke's law, where Young's modulus is the constant of proportionality between stress and strain. However, biological tissues differ in several ways from this behaviour. Take the gastrointestinal tract as an example. The geometry is complex in most parts of the tract, the structure is layered and heterogeneous, the mechanical properties depend on the direction – that is, the behaviour is anisotropic – and the deformation is large, resulting in non-linear stress–strain curves. Due to the high water content in tissue it not only has the properties of solids but also those of fluids; i.e., the behaviour is viscoelastic. On top of this, the gastrointestinal tract contains muscle cells with the ability to contract and with some degree of control from nervous structures intrinsic and extrinsic to the gastrointestinal wall. This complexity makes the mechanical analysis much more difficult than for common engineering materials. It is therefore clear that simple measures such as compliance based on pressure–volume measurements are insufficient. Even for the stress–strain analysis, we need simplifying assumptions to reduce the number of experiments and the complexity of the analysis. Such assumptions, however, can not be chosen randomly, they must relate to the scientific problem and the validity should be tested during the process.

Biomechanics takes advantage of engineering principles and requires understanding of mathematics, mechanics, statistics, and biology. When applied to gastroenterology, it also requires a thorough understanding of gastrointestinal structure, anatomy, function, symptomatology and pathophysiology. The complex features of the gastrointestinal tract must be tackled by a multidisciplinary effort through the use of experimental, analytical and numerical methods. A biomechanical approach is greatly needed to enhance our understanding of mechanoreceptor behaviour, smooth muscle function, perception, bolus transport mechanics, tissue growth and remodelling during intervention and disease.

Fascinating and useful information about gastrointestinal mechanics has only begun to penetrate to the level of common knowledge in science. The unfortunate consequence of this deficit relates to the prominence of disordered gastrointestinal function in medicine. Motor disorders constitute the root problem in most patients with gastrointestinal complaints. Nearly all oesophageal disease, for example, represents abnormal mechanical function in that organ. Mechanical dysfunction including the motor disorders often complicates connective tissue diseases, diabetes, and neurological disorders. Why have physiology and medicine neglected the study of gastrointestinal mechanics? The gut is primarily a mechanical device. Physics, mechanical engineering, and mathematics, the "basic sciences" of mechanical processes, have never held an important place in the fundamental education of biologists.

1.3 History and Background

The concepts of biomechanics can be traced back to before Christ (see Fung (1993) for a thorough review of the history). However, it was not until the seventeenth century that biomechanics became a discipline with significant contributions from Galileo Galilei, William Harvey, Robert Hooke, and Isaac Newton, among others. Borelli was one of the first to treat the gastrointestinal tract as a biomechanical system, also in the seventeenth century. Thorough studies of the motor function first came about in the nineteenth and twentieth centuries. Special attention should be paid to Bayliss and Starling's formulation of the law of the intestine in 1899–1901 with the discovery of the polarised intestinal reflexes, though Nothnagel in 1882 and Mall in 1896 already had suggested the existence of these reflexes. Trendelenburg's model for studying peristalsis (1917) was also an important development for further studies and modifications. Walter B. Cannon's book, *The Mechanical Factors of Digestion*, from 1911 was also a step on the road as was the extensive work published by Yamada in 1970 on mechanical properties of almost all kinds of biological tissue. Nevertheless, gastroenterology largely escaped the developments that took place in mechanical studies of other tissues, especially in orthopaedics, urology and the cardiovascular field. James Christensen supervised several PhD students in the 1970s, primarily on various aspects of gastrointestinal flow and the relation to contractile activity. Only some of this work was published in journals, the rest remains in the theses. Several new methods have been developed, perhaps the one with largest impact in vitro thus far is the uniaxial strips technique for studying the active and passive muscle mechanical behaviour and the response to pharmacological intervention. Balloon distension techniques have been developed for in vivo studies. However, from a mechanical point of view, only impedance planimetry has attract attention due to the thorough validation of this technique and the results obtained in gastroenterology as well as in urology (several papers by Lose, Colstrup and co-workers) and the cardiovascular field (several papers by Frøbert, Tanko and co-workers). New techniques based on non-invasive imaging techniques using ultrasound, MR-scanning and multi-slice CT-scanning in combination with balloon distension are, however, emerging as the techniques for future in vivo studies.

Treating the gastrointestinal tract from the bioengineering point of view has been done by very few. Sporadic (but very good) publications came from Fung, Tozeren, Srivastava and Srivastava, and Metry among others. Only a few groups have seriously entered the field. James Brasseur (now professor at Pennsylvania State University) has published a number of publications on flow and contractility in the oesophagus during the past two decades. Miftakhov (now at University of Iowa Hospitals and Clinics) has in several papers treated gastrointestinal behaviour using bioengineering principles. At last, my group has focused on the development of new biomechanical equipment, animal and human experimental work, and lately on the residual stresses and strains in the gastrointestinal tract.

1.4 Applications of Biomechanics

The gastrointestinal tract acts as a mechanical system and serves primarily mechanical functions. Therefore, biomechanical and bioengineering principles can be applied to almost any problem related to gastrointestinal function and pathophysiology. In the gastrointestinal tract, mechanical data and analysis are important for an understanding of passive viscoelastic properties, peristaltic reflexes, perception, mechano-receptor responses, tone, bolus transport mechanics, growth, and remodelling in experimental and clinical studies. These fields of investigation are in many ways related and will be treated in detail later in this book.

1.4.1 Mechanoreceptors, Peristaltic Reflex Activity and Sensory Function Testing

Distension of the gastrointestinal tract elicits reflex-mediated inhibition and stimulation of motility via intrinsic or extrinsic neural circuits and sensations such as pain. The sensing elements are afferent neurons with the mechano-receptors located in the gastrointestinal wall. The mechano-receptors have been classified on the basis of their pressure or volume thresholds to balloon distension, which in turn relates to whether the afferents project to the vagal nerve or to the spinal cord. However, it is a common mistake to believe that mechano-receptors are sensitive to variations in pressure or volume. A large variation in the pressure and volume thresholds needed to trigger secondary peristalsis and perception have been found in various studies and species, thus suggesting that pressure is not the direct stimulus. Instead, the receptors are stimulated by forces and deformations acting in the intestinal wall due to changes in transmural pressure. The same mechanism accounts for the baroreceptors in the cardiovascular system. Despite the name (baro = pressure), the baroreceptor nerve activity is associated with wall deformation, because restricting wall distension abolishes baroreceptor-mediated reflexes. In a similar fashion, data obtained by Ginzel (1959) and Kosterlitz and Robinson (1959) indicate that stretch is a more potent stimulus for gastrointestinal mechanoreceptors than pressure is. Yuan et al. (1991) pointed out that different mechanoreceptors respond to different kinds of mechanical stimuli, e.g., mucosal receptors respond to distortion and pinching but are relatively insensitive to contraction and distension. Since some of these studies have been available for more than 40 years it is strange that many researchers still consider pressure or volume to be the determinant of evoked reflex responses.

Since pressure is not the direct stimulus for the mechanoreceptor, a large variation arises when expressing the receptor response as a function of pressure, making it difficult to draw conclusions. Despite the vast number of studies in this area, the essential stimulus has not yet been defined. Further analysis must consider the fact that mechanical forces imposed on the tissue induce stresses and strains in various directions. Thus, distension will, in addition to circumferential tensile stress and strain, induce longitudinal tensile stress and strain and radial compressive stress and strain. For the study of mechanoreceptor kinematics, it is therefore important to express the stimulus in terms of magnitude and direction of stress, strain, and strain rate. Unfortunately, it is impossible to measure stresses and strains exactly at the receptor sites so theoretical solutions based on experimentally determined mechanical properties in different layers of the gastrointestinal wall will be necessary. By determining the nature of these mechanical stimuli we will learn more about the mechanoreceptors and how they adapt to the physical environment. This will allow us to establish the relation between the activity in the neurons, electrical responses in neurons, and the mechanical behaviour of the gut. If the appropriate mechanical stimulus is not known, the dependent variable data are merely to be considered qualitative. Miftakhov has, during the past ten years, done extensive work on gastrointestinal peristalsis and the regulation of it.

1.4.2 Smooth Muscle Tone

Tone exists in visceral muscle as it does in somatic muscle. Tone in gut muscle, however, has received comparatively little attention until recent decades when its significance in the motility of the gastrointestinal tract became evident. Tone, as a form of active muscle contraction, is commonly contrasted to what is called phasic or rhythmic contraction on the basis of temporal characteristics. The distinct differences between these two forms of gross behaviour preclude the need for precise quantitative criteria to distinguish them.

The demonstration of tone in gastrointestinal muscle in vitro is relatively straightforward, but the measurement of its force in vivo is not (Gregersen and Christensen, 2000). The conventional method to record gastrointestinal motor function – intraluminal manometry – cannot detect tone in the intact gut except in sphincteric regions. Most studies of tone in vivo define tone in terms of changes in volume at a constant predefined pressure, or a change in the slope of the compliance curve, $\Delta V/\Delta P$ without knowledge of the nature of the active and passive stresses but this can lead to confusion due to the use of insufficient methods and protocols. There are several major problems in the use of pressure–volume measurements in balloons to record tone. Many investigators seem to be unaware of the impediments they face both from the obscurities of gastrointestinal physiology and from the unfamiliarity of the mechanical principles that apply to the distension of the balloon within a deformable cylinder like the gastrointestinal tract.

Physiologists must deal with the mechanical concepts of stress and strain in treating tone as the mechanical phenomenon it is. This demands consideration of the tissue mechanics and geometry of the various sites and organs under scrutiny. The gastrointestinal tract is a deformable tube and its smooth muscle mostly operates under essentially isotonic conditions in vivo, being allowed to shorten with little resistance. Therefore, it makes geometric sense to define tone as a sustained active stress tending to produce strain that leads to a reduction in luminal cross-sectional area and circumference, or surface area (see Chapter 5). This stress represents the activation of the smooth muscle in the walls of the cylinder.

1.4.3 Bolus Transport Mechanics

Many factors influence flow and bolus transport in distensible organs such as the gastrointestinal tract. The driving forces are the pressure generated by the contractile peristaltic forces and, to a lesser extent, the hydrostatic force of gravity. Biomechanical models predict that factors such as the shape and size of the luminal cross-section, bolus size and viscosity, and tissue viscoelastic properties such as the elastic modulus, strain rate and shear properties of the wall are important determinants of flow. The gastrointestinal tract is stretched in the vicinity of a bolus; i.e., the contents propelled in front of a peristaltic contraction bulge out the intestinal wall. This supports the hypothesis that the intestinal reflexes initiated by localised stretch are of physiological significance for flow. Distensibility studies may help us understand the complex fluid mechanical behaviour of the gastrointestinal tract.

Oesophageal flow properties have been studied by Brasseur and co-workers while gastric emptying, antroduodenal mechanics, and intestinal flow properties were described primarily by Christensen, Meyer, Schulze-Delrieu and their co-workers. Weems reviewed the literature on flow properties in the gastrointestinal tract. Mathematical modelling of peristaltic transport in distensible tubes has also been attempted (Stavitsky et al., 1981; Srivastava and Srivastava, 1983; Fung and Yih, 1968; Li and Brasseur, 1993; and several papers by Miftakhov). The mathematical modelling of peristaltic transport is based on the premise that the interaction between gastrointestinal elasticity and bolus transport is best understood by examining the fluid dynamics equations (equations describing bolus transport) and equations describing gastrointestinal deformation (the constitutive equations). The interaction between these equations is governed by the boundary conditions.

1.4.4 Mechanical Forces, Growth, Remodelling and Morphology

Mechanical forces are a major determinant of tissue behaviour and are likely the single most important factor influencing growth and remodelling of tissue. An understanding of this relationship is important in deciphering how the gastrointestinal tract responds and adapts to changes in the physical environment. The passive elastic properties of the intact organ must be studied since these properties reflect the state of the structural components determining a specific function or malfunction. Furthermore, the passive properties determine the active muscle behaviour as predicted by the Hill model (see Chapter 5). It is desirable to study the biomechanical and morphological properties of the tissue concurrently. The literature is relatively sparse in this regard. Gabella evaluated the structural apparatus for force transmission in smooth muscles, and Orberg et al., Fackler et al., and Sacks and Gloeckner described the biaxial orientation of the collagen fibrils and the pattern of uncrimping in response to stress. Collagen is likely the most important determinant of the passive biomechanical properties of the gastrointestinal tract since large amounts are found in its wall. Further studies are needed to clarify the relation between the biomechanical properties of the gut and its morphology. Such studies should also utilise molecular techniques. The advent of molecular techniques provides an opportunity to study the effect of mechanical stimuli on the gene expression of various structural proteins such as collagens, fibronectin and stress proteins on the transcriptional level. This approach is now widely used in cardiovascular research and will undoubtedly provide us with important new insights into the mechanisms of structural remodelling of tissue in health and disease.

1.4.5 Clinical Perspectives

The clinical aspects of gastrointestinal biomechanics are mainly related to bolus transport and remodelling and to the development of new diagnostic tools based on a rational bioengineering approach. Since the mechanical properties are important for the physiologic function of the gut, it also plays an important role in diseases that affect wall structure and motility. Several diseases cause tissue remodelling that is associated with changes in the mechanical properties. Obstructive diseases, as an example, are characterised by proximal dilatation and wall thickening. This is perhaps caused by the repeated stretch during passage of foods due to the increased resistance to flow and the repetitive high-pressure contractions proximal to the site of obstruction. Increased wall stress due to progressive dilatation and strain are likely responsible for the development of collagenous infiltration. Hirschsprung's disease, ulcerative colitis, diabetes, systemic sclerosis and sclerotherapy-induced damage are some examples of diseases and treatments associated with remodelled morphology and biomechanical properties. Biomechanics is relevant in studying dysmotility and perception and thereby also for studying functional diseases such as non-cardiac chest pain and irritable bowel syndrome where many previous tests have proved obsolete.

1.5 The Bioengineering Approach

This book emphasises a rational approach adopted and modified from physics and engineering for studying problems in biomechanics (Fung, 1993). A bioengineering plan is needed for solving important problems in gastrointestinal physiology and pathophysiology. It emphasises a thorough consideration of geometry, structure, biological factors, mechanical properties, and determination of constitutive equations. Then boundary-value problems can be formulated and solved, further testing can be done and the data compared to the theory. Then the theory can be used to predict the behaviour during disease or medical and surgical intervention. The application of the biomechanical approach to the study of the gastrointestinal structure and function represents an important interdisciplinary approach that can add rigorous insight to our understanding of gastrointestinal design and function. The bioengineering approach can be formulated in more detail as follows (summary from Fung, 1993).

- First, we need to study the geometric configuration; i.e., the morphology of the organism, anatomy of the organ, histology of the tissue, and the structure and ultrastructure of the materials.
- The biophysical properties of the materials and tissues that are involved in a problem must be determined.
- The governing differential of integral equations must be derived on the basis of fundamental laws of physics (conservation of mass, momentum, and energy, Maxwell's equations, etc.) and the constitutive equations of the materials.

- The environment in which an organ works must be understood in order to obtain meaningful boundary conditions.
- Based on the approach mentioned above, the boundary-value problems (differential equations with appropriate initial and boundary conditions) can be solved by experiments, analytically or numerically.
- Physiological experiments that will test the solutions of the boundary-value problems must be performed. The mathematical problem must be reformulated and resolved if necessary, to make sure that the theory and experiment correspond to each other.
- The experimental results must be compared with the corresponding theoretical results. If the hypotheses made in the theory are justified, then the numerical values of the undetermined coefficients in the constitutive equations can be determined.
- The outcome of other boundary-value problems associated with the same basic equations can then be predicted using the validated theory. With this basic information, biomechanics can be used for the invention of new devices and methods of intervention and treatment.

A number of publications exist on the anatomy and structure of the gastrointestinal tract (Gabella, 1987), though the data often have been collected on an empirical basis rather than with the above approach in mind. What work needs to be done next in gastroenterology is quite clear from a bioengineer's view: the constitutive equations, i.e., the three-dimensional stress–strain relationships of the gastrointestinal tissues, must be determined. With the constitutive equations, the distribution of physical stresses and strains in the organs in vivo, and the function of bolus transport and fluid movement in the gastrointestinal tract can be analysed by methods of continuum mechanics. The analytical results will relate the stress and function of the gastrointestinal tract with the structure represented by the geometric parameters. The job of determining the constitutive equations is not as daunting as it seems because the mathematical form of most soft tissues are known. If the hypothesis that the constitutive equations of the gastrointestinal tissues are similar to those of other soft tissues can be confirmed, then it remains only to identify the mathematical constants of the gastrointestinal tissues. Instruments are available and this task is currently being pursued. Determination of the geometry and the solid mechanical properties are, however, only steps in the direction of understanding the flow in the gastrointestinal tract.

1.6 Literature

The literature list contains papers mentioned in the text as well as other papers of interest.

Bayliss, VM, Starling, EH. 1899. The movements and innervation of the small intestine. *J Physiol*, **24**, 99–143.

Bayliss, VM, Starling, EH. 1901. The movements and innervation of the small intestine. *J Physiol*, **26**, 125–38.

Brasseur, JG. 1993. Mechanical studies of the esophageal function. *Dysphagia*, **8**, 384–6.

Cannon, WB. 1911. *The Mechanical Factors of Digestion*. Edward Arnold: London.

Colstrup, H, Mortensen, J, Kristensen, K. 1983. A probe for measurement of related values of cross-sectional area and pressure. *Urology Res*, **11**, 139–43.

Fackler, K, Klein, L, Hiltner, A. 1981. Polarizing light microscopy of intestine and its relationship to mechanical behaviour. *J Microscopy*, **124**, 305–11.

Frøbert, O, Gregersen, H, Bagger, JP. 1996. Mechanics of porcine coronary arteries ex vivo employing impedance planimetry: a new intravascular technique. *Ann Biomed Eng*, **24**, 148–55.

Frøbert, O, Mikkelsen, EO, Gregersen, H, Nyborg, NCB, Bagger, JP. 1996. Porcine coronary artery pharmacodynamics in vitro evaluated by a new intravascular technique. Relation to axial stretch. *J Pharmacol Toxicol Methods*, **36**, 13–19.

Frøbert, O, Schiønning, J, Gregersen, H, Baandrup, U, Petersen, JAK, Bagger, JP. 1997. Impaired human coronary artery distensibility by atherosclerotic lesions. A mechanical and histological investigation. *Int J Exp Pathol*, **78**, 421–8.

Fung, YC, Yih, CS. 1968. Peristaltic transport. *J App Mech*, 669–75.

Fung, YC. 1993. *Biomechanics. Mechanical Properties of Living Tissues.* Springer-Verlag: New York.

Gabella, G. 1984. Structural apparatus for force transmission in smooth muscles. *Physiol Rev*, **64**, 455–77.

Gabella, G. 1987. Structure of muscles and nerves in the gastrointestinal tract. In: *Physiology of the Gastrointestinal Tract*, Second Edition, ed. Johnson, LR, Christensen, J, Jackson, MJ, Jacobson, ED, Walsh, JH, pp. 335–82. New York: Raven Press.

Ginzel, KH. 1959. Investigations concerning the initiation of the peristaltic reflex in the guinea-pig ileum. *J Physio*l, **148**, 75–6.

Gregersen, H, Andersen, MB. 1991. Impedance measurement system for quantification of cross-sectional area in the gastrointestinal tract. *Med Biol Eng Comput*, **29**, 108–10.

Gregersen, H, Kassab, GS. 1996. Biomechanics of the gastrointestinal tract. *Neurogastroenterol and Motil*, **8**, 277–97.

Gregersen, H, Christensen, J. 2000. Gastrointestinal tone. *Neurogastroenterol and Motil*, **12**: 501–8.

Kosterlitz, HW, Robinson, JA. 1959. Reflex contractions of the longitudinal muscle coat of the isolated guinea-pig ileum. *J Physiol*, **146**, 369–79

Li, M, Brasseur, JG, Kern, MK, et al. 1992. Viscosity measurements of barium sulfate mixtures for use in motility studies of the pharynx and esophagus. *Dysphagia*, **7**, 17–30.

Li, M, Brasseur, JG. 1993. Non-steady peristaltic transport in finite-length tubes. *J Fluid Mech*, **248**, 129–51.

Macagno, E, Melville, J, Christensen, J. 1975. A model for longitudinal motility of the small intestine. *Biorheology*, **12**, 369–76.

Macagno, EO, Christensen, J. 1980. Fluid mechanics of the duodenum. *Ann Rev Fluid Mech*, **12**, 139–58.

Mall, F. 1896. A study of the intestinal contraction. *Johns Hopkins Hosp Rep*, **1**, 37–75.

Melville, J, Macagno, E, Christensen, J. 1975. Longitudinal contractions in the duodenum: their fluid-mechanical function. *Am J Physiol*, **228**, 1887–92.

Metry, S, Arhan, P, Chauvet, G. 1995. A mathematical analysis of intestinal peristaltic waves. *Med Eng Phys*, **17**, 204–14.

Meyer, JH, Gu, Y, Elashoff, J, Reedy, T, Dressman, J, Amidon, G. 1986. Effects of viscosity and fluid outflow on postcibal gastric emptying of solids. *Am J Physiol*, **250**, G161–4.

Miftakov, RN. 1994. Mathematical modeling of the peristaltic reflex: a numerical experiment. *J Math Sci*, 71, 2775–89.

Miftakhov, RN, Wingate, DL. 1994. Numerical simulation of the peristaltic reflex of the small bowel. *Biorheology*, **31**, 309–25.

Miftakhov, RN, Abdusheva, GR, Christensen, J. 1999. Numerical simulation of motility patterns of the small bowel. Part I – formulation of a mathematical model. *J Theor Biol* **197**, 89–112.

Nothnagel, H. 1882. Zur chemischen reizung der glatten muskeln, zugleich als beitrag zur physiologie des darms. *Arch Pathol Anat Physiol Klin Med*, **88**, 1–11.

Orberg, J, Baer, E, Hiltner, J. 1983. Organization of collagen fibers in the intestine. *Connect Tissue Res*, **11**, 285–97.

Ren, J, Massey, BT, Dodds, WJ, et al. 1993. Determinants of the bolus pressure during esophageal peristaltic bolus transport. *Am J Physiol*, **264**, G407–13.

Schulze-Delrieu, K, Wall, JP. 1983. Determinants of flow across isolated gastroduodenal junctions of cats and rabbits. *Am J Physiol*, **245**, G257–64.

Schulze-Delrieu, K. 1992. Clearance patterns of the isolated guinea pig duodenum. *Gastroenterology*, **102**, 849–56.

Srivastava, LM, Srivastava, VP, Sinha, SN. 1983. Peristaltic transport of a physiological fluid. Part I. Flow in non-uniform geometry. *Biorheology*, **20**, 153–66.

Stavitsky, D, Macagno, EO, Christensen, J. 1981. Finite-element analysis of flow induced by contractions like those of the intestine. *J Biomech*, **14**, 183–93.

Tanko, L, Mikkelsen, EO, Frøbert, O, Bagger, JP, Gregersen, H. 1998. A new method for combined isometric and isobaric pharmacodynamics studies on porcine coronary arteries. *Clin Exp Pharmacol Physiol*, **25**, 919–27.

Tanko, LB, Simonsen, U, Frøbert, O, Gregersen, H, Bagger, JP, Mikkelsen, EO. 2000. Vascular reactivity to nifedipine and Ca^{++} in vitro. The role of preactivation, wall tension and geometry. *Eur J Pharmacol*, **387**, 305–14.

Tanko, LB, Simonsen, U, Matrougui, K, Gregersen, H, Frøbert, O, Bagger, JP, Mikkelsen, EO. 2001. Axial

stretch modifies contractility of porcine coronary arteries by a protein kinase C-dependent mechanism. *Pharmacol Toxicol*, **88**, 89–97.
Tözeren, A, Özkaya, N, Tözeren, H. 1982. Flow of particles along a deformable tube. *J Biomech*, **15**, 517–27.
Trendelenburg, P. 1917. Physiologische and pharmacologische versuche über die dünndarmsperistaltik. *Naunyn Schmiedebergs Arch Pharmacol*, **81**, 55–129.
Weems, WA. 1987. Intestinal fluid flow: Its production and control. In: *Physiology of the Gastrointestinal Tract*, ed. Johnson, LR, Christensen, J, Jackson, MJ, Jacobson, ED, Walsh, JH, pp. 571–93. New York: Raven Press.
Yamada, H. 1970. *Strength of Biological Materials*. Baltimore: The Williams & Wilkins Company.
Yuan, SY, Furness, JB, Bornstein, JC, Smith, TK. 1991. Mucosal distorsion by compression elicits polarized reflexes and enhances responses of the circular muscle to distension in the small intestine. *J Autonom Nervous System*, **35**, 219–26.

2. Geometry, Structure and Motor Function of the Gastrointestinal Tract

2.1 Introduction

The gastrointestinal tract is a continuous channel through the body with oral and anal orifices for intake and output (Figure 2.1). It consists of a series of organs with the biliary tract as a major side-branch. These organs, named as though they are distinct and separate constructions, resemble one another in constitution, the elements being variously arranged in such a way as to produce cylinders, spheroids, or intermediate forms. This chapter deals primarily with the basic structure of the gastrointestinal wall and with the organ-specific variations.

The discussion of gastrointestinal anatomy deals with the pharynx and following parts of the system. Organ-specific variations from the general pattern in humans are described but the reader should be aware of large inter-species variations. The structure and function of the gastrointestinal tract differ enormously among species according to nutritional mechanisms and practices. Different kinds of diets demand

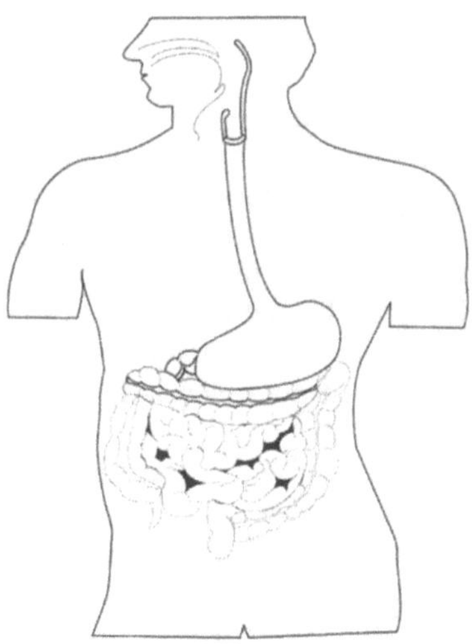

Figure 2.1. An overview of the gastrointestinal tract with its organs in series extending from the mouth to the anus.

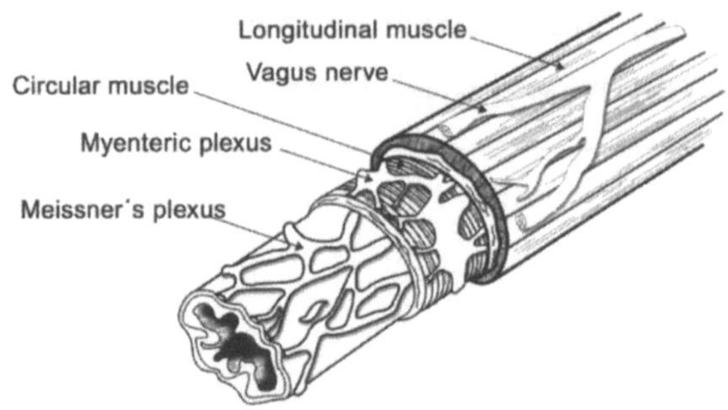

Figure 2.2. The laminated composite structure of the oesophagus. A muscle layer with the long axis of the cells lying in circumferential direction is surrounded by a muscle layer with the long axis of the cells lying in longitudinal direction. The myenteric plexus of nerve cells is found in between these muscle layers. The circular muscle layer surrounds the submucosal layer and in between these layers the submucosal nerve plexus is located. Another muscle layer is located on the inside of the submucosa (not labelled in the figure).

different physical treatments to prepare foods for enzymatic degradation. For that reason, the motor apparatus responsible for the necessary mechanical processing varies from the basic pattern according to the diet of the species.

The wall of the gastrointestinal tract is a laminated composite structure, as exemplified by the oesophagus in Figure 2.2. Though its various layers necessarily operate in unity because they are bound together, they possess dissimilar properties, both physical and physiological, that contribute differently to the net motor function of the system.

2.2 The Structure and Composition of the Muscle Layers

2.2.1 The Main Muscle Coat

The main muscle coat is also called the muscularis propria and consists of two layers. A network of collagen fibres forms the basic structure or stroma in each layer. The muscle cells lie within the interstices of this web, tightly attached both to one another and to the fibrous elements of the mesh. The muscle cells themselves are cylindrical. They are capable of shortening separately in the contraction process. The cells of one muscle layer all lie with their axes in the same direction. Their union through their attachment to the stroma causes the summation of their separate actions. Since they usually contract essentially simultaneously, the muscle layer seems to move as a unit.

The two layers of muscle differ in the orientation of their muscle cells. The cylindrical cells in the outer layer of the muscularis propria are oriented with their axes in the direction of the axis of the cylindrical conduit. This layer is called the longitudinal muscle layer. In the inner layer of muscle, the cells lie in the direction of the circumference of the cylinder. This layer is called the circular muscle layer.

The circular muscle layer contains the interstitial cells of Cajal, a special set of modified muscle cells. These are stellate, with long processes that extend widely to contact many muscle cells. The axons that provide the neural regulation of the muscle are concentrated in an intimate relationship with these interstitial cells. The interstitial cells seem to be derived from muscle cells, modified for the transmission of information both from nerve to muscle and from cell to cell within the muscle mass.

2.2.2 The Muscle of the Mucosa

A third layer of muscle, the muscularis mucosae, exists throughout almost the entire tract. The submucosa separates it from the main muscle coat. In general, its structure is the same as that of the other muscle layers, a collagen mesh forming a stroma in which the muscle cells are embedded. The muscle cells are oriented with their axes in all directions. It forms a part of the mucosa, together with two other layers, the lamina propria and the epithelium.

2.2.3 Organ-specific Variations in the Muscle Layers

Pharynx and Upper Oesophageal Sphincter

The pharyngeal musculature differs greatly from the general pattern. The cells are striated (somatic) muscle cells and they follow a circular rather than an axial direction. The musculature of the oropharynx is arranged to form three overlapping layers, the upper, middle, and lower pharyngeal constrictors. Each constrictor has a right and left part. The two halves are physically continuous, fusing at the midline so that a constrictor functions as a single encircling unit. The inferior constrictor thickens at the junction with the oesophagus to form the upper oesophageal sphincter; that is, this sphincter is continuous with the pharyngeal musculature, but distinct from it in its thickness and its gross anatomical relationship to the airway. Instead of encircling the pharynx like the rest of the constrictors, the sphincter muscle inserts at both ends to the ends of the cricoid cartilage so that it closes the cephalic end of the oesophageal lumen by compressing it against the trachea when contracting. The pharyngeal mucosa throughout this region contains no muscle.

The Oesophagus and the Lower Oesophageal Sphincter

The muscle of the oesophagus differs from the general description given above mainly in the nature of the muscle cells. The musculature is striated (somatic) muscle, like that of the pharynx, from the cephalic end of the organ to a point about one-third of the way to the stomach. At this level, the muscle becomes visceral (smooth) muscle, which continues throughout the remainder of the gastrointestinal tract. This pattern is seen in humans and in some marsupials such as opossums, whereas most other species have either striated or smooth muscle in the whole oesophagus. The two kinds of muscle cells, striated and smooth, are intermingled for only a very short distance within both the outer and inner layers. The striated muscle also contains sensory structures - muscle spindles - like those found in somatic muscle everywhere.

The lower oesophageal sphincter, evident grossly as a ridge, represents a thickening of the circular muscle layer shortly above the stomach. The ultrastructure of this muscle differs from that of the body of the oesophagus in the size and number of mitochondria, and it differs from that muscle in the oxygen-dependence of its contraction. The two intramural plexuses at the level of the sphincter resemble those in the remaining part of the oesophagus, except for the lower distribution density of nerve cells in the myenteric plexus at this level.

The prominent mucosal muscle consists of smooth muscle throughout the oesophagus in all mammals. The muscle of the mucosa extends from a level just below the cricopharyngeus to the stomach where it merges into the gastric mucosal muscle. The muscle of the mucosa in the oesophagus differs from the general description provided above in two ways. It is very thick (fully as thick as the inner layer of the main muscle coat) and its cylindrical muscle fibres (visceral muscle throughout the organ) all lie with their axes parallel to the axis of the tubular organ. This implies that the contraction of this muscle layer tends to shorten the oesophagus, along with contractions of the longitudinal muscle layer of the main muscle coat.

The interstitial cells of Cajal in the circular smooth muscle layer are distributed diffusely throughout that layer. There are none in the striated muscle of the cephalic part of the oesophagus.

The Stomach and Pyloric Sphincter

The asymmetrical geometry of the stomach arises from a widening at the cephalic end of the basic form of a collapsible cylindrical tube, an enlargement that occurs only on its left side. This bulge becomes a saccule, the fundus of the stomach. It serves to store ingested materials, so that there is no need for the organism to take in nutrients continuously. The fact that this part of the stomach possesses a third muscle layer, the oblique muscle layer, probably reflects this storage function. The oblique muscle layer lies on the luminal side of the inner (circular) layer of muscle. These two muscle layers are separate throughout the fundus but they fuse in the region that separates the fundus from the caudal part of the stomach, the antrum. Thus, the oblique muscle layer seems to arise from the inner circular muscle layer.

The circular muscle layer thickens progressively toward the junction of the stomach with the duodenum. A particular thickening characterises the most distal segment, the pylorus or the pyloric sphincter.

The outer longitudinal muscle layer enters into a special relationship with the pylorus. Most of this muscle layer inserts into the thickened muscle of the pyloric loop, while virtually none passes outside this loop to join with the longitudinal muscle layer of the duodenum. This arrangement implies that this muscle layer shortens the stomach as it contracts.

The interstitial cells of Cajal in the stomach lie in two layers, in the plane of the myenteric plexus and in the circular muscle layer. A pronounced difference exists between the proximal stomach and the distal stomach in their distribution density. They are virtually absent from both locations in the cardia and fundus, rising to a maximal density at the level of the pylorus.

The Small Intestine and the Ileo–caecal Sphincter

The basic geometry of the gastrointestinal tract, as described above, essentially describes the whole of the small intestine. The outer longitudinal muscle layer is somewhat thinner than the circular muscle layer as can be seen in the histological section in Figure 2.3. The mucosal muscle layer is much thinner. Its fibres lie with no dominant orientation. It gives off slips that extend into each villus of the mucosa. The organ is not uniform throughout its length. A slight gradient in the thickness of the main muscle coats exists so that the muscle layers appear to be a little thicker at the cephalic end of the organ. The interstitial cells of Cajal lie concentrated in a single plane within the substance of the circular muscle layer. They appear to be a little more numerous in the duodenum than they are in the ileum.

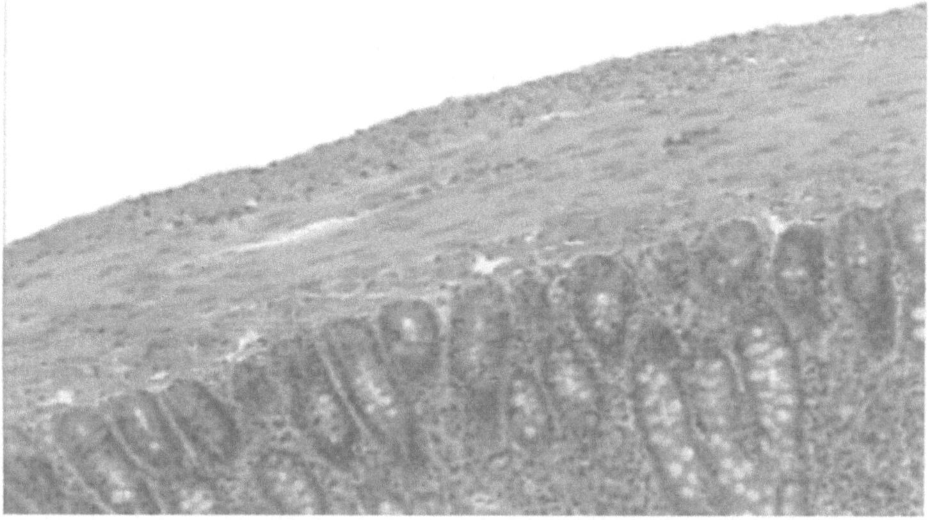

Figure 2.3. H-E-stained histological section of rat intestine.

The Large Intestine and Anal Sphincters

The outer longitudinal muscle layer forms three thick bundles, the taeniae of the colon, that extend throughout the length of the organ from the tip of the caecum to the proximal end of the rectum. A very thin layer of the longitudinal musculature covers the surface of the colon between these taeniae.

The circular muscle layer forms narrow thickenings, the haustral markings, at fairly regular intervals along the organ. It also bulges between the three taeniae. The whole colon, except for the rectum, is consequently sacculated in appearance, the saccules being defined both by the haustral markings and by the restriction that the taeniae impose. The circular muscle layer thickens at the caudal limit of the colon to form the internal anal sphincter. The external anal sphincter is not a part of the gastrointestinal tract but a somatic muscular structure deriving from the pelvic floor, whose position and function, however, give it important regulatory control over defecation.

The interstitial cells of Cajal occupy two planes in the colon, the submucosal surface of the circular muscle layer and the plane of the myenteric plexus in the intermuscular cleft. Those interstitial cells associated with the myenteric plexus are relatively sparse. The dense plexus of interstitial cells on the submucosal surface of the circular muscle layer contains abundant axons derived from ganglion cells in the deeper part of the two divisions of the colonic submucosal plexus.

The Biliary Tract

The bile duct and its branches and the pancreatic duct lack organised intrinsic musculature except for the thick and short segment adjacent to the duodenum where a circular muscle layer forms the sphincter of Oddi. The sphincter of Oddi contains thickened circular muscle at the proximal margin, the sphincter choledochus and at the distal margin, the sphincter ampullae. The bile duct consists of an outer layer of loose connective tissue, a middle sub-epithelial layer of collagen with a few scattered longitudinal muscle fibres. The wall of the gall bladder contains loosely organised muscle bundles within a loose connective tissue stroma.

2.3 The Structure and Composition of the Connective Tissue Layers

The term connective tissue refers to a material made up principally of collagen and elastin fibres forming a loose mesh in which many kinds of small cells with general maintenance functions and immune functions are scattered. Collagen is a structural protein comprised of 3 α-chains arranged in a triple helix. The primary amino acid structure consists of repeating units of Glycine-X-Y where X and Y often are proline and hydroxyproline. The latter comprises approximately 14% of collagen and is often used as a biochemical marker for the collagen content of tissues due to its relative specificity to collagen. More than 15 types of collagen have been discovered, with the fibrillar types 1 and 3 being the most frequent in the gastrointestinal tract. Once a fibril is formed and the collagen cross-linked, the fibril becomes strong and stable. Thus, collagen fibrils have a high stiffness and are often abundant in the extracellular matrix. Therefore, they contribute significantly to the passive mechanical properties of tissue. However, its arrangement is important and in the gastrointestinal tract it appears somewhat disorganised compared to structures such as tendons. Figure 2.4 shows collagen in the rabbit submucosa of the oesophagus. It can be seen that the fibres run in several directions with much variation in size.

The picrosirius red stain is specific for collagen. When viewed with polarised light, picrosirius red enhances the birefringence of collagen molecules and results in different birefringence colours as a function of fibril size and orientation (Figure 2.5).

2.3.1 The Submucosa

A broad zone of connective tissue lies between the main muscle coat and the mucosa (Figure 2.5). Most of the space in the submucosa is occupied by water. Both the thickness and the structural laxity of this lamina allow the mucosa to move freely and widely over the inner surface of the muscle coat. The submucosa also contains a network of nerves called the submucosal plexus.

The submucosa has been called the skeleton of the intestine due to its high content of collagen. Both Fackler et al. (1981) and Orberg et al. (1982; 1983) have demonstrated that the collagen fibres in the submucosa of the rat intestine, when exposed to stress, biaxially orient at approximately –30 and +30 to the longitudinal axis.

2.3.2 The Subserosa

The main muscle coat is covered on the outside by the serosa, a layer of epithelial cells that envelops the whole of the intra-abdominal gastrointestinal tract and lines the abdomen as a membrane called the peritoneum. This epithelium is tightly attached to

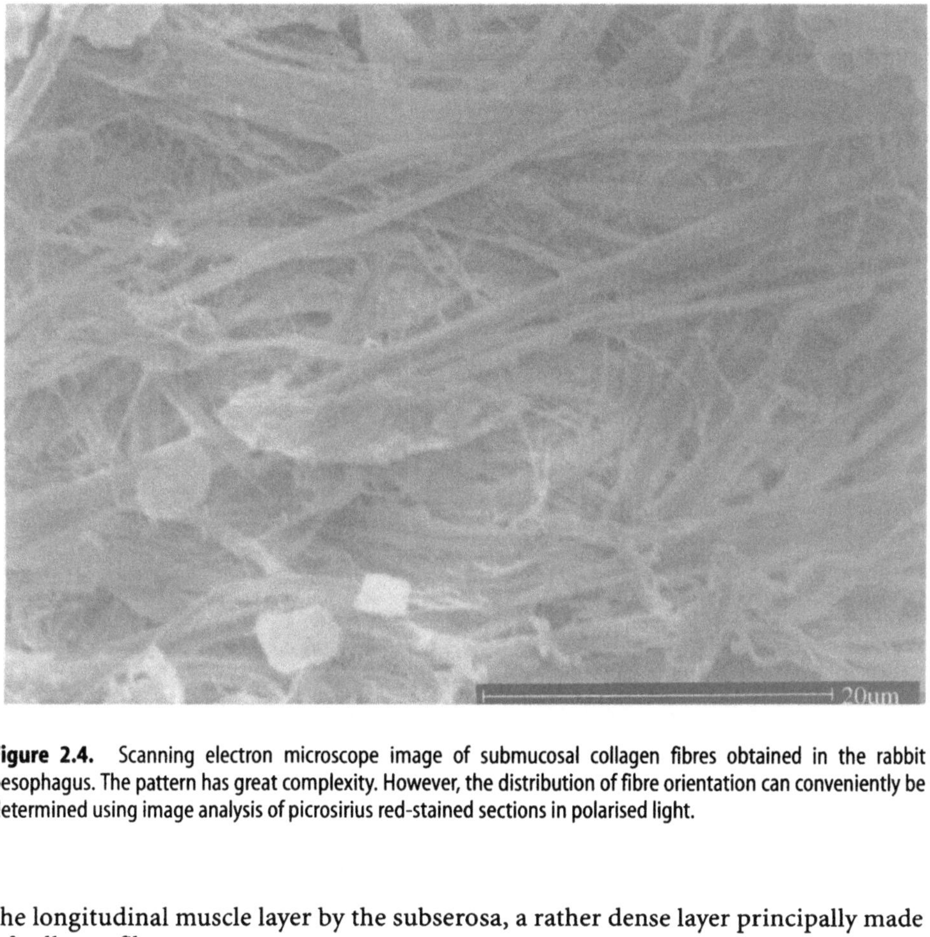

Figure 2.4. Scanning electron microscope image of submucosal collagen fibres obtained in the rabbit oesophagus. The pattern has great complexity. However, the distribution of fibre orientation can conveniently be determined using image analysis of picrosirius red-stained sections in polarised light.

the longitudinal muscle layer by the subserosa, a rather dense layer principally made of collagen fibres.

2.3.3 The Lamina Propria

Another layer of connective tissue, the lamina propria, lies between the muscularis mucosae and the layer of cells that lines the entire gastrointestinal tract, the gastrointestinal epithelium. The lamina propria appears histologically to be somewhat denser in its composition than the submucosa. Fibres of collagen and elastin, lacking any obvious organisation, constitute the principal matrix elements in this layer. The layer also contains many kinds of scattered small cells as well as a few nerve fibres. It firmly attaches the epithelium to the muscularis mucosae.

2.3.4 Organ-specific Variations in the Connective Tissue

Pharynx and Upper Oesophageal Sphincter

The pharynx is covered on the outside by a tough collagenous muscle sheath that separates it from other structures in the neck and thus allows the muscle to move freely without tethering. A similar sheathing exists between the constrictors as they

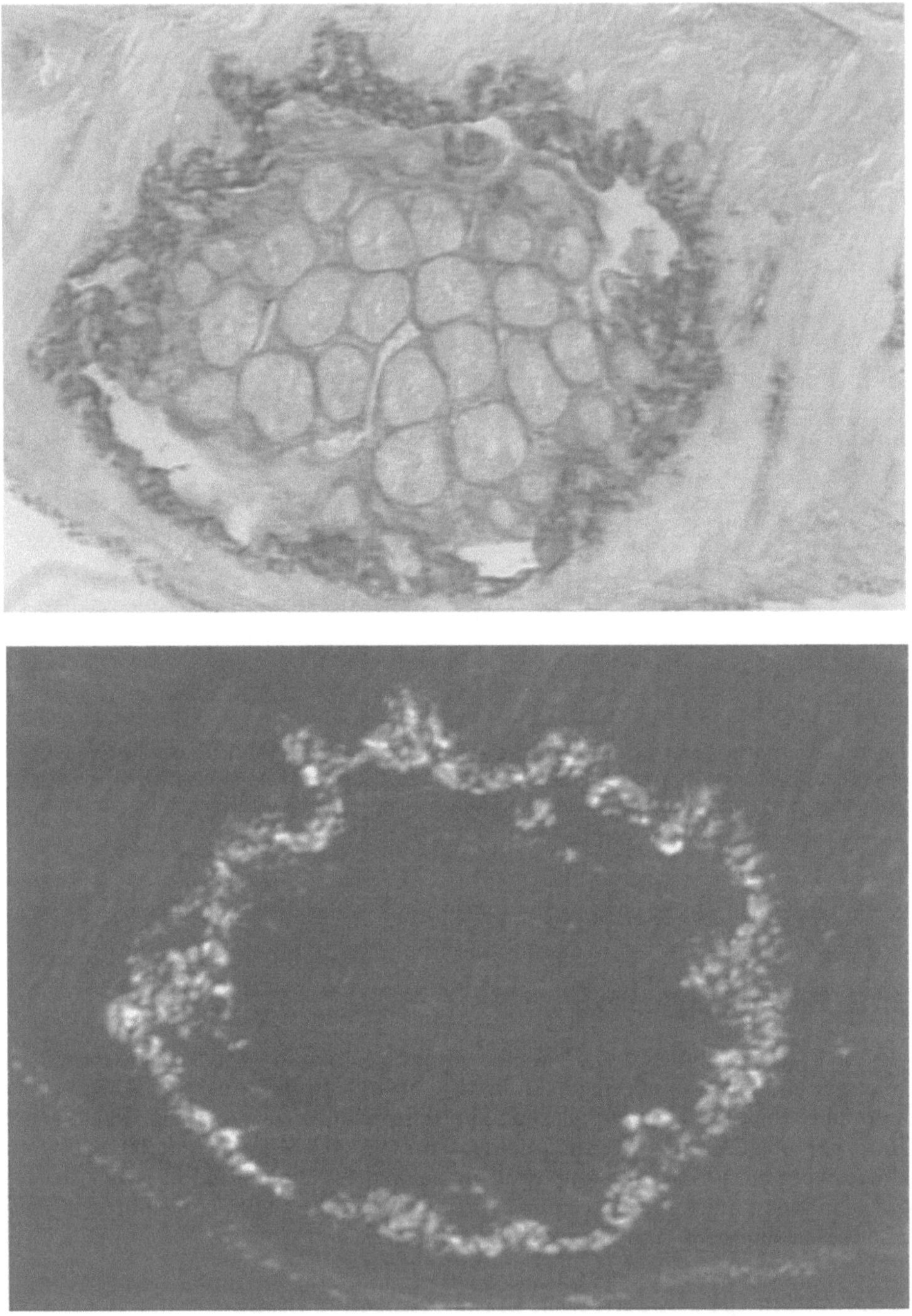

Figure 2.5. Two images of the same section obtained from the small intestine and stained with picrosirius red that stains collagen red and muscle yellow in brightfield microscopy (top) and collagen red and green using polarised microscopy (bottom). The cutting is made along the longitudinal axis and penetrates from the circular muscle layer into the submucosa and mucosa in the centre of the images. The submucosal collagen is easy to identify. Some collagen can also be identified in the subserosa (see also Plate I).

overlap. Such a connective tissue sheath also covers the inner surface of the musculature to make a base for the attachment of the epithelium.

The Oesophagus and the Lower Oesophageal Sphincter

The oesophagus possesses a special connective tissue structure in the form of the phreno-oesophageal ligament, a collar made of collagen and elastin fibres that seals the organ within the oesophageal hiatus of the diaphragm. This ligament has two layers, one that inserts on the organ at about the cephalic edge of the lower oesophageal sphincter and one that inserts at about the caudal border of the sphincter. The two layers also pass to the upper and lower surfaces, respectively, of the diaphragm. The phreno-oesophageal ligament is a tough structure, as the term "ligament" implies, but it is generous, long and elastic enough to allow considerable axial movement of the oesophagus relative to the diaphragm, as must obviously occurs in breathing and swallowing. Since the whole organ is tethered at both ends (to the cricoid cartilage and to the diaphragm), it remains rather well fixed in a cephalo-caudal orientation.

The Stomach and Pyloric Sphincter

The stomach hangs in a mesentery, a fold of the peritoneum encasing the organ and attaching it to the posterior wall of the abdomen. From the point of view of its mechanics, this arrangement means that the organ is essentially free of any encumbrances to its movements. The phreno-oesophageal ligament fixes it loosely to the diaphragm at its diaphragmatic end. The retroperitoneal position of the duodenum ties the pyloric end of the stomach to the posterior abdominal wall at the other end. A thin sheet of connective tissue separates the oblique and circular muscle layers in the gastric fundus.

The very thick submucosa of the stomach allows the thick mucosa be thrown into folds except when the organ is maximally distended.

The Small Intestine and the Ileo–caecal Sphincter

The duodenum is relatively fixed in position, being retroperitoneal while the rest of the intestine, like the stomach, is suspended from the posterior abdominal wall by a mesentery. The shortness of the intestinal mesenteric attachment at the ileo–caecal junction tends to fix that end in position as well. Between the intestinal extremities, the generous mesentery allows considerable movement of the intestine. Thus, these attachments do not affect the freedom of the intrinsic movements of the intestinal wall in its contractions. They seem, however, to keep the long and pliable intestine oriented in such a way as to prevent twists and other such gross deformations. Gabella, Fackler and Orberg provide many useful data on the submucosa of the small intestine.

The Large Intestine and Anal Sphincter

The colon is fixed in position at several points, but this fixation seems not to impede the movements that result from contractions of its intrinsic musculature. Rather, they serve to keep the organ loosely in place. The shortness of the colonic mesentery in general serves, in part, to do this. The mesentery is particularly short at the ileo–caecal junction, at the hepatic flexure, and at the splenic flexure where they are sometimes referred to as ligaments. At the rectosigmoid junction, the organ becomes extraperitoneal. The other pelvic viscera closely opposes it in the pelvis.

The Biliary Tract

The gall bladder has no submucosa. In the biliary tract itself there is a submucosa of looser connective tissue.

2.4 The Structure and Composition of the Epithelial Layers

2.4.1 The Gastrointestinal Epithelium

The innermost surface of the gastrointestinal tract is covered by the gastrointestinal epithelium, a layer of cells that differs greatly in structure and function in the various parts of the tract. These cells are closely attached to one another to form a continuous and quite homogeneous sheet. There is little or no collagen within such epithelial sheets, but they tend to be held together in part through their adherence to the basal lamina, a sheet of amorphous material to which epithelial cells are attached.

2.4.2 The Serosa

The serosa is a sheet of epithelial cells that encloses the intra-abdominal parts of the gastrointestinal tract, the peritoneum. Its cells are adherent to one another both directly and through their position on the collagen mesh of the subserosa.

2.4.3 Organ-specific Variations in the Epithelial Layers

Pharynx and Upper Oesophageal Sphincter

The epithelium lining the pharynx is squamous epithelium, like the skin but devoid of the keratinised outer layer that is found in the skin. As a squamous epithelium, it is tougher and less permeable than the epithelium of the more caudal parts of the gastrointestinal tract. Serosa does not surround the pharynx because it does not lie within a serosal cavity.

The Oesophagus and the Lower Oesophageal Sphincter

The oesophageal epithelium is similar to that of the pharynx, a non-keratinised squamous epithelium. As such, it is relatively tough and impermeable. It ends almost exactly at the point where the oesophagus and stomach join in terms of the musculature; that is, the very different epithelia of the oesophagus and stomach seem to be very strictly confined to their respective organs. In a common disease state like chronic gastro-oesophageal reflux, however, the epithelial junction is often found well above the muscularly defined union of the two organs. Thus, the location of the epithelial junction is defined physiologically rather than being fixed anatomically.

The Biliary Tract

The submucosa consists of looser connective tissue, over which the cuboidal epithelium lies.

2.5 Intramural Nerves

A system of nerves extends throughout all layers of the gastrointestinal wall. Its operation is most evident in the motions of the muscular walls, although it serves other functions, such as the regulation of secretion and absorption. It is usually thought of in terms of its regulation of muscular contraction and relaxation. In this role, which prominently includes autoregulation, the intramural system of nerves must incorporate the ability to sense motions through sensory structures called mechanoreceptors. Also, it must regulate the actions of the three muscle layers. These different functions dictate the geometry of the system.

2.5.1 The Myenteric Plexus

A network of nerves, the myenteric plexus, is embedded in the loose collagen matrix of the intermuscular plane (Figure 2.2). This set of nerves is essential in the regulation of the contractions of the adjacent musculature, the two layers of the muscularis propria. The basic form of nerve cells, a single rounded cell body that gives off one or more long processes – axons – gives this network its form. The nerve cell bodies form clusters or nodes, the ganglia, joined together by bundles of axons to make a two-dimensional rhomboidal mesh.

Unlike the brain, which is essentially immobile, the gastrointestinal wall moves a great deal and the axons, being forced to move as well, might be stretched and broken without safeguard. The geometry of the myenteric plexus contributes the necessary protection. The whole plexus floats in the loose collagen stroma. Also, the mesh can be considerably deformed without damage because its polygonal structure can accommodate great change in either the axial or the circumferential direction as long as a change in the one dimension is compensated by a change in the other.

Several different kinds of nerve cells, having different functions, make up the myenteric plexus. One can postulate that they constitute (i) sensory cells, (ii) motor cells and (iii) internuncial cells, cells connecting sensory to motor nerve cells. These forms cannot be distinguished in structural or geometric terms.

Sensory structures, the mechanoreceptors essential to the self-regulatory function of the gastrointestinal musculature, must be present in the gastrointestinal wall. Some of the necessary mechanoreceptors have been identified as leaf-like elements in ganglia of the myenteric plexus, the intraganglionic laminar endings.

2.5.2 The Submucosal Plexus

The thick and gelatinous submucosa contains another plexus of nerves that resembles the myenteric plexus. In this position, this plexus seems to serve to regulate the whole mucosa, especially the mucosal muscle, as well as the circular muscle layer of the muscularis propria. Its ganglia and interconnecting nerve bundles, however, are less densely distributed than those of the myenteric plexus and the mesh they form has a less uniform pattern. Its arrangement seems to offer protection from damage that might ensue from the motions of the muscular wall of the gastrointestinal tract.

2.5.3 The Nerves of the Lamina Propria and Serosa

The lamina propria contains many nerves but only as fibres, axons connecting the submucosal plexus to the epithelium. For the most part, these seem to extend mainly in the radial axis, passing only a short distance in the axial and circumferential directions. Still, they tend to be convoluted, thus accommodating to the motions of the mucosa.

The serosa contains particular neural structures known to be mechanoreceptive, the Pacinian corpuscles. They presumably constitute the endings of sensory nerves whose cell bodies lie in the myenteric plexus.

2.5.4 Organ-specific Variations in the Intramural Nerves

Pharynx and Upper Oesophageal Sphincter

No nerve plexus exists between these muscle layers. Their nerve supply comes instead directly from the central nervous system through somatic motor fibres contained in the glossopharyngeal nerve. They activate the muscle through the motor end plates, the classical neuromuscular junction that characterises striated muscle everywhere. In addition, sensory nerve fibres end in muscle spindles, mechanoreceptors, that appear to be identical to those found in somatic muscle.

The Oesophagus and the Lower Oesophageal Sphincter

The myenteric plexus of the oesophagus differs from the general pattern with its relatively few ganglia and in its lack of a regular polygonal pattern. This plexus can be seen even in the striated muscle part of the organ where it seems likely to serve mainly mucosal functions. The interganglionic bundles of axons extend mainly in the direction of the axis of the organ, and there are many large nerve bundles, branches from the extrinsic innervation – the vagus nerves – running through the plexus parallel to the axis of the organ. This prominence of axially oriented bundles of nerve fibres in the wall of the oesophagus is consistent with the lack of axial stretching of the organ in its normal operation.

The submucosal plexus differs in that there are no ganglia, and the few nerve fibre bundles to be found there all contain axons that extend there from the myenteric plexus or the extrinsic nerves.

Two apparently sensory structures characterise the oesophagus. The ganglia of the myenteric plexus contain laminar stellate structures, the intraganglionic laminar endings. These terminals of the bipolar nerve cells of the vagal nodose ganglia may well be stretch receptors. Also, the squamous epithelium of the oesophagus contains solitary nerve fibres that depart from the submucosal plexus to extend almost all the way to the luminal surface, often ending in small knobs. They are probably sensory fibres of the thoracolumbar system and possibly modulate the pain of oesophageal mucosal inflammation.

The Stomach and Pyloric Sphincter

The gastric myenteric plexus lies only beneath the longitudinal muscle layer. There is none between the circular and oblique muscle layers. The plexus consists of large ganglia connected by thick fascicles to form a very irregular pattern of polygons. Thick fascicles, extensions of the vagal innervation, ramify in the plane of the myenteric plexus distributing nerve fibres of extrinsic origin especially to the proximal stomach, sparing the antrum. The distribution density of nerve cells in the myenteric plexus rises somewhat from the proximal stomach to the pylorus. The intramural extensions of the vagal nerves that enter the stomach from the oesophagus tend to lie in the direction of the axis of the organ, minimising their possible damage from the stretching effect of gastric filling. The submucosal plexus is sparse.

The Small Intestine

The myenteric plexus contains large, uniformly spaced ganglia linked by fascicles to form a regular polygonal plexus. The submucosal plexus, where the distribution density of nerve cells far exceeds that in this plexus anywhere else in the gut, actually constitutes a bilayered structure, one layer (Meissner's plexus) lying just beneath the mucosa with the other (Henle's plexus) closer to the circular muscle layer.

The Large Intestine and Anorectum

The myenteric plexus closely resembles that of the small intestine. The distribution density of its ganglion cells declines throughout the distal colon to reach a nadir at the level of the internal anal sphincter and just above it. The myenteric plexus contains conspicuous large fascicles of nerves, intramural extensions of the extrinsic pelvic innervation, which arise from colonic branches of the pelvic plexus that penetrate the longitudinal muscle layer at the rectosigmoid junction. From here, they extend to about the level of the hepatic flexure and to the internal anal sphincter, distributing fibres of extrinsic origin to the ganglia of the plexus (see the section about extramural innervation).

The submucosal plexus of the large intestine closely resembles that of the small intestine. It has the same bilayered structure with a Meissner's plexus just beneath the mucosa and a Henle's plexus next to the circular muscle layer.

The Biliary Tract

No intramural innervation exists at all, the intrinsic innervation simply lying on the outside of the system of ducts, but seeming to be an extension of the intestinal myenteric plexus.

2.6 Extramural Nerves

The extrinsic nerves clearly possess regulatory properties in gastrointestinal motility, but the strength of this effect varies greatly among organs. In the oesophagus and the distal colon, normal motor function requires an intact extrinsic innervation, but motility in the small intestine seems to require less control from extrinsic nervous structures.

Although the intramural system of nerves acts with such autonomy as to be called the little brain, it must interact to some extent with regulatory centres in the brain and spinal cord, for major nerve trunks connect the gastrointestinal tract with the relevant parts of the central nervous system. The geometry of the connections leads to the designation of two separate extramural systems that act separately.

2.6.1 The Craniosacral Innervation

The craniosacral innervation links the gastrointestinal tract to two parts of the central nervous system, the brainstem in the cranial part of the central nervous system and the sacral part of the spinal cord. Motor fibres of the vagus nerves originate in the two vagal motor nuclei in the floor of the fourth ventricle, the nucleus ambiguus and the dorsal motor nucleus. The somas of the sensory fibres in the vagus nerves lie in the ganglion nodosum whose unipolar cells possess a branched process, one branch receiving signals from the gut and the other carrying them on to the brain. A major part of the sensory input from the gut goes to the nucleus of the tractus solitarius.

The two vagi exit the skull through the jugular foramina and follow the oesophagus to about its midpoint where they break up into branches that coalesce, divide, and fuse again, equally mixing fibres from the right and left vagi into the oesophageal plexus. This coarse network surrounds the organ from about the level of the aortic arch to the stomach. Branches from the vagi, both offshoots from the recurrent laryngeal nerves and specific oesophageal branches from the oesophageal plexus, enter the oesophagus, supplying fibres to the ganglia of the myenteric plexus. The bundles of the oesophageal plexus coalesce just above the diaphragm to form the vagal trunks, four or five thick fascicles that pass through the oesophageal hiatus in the diaphragm to reach the stomach. They break up below the diaphragm into branches to the various viscera. Branches of the vagi to the other parts than the stomach of the tract cannot be traced readily beyond the first part of the duodenum, but physiological evidence indicates that vagal influences extend to the middle of the colon.

The vagus nerves initiate the oesophageal motor responses to deglutition. Swallowing initiates a coordinated discharge of the motor nerve cells in the vagal nuclei. The organisation of the muscular actions that occur after a swallow indicate the existence of a swallowing centre that lies close to these nuclei. Its sensory inputs include signals from chemoreceptors and mechanoreceptors traversing the glossopharyngeal nerves. Its output regulates the discharge of the somatic motor fibres that pass through this nerve to supply the somatic musculature of the pharynx and proximal oesophagus, as well as the preganglionic parasympathetic fibres to oesophageal visceral muscle.

The pelvic nerves emanate from the roots of the second through fourth sacral segments of the spinal cord. Following the general rule for spinal nerves, the sensory nerve fibres that carry impulses from the gut originate from nerve cells in the dorsal root ganglia that also send fibres to cells in the dorsal horn of the cord. Motor fibres in the sacral outflow originate in cells of the ventral horn; the ventral horn cells of this region are sometimes referred to as Onuf's nucleus. Bundles of nerve fibres from these sacral roots enter into the formation of the pelvic plexus, a coarse network of fascicles that lies among the pelvic viscera. The pelvic plexus contains scattered small ganglia whose relation to the colonic innervation remains unclear. Colonic branches leave this plexus to enter the colon at the level of the rectosigmoid junction and pass from there in both directions principally in the direction of the axis of the organ.

2.6.2 The Thoracolumbar Innervation

The second extramural system of nerves is called the thoracolumbar innervation because its motor and sensory components connect the gastrointestinal tract to the thoracic and lumbar segments of the spinal cord. Much of the gastrointestinal tract receives these fibres by way of the splanchnic nerves. These nerve trunks constitute a fusion of the segmental nerves from the thoracic part of the spinal cord and pass to the prevertebral ganglia. The three ganglia here – the celiac, superior mesenteric, and inferior mesenteric ganglia – all send fibres to the gut by way of perivascular nerves and the hypogastric nerve to enter the gastrointestinal wall. The fibres terminate in the myenteric plexus where they supply ganglion cells, as well as in the vasculature.

The three ganglia contain the nerve cell bodies of neurons of two classes, the principal ganglionic neurons and the small, intensely fluorescent (sif), cells, as well as chromaffin cells and mast cells.

The thoracolumbar pathways most notably transmit nerve impulses that induce painful sensations of gastrointestinal origin. Also, organ-to-organ reflexes make use of pathways through the prevertebral ganglia as a part of the self-regulatory mechanism in gastrointestinal motility. The chromaffin cells and mast cells may serve some modulatory function in the neurohumoral transmission that occurs at the principal ganglionic neurons through the release of the bioactive substances they contain.

2.7 The Mechanical System: Motor and Flow Patterns in Organs

2.7.1 Introduction

The motions of the gastrointestinal tract must, at all times, keep the contents of the tract in their intended place. This means not only that they accomplish net antegrade flow but also that they mix the contents, move it across the surface where absorption occurs, and, critically, keep it out of places where it does not belong. Much disease results from the movement of the contents of one region into another where it is not tolerated. The sphincters of the tract critically serve this barrier function. Transit, the end result of gastrointestinal motility, varies greatly from one part of the tract to another. The first physiologists recognised the existence of transit in the gut but they had no way to examine it. Even yet, knowledge of the various patterns of transit remains incomplete because of methodological limitations. The following sections briefly describe the known patterns of transit that occur throughout the gastrointestinal tract.

Quite often, one meets the view that all aspects of gastrointestinal contractions represent the effects of nerves acting on an intrinsically passive musculature. However, many aspects of gastrointestinal motility arise from the special properties of the visceral muscle itself. These so-called myogenic factors in motility bear especially upon two matters: the rhythmicity of phasic contractions – those lasting a matter of seconds with a rapid onset and offset, and capacity of the muscle to generate tone – a sustained contraction. The former has been extensively studied whereas the latter largely has escaped investigation (see Chapter 5).

Rhythmic contractions, phasic contractions occurring at regular intervals, occur in the distal stomach, small intestine, sphincter of Oddi, and colon, but not in the oesophagus, proximal stomach, gall bladder, rectum, and the other sphincters of the gut. Some degree of tone occurs in the muscle throughout the gut, but it varies greatly in magnitude from one part to another. For example, the smooth muscle of the body of the oesophagus exhibits no tone at all, while, in contrast, tonic contraction accounts entirely for the closure of the lumen that defines sphincters.

In order to understand contractile activity and flow in the gastrointestinal tract and why it varies so much between different parts of the gastrointestinal tract, motility control will now be described.

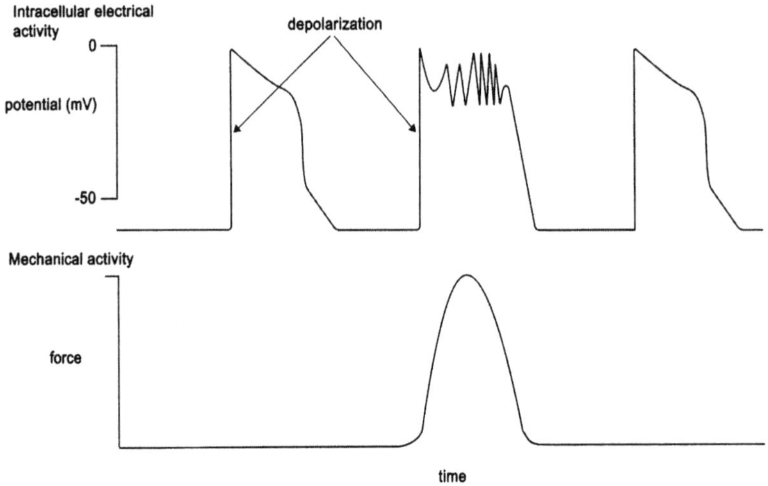

Figure 2.6. Top: Intracellular recording showing rhythmic fluctuations of the membrane potential without and with spike potentials. Bottom: Contraction of the smooth muscle when spike potentials are present.

2.7.2 Origin of Rhythmicity

Rhythmic contractions have a rapid onset and offset, a duration that is measured in seconds, and a pattern of regular recurrence at intervals measured in seconds. The regions of the gut that exhibit rhythmic contractions also display an electrical phenomenon called electrical slow waves. When recorded from a series of electrodes placed at close intervals along the intestine or distal stomach, they occur in a spatio-temporal sequence such that they appear to migrate along the organ. Similar signals can be recorded from the colon and the sphincter of Oddi, but only if the electrodes are placed deep in the wall. Intracellular microelectrode recordings from smooth muscle cells in rhythmically active parts of the gut show electrical slow waves as a signal with a relatively rapid depolarisation, a plateau, and a slower repolarisation (Figure 2.6). The parts of the gastrointestinal tract that do not generate rhythmic contractions lack such signals. The frequency of these electrical slow waves at any location corresponds to the fundamental frequency of the rhythmic contractions that occur at that site, and the direction and velocity of their migration conform to those characteristics of the peristaltic contractions of that region. They are considered to be pacemaker signals establishing the rhythmicity of contractions in those regions.

These electrical signals originate in close relationship to the interstitial cells of Cajal. The generation of slow waves requires the interaction of interstitial cells with smooth muscle cells but the exact nature of that interaction remains unknown. Interstitial cells resemble muscle cells in structure and can be viewed as specialised muscle cells.

When the muscle contracts in the neighbourhood of a gross recording electrode, the electrical record often shows a distinctive signal, a burst of one or more rapid electrical transients called the spike burst, superimposed upon the plateau of the electrical slow wave (Figure 2.6). This signal cannot be considered to be a true action potential because small rhythmic contractions in the intestine can occur without an accompanying spike burst on the corresponding slow waves, and even strong contractions regularly lack associated spike bursts in some places, especially in the distal stomach and colon. The depolarisation partly represents an inward current from the entry of calcium into the cell through the voltage-dependent and calcium-dependent L- and T-type calcium channels of the muscle. The magnitude of the depolarisation is determined by the modifications in the activity of those channels brought about by the changes in the membrane voltage and in the intracellular calcium concentration, and by the activation of calcium-dependent potassium channels that open to create an outward current. The plateau reflects a balance in the inward and outward currents. Repolarisation comes with the further rise in intracellular calcium concentration produced by calcium-induced calcium release from intracellular stores. This fully inactivates the calcium channels and fully activates the potassium channel. Smooth muscle cells also possess voltage-dependent potassium channels and chloride channels. The interactions among these five channels can fully account for the generation of both slow waves and spikes.

The gap junctions that link smooth muscle cells together ensure that the ion movements across membranes occur simultaneously within all muscle cells over a relatively large area, thus providing for synchronicity of slow waves and contractions; that is, the muscle constitutes a functional syncytium. It is not clear, however, what factors establish the frequency of slow waves. Also, there is no theory to explain the mechanism that determines the way the electrical signals progress through an organ.

2.7.3 Patterns in Various Parts of the Gastrointestinal Tract

The patterns fall into three categories; flow (transit), the movements of the muscular walls, and the control systems that regulate the movements. The concept of the former two categories is the subject in the next sections. Since this book is about mechanics rather than about regulation, the third category is not described in detail in this book.

Oropharynx

The oropharynx serves as a conduit for respiration through its rostral connections to the nasopharynx and the oral cavity and its caudal conjunction with the laryngeal inlet. Its intraluminal pressure varies during respiration, between about +1.4 and about −0.2 mmHg relative to atmospheric pressure. At rest the pharyngeal constrictors maintain a state of partial contraction. This muscle tone is low enough to allow the pharynx to remain open as a part of the airway, but great enough to reduce the size of the organ so that no significant pooling of fluids occurs there. The stability of the tone reflects the stability of the neural discharge. This firing rate seems most likely to depend, in part, upon the mechanoreceptors located in this musculature. As a part of the inferior constrictor of the pharynx, the upper oesophageal sphincter must possess the same regulatory mechanisms as that constrictor.

After mastication, food enters the pharynx intermittently in boli, delivered there by the movements of the tongue and palate that precede a swallow. Propelled by the contractions that a swallow induces in the pharyngeal musculature, each bolus moves rapidly through the pharynx and the upper oesophageal sphincter into the oesophagus. The apposition of the soft palate to the posterior pharyngeal wall produces a pressure of about 160 mmHg and initiates pharyngeal peristalsis. The bolus usually clears the pharyngeal lumen within about one second, leaving no residual. This pattern of transit characterises every bolus, regardless of the size and physical properties of the swallowed material. The fact that a swallow cannot be repeated at intervals of less than about two seconds provides for an upper limit to the frequency with which food can be delivered to the oesophagus. The moving front of contraction passes through the pharyngeal constrictors in sequence, traversing the pharynx and hypopharynx at about 15 cm per second to reach the upper oesophageal sphincter in about one second. If the bolus is a thin liquid, it may flow so fast under hydrostatic force that it reaches the sphincter almost before the pharyngeal contraction has begun. It will, however, arrive after the sphincter has relaxed because the relaxation in this complex of muscles always precedes the contraction. If the liquid bolus arrives too soon, it enters the larynx, an abnormal event called aspiration or laryngeal penetration. The duration of events in the pharyngeal muscles is such that solid masses that move through the organ only under the influence of the pharyngeal contraction also reach the sphincter while it is still relaxed. The initial response of the constrictors is a very brief loss of the stable tone that they manifest at rest. This brief and inapparent inhibition, however, is followed by a powerful discharge of the nerves, producing a contraction. The contraction seems to represent a rebound of the nerves to the antecedent inhibition. Although the neural inhibition that precedes the swallowing-induced contraction in the pharynx has no obvious effect on the mechanics of the main body of the pharynx, it has a crucial role in the function of the upper oesophageal sphincter. The tone of the sphincter accomplishes the critical purpose of maintaining closure of the lumen into the oesophagus, thus keeping the airway separated from the gastrointestinal tract. When the muscle of the sphincter relaxes with the initiation of a swallow, any hindrance the sphincteric muscle offers to the opening of the lumen disappears. A bolus of swallowed material approaching from above under the influence of a progressive contraction of the pharyngeal constrictors thus meets no resistance.

Oesophagus

The oesophagus and its sphincter remain stable at rest. The flaccid walls of the oesophageal body exhibit only passive movements, a result of the motions of the chest in breathing and of cardiac events. Intraluminal pressure in this part of the organ, essentially the same as intrathoracic pressure, varies at rest between a minimum of -15 (in inspiration) and a maximum of +5 (in expiration) mmHg, according to the respiratory effort. Both the striated muscle of the upper third of the organ and the visceral muscle of the rest – except the sphincter – exhibit no tone. The thickened zone in the circular muscle layer just above the junction to the stomach exhibits enough tone that the lumen at this level is occluded. The normal resting pressure recorded by conventional manometry in the lower oesophageal sphincter generally lies between 10 and 40 mmHg above atmospheric pressure. As for the upper oesophageal sphincter, this pressure partly represents an artefact from the reflex contraction of the muscle induced by the presence of the catheter in the lumen. The actual intraluminal pressure in the sphincter must be the same as that within the tissue of the sphincter, and this cannot exceed the vascular perfusion pressure of the tissue. The similarly closed upper oesophageal sphincter and this sphincter at the stomach seal the lumen of the flaccid oesophageal body from the atmosphere and from the stomach respectively. Thus, the fluctuating pressure in this empty potential space is sub-atmospheric, reflecting intrathoracic pressure. There are normally no contents in the lumen of this segment except for a very small quantity of saliva, enough simply to coat the surface.

Transit through the upper oesophageal sphincter occurs only in those brief intervals when it is open, about 1.5 seconds after each initiation of a swallow. The rapid and powerful contraction of the pharyngeal musculature induces a rapid movement of the pharyngeal contents through the opening – essentially a jet.

The slower and more variable transit through the oesophageal body depends upon the physical properties of the bolus and the position of the thorax when the swallow occurs. People swallow quite fluid boli most of the time when normally seated with the trunk upright. Under these conditions, the bolus passes in two or three seconds to the level of the lower oesophageal sphincter, moving mainly under the influence of gravity. When either or both the consistency of the bolus and the position of the thorax reduce the effect of gravity, the intrinsic peristaltic mechanism in the oesophagus must carry the bolus. In this case, it moves at about 2–4 cm per second, requiring 5–7 seconds to traverse the distance to the lower oesophageal sphincter. The magnitude of the intraluminal pressure induced by this peristaltic contraction, as measured by oesophageal manometry, varies following the same general pattern along the conduit as the velocity does. They both rise slightly from a nadir at the junction of the two kinds of muscle and then both decline slightly close to the lower oesophageal sphincter. The force of the contraction front may be required to carry more solid boluses throughout most or all of the organ. Even very solid things, such as large tablets, or artefacts such as coins, are transported to the stomach readily by the oesophageal contraction. The distensibility of the organ throughout its entire length allows it to accommodate and pass objects of large diameter. The outer longitudinal muscle layer and the parallel mucosal muscle layer of the oesophagus participate in this event, tending to draw the gastric end of the organ, which is relatively mobile, toward the pharyngeal end where the apparatus of bones and cartilages associated with the parapharyngeal muscles anchor it to the skull. The mechanical function of such axial oesophageal contraction remains obscure. The rigidity of the wall created by the contraction may facilitate the flow of the swallowed bolus. The lubrication that the saliva provides critically facilitates the oesophageal transport of solids. Saliva, swallowed throughout the day and night, bathes the oesophageal mucosa with a considerable volume of a weak buffer solution. This process opposes the erosion of the oesophageal mucosa by refluxed gastric acid and pepsin and removes bile salts and any corrosive drugs that may be present in the contents of the stomach.

Relaxation of the lower oesophageal sphincter takes place at about the same time that the upper oesophageal sphincter contracts, so that the oesophageal lumen is open to that of the stomach for the full duration of the propulsive contraction in the oesophageal body. This sphincter remains relaxed for several seconds, tone returning only when the front of the contraction travelling along the oesophageal body reaches it. The front simply moves forward into the sphincter and makes it contract, the force rising well above that characteristic of the resting state for a few seconds before it declines to that level.

Filling and Emptying the Stomach

The need of the animal for the digestive tract to prepare food before its delivery to the intestine and for it to be delivered in a controlled manner led to the evolution of the stomach. Boli usually enter the stomach at least several seconds apart because

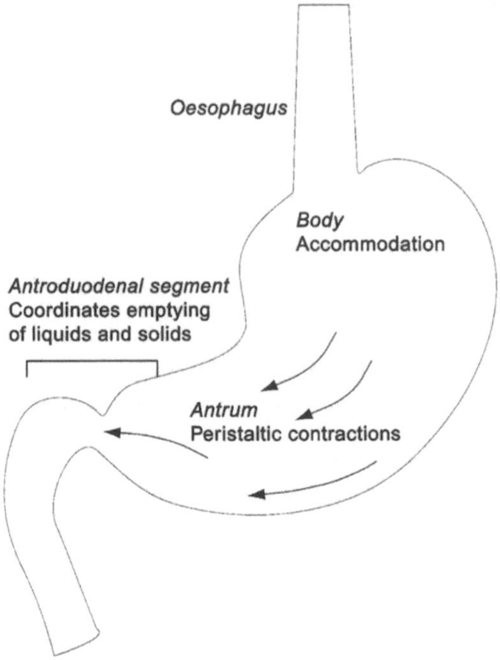

Figure 2.7. Geometry of the stomach. See the text for further explanation.

effective swallowing cannot be induced more rapidly. After a bolus first enters the stomach, the solid part remains in the fundus, layered at the periphery over the mucosa, while the more fluid parts flow more toward the antrum (Figure 2.7). With subsequent boli the solid components tend to move into the proximal stomach as well, each coming to lie in a layer on the luminal side of the solids from the last bolus. Liquids pass around these solids to enter the distal stomach directly.

Most species eat rapidly and at long intervals so that a large volume of food enters the stomach over a short time. The stomach relaxes to accommodate this increase in volume with little or no increase in pressure. This process is called receptive relaxation or accommodation. The term receptive relaxation refers to the fact that the muscle of the fundus of the stomach relaxes as the stomach fills, expanding luminal volume so that intragastric pressure remains constant over a very large volume. This is somewhat similar to what happens in the urinary bladder. This process seems to characterise especially the oblique muscle layer because the increase in volume produces an expansion especially in the axis of the fundus, the direction of the fibres of that muscle layer. Of course, some relaxation must occur in the other layers as well. As a result, the human stomach increases in size from its "empty" volume (the volume aspirated after a prolonged fast) of about 50 ml to a "filled" volume of 1500 ml or more. It can expand at a rate of at least 100 ml per min.

The storage of the ingested solids in the proximal stomach lasts for a long time relative to the time required for their entry. Almost immediately after entry begins, the secretion of gastric juice and mucus starts. This occurs especially from the mucosa of the proximal stomach, so that the layering of one bolus over another must take place at the same time that gastric juice and mucus are beginning to spread over and around the solid mass to flow into the distal stomach. The volume of gastric fluids secreted after a meal often exceeds that of the ingested material, though their production takes a relatively long time. As a result, the quantity of fluid presented to the duodenum following a meal can greatly surpass the volume ingested.

The emptying of the stomach must be closely regulated both to prevent the duodenum from receiving more fat than the slow process of intestinal lipolysis can handle and to protect the duodenum from the damaging effect of excessive amounts of gastric acid. Contraction of the muscle of the proximal stomach provides most of the force required to empty the stomach. Having relaxed in the process of gastric filling, this muscle then begins to regain its tone. Its slow contraction presses the material stored in the saccular proximal stomach into the conical distal stomach where antral peristalsis kneads and stirs it. The antral contractions begin as shallow circumferential indentations at the junction between the proximal and distal stomach. They migrate through the conical antrum to the pylorus, deepening and accelerating as they progress. They occur at intervals of 20 seconds or integral multiples of 20 seconds in humans. Their force can produce an intraluminal pressure of more than 100 mmHg. They appear to deepen as they approach the pylorus, in part because of the underlying conical configuration of the antrum. Antral peristalsis reduces the solid material, suspended in large particles in the stored gastric content, to smaller and smaller particles. The larger particles move at first toward the gastroduodenal junction, carried in that direction by the peristaltic antral contractions, only to be pushed back into more proximal regions because of the geometry of the antrum. The progressive narrowing of the conical antral lumen leads finally to the development of a lumen ahead of the solid mass that is too small to receive it, and so the bolus must move back as the antrum narrows further. This back-and-forth transit of large solid masses continues until grinding has reduced the masses to particles with a diameter of about 1–2 mm, a dimension that allows them to escape through the pylorus into the duodenum, suspended in fluid. A special term, gastric sieving, refers to this process of effective filtration. Thus, the stomach empties a fluid that constitutes a suspension of solid particles 1 to 2 mm in diameter through the pylorus. The fluid escapes from the stomach in boluses because of the rhythm of antral peristalsis.

After an initial lag time, bulk emptying follows a single exponential curve with a half time of about 45 minutes for a standard meal. The chemical composition of the gastric contents regulates the bulk emptying rate, the flow being slowed especially by lipids. Excessive acidity and osmotic pressure in the gastric effluent can also slow emptying. The responsible chemoreceptors lie in the duodenal mucosa and excite an inhibitory reflex that traverses vagal pathways. The reflex probably retards both the restoration of tone in the proximal stomach and antral peristalsis.

No complete luminal occlusion exists at the pylorus when the stomach is quiet. The muscle of the pylorus does, however, show the high degree of myogenic tone characteristic of other sphincters. Thus, it is probably tonically contracted in vivo to some degree but not enough to fully occlude the lumen. The pylorus does, however, close forcefully when a peristaltic contraction reaches it. Because of the acceleration of peristalsis, each wave ends with the terminal antral contraction. This event includes the pylorus. Thus, the pylorus closes with each terminal antral contraction, remaining closed until after the contraction has ended. Then it opens again after a second or so.

The Small Intestine

The small intestine receives about 6 to 12 litres of fluid daily and it delivers 1.0 to 1.5 litres daily to the colon. Flow in the intestine seems to exhibit patterns of great complexity between the extremities of the organ. The limitations of available methods have prevented their careful examination. The small intestine exhibits distinct fasting and postprandial patterns. The pattern of intestinal flow is different during the migrating motor complex seen during fasting than it is in the postprandial state. Since the intestine is nearly empty in fasting, however, the quantity of flow of the intestinal content during the migrating motor complex must be small.

Net transit through the intestine is antegrade but transient retrograde movement occurs as well; a single particle can pass backward 50 cm or more. The rate of net antegrade transit is variable. Clearly, a great deal of bulk mixing takes place in the small intestine; likewise, an intense microcirculation of intestinal contents across the mucosal surface occurs, for the stirring of the unstirred layer is essential for the optimal absorption of nutrients.

Digestion and absorption in the intestine require not only the constant mixing and antegrade propulsion of the intestinal content but also its local microcirculation across the absorbing surface of the epithelium. These three kinds of flows result from the combined contractions of the three muscular layers, the longitudinal and circular layers of the tunica muscularis and the muscle of the mucosa.

Although the structure of the small intestine looks the same all along its length, its motility varies quantitatively from one region to another. Transit slows greatly from the duodenum to the ileum. The volume of the fluid load also falls greatly. Both the viscosity and the chemical composition of the contents change. These changes correlate with quantitative differences in motility.

The two layers of the tunica muscularis seem to act sometimes together and sometimes independently. Contractions of both layers move the contents of the intestine. Theoretical considerations suggest that contractions of the circular layer principally cause antegrade transit while those of the longitudinal muscle layer induce mixing movements. The usual assumption of symmetry, both radial and axial, in the peristaltic contractions of the intestine does not accord with the facts. Both radial and axial asymmetries characterise the contractions of these muscle layers in the intestine. They undoubtedly affect flow patterns. These asymmetries result both from the nature of the coordination of the two muscle layers and from the patterns of spread of excitation within the muscle layers.

Some contractions of the circular muscle layer spread smoothly along the intestine in the pattern called peristalsis. Others involve the simultaneous contraction of short segments of the intestine that are separated by uncontracted segments of similar length (segmentation pattern). A third pattern of contractions, called pendular movements, manifests itself as swaying motions seen in isolated loops of intestine, probably reflecting rhythmic contractions confined to the longitudinal layer of muscle.

Contractions of the longitudinal muscle layer shorten the intestine. Theoretical considerations suggest that the rhythmic alternated shortening and lengthening of the intestine promote the mixing of the content. Peristaltic contractions of the circular muscle layer, in contrast, mainly produce forward transit with relatively little mixing. Since the two kinds of contractions occur together most of the time, the highly complex flows they induce must be of both kinds.

The function of the muscle of the mucosa may not be entirely independent of the actions of the other muscle layers, but the consequences of its action are distinctly different. The function of the contractions of the muscle of the mucosa cannot easily be investigated directly. This layer shifts the pliable mucosa back and forth over the somewhat stiffer tunica muscularis. Also, the fingers of mucosal muscle that extend into the mucosal villi make them move rhythmically. Both functions must mix the layer of fluid at the absorbing epithelial surface. Since diffusion through this layer is the rate-limiting step in the intestinal absorption of many nutrients, the mucosal muscle, by stirring this otherwise unstirred layer of fluid, may have a critical place in nutrition.

The Large Intestine and Anorectum

The movement of intestinal contents through the ileo-caecal valve rarely stops for very long except in prolonged fasting, but it fluctuates widely, rising in phase with meals. The actual daily volume is 1.0 to 1.5 litres. The minimal daily output from an established ileostomy is most often more than 800 ml. This may be considered the obligatory volume of input to the large intestine. In diarrhoea due to defective absorption by the small intestine, the volume of flow into the colon may exceed the normal by a factor of 10 to 15.

Just as we lack methods to describe transit in any detail in the small intestine, so are we unable to describe that in the large intestine except in terms of bulk flows. Fluid remains in the caecum and ascending colon for a prolonged period while its volume is reduced by the absorption of water. Extensive mixing of the content occurs during its prolonged stay in this region of the colon. However, flow is faster on the average beyond about the level of the hepatic flexure. From here to the rectum, movement is highly intermittent, the faecal mass remaining stationary for tens of minutes to hours and then rapidly moving antegrade for a variable distance. Such an episode of rapid transit is called the mass movement. The faecal mass thus comes to the rectum in boli. The rectum and a variable length of the left colon above it can store the faecal mass in a variable volume for a variable time.

The fundamental patterns of contractions vary in the different parts of the large intestine. These different patterns define three parts of the organ in series, the right colon (ascending colon and caecum), the left colon (transverse, descending and sigmoid colons), and the anorectum (rectum and anal canal).

1. **Right Colon.** The characteristic intermittent antiperistalsis in this part of the large intestine consists of indenting rings, formed by rhythmic contractions of the circular muscle layer, that migrate toward the caecal tip. During antiperistalsis, the contractions are intermittent, occurring repeatedly in periods lasting tens of minutes with intervening rest periods of similar length. They begin at about the hepatic flexure and pass without interruption to the caecal tip, driving the colonic contents towards the caecum, from which it cannot escape, with consequent stirring of that material. Of course, orthoperistaltic contractions also occur, emptying the region. They are very similar to antiperistalsis except in the direction of their propagation. The controls of these rhythmic phasic contractions remain partly unknown. The only control mechanism described is a system of electrical slow waves very much like those that occur in the distal stomach and small intestine. In antiperistalsis, these begin at a pacemaking site at about the level of the hepatic flexure and migrate to the caecum, driving the contractions of

the circular muscle layer. Their direction of migration changes at intervals when the pacemaker leaps to the caecal tip to produce orthoperistalsis.

2. **Left Colon.** In the colon beyond the hepatic flexure, the same investigators who first described antiperistalsis in the right colon in experimental animals observed only orthoperistalsis, rhythmic ring contractions that move always toward the anus. These contractions occur in temporal sequences like those of antiperistalsis and orthoperistalsis in the right colon, and with a similar frequency, and they similarly push colonic content before them. Electrical slow waves like those of the right colon drive these contractions, differing from those of the right colon only in the constancy of the direction of their migration. Such orthoperistalsis has not been described from direct observation in man, but its existence is implied by manometric studies.
3. **Anorectum.** The third part of the colon exhibits no spontaneous rhythmic contractions, but it contracts powerfully on stimulation of the pelvic nerves. The rectum behaves like this; it receives and stores faecal material, contracting only in the complex operation of defaecation. This part of the colon generates no electrical slow wave activity, consistent with the absence of rhythmic contractions. The anal canal, a ring of the circular muscle layer at the end of the rectum, maintains a state of constant contraction, shortening sufficiently to create a barrier to the transit of the solid material stored in the rectum.

The Biliary System

Bile passes continuously from the liver, into the biliary system at about one litre per day. During fasting, the rate of bile production fluctuates with the migrating motor complex, as does the volume of pancreatic fluid flow. About one-fourth of the bile secreted in fasting passes directly into the duodenum through the sphincter of Oddi. The rest enters the gall bladder. This division results from the very small pressure gradient produced by the slight resistance to flow through the sphincter of Oddi. The pressure gradient across the sphincter rarely exceeds 4–5 mmHg. The gall bladder thus fills slowly under low pressure, probably expanding in a process resembling that of receptive relaxation in the stomach, so that its intraluminal pressure rises negligibly with volume. With a meal, the flow pattern is altered since bile production from the liver doubles in volume, resistance to transit at the sphincter of Oddi falls, and the gall bladder contracts. The muscle of the filled gall bladder contracts very slowly to reduce its volume (usually about 35 ml) by 50 to 75% in 30 to 60 minutes after the meal. The organ never empties completely, however, about 10% of the volume remaining in the organ. The pressure produced in the common bile duct by the contraction of the gall bladder never exceeds the maximal secretion pressure of the liver, about 8 mmHg. After emptying, the flow pattern slowly returns to that which characterises fasting. The rate of transit of bile through the common duct in fasting is 0.5 to 1.0 ml per minute. It can double or triple under the influence of meal-enhanced bile secretion and postprandial gall bladder contraction.

The contraction that empties the filled gall bladder after a meal can take 30 minutes or more for completion, and generates a pressure of less than 20 mmHg. During this period, the organ shows little consistent electromyographic activity, but intermittent rhythmic bursts of low amplitude spikes sometimes occur. They probably represent transient rises in tone. An underlying basal level of tone prevails, however, which seems to be chiefly responsible for emptying.

After the gall bladder has emptied during fasting and after the gastrointestinal migrating motor complex has developed, the gall bladder contracts and relaxes periodically in rhythm with the complex. During the last part of phase II of the migrating motor complex, these rhythmic contractions achieve a magnitude that is as much as 40% of the magnitude of the contractions that empty the gall bladder postprandially. The rhythmic contractions of the gall bladder cease just before phase III of the complex enters the duodenum. Thus, they enhance the delivery of bile to the intestine just before maximal contractile activity takes place in the duodenum. They also partially empty the gall bladder of viscous concentrated bile, microcalculi, bacteria, and debris accumulated during fasting, allowing it to continue to fill with dilute hepatic bile in the intervals between migrating motor complexes.

The filling of the gall bladder, which occurs in the 90-minute intervals between one episode of the migrating myoelectric complex and the next, reflects the net effect of several variable processes. These include the resistance to the flow of bile through the sphincter of Oddi, the resistance to its flow through the cystic duct, and a probable receptive relaxation of the gall bladder produced by a local reflex in which stretch of the organ induces its relaxation. Relaxation of the gall bladder also represents the fall in the level of cholecystokinin circulating in the blood. The normal net effect of these factors is the diversion of 75% of hepatic bile to the gall bladder in fasting, with 25% passing directly into the duodenum.

The structure of the cystic duct, with its spiral valve of Heister, suggests its importance in directing the movement of bile. The cystic duct wall contains smooth muscle throughout, and bursts of spikes moving peristaltically towards the common duct have been recorded electromyographically. Thus, the duct seems to have a limited function as a peristaltic pump. Small changes in the diameter of the duct produce large changes in resistance because of its long and narrow lumen.

In man, the common duct acts as a passive conduit. The longitudinal muscle in the wall of the duct may not contract. In any case, its contraction cannot influence the flow of bile very much because of its sparsity. Other species, however, possess somewhat more muscle in the common duct and bursts of spikes have been detected in such species.

The resistance to transit produced by the tonic closure of the sphincter of Oddi raises the pressure in both the common and pancreatic ducts because of the constancy of hepatic and pancreatic secretion. The mean pressure in fasting in the common duct can exceed that in the duodenum by up to 12 mmHg, and the mean pressure in the pancreatic duct can be up to about 15 mmHg higher than intraduodenal pressure. Manometry of the sphincter, about 6 mm long, reveals an intraluminal pressure 4 mmHg above that in the common duct. As in other sphincters, a closure of the lumen at this point sufficient to produce an adequate resistance to transit does not require an intraluminal pressure much above that in the common bile duct. The two discrete muscular components of the sphincter identified anatomically cannot be distinguished manometrically. The sphincter seems to act as a single functional unit. Periodic transient contractions occur in the sphincter at four per minute, especially prominent in the last part of phase II and in phase III of the migrating motor complex. They average about 130 mmHg in amplitude and last about four to five seconds. These contractions are peristaltic, propagating toward the duodenum for 60% of the time and toward the common bile duct for 14% of the time. They occur simultaneously throughout the sphincter 26% of the time. These phasic contractions can both enhance and slow transit through the sphincter, for they make the sphincter act as a peristaltic pump capable of directing transit in either direction as their movement dictates.

The reciprocal relationship between gall bladder contraction and sphincter contraction is currently attributed to the differential effect of cholecystokinin on the two organs. Its ability to inhibit the muscle of the sphincter is the result of its excitation of inhibitory nerves. Other autonomic nerves may also affect the sphincter, but evidence on this point remains incomplete.

2.8 Verification of Simplifying Geometric Assumptions and Error Considerations

Biomechanical analysis is often based on simplifying geometric assumptions, where the organ under study is assumed to conform to certain shapes such as spheres or tubes. The assumptions simplify the analysis but verification should be done and possible errors computed. The next chapter deals with biomechanical analysis but it is convenient already in this chapter on geometry and structure to present to the readers the assumptions regarding geometry, whether these assumptions are valid and how big the errors might be. Other assumptions about homogeneity, isotropy, thin- versus thick-wall, compressibility, etc. will be treated elsewhere in this book.

The oesophagus and intestines are often assumed to be straight cylindrical tubes, meaning that the cross-sections of these organs both at unloaded and loaded states are circular (both the inner and outer surface) and straight. The oesophagus appears to be cylindrical at a glance but what about the inner surface? It is well known that buckles (mucosal folds exists) at low pressures. Then at what pressure can we regard it as circular? Such questions have implications for the stress–strain analysis and for a flow analysis. Though medical imaging techniques such as MR-scanning and ultrasound have improved tremendously during the past decades, it can still be quite difficult to obtain the geometric validation data in vivo and the correct reference state (the zero-stress state; Chapter 6) cannot be obtained unless the tissue is excised. In vitro experiments, on the other hand, can give much more accurate data under homeostatic conditions and close to the zero-stress state.

Error considerations and an error propagation analysis should be done for each scientific problem, unless previous studies have verified the assumptions used. Below is an example of how to compute the errors in circumference when an elliptical cross-section is assumed to be circular (modified from Gregersen and Kassab, 1996). Clearly, an error in circumferential dimension will propagate and give errors in the strain and stress. If Kirchhoff stress and Green strain (see next chapter for definitions) are computed, the error in these parameters will depend both on the possible errors in the reference state and on the loaded state circumferences. Measurement uncertainty also needs to be taken into account in the error analysis. Furthermore, an example with the introduction of a correction factor based on experimental data on casts is provided (Gregersen et al., 1997). Finally, some directions are given for slightly more complex geometries. One can imagine however, that an organ like the stomach is difficult to deal with and that its geometry is far from the spherical shape some investigators claim.

2.8.1 Determination of the Error when an Elliptical Cross-section is Assumed to be Circular

Consider a segment of a tubular part of the gastrointestinal tract with a fixed length and two special cross-sectional geometries of the organ: a circle with radius r and an ellipse with major and minor radii a and b, respectively. According to Gregersen and

Kassab (1996) the equations below relate the circumference and luminal cross-sectional area (or volume if the length is constant) to the radii for the two geometries.

Elliptic shape

Cross-sectional area = $\pi \cdot a \cdot b$ (2.1a)

$$\text{Circumference} = \int_0^{2\pi} \left[a^2 \sin^2\theta + b^2 \cos^2\theta\right]^{1/2} d\theta \tag{2.1b}$$

Circular shape (a = b = r)

Cross-sectional area = $\pi \cdot r^2$ (2.2a)

Circumference = $2 \cdot \pi \cdot r$ (2.2b)

The error in the area (E_A) when an elliptical cross-section is assumed to be circular is

$$E_A = \frac{\pi r^2 - \pi ab}{\pi r^2} = 1 - \left(\frac{ab}{r^2}\right) \tag{2.3}$$

Assuming a constant circumference, then

$$r = \frac{1}{2\pi} \int_0^{2\pi} \left[a^2 \sin^2\theta + b^2 \cos^2\theta\right]^{1/2} d\theta \tag{2.4}$$

Hence, substitution of r from Equation 2.4 into Equation 2.3 yields

$$E_A = 1 - \frac{4\pi^2 ab}{\left[\int_0^{2\pi} \left[a^2 \sin^2\theta + b^2 \cos^2\theta\right]^{1/2} d\theta\right]^2} = \tag{2.5}$$

or

$$E_A = 1 - \frac{4\pi^2}{I^2}, \tag{2.6}$$

$$\text{where } I = \int_0^{2\pi} \left[\left(\frac{a}{b}\right) \sin^2\theta + \left(\frac{a}{b}\right)^{-1} \cos^2\theta\right]^{1/2} d\theta$$

The dashed curve in Figure 2.8 illustrates the error in cross-sectional area as a function of the *a/b* ratio when the geometry deviates from a circle. It is clearly shown that the error in the CSA or volume can be considerable at a constant circumference (equivalent to no variation in tone). Thus, the circular assumption may introduce large errors when the cross-section is elliptical. The *a/b* ratio is high in the resting state since the intestines are collapsed and become more circular as the distension pressure is increased.

In a similar fashion, we can compute the error in the circumference (E_C) when an elliptical section is assumed to be circular as follows:

$$E_C = \frac{2\pi r - \int_0^{2\pi} \left[a^2 \sin^2\theta + b^2 \cos^2\theta\right]^{1/2} d\theta}{2\pi r} = 1 - \frac{1}{2\pi} \frac{\int_0^{2\pi} \left[a^2 \sin^2\theta + b^2 \cos^2\theta\right]^{1/2} d\theta}{r} \tag{2.7}$$

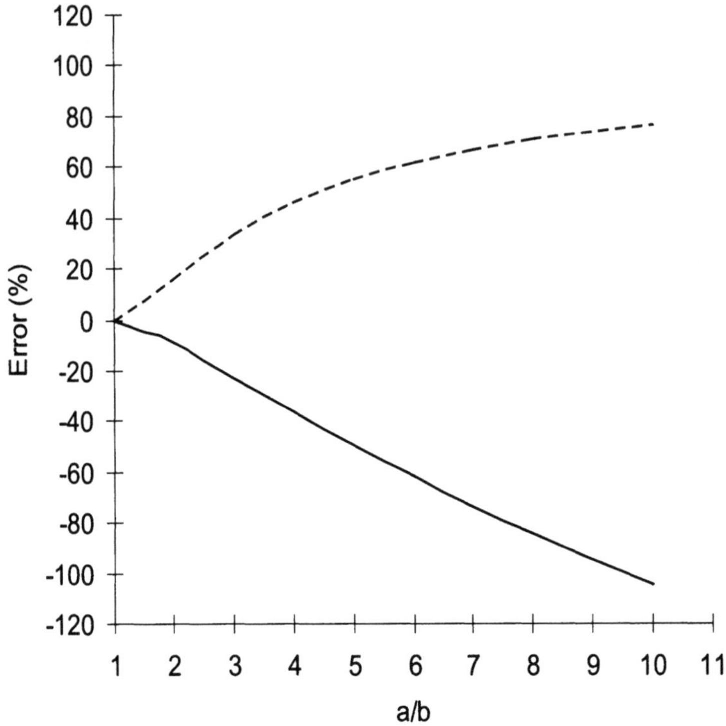

Figure 2.8. The graph shows the resulting error when an elliptical cross-section is assumed to be circular as a function of the major-to-minor axis ratio (a/b). The dashed line shows the error in the CSA for a constant circumference while the solid line shows the error in the circumference at constant CSA. The curves are based on the equations in this chapter. When the a/b ratio increases, the cross-section becomes more elliptical. This plot shows that without any change in circumference (tone), changes in geometry can cause the CSA or volume to vary considerably. The a/b ratio is high in the intestines at low pressures because they are collapsed in the resting state. Thus, unless geometry is taken into account, the estimation of circumference (tone) is at best qualitative (With permission, Gregersen and Kassab. Biomechanics of the GI tract; *Neurogastroenterology and Motility,* 1996; 8: 277–297, Blackwell Science).

Assuming a constant cross-sectional area:

$$\pi r^2 = \pi ab, \text{ or } r = \sqrt{ab}, \tag{2.8}$$

thus

$$E_C = 1 - \frac{I}{2\pi} \tag{2.9}$$

The solid line curve in Figure 2.8 illustrates the error in circumference as a function of the *a/b* ratio.

Equations 2.5 and 2.9 can be used to relate the error in circumference to that in area as follows

$$E_C = 1 - \frac{1}{\sqrt{1-E_A}} \tag{2.10}$$

These equations only consider the case when an elliptical cross-section is assumed to be circular. The evaluation of errors (in, for example, volume or surface area) is considerably more difficult for organs with more complex geometry, e.g., the stomach (see also Section 2.8.4).

2.8.2 Correction of Data using Casts of the Guinea Pig Duodenum

Morphometric data can be obtained from photographs of intestinal segments in the zero-stress and no-load states. The loaded state can be measured on photographs of the intestinal segment with a silicone elastomer cast inside (provides data on the outer wall of serosa) and of photographs of the cast itself (provides data on the inner wall of mucosa). The cross-sections may not be exactly round. The photographs allow measurements of the major (D_{ma}) and minor (D_{mi}) diameters of the cross-section of the loaded state (these diameters are related to the major and minor radii in the analysis above). If one could assume the cross-section to be elliptic, then it is possible to compute the circumferences of the mucosa and serosa, C_l-mucosa and C_l-serosa, respectively. According to Gregersen et al. (1997) the exact formula is:

$$C_l = 2D_{ma}E(k), \tag{2.11}$$

in which E(k) is the complete elliptic integral of the second kind, and k is a parameter defined by

$$k^2 = 1 - \left(\frac{D_{mi}}{D_{ma}}\right)^2 \tag{2.12}$$

If the ratio D_{mi}/D_{ma} is very close to 1, then it can be shown that Equation 2.11 may be approximated by

$$C_l = \frac{D_{ma} + D_{mi}}{2}\pi \tag{2.13}$$

For example, if $D_{ma}/D_{mi} = 1.15$, then the C_l calculated by Equation 2.13 is 0.12% of that given by Equation 2.11. The use of Equation 2.13 involves the assumptions that the cross-section is elliptical and that D_{mi}/D_{ma} duodenum is quite close to 1. To test these assumptions, casts obtained from the guinea pig duodenum were sliced. From images of these slices we measured D_{ma}, D_{mi} and C_l. Figure 2.9 shows the circumference of the cast slices measured directly from photographs and the circumference computed from measurements of the major and minor radii from the same photographs using Equation 2.13. The regression line of Figure 2.9 can be expressed in the form

$$C_{l\text{-mucosa}} = \left[\frac{D_{ma} + D_{mi}}{2}\pi \cdot A\right] + B \tag{2.14}$$

where A and B are empirical constants, 1.17 and 0.297 mm, respectively. The correlation coefficient was 0.98. Equation 2.14 is an improvement of Equations 2.11 and 2.13 for a duodenum whose cross-section is not exactly elliptical and whose D_{ma}/D_{mi} is not equal to 1. The serosal circumferences also need correction and can be calculated on the basis of Equation 2.4 and the wall thickness. First, the mean wall thickness (h)

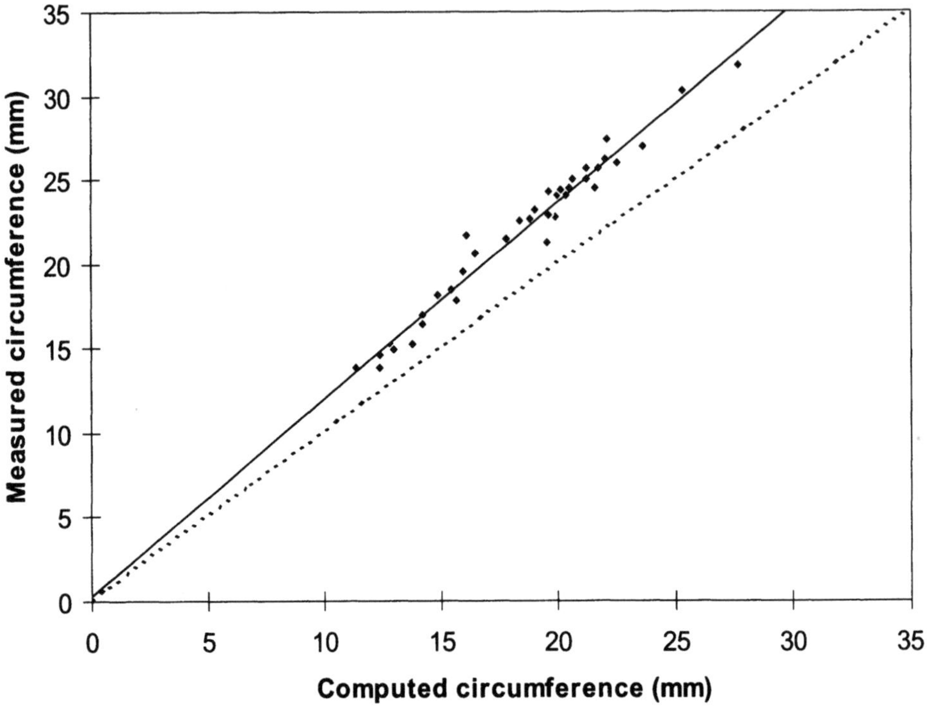

Figure 2.9. Correction factor for mucosal circumference measurements. In these experiments the circumference of cast slices was measured directly from photographs (Y-data) and also computed from measurements of the major and minor radii from the same photographs (X-data). Since the measured and computed circumferences were not identical, a correction factor was introduced in all other experiments based on the linear relationship (Y = 1.17X + 0.297, r = 0.98) found in these experiments by least square fitting (the solid line). The dashed line is the line of identity between X and Y (With permission from Gregersen et al; *Am J Physiol*, 1997: 237; G868).

in the loaded state can be calculated as

$$h = \frac{1}{2}\left(\frac{OD_{ma} - ID_{ma}}{2} + \frac{OD_{mi} - ID_{mi}}{2}\right), \tag{2.15}$$

where OD and ID denote the outer (serosal) and inner (mucosal) diameters, respectively. The serosal circumference based on the correction factor can be computed as:

$$C_{l\text{-serosa}} = B + 2\pi h - 2\pi h A + A\frac{OD_{ma} + OD_{mi}}{2}\pi . \tag{2.16}$$

After appropriate corrections are done, homeostatic strains can be computed according to the equations given in the next chapter and in Gregersen et al. (1997).

2.8.3 Verification of the Circularity Assumption for the Rat Oesophagus and Small Intestine

There are several ways to verify the circularity assumption. Imaging using B-mode ultrasonography has already been mentioned. Most image analysis programs have tools for measurement of circumferential length C and area A. From these measurements the circularity can be computed as

$$Circularity = \frac{C^2}{A}. \tag{2.17}$$

For a circle, this measurement gives a value of 4π or 12.57. Any other geometry will give a higher value. Table 2.1. provides data obtained from rat gastrointestinal tissue fixed in formalin at two different pressures (n = 12 for each segment and pressure). As it can be seen the values obtained are not far from the value for a circle.

Table 2.1. Data obtained from rat gastrointestinal tissue fixed in formalin at two different pressures (n = 12 for each segment and pressure)

	Duodenum 0.2 kPa	Duodenum 1 kPa	Jejunum 0.2 kPa	Jejunum 1 kPa	Ileum 0.2 kPa	Ileum 1 kPa	Oesophagus 0.2 kPa	Oesophagus 2 kPa
Circularity of inner surface	13.3±0.2	14.1±0.3	13.7±1.4	13.4±0.4	13.7±0.6	13.4±0.3	12.6±0.1	12.6±0.0
Circularity of outer surface	13.1±0.2	13.5±0.0	13.4±0.9	13.1±0.3	13.4±0.6	13.1±0.3	12.7±0.1	12.6±0.0

The oesophagus is remarkably circular both at the inner and outer surfaces. The fixation procedures may have influenced the results but the data are in agreement with results obtained in snap-frozen sections. To avoid the effect of fixation, additional experiments were done where outer diameters were measured from images of the specimens from several angles around the circumference. Since the area can not be measured using this method, the outer diameters obtained at different angles around the circumference must be compared statistically. Table 2.2 provides mean and SE data from four parts of the rat gastrointestinal tract. The magnitude of the SE values are mainly due to differences in size between animals. Statistical analysis revealed that at all pressures, the diameter did not depend on the angle the photograph was taken from.

2.8.4 The Surface Area and Volume of a Prolate Spheroid

The prolate spheroid is a slightly more complex geometric structure than the cylindrical tube and the sphere. It may serve as a better geometric model for organs like the rectum and gall bladder than the simple geometries considered above. A prolate

Table 2.2. Outer diameters as function of angles of rotation and applied pressure. The unit for pressures is kPa. Angle 0 is where the mesenterium is attached in the small intestine or the posterior of the oesophagus. The data were obtained by Dr. Jingbo Zhao

Duodenum Angle (degrees)	P=0	P=0.2	P=0.4	P=0.6	P=0.8	P=1
0	4.51±0.14	5.09±0.09	5.31±0.09	5.41±0.14	5.42±0.14	5.54±0.13
30	4.57±0.15	5.02±0.22	5.32±0.15	5.48±0.21	5.58±0.23	5.56±0.23
60	4.56±0.33	5.31±0.02	5.49±0.23	5.52±0.26	5.59±0.19	5.55±0.21
90	4.56±0.28	5.22±0.11	5.44±0.22	5.52±0.15	5.55±0.11	5.52±0.16
Jejunum Angle (degrees)	P=0	P=0.2	P=0.4	P=0.6	P=0.8	P=1
0	4.30±0.24	4.57±0.15	4.71±0.21	4.79±0.19	4.86±0.01	4.85±0.15
30	4.29±0.29	4.60±0.19	4.73±0.12	4.84±0.12	4.89±0.06	4.91±0.04
60	4.31±0.30	4.64±0.21	4.81±0.08	4.91±0.08	4.89±0.05	4.97±0.02
90	4.33±0.24	4.62±0.23	4.87±0.24	4.92±0.17	4.90±0.06	4.96±0.11
Ileum Angle (degrees)	P=0	P=0.2	P=0.4	P=0.6	P=0.8	P=1
0	4.81±0.39	5.24±0.54	5.47±0.60	5.55±0.62	5.67±0.62	5.68±0.66
30	4.89±0.49	5.35±0.59	5.54±0.72	5.76±0.74	5.79±0.71	5.89±0.69
60	4.85±0.51	5.35±0.55	5.56±0.64	5.59±0.66	5.71±0.55	5.81±0.67
90	4.76±0.38	5.30±0.45	5.50±0.68	5.60±0.61	5.69±0.62	5.74±0.65
Oesophagus Angle (degrees)	P=0	P=0.5	P=1	P=1.5	P=2	P=2.5
0	2.92±0.27	3.48±0.06	3.75±0.08	3.95±0.08	4.08±0.06	4.11±0.05
30	2.97±0.15	3.49±0.06	3.76±0.08	3.94±0.08	4.07±0.06	4.11±0.05
60	3.10±0.15	3.47±0.04	3.74±0.10	3.94±0.10	4.07±0.08	4.12±0.06
90	3.04±0.11	3.48±0.05	3.76±0.11	3.95±0.10	4.07±0.09	4.12±0.06

spheroid is a spheroid that is "pointy"; that is, the polar radius c is greater than the equatorial radius a. An egg is close to this shape if it has the same shape at both ends. It is obtained by rotating an ellipse about its major axis (in the direction of c). The surface area is:

$$S = 2\pi a^2 + 2\pi \frac{ac}{e}\sin^{-1} e, \tag{2.18}$$

where e is the ellipticity of the prolate spheroid. The volume is:

$$V = \frac{4}{3}\pi a^2 c. \tag{2.19}$$

The interested reader can find the derivation in geometry books or at www://mathworld.wolfram.com/ProlateSpheroid.html. Knowing the surface area and volume of a sphere, the reader can now do an error analysis similar to the one provided in Section 2.8.2. Regen (1996) provides formulations for tension calculations in spheroid structures. The stomach is more complex than spheroidal geometry and may require numerical analysis due to the geometric complexity.

2.9 Literature

The literature list contains papers mentioned in the text as well as other papers of interest.

Christensen, J. 1994. The motility of the colon. In: *Physiology of the Gastrointestinal Tract*, third edition, ed. Johnson, LR, Alpers, DH, Christensen, J, Jacobson, ED, Walsh JH, pp. 991–1024. New York: Raven Press.

Conklin, JL, Christensen, J. 1994. Motor functions of the pharynx and esophagus. In: *Physiology of the Gastrointestinal Tract*, third edition, ed. Johnson, LR, Alpers, DH, Christensen, J, Jacobson, ED, Walsh, JH., pp. 903–28. New York: Raven Press.

Fackler, K, Klein, L, Hiltner, A. 1981. Polarizing light microscopy of intestine and its relationship to mechanical behaviour. *J Microscopy*, **124**, 305–11.

Furness, JB, Costa, M. 1987. *The Enteric Nervous System*. Edinburgh: Churchill Livingstone.

Gabella, G. 1994. Structure of muscles and nerves in the gastrointestinal tract. In: *Physiology of the Gastrointestinal Tract*, third edition, ed. Johnson, LR, Alpers, DH, Christensen, J, Jacobson, ED, Walsh, JH., pp. 381–422. New York: Raven Press.

Gregersen, H, Kassab, GS, Pallencaoe, E, Lee, C, Chien, S, Skalak, R, Fung, YC. 1997. Morphometry and strain distribution in guinea pig duodenum with reference to the zero-stress state. *Am J Physiol*, **273**, G865–74.

Gregersen, H, Kassab, GS. 1996. Biomechanics of the gastrointestinal tract. *Neurogastroenterol and Motil*, **8**, 277–97.

Hogan, WJ, Dodds, WJ, Geenen, JE. 1988. The biliary tract. In: *A Guide to Gastrointestinal Motility*, ed. Christensen, J, Wingate, DL., pp. 157–97. Bristol: John Wright & Sons Ltd.

Kosterlitz, HW, Robinson, JA. 1959. Reflex contractions of the longitudinal muscle coat of the isolated guinea-pig ileum. *J Physiol*, **146**, 369–79

Orberg, JW, Baer, E, Hiltner, A. 1983. Organization of collagen fibers in the intestine. *Connect Tissue Res*, **11**, 285–97.

Orberg, JW, Klein, L, Hiltner, A. 1982. Scanning electron microscopy of collagen fibers in the intestine. *Connect Tissue Res*, **9**, 187–93.

Regen, DM. 1996. Tensions and stresses of ellipsoidal chambers. *Ann Biomed Eng*, **24**, 400–17.

Wood, JD. 1994. Physiology of the enteric nervous system. In: *Physiology of the Gastrointestinal Tract*, third edition, ed. Johnson, LR, Alpers, DH, Christensen, J, Jacobson, ED, Walsh, JH, pp. 423–82. New York: Raven Press.

3. Basic Mechanical Theory

3.1 Introduction

Mechanics of a material deals with the behaviour of a solid body subjected to various types of loading. Mechanics is the study of motion of matter and the forces that cause such motion. It is applied to the analysis of any dynamical system ranging from atoms to solar systems. The analysis of stress, deformation and stability of thin-walled tubes is a classical subject of physics and engineering. Theories have been developed by Bernoulli, Cauchy, Euler, Flügge, Kirchhoff, Reissner, Timeshenko and many other famous scientists. All theories of thin-shell structures have the common objective to represent the three-dimensional structure by a two-dimensional surface. In the classical theory, this is accomplished through the Bernoulli–Kirchhoff hypothesis, which states that all points lying on a normal of the neutral surface before deformation do the same after deformation, that for all kinematical relations, the coordinate z of a point (distance from the neutral surface, positive outwards) is unaffected by the deformation of the shell, and for all considerations of the stress system, the stress σ_z may be ignored (Fung and Liu, 1995). The classical theories use four additional hypotheses: 1) that the material is homogenous, 2) that the stress–strain relationship is linear, 3) that the deformation is so small that the strain–displacement relationship is also linear, and 4) that the shell is stress-free when all the external loads are removed. In mechanics of structures like biological tissues, all four additional hypotheses needs relaxation but the Bernoulli–Kirchhoff hypothesis can be retained. Hypothesis number four has been thought to be so natural that it was not questioned for a long time. Fung, and Vaishnav and Vossoughi independently found in 1983 that this "facit" hypothesis was wrong for blood vessels. Gregersen and Kassab confirmed this in 1996 for the gastrointestinal tract. The general consensus today is that most hollow organs have residual stresses and strains (see Chapter 6).

Biomechanics aims to understand the mechanics of living tissues. The principal objective of biomechanics is to determine the stresses and strains in biological structures when forces are acting on them. If we can determine these quantities from close to the unloaded state up to loads causing failure, we will obtain a complete picture of the biomechanical behaviour. Forces applied to solids cause deformation while forces applied to fluids cause flow. Biological tissues such as the gastrointestinal tract express mechanical properties in between the solid and fluid properties; i.e., the deformation is finite, the stress–strain relation is non-linear, there is a pronounced viscoelastic component and anisotropy prevails due to the heterogeneous laminated structure. Due to the high water content in biological tissues, they express mechanical

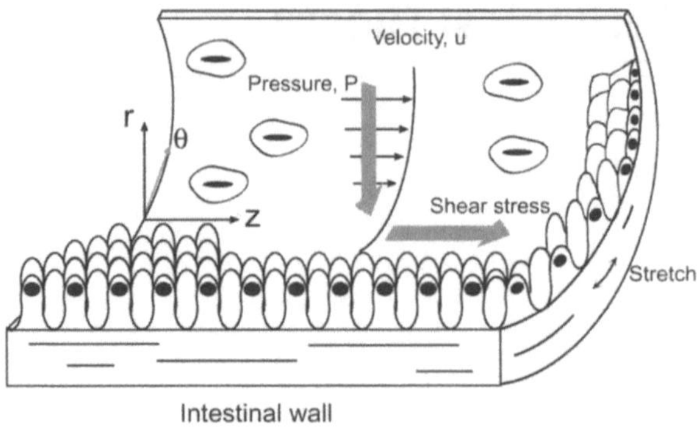

Figure 3.1. A segment of the intestinal wall. The longitudinal, radial and circumferential directions are symbolised by z, r, and q. The pressure P from a bolus or a distending balloon induces a normal stress that will stretch the tube in circumferential directions and radial compression (a thinner wall). Longitudinal extension may also occur. A moving bolus will also cause shear stresses in the mucosa.

properties of both an elastic solid and viscous fluid. The mechanical properties are time-dependent in that the stress–strain response does not occur instantly. When the material is suddenly strained and the strain is maintained constant, the corresponding stresses induced in the wall decrease with time. This phenomenon is called stress relaxation. If the material is suddenly stressed and the stress is maintained constant, the material will continue to deform. This phenomenon is called creep. If the material is subjected to a cyclic loading, the stress–strain relationship in the loading process is different from that in the unloading process and the phenomenon is called hysteresis (Dobrin, 1978; Fung, 1993; Fung and Yih, 1968). Stress relaxation, creep and hysteresis are features of viscoelasticity. Viscoelastic properties are described in this chapter and experimental data are given in Chapter 6. Furthermore, other history-dependent properties such as the preconditioning behaviour and strain softening are treated in greater detail in Chapter 7.

When dealing with the mechanical properties of a biological structure like the gastrointestinal tract, we must consider the structure and geometry. These aspects were dealt with in Chapter 2, where data on the normal motor function, primarily in humans, were also reviewed. This chapter serves to provide the basic theory for treating the gastrointestinal tissue from a mechanical point of view. This theory will in the first place help us to understand the passive properties; i.e., the gastrointestinal tract as a passive structure. It is important that we understand the passive properties before we use the same equations and other equations to describe the active muscle contraction and flow (see Chapter 5). A number of the terms used in this chapter and the next chapters are defined in Table 3.1.

In the gastrointestinal tract, we must consider the external forces from the environment the gastrointestinal tract is a part of and the forces it generates by itself. When external forces (from outside or from the lumen) are applied to a gastrointestinal segment, it deforms, resisting the forces. It is common practice to use distensibility and stiffness to describe the deformation and the resistance to deformation, respectively. However, definition of these parameters for the gastrointestinal tract can be difficult since no single parameter can describe the complex mechanical behaviour of the gastrointestinal system. To arrive at useful approximations, it is important to understand the relationship between stress and strain; i.e., the constitutive equation.

Table 3.1. Reference guide to mechanical terms

Anisotropy	Materials whose mechanical properties depend on directions are said to be anisotropic. Biological tissues are usually anisotropic, mainly due to their heterogeneous, layered structure.
Buckling	Some materials are strong in tension but offer little resistance to deformation when uniaxially compressed. The large deformation in compression is called buckling in engineering literature. Buckling is a common phenomenon in organs. Tissue is in tension when the stretch ratio $\lambda > 1$ and in compression when $\lambda < 1$. Tissue buckling occurs when $\lambda < 1$.
Compliance	A measure of a hollow organ to change its volume, i.e. the ratio between volume change and pressure change. The inverse of compliance is elastance.
Composite materials	A structural composite is a material system consisting of two of more phases on a macroscopic scale, whose mechanical performance and properties are designed to be superior to those of the constituent materials acting independently. Many types of composites exist, e.g. laminated structures or fibres embedded in a matrix. The properties of a composite material depend on the properties of the constituents, geometry, and distribution of the phases.
Constitutive equation	A constitutive equation describes a physical property of a material such as the mechanical properties of a material (the stress-strain relation). Since a wide range of materials exist, it is not surprising that many constitutive equations exist. A constitutive equation is a tensor equation where every term in the equation must be a tensor of the same rank.
Elastic deformation	Deformation that returns to its initial state when the stress is removed. If it does not return to the initial state, the deformation is plastic.
Elastic modulus	The proportionality constant between stress and strain. Hooke's law applies for a homogenous, isotropic, linearly elastic material implying that a single elastic modulus describes the stiffness. Since the mechanical behaviour in biological tissues is non-linear, an incremental approach must be used.
Incompressibility	If a material is incompressible it means that the volume is conserved under various conditions.
Isotropy	A material is isotropic when its properties are the same in all directions or are independent of the orientation of the reference axis. A material is called transversely isotropic when one of its principal planes is a plane of isotropy, i.e. at every point there is a plane on which the mechanical properties are the same in all directions.
Newtonian fluid	For a Newtonian fluid, the viscous stresses are proportional to the element strain rates and the coefficient of viscosity. Most fluids do not behave as Newtonian fluids.
Plastic deformation	Deformation that does not return to its initial state when the stress is removed.
Preconditioning	In mechanical testing of living tissues in vitro, the loading and unloading processes are repeated for a number of cycles until the stress-strain relation become stabilised and repeatable results are obtained.
Pressure	A measure of the force per unit area exerted, for example by a fluid. The SI unit for pressure is $N\ m^{-2}$ or pascal.
Pseudoelastic	The loading and unloading curves of the stress-strain relation are considered to represent properties of two materials with different elasticity.
Residual strain	The strain between the stress-free (zero-stress) state and the no-load state (where all external forces are absent).

Table 3.1. Reference guide to mechanical terms (Continued)

Strain	Forces applied to solids cause deformation or strain. Consider a string with initial length L_0 and stretched length L. It is useful to describe the change by dimensionless ratios such as L/L_0 or $(L-L_0)/L_0$ since it eliminates the absolute length from consideration. Elongation causes tensile (positive) strain while shortening causes compressive (negative) strain. Strain is a tensor quantity with six independent components (ε_{11}, ε_{22}, ε_{33}, $\varepsilon_{12} = \varepsilon_{21}$, $\varepsilon_{13} = \varepsilon_{31}$, and $\varepsilon_{23} = \varepsilon_{32}$). The components ε_{11}, ε_{22}, and ε_{33} are called normal strains, and the remaining components are called shear strains. A normal strain is perpendicular to the surface while shear strain is parallel to the surface.
Stress	Force per unit surface area that the part lying on the positive side of a surface element (the side on the positive side of the outer normal) exerts on the part lying on the negative side. Stress is a tensor quantity with six independent components (τ_{11}, τ_{22}, τ_{33}, $\tau_{12} = \tau_{21}$, $\tau_{13} = \tau_{31}$, and $\tau_{23} = \tau_{32}$). The components τ_{11}, τ_{22}, and τ_{33} are called normal stresses, and the remaining components are called shear stresses. A normal stress is perpendicular to the surface while shear stress is parallel to the surface.
Viscoelasticity	Time dependence of the response to stress or strain. Stress relaxation, creep and hysteresis are features of viscoelasticity.
Viscosity	The property of resisting deformation in a fluid.
Zero-stress state	The tissue configuration where neither external nor internal forces are present. For a tubular organ, the zero-stress state is obtained by making radial cuts in a ring of tissue so that it springs open into a sector.

Figure 3.1 illustrates the basic geometry of a cylindrical segment of the gastrointestinal tract. The three principal directions, longitudinal (z), radial (r) and circumferential (θ) are illustrated. The pressure P from a bolus or a distending balloon will induce a normal stress that will stretch the tube in circumferential directions and will probably also cause longitudinal extension and radial contraction, meaning that the wall becomes thinner. Assuming that the pressure is generated by a moving bolus, forces will also occur in a longitudinal direction, causing a shear stress in the mucosa. The normal stresses and shear stresses together with corresponding strains will be dealt with below and definitions are given in Table 3.1.

3.2 Stress

Consider a large and a small specimen. The large specimen can sustain a large force where the small specimen can sustain a much smaller force. Therefore, the force relative to the size is important. Stress is force per unit cross-sectional area ($\sigma = F/A$) with units of Pa or N m^{-2} (the SI unit for force is the Newton, being the force required to give a mass of one kilogram an acceleration of 1 m s^{-2}). On any surface, the force may be applied perpendicular to the surface, such as the bolus pressure (normal stress) exerted on the wall, or parallel to the surface, such as the force exerted by the fluid flow (shear stress) on the wall. Normal stresses may be either compressive or tensile. A force may be applied in any direction and can induce stresses and strains in various directions. At any given point in the body, the state of stress is described by a stress tensor that consists of three normal stresses and six shear stresses, three of which are independent.

To understand the concepts of tensors it is useful first to consider a vector. A vector is a quantity having both magnitude and direction such as velocity, force and acceleration.

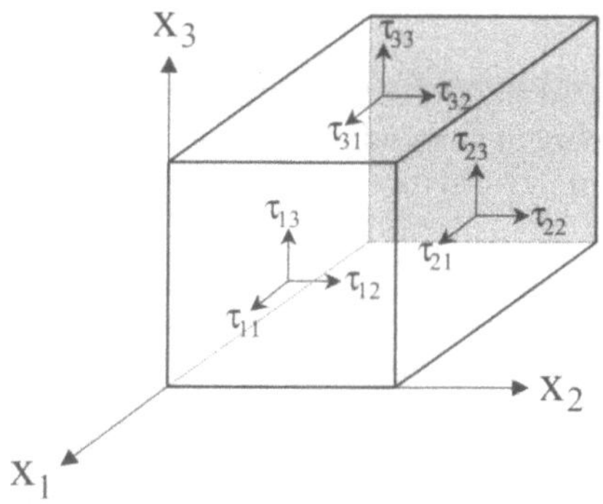

Figure 3.2. Notation of stress components. Modified from Fung (1993).

The essential feature of a vector is that in any rectangular Cartesian coordinate system it is represented by three components, each associated with a particular axis. These components depend only on the orientation of the axis and transform when the axes are rotated. Tensor analysis can be regarded as a generalisation of vector analysis to certain entities known as tensors that require more than three components for their complete specification. Tensor analysis is used as a mathematical tool to make physical laws independent of any particular coordinate system. A vector is a first-order tensor with three components. In the example of stress, an elastic body that deforms can be considered. At any particular point within the body, consider a small plane area at P whose normal is in the direction of the x_1 axis of a rectangular Cartesian coordinate system. The force per unit area on this element is a vector having three rectangular Cartesian components which may be denoted by τ_{11}, τ_{12}, and τ_{13} (Figure 3.2). In the same way, the forces per unit area on small plane areas at P whose normals are in directions $x2$ and $x3$ will have components that can be denoted τ_{21}, τ_{22}, τ_{23}, and τ_{31}, τ_{32}, τ_{33}, respectively. These nine components (τ_{ij} where i, j = 1, 2, 3) are sufficient for fully specifying the state of stress at the point P. This means that stress is a second-order tensor. Similar considerations can be made for strain; i.e., strain also consists of nine components. Since both stress and strain are second-order tensor quantities, the stress–strain relation will be of the fourth order; i.e., will contain 81 components. Thus, in general, it would require 81 elastic constants to characterise a material fully. Assumptions such as the isotropy assumption can reduce the number of components considerably. For example, if the stress and strain tensors are symmetric, the number of constants is reduced to 36.

Table 3.2 gives the notation for the stress tensor in a square matrix form.

Table 3.2. Notation of stress components

	Stress component 1	Stress component 2	Stress component 3
Surface normal to x_1	τ_{11}	τ_{12}	τ_{13}
Surface normal to x_2	τ_{21}	τ_{22}	τ_{23}
Surface normal to x_3	τ_{31}	τ_{32}	τ_{33}

Here, τ_{11}, τ_{22}, and τ_{33} are the normal stresses and the remaining six components are shear stresses.

It is not the intention to make much use of tensor analysis in this book but it is important to keep the concept in mind. This book is primarily based on derivation of equations for simple geometric structures based on equilibrium equations. Laplace's law is such an example.

3.2.1 Stress in Thin-walled Cylindrical Pressure Vessels

Cylindrical pressure vessels having a circular cross-section are found in many industrial settings. Pressurised pipes in water supply systems are an example. In the following we will assume that the gastrointestinal tract is a thin-walled cylindrical pressure vessel and that the weight of the pressure vessel and its contents can be neglected (argument for studying the tissue when it is immersed in a fluid-filled organ bath). In a cylindrical tube, we deal with radial, circumferential and longitudinal components of stress in the respective directions. These are the normal components of stress in the wall of the cylinder (Figure 3.3). There are also three additional shear components. In the following we shall consider the circumferential stress, also called the hoop stress, and the longitudinal stress, also called the axial stress.

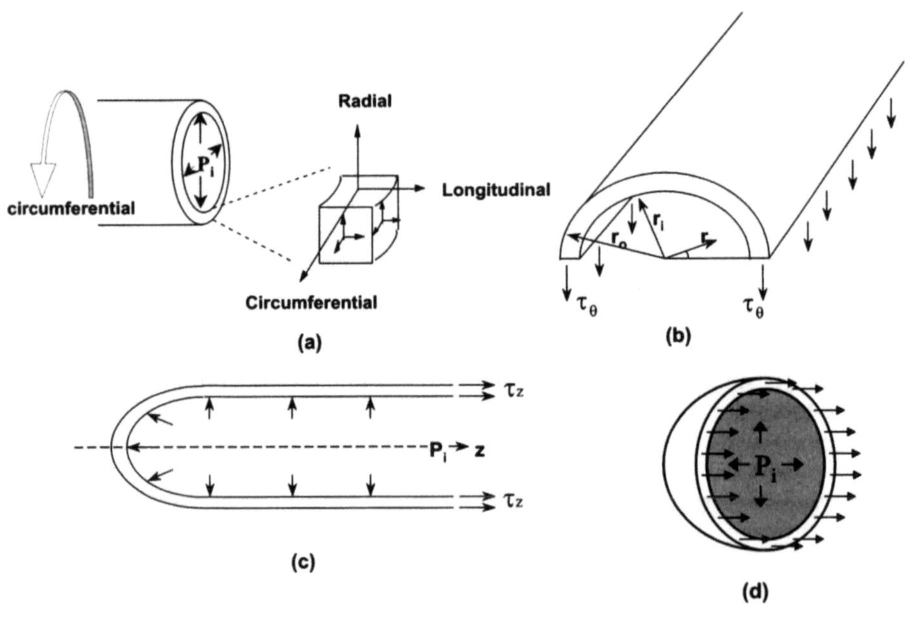

Figure 3.3. A pressurised cylindrical tube and sphere. (a) An infinitesimal element of the cylindrical tube showing the radial, longitudinal and circumferential directions. (b) A free-body diagram of half of the tube cut parallel to the central axis. (c) A free-body diagram of the tube cut perpendicular to the central axis. (d) A free-body diagram of the pressurised spherical shell. See the text for explanation of symbols.

Circumferential Stress

In tubular organs the major tensile stress induced by distension is in the circumferential direction (Dobrin, 1978). During luminal pressure loading, the equilibrium condition requires that the force in the intestinal wall in the circumferential direction be balanced by the force in the intestinal lumen contributed by the inflation pressure. When the geometry is cylindrical, it can be shown that the average circumferential wall stress is

$$\tau_\theta = \frac{Pr_i}{h}, \tag{3.1}$$

where P, r_i, and h are the pressure, internal radius, and wall thickness, respectively. This formula is commonly known as Laplace's law. This law can be derived from consideration of equilibrium as follows.

Consider a cylindrical intestine subjected to an internal pressure P_i, as shown in Figure 3.3a. The pressure in the intestine induces stress in the intestinal wall. Under equilibrium conditions, the force in the intestinal wall in the circumferential direction $2\tau_\theta(r_o - r_i)L$, is balanced by the force in the intestinal lumen contributed by the pressure $2Lr_iP_i$, as shown in Figure 3.3b. Hence, under equilibrium conditions, $2\tau_\theta(r_o - r_i)L = 2Lr_iP_i$ and the equilibrium equation in the circumferential direction can be expressed as

$$\tau_\theta = \frac{Pr_i}{r_o - r_i}, \tag{3.2}$$

where τ_θ is circumferential stress, r_i is internal radius and r_o is outer radius. Because $r_o - r_i = h$, then Equation 3.2. is simplified to Equation 3.1. It should be noted that the stress in Equation 3.1 is averaged over the thickness of the segment and does not describe any regional distribution of stress across the wall thickness. Furthermore, the Laplace stress may refer to either the wall thickness in the undeformed state (h_0) (named the *engineering stress*) or to the thickness in the deformed state *(true stress)*.

If the segment geometry is conical, the *circumferential stress* is given as (Nash, 1994)

$$\tau_\theta = \frac{pr_i}{h\cos\alpha}, \tag{3.3}$$

where α is the angle between the wall of the cone and the centre line axis. If α is small, Equation 3.3 reduces to Equation 3.1. If the deformation is expressed in terms of Green strain (large deformation theory with reference to the initial state (Lagrangian description), see later), the stress must be expressed as a commensurate measure and should in that case be expressed in the sense of Kirchhoff stress:

$$S_\theta = \frac{Pr_i}{h\,\lambda_\theta^2} \tag{3.4}$$

The Kirchhoff stress is especially useful in biaxial strain energy functions with uniform $S_{\theta\theta}$ *and* S_{zz}. Stress in radial directions is ignored.

Longitudinal Stress

The longitudinal stress in an intestinal wall can be determined in a similar way to the circumferential stress based on the equilibrium of forces in the longitudinal direction. The product of the longitudinal stress and the cross-sectional area of the intestinal wall is the force that balances the total longitudinal force acting on the intestine as shown in Figure 3.3c. The longitudinal force in the intestinal wall $\tau_z\pi(r_o^2 - r_i^2)$ is balanced only by the pressure component $P_i\pi r_i^2$ since the external pressure is assumed to be zero. Thus, $\tau_z\pi(r_o^2 - r_i^2) = P_i\pi r_i^2$ and the longitudinal stress τ_z can be expressed as:

$$\tau_z = \frac{Pr_i^2}{r_o^2 - r_i^2}. \tag{3.5}$$

If the wall thickness-to-radius ratio is small, so that $r_o = r_i = r$ and $r_o - r_i = h$, then the equation for longitudinal stress is simplified to

$$\tau_z = \frac{Pr_i}{2h}. \tag{3.6}$$

Stress in a Thin-walled Spherical Shell

The free-body diagram for the geometry is shown in Figure 3.3d. In a similar way to the longitudinal direction of a pressurised pressure vessel, it can be shown that the force in the wall of a sphere $\tau_z\pi(r_o^2 - r_i^2)$ is balanced only by the pressure component $P_i\pi r_i^2$. With the same assumptions as above, Equation 3.6 also accounts for the thin-walled pressurised spherical shell.

3.2.2 Additional Notes on Laplace's Law

The law of Laplace has been used extensively in gastroenterology and the cardiovascular field due to the simplicity in explaining why rupture occurs when segments are excessively distended. Another important implication of the law is that the wall stress is related to pressure and the radius-to-wall thickness ratio. Hence, the conventional comparison of wall properties such as the distensibility and tone at the same pressure and threshold for mechanoreceptor response regardless of the radius-to-wall thickness ratio is often misleading (see later).

Using Laplace's law, it is important to recognise the assumptions upon which it is based. In fact, it was probably never meant to be used in physiology. The following assumptions apply: the geometry of the segment must be circular cylindrical with equal thickness along the circumference, the material properties must be isotropic and a static equilibrium of forces is required; i.e., acceleration must not occur. The question arises whether the gastrointestinal tract can be regarded as being thin walled or thick walled. This is important because the stress cannot be assumed uniformly distributed through the wall thickness in a thick-walled cylinder. The limit is often said to be at a thickness-to-radius ratio of 10% (if so, there is less than a 5% difference in the stress distribution from the inner to outer surface of the wall). The heart, arteries and arterioles have thickness-to-radius ratios of 0.25, 0.20 and 1.0 and must be modelled as thick-walled shells. The corresponding value for veins is 0.03. Hence, only veins can strictly be regarded as being thin walled. However, theory for thin-walled structures have often been employed for other parts of the cardiovascular system and justified by the fact that the average stress is a good approximation. Considering the gastrointestinal tract, only few data on the thickness-to-radius ratio appear in the literature. Figure 3.4 provides new data from the rat oesophagus and small intestine. The oesophagus is thick walled, especially at low pressures where the lumen is small. Even the duodenum and jejunum must be considered as thick-walled organs whereas the ileum approximates the thin-walled structure. Gao and Gregersen (2000) provided data to support the fact that the large intestine can be considered a thin-walled organ

3.2.3 The Thick-walled Cylinder

If the organ is thick walled, it is not appropriate to average the stress over the thickness. In that case the circumferential distribution of stress across the circular cylinder can be computed according to Fung (1994) as

$$\sigma_\theta = \frac{r_o^2 r_i^2 (p_o - p_i)}{r^2 (r_o^2 - r_i^2)} + \frac{p_i r_i^2 - p_0 r_o^2}{r_o^2 - r_i^2}, \tag{3.7}$$

where r_0 and r_i are outer and inner radii whereas r is the radial location of a point in the wall. p_0 and p_i are the outside and inside pressures. Consider an example with an inner and outer pressure of 10 and 0 kPa and outer and inner radii of 10 and 7 mm. Then the circumferential stress can be computed as a function of the radius in the wall as shown in Figure 3.5. In a similar fashion it is possible to compute the radial stress distribution (Fung, 1994). The thick-walled approach is better than the thin-walled one if we want to determine the stress in the vicinity of the mechanoreceptors located in the wall of the gastrointestinal tract, especially in the thick-walled oesophagus. However, at least two important considerations must be taken into account. First, a physiological mechanism exists to reduce the gradient in stress through the wall. Residual stress and strain (prestress) exist in biological tissues including the gastrointestinal tract as shown by a segment opening into a sector when cut radially. A cylinder that is not prestressed will bear the highest tensile stress at the inner edge (as shown in Figure 3.5), which will further increase during loading since the radius-to-wall thickness ratio increases. However, the residual strain in biological tissues serves as a mechanism for reducing the stress gradient from the inner to the outer wall under homeostatic conditions. Hence, residual strain may prevent damage to the mucosa in the gastrointestinal tract during large bolus transport (see the notes on residual stress and strain in Chapter 6). Second, the gastrointestinal tract is a layered structure where the mechanical properties differ between the layers. This fact greatly complicates mechanical studies and puts emphasis on development of composite models (see Chapters 4 and 6)

3.2.4 Computation of Tension in Membranes

The following formulas may be better known to the readers as Laplace's law. In the following case the organ is considered to be very thin walled; hence, membrane theory must be considered rather than the shell theory presented above, and tension is computed rather than stress. This approach is often used in physiology when the wall thickness is immeasurable.

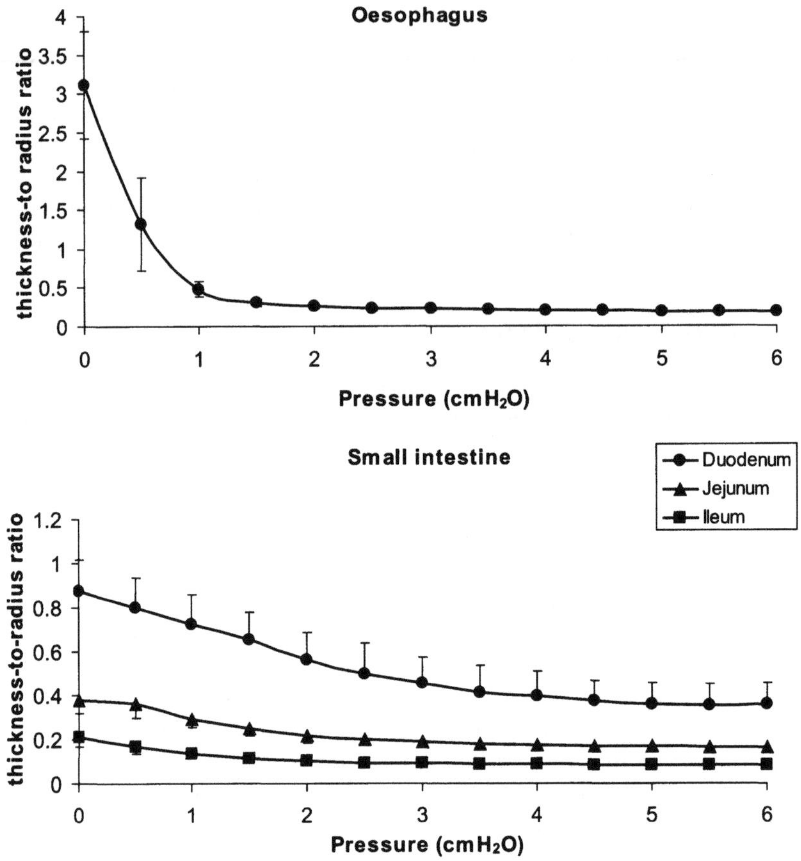

Figure 3.4. Data on thickness-to-radius ratio for the rat oesophagus and small intestine as function of pressure. Only the ileum can be considered strictly thin-walled at pressures above 2 cmH_2O.

Laplace's law originally refers to the relationship between the pressure difference, the wall tension, and the curvature of the membrane surface. Consider a thin-walled membrane surface and assume that the wall tension is constant everywhere. Then Laplace's law reads:

$$T = \Delta P \left(\frac{1}{r_1} + \frac{1}{r_2} \right)^{-1}, \tag{3.8}$$

where r_1 and r_2 are the principal radii of curvature of the surface, T is the total tension per unit length of the mid-surface of the membrane, and ΔP is the transmural pressure difference. ΔP is often assumed equal to the pressure inside the membrane due to the assumption that the external pressure is zero. The equation reduces to $T = \Delta Pr$ in the case of a cylinder since one of the radii tends to infinity, and to $T = \Delta Pr/2$ for a sphere since the two radii in that case are equal. Laplace's law is often used for balloon

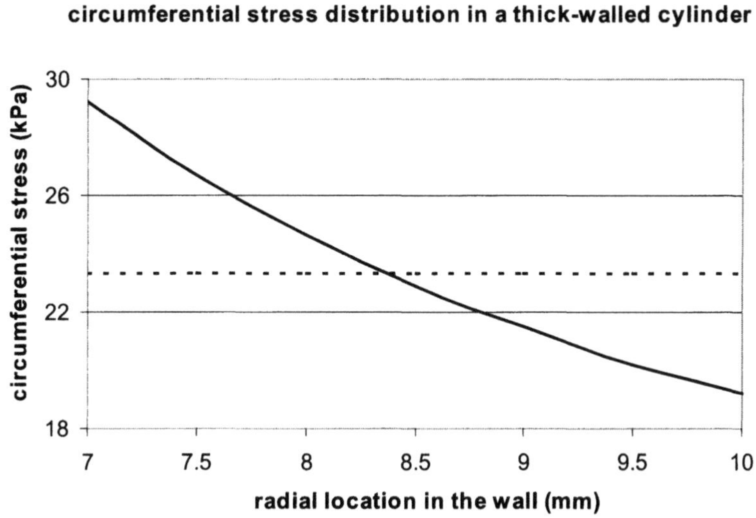

Figure 3.5. The distribution of circumferential stress computed according to thick-walled theory (Equation 3.7, solid line) and using thin-walled theory (Equation 3.1, dotted line). The inner pressure is in this example 10 kPa and the inner and outer radii are 7 and 10 mm. The thick-walled approach considers the non-linear distribution in the wall and shows that the highest stress is found at the inner surface. The thin-walled approach, however, approximates the average stress in the wall.

distension studies in tubular organs if the wall thickness cannot be measured. More general equations have to be considered if the luminal cross-sectional area is elliptical rather than circular.

Equation 3.8 is valid as long as the membrane is so thin that bending rigidity can be neglected. Furthermore, there are several underlying assumptions: first, a cylindrical lumen must be assumed, and second, the material property is assumed to be isotropic. Although the isotropy assumption is not valid for many biological tissues, the parameter is still useful. Finally, the analysis requires a static equilibrium of forces and consequently zero inertial forces. The circumferential tension is equal to the product of the average circumferential stress (Equation 3.1) and wall thickness.

3.3 Deformation

The term deformation refers to a change in the shape of a continuum (a continuous distribution of matter in space) between some initial (undeformed) configuration and a subsequent (deformed) configuration. A force may be applied in any direction and can induce strains in various directions. On any surface, the strain may be perpendicular to the surface (normal strain), or parallel to the surface, such as the strain exerted by the fluid flow (shear strain) on the wall. At any given point in the body, the state of strain is described by a strain tensor that consists of three normal strains and six shear strains, three of which are independent (as shown in Figure 3.2 and Table 3.2 for stress). Tensor analysis is a generalisation of vector analysis to certain mathematical and physical entities known as tensors that require more than three components for their complete specification.

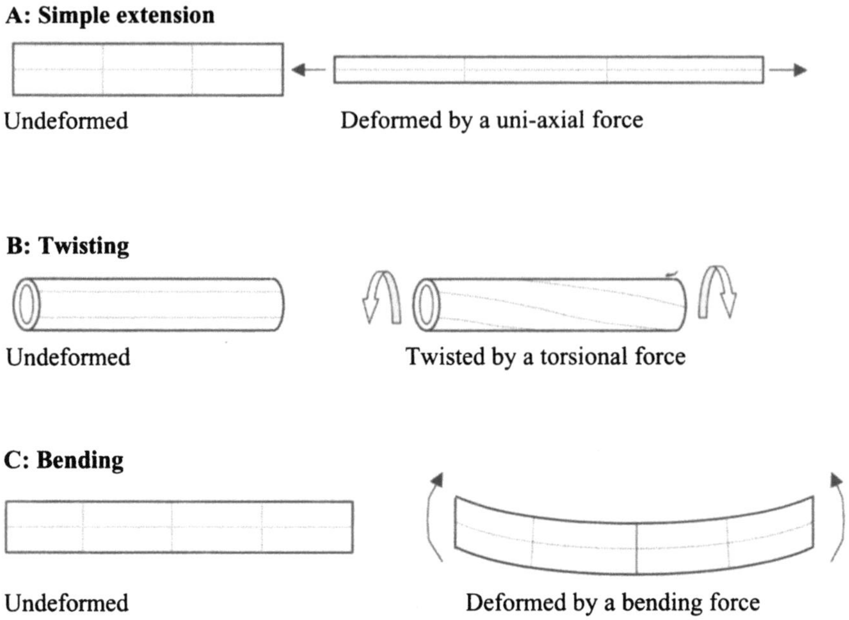

Figure 3.6. A) Extension of a bar with undeformed length (L_0) at resting state and length L at deformed state caused by a uniaxial force. The force induces a positive (tensile) strain. A force in the opposite direction will induce a negative (compressive) strain. B) The effect of twisting a tissue strip. C) The effect of bending a specimen. No strain occurs at the horizontal dotted line. Therefore, this line is called the neutral axis. On the convex side the tissue is in tension, on the concave side the tissue is in compression.

How should we quantify deformation? Deformation can take many forms, e.g., if we pull a tissue strip, it stretches, and if we inflate a tubular organ, it distends.

To be able to describe such deformations quantitatively, the strain measures must be introduced. For a continuum subjected to deformation, strains can be defined in several different ways in relation to the deformation gradient. For simplicity, consider a tissue strip of initial length L_0 (Figure 3.6a). If we stretch it to length L, the change in length can be described by several dimensionless ratios such as:

$$\text{Stretch ratio } \lambda = \frac{L}{L_0}, \tag{3.9}$$

$$\text{Cauchy strain } \in = \frac{L - L_0}{L_0}, \tag{3.10}$$

$$\text{Green's strain } \mathrm{E} = \frac{L^2 - L_0^2}{2L_0^2}. \tag{3.11}$$

Thus, for a continuum subjected to finite deformation, strains can be defined in several different ways in relation to the deformation gradient (Fung, 1990). The selection of proper strain measures is dictated primarily by the stress–strain relationship. The strains can be computed for all surfaces or interfaces between layers where the geometric data can be obtained. The measures are dimensionless This is advantageous since it eliminates the absolute length and any system of units from consideration. This makes comparison possible between specimens of various sizes. In Equations 3.9–3.11, the strain measures are expressed as a fraction of the initial length (Lagrangian strains). However, they may also be expressed as a fraction of the final length (Eulerian strains). Either of these strain measures is useful. In infinitesimal elongations, the strain measures are equal. However, in finite elongations, they are different, which easily can be shown by examples. The Cauchy strain is especially useful in linearised theory of elasticity, which is valid when $\in$ is infinitesimal. Hence, it is usually called the "infinitesimal strain" or "engineering strain". For finite deformations, strain defined by Green is more conveniently related to stress. One strain measure can readily be transformed into another as shown below:

$$\in = \lambda - 1 \text{ and } E = \frac{(\in + 1)^2 - 1}{2}. \tag{3.12}$$

At this point, we may learn more about deformation from the tissue strip stretch experiment. Due to the incompressible nature of tissue, stretching the strip causes a lateral contraction (radial narrowing). The ratio of lateral strain to longitudinal strain in a body under tensile or compressive forces is called Poisson's ratio (see text in relation to Equation 3.19). For biological materials, which are regarded as incompressible in the physiological range, the Poisson's ratio is often regarded as 0.5. Thus, deformation in one direction causes deformation in other directions.

In some cases, one strain measure may have advantages over others. As stated before, Green strain and Kirchhoff stress are commensurate measures useful in strain energy functions. On the other hand, the stretch ratio is a convenient measure when the tissue can be considered incompressible. In the case of incompressibility, the product of the stretch ratios in the three principal directions equals 1. Thus, if two stretch ratios are known, it is possible to compute the last one. The proof for this relation is simple. Consider a rectangular block of tissue with length L, width W and height H. In the undeformed state we use the subscript 0. In the undeformed state the volume $V_0 = L_0 W_0 H_0$ and in the deformed state $V = L W H$. If the tissue is incompressible, $V = V_0$. Therefore, $L_0 W_0 H_0 = L W H$ or

$$\frac{L W H}{L_0 W_0 H_0} = 1 \text{ or } \lambda_L \lambda_W \lambda_H = 1. \tag{3.13}$$

It is a characteristic of soft biological tissues that they can undergo large deformations with a very low degree of compressibility. Thus, these tissues resist volume changes much more than shape changes. For the majority of practical applications within the physiological range, soft tissues can be considered incompressible. This has been verified for arterial tissue but is as yet an assumption in gastroenterology.

In this book, focus is primarily on deformations caused by applying luminal forces (pressure) in tubular organs such as the oesophagus and the intestines. This approach obviously is more physiological than using tissue strip specimens. The rationale will be explained further in the chapters about smooth muscle function and residual strain. The strain measures can readily be used for studying organs of complex configuration as long as the geometric data can be obtained. In the case of the gastrointestinal tract, let us for simplicity consider a straight circular cylindrical tube of homogenous material. We may refer to radial, circumferential and longitudinal components of strain in the respective directions as defined before. These are the normal components of strain in the wall of the cylinder. During luminal pressure loading (distension) the circumferential length usually increases (tensile circumferential strain), the wall thickness decreases (compressive radial strain) and compression or elongation may occur in the longitudinal direction (dependent on the material properties).

One issue that has not yet been touched upon is the determination of the initial (reference) length. It is apparent from Equations 3.9–3.11 that the strain measure depends on the correct determination of the initial length. This is a difficult task for several reasons. First, it may be difficult to suppress smooth muscle activity. Second, strain may vary throughout the intestinal wall. Finally, as outlined in Chapter 6, the zero-stress state can only be determined when the tissue is cut open, which is obviously only possible in vitro. When the strain is referred to the zero-stress state, the tissue is in tension when $\lambda > 1$. It is in compression when $\lambda < 1$. Tissue buckling occurs when $\lambda < 1$. Our assumption is that for very soft tissues in the neighbourhood of a state of zero stress or no load, the critical buckling stress tends to zero. Oesophagus, intestine, and blood vessels probably belong to this category. Buckling is a global or macroscopic phenomenon. Hence, the edge lengths E should be measured with a scale commensurate with the wall thickness of the tube, and not by fractal considerations.

In the case where the geometry of the zero-stress state cannot be obtained, other methods are needed. For comparison of strains (and tension–strain relations) between different segments or before and after intervention, r_0 can be determined at the same level of tension rather than at the same pressure level in accordance with Laplace's law. Thus, a plot of tension versus radius for each set of data can be extrapolated to give the lowest tension value. At this specific tension, it is easy to determine r_0 graphically for the different specimens. For ease of measurements, r and r_0 can be replaced by c and c_0, the corresponding circumferences. Usually, some kind of curve fitting is used and in that regard it is important to obtain numerous valid measurements points as close to the unloaded state as possible. However, it is important to realise that the curves may be translated from a curve referring to the true zero-stress state on the strain axis. It is also emphasised that in determining average stress–strain relations such as Cauchy strain and Cauchy stress (according to Laplace's law), the mid-wall radius needs to be determined. This is often impossible in vivo and, therefore, in many studies the luminal radius is used.

There are several ways to obtain strain measures. In vitro, strains are often computed from measurements of changes in the total length of a strip, as the change in distance between markers located on the surface or embedded in the tissue, or from the measurement of changes in diameter in intact specimens. In vitro, it will also often be possible to determine the zero-stress state as the reference. In vivo, we often rely on balloon distension techniques with measurement of volume or cross-sectional area along with the balloon pressure. Recently, Jørgensen et al. (1995) made a promising development by combining impedance planimetry with high-frequency endoluminal ultrasonography. Hereby, it is possible to correct for variation in wall thickness to obtain a measure of stress–strain distributions in vivo. These methods are treated in the next chapter. Several ultrasound techniques also seem promising. When comparing in vivo experiments with in vitro experiments, it is important to notice that the segments are often free to lengthen in vitro, but they may be tethered to the surrounding structures in vivo. An important observation is that the oesophagus is always under considerable longitudinal stress in vivo. When removed from the body, it shortens by up to 50% as observed in guinea-pigs, with a corresponding increase in diameter. This large residual longitudinal strain may have important physiological implications during bolus transport in the oesophagus.

The effect of bending a beam is illustrated in Figure 3.6c. This kind of experiment is useful because it brings one part of the specimen in tension whereas the part on the other side of the neutral axis is in compression. Such experiments have been used to determine the stress–strain properties close to the zero-stress state. Unfortunately, no data has yet been presented for the gastrointestinal tract. The interested reader must therefore turn to cardiovascular literature.

3.3.1 Calculation of 2D Surface Strains

In experiments in vitro it is possible to obtain surface strains in the two principal directions (circumferential and longitudinal) and a circumferential–longitudinal shear strain from the location of markers on the surface. These local 2D surface strains are computed from displacement of microbeads or other markers located on the surface. Three microbeads must be chosen at each region of interest and their locations recorded in the initial (i) and final (f) state. The triplet of microbeads should always be selected from the middle of each segment and they must be closely spaced in order to minimise the effect of curvature. Errors in such analysis are given by Yu and co-workers. The change in length for each side of the triangle can be calculated using Pythagoras' Theorem:

$$ds_f^2 - ds_i^2 = dx_{1f}^2 + dx_{2f}^2 - dx_{1i}^2 - dx_{2i}^2, \tag{3.14}$$

where ds is the length of the side and dx_1 and dx_2 are the components in the x_1 (circumferential) and x_2 (longitudinal) directions in the final and initial state. The change in the lengths of the sides can then be equated to the strain tensor (E) as follows:

$$ds_f^2 - ds_i^2 = \sum_{l=1}^{2} \sum_{m=1}^{2} 2E_{lm}\, dx_l dx_m. \tag{3.15}$$

By equating the changes in the lengths of each of the three sides to the components of the strain tensor, a system of linear equations is constructed. This system can be solved to yield the three independent strain components E_{11}, $E_{12}=E_{21}$ and E_{22}.

If we have more than three pairs of points, then we use the method of least squares of errors to determine a set of E_{ij} that minimises the errors. Figure 3.7 shows an intestinal segment from a rat at various applied pressures. Several markers can be identified on the photographs and some of these are used in triplets to draw triangles to be used in the analysis above.

3.4 The Stress–Strain Relation

The proportionality constant between stress and strain is called the *elastic modulus* and for a linear, Hookean material it is called Young's modulus. For such material the mechanical properties are elastic and the constitutive equation is simplified to Hooke's law (see below). However, in biological tissues the relation between stress

P = 0 kPa

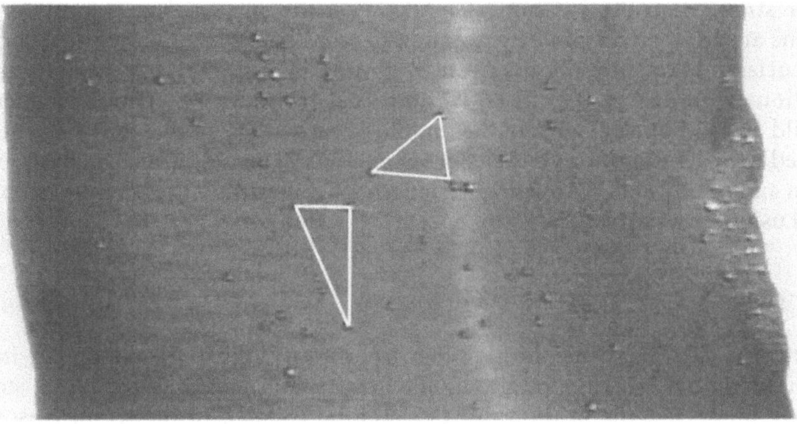

P = 0.5 kPa

P = 2 kPa

Figure 3.7. An intestinal segment distended with various luminal pressures. The markers are used for 2D-strain determinations according to the analysis in the text. The arrow indicates the axial direction.

and strain is non-linear and the strain is usually large (finite deformation). Examples of non-linear stress–strain curves for gastrointestinal tissue are found elsewhere in this book. The non-linear (usually exponential-like) mechanical behaviour is likely due to the properties of collagen and facilitates stretch in the physiological pressure range and prevents overstretch and damage to the tissue at higher stress levels. Overstretch can induce a plastic deformation whereby the tissue can no longer return to its original state when unstressed. Due to the non-linearity it is necessary to compute an incremental elastic modulus. The gastrointestinal wall is similar to other biological tissues in that it has complex three-dimensional structures that have different material properties in different directions. This important feature is called anisotropy and it implies that a large set of material constants have to be specified in order to completely describe the mechanical behaviour. The so-called constitutive equation relates stress and strain through a set of material constants.

The constitutive equations of a solid that consists of a homogeneous, isotropic, linearly elastic material contain only two material constants (Gregersen and Kassab, 1996):

$$\varepsilon_{ij} = \frac{1}{2\mu}\left[\frac{-\lambda}{3\lambda+2\mu}\tau_{\alpha\alpha}\,\delta_{ij}+\tau_{ij}\right], \tag{3.16}$$

where i and j are indices ranging from integers 1 to 3. The ith index denotes the component in the ith direction whereas the jth index denotes the surface perpendicular to the jth direction. The repetition of an index in a term denotes a summation with respect to that index over its range.

Several special cases are considered below.

a) A uniaxial state of stress with the non-zero stress component τ_{11} corresponding to the x_1-direction. From Equation 3.16, we obtain

$$\varepsilon_{11} = \frac{\lambda + \mu}{\mu(3\lambda + 2\mu)}\tau_{11}, \varepsilon_{22} = \varepsilon_{33} = \frac{-\lambda}{2\mu(3\lambda + 2\mu)}\tau_{11}. \tag{3.17}$$

The coefficient

$$E = \mu\,\frac{(3\lambda + 2\mu)}{\lambda + \mu} \tag{3.18}$$

is called the *elastic modulus* or *Young's modulus*, which can be measured in a uniaxial tension test. The ratio $\frac{-\varepsilon_{22}}{\varepsilon_{11}} = \frac{-\varepsilon_{33}}{\varepsilon_{11}}$ is given by

$$\upsilon = \frac{\lambda}{2(\lambda+\mu)}, \tag{3.19}$$

which is known as *Poisson's ratio*. It is a measure of the lateral contraction (extension) produced by an axial tension (compression). The physical dimension of E is force divided by squared length, while υ is a dimensionless quantity.

b) Consider a state of plane stress in pure shear in which the only non-zero component of the stress tensor is $\tau_{12}=\tau_{21} \neq 0$. The corresponding non-zero strain component is given by (Gregersen and Kassab 1996)

$$\varepsilon_{12} = \frac{1}{2\mu}\tau_{12} \tag{3.20}$$

The ratio of shearing stress τ_{12} and the corresponding change $2\varepsilon_{12}$ of an initially right material angle is known as the shear modulus. In engineering literature, the symbol *G* is widely used to describe the shear modulus.

Consider a hydrostatic state of stress, i.e.,

$$\tau_{ij} = -p\delta_{ij}, \tag{3.21}$$

Combining this result with Equation 3.16 yields

$$\varepsilon_{ij} = \frac{-1}{3\lambda+2\mu} p\delta_{ij} \tag{3.22}$$

We obtain by setting i = j, and summing on j

$$-p = K\varepsilon_{\alpha\alpha}, \tag{3.23}$$

where $\varepsilon_{\alpha\alpha} = \varepsilon_{11} + \varepsilon_{22} + \varepsilon_{33}$ and the constant

$$K = \lambda + \frac{2}{3}\mu \tag{3.24}$$

is the bulk modulus. This modulus is known as a measure of the compressibility of the solid. K and, hence, λ are unbounded ($\rightarrow\infty$) for an incompressible material. The bulk modulus K must be positive if a hydrostatic pressure is to be accompanied by a volume decrease. Moreover, the shearing should occur in the direction of the shearing stress. Therefore, from a purely physical consideration, we expect that

$$\infty \geq (\lambda + \frac{2}{3}\mu) > 0, \ \infty > \mu > 0$$

The corresponding bounds on E and υ are

$$0 < E \leq 3\mu, \ -1 < \upsilon \leq \frac{1}{2} \tag{3.25}$$

Hence, a simple tension is always accompanied by an extension. On the other hand, in a direction normal to the direction of this tension, a contraction may take place.

Often, material constants are derived using exponential or polynomial laws. The stress–strain data are plotted and can be fitted using a least-square method with an exponential stress–strain relationship such as of the form (Fung, 1990):

$$\sigma = \beta\left[\exp(\alpha\varepsilon) - 1\right] \tag{3.26}$$

for the circumferential and longitudinal directions, respectively. Since the strain is referred to the zero-stress state, we must have $\sigma = 0$ when $\varepsilon = 0$, as satisfied by Equation 3.27. A least-square fit is used to determine the values of α and β for the circumferential and longitudinal directions at the various times. The Young's modulus is the slope of the stress–strain relationship and can be computed analytically from Equation 3.26 as

$$E = \frac{d\sigma}{d\varepsilon} = \alpha[\sigma + \beta]. \tag{3.27}$$

Normally, a linear relation will be found between Young's modulus and the stress as a result of the exponential nature of the stress–strain relation. This equation was originally proposed for uniaxial experiments but can be used independently for circumferential and longitudinal data obtained from gastrointestinal distension experiments. A biaxial approach is given later in this chapter.

Since the behaviour is non-linear, incremental elastic moduli can also be computed. In order to compare these moduli, though, they must be measured under conditions of constant strain. Bergel (1972) and Dobrin (1978) provide examples of the error introduced in expressing such moduli as a function of pressure and different length settings in the cardiovascular system.

3.4.1 Computation of Morphometric Parameters in a Cylindrical Structure based on Experimental Data in Vitro

From biomechanical experiments conducted in vitro on gastrointestinal segments, the pressure difference across the wall (the distension pressure), the outer dimensions such as the diameter and the length are usually known. Often, images of segments loaded at various distension pressures are frame grabbed. It is important to avoid edge effects. The internal diameter and wall thickness can, by means of video imaging, not be measured directly but can be calculated with the following assumptions: 1) the wall is incompressible; i.e., the volume of the wall does not change during distension, and 2) the shape of the segment is cylindrical. Furthermore, the wall thickness and wall area at no-load conditions and the circumferential lengths at zero-stress state must be measured. The mid-wall circumference must be measured or computed from the zero-stress images. The total wall volume (V) is defined as

$$V = A_0 l_0 , \tag{3.28}$$

where l_0 and A_0 are the length and wall area at the no-load state, respectively.

The assumptions of incompressibility and cylindricity lead to the equation:

$$V = A_0 l_0 = \pi(r_o^2 - r_i^2)l, \tag{3.29}$$

where r_0 and r_i are the outer and inner radius at the pressurised state, respectively, and l the length at pressurised state. The wall thickness can be computed at the pressurised state as

$$h = r_o - r_i = r_o - \sqrt{r_o^2 - \frac{A_o}{\pi\lambda}}, \text{ where } \lambda = \frac{l}{l_o}. \tag{3.30}$$

Knowing these morphometric data, one can now proceed to compute circumferential stresses and strains according to the equations stated previously in this chapter and the stress–strain data can be fitted to constitutive functions (for example, to the uniaxial function given above or the biaxial approach below).

3.4.2 An Approach for Analysis of Strain and Stress in the Wall of the Gastrointestinal Tract

In order to quantify wall stresses, it is necessary to have an accurate measurement of the strain field to which the gastrointestinal tract is subjected and to have a reliable constitutive equation that relates the strains to stresses. Based on previous data, we assume the wall to be an incompressible, non-linearly elastic orthotropic material subjected to finite deformation. This implies the use of a strain energy function for which the strains must be given with reference to the zero-stress state. The strain energy function represents stored energy per unit volume of the gastrointestinal wall. One of the forms for the strain energy function in a two-dimensional analysis is expressed as follows (Fung, 1981):

$$\rho_0 W = \frac{C}{2} e^Q \tag{3.31}$$

$$Q = a_1(E^2_{\theta\theta} - E^{*2}_{\theta\theta}) + a_2(E^2_{zz} - E^{*2}_{zz}) + 2a_4(E_{\theta\theta} E_{zz} - E^*_{\theta\theta} E^*_{zz})] , \tag{3.32}$$

where ρ_0 is the material density of the artery (mass per unit volume). W is the strain energy per unit mass, $\rho_0 W$ is the strain energy per unit volume, $E_{\theta\theta}$ and E_{zz} are the circumferential and longitudinal Green's strains, respectively. $E^*_{\theta\theta}$ and E^*_{zz} are reference strains measured at a physiological pressure, and C, a_1, a_2, and a_4 are material coefficients. Under assumptions that materials in a wall are homogeneous and pseudoelastic (i.e., the loading and unloading curves of the stress–strain relation are considered to represent properties of two materials with different elasticity), the strain energy function can be applied to loading and unloading processes separately and stress components expressed as

$$S_{ij} = \frac{\partial(\rho_0 W)}{\partial E_{ij}}, \tag{3.33}$$

where S_{ij} and E_{ij} are components of Kirchhoff's stress and Green's strain, respectively. The assumption on homogeneity in the wall can in part be tested in the case of the oesophagus by separating the layers of the oesophagus and testing the layers separately. By combining Equations 3.31 and 3.33, with the radial component and all shear components neglected, the stress–strain relations of a gastrointestinal segment in both circumferential and longitudinal directions can be obtained. The stress and strain components can be determined experimentally and the coefficients of the strain energy function, C, a_1, a_2, and a_4 can be determined by using a non-linear, curve-fitting method.

The zero-stress state of the gastrointestinal segment, at which the diameter and length are measured in the mechanical test, can be determined as described in Chapter 6. The circumferential length of the inner and outer walls of the segment must be measured at the zero-stress state with the aid of an image analysis system. The mid-wall circumferential length of the segment is calculated as

$$L_\theta = \frac{L_\theta^i + L_\theta^o}{2}, \tag{3.34}$$

where L_θ, L^i_θ and L^o_θ are the mid-, inner- and outer-wall circumferential lengths of the segment at zero-stress state, respectively.

The mid-wall strain and average stress of the segment (or its sublayers after separation) can be determined under assumptions that the materials in the wall are homogenous and the shape is cylindrical. The mid-wall circumferential strain can be calculated on the basis of experimentally measured outer diameters at varying inflation and deflation pressures and reference mid-wall circumferential length at zero-stress state. With an assumption that materials in the wall are incompressible, the mid-wall circumferential length of the segment at a given inflation (or deflation) pressure can be computed based on the outer diameter and longitudinal length of the pressurised segment, and the inner- and outer-wall circumferential and longitudinal lengths at zero-stress state. The mid-wall circumferential stretch ratio of the segment, λ_θ, at a given pressure can be computed with respect to zero-stress state as the ratio between the mid-wall circumferential length at a given pressure and the middle-wall circumferential length at zero-stress state. Similarly, the longitudinal stretch ratio, λ_z, at a given pressure can be computed with respect to the zero-stress state as the ratio between the longitudinal length of the segment at a given pressure and the longitudinal length at zero-stress state. The circumferential and longitudinal strains of the segment at a given pressure can be computed with the following equations:

$$E_{\theta\theta} = \frac{\lambda_\theta^2 - 1}{2}, \tag{3.35}$$

$$E_{zz} = \frac{\lambda_z^2 - 1}{2}, \tag{3.36}$$

where $E_{\theta\theta}$ and E_{zz} are circumferential and longitudinal Green's strains, respectively. At an equilibrium condition, the average circumferential stress in the wall at a given pressure can be computed with an assumption that the shape is cylindrical:

$$S_{\theta\theta} = \frac{\mathrm{Pr}_i}{h\lambda_\theta^2}, \tag{3.37}$$

where $S_{\theta\theta}$ is the circumferential Kirchhoff's stress, P is the inflation pressure, r_i is the inner radius, and h is the wall thickness at the given pressure. These parameters can be determined experimentally as described earlier. With a similar approach, average longitudinal stress in the wall at a given pressure can be computed using the following equation:

$$S_{zz} = \frac{\mathrm{Pr}_i^2}{h\lambda_z^2(r_o + r_i)}, \tag{3.38}$$

where S_{zz} is longitudinal Kirchhoff's stress and r_o is outer radius at a given pressure.

By substituting Equation 3.31 into Equation 3.33, the following stress–strain relations for the gastrointestinal segment in both the circumferential (θ) and longitudinal (z) directions can be obtained:

$$S_{\theta\theta} = C(a_1 E_{\theta\theta} + a_4 E_{zz})e^Q, \tag{3.39}$$

$$S_{zz} = C(a_2 E_{zz} + a_4 E_{\theta\theta})e^Q, \tag{3.40}$$

Reference strains $E^*_{\theta\theta}$ and E^*_{zz} are selected as the strain components at physiological pressures in the circumferential and longitudinal directions, respectively. A Marquardt's non-linear, least-squares algorithm can be used to fit the experimental data, and the coefficients C, a_1, a_2, and a_4, can be determined by minimising the sum of the squares of the differences between experimental and theoretical data.

3.5 Viscoelasticity

Biological tissues reveal properties of both elastic solid and viscous fluid. Thus, the stress depends not only on the applied strain as in a solid, but also on the rate of strain as in a viscous fluid. In other words, the response is time-dependent in that the stress–strain response does not occur instantly. When the material is suddenly strained and the strain is maintained constant, the corresponding stresses induced in the wall decrease with time. This phenomenon is called stress relaxation (Figure 3.8).

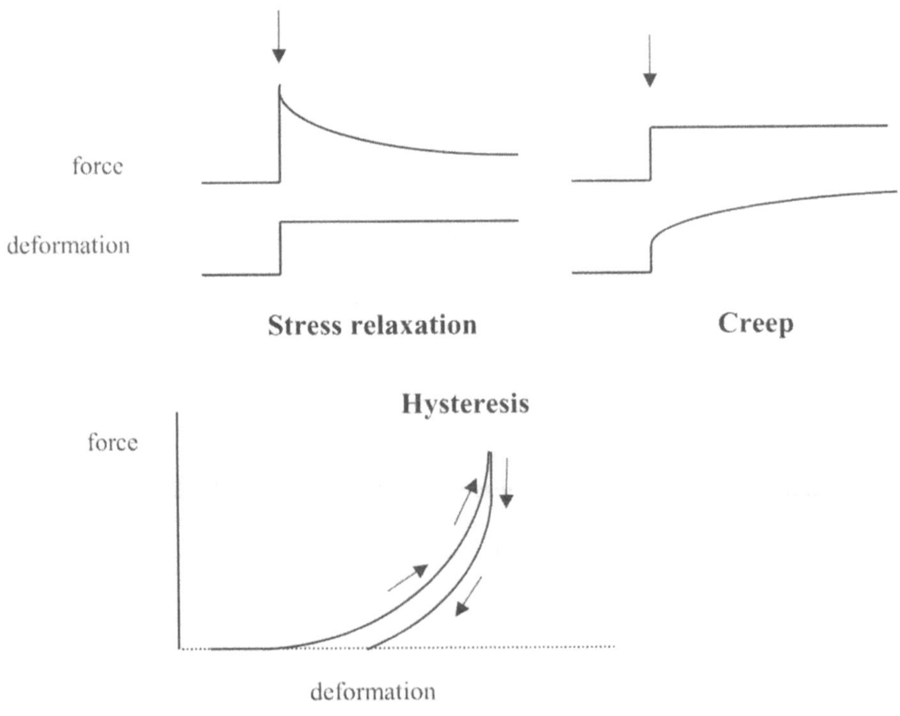

Figure 3.8. Illustration of stress relaxation, creep and hysteresis.

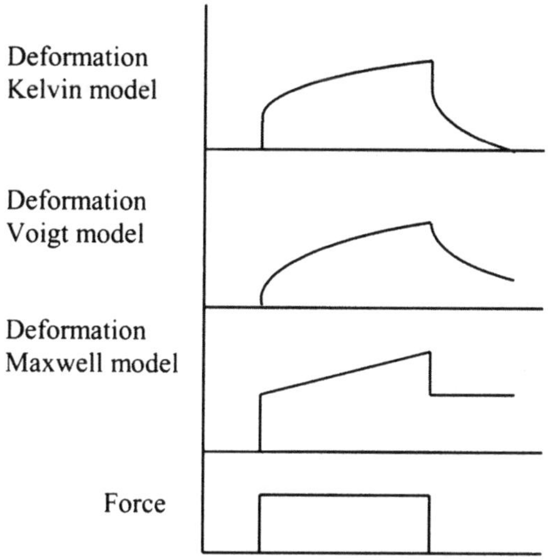

Figure 3.9. Illustration of the creep behaviour in the three viscoelastic models.

If the material is suddenly stressed and the stress is maintained constant, the material will continue to deform. This phenomenon is called creep. If the material is subjected to a cyclic loading, the stress–strain relationship in the loading process is somewhat different from that in the unloading process (Figure 3.8) and the phenomenon is called hysteresis (Dobrin, 1978; Fung, 1993; Fung, 1968). Stress relaxation, creep, and hysteresis are features of viscoelasticity.

Viscoelastic properties in terms of pressure relaxation curves have been described for the normal and diseased human rectum (Arhan et al., 1976; 1978). Often, the viscoelastic behaviour is described in terms of models. Three simple models are often used: the Maxwell model, the Voigt model and the Kelvin model (standard linear solid, see also Chapter 7). The models combine linear springs (constant μ) and dashpots with coefficient of viscosity (η). The spring is supposed to produce an instant deformation proportional to the load whereas the dashpot produces a velocity proportional to the load at any instant. The relationship $F = \mu u$, where F is a force acting on the spring and u is the extension of the spring describes the spring. For the dashpot we have the relationship $F = \eta u$, where u is the velocity of deflection. The Maxwell body is the combination of a spring and dashpot in series. The Voigt body is the spring and dashpot in parallel whereas the Kelvin model has a Maxwell body in parallel with a dashpot. On the basis of the equations for the dashpot and spring, creep functions and relaxation functions can be derived for these models (Fung, 1993). Figure 3.9 illustrates the creep behaviour for the three models. More complex functions exist. However, viscoelastic models do not account for all history-dependent mechanical behaviour (see Gregersen et al., 1998 and Chapter 7 about preconditioning and strain softening).

3.6 Numerical Methods

The complexity of geometry and mechanical problems often makes it impossible to use methods as outlined previously in this chapter. It that case we must turn to computational methods, of which the finite element method has gained most interest. In finite element analysis, the domain to be analysed is divided into elements and the displacement field within each element is interpolated in terms of displacements at a few points around the element boundary. These points are called nodes. The interpolation is done so that the displacement field is continuous across element boundaries. The strain at every point can then be expressed in terms of nodal displacements. It is then required that the associated stresses, through the stress–strain relation, satisfy the principle of virtual work for arbitrary variations of the nodal displacements. This generates numerous equations to be solved by computer.

3.7 Flow Considerations

This book does not intend to provide an analysis of gastrointestinal flow, despite the fact that flow is the end result of the motion of the gastrointestinal tract. I merely wish to point out the most important concepts for the study of fluids in motion. Three significant concepts in fluid flow are:

- The principle of conservation of mass
- The principle of kinetic energy
- The principle of momentum

from which the equations of continuity, flow equations and equations evaluating dynamic forces exerted by flowing fluids are developed.

Chapter 2 provides a summary of data primarily obtained in transit studies in humans in vivo. Compared to the work on muscle contractility, the work on gastrointestinal flow seems more scattered, probably due to the difficulties in studying it and to the great complexity in geometry, composition and properties of the fluids, and the kind of flows. Several studies, more or less theoretical by nature, have been published on flow and the consequences of various types of contractions. Brasseur and coworkers applied bioengineering principles in their extensive work on bolus flow in the oesophagus (see, for example, Brasseur, 1987; 1993). In addition, the reader should consult the work by Bertuzzi, Denli, Stavitsky, Macagno, Fung, Tözeren, and Miftakhov (some references are given in the literature list). Though the complexity is great in vivo, fluid flow is a factor that can be controlled in vitro in experimental setups. For example, it is possible to impose predetermined shear stresses and shear rates to cells grown in culture to study their responses.

3.8 Some Notes on other (Commonly Used) Mechanical Parameters

Such complexities of the gastrointestinal wall as anisotropy, non-linear behaviour, finite deformation, and viscoelasticity make simple estimates of stiffness unsound. In fact, even techniques with far better resolution than those currently available would not permit complete characterisation of the three-dimensional mechanical behaviour of the gastrointestinal tract in vivo. However, some general expression of the overall stiffness is useful. Several commonly used elasticity parameters in gastroenterology are defined and discussed below (summary of Gregersen and Kassab, 1996).

Compliance

Compliance is defined as the change in luminal dimension (volume or cross-sectional area) divided by the corresponding change in pressure:

$$C = \frac{\Delta V}{\Delta P}, \text{ or } C = \frac{\Delta CSA}{\Delta P}, \tag{3.41}$$

where V, CSA and P are volume, cross-sectional area, and pressure, respectively. In this respect, compliance is a reciprocal measure of stiffness. Usually, it is given either as a single averaged value or as a function of pressure. The compliance parameter merely expresses the differences in luminal dimensions between pressure steps. Hence, the parameter does not take into account the actual degree of stretch that occurs under luminal pressure loading, the variations in the unstressed basal luminal diameter, or the wall thickness. Consequently, compliance should be avoided. Despite these convincing arguments, it is often used in gastroenterology literature.

Pressure Elastic Modulus

The pressure elastic modulus is a measure of stiffness and is defined as

$$E_P = \frac{\Delta P}{\left(\frac{\Delta r}{r}\right)}, \tag{3.42}$$

where ΔP and Δr denote changes in internal pressure and radius. This parameter can be used to compare segments with different pressures. It is a better measure than compliance because it considers the degree of stretch but still does not account for the wall thickness. Although this variable has units of pressure similar to an elastic modulus, it is different from the elastic modulus of the wall because the pressure in the lumen is not the same as the tensile stress in the wall.

Cross-sectional Area Distensibility

Volume is not a convenient measure in a cylindrical organ due to the effect of elongation of the distending balloon. Cross-sectional areas and circumferences are much more optimal. It is often advantageous to express the distensibility rather than stiffness in terms of cross-sectional area and transmural pressure. Thus, the distensibility is defined as the ratio of fractional change of cross-sectional area (CSA) to the change in transmural pressure (P_{tm}):

$$D = \left(\frac{1}{CSA_0}\frac{\Delta CSA}{\Delta P_{tm}}\right), \tag{3.43}$$

where CSA_0 is the reference cross-sectional area. This parameter can be directly computed from a P–CSA curve. However, it is useful to convert D into an incremental Young's modulus for circumferential extensions of the whole intestinal wall, modelled as a uniform cylinder with homogenous, isotropic walls. The use of an incremental Young's modulus is made necessary by the non-linearity of the relation between circumferential stress and cross-sectional area. Hence, a single elastic parameter can be defined by considering small departures from a mean, prestressed, in vivo state, and linearising the stress–strain curves. This may be useful if the amplitude of the pressure is small. For an increase in pressure within a thin-walled isotropic intestinal segment whose length is held constant, Young's modulus E is related to the distensibility D by:

$$D^{-1}=\frac{\mathrm{E}}{1-\alpha^2}\frac{h}{d}, \qquad (3.44)$$

where h is the wall thickness, d is the intestinal diameter and α is Poisson's ratio (equal to 0.5 if the material is incompressible). This result follows from the classical elasticity of shells and is referenced by Bergel (1972) in a more general form for thick-walled tubes.

3.9 Literature

The literature list contains papers mentioned in the text as well as other papers of interest.

Arhan, P, Devroede, G, Denis, K et al. 1978. Viscoelastic properties of the rectal wall in Hirschsprung's disease. *J Clin Invest*, 62, 82–7.

Arhan, P, Faverdin, C, Persoz, B et al. 1976. Relationship between viscoelastic properties of the rectum and anal pressure in man. *J Applied Physiol*, 41, 677–82.

Bergel, DH. 1972. The properties of blood vessels. In: *Biomechanics: Its Foundations and Objectives*, ed. Fung, YC, Perrone, N, Anliker, M. Englewood Cliffs, NJ: Prentice Hall.

Bertuzzi, A, Salinari, S, Mancinelli, R, Pescatori, M. 1983. Peristaltic transport of a solid bolus. *J Biomech*, 16, 459–64.

Brasseur, JG. 1987. A fluid mechanical perspective on esophageal bolus transport. *Dysphagia*, 2, 32–39.

Brasseur, JG. 1993. Mechanical studies of the esophageal function. *Dysphagia*, 8, 384–6.

Denli, N. 1975. An analytical model of flow induced by longitudinal contractions in the small intestine. Thesis, University of Iowa.

Dobrin, PB. 1972. Vascular mechanics. In: *Handbook of Physiology - The Cardiovascular System*, pp. 65–102. Englewood Cliffs, NJ: Prentice-Hall.

Dobrin, PB. 1978. Mechanical properties of arteries. *Physiol Rev*, 58, 397–460.

Fung, YC, Liu, SQ. 1995. Determination of the mechanical properties of the different layers of blood vessel in vivo. *Proc Natl Acad Sci U S A*, 92, 2169–73.

Fung, YC, Yih, CS. 1968. Peristaltic transport. *J Applied Mechanics*, 35, 669–75.

Fung, YC. 1967. Elasticity of soft tissues in simple elongation. *Am J Physiol*, 28, 1532–44.

Fung, YC. 1968. Biomechanics. Its scope, history, and some problems of continuum mechanics in physiology. *Applied Mechanics Reviews*, 21, 1–20.

Fung, YC. 1981. *Biomechanics. Mechanical Properties of Living Tissues*. New York: Springer-Verlag.

Fung, YC. 1983. What principle governs the stress distribution in living organs? In: *Biomechanics in China, Japan and USA*, ed. Fung, YC, Fukada, E, Junjian, W, pp. 1–13. Beijing, China: Science.

Fung YC. 1990. *Biomechanics, Motion, Flow and Growth*. New York: Springer Verlag.

Fung, YC. 1993. *Biomechanics. Mechanical Properties of Living Tissues*, second edition. New York: Springer-Verlag.

Fung, YC. 1994. *A First Course in Continuum Mechanics*. Englewood Cliffs, NJ: Prentice Hall.

Gao, C, Gregersen, H. 2000. Biomechanical and morphological properties in rat large intestine. *J Biomech*, 33, 1089–97.

Gregersen, H, Emery, J, McCulloch, AD. 1998. History-dependent mechanical behavior of the guinea-pig small intestine. *Ann Biomed Eng*, 26, 1–9.

Gregersen, H, Kassab, GS. 1996. Biomechanics of the gastrointestinal tract. *Neurogastroenterol Motil*, 8, 277–97.

Jørgensen, CJ, Dall, FH, Jensen, SL, Gregersen, H. 1995. A new combined ultrasound-impedance planimetry measuring system for quantification of organ wall biomechanics in vivo. *J Biomech*, 28, 863–7.

Macagno, EO, Christensen, J. 1981. Fluid mechanics of gastrointestinal flow. In: *Physiology of the Gastrointestinal Tract*, ed. Johnson, LR et al., ch. 10. New York: Raven Press.

Miftakhov, RN, Abdusheva, GR, Christensen, J. 1999. Numerical simulation of motility patterns of the small bowel. Part I - Formulation of a mathematical model. *J Theor Biol*, 197, 89–112.
Miftakhov, RN, Wingate, DL. 1994. Numerical simulation of the peristaltic reflex of the small bowel. *Biorheology*, 31, 309–25.
Miftakov, RN. 1994. Mathematical modeling of the peristaltic reflex: A numerical experiment. *J Math Sci*, 71, 2775–89.
Nash, W. 1994. *Theory and Problems of Strength of Materials*, third edition. USA: McGraw-Hill.
Ren, J, Massey, BT, Dodds, WJ, Kern, MK, Brasseur, JG, Shaker, S et al. 1993. Determinants of the bolus pressure during esophageal peristaltic bolus transport. *Am J Physiol*, 264, G407–13.
Singerman, RBJ. 1974. Fluid mechanics of the human duodenum. Thesis, University of Iowa.
Stavitsky, D. 1979. Flow and mixing in a contracting channel with applications to the human intestine. Thesis, University of Iowa.
Tözeren, A, Özkaya, N, Tözeren, H. 1982. Flow of particles along a deformable tube. *J Biomech*, 15, 517–27.
Vaishnav, RN, Vossoughi, J. 1983. Estimation of residual strains in aortic segments. In: *Biomedical Engineering II. Recent Developments*, ed. Hall, CW, pp. 330–33. New York: Pergamon Press.
Weems, WA. Intestinal fluid flow: Its production and control. In: Johnson LR, Christensen J, Jackson MJ, Jacobson ED, *Physiology of The Gastrointestinal Tract*. New York: Raven Press. 1987, Chapter 17, pp571-593.
Yu, Q, Zhou, J, Fung, YC. 1993. Neutral axis location in bending and Young's modulus of different layers of arterial wall. *Am J Physiol*, 265, H52–60.

4. Biomechanical Methods and Analysis

As Walter Cannon published his book entitled *The Mechanical Factors of Digestion* almost 100 years ago (Cannon, 1911), one might expect the knowledge base about gastrointestinal mechanics to be very large today. Unfortunately, this is not the case. The development in methodology has indeed been slow and unstructured. Today, basic research in gastroenterology and consequently also the diagnostics of "mechanical" diseases suffer greatly from these limitations. Multidisciplinary efforts and use of bioengineering principles are needed to overcome the limitations and to make new progress.

This chapter outlines a number of useful biomechanical methods and analyses, some old and some so new that they have not been published before (cardiovascular physiologists may find the layered models interesting and may wish to evaluate the usefulness in their own research field). The three main themes are conventional motility recording techniques, balloon distension and other mechanical techniques, primarily for mechanical testing in vitro. The basic theory of mechanics is described in Chapter 3. Furthermore, other methods and analyses related to testing of smooth muscle (in particular, membrane properties, stretching and shearing cells grown on membranes), the zero-stress state, and viscoelastic analysis are given in the respective chapters rather than in this chapter.

4.1 Conventional Motility Recording Techniques

Several techniques exist for measurement of motility in vivo including electromyography, manometry, transit time measurements, non-invasive techniques and methodologies for recordings of distensibility. It is beyond the scope of this book to describe all these techniques in detail, therefore, only the techniques useful in mechanical sense are briefly described with respect to methodology and applicability in animal experimental and human studies.

4.1.1 Manometry

Manometry is commonly used for recording of mechanical activity in the gastrointestinal tract. It detects changes in luminal pressure, most often caused by contractions. Essentially, two different ways of measuring luminal pressure exist; i.e., luminal pressure-sensitive strain gauges and perfused low-compliance manometric technique.

The strain-gauge technique is the simplest with the sensor placed in the organ in question. The technique provides good dynamic characteristics and it can be used in portable systems for long-term recordings. Widespread use of these miniature transducers is, however, limited by a high cost.

The perfusion technique is more prone to sources of error than the strain-gauge technique but is nevertheless widely used, due mainly to the low costs and the fact that the probes are tolerable for human subjects and animals. Furthermore, the dynamic characteristics are sufficient for most types of recordings in the gastrointestinal tract. The use of the technique can be traced back at least to the work in the late 1960s by Pope (1967) and others. The probes are single- or multi-lumen with up to and above 12 channels and side holes. Most probes have a diameter of less than 5 mm. The perfusion channels usually have a diameter from 0.4 to 1.0 mm. Each channel is perfused at a low perfusion rate. This low rate is maintained by a high feeding pressure in the fluid reservoir and by inserting high-resistance steel capillaries between the fluid reservoir and the transducer. The fluid flows from the transducer chamber via the channels in the probe and into the gut. When a lumen-occluding contraction occurs in the gastrointestinal tract, pressure increases in the vicinity of the side hole. The obstruction of the flow in the channel results in counter-pressure in the fluid in the channel (amounting to the same pressure as the luminal pressure). The pressure is transmitter via the fluid in the channel to the transducer. The analogue signal from the transducer is amplified and online recordings are visualised on chart recorders and on computer monitors after analogue-to-digital conversion. In gastrointestinal sphincters, phasic and tonic activity can be recorded by perfusion either by stationary or pull-through techniques. In order to obtain more reliable recordings, a cuff or a sleeve may cover the perfused side hole. This is of special importance in the lower oesophageal sphincter where respiratory movements tend to move the probe forwards and backwards.

Evaluation of the recording fidelity of the low-compliance perfusion system must consider several factors such as the characteristics of the pressure events and performance of the recording system. Often, reliable recordings can be obtained with infusion rates in the range of 0.1–0.5 ml min^{-1} per channel. The compliance of the recording system is important. Due to the use of steel capillaries and transducers with small rigid chambers and little displacement of the membrane, the probe itself becomes the major compliance source of the system. Therefore, thick-walled, inelastic catheters of the shortest length and smallest internal diameter should be used in order to improve the dynamics of the system. A rise rate of 150 mmHg per second and a frequency response of 1.0–2.0 Hz are sufficient for gastrointestinal pressure recordings. Also, the presence of air bubbles in the fluid decrease the recording fidelity. It is important to emphasise that the technique in non-sphincteric regions only measures phasic contractions from the circular muscle layer and only those contractions that occlude the lumen are detected. Thus, phasic contractions and tone, which do not constrict the lumen to the diameter of the probe, will not be recorded.

Comparing these two manometry techniques, Valori and co-workers (1986) found that they recorded the same number of contractions in the canine duodenum. However, these two techniques only recorded between 87% and 90% of vagal- and field-stimulated activity recorded by strain gauges sewn onto the serosa. Possible radial duodenal pressure variations and artefacts from movements and respiration must also be considered. Furthermore, calculated propagation velocities of contractions should be interpreted with caution because the gut may telescope onto the probe.

4.1.2 Serosal Strain-gauge Transducers

Extra-luminal strain-gauge transducers for the recording of gastrointestinal motility in laboratory animals were also developed in the 1960s. Many companies manufacture these strain gauges and they are widely used for industrial purposes. For use in the gastrointestinal tract they must be processed into force transducers, which can be done by gluing one or two strain gauges on a thin sheet of stainless steel, attaching lead wires and embedding them between silicone plates. For this procedure the use of curved moulds is advantageous. A major problem is the sealing of the transducers and the lead wires against saline and body fluids. Since watertight strain gauges are available, their use is of great advantage. Various materials such as special preparations of epoxy are used to protect the soldering spots. The lifetime of the transducers range from weeks to several months. If serosal strain-gauge transducers are used to investigate intestinal motor patterns, it is important that all the transducers operate over a long period of time. The transducers are sutured via the silicone sheets onto the serosal surface of the stomach or gut to record circular muscle contraction. Before suturing it is useful to administer atropine and to mark the sites of implantation in order to avoid shortening of the gut due to handling and to define exactly the inter-transducer distances. When multiple transducers are implanted at short distances, adhesions with the omentum or with intestinal loops and encapsulation by connective tissue might occur; therefore, the number of transducers should be restricted. Adhesions and encapsulation might reduce the signals of the transducers. In a long-lasting study, decreasing sensitivity can be compensated for by comparing the signals of phase III of the migrating motility complex over the period of the study.

Serosal strain-gauge transducers work as force transducers. The transducers represent a quarter or half bridge of a Wheatstone amplifier. In general, they record contractile events as luminal manometry does. However, serosal strain gauge transducers are more sensitive and record even small phasic contractions of the circular muscle layer. Changes of gastrointestinal tone are also recorded if they are associated with an increase in wall tension. In contrast, simultaneous video-fluoroscopy indicated that changes in intestinal diameter often produced no clear-cut fluctuation of the baseline of the strain-gauge signals. Besides their advantage of high sensitivity to phasic contractions, serosal strain-gauge transducers do not impede luminal flow and produce no or only negligible artefacts. In comparison with electromyography, the signals of the serosal force transducers are more clear-cut and have a closer relationship to the flow of luminal contents. Strain-gauge transducers can be calibrated in units of milliNewtons before implantation.

4.1.3 Measurement of Flow of Luminal Contents along the Gut

Two parameters are usually used to determine the flow of luminal contents along the gut in vivo: the flow rate and the transit time. The flow rate measures the volume flow along the gut in units of ml min^{-1} whereas the transit time measures the velocity of the aboral movement of a luminal bolus in units of cm min^{-1}. For the measurement of both parameters, markers are necessary which should be inert and non-absorbable. There are markers for both liquids and solids. A combination of both types of marker is often used for gastric emptying studies, whereas the flow of luminal contents in the small intestine is usually measured with liquid markers. Liquid markers are substances that are soluble and flow along the gut together with the liquid phase of the luminal contents. Several types of liquid markers are commonly used: dye markers like phenol red, polyethylene glycol, chelates of heavy metal cations like CrEDTA or CoEDTA or soluble isotopes. They can be measured photometrically, nephelometrically, by atomic absorption spectrometry and by isotopic methods. To measure the flow rate, the marker is continuously infused into the gut and samples are taken further distally at definite time intervals. To measure the transit time the marker is injected as a bolus and samples are taken distally at short intervals of time. The transit time can be defined: a) as the period between injection and the first appearance of the marker, b) as the period until the maximum concentration of the marker is found, and c) as the period until the marker recovered reaches the value of 50%. These measurements of flow rate and transit time in the small intestine are invasive because catheters are necessary for the administration of the marker and for sampling of luminal contents. However, intubation of the gut is not necessary when isotopes are administered with a meal and the radioactivity is measured with a gamma-camera in the regions of interest, e.g., at the terminal ileum or caecum. With this kind of scintigraphic measurement, profiles of gastric emptying and colonic filling can be obtained and consequently the small bowel transit time can be determined.

Another non-invasive test was developed to measure the mouth-to-caecum or the duodenal-to-caecum transit time. When non-absorbable carbohydrates like lactulose are administered with a meal or injected via catheters into the duodenum, they increase the fermentation in the large intestine associated with hydrogen production and absorption. The hydrogen is exhaled in the breath and can be determined with special detectors. The rise in hydrogen concentration indicates the arrival of non-absorbable carbohydrates in the caecum and, therefore, is a measure for the mouth-to-caecum or duodenal-to-caecum transit time. This breath hydrogen test is most frequently used in clinical studies because it is non-invasive and free of hazards by radiation.

It is clear from the above discussion that these methods provide global measures of flow. Information about localised flow, e.g., flow related to individual contractions, cannot be obtained by these techniques. However, in the oesophagus, radiographic techniques provide some information about the flow properties in relation to primary peristalsis (see also the section on ultrasonography).

4.1.4 Non-invasive Methods

According to Werner Heisenberg's principle of uncertainty, it is impossible to obtain an exact measure of a phenomenon without changing the conditions one wishes to measure. The introduction of non-invasive methods, however, seems to invalidate this principle to some degree. Non-invasive studies of motility are becoming increasingly popular, especially for studies of gastric emptying in humans. The popularity stems from the fact that the methods are safe and seem not to influence the organ under study.

Both static and real-time ultrasonography can be used to visualise some areas of the gastrointestinal tract such as the stomach, pylorus and duodenum. Usually, a 3–7.5 MHz ultrasound probe is used, providing fairly good resolution and a sufficient penetration range. The principle is that short pulses of ultrasound will pass through a liquid but are partially reflected by solid interfaces, producing echoes. The depth of the reflecting interface determines the time taken for an echo to return to the ultrasound transducer. In real-time ultrasonography a continuous, moving, two-dimensional image of a slice of tissue is obtained. The technique is useful for the study of gastric emptying and antroduodenal motility and antral size. In addition, movements of luminal contents across the pylorus can be studied by duplex sonography, which is real-time scan combined with Doppler technology (Hausken et al., 1992). Some problems may, however, arise from individual anatomical variations that may prevent visualisation in a single plane. Furthermore, subcutaneous fat and bowel gas may blur the images.

Radiographic methods based on conventional X-rays and CT scans mainly provide data on transit of contents. Visualisation of gastric emptying is based on radiological visualisation of emptying of a contrast medium or capsules mixed with a meal, or evaluation with a computer-assisted gamma-camera of the decrease of radioactivity in the gastric area after administration of a radio-labelled meal. Disadvantages of most radiographic techniques are radiation hazards and the fact that the studies can only be performed with certain kinds of meals.

4.2 Balloon Distension

Balloon distension is a commonly used technique in visceral organs. Research studies take advantage of this technique for studying organ physiology; e.g., for investigating the force–deformation relationship using various mechano-sensitive receptors in the gastrointestinal wall. Balloon distension is also used for diagnostic purposes, e.g., in the diagnostics of non-cardiac chest pain (Rao et al., 1996) and for treatment of diseases such as atherosclerosis, bleeding oesophageal varices caused by liver disease and lower oesophageal sphincter occlusion caused by achalasia.

Balloon distension can be done in vivo and in vitro with concomitant measurements of pressure, volume and cross-sectional area alone or in combination. Furthermore, the balloon distension technique can be combined with various imaging techniques such as B-mode ultrasonography to obtain geometric data about the geometry of the cross-section and wall thickness. Basic considerations regarding balloon distension as a mechanical tool are provided below. Balloon distension will be treated in more detail later in this chapter.

4.2.1 Some Notes on Distension Protocols

Distensions can be performed in numerous ways. If pressure is controlled, then it is called an isobaric protocol. If volume is controlled, then it is an isovolumetric protocol. The distension test can be done as a step test, a staircase test or as a ramp test (Figure 4.1) where pressure, volume, diameter, cross-sectional area, tension and strain can be controlled, dependent on the experiment. Selection of the test must depend on the purpose of the study but it is important that certain considerations are made. Reproducibility is important, but is often not considered in studies in vivo. In order to obtain reproducible results, the tissue must be preconditioned; i.e., the test must be repeated until the various responses become repeatable (see Chapter 7). The duration of loading and the magnitude of loading are also important for physiological processes in the tissue and for whether irreversible or reversible tissue damage occur. The staircase procedure may be especially troublesome in relation to the duration of the stimulus. It is also important to realise that the step test and the staircase test are especially suited for viscoelastic studies because time-dependent properties have a major influence on the results. The ramp test, on the other hand, provides a

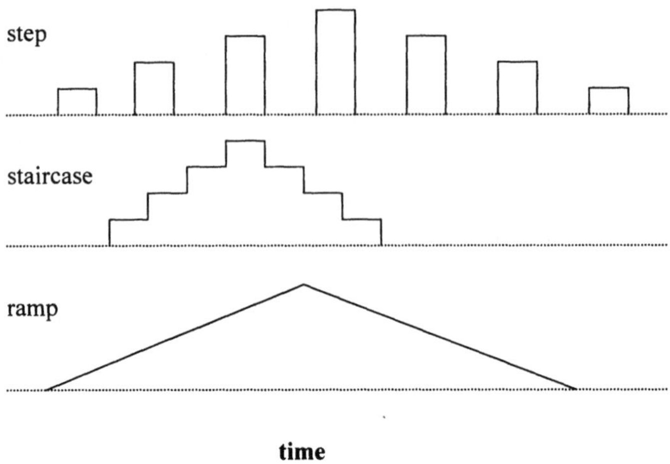

Figure 4.1. Examples of step, staircase and ramp testing schemes. The time scale is arbitrary. Normally, the ramp test is by far the fastest to run because equilibrium is not required.

continuously increasing load where time-dependent properties are not allowed to become significant.

4.2.2 Balloon Techniques. General Consideration of Geometric Factors

Since available in vivo distensibility data are based mainly on balloon distension methods, we will consider these methods in greater detail. The most important aspects relate to the experimental design, the probe design, the method of measurement, the geometry and mechanical properties of the balloons, and the assumptions for the analysis; e.g., the analysis depends on the mode of filling (continuous or stepwise).

In general, the boundary of a geometrically measurable parameter must be in complete contact with the object being measured; i.e., when measuring the cross-sectional area of a tube or the volume of any organ, the whole circumference and surface area of the balloon must be in contact with the tissue. This is never completely attained in volume measurements in the intestines and the stomach. It is important to ensure that the applied pressure is properly transmitted to the tissue; i.e., that the balloon is sufficiently large so that it does not contribute its own elastic properties to the results (here, the word balloon is in some ways misleading. Bag is a more appropriate word), otherwise, correction has to be done. Thus, the pressure drop must be confined to the wall itself (as long as the surrounding structures do not contribute, e.g., by tethering). Hysteresis of the balloon and infusion system is usually negligible but should always be tested. Friction between the tissue and balloon wall is by experience not negligible; i.e., the longitudinal strain in the wall depends on whether the segment is distended by a balloon or by filling the lumen with fluid, even at comparable loadings.

To avoid difficulties in the interpretation of results in studies of the intestines and oesophagus, the balloon must be rather short in order to provide a localised distension to imitate the size and shape of a bolus. In some studies of the intestine, very long balloons were used (see, for example, Rouillon et al., 1991, and Bell et al., 1991). There is little physiological significance in the use of long balloons and data on tone and distensibility may be difficult to interpret since reflex stimulation and inhibition of phasic activity are likely to occur simultaneously at the proximal and distal end of the balloon. Consequently, volume measurements in a long balloon may show no change even though the shape changes are considerable and vice versa. In other words, the results depend on the level of phasic contractility in the segment under study. Therefore, the definition of specific tone (see Chapter 5) does not apply for studies using the barostat (a method for combined pressure–volume measurement) since this technique provides a three-dimensional measure of tone that is difficult to differentiate from phasic contractility (tetanic tone is a summation of phasic contractions). The dynamic characteristics of the measurement system also make it difficult to compare results from different studies, e.g., short balloons encounter only a few contractions while long balloons sum up phasic contractions in a tetanic tone pattern. Consequently, due to these limitations with balloon techniques, it is essential that measurements are obtained at one location; i.e., the luminal cross-sectional area must be measured rather than volume, and in any case, the balloon should be short. This is further emphasised by the fact that the balloon will tend to elongate rather than distend the organ wall due to resistance to deformation in the radial direction. The degree of elongation is obviously dependent on the balloon material but will inevitably occur with any balloon (Figure 4.2). This has two very important implications. First, parameters such as wall tension and compliance are overestimated. Furthermore, the large variation in intestinal size between subjects gives rise to large variations in pressure–volume relations and compliance. Second, even a small error in volume in the low-pressure range, as shown in Figure 4.2, will affect the results of a tension–strain analysis since the reference length for the strain measurement is determined in this range. To overcome these invalidating problems, impedance planimetry and endosonographic methods for measurement of luminal cross-sectional area have been developed. Cross-sectional area as a function of pressure provides important information on luminal dimensions during loading but gives little information about material properties without further mechanical analysis.

Biomechanical analysis is dependent on geometrical factors. Most of the equations used in biomechanics are based on geometric assumptions. As an example, the Laplace equation is valid for a circular tube. Chapter 2 demonstrates the magnitude

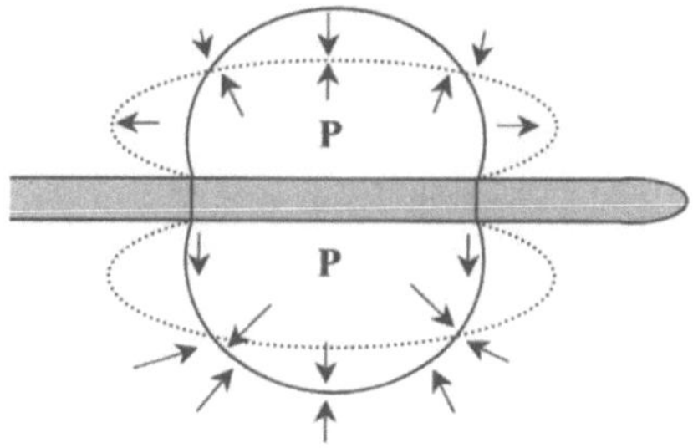

Figure 4.2. A schematic illustration of a commonly used bag type for oesophageal and intestinal studies. Two degrees of distension are shown for a fairly rigid bag. The solid line shows when the pressure is high, resulting in a large diameter and an almost spherical bag geometry. The dotted line illustrates a lower pressure where the radial stretch is less. Due to the finite dimension of the bag, the bag will appear longer and bulge out in both ends. For this reason, volume measurements are not usable in calculations of compliance, tension and strain. The arrows indicate the direction of forces for the balloon pressurised with the lower pressure (top) and for the higher pressure (bottom).

of the error induced when an elliptical geometry is considered to be circular. In more complex geometries, the expressions for possible errors are much more complex. Obviously, this has consequences for interpretation of data on tone and distensibility, yet very few studies consider the proper geometry and provide measures of whether the tubular sections of the gastrointestinal tract are circular and whether the stomach is spherical. The current knowledge, therefore, is based on qualitative data and more sophisticated analysis is greatly needed.

Due to the arguments provided above, the balloon distension techniques that are based purely on simultaneous pressure–volume measurements will not be further considered as a biomechanical tool. The measurements of pressure and volume may be accurate but any derived parameter such as tension is invalid due to the numerous assumptions made and the errors caused by elongation and contractions. The following is a description of impedance planimetry, a technique based on fewer assumptions and one which is much better validated.

4.2.3 Impedance Planimetry

Impedance planimetry is a technique for measurement of luminal cross-sectional area (CSA) in hollow organs, especially cylindrical tubes. Other names such as the field gradient principle and the four-electrode technique are still in use, especially in the field of urology. Here, the term impedance planimetry will be used as suggested by Gregersen and Andersen (1991). Professor James Christensen from Iowa City originally suggested the term. It is informative since the technique measures the cross-sectional area of a plane by means of electrical impedance measurements. Impedance planimetry utilises a simple electrical principle based on Ohm's law. However, as with any technique, sources of errors and the system's static and dynamic properties must be considered. Before introducing the principle of CSA measurement, the sources of error, system properties in further detail, and a description of the major historical landmarks in the development of impedance planimetry are given.

Table 4.1. Landmarks in the impedance planimetry story

Year	Authors	Landmark
1971	Harris et al.	Described the field gradient principle
1978	Fisher	First results on flow in oesophagus
1979	Axenborg	Development and application in vascular research
1983	Colstrup et al.	First paper introducing the balloon method
1988	Gregersen et al.	Modification of the balloon method for use in gastroenterology
1991	Gregersen and Andersen	Introduction of the term impedance planimetry
1991	Silny et al.	Modification of the non-balloon system. Application in oesophagus and small intestine
1995–96	Villadsen et al., Rao et al.	Provided clinical data on oesophageal function in patients with systemic sclerosis and non-cardiac chest pain
1999	Gregersen et al.	Development of tension- and strain-controlled system (tensiometer and tensiostat)

The major historical landmarks of the development are outlined in Table 4.1. More than two decades ago Harris et al. (1971) described the principles of a method using impedance measurements based on the field gradient principle for the assessment of CSA in the ureter. The measurement system consisted of four or more electrodes placed in a linear array on a thin probe. When a voltage was applied to the two end electrodes, it caused a current to flow when a urine bolus passed the electrodes. The sensing elements (the detection electrodes) measured the potential difference between them. Since an inverse relation exists between CSA and the potential difference, CSA could be deduced from measurements of voltage values (see later). The system allowed measurements of CSA, contraction velocity and bolus shape. In the late 1970s, two- and four-electrode techniques were combined with hot-film anemometry for the study of peristalsis and urine flow in the porcine ureter (several papers by Djurhuus and co-workers). Later, it was combined with manometry and with flow measurements by Mortensen for the study of urethra during micturition. Apparently without knowing about the papers by Harris' and Djurhuus' groups, Fisher et al. (1978) and Fass et al. (1989) applied a variant of the method for the determination of bolus velocity and clearance in the human oesophagus using 8- and 20-electrode systems, respectively.

The method was further developed in 1983 by Colstrup, who introduced a system with the electrodes placed inside a fluid-filled, non-conductive latex balloon for the measurement of biomechanical wall parameters such as compliance, defined as the change in cross-sectional area divided by the change in balloon pressure. Their system had the obvious merit that the current was confined to the fluid in the balloon. Furthermore, the system proved sufficient for studies of active and passive wall properties in the urethra of healthy females and stress-incontinent females (several papers by Lose and Colstrup).

The balloon method was introduced in 1988 for studies of the gastrointestinal tract by Gregersen. Due to the large dimension of the gastrointestinal tract, however, further development of the measuring system took place and the technique was termed impedance planimetry in 1991 since it is based on measurements of a two-dimensional variable from electrical impedance recordings. The next major technological step forward was the development of a tensiometer and tensiostat in 1999 by Gregersen, Barlow and Thompson (see later in this chapter).

Principle of Cross-sectional Area Measurement

Consider a linear array of four electrodes mounted on a catheter. When a constant alternating current I is induced in an electrical conductor by the two outer (excitation) electrodes, the potential difference V between the two inner (detection) electrodes according to Ohm's law is

$$V = I\,Z. \tag{4.1}$$

If the two excitation electrodes are sufficiently far away from the detection electrodes to create a uniform electrical field in the area around the two detection electrodes, Z, the electrical impedance of the fluid, can also be expressed as $d\ c^{-1}$ CSA^{-1}, with d being the distance between the detection electrodes and δ the conductivity of the fluid. Thus, the potential difference can be expressed as

$$V = I\,d\,\delta^{-1}\,CSA^{-1}, \tag{4.2}$$

Table 4.2. Survey of literature on intraluminal impedance measurements

Study Object	Number of Electrodes	Balloon Mounted	Derived Parameters	Literature
Theoretical approach rabbit urethra	4	No	flow velocity, bolus shape	Harris et al., 1971
pig ureter	2, 4	No	none, peristaltic rate	Rask-Andersen and Djurhuus, 1976
male urethra (normal)	4	yes/no	none, compliance distensibility	Mortensen et al., 1983; Harada, 1985; Mortensen, 1989
female urethra (normal)	4	yes	opening pressure, rigidity, stress relaxation hysteresis, compliance, work power generation, elastance	Colstrup et al., 1983 Colstrup, 1984 Lose and Colstrup, 1991; Lose et al., 1988; Thind et al., 1991; 1992; Thind, 1992; Lose and Schroeder, 1990; Colstrup, 1985
female urethra (stress incontinence)	4	Yes	elastance, hysteresis, stress relaxation power generation, relaxation, closure pressure	Lose and Colstrup, 1990; 1991; Lose, 1990; 1991; Thind and Lose, 1992
in vitro validation	4	Yes	none, balloon hysteresis	Lose et al., 1986; Gregersen et al., 1988; 1991; Colstrup et al, 1983; Jørgensen et al., 1995
human oesophagus	4, 8, 20 variable	yes/no no	none, bolus velocity tension, elastic modulus	Fisher et al., 1978; Silny, 1991; Orvar et al., 1993; Fass et al., 1989 Villadsen et al., 1995; 1997; 2001; Rao et al., 1996; Gregersen et al., 1999
rabbit, pig, guinea pig and opossum oesophagus	4	yes	compliance, hysteresis, wall tension	Gregersen et al., 1988; 1991; 1992; 1996 Juhl et al., 1994; Gregersen and Loft, 1996; Vinter-Jensen et al., 1994; Assentoft et al., 2000; Pedersen et al., 1998; 1999
human ileum	variable	no	bolus velocity	Silny, 1991
human duodenum	2	yes	tone	Gregersen et al., 1990; 1992 Gregersen, 1994
porcine ileum	4	yes	tone	Harling et al., 1990
porcine duodenum	4, 5	yes	wall tension, stress, elastic modulus, compliance creep, strain	Gregersen et al., 1991; 1992 Jørgensen et al., 1991
rat and guinea pig small intestine	4	yes	tension, strain	Storkholm et al., 1994; 1998; Duch et al., 1996; Vinter-Jensen et al., 1996; Jørgensen et al., 2001
porcine and mouse rectum	4	yes	compliance	Dall et al., 1991; Lundby et al., 1999
human rectum	4	yes	compliance, tension	Dall et al., 1993; Krogh et al., 2001
human anal canal			rigidity, hysteresis, compliance, opening pressure, closing pressure	Rasmussen et al., 1990 Gregersen et al., 1991
porcine bile duct	4	yes	tension, strain	Duch et al., 1998
Arteries	4	yes	tension, strain	Frøbert et al., 1995; 1996; 1997; 1998 Storkholm et al., 1997 Tanko et al., 1999; 2000; 2001

where I, d, and δ are constant within an experiment (see the section about possible errors). Thus, V is inversely proportional to the CSA:

$$V = k\ CSA^{-1}. \tag{4.3}$$

Direct proportionality between the output V and the CSA can be obtained by means of reciprocating software. After reciprocation, Equation 4.3 can be written as

$$V = k\ CSA, \tag{4.4}$$

where k is a calibration constant.

The electrical field created by the current from the excitation electrodes is either confined to the conducting fluid inside a balloon embracing the electrodes on the probe or it can be conducted through the fluid in the lumen of the gastrointestinal tract (and likely also through the wall) if a balloon system is not used. The balloon system is useful for distension studies. To control the pressure or volume inside the balloon, a channel connects the balloon with a fluid container that can be raised to increase the balloon pressure or to a pump (Figure 4.3). The system without the balloon is useful for studying CSA during bolus transport and flow.

The four-electrode system can be extended to include many sets of detection electrodes (multi-electrode systems) so that serial cross-sections can be obtained on the longitudinal axis (Figure 4.3). This has the advantage that spatial variations in CSA can be evaluated since a third dimension is added, thus, the spatial and temporal distribution of contractions can be described in more detail. If no detection electrodes are in common, the numbers of CSA measurement (N) equal n/2, where n denotes the number of detection electrodes. However, it is possible to use electrodes in common. In that case $N = n-1$ if CSA measurements are allowed only from adjacent electrodes and $N = (n(n-1))/2$ if the bound of adjacent electrodes is eliminated.

A two-electrode system can also be utilised since the two excitation electrodes can also serve as detection electrodes to measure the potential difference (Figure 4.3). The two-electrode system is advantageous if the balloon is straight in the area between the two electrodes, because the electrode distance can be shorter than in the four-electrode system. However, errors, in particular due to the slope of the wall between the detection electrodes, may influence the CSA measurements more easily with the two-electrode technique. Therefore, the two-electrode system should only be used with caution.

Signal Conditioning System

Each channel in the impedance planimeter consists of an excitation unit with a sine wave generator and alternating current generator, and a signal processing unit with a differential amplifier, rectifier and filter. The sine wave generator generates a sine voltage with low harmonic distortion. The signal is fed into the alternating current generator where it is converted into a constant AC current. The current is delivered to the two excitation electrodes which generates an electrical field between them. The potential difference between the two detection electrodes is amplified, filtered, analogue-to-digital converted and fed to the computer. The advantages of digitalisation and computer analysis are many, e.g., lower noise level, better filtration possibilities, extended data analysis such as fast Fourier analysis, and better display options.

Figure 4.3 illustrates a four-electrode probe utilising the balloon system and the design of the signal processing system used in our lab.

The excitation current is 100 microamperes in the clinical impedance planimeters and 250 microamperes in the impedance planimeters for animal experimental use. Increasing the current is advantageous because it increases the magnitude of the potential difference that provides a better signal-to-noise ratio. Furthermore, the "linear" range of the calibration curve increases. However, patient security must be considered and this obviously puts limits to the magnitude of the current and the AC frequency.

Impedance Planimetric Calibration Procedures

Calibrations are done in rigid tubes or in a plastic block with holes of known luminal CSAs (Figure 4.4). If an impedance planimetric system without balloon is used, it is important that the calibration tube material is electrically non-conducting. Hard plastic materials are a good choice and drilled holes consecutively increasing in size by, for example, 1 mm in diameter from a few mm to 45 mm in diameter will cover most biological applications except for the large bowel in humans. The calibration constant depends on current intensity, conductivity of the fluid, temperature, and the distance between the excitation electrodes as pointed out in Equations 4.2–4.4. However, since the calibration curve is non-linear, a single constant cannot be used. Previous studies have used either an initial two-point calibration and only done measurement within the "linear" CSA range or multi-point calibration data using up to 16 calibrations points have been obtained to linearise the data. The multi-point calibration has the advantage that it increases the resolution, accuracy and range of measurement when compared to the two-point calibration system. However, future software solutions should implement the equations derived from the modelling section (see later) to further increase the accuracy.

Errors and Optimisation Procedures

The paper by Harris from 1971 on application of the field gradient principle to ureteral physiology still stands as one of the most important methodological papers. Most of the other papers in the 1970s and 1980s merely show experimental validation of the errors pointed out by Harris. First of all, it was demonstrated that Equation 4.2 is valid for a straight tube as long as the slope of the wall between the detection electrodes is small. Further, an approximation of the potential distribution in a conical tube is provided, which leads to an estimation of the error associated with the slope of the wall between the detection electrodes and for the error due to asymmetric probe positioning; i.e., radial displacement of the probe.

Clearly, it is advantageous to keep the distance between the detection electrodes small and to keep the probe in a central position. Several subsequent studies by other authors have added experimental data to the theoretical error analysis by Harris and co-workers. Since the paper by Harris uses the technique without implementation of balloons, the electrophysiological effect of inducing current in ureteral tissue, shunting of current in the tissue, and flow measurement are also considered.

A number of considerations concerning the accuracy of the probe measurement that affect probe design was provided by Harris and co-workers in 1971.

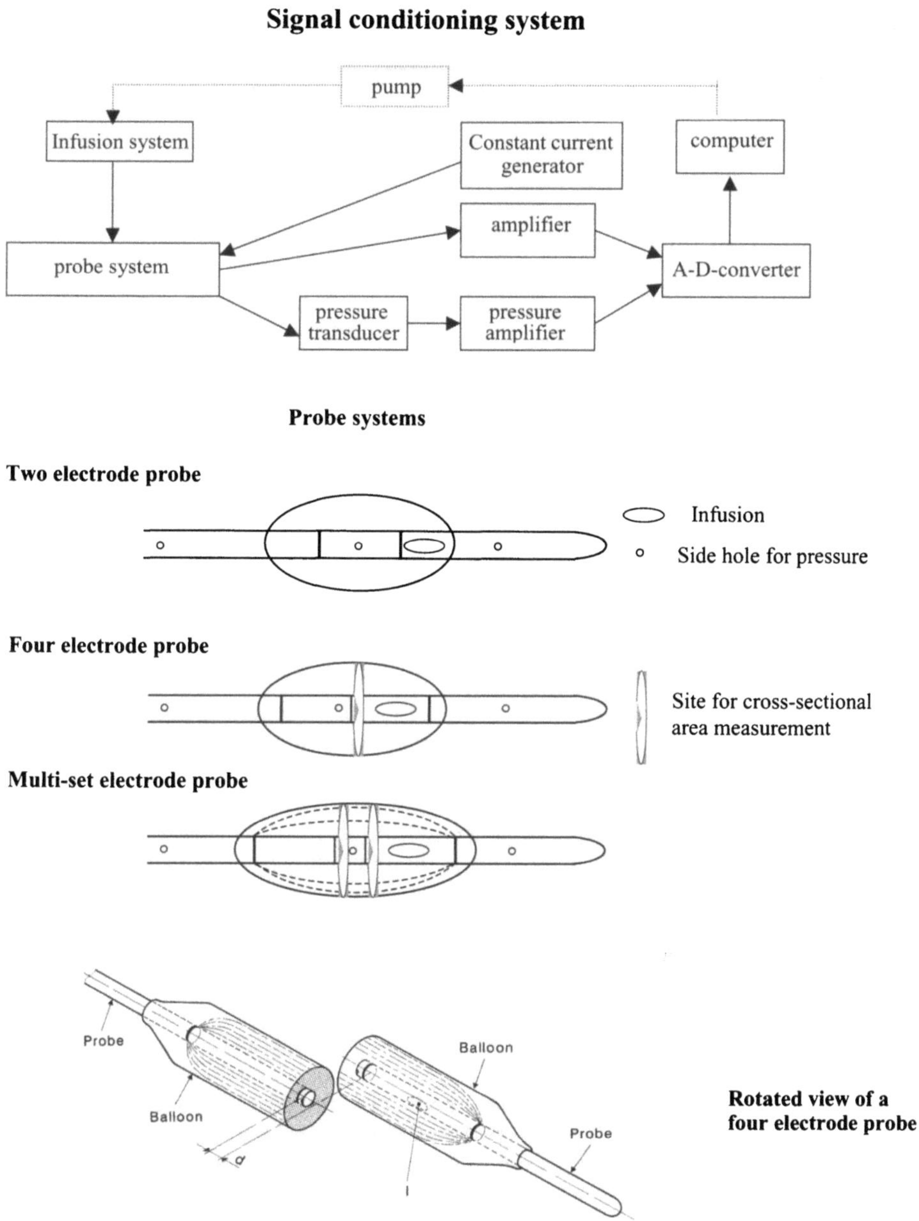

Figure 4.3. Schematics of the impedance planimetry and tensiostat system. Three different probe designs are illustrated (the four-electrode system is shown from two angles). The system renders possible pressure–cross-sectional area measurements, evaluation of peristaltic reflex responses and standardisation of the stimulation stimulus in terms of tension or strain. The group of dashed lines (shown only inside the two bottom balloons), forming a field between the two outer excitation electrodes, represents the flow of constant AC current. In the tensiometer system a computer-controlled pump is added to the system as shown by the dotted lines. The computer calculates tension in real time from measurements of balloon pressure and CSA and remote controls the pump to infuse/withdraw fluid from the balloon in order to change tension. I and d in the bottom figure denote the infusion channel (shown by other symbols in the upper figures) and the distance between the detection electrodes.

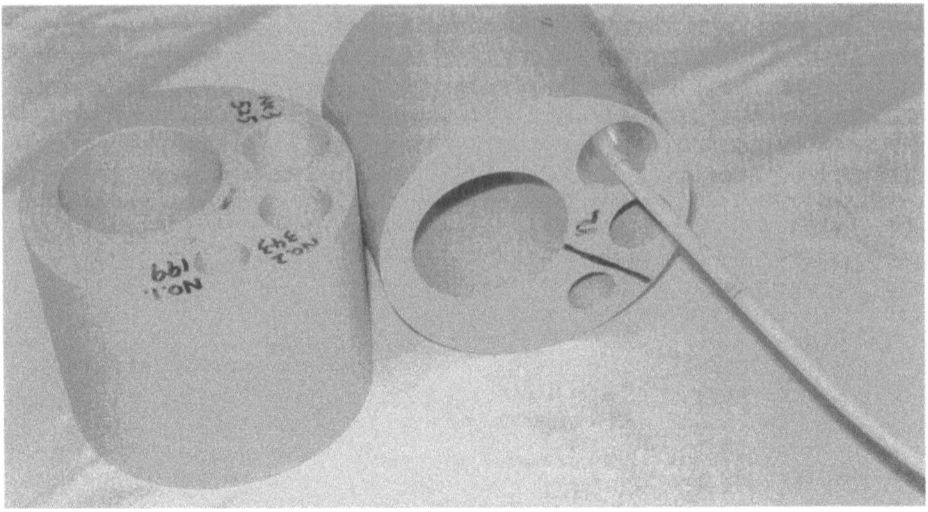

Figure 4.4. Two calibration blocks made from a hard plastic material. The exact size of each calibration hole is known. The balloon on a 14F probe is distended inside one hole.

1. The distance between the excitation and detection electrodes must be finite and large enough to provide a uniform current flow in the region of the detection electrodes. The distance should be approximately equal to the radius of the bolus (or balloon) at the excitation electrode.
2. The distance between the excitation and detection electrodes can be decreased somewhat by using relatively large excitation electrodes. Low-impedance electrodes of large area serve to distribute the current more uniformly than point electrodes do.
3. The distance between the detection electrodes must be small (<< the expected radius) so that the variation in cross-section from one electrode to the next is small compared to the cross-section itself.
4. The detection electrodes must be as small as possible. Further, they must be high-impedance elements to provide a minimum scattering effect.

Other papers also deal with possible sources of error related to impedance planimetric measurements (several papers by Djurhuus, Mortensen, Lose and Gregersen). The errors can be divided into those errors directly associated with the CSA measurement and those errors related to the derived biomechanical parameters (see Chapters 2 and 3). The following paragraphs consider the sources of error on CSA and pressure measurements, and how to optimise the measurements.

Potential Difference (Conductivity, Distance Between Detection Electrodes, and Current)

According to Equations 4.2–4.4, the distance between the detection electrodes, the conductivity of the fluid, and the current must be constant during an experiment. When using a reliable constant current generator, a probe with fixed electrodes and the balloon system, and where the current is confined to the fluid in the balloon, this assumption is valid. Any error related to these factors will be linear if they were variables rather than constants. It can be seen from Equation 4.2 that the potential difference increases when: the current increases, the distance between the detection electrodes increases, and when the conductivity decrease. If the magnitude of the potential difference between the detection electrodes increases by spacing the electrodes further apart or by increasing the current, the signal-to-noise ratio will increase. However, a large distance between the detection electrodes may induce errors due to the slope of the wall and there are bounds on the current used due to patient security matters. Therefore, the best way to increase the potential difference is by using a fluid with low electrical conductance. Most studies done in small species use physiological saline solution (0.9% NaCl) as the fluid in the balloon. However, when a larger CSA range is required, it is common to dilute the saline by 10–1000 times with distilled water. Equations 4.2–4.4 emphasise the difficulties when a system without balloon is used since the conductivity may not be known, and most likely is not constant. Shunting of current through the gastrointestinal wall may be significant and has to be corrected for, if possible. Thus, it is difficult to provide valid CSA measurement from the small intestine whereas the problem may be more easily overcome for the oesophagus where a sample of the swallowed fluid can be sampled and its conductivity measured.

An error may be introduced due to the temperature-dependence of conductivity. It is, therefore, important to maintain the temperature constant during both calibration and experiment. However, from physics we know that this dependency is linear and consequently a correction factor can be introduced if the temperature is measured.

Electrode Distances and Calibration Curves

The measurements must be within the range of the calibration curve that approaches an asymptotic constant at high CSAs. The calibration curve is highly dependent on the distance between the excitation and detection electrodes; measurements can therefore be optimised by selecting a proper distance between the electrodes. With the current system, the rule of thumb is that the highest measurable CSA corresponds to a diameter of 4–5 times the distance between the detection and excitation electrodes. However, this is only a guideline since the linearity of the calibration curve likely also depends on electrode materials, electrode size, current, etc. The range of the calibration curve also depends on the impedance of the fluid and on the signal-conditioning system (gain setting, etc). The model presented later in this chapter is prepared for considering these factors.

Dislocation of the Probe from the Centre Axis and Balloon Properties

The error due to dislocation of the probe from the centre of the longitudinal axis to an eccentric position in the balloon can be calculated (Harris et al., 1971). Experimental validation studies in vitro, where the probe was forced to an eccentric positions, have shown that the error can be large. The error depends on electrode distances, current, and balloon size. The errors found in systems used in the gastrointestinal tract are remarkably larger than with systems used in urological studies. The dislocation error can easily be avoided by using a large balloon that unfolds without stretching during distension of the organ under study. Further, the balloon must be mounted loosely in the axial direction. A properly mounted and sized balloon will straighten the probe, otherwise, the probe will bend and dislocate to an eccentric position with subsequent errors in CSA measurements. This can easily be evaluated by vision during inflation of the balloon. Ultrasonography and X-ray can be used to validate the position of the probe in studies in vivo.

The balloon material is also important. A thin, non-stretchable and non-conducting material with low bending rigidity such as 30–50 micrometer-thick polyurethane is optimal. Stretchable materials such as latex deform primarily in the longitudinal direction during distension due to the resistance to stretch in the radial direction. The balloon elongation does not directly affect the CSA measurements (but it severely affects any volume measurement). Consequently, the size of the distension area is increased and quantification of evoked responses such as secondary peristalsis, afferent nerve activity and pain is more difficult.

Slope of the Wall Between the Detection Electrodes and Configuration of the Balloon Outside the Detection Area

Harris devised how to compute the error due to the slope of the wall between the detection electrodes. In theory, this error can be rather large. Experimental validation of the error demands visualisation by means of X-rays or ultrasonography. Most gastrointestinal organs studied have been fairly straight; i.e., this error can be neglected. However, in sphincteric regions, one of the detection electrodes may be placed in the sphincter region while the other is outside the sphincter during distension. Furthermore, propagating contractions will inevitably reach one detection electrode before the other. The detection electrodes must therefore be placed as close as possible to each other.

The shape of the balloon in the region outside the detection electrodes is also important (see the model later in this chapter).

Irregular (Non-circular) Shape of the Cross-section

Many biological organs have irregular cross-sections, e.g., the small intestine is collapsed and the oesophageal wall is buckled at rest and at low luminal pressures. Though the lumens of these organs become circular at higher pressures, we must consider the error due to irregularity. Many validation experiments have been done where rods have been inserted to make the lumen of a calibration tube irregular. It can be concluded from these studies that this error is insignificant.

Hysteresis and Dynamic Properties of the Measurement System

The dynamic properties of a measurement system must be characterised. If the upper frequency limit is too low, it will result in amplitude reduction and phase shift. Considering impedance planimetry, the properties of the CSA measurement system, the pressure measurement system, and the infusion/balloon system must be evaluated. Since the CSA measurements are based on electrical measurements rather than on a mechanical system, the dynamic characteristics of the CSA measuring system is of no concern. The upper frequency limit is by far higher than the frequency of contraction observed in gastrointestinal motility recordings. Pressure measurements, in general, do not impose problems for measurements in the gastrointestinal tract, however, it is important with the perfused manometry system to check for air bubbles and kinking of the probe. The infusion system gives rise to the biggest concerns. Only if a pump controls the infusion or the volume is changed in other controlled ways, or if recordings of passive tissue properties (i.e., without muscle contractions) are done, is the infusion time not important. However, if a level container controls pressure, fluid is allowed to move in and out of the balloon during the distension. When phasic contractions are present, fluid will flow between the level container and the balloon. This fluid flow depends on a number of factors basically determined by Poiseuille's law. To reduce the resistance to flow, the infusion channel should be as short as possible with the largest possible luminal diameter. Reduction can also be obtained by using a fluid with a lower viscosity than physiological saline. Furthermore, the smaller the balloon the better; i.e., the length of the balloon should be as short as possible to reduce the fluid volume in the balloon.

Hysteresis must also be considered. This is a measure of the difference in pressure–CSA curves between inflation and deflation of the balloon. Several validation studies have shown that hysteresis can be neglected with the probes currently in use.

Procedures Related to Studying the Human Oesophagus with Impedance Planimetry

This section describes the most common procedures for oesophageal measurements. The procedures are, with slight modifications, also useful for studying other organs. The descriptions consider investigations of oesophagus in healthy volunteers but clearly there may be specific problems related to various diseases.

Oesophageal studies are usually performed after an overnight fast. The 0.5 m-long 14F probe contains side holes for pressure measurements distal to, proximal to and inside the balloon. After proper calibration of the impedance planimeter at body temperature and calibration of the pressure sensors, the probe is passed down to the oesophagus via either the oral cavity or via the nostrils with the patient in an upright position. Before passage through the oral cavity, an anaesthetic drug can be used for local oropharyngeal anaesthesia. The patient is for the rest of the investigation in the supine position with the head slightly elevated. The distal side hole is used to locate the lower oesophageal sphincter (or the pressure inversion point in case of non-measurable sphincter pressure). At this point, ten wet swallows and ten dry swallows are done to evaluate oesophageal peristalsis. Then the probe is retracted to place the middle of the balloon either 7 or 15 cm above the sphincter. The axial location is important since the distribution of smooth and striated muscle varies in the axial direction and the phrenic ampulla is located distally in the oesophagus. Several ways exist for determination of the oesophageal resting pressure that serves as the reference for the distension series. One way is to carefully measure the end-expiratory pressure at the balloon distension site. Since that pressure will serve as the zero pressure, the level of the fluid container has to be changed accordingly. However, it has to be emphasised that the oesophagus is difficult to study due to respiration and the pronounced secondary peristalsis during balloon distensions. The distension series can be performed as a step test, where the pressure is changed by equilibrating the pressure in the balloon by changing the height of the fluid container above the zero level or by infusing a certain volume for some time. Usually, the pressure is changed stepwise to 0.5, 1, 1.5, 2, 3, 4, and 5 kPa (1 kPa = 10 cmH_2O). Higher pressures may cause pain in volunteers and several patient groups experience pain at very low pressures. A different way of distension is the ramp test, where the pressure or volume rate is controlled, e.g., 2.5 kPa per minute or 25 ml per minute may be infused until pain appears. Perception data are important in these studies. The study subjects are asked not to swallow during the distension period. After steady state is reached at each step, the pressure is lowered to below zero for 1–2 minutes in order to completely empty the balloon. Steady state may be hard to define due to the numerous contractions, especially in patients with hyperperistalsis. In that case, the CSA value to be used for further analysis may be obtained after 2 minutes distension or found as the average last 30 seconds of distension. The uncertainty of defining steady state may make comparison between studies difficult. One way to reduce peristalsis is to administer 20 mg butylscopolamine intravenously. This can be repeated when peristalsis returns or the drug may be infused on a continuous basis. This facilitates studies of passive tissue properties though complete muscle relaxation is impossible to obtain in humans. On the other hand, in many physiological and clinical studies, the issue of interest is the evoked secondary peristalsis. The probe with the emptied balloon is slowly withdrawn after the biomechanical test is completed.

4.2.4 Techniques Based on Impedance Planimetry

The tensiometer–tensiostat and combined impedance planimetry–ultrasound techniques are treated in more detail in this section. The multi-modal stimulation technique, where the mechanical stimulus is based on impedance planimetry, is described in Chapter 5.

Tensiometer–Tensiostat

Impedance planimetry has been further developed to a tensiometer and tensiostat for measurement and control of circumferential wall tension and strain in the gastrointestinal tract (Gregersen et al., 1999). The innovation was facilitated by the need for better control of the distension stimulus since the mechanoreceptors in the intrinsic and extrinsic pathways respond to changes in stress and strain rather than to changes in pressure or volume. The main assumptions on which this technique is based are that 1) pressure and CSA are reliably measured, 2) the use of Laplace's law is valid, and 3) the dynamics of the measurement system are adjusted to the organ under study.

Mechanical Set-up and Signal Processing

A diagram of the mechanical set-up is included in Figure 4.3. The system consists of a probe with four ring electrodes and a 4-cm-long and 5-cm-diameter balloon, an impedance planimeter, pressure transducer, pressure amplifiers, and a pump with interface to a Pentium PC with dedicated software. The peristaltic pump has a maximum flow rate of 1.3 L min^{-1} with a silicone tube with ID = 10 mm. The pump is remotely controlled by a PC through the RS232 port (baud rate 9600 (approximately 100 commands per second), ASCII code communication) and infuses and withdraws fluid from a reservoir into the balloon mounted on a multi-lumen 14F PVC-probe. The pressure is measured inside the balloon using a low-compliance perfused pneumohydraulic system connected to the external transducer. The resolution of the CSA and pressure measurements was 0.1 mm^2 and 0.2 mmHg in the first system developed. The dynamic characteristics of the pressure system were 0.15 seconds rise time, damping ratio of 0.75, and 2.4 Hz undamped resonance frequency. The delay in sending commands from the computer to the pump response was in the range of 5–30 ms.

Balloon pressure and CSA are measured by means of manometry and impedance planimetry and the signals are amplified, analogue-to-digital converted and transferred to the computer for real-time display and tension computation. The pump is not used in open-loop feedback mode (tension measurement mode, pressure or volume control mode). Hence, the system records variations in tension when the balloon is inflated to a given pressure, volume or CSA. In closed-loop feedback mode (tension control mode), the PC regulates the pump to infuse and withdraw fluid to obtain the selected tension. Calibrating the pressure transducer and impedance planimeter separately calibrates the tensiometer. The former is done by applying pressures of 0 and 50 mmHg, the latter by distending the balloon in a range of PVC calibration tubes of known CSA. The infusion rate of the pump is regulated by a software subroutine. The volume rate regulation is implemented in the software to ensure a fast response of the system as well as minimisation of overshoot. The setting is variable with up to fourteen regulator steps. The first step covers the 2% tension range closest to the preset value where the pump flow rate was zero. The next step covers the range from ±2–5% where the pump only was allowed to work at a low volume rate. The next step covers the range deviating 5–10% from the preset tension with a slightly higher maximum volume rate, and so on with the following steps. When the computed tension is far from the preset tension, fluid is infused and withdrawn at a high volume rate. When tension reaches values closer to the preset tension, the volume rate is lowered.

Safety Features for Human Studies

Safety features are implemented both in the software and hardware. The software has variable settings for maximum pressure, CSA and tension. If one of these thresholds is exceeded, the PC commands the pump to stop infusion instantly. Failure of communication between the PC and pump can potentially result in a continuous infusion of fluid into the balloon. This is avoided by using a syringe as a reservoir and by keeping the fluid volume in the reservoir low (less than 60 ml in oesophagus studies). The low-friction piston in the syringe ensures that infusion stops when the reservoir empties. An off-device that allows the patient to press a button to stop infusion by sending a command to the software is also implemented.

Validation Data

The system has been validated in vitro by using various-sized PVC tubes and pressure-test chambers. Furthermore, instant changes in pressure and CSA were imposed to test how the system reacted in tension measurement mode and tension control mode. The tension error is a function of pressure and radius. Error propagation analysis showed in the pressure and radius ranges from 10–100 mmHg and 5–20 mm that the tension error was 1.1–5.6%. The tension resolution was approximately 0.1 mm mmHg. Physiological data are given in Gregersen et al. (1999).

Combined Impedance Planimetry–High-frequency Ultrasonography

Ultrasound in the frequency domain up to 30 MHz is a method that can be combined with impedance planimetry since it gives useful data on the geometry of the cross-section and on the wall thickness. Hence, it is, under certain assumptions, possible to obtain stress estimates rather than tension estimates in vivo. The ultrasound transducer can be placed either on the probe inside the balloon or on the balloon. Furthermore, transabdominal scans can be made. All techniques have been in use to date, though not always with biomechanical purposes in mind.

4.2.5 Ultrasonographic Studies in Vivo

The combination of impedance planimetry and ultrasound mentioned above has been followed up by studies based entirely on ultrasonography. Two basic techniques

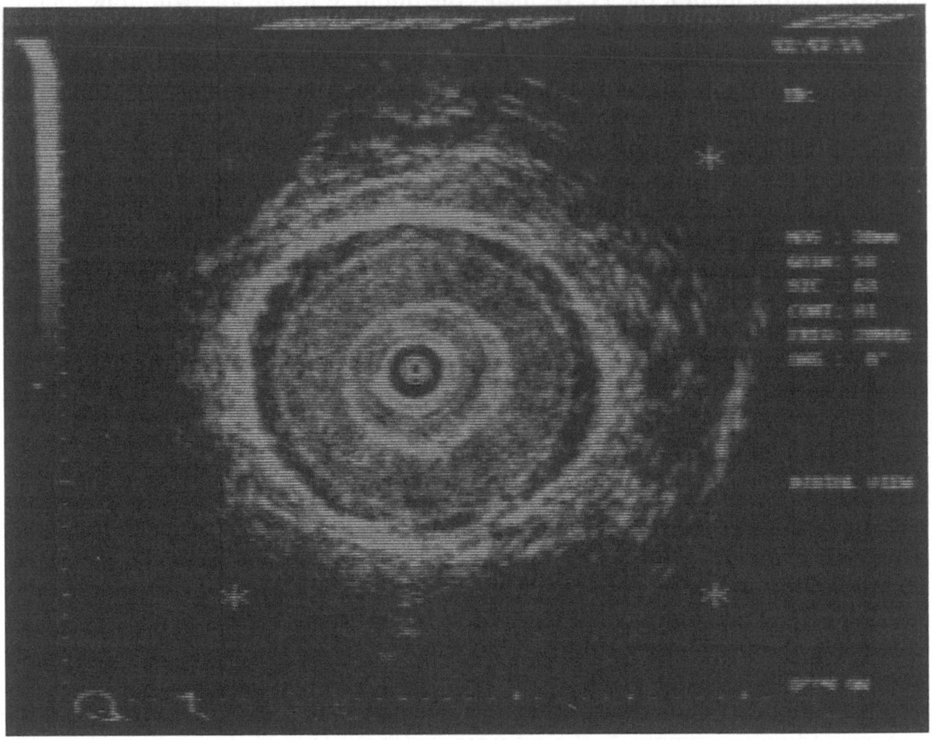

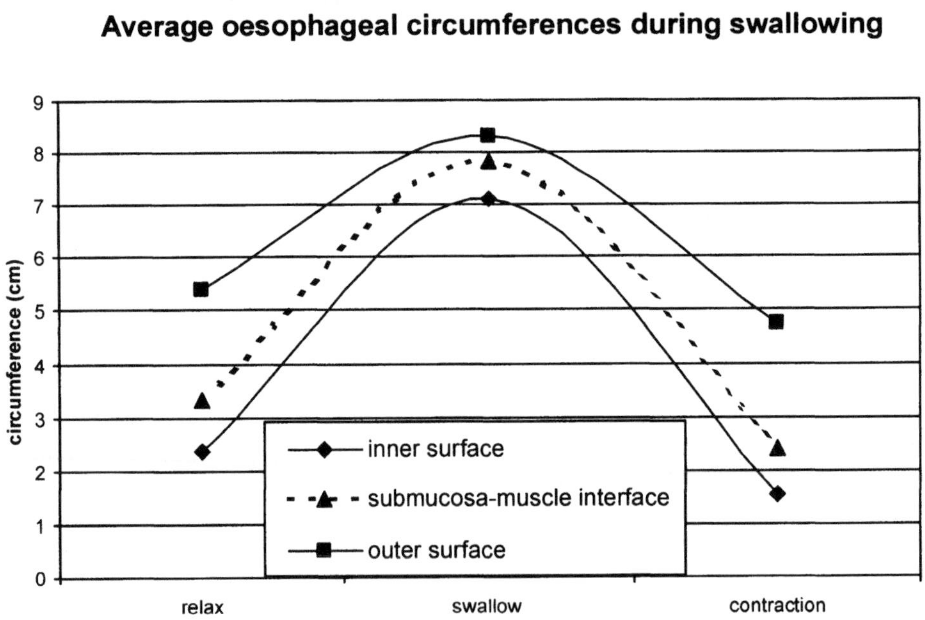

Figure 4.5. Top. Ultrasonographic image of oesophageal wall during a contraction (primary peristalsis). The image is obtained using a 15 MHz endoscopic ultrasound probe in a young volunteer at Haukeland Hospital in Bergen, Norway. Several layers and their interfaces are visible. Bottom. Measurements are obtained at the surfaces and interfaces. The data indicates that the largest strains during swallowing occur in the inner layers.

are in use, namely endoscopic ultrasound and transabdominal B-mode ultrasonography. Ultrasound has also been used for years in studies of tissue structure and diagnostics. Furthermore, it has been used for functional studies, especially in the anorectum but only rarely for biomechanical purposes.

An example of the endoscopic approach is shown in Figure 4.5 (top) where the geometry of the layers of the oesophageal wall are clearly seen during a contraction. Figure 4.5 (bottom) shows averaged circumferences from four swallowing sequences in a human volunteer. The data are obtained when the oesophagus is at rest before the swallow, during bolus passage (when the largest cross-sectional area was recorded) and during the contraction (at the point where pressure was at its maximum).

In some regions of the gastrointestinal tract, such as in the antrum, it is possible to obtain images using transabdominal ultrasonography (Figure 4.6). If combined with balloon distension where the pressure is measured and can be controlled, it is possible to compute stress–strain relations in vivo. Data based on multislice-CT scanning and MR-scanning will likely also appear soon in the literature.

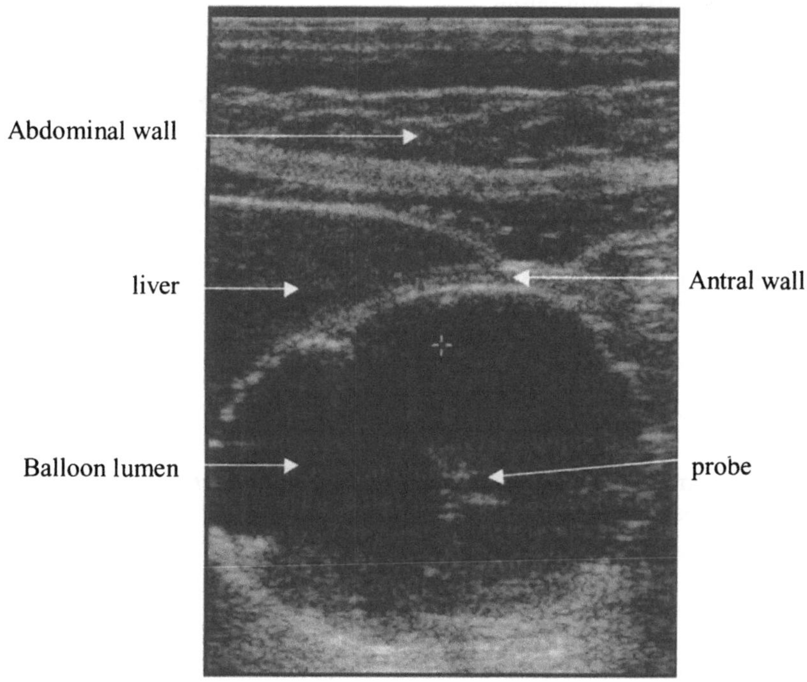

Figure 4.6. B-mode ultrasonographic image of the antrum (cross-section) during balloon distension. The image is obtained with an 8 MHz scanner. The probe the balloon is mounted on is visible in the centre of the balloon. Several layers can be identified in the antrum wall. This is especially evident in the part of the antrum closest to the transducer.

4.2.6 Non-linear Statistical Method for Analysis of Tension–Strain Data Obtained by Impedance Planimetry or Other Methods

Data sets where both the X- and Y-variable show variation are difficult to analyse statistically. It is clear that simple tests such as t-test and ANOVA cannot be used directly. Tension–strain data obtained by impedance planimetry is an example of such data because both tension and strain are computed. Traditionally, curve fitting has been done followed by determination of specific values that can be compared by traditional statistical tests or by using a simple linear regression model. This, however, does not reflect the way the measurements are obtained. A solution based on a non-linear statistical model is given below (Andersen et al., 2000).

Consider a series of pairs of measurements (p_i, r_i), $i = 1,\ldots,n$ of change in pressure (p) and radius (r) obtained from the same experiment. The pressure measurement error is assumed negligible and the radii are considered independent observations from normal distributions with common variance σ^2 and means ξ_i satisfying the relation

$$\ln\left(p_i \xi_i\right) = \alpha + \beta \cdot \xi_i, i=1,\ldots,n \tag{4.5}$$

The parameters (α, β) with the pressure determine the mean ξ_i ; i.e., we have an implicitly defined non-linear model. To investigate the implications of the model assumptions consider the function

$$g(\xi) = \ln(p\xi) - \alpha - \beta \cdot \xi \tag{4.6}$$

radius and also the tension increase with pressure, hence, $\beta>0$. G attains its maximum for $\xi = \beta^{-1}$ and tends to $-\infty$ for ξ going to 0 or $+\infty$. The relation has no solutions if $p<\beta\exp(\alpha+1)$. There is one solution if $p=\beta\exp(\alpha+1)$ and two solutions if $p>\beta\exp(\alpha+1)$. In the latter case, the solution satisfying $\xi = \beta^{-1}$ is the relevant solution. To find the maximum likelihood estimates the log-likelihood function:

$$l\left(\alpha, \beta, \{\xi_i\}_{i=1}^{n}, \sigma^2\right) = -\frac{n}{2}\ln(2\pi\sigma^2) - \frac{1}{2\sigma^2}\sum_{i=1}^{n}\left(r_i - \xi_i\right)^2 \tag{4.7}$$

must be maximised subject to the constraints

$$g(\alpha, \beta; \xi_i) = \ln(p_i \xi_i) - \alpha - \beta\xi_i = 0, \; i = 1,\ldots,n \; . \tag{4.8}$$

This can be accomplished by applying the Lagrange multiplier method. The common variance can be estimated by

$$\hat{\sigma}^2 = \sum\left(r_i - \hat{\xi}_i\right)^2 / (n-2), \tag{4.9}$$

where ξ_i is obtained from Equation 4.5 with (α, β) replaced by the maximum likelihood estimates $(\hat{\alpha}, \hat{\beta})$. Under mild regulatory conditions, ensuring that the information tends to infinity, the maximum likelihood estimates (α, β) are asymptotically distributed as a two-dimensional normal distribution:

$$\sqrt{n}\left\{\begin{pmatrix}\hat{\alpha}\\ \hat{\beta}\end{pmatrix} - \begin{pmatrix}\alpha\\ \beta\end{pmatrix}\right\} I\begin{pmatrix}\hat{\alpha}\,\hat{\beta}\end{pmatrix}^{0.5} \rightarrow N_2\left(\begin{pmatrix}0\\ 0\end{pmatrix}, \begin{pmatrix}1 & 0\\ 0 & 1\end{pmatrix}\right), \tag{4.10}$$

where

$$I(\alpha,\beta)=\frac{1}{n}E\{-D^2_{\alpha\beta}l\}=\frac{1}{n\sigma^2}\sum_{i=1}^{n}\frac{1}{(\xi_i^2-\beta)^2}\begin{pmatrix}1 & \xi_i\\ \xi_i & \xi_i^2\end{pmatrix} \tag{4.11}$$

The estimates of α and β are asymptotically independent of the estimate of σ^2. The validity of these estimates and comparison with other models are covered by Andersen et al. (2000).

4.2.7 Finite Difference Model of the Impedance Planimetry Technique

Impedance planimetry is a relatively inexpensive and simple technique for cross-sectional area (CSA) measurements. As explained before, this technique is based on voltage-drop measurement in a saline solution volume conductor, with geometry dependent on the unknown CSA, crossed by a known constant current. The measured voltage depends on the volume conductor impedance, implicit the CSA, for a constant current. In the idealised case, Ohm's law should give this dependency, but in practice this does not hold because the current is injected through a surface of the volume conductor while the voltage drop is measured, theoretically, in two points. A mathematical description of the injected current–voltage drop should take into account the potential distribution along the current's entry and exit surfaces. The modelling presented in this section was done by Romulus Lontis.

It is difficult to know in practice the potential distribution in the volume conductor within the balloon. For a better design, a model must be created of the balloon catheter that allows the computations of the potential. This model should give insight in:

- CSA–voltage dependency (the model transfer function), with a constant current injection,
- voltage–angle dependency, for a constant CSA, where the angle is the slope made by the tube wall,
- voltage–measurement electrodes distance dependency, for a constant CSA,
- voltage–excitation electrodes distance dependency, for a constant CSA

Initial Assumptions on the Model

The model is represented in half cross-section and a cylinder tube gives the CSA. The basic shape of the balloon is an ellipse arc of constant length, with a linear constrain given by the cylinder wall (Figure 4.7). The *initial shape* is half of an ellipse, the arc EGD, with the axis b_{max}, a_{min}, centred on O_{max}. This corresponds (theoretically) to the maximal cross-sectional area measurement set-up ($b_{max} = R_{max}$, a_{min} is the catheter half length), respectively to the unconstrained, full extension of the balloon.

The *default shape* corresponds to a cylinder radius of 13 mm (arbitrarily taken, see default values below). The wall constrain is the segment AIB on the ellipse arc EHD. The axis b_d, a_d are computed. Thus, the ellipse arc EHD, centred on O_d, has the same length as the initial ellipse arc EGD.

- The EHD arc length is computed as the sum between the FHC arc length and FE, CD segments length, which represent an approximation of the arc FE and CD respectively.
- It is imposed that a percent of the sum b_d + *vertical* O_d *centre coordinate* (which is the length of the segment HO_{max}) must be equal to the cylinder radius R_d

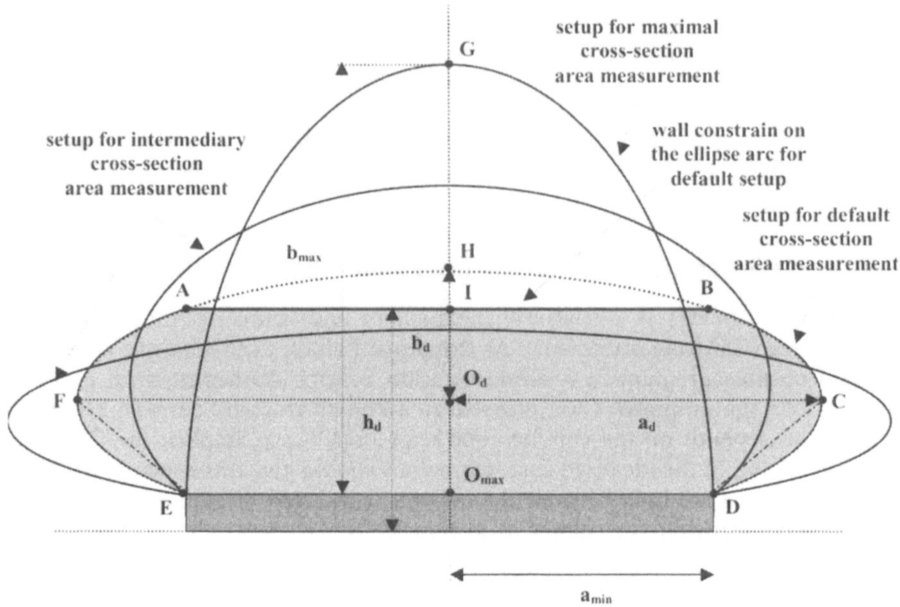

Figure 4.7. The balloon shape for different values of the cylinder radius. The ellipse arc between ED points is of constant length (an approximation, see text). The wall constrains this shape by the segment AIB (default setup), whose high, h_d, is a percent (default value, 90%) of the segment HO_{max}.

(marked as h_d). This relation establishes the position of the wall constriction, the segment AIB, relative to the ellipse arc EHD.

- a_d is iterative increased, b_d is computed in order to fulfil the previous constriction, the EHD arc length is computed according to the a_d, b_d new values. The iteration is done when the EHD arc length is at least equal to the EGD initial arc length.

The *intermediary shape* corresponds to the cylinder radius range 7–25 mm. It is computed in the same way as the default shape.

The segment AB represents the wall constriction on the ellipse. The angle effect can be simulated by rotating the AB segment around the I point, with the α angle. In 3D, this is analogue to cylinder transformation into a cone.

The following should be noted.

- The length of the balloon, according to the model, depends on the values of the H_max (maximum cylinder radius = b_{max}) and the catheter length (= $2*a_{min}$). The length of the balloon should be taken into account in measuring the cross-sectional area.
- The model presented above has constant length in cross-section. This can be seen in different phases corresponding to the value of the measured cylinder radius.
- The balloon surface folding effect as the radius of the measured cylinder decreases can be simulated by proper weighting of the balloon cross-section length as the cylinder radius under measure decreases. A possible weighting function can be linear, convex or concave hyperbolic.

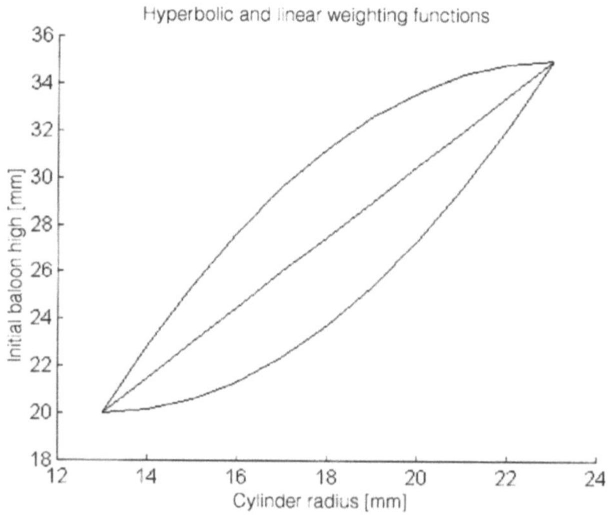

Figure 4.8. Hyperbolic and linear weighting function for default model.

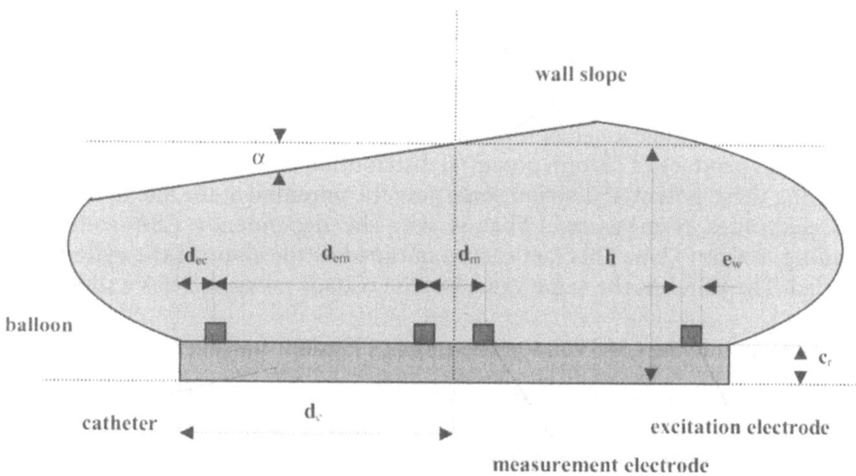

Figure 4.9. The balloon modelled as an ellipse arc with the wall constriction as linear segments of different slope. Model geometric parameters.

Model Parameters

Simulation analysis is performed in order to understand better the mechanism underlying the voltage measurement. This should lead to the balloon catheter design optimisation. Model parameters to be optimised are the geometric and material conductivity properties. Default values refer to the size of the balloon used. The minimum and maximum in Table 4.3 gives the range within which can be found an optimal value for the parameter.

Table 4.3.

	Symbol	Unit	Min	Default	Max
Electrode width	e_w	[mm]	0.1	0.8	1
Measurement electrode distance	d_m	[mm]	1	5	10
Excitation electrode distance	d_e	[mm]	30	30	60
Cylinder cross-sectional area	CSA	[mm^2]	78	530	1965
Cylinder radius	h	[mm]	5	13	25
Excitation electrode–catheter end distance	d_{ec}	[mm]		4	
Catheter radius	c_r	[mm]		3	
Catheter length	$2*d_c$	[mm]	38	38	68
Wall slope	α	[degrees]	−15	0	15
Catheter conductivity	σ_c	[S/m]		0.0001	
Electrode conductivity	σ_e	[S/m]		15	
Balloon conductivity	σ_b	[S/m]		0.0001	
Saline solution conductivity (0.9% NaCl at T=25 °C, dilution=100)	σ_s	[S/m]		1.5/dilution	

Experimental Setup

A constant current of 100 μA, typically in the range of 20–200 μA, is injected through the excitation electrodes. The voltage drop along the detection electrodes is measured. The volume conductor defined by ABCD (see Figure 4.10) has the impedance dependent on the CSA, if the distance between the detection electrodes is constant as the CSA varies. As mentioned before, the current has a distribution on the BD and AC surfaces and produces a certain potential distribution along these surfaces. Instead of measuring these potential distributions, only the potential at the site of the measurement electrodes is measured. That is why the dependency CSA-voltage is not according to Ohm's law. This fact can be outlined as the slope of the cylinder wall is modified. The CSA has the same value but the voltage measured has a slight change.

Characterisation of the CSA–Voltage Dependency (System Transfer Function)

This dependency has a similar shape as a decaying exponential. Below is an example of a set of transfer functions for different values of the wall slope (Figure 4.11). The left side represents experimental measurements. The angle values are: 0, 1, 3, 5, 7, 9, 11 degrees. The right side represents simulations, with the angles 0, 1, 2, 3, 4, 5, 6, 7, 8, 9, 10, 11 degrees. The angle effect is obvious, but the errors introduced are relatively small.

Potential Distribution

Figure 4.12 presents the potential distributions for the default model (A, C for angle = 0° and B, D for angle=14°). In the contour plots C, D, given by 60 equally spaced isopotential lines, can be seen how the potential is distributed around the excitation electrodes, internal balloon surface, detection electrodes and, most importantly, along the verticals above the detection electrodes. The isopotential lines are normal to the internal balloon surface, meaning that the current is tangential to this surface and it is preserved within the balloon volume. The verticals above the detection electrodes represent the entry and exit surfaces for the current in and from the volume

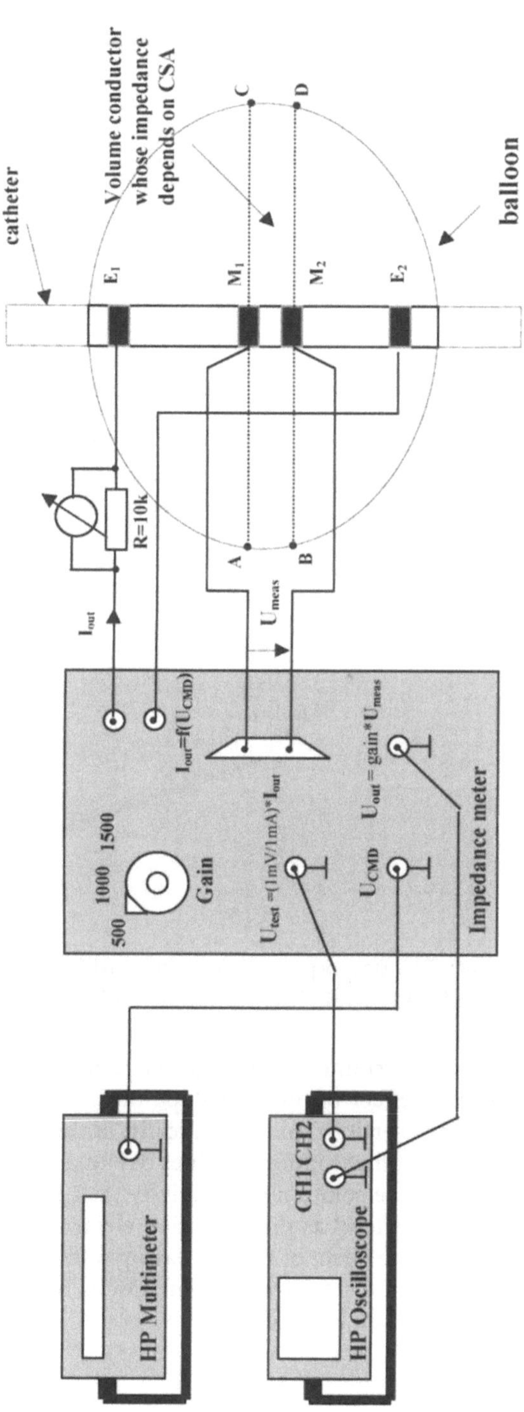

Figure 4.10. Experimental set-up.

a

b

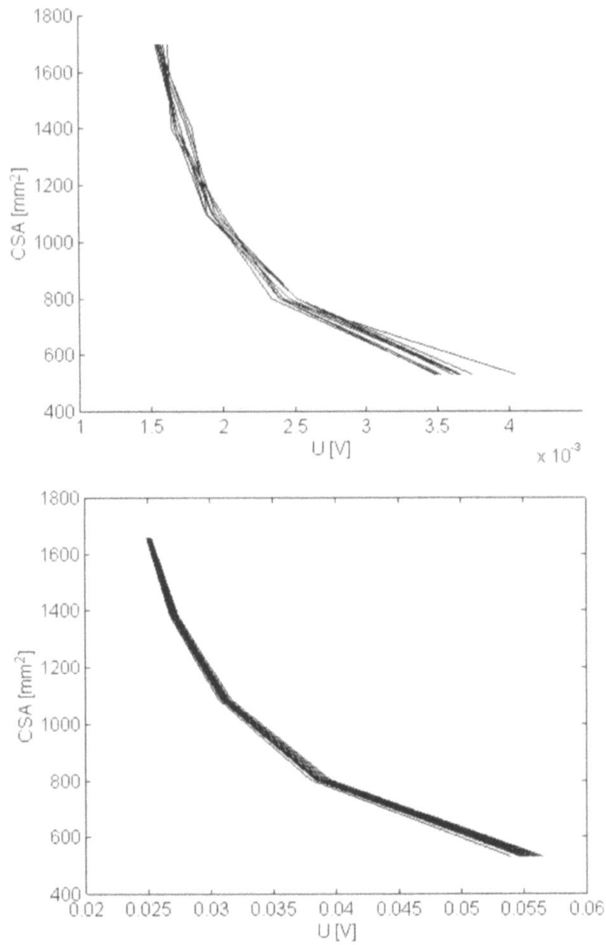

Figure 4.11. The transfer function: CSA-voltage dependency for default model, for angles 0 to 11°. **a:** experimental measurements, **b:** simulations. The projection on the x-axis differs due to difference in saline conductivity.

conductor whose impedance determines the voltage between the detection electrodes (Figure 4.10). The deforming effect of the wall slope on the potential distribution along the detection electrodes verticals and, specifically, at the detection electrodes site can be seen. Deformation of the isopotential lines due to a higher detection electrode conductivity than the saline solution conductivity (at 2.5 mm at the right and left from the centre, at the same level as the excitation electrodes) is also observed. The surface plots C, D outline the effect of the increasing wall slope on the potential distribution above the excitation electrodes. The left side is more negative and the right side is less positive and the angle increases.

Computational Technique

Model representation is a half cross-section in cylindrical coordinates due to rotational symmetry along the z axis (segment AF in Figure 4.13).

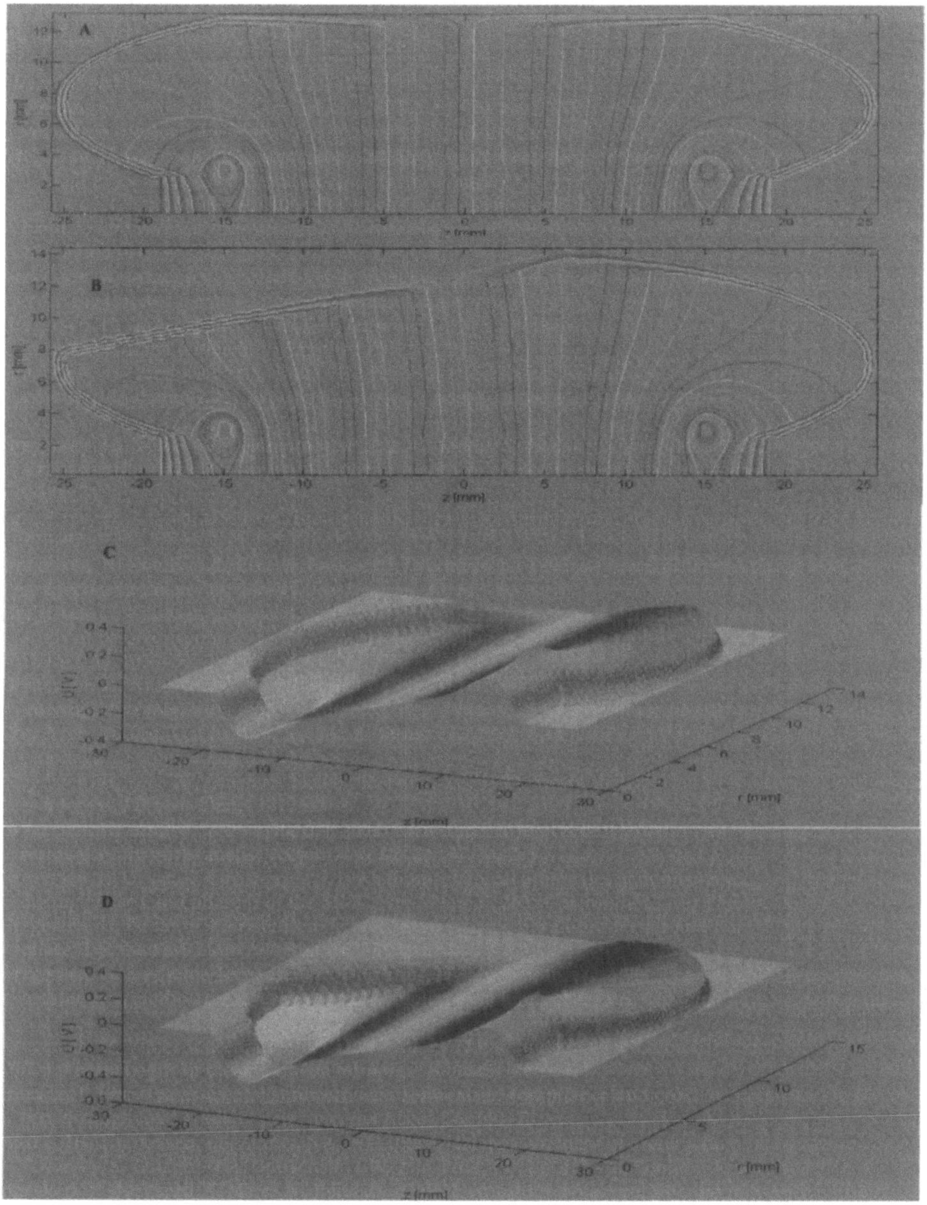

Figure 4.12. The potential distributions for default model with angle=0° (A, C) and angle=14° (B, D) (see also Plate II).

A constant current I is injected through the excitation electrodes E_{e1}, E_{e2}. This will produce a potential distribution in the volume conductor within the balloon. Ohm's law gives the relation between the current density and the electric field intensity as

$$J = \sigma E \,, \tag{4.12}$$

where the electric field intensity can be expressed by the scalar electric potential as

$$E = -\nabla V \tag{4.13}$$

This will yield

$$J = -\sigma \nabla V \tag{4.14}$$

Applying the divergence operator on the equation gives the relation between the current source and the potential distribution:

$$\nabla * J = -\nabla * (\sigma \nabla V) \tag{4.15}$$

According to the charge continuity law the current distribution should have divergence zero. In the present model only in two points, the centre of the excitation electrodes, is the divergence not zero because of the current injection. They are the source points.

The finite difference method is used for solving Equation 4.15. The lambda operator is expressed in 3D Cartesian and cylindrical coordinates as:

$$\nabla = \frac{\partial}{\partial x}\overline{e}_x + \frac{\partial}{\partial y}\overline{e}_y + \frac{\partial}{\partial z}\overline{e}_z \tag{4.16}$$

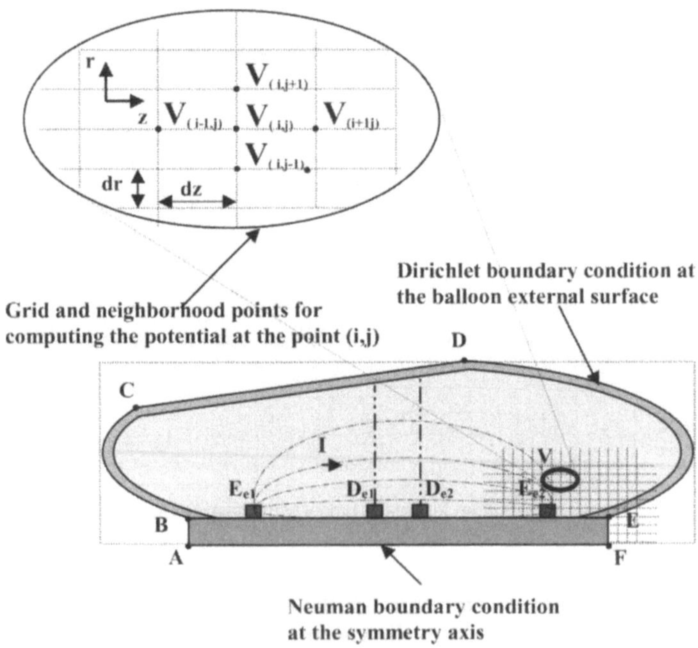

Figure 4.13. Model in half cross-section, cylindrical coordinates representation, with example of mesh generation and boundary conditions.

$$\nabla = \frac{\partial}{\partial r}\overline{e}_r + \frac{1}{r}\frac{\partial}{\partial \varphi}\overline{e}_\varphi + \frac{\partial}{\partial z}\overline{e}_z \tag{4.17}$$

The right term of Equation 4.15 becomes in 3D:

$$\nabla * (\sigma \nabla V) = \frac{1}{r}\frac{\partial}{\partial r}\left(r\sigma\frac{\partial V}{\partial r}\right) + \frac{1}{r^2}\frac{\partial}{\partial \varphi}\left(\sigma\frac{\partial V}{\partial \varphi}\right) + \frac{\partial}{\partial z}\left(\sigma\frac{\partial V}{\partial z}\right) \tag{4.18}$$

Due to rotational symmetry, the differential along the φ dimension is zero. Thus, Equation 4.18 can be reduced to 2D:

$$\nabla * (\sigma \nabla V) = \frac{1}{r}\frac{\partial}{\partial r}\left(r\sigma\frac{\partial V}{\partial r}\right) + \frac{\partial}{\partial z}\left(\sigma\frac{\partial V}{\partial z}\right) \tag{4.19}$$

where the differential of a product can be expressed as:

$$\frac{1}{r}\frac{\partial}{\partial r}\left(r\sigma\frac{\partial V}{\partial r}\right) = \frac{1}{r}\frac{\partial r}{\partial r}\sigma\frac{\partial V}{\partial r} + \frac{1}{r}r\frac{\partial}{\partial r}\left(\sigma\frac{\partial V}{\partial r}\right) = \frac{1}{r}\sigma\frac{\partial V}{\partial r} + \frac{\partial}{\partial r}\left(\sigma\frac{\partial V}{\partial r}\right) \tag{4.20}$$

Equation 15 becomes:

$$\nabla * J = \frac{1}{r}\sigma\frac{\partial V}{\partial r} + \frac{\partial}{\partial r}\left(\sigma\frac{\partial V}{\partial r}\right) + \frac{\partial}{\partial z}\left(\sigma\frac{\partial V}{\partial z}\right) \tag{4.21}$$

The mesh is performed with equal steps in both directions, r respective z, within the rectangle ADF. In its discrete form the equation relates the potential in a point (i, j) with the values of the potential in its close neighbourhood, the points (i+1,j) (i–1,j) (i,j+1) (i,j–1). Each point has associated a value for the conductivity $\sigma_{i,j}$, as follows:

$\sigma_{i,j}$ corresponds to the segments (i,j) ↔ (i+1,j), respective (i,j) ↔ (i,j+1)
$\sigma_{-1,j}$ corresponds to the segment (i–1,j) ↔ (i,j)
$\sigma_{i,j-1}$ corresponds to the segment (i,j–1) ↔ (i,j)

The terms in Equation 4.21 can be written in the discrete form as follows

- The first order derivative of the potential along the r axis is computed as the mean between the first order derivatives in the points (i,j) and (i,j–1), in order to obtain a symmetry around the point (i, j)

$$\frac{1}{r}\sigma\frac{\partial V}{\partial r} \Rightarrow \frac{1}{2r_{i,j}}\left(\frac{\sigma_{i,j}\left(V_{i,j+1} - V_{i,j}\right) + \sigma_{i,j-1}\left(V_{i,j} - V_{i,j-1}\right)}{\Delta_r}\right) \tag{4.22}$$

The second order derivative of the potential along the r axis is

$$\frac{\partial}{\partial r}\left(\sigma\frac{\partial V}{\partial r}\right) \Rightarrow \frac{\sigma\left.\frac{\partial V}{\partial r}\right|_{i,j} - \sigma\left.\frac{\partial V}{\partial r}\right|_{i,j-1}}{\Delta_r} \Rightarrow \frac{\sigma_{i,j}\left(V_{i,j+1} - V_{i,j}\right)\frac{1}{\Delta_r} + \sigma_{i,j-1}\left(V_{i,j} - V_{i,j-1}\right)\frac{1}{\Delta_r}}{\Delta_r}$$

(4.23)

$$\frac{\partial}{\partial r}\left(\sigma\frac{\partial V}{\partial r}\right)\Rightarrow\frac{\sigma_{i,j}V_{i,j+1}-\left(\sigma_{i,j}+\sigma_{i,j-1}\right)V_{i,j}+\sigma_{i,j-1}V_{i,j-1}}{\Delta_r^2} \tag{4.24}$$

- Similarly, the second order derivative of the potential along the z axis is

$$\frac{\partial}{\partial z}\left(\sigma\frac{\partial V}{\partial z}\right)\Rightarrow\frac{\sigma_{i,j}V_{i+1,j}-\left(\sigma_{i,j}+\sigma_{i-1,j}\right)V_{i,j}+\sigma_{i-1,j}V_{i-1,j}}{\Delta_z^2} \tag{4.25}$$

- The source term is represented in the discrete form as

$$\nabla * J\Rightarrow S_{i,j} \tag{4.26}$$

The mesh is uniformly generated, with the same step on both directions

$$\Delta_r=\Delta_z=\Delta \tag{4.27}$$

Equation 4.21 becomes, in the discrete form

$$-S_{i,j}=\frac{\Delta}{2r_{i,j}}\left(\frac{\sigma_{i,j}\left(V_{i,j+1}-V_{i,j}\right)+\sigma_{i,j-1}\left(V_{i,j}-V_{i,j-1}\right)}{\Delta^2}\right)+\frac{2r_{i,j}}{2r_{i,j}}\left(\frac{\sigma_{i,j}V_{i,j+1}-(\sigma_{i,j}+\sigma_{i,j-1})V_{i,j}+\sigma_{i,j-1}V_{i,j-1}}{\Delta^2}\right)+$$
$$+\frac{2r_{i,j}}{2r_{i,j}}\left(\frac{\sigma_{i,j}V_{i+1,j}-(\sigma_{i,j}+\sigma_{i-1,j})V_{i,j}+\sigma_{i-1,j}V_{i-1,j}}{\Delta_r^2}\right)$$

and can be rewritten as:

$$\frac{2r_{i,j}\left(2\sigma_{i,j}+\sigma_{i,j-1}+\sigma_{i-1,j}\right)+\Delta\left(\sigma_{i,j}-\sigma_{i,j-1}\right)}{2r_{i,j}\Delta^2}V_{i,j}=\frac{\left(2r_{i,j}-\Delta\right)\sigma_{i,j-1}}{2r_{i,j}\Delta^2}V_{i,j-1}+\frac{\left(2r_{i,j}+\Delta\right)\sigma_{i,j}}{2r_{i,j}\Delta^2}V_{i,j+1}$$
$$+\frac{2r_{i,j}\sigma_{i-1,j}}{2r_{i,j}\Delta^2}V_{i-1,j}+\frac{2r_{i,j}\sigma_{i,j}}{2r_{i,j}\Delta^2}V_{i+1,j}+S_{i,j}$$

Further, it can be generalised as

$$\left(A_{i,j}+\mathbf{B}_{i,j}+C_{i,j}+D_{i,j}\right)V_{i,j}=A_{i,j}V_{i,j-1}+\mathbf{B}_{i,j}V_{i,j+1}+C_{i,j}V_{i-1,j}+D_{i,j}V_{i+1,j}+S_{i,j} \tag{4.31}$$

with the following expression for the coefficients

$$A_{i,j}=\frac{\sigma_{i,j-1}}{\Delta}\left(\frac{1}{\Delta}-\frac{1}{2r_{i,j}}\right), \tag{4.31}$$

$$B_{i,j}=\frac{\sigma_{i,j}}{\Delta}\left(\frac{1}{\Delta}+\frac{1}{2r_{i,j}}\right), \tag{4.32}$$

$$C_{i,j} = \frac{\sigma_{i-1,j}}{\Delta^2}, \tag{4.33}$$

$$D_{i,j} = \frac{\sigma_{i,j}}{2} \tag{4.34}$$

and the source term, with the expression only in the points corresponding to the centre of the excitation electrodes (source with $I>0$ and the sink with $I<0$):

$$S_{i,j} = \frac{I}{2\ {}^{2}\pi R_e}, \tag{4.35}$$

where R_e is the mean radius formed by the excitation electrode around the catheter, and I the current through the electrode.

A very important issue in numerical computations is to establish the boundary conditions. Thus, at the external surface of the balloon, it is reasonable to assume $V=0$, which represents a Dirichlet boundary condition. At the symmetry axis, the segment AF in figure 4.13, a Neumann condition is imposed: the first derivative of the potential normal to the symmetry axis is zero $\frac{\partial V}{\partial r} = 0$. This expresses that no current crosses the symmetry axis.

The computational technique is iterative, known as Liebmann's method, which is much faster than a general simultaneous equation solver. In each step are used either odd index sum points or even index sum points. This ensures a faster convergence and less computational cost.

The initial value for the potential is usually chosen as zero. After each step, due to the source points, the potential will increase toward the solution. The difference between the potential values in the same point between consecutive iterations is the residual $R_{i,j}$. The values of this residual will give the error that is accepted at the end of the iteration algorithm between the real solution and the value of the potential obtained. The convergence can be improved by weighting the residual with the over-relaxation factor, β. After each iteration the new value of the potential in a point can be computed according to the relation

$$V_{i,j}^{new} = V_{i,j}^{old} + \beta R_{i,j}. \tag{4.36}$$

The over-relaxation factor has the usual value $1 < \beta < 2$. A value greater than 2 leads to instability.

4.2.8 Determination of Membrane Biaxial Tension During Balloon Distension

In a biomechanical analysis it is important to know the force–deformation (stress–strain) relationship of the tissue. If the wall thickness is measurable and the intestine remains in cylindrical shape, then the average stress in the tissue can be determined according to Laplace's law. However, often the thickness is not measurable and the shape of the balloon-distended intestine is not exactly a cylinder; then the force exerted on the tissue is expressed in terms of tension, and Laplace's law must be improved to remove the cylindrical shape restriction. The distribution of tension in the circumferential and axial directions everywhere in the balloon-distended area would be of interest to know. The equations below develop the path for determination of the circumferential and longitudinal components of tension during balloon distension. When we know the shape of the intestine, we can compute the circumferential and longitudinal components of tension everywhere. The theory will help to provide insight into the stimulus exerted by balloons during distension.

Development of Theory and Principles for Determination of Tensions During Balloon Distension

The theory can be used in any hollow visceral organ in vitro or in vivo as long as the geometric data and the balloon pressure can be measured. Bending forces and moments are neglected. Assume that the balloon and intestine form a tube of revolution, that the structure and deformation are axisymmetric, that the zero-stress state of the balloon has a diameter larger than that of the distended intestine and that its length is sufficiently greater than the intestinal segment in contact. The intestine resists the pressure imposed through the balloon. Assume also that the shear stress (friction) between the balloon and intestine is negligible. At places where the balloon is in contact with the intestine during distension, the tissue and balloon together are considered a membranous structure. Coordinates as shown in Figure 4.14 are used, with z as the longitudinal axis, r as the radial axis, θ the polar angle and ϕ the latitudinal angle. θ and ϕ form a set of curvilinear orthogonal coordinates. Membrane stress resultants are N_θ, N_ϕ, the circumferential and longitudinal membrane stress, respectively. Definitions of these coordinates and stress resultants are given in the appendix. We assume for an oversized balloon that $N_\theta = N_\phi = 0$ for the balloon.

The boundary condition for stress in the intestine is $N_\theta = 0 = N_\phi$ when $z \rightarrow \infty$. The shear $N_{\theta\phi}$ is zero for axisymmetric deformation. The radius of curvature of the longitudinal section is r_1. The radius of curvature of the intestinal surface in a normal section orthogonal to the longitudinal direction is r_2. Figure 4.14 shows the following geometric relationships (yet unpublished work by Gregersen, Kassab and Fung)

$$r = r_2 \sin \phi, \; ds = r_1 \, d\phi \tag{4.37}$$

$$dz = r_1 \sin \phi \, d\phi, \; dr = r_1 \cos \phi \, d\phi \tag{4.38}$$

The equations of equilibrium are

$$\frac{N_\phi}{r_1} + \frac{N_\theta}{r_2} = p, \tag{4.39}$$

$$\frac{d}{d\phi}(rN_\phi) - r_1 N_\theta \cos\phi = 0, \tag{4.40}$$

where p is the pressure in the balloon.

By solving Equation 4.39 for N_θ and substituting the result into Equation 4.40, we obtain after reducing:

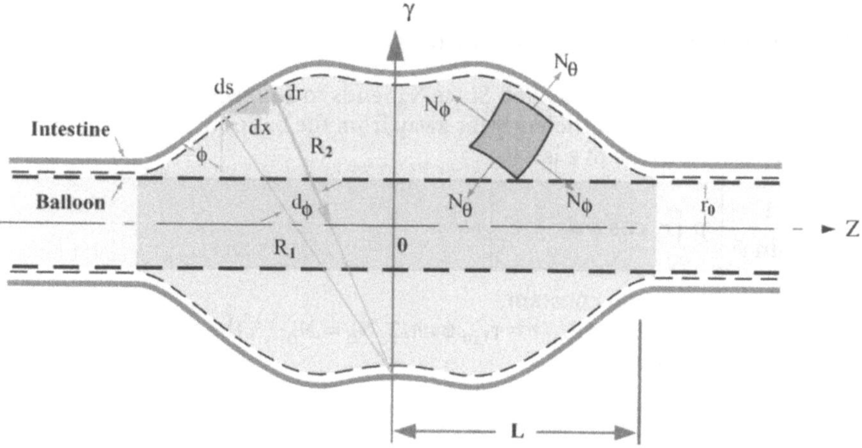

Figure 4.14. Schematic drawing of the distended segment of an intestine. N_ϕ and N_θ denote the tension in longitudinal and circumferential directions. To analyse the tension in the intestine as the balloon is inflated, we have to describe the geometry and define the notions of the stresses. Two systems of coordinates are convenient. (1) the 3-dimensional cylindrical polar coordinates (r, θ, z). (2) the two-dimensional latitude–longitude coordinates (ϕ, θ) on the surface of the intestine, similar to the one we use for our globe in geography. The latitudinal angle ϕ is defined as the angle between the normal vector to the intestinal surface and the axis of symmetry z. The longitudinal angle θ is the angle of rotation of the meridian cross-section about the z axis. Now there exists a normal vector at any point P on the surface with coordinate ϕ and θ. At a point neighbouring to P, with coordinates (ϕ+dϕ and θ), there is another normal vector. These two normal vectors will intersect at a distance r_1 from the surface. This r_1 is a principal radius of curvature of the surface in the longitudinal cross-section. On the other hand, the normals at (ϕ, θ) and (ϕ+dθ and θ) will intersect at a distance r_2 from the surface. r_2 is also a principal radius of curvature of the surface. The radius of curvature of the curve of intersection of any other normal plane at P lies between r_1 and r_2. So the curvature of the intestinal surface is defined by r_1 and r_2.

The stresses of interest in an axisymmetric tube subjected to axisymmetric load are those lying in the wall, acting on normal cross-sections of the wall. In this figure, a rectangular element with sides df and dq is shown, similar to a sphere with a small latitudinal and longitudinal difference. The stresses s_{ff} and s_{qq} can be much larger than the pressure in the balloon (p). When the shear stress between the balloon and intestine is ignored, and the bending stresses are ignored because the intestinal wall thickness is small compared with the radii of curvature, then s_{ff} and s_{qq} are uniformly distributed throughout the intestinal wall. Multiplying the uniform stresses s_{ff} and s_{qq} by the wall thickness, h, we obtain $N_f = s_{ff}$ h and $N_q = s_{qq}$ h, which are called membrane tensions in cross-sections whose normal vectors lie in the direction of f and q axes, respectively. The units of N_f, N_q are force per unit length, N/m, like those of surface tension, while the stresses have units N/m^2. N_f and N_q are by some called stress resultants or membrane stress resultants to recognise the fact that they are the integrals of stresses throughout the thickness of the membrane wall. (see also Plate III.)

$$\frac{d}{d\phi}(rN_\phi) + r_2 \cos\phi N_\phi - pr_1 r_2 \cos\phi = 0\,. \qquad (4.41$$

Multiplying by sin ϕ and using Equation 4.37, we obtain after rearrangement:

$$N_\phi = \frac{1}{r_2 \sin^2\phi}\int pr_1 r_2 \cos\phi \sin\phi \, d\phi\,. \qquad (4.42)$$

We obtain using the geometric relationships in Equations 4.37 and 4.38:

$$N_\phi = \frac{1}{r \sin \phi} \int_{r_o}^{r} pr\, dr = \frac{p(r^2 - r_o^2)}{2r \sin \phi}, \tag{4.43}$$

where r_0 is an integration constant. Since N_ϕ tends to zero when $z \to \infty$, we see that r_0 is the radius of intestine sufficiently far away from the balloon. An alternative expression for N_ϕ as a function of z is:

$$N_\phi = \frac{1}{r \sin \phi} \left[p \int_0^z r_2 \cos \phi \, dz + C \right], \tag{4.44}$$

where C is an integration constant.

If, at a large distance, $z = L_0$, $r = r_{L0}$, $\phi = \pi/2$, $N_\phi = N_\phi^{(\infty)}$, the total longitudinal force $= 2\pi r_\infty N_\phi^{(\infty)} = T$, then

$$T = 2\pi \left[p \int_0^{L_0} r_2 \cos \phi dz + C \right]. \tag{4.45}$$

Of special interest is the case in which T = 0; i.e., the longitudinal tension in the intact intestine is zero. Then:

$$C = -p \int_0^{L_0} r_2 \cos \phi dz . \tag{4.46}$$

With this C, we can then compute the value of $N_\phi^{(0)}$. This gives the longitudinal tensile stress resultant N_ϕ at the centre section z = 0.

Using Equation 4.43 or 4.44 in Equation 4.39, we have

$$N_\theta = r_2 (p - \frac{N_\phi}{r_1}) . \tag{4.47}$$

Combining with Equation 4.43 or 4.44, we obtain:

$$N_\theta = r_2\, p\, (1 - \frac{r^2 - r_o^2}{2rr_1 \sin \phi}) . \tag{4.48}$$

Hence, if the radii of curvature r_1, r_2 are measured from a photograph of the intestinal profile, then the membrane stress resultants N_θ, N_ϕ can be computed everywhere in the intestine in contact with the balloon. From the images of the intestinal specimens we measured the radii along the z-axis. A cubic polynomial was fitted to the measured radii using a least square fit.

Figure 4.15 shows a porcine intestinal segment during balloon distension with a pressure of 10 cmH_2O (1 kPa). The radii obtained at three pressure levels as a function of z is shown in Figure 4.16 for one pig experiment. Figure 4.16 also shows N_ϕ, N_θ and T_L as functions of z at the same pressure level (the tension T_L was computed according to Laplace's law, $T_L = p\, r$ for comparison with N_θ). N_ϕ increased with the pressure, was highest at the mid-balloon location and fell towards the end of the balloon. N_θ increased with the pressure applied and was in general 2–3 times higher than N_ϕ, At the highest pressure applied in the pigs, the maximum N_θ was located approximately 25% away from the middle of the balloon (the middle and end of the balloon being defined as 0% and 100%, respectively). T_L exceeded N_θ from 0% to about 70% away from the middle of the balloon. Only at the location of maximum N_θ

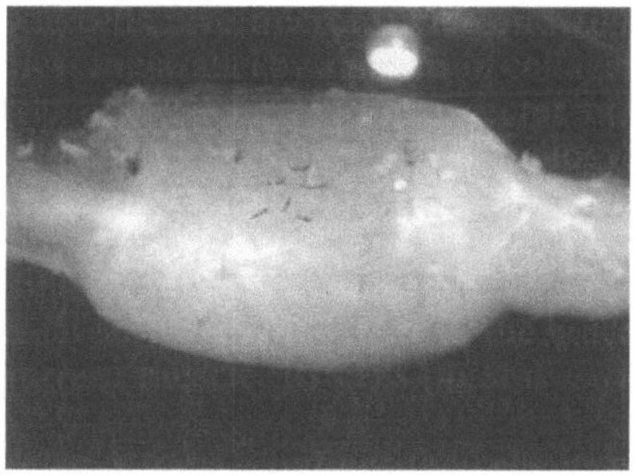

Figure 4.15. A photograph of an intestinal segment during balloon distension with a pressure of 10 cmH_2O. z is in the direction of the longitudinal axis of the segment (approximately horizontal in this image). The radii normal to the z-axis were measured from the images and used for the computation of tensions.

were N_θ and T_L similar. At about 80% away from the middle of the balloon, N_θ became higher than T_L. Figure 4.15 shows the comparison of N_θ and T_L from one pig experiment at a pressure of 15 cmH_2O. The same pattern was found at lower pressures and when using the smaller balloon in guinea pigs. The analysis implicates that tension computed directly from Laplace's law does not provide very accurate measures of circumferential tension.

In the data mentioned above the radii were measured on the exterior surface of the intestine. For a fairly thin-walled organ there is not much difference between the internal and external radii but for a thick-walled organ, this may be a matter of issue. The intestine is regarded as a thin-walled organ during distension (thickness-to-radius ratio $<<$10%, see Chapter 3 for values obtained in rats). The theory presented can be applied to the in vivo situation as long as the shape and the pressure can be obtained. Balloon distension with pressure–volume measurement cannot provide such data. However, there are several other ways to obtain the data in vivo. Impedance planimetry can provide measures of the balloon cross-sectional area at various locations on the z-axis (Scheel et al., unpublished data) and assuming circularity, the internal radii can be computed. X-ray is another way to provide the geometric data (see, for example, Figure 2 in Villadsen et al., 1995). New imaging techniques call for attention. B-mode ultrasound, 3D-ultrasound, multi-slice CT-scanning and NMR scanning can all provide a three-dimensional profile of the distended area. Hence, data obtained with these technique can be used.

4.3 Other Mechanical Techniques and Analytical Approaches

The question of mechanical properties can be formulated in many different forms, varying in degree of generality and difficulty. A number of methods and experiments have been proposed for studying the mechanical properties. These methods range from techniques for investigating cell properties to whole organs in clinical settings. Although most of these experiments are in vitro, they may contribute significantly to our understanding of the complex behaviour of the gastrointestinal tract in health and disease. It is to reach too far to describe all these methods in detail. Information about

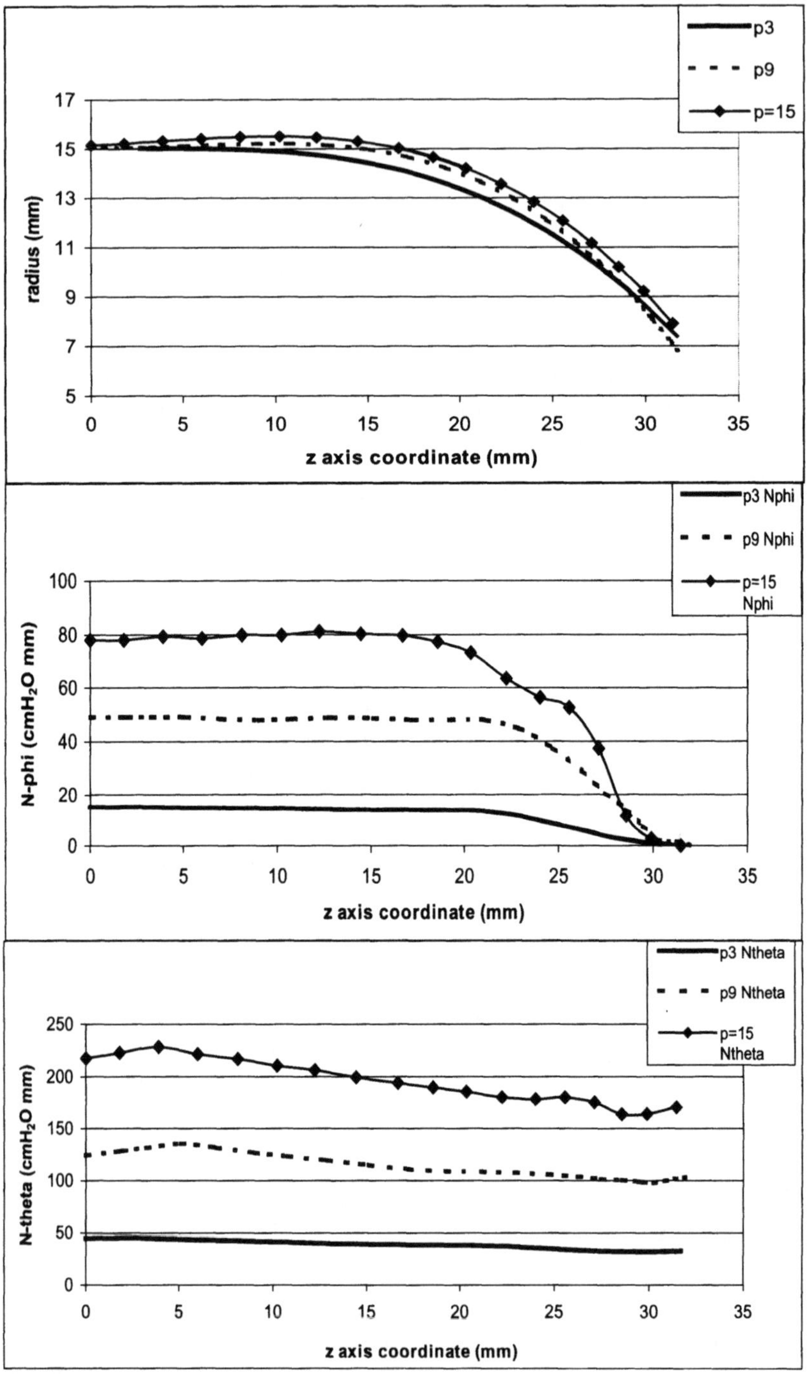

Figure 4.16. Representation of radii, N_ϕ, N_θ and T_L as functions of z obtained at pressure levels up to 15 cmH_2O for one of the pig experiments. The data are obtained in one quadrangle of the balloon and only represent a limited number of the pressure levels. The intestinal segments were preconditioned before testing. N_ϕ and T_L resemble the shape of the radii whereas N_θ has a local maximum 20–30% away from the middle of the balloon.

techniques for evaluating the active mechanical behaviour in vivo such as manometry and radiography may be obtained from other sources. Rather, this section aims to give the reader a review of some very useful techniques and to provide some guidelines as to which measures are useful when studying specimens from the gastrointestinal tract in vivo and in vitro. Techniques for studying the zero-stress state are reviewed in detail in Chapter 6. Techniques and analyses are also provided in other chapters.

4.3.1 Uniaxial Testing

Resected specimens are placed in physiological solution. After cleaning, further dissection and an equilibration period, the strip is mounted between two hooks using silk thread. The tissue strips may contain several layers of the wall or a single layer may be studied after separation from other layers. One hook is attached to a force transducer whereas the other hook is attached to a slider so the length of the strip can be changed. The solution is oxygenated, pH- and temperature-controlled to maintain physiological conditions. The length of the specimen is often between 1 and 5 mm and the resting length is measured using a stereomicroscope, often after a small longitudinal tension is applied. By increasing the length stepwise between the two knots with 10% or 20% of the resting length and recording corresponding values of passive, spontaneous active and stimulated (often by carbachol) tension, length–tension curves for each type of preparation can be obtained. From such isometric experiments the optimum active tension can be determined. In the setup for isotonic studies, the tension is fixed and the length of the strip is allowed to change.

Comparison of Data Obtained in Intact Segments with those from the Uniaxial Strip Method

Uniaxial testing of tissue strips has been used extensively in pharmacological studies of smooth muscle behaviour. It is a standardised technique that has certain advantages. Mechanical data obtained from intact segments are generally not directly comparable to data obtained in isolated smooth muscle strips. This is due to the many differences between the two techniques. In strip studies it is common to determine the active and passive length–tension relations, force–velocity relations, and stress relaxation (Meiss, 1971; Tøttrup et al., 1990). Furthermore, it is possible to study the behaviour under either isotonic or isometric conditions in uniaxial isolated strip experiments. Unfortunately, in the isolated strip studies, the structural integrity of the organ wall is not preserved and the strip is not necessarily in its zero-stress state at resting conditions. Hence, some strain may be imposed during the mounting of the gastrointestinal specimens. The zero-stress state must be considered in the analysis of stress distribution in the organ and in the determination of the resting length. The isolated strip technique, however, is useful for studying the effect of pharmacological agents on smooth muscle and it may provide useful data on the mechanical properties of individual layers in the wall by studying circular and longitudinal strips. In contrast to strip investigations, it is more difficult to compute stress in vivo where purely isometric or isotonic conditions are difficult to obtain. On the other hand, the advantage of studying intact organs is that one can measure their biomechanical properties in conditions closer to the in vivo condition (Fung, 1993) with proper geometry and boundary conditions.

4.3.2 Biaxial and Triaxial Testing

Uniaxial loading experiments such as the conventional strip experiments cannot provide the full relationship between all the components of stress and strain. Biaxial and triaxial tests are required to provide further information for the constitutive equation of a material. Biaxial testing can be performed in various ways. Well-known examples are biaxial testing of flat sheets and inflation-axial stretching experiments in tubular segments. In the former, the sheet of tissue is mounted in a biaxial testing machine. Forces are applied in the two orthogonal directions of the plane and the corresponding strains measured. Biaxial stretching of sheets has not been used much in gastroenterology. The latter is in part described together with a two-dimensional strain energy function in Chapter 3. A 2D strain analysis based on displacement of surface markers (also applicable to embedded particles) is also presented in Chapter 3 and biaxial stretch of smooth muscle cells in culture is described in Chapter 5. Balloon distension, as a specialised form for multiaxial test is discussed in this chapter. This section is therefore confined to the description of triaxial experiments; i.e., a device for inflation-stretching and torsion of tubular organs.

Three-dimensional modelling requires a tri-dimensional constitutive equation that incorporates information in the axial, circumferential and shearing directions. For this purpose a new triaxial mechanical test equipment has been developed (Figure 4.17). A tubular tissue sample is mounted on cannulas, submersed in an organ bath and attached to the test machine. The test machine utilises a linear stage and two step motors controlled by a computerised controller to provide longitudinal stretch and twisting of tubular tissue samples. The lumen can be pressurised by a pump connected to the cannulas. A pressure transducer measures luminal pressure. Torsion is measured using a sin/cos optical encoder in conjunction with flexural pivot bearings. Outer dimensions of the segment are measured using a video camera. Transducer signals are conditioned and then converted to digital format by an A/D card. The system and data acquisition are controlled using software written in the Labview programming language and the data analysis performed using custom Visual Basic macros. Use of the triaxial test machine allows versatility in study design and the examination of multiple sample properties. Extension, pressure, and twisting

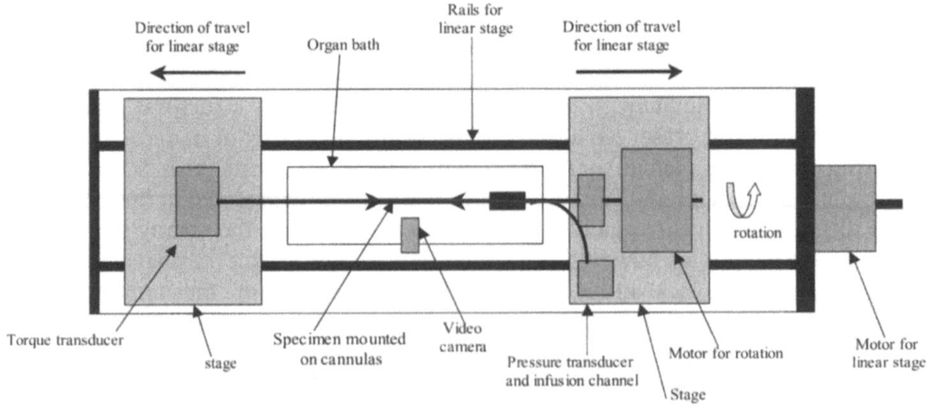

Figure 4.17. Schematic of the triaxial test equipment for inflation, stretching and torsion experiments (original design by Judd Day).

can each be varied independently while monitoring sample geometry. This facilitates the study of material properties over a wide range of physiological conditions.

In the uniaxial and biaxial experiment with determination of changes in diameter and length, we only consider normal stresses and strains in the circumferential and longitudinal directions. Shear components, however, are also important in the longitudinal–circumferential directions. Hence, the data on the shear modulus and its dependence on the circumferential and longitudinal stresses must be included in the constitutive equation. Recently, Deng et al. (1994) presented an experimental method to determine the shear modulus of arteries. They applied torsion to an arterial segment in vitro and subsequently measured the torque and shear deformation to solve for the shear modulus G from the equilibrium of forces equation:

$$\mathrm{T} = GJ\frac{\theta}{L} \tag{4.49}$$

where T is the torque, J is the polar moment of inertia of the vessel cross-section and $\frac{\theta}{L}$ is the angle of twist per unit length. The method is equally applicable to the gastrointestinal tract. Figure 4.18 shows data obtained in the duodenum of guinea pigs. The top figure shows the torque as a function of the twist angle (from a ramp test to and from ±25 degrees). The data are obtained after a series of preconditioning cycles. Therefore, the hysteresis is small. Within the range studied, a good relationship was found. From the bottom figure, it appears that the shear modulus is linearly related to the distension pressure.

4.3.3 Some Notes on Diameter Measurements and Analysis

Diameter measures can be obtained in several ways. Diameters of the gastrointestinal tract in vivo can be obtained by diagnostic imaging techniques. In vitro, outer diameters are often obtained from video images of the specimens. The inner diameters can be computed assuming tissue incompressibility if the no-load dimensions can be measured. The inner diameters can also be obtained by making a cast of the lumen in vivo or in vitro. In this case, cannulas are inserted in the proximal and distal ends of the segment that is perfused with a catalysed silicone elastomer solution at a preselected inlet pressure. The use of silicone elastomer renders possible the reconstruction of organ geometry. After perfusion for several minutes, the outlet is clamped and the silicone elastomer hardens in approximately 20 minutes. Then the segment can be excised and photographed with and without the cast inside. Photographs of the cast will provide the inner diameters and other relevant geometry.

Pressure–Diameter Experiments

The pressure–diameter relation is popular among cardiovascular physiologists since it plays an important role in the pressure–flow relationship of blood through an organ and because diameters have been measurable with ultrasound. The pressure–diameter relation can also be used to understand the distensibility of the intestine (Gregersen and Kassab, 1996). The compliance of an intestine as expressed by the percentage change of diameter versus change in transmural pressure can be shown to be a function of the ratio of the wall thickness to intestinal diameter and on the

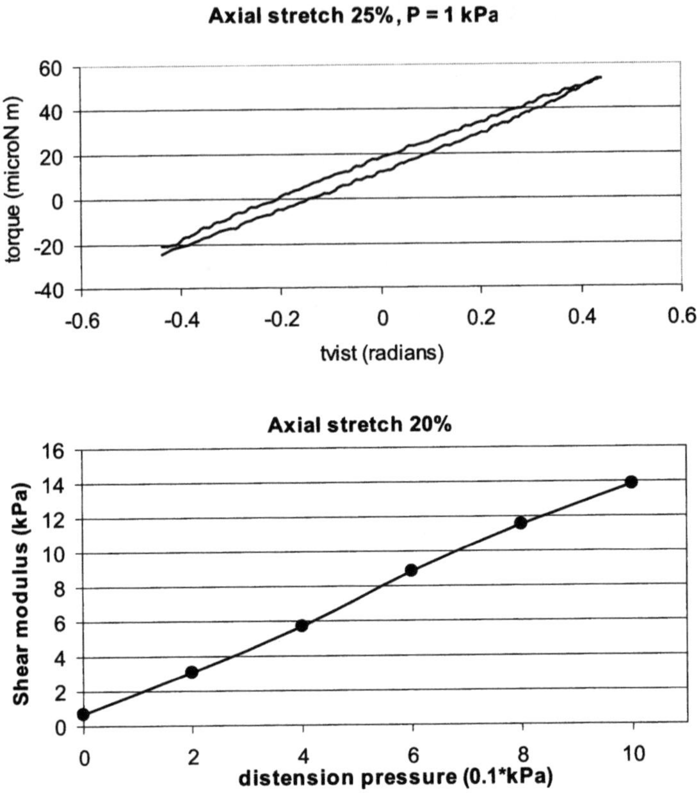

Figure 4.18. Data obtained from triaxial experiments in the guinea pig duodenum. Torque as a function of twist angle at 25% stretch (top) and shear modulus as a function of pressure at 20% stretch (bottom).

Young's modulus of the wall material. A simple estimation can be made based on the assumption of thin-walled tube. The tensile stress in the wall of a tube of diameter D is, according to Laplace's formula, given by

$$\tau_\theta = \Delta P_{tm} \frac{D}{2h} \tag{4.50}$$

where h is the wall thickness. The strain $\frac{\tau_\theta}{E}$, which is also the change of diameter divided by the diameter can be written as

$$\frac{\Delta D}{Do} = \frac{D}{2hE} \Delta P_{tm}. \tag{4.51}$$

Hence, the compliance $\frac{D}{2hE}$ is proportional to the diameter–wall thickness ratio, $\frac{D}{h}$, and inversely proportional to the material constant E.

The reference diameter D_0 cannot be taken as the diameter at zero pressure since the cross-sectional area of the intestine is not circular at that pressure. Upon luminal pressure loading, the intestine becomes circular and hence it is useful to analyse the tendency for the cross-section to become circular as the pressure increases. The complex differential equations for a cylindrical shell (Flügge, 1960; Prescott, 1946) can be simplified considerably if the deformation is taken to be the same along each generator line, so that a cylinder deforms into another with a different cross-section. The following assumptions are necessary: the wall thickness-to-diameter ratio is sufficiently small, the material is isotropic and obeys Hooke's law with a Young's modulus E and Poisson's ratio ν, and the cylinder has an elliptical cross-section in the no-load condition so that its curvature can be expressed by the equation

$$C_0 = \frac{1}{a} + \frac{\varepsilon}{a}\cos 2\theta \tag{4.52}$$

where a is the mean initial radius, θ is the polar angle, and ε is a measure of the ellipticity of the cross-section. Yen et al. (1980) have shown that a simple solution can be analytically determined if the initial ellipticity is small ($\varepsilon \ll 1$). The solution can be expressed as the change of curvature ΔC as follows

$$\Delta C = \frac{-\frac{\varepsilon}{a}\cos 2\theta}{1 + \frac{Eh^3}{3(1-\upsilon^2)a^3\Delta P_{tm}}}. \tag{4.53}$$

This solution shows that when the transmural pressure ΔP_{tm} is zero, the denominator tends to infinity and $\Delta C = 0$, so that the original elliptical cross-section is obtained. When the transmural pressure tends to infinity, the solution becomes

$$\Delta C = \frac{-\varepsilon}{a}\cos 2\theta. \tag{4.54}$$

This implies a circular cross-section.

The transmural pressure at which the ellipticity is reduced to 5% of its initial value can be obtained by setting the denominator of the fraction on the right hand side of Equation 4.53 to 1.05

$$\Delta P_{tm} = \frac{E}{0.15(1-\upsilon^2)}\left(\frac{h}{a}\right)^3. \tag{4.55}$$

Hence, assuming that the material is incompressible ($\nu = 0.5$), the value of the transmural pressure required to reduce the ellipticity to 5% of its initial value can be computed given the material's Young's modulus and thickness-to-radius ratio.

The cross-sections may not be exactly round. Assume that measurements of the major (D_{ma}) and minor (D_{mi}) diameters of the cross-section in the loaded state can be obtained ad modem Gregersen (1997). If, assuming the cross-section to be elliptic, then it is possible to compute the circumferences of the mucosa and serosa, C_l-mucosa and C_l-serosa, respectively. The exact formula is

$$C_l = 2D_{ma}E(k) \tag{4.56}$$

in which E(k) is the complete elliptic integral of the second kind, and k is a parameter defined by

$$k^2=1-\left(\frac{D_{mi}}{D_{ma}}\right)^2. \tag{4.57}$$

If the ratio D_{mi}/D_{ma} is very close to 1, then it can be shown that Equation 4.56 may be approximated by

$$C_l=\frac{D_{ma}+D_{mi}}{2}\pi. \tag{4.58}$$

For example, if D_{ma}/D_{mi} = 1.15, then the C_l calculated by Equation 4.58 is 0.12% that given by Equation 4.56. The use of Equation 4.58 involves two assumptions: that the cross-section is elliptical, and that D_{mi}/D_{ma} is quite close to 1. To test these assumptions, casts of the intestinal lumen must be sliced. From photographs of these slices, D_{ma}, D_{mi} and C_l can be measured. The regression line of the circumference of the cast slices measured directly from photographs and the circumference computed from measurements of the major and minor radii from the same photographs using Equation 4.58 can be expressed in the form:

$$C_{l\text{-mucosa}}=\left[\frac{D_{ma}+D_{mi}}{2}\pi\cdot A\right]+B \tag{4.59}$$

where A and B are empirical constants. Equation 4.59 is an improvement of Equations 4.56 and 4.58 for segments whose cross-section is not exactly elliptical and whose D_{ma}/D_{mi} is not equal to 1. If the serosal circumferences also need correction, they can be calculated on the basis of Equation 4.59 and the wall thickness. First, the mean wall thickness (h) in the loaded state can be calculated as

$$h=\frac{1}{2}\left(\frac{OD_{ma}-ID_{ma}}{2}+\frac{OD_{mi}-ID_{mi}}{2}\right) \tag{4.60}$$

where OD and ID denote the outer (serosal) and inner (mucosal) diameters, respectively. The serosal circumference based on the correction factor can be computed as:

$$C_{l\text{-serosa}}=B+2\pi h-2\pi hA+A\frac{OD_{ma}+OD_{mi}}{2}\pi. \tag{4.61}$$

4.3.4 Bending Experiments. Determination of the Elastic Properties in a Two-layer Model

Experimental results on gastrointestinal elasticity have thus far been analysed with the gastrointestinal wall treated as a homogeneous material. Since the structure of the wall in the gastrointestinal tract is non-uniform, we must evaluate the stress–strain relations of different layers of the wall. Hence, data on the force–deformation relation in the vicinity of mechanoreceptors at various locations can be obtained. Experiments involving pressurisation and longitudinal stretching yield a single strain. Hence, non-axisymmetric deformation must be studied so that the different layers and locations along the wall deform to different degrees. Bending is one way of imposing a non-axisymmetric deformation on the wall of a tubular organ. In experiments based on bending theory, Yu et al. (1993) showed a large difference between Young's modulus of the intima-media and the adventitia layers in the pig thoracic aorta. The method is applicable to the gastrointestinal tract where differences in elasticity between individual layers are expected due to the differences in the structure, e.g., the orientation and configuration of collagen fibres.

4.3.5 Elastic Stiffness Determined with Scanning Acoustic Microscopy

The advancement of science in recent years follows a trend of wishing to know greater details in physiology, medicine and tissue engineering. Biomechanical models are often based on the assumption that the tissue is homogenous; i.e., consisting of one layer only. Most internal organs, however, are multi-layered. Obtaining data from several layers is a step towards the next level in the hierarchy of the structure of living tissue. For the cardiovascular system, Yu et al. (1993) presented a two-layer model based on bending experiments for determination of stress–strain properties. Gregersen et al. (1999) studied the stress-free state in the guinea pig oesophagus and found significant differences in residual strain between the muscle and submucosa layers. Data relating microscale biomechanical properties and gastrointestinal histology are sparse although tissue components for force transmission; i.e., smooth muscles and collagen, have been identified. Scanning acoustic microscopy (SAM) can provide such data in vitro by means of ultrasound in the gigahertz range and the use of elementary theory of elasticity. The following shows how advanced high-frequency ultrasound and elementary theory of elasticity can be utilised to quantify the intestinal properties on the micrometer scale (Assentoft et al., 2001).

SAM Microscope

The SAM microscope utilises ultrasound signal frequencies (f) in the gigahertz range to image and measure elastic properties in sectioned tissue specimens. In both imaging and measurement mode, operation is by way of a focused acoustic lens with a piezoelectric transducer transmitting and receiving ultrasound to and from the tissue. The microscope may be operated at $f = 500$ MHz using a lens with a numerical aperture (NA) of 0.98 to yield a resolution (w) of approximately 1.8 μm:

$$w = \frac{0.51\ C_{couplant}}{NA\, f} \tag{4.62}$$

where $C_{couplant}$ is the ultrasound propagation speed in the acoustic couplant. Keeping the temperature constant at 37 °C using a microscope stage heater and using degassed saline 0.9% as acoustic couplant. $C_{couplant}$ is 1,532.9 m s^{-1} (Chen and Millero, 1977).

In image mode, the lens x-y-scans fields of interest, magnifying between 125× and 600×. In this mode, the reflected ultrasound is converted to 512 by 512 pixel, 8-bit greyscale images. Once an image is obtained, the microscope is switched to the time-resolved measurement mode where it scans the lens along a programmed scan line while digitising and recording individual reflected ultrasound waves.

Calculation of Elasticity

Elementary elasticity theory can be applied to SAM measurements for calculation of c_{11}, which expresses the elastic stiffness. In acoustics, the speed C and acoustic impedance Z, where the latter is the ratio of stress or pressure to particle displacement velocity, are the parameters necessary to calculate c_{11} (Briggs, 1992). The recorded ultrasound waves contain echoes in the form of voltage amplitude signals originating from the upper tissue surface and the tissue–substrate interface. The method of detecting the relevant echoes by way of computer-based signal deconvolution and waveform recognition is described in detail by Briggs et al. (1993). For determination of C and Z, we need individual wave amplitudes (A_1 and A_2) and timings (t_1 and t_2), reference amplitude and timing of an echo in a wave received from a substrate without tissue (A_0 and t_0), standard C and Z values of the couplant (1,532.9 m s^{-1} and 1.53 MPa s m^{-1}, respectively) into:

$$C = C_{couplant} \frac{t_0 - t_1}{t_2 - t_1} (\text{m s}^{-1}) \tag{4.63}$$

and

$$Z = Z_{couplant} \frac{A_0 + A_1}{A_0 - A_1} (\text{Pa s m}^{-1}). \tag{4.64}$$

In the analysis we neglect absorption in the thickness of the sample, and take the variables in Equation 4.64 to be real-valued.

c_{11} is computed from

$$c_{11} = C\,Z\ (\text{Pa}). \tag{4.65}$$

According to Briggs (1992), c_{11} is related to Young's modulus (E) by the relationship

$$c_{11} = \frac{(1-\sigma)E}{(1+\sigma)(1-2\sigma)} \tag{4.66}$$

where σ is Poisson's ratio. Because σ is only slightly less than 0.5 for soft tissue, c_{11} may be considerably higher than E.

4.3.6 Propagation Speed of Sound Assessment in the Layered Wall in Vitro by Means of Acoustic Microscopy

In the early 1930s, Sokolov explored the usefulness of ultrasound for imaging internal structures in optically opaque objects and he was the first to suggest ultrasound at 3 GHz for imaging small objects. This technique is now known as *acoustic microscopy*, which is defined as a general term for high resolution, high frequency ultrasonic inspection techniques that produce images of features beneath the surface of the sample. Further developments led to the Scanning Laser Acoustic Microscope (SLAM), which was introduced by Korpel and co-workers (1971). The SLAM is a transmission mode instrument that creates real-time acoustic images of a sample throughout its entire thickness. A collimated continuous-wave ultrasound beam at frequencies from 10 to 500 MHz is produced by a piezoelectric transducer located beneath the sample. When the ultrasound wave propagates through the sample, the wave is affected by mechanical inhomogeneities in the material. A scanned laser beam is used as the ultrasound detector. The ability of the SLAM to produce simultaneously optical and acoustic images from which the acoustic properties of the specimen can be calculated facilitates its use in this field of biology. The ultrasonic attenuation and propagation speed can be estimated from the obtained information. Conventional tissue fixation and staining are not required for the SLAM imaging. Living cells and tissues can be studied.

Scanning Laser Acoustic Microscopy

The technical details and operating principles of the SLAM were described in detail by Kessler et al. (1972). Three SLAM modes produce three different images. For all modes, the sample is located between the SLAM stage and plastic coverslip. The coverslip is coated with a partially reflecting optical layer. In the *optical mode*, a focused laser beam scans the specimen from above, and is transmitted through the coverslip and specimen to a photodiode at the base of the stage. The received photodiode signal is electronically processed and displayed to a television monitor; the SLAM's optical image is comparable to that of conventional optical microscopy at a magnification of 100×, but is not comparable in that the light source is that of a laser. In the *acoustic mode*, the specimen is insonified with an ultrasonic wave generated by a piezoelectric transducer located below the specimen. The sound wave traverses the specimen and is incident on the lower surface of the coverslip – the surface with the coating. The acoustically generated deflections on this surface of the coverslip are scanned by the laser beam that in turn are reflected to a photodiode. The laser signal is then processed into an acoustic-mode image and displayed in real time on the TV monitor. The ultrasonic attenuation of the specimen can be calculated from this acoustic image ad modem Tervola (1985). In the *interference mode*, the laser beam is detected by the same photodiode as in the acoustic mode and it is then mixed with a reference signal to produce an interference image displayed on the TV monitor. From the interference image, the acoustic propagation speed is calculated from the lateral (horizontal) shift of the vertical interference lines. The lines shift to the right when the sound waves enter an object having a higher speed relative to the coupling reference medium. Quantitative speed profiles can be obtained from several image regions in different loci. The propagation speed of the specimen is calculated ad modem Goss and O'Brien in relation to that of the reference medium according to the following expression

$$C_X = \frac{C_O}{\sin\theta_O}\left[\tan^{-1}\left(\frac{1}{\dfrac{1}{\tan\theta_O} - \dfrac{\lambda_O N}{T\sin\theta_O}}\right)\right] \text{ m / s} \tag{4.67}$$

where C_X is the propagation speed in the specimen of interest, C_O is the propagation speed in the reference medium, λ_O is the wavelength of sound in the reference medium, T is the specimen thickness, N is the measured normalised lateral fringe shift, θ_O is the angle between the direction of sound propagation in the reference medium and the normal to the stage surface, and is determined from Snell's Law:

$$\theta_O = \sin^{-1}\left[\frac{C_O}{C_S}\sin\theta_S\right] \tag{4.68}$$

where C_S is the propagation speed in the fused silica stage (5968 m/s), and θ_S is the angle the sound wave travels through the stage (45°). Measurements of the propagation speed can be done along the vertical line in each layer of the wall to yield a speed profile. Preliminary data obtained in the guinea pig oesophagus are given by Assentoft and co-workers (1996).

4.3.7 Model for Determination of Homeostatic Elastic Moduli in Two Layers of the Oesophagus

To understand gastrointestinal function it is necessary to know how the stress and strain in the gastrointestinal tract can be computed, and how to determine the stress–strain relationship of the wall materials. This section is devoted to the issue of determining the incremental elastic moduli in the layers of the oesophagus under homeostatic conditions.

The basic theory is borrowed from a previous paper on the mechanical properties of blood vessels by Fung and Liu (1995). Due to the similarity in composite tissue structure between the oesophagus wall and blood vessel wall, the theory and experimental design can largely be used in the research of oesophagus. The most useful equipment for testing the mechanical properties of tubular structures is the triaxial test machine as described elsewhere in this chapter. This machine imposes longitudinal stretch, internal pressure, and twisting on an excised isolated test specimen. Fung and Liu (1995) describe the method of measuring incremental elastic moduli of cylindrical tubes in vivo. Data from testing machines can be interpreted in the multi-layered sense in spite of the fact that the testing presumes a homogeneous wall material.

The oesophagus has mucosa–submucosa and muscle layers that can be analysed as elastic cylindrical shells. Each layer has its own zero-stress state, and its own elastic constants. The whole wall is composed of these layers. The incremental elastic moduli in the individual layers at in-vivo (homeostatic) conditions can be computed from the stress–strain relation and zero-stress state of the whole tube with the method of analysis presented below. Four types of experiments including inflation, axial stretching, axial bending and circumferential bending must be applied to obtain the material parameters of the two-layered oesophagus at homeostatic conditions.

The tissue samples must be mounted in the test machine. When a tube is tested in these machines, its geometry is shown in Figure 4.19. Before running the actual tests, the segments must be preconditioned by loading and unloading the segment until the stress–strain relation becomes stabilised. The transmural pressure difference and axial stretch can be controlled and the axial force measured. Furthermore, the outer diameter can be determined optically by using a CCD camera and a VCR. After running the test on the oesophageal specimen, the zero-stress state must be determined as described in Chapter 6.

Biomechanical Preliminary and Definitions of Incremental Elastic Moduli

According to Fung et al. (1993) the relationship between the Kirchhoff stresses, S_{ij}, and the Green strains, e_{ij}, of a multi-layered organ can be expressed by the following strain energy function

$$\rho_0 w = (c/2)[\exp(Q) - Q - 1] + (q/2) \tag{4.69}$$

where

$$Q = a_1 e_{\theta\theta}^{\ 2} + a_2 e_{xx}^{\ 2} + a_3 e_{\theta x}^{\ 2} + 2a_4 e_{\theta\theta} e_{xx} \tag{4.70}$$

$$q = b_1 e_{\theta\theta}^{\ 2} + b_2 e_{xx}^{\ 2} + b_3 e_{\theta x}^{\ 2} + 2b_4 e_{\theta\theta} e_{xx} \tag{4.71}$$

and c, $a_{i,}$ b_i (i = 1, 2, 3, 4) are characteristic material constants, different for each layer.

According to the general theory of continuum mechanics, the stress components S_{ij} are related to the strain components e_{ij} by the formula

$$S_{ij} = \partial(\rho_0 w) / \partial e_{ij}, \qquad (i, j = 1,2,3). \tag{4.72}$$

If we treat the oesophagus as a circular cylindrical shell and consider small perturbations of stress and strain from a homeostatic in vivo state, defined by stress S_{ij} and strains e_{ij}, then the perturbations may be written as

$$S_{ij} = S_{ij}^0 + \delta S_{ij}, \qquad e_{ij} = e_{ij}^0 + \delta e_{ij} \tag{4.73}$$

in which δS_{ij} and δe_{ij} are infinitesimal compared with S_{ij}^0, e_{ij}^0, respectively. From Equation 4.72, we have, by omitting higher order terms,

$$\delta S_{ij} = \frac{\partial^2 \rho_0 W}{\partial e_{km} \partial e_{ij}} \delta e_{km} = C_{ijkm} \delta e_{km} \tag{4.74}$$

where C_{ijkm} are the values of the second partial derivatives of $\rho_0 W$ evaluated at the homeostatic state. Research work on the zero-stress state of the oesophagus has shown that e_{ij}^0 are uniform throughout the wall in the homeostatic state. Therefore, C_{ijkm} are constants in each layer. Hence, while the non-linear Equations 4.69–72 are valid for finite strains, Equation 4.74 is a linear incremental stress–strain relationship. The incremental Cauchy stress and Cauchy strain can also express this relationship

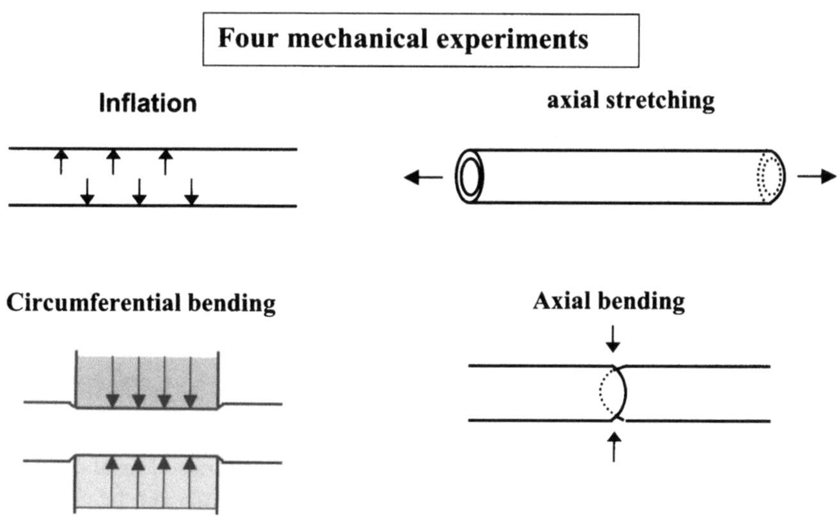

Figure 4.19. Four types of mechanical experiments.

$$\sigma_{11} = Y_{11}\varepsilon_{11} + Y_{12}\varepsilon_{22}, \qquad \sigma_{22} = Y_{21}\varepsilon_{11} + Y_{22}\varepsilon_{22}, \qquad \sigma_{12} = 2G\varepsilon_{12} \tag{4.75}$$

Y_{11} and Y_{22} are defined as the incremental Young's modulus in the circumferential and longitudinal directions, respectively. G is the incremental shear modulus, Y_{12} and Y_{21} are defined as cross modulus.

Principles for Experimental Determination of the Elastic Constants of Two Tissue Layers

To determine the three elasticity constants Y_{11}, $Y_{12} = Y_{21}$, Y_{22} in each of the two layers of the oesophagus, we must first derive six independent equations from the basic principles of conservation of mass, momentum, and energy, the fundamental equations describing the mechanical properties of the materials, (Equations 4.69, 4.72, 4.74, 4.75, called constitutive equations), and a sufficient number of test protocols. The process of deriving the needed equations is said to be the solution to boundary-value problems because each protocol specifies force and displacements on the boundaries. We assume that the oesophagus is a shell. Four types of experiments must be done as sketched in Figure 4.19; namely, the inflation or deflation of the oesophagus by changing internal pressure, the axial lengthening or shortening, the circumferential bending by compressing the oesophagus between diametrically opposite knife edges, and the axial bending by applying a ring of compressive force. The basic theory presented by Fung and Liu (1995) is used. The principal hypothesis and results are listed below.

Bernoulli–Kirchhoff Hypothesis

According to this hypothesis, in shell deformation all points lying on a normal of the neutral surface before deformation do the same after deformation and the coordinate z of a point is unaffected by the deformation of the shell. Hence, the deformation of the shell can be described by the deformation of any of these normal planes. It consists of a *stretching* (translation of the plane) and a *bending* (rotation of the plane). Changes in the direction perpendicular to the neutral surface can be ignored.

Coordinate System

A local rectangular Cartesian frame of reference o-xyz is used to describe the shell. The origin o is located on the neutral surface of the shell. The x-axis points in the direction of the longitudinal axis of oesophagus. The y-axis points in the circumferential direction. The circumferential location is conveniently described by polar coordinates θ, so that $y = a\theta$, a being the radius of the neutral surface; i.e., the y-axis points in the θ direction. The z-axis is perpendicular to the neutral surface, positive outward. The origin can be chosen to be located on the neutral surface to get a better representation of residual stress. If every step was consistently done, the final equations are the same as the alternative location on the neutral surface.

Membrane Tensions and Bending Moments (Also Called Stress Resultants and Stress Moments)

According to the Bernoulli–Kirchhoff hypothesis and the choice of local coordinates, the stress system in the shell (oesophagus) for the experiments can be described by the stress resultants N_θ, N_x and stress moments M_θ and M_x. They are defined by the equations

$$N_\theta = \int_{h_i}^{h_m} \sigma_{\theta\theta}^{(sl)} dz + \int_{h_m}^{h_a} \sigma_{\theta\theta}^{(ml)} dz \; N_x = \int_{h_i}^{h_m} \sigma_{xx}^{(sl)} (1+z/a)dz + \int_{h_m}^{h_a} \sigma_{xx}^{(ml)} (1+z/a)dz,$$

$$N_\theta = \int_{h_i}^{h_m} \sigma_{\theta\theta}^{(sl)} dz + \int_{h_m}^{h_a} \sigma_{\theta\theta}^{(ml)} dz$$

$$M_x = \int_{h_i}^{h_m} \sigma_{xx}^{(sl)} z(1+z/a)dz + \int_{h_m}^{h_a} \sigma_{xx}^{(ml)} z(1+z/a)dz \tag{4.76}$$

where h_i, h_m and h_a are the z coordinates of the inner wall, the interface of mucosa-submucosa and muscle and the outer wall of muscle, respectively. The superscripts (sl) and (ml) indicate the mucosa–submucosa layer and muscle layers, respectively, and a is the radius of the neutral surface. The solution method of a, h_i, h_m and h_a is similar to that in Chuong and Fung (1986). The value of a, h_i, h_m and h_a are obtained from the following equations

$$R_i = L_i / 2\theta_0, \qquad R_e = L_e / 2\theta_0 \tag{4.77}$$

$$\theta_0 (R_e^2 - R_i^2) = \pi\lambda_z (r_e^2 - r_i^2) \tag{4.78}$$

$$\theta_0 (R_e^2 - (R_i + h_{im}^0)^2) = \pi\lambda_z (r_e^2 - (r_i + h_{im})^2) \tag{4.79}$$

where r_e is the outer radius of the oesophagus in the in vivo state, which can be measured during experiment. Therefore, the inner radius, r_i , the inner layer thickness h_{im} can be obtained from Equations 4.77–79. R_e, R_i, L_e, L_i, θ_0, h_{im}^0, are, respectively, the outer radius, inner radius, outer arc length, inner arc length, opening angle, and inner wall thickness in zero-stress state. λ_z is the longitudinal stretch ratio in the in vivo state.

The stress and strain in the radial direction can not be neglected since the oesophagus is thick-walled. In order to get the location of the neutral surface, the assumption is made that the relative location of the neutral surface does not change during the whole deformation period. According to this assumption, the location of the neutral surface in the in vitro state can be calculated from:

$$\begin{cases} \theta_0 (R_e^2 - R_n^2) = \pi (r_{enoload}^2 - r_{nnoload}^2) \\ \lambda_\theta = (\dfrac{\pi}{\theta_0}) \dfrac{r_{nnoload}}{R_n} = 1 \end{cases} \tag{4.80}$$

$$\theta_0 (R_e^2 - R_i^2) = \pi (r_{enoload}^2 - r_{inoload}^2)$$

where $r_{enoload}$, $r_{inoload}$ and $r_{nnoload}$ are, respectively, the outer radius , the inner radius and the neutral surface radius in the in vitro state; R_n the neutral surface radius in the zero-stress state.

Then the radius of the neutral surface, a, the location of the mucosa-submucosa layer, h_i, the location of the interface between the mucosa-submucosa and muscle layers, h_m, and the location of the muscle layer h_a in the in vivo state can be obtained from

$$\frac{a - r_i}{r_e - r_i} = \frac{r_{nnoload} - r_{inoload}}{r_{enoload} - r_{inoload}}$$

$$h_i = r_i - a, \qquad h_m = r_i + h_{im} - a, \qquad h_a = r_e - a. \tag{4.81}$$

Relationship Between Membrane Tensions, Bending Moments, and Shell Deformation

The deformation of the shell (oesophagus) is described by the displacements in the x, θ, z directions of a point on the neutral surface , u, v and w. The displacements of any other points in the shell can be computed according to the Bernoulli–Kirchhoff hypothesis, then the strains can be computed, and stresses are given by Equation 4.75. Substituting these results into Equation 4.76, we obtain a full set of equations relating N_x, N_θ, M_x, M_θ, with u, v, w and their derivatives.

$\frac{\partial u}{\partial x}$, $\frac{\partial v}{\partial \theta}$, $\frac{\partial^2 w}{\partial x^2}$, $\frac{\partial^2 w}{\partial \theta^2}$ and the elastic constants Y_{xx}^{sl} , $Y_{x\theta}^{sl}$, $Y_{\theta\theta}^{sl}$ for the inner, submucosa layer (sl) of the oesophagus, and Y_{xx}^{ml} , $Y_{x\theta}^{ml}$, $Y_{\theta\theta}^{ml}$ for the outer, muscle layer (ml) of the oesophagus.

In these experiment, axisymmetric perturbations are imposed on the oesophagus by using incremental changes of pressure, axial force and a ring of uniform transverse load. With the axisymmetric load, the membrane shear force $N_{x\theta}$ and transverse shear force Q_θ, and the moment $M_{x\theta}$ are zero, whereas the displacement v=0, and u and w are functions of x only, the circumferential membrane tension is the function of x only. So the expression of membrane tension N_θ, N_x and membrane moment M_x under the axisymmetric loads are:

$$N_\theta = a(\partial u/\partial x)\left(Y_{\theta x}^{(im)}C_1^{(im)} + Y_{\theta x}^{(ad)}C_1^{(ad)}\right) + w\{Y_{\theta\theta}^{(im)}\left[C_1^{(im)} - C_2^{(im)} + C_3^{(im)}\right] + Y_{\theta\theta}^{(ad)}\left[C_1^{(ad)} - C_2^{(ad)} + C_3^{(ad)}\right]\}$$
$$-a^2(\partial^2 w/\partial x^2)\left(Y_{\theta x}^{(im)}C_2^{(im)} + Y_{\theta x}^{(ad)}C_2^{(ad)}\right) \tag{4.82}$$
$$N_x = a(\partial u/\partial x)\{Y_{xx}^{(im)}(C_1^{(im)} + C_2^{(im)}) + Y_{xx}^{(ad)}(C_1^{(ad)} + C_2^{(ad)})\} + w\{Y_{x\theta}^{(im)}C_1^{(im)} + Y_{x\theta}^{(ad)}C_1^{(ad)}\}$$
$$-a^2(\partial^2 w/\partial x^2)\{Y_{xx}^{(im)}(C_2^{(im)} + C_3^{(im)}) + Y_{xx}^{(ad)}(C_2^{(ad)} + C_3^{(ad)})\} \tag{4.83}$$
$$M_x = a^2(\partial u/\partial x)\{Y_{xx}^{(im)}(C_2^{(im)} + C_3^{(im)}) + Y_{xx}^{(ad)}(C_2^{(ad)} + C_3^{(ad)})\} + aw\{Y_{x\theta}^{(im)}C_2^{(im)} + Y_{x\theta}^{(ad)}C_2^{(ad)}\}$$
$$-a^3(\partial^2 w/\partial x^2)\{Y_{xx}^{(im)}C_3^{(im)} + Y_{xx}^{(ad)}C_3^{(ad)}\} \tag{4.84}$$

in which

$$C_1 = (1/a)\int dz, \qquad C_2 = (1/a)\int (z/a)dz, \qquad C_3 = (1/a)\int (z/a)^2 dz. \tag{4.85}$$

They are dimensionless functions of the limits of integration. If the limits are from h_i to h_m, we write these Cs as $C_i^{(im)}(i = 1,2,3)$. If the limits were from h_m to h_a, we write them as $C_i^{(ad)}(i = 1,2,3)$, we write the modulus of the inner layer as $Y_{ii}^{sl}, (i, j = x, z, \theta)$, and the modulus of the oul2ter layer as $Y_{ij}^{ml}, (i, j = x, z, \theta)$, u, w and a are, respectively, the displacement in the x direction, in the radius direction and the radius of neutral surface.

Inflation and Axial Stretch Experiments

In the case of uniform inflation, an incremental pressure Δp_i induces a measurable change of radius without a change of axial length, so the boundary conditions are:

$$u = 0, \qquad w = const.$$

Then, by Laplace equation, $N_\theta = a\Delta P_i$, so Equation 4.82 becomes

$$a\Delta p_i / w = Y_{\theta\theta}^{(im)}\left[C_1^{(im)} - C_2^{(im)} + C_3^{(im)}\right]) + Y_{\theta\theta}^{(ad)}\left[C_1^{(ad)} - C_2^{(ad)} + C_3^{(ad)}\right]. \quad (4.86)$$

The inflation plus stretching experiment can be divided into three sub-experiments as follows.

Sub-experiment 1, uniform stretch without the pressure, so the boundary conditions in this case are:

$$w = const\ , \qquad \partial u / \partial x = const \quad (4.87)$$

These constants can be measured and Equation 4.83 can be written as:

$$F/(2\pi(a+w) = a(\partial u/\partial x)\{Y_{xx}^{(im)}(C_1^{(im)} + C_2^{(im)}) + Y_{xx}^{(ad)}(C_1^{(ad)} + C_2^{(ad)})\} + w\{Y_{x\theta}^{(im)}C_1^{(im)} + Y_{x\theta}^{(ad)}C_1^{(ad)}\} \quad (4.88)$$

where F is the stretch force which can be measured.

Sub-experiment 2. uniform stretch with constant pressure P_i, then by Laplace equation, $N_\theta = P_i w$, so Equation 4.82 can be written as:

$$p_i w = a(\partial u / \partial x)\left(Y_{\theta x}^{(im)}C_1^{(im)} + Y_{\theta x}^{(ad)}C_1^{(ad)}\right) + w\{Y_{\theta\theta}^{(im)}\left[C_1^{(im)} - C_2^{(im)} + C_3^{(im)}\right]) + Y_{\theta\theta}^{(ad)}\left[C_1^{(ad)} - C_2^{(ad)} + C_3^{(ad)}\right]\} \quad (4.89)$$

in which w and $\partial u/\partial x$ can be measured from experiment.

Sub-experiment 3. uniform stretch with constant radius by a controlled change of pressure, then $N_x = F / 2\pi a$, so Equation 4.83 in this case is

$$F / 2\pi a = a(\partial u / \partial x)\{Y_{xx}^{(im)}(C_1^{(im)} + C_2^{(im)}) + Y_{xx}^{(ad)}(C_1^{(ad)} + C_2^{(ad)})\} \quad (4.90)$$

in which F is axial force and $\partial u/\partial x$ is uniaxial strain. Both can be measured from experiment.

Axial Bending Analysis

In the axial bending experiment, the oesophageal sample is mounted between two connectors and stretched to the in vivo length. In an oesophagus, internal pressure need not be applied. A thin nylon surgical suture is wound around the sample as shown in Figure 4.19. One end of the suture is firmly attached to a plate above the sample in order to avoid shearing of the sample during the experiment. Small weights for preventing wrinkling can be hung on the other end of the suture to impose force on the sample. The force–deformation relationship can be obtained from these experiments where the deformation is evaluated in terms of changes in diameter. The zero-stress state of the oesophagus can be obtained as described in Chapter 6.

The boundary conditions in the axial bending with a ring of load of uniform intensity Hkg/m experiment are:

$$\left.\frac{\partial w}{\partial x}\right|_{x=0,x=\pm\infty} = 0, \qquad \left.\frac{\partial^2 w}{\partial x^2}\right|_{x=\pm\infty} = 0, \qquad \left.Q_x\right|_{x=0} = H/2. \quad (4.91)$$

Finally, the result of this experiment can be written as the equation

$$w = -Ha/(2D^{1/4}F^{3/4}) \tag{4.92}$$

where w can be measured during the experiment, and D and F , respectively, are

$$D = a^4\{Y_{xx}^{(im)}C_3^{(im)} + Y_{xx}^{(ad)}C_3^{(ad)}\},$$
$$F = \{Y_{\theta\theta}^{(im)}\left[C_1^{(im)} - C_2^{(im)} + C_3^{(im)}\right]) + Y_{\theta\theta}^{(ad)}\left[C_1^{(ad)} - C_2^{(ad)} + C_3^{(ad)}\right]\}. \tag{4.93}$$
$$-\left(Y_{\theta x}^{(im)}C_1^{(im)} + Y_{\theta x}^{(ad)}C_1^{(ad)}\right)^2 / \left(Y_{xx}^{(im)}C_1^{(im)} + Y_{xx}^{(ad)}C_1^{(ad)}\right)$$

Circumferential Bending Experiments

The oesophageal sample must be mounted in an organ bath between two connectors and the sample stretched to the in vivo length. The cross-section of the sample maintains a circle. A thin metal plate (e.g., length 8 mm and height 0.2 mm) can be used to impose small forces on the mid-portion of the sample. A similar plate placed on the other side of the sample measures the force (see details in Figure 4.16). By placing a prism next to the sample, the sample can be viewed from the vertical position using the video camera and from the horizontal position using the video camera and the prism. The deformation is evaluated as the change in diameter from the two views. The force-deformation data is used in the analysis. The circumferential bending experiment as described here differs from the micropipette bending experiments done previously by Fung and Liu (1995) in the sense that compression rather than suction is applied to the sample. Furthermore, the metal plate is much thinner than the micropipette used previously and the deformation can more readily be measured.

Note that Fung and Liu's (1995) directions were not followed in one important aspect. We did not use micropipettes for the circumferential bending experiment. We used knife edges (i.e., thin plates) to compress the oesophagus at the opposite ends of a diameter in the circumferential bending experiment. This simplifies the experiment and formulas. The extra degree of freedom of sucking on the micropipette was not practical in the oesophagus. A pipette whose length is at least 10 times its width and whose width is smaller than the vessel diameter was used to press on the oesophagus in vivo symmetrically, therefore, the deformation of the oesophagus can be viewed as in a state of plane strain under this boundary condition, so $u, \partial u/\partial x, \partial v/\partial x, \partial w/\partial x, \partial^2 w/\partial x^2$ are zero, and, furthermore, with the hypothesis in shell theory, the stretching of the neutral surface is negligible, so $\partial v/\partial\theta = 0$. With these two assumptions, Equation 4.84 can be written as

$$M_\theta = aw\{Y_{\theta\theta}^{(im)}(C_2^{(im)} - C_3^{(im)}) + Y_{\theta\theta}^{(ad)}(C_2^{(ad)} - C_3^{(ad)})\} - a\frac{\partial^2 w}{\partial\theta^2}\{Y_{\theta\theta}^{(im)}C_3^{(im)} + Y_{\theta\theta}^{(ad)}C_3^{(ad)}. \tag{4.94}$$

Let

$$A = Y_{\theta\theta}^{(im)}C_3^{(im)} + Y_{\theta\theta}^{(ad)}C_3^{(ad)}, \qquad B = Y_{\theta\theta}^{(im)}(C_2^{(im)} - C_3^{(im)}) + Y_{\theta\theta}^{(ad)}(C_2^{(ad)} - C_3^{(ad)}), \qquad k^2 = B/A. \tag{4.95}$$

Regarding Equation 4.78 as a differential equation of w(θ), we obtain the solution of w as:

If k=0, then

$$w(\theta) = \beta_1 + \beta_2\theta + \frac{1}{aA}\int_0^{\theta}(\theta - \xi)M_\theta(\xi)d\xi\,, \tag{4.96}$$

otherwise

$$w(\theta) = \beta_1 \sin k\theta + \beta_2 \cos k\theta + \frac{1}{akA}\int_{-\pi/2}^{\theta} \sin k(\theta - \xi)M_\theta(\xi)d\xi \tag{4.97}$$

the boundary conditions are

$$\left.\frac{\partial w}{\partial\theta}\right|_{\theta=0} = 0, \qquad \left.\frac{\partial w}{\partial\theta}\right|_{\theta=\pm\frac{\pi}{2}} = 0, \qquad \int_0^{2\pi} w(\theta) = 0\,. \tag{4.98}$$

The bending moment $M(\theta)$ in this case can be written as

$$M_0(\theta) = 0.5Fa\sin(\theta) \tag{4.99}$$

where F is the force on the thin plate. This formula is more concise compared to Fung and Liu (1995). F is the press force on the pipette, w is the deformation in radius, and the w value at $\theta=\pi/2$ can be measured from experiment. Because C_2, C_3 are small numbers of the order of 2, 3, respectively, so an assumption of $(C_2^{(im)} - C_3^{(im)}) = (C_2^{(ad)} - C_3^{(ad)}) = 0$ is used in this experiment. The solution of constants β_1, β_2 can be obtained from Equation 4.98.

Use of the Equations in Experiments

In each experiment the values of N_x, N_θ, M_x, M_θ from statics and shell theory, and the deformation u, v, w, $\partial u/\partial x$, $\partial v/\partial\theta$, $\partial^2 w/\partial x^2$, $\partial^2 w/\partial\theta$ from recorded images are determined. Then the derived equations can be used to compute the elastic moduli $Y_{xx}^{sl}, Y_{xx}^{ml}, Y_{x\theta}^{sl}, Y_{x\theta}^{ml}, Y_{\theta\theta}^{sl}, Y_{\theta\theta}^{ml}$.

These equations are solvable when the thicknesses of the two layers are not equal. When the thicknesses of the two layers are equal, the equations are ill-conditioned. In the latter case, other ways must be devised to measure the elastic constants of the two layers, for example, by dissecting the oesophagus into two separate layers to be tested separately. Then we have the task of showing that the dissection did not cause any change in the mechanical properties of the tissue.

Based on the analysis, we find that elastic moduli are highest in the longitudinal direction (approximately 160 kPa for both layers) and that the submucosa layer and the muscle layer have different elastic moduli and different residual strains in the circumferential direction. In the circumferential direction the incremental modulus is a factor 7 higher in the submucosa layer (approximately 6 kPa) compared to the muscle layer 0.9 kPa).

4.3.8 The Stress Distribution in the Layered Wall of the Oesophagus

The previous section presented a method to determine the homeostatic elastic moduli of two layers of the oesophagus. In the multi-layered composite oesophageal structure, the material constants and the stress distribution across the whole wall is non-homogeneous. Each layer has its own zero-stress state as well as its own mechanical properties. Therefore, data on the strain and stress distribution in different layers will help us understand the relationship between the stress and remodelling of the tissue. In this section a method will be presented for determination of the stress distribution in the layered oesophageal wall.

Using this method, the muscle layer, submucosa layer, as well as the intact layer are simulated individually from multiaxial inflation experiment. An anisotropy exponential pseudo-strain energy density function is used as the constitutive equation to fit to the experimental loading curve, and is used to calculate the stress through the wall of each layer. With the present results, a reasonable reference state to the multi-layered tissue can be determined and further research about the layered oesophagus, such as the numerical simulation of the remodelling in oesophagus, can be initiated.

Preliminary Work

The muscle layer, submucosa layer, as well as the intact layer of oesophagus wall, are characterised by a pseudo-strain energy density function of the exponential type (Chuong and Fung, 1983; 1986; Fung et al., 1993):

$$\rho_0 w = (c/2)[\exp(Q)] \tag{4.100}$$

where $Q = a_1 E_{\theta\theta}{}^2 + a_2 E_{zz}{}^2 + a_3 E_{rr}{}^2 + 2a_4 E_{\theta\theta} E_{zz} + 2a_5 E_{rr} E_{zz} + 2a_6 E_{\theta\theta} E_{rr}$. All E_{ij} (i,j=θ,r,z) terms are Green strains referred to the zero-stress state, θ, r, z are the circumferential, radial and axial directions, respectively. The parameter c expresses a scale to all stress components, and the parameters a_i are the material constant, which affect the non-linearity and anisotropy of the material. Each layer of oesophagus is modelled as different cylinders in the no-load state and with opening angles in the zero-stress state. With the unloaded dimensions and the opening angle in the zero-stress state, the dimensions in the zero-stress state can be calculated by requiring the same value in both states. The strain referenced to the zero-stress state is expressed as

$$\lambda_r = \frac{\partial r}{\partial R}, \tag{4.101}$$

$$\lambda_\theta = \frac{\pi r}{R\theta}, \tag{4.102}$$

$$\lambda_z = \frac{l}{L} \tag{4.103}$$

where $\lambda_r, \lambda_\theta, \lambda_z$ is the stretch ratio in the radial, circumferential and longitudinal direction, respectively. r, l and R, L are the radii and the length in the loaded and the zero-stress state, respectively; θ is the opening angle in the zero-stress state. The Green strain in the three directions can be calculated from

$$E_i = \frac{1}{2}(\lambda_i^2 - 1), \qquad i = r, \theta, z. \tag{4.104}$$

The Cauchy stress, σ, calculated from the partial derivatives of W, can be written for the three directions as

$$\sigma_\theta = C(1+2E_\theta)(a_1E_\theta + a_4E_z + a_6E_r)e^Q \quad \sigma_z = C(1+2E_z)(a_4E_\theta + a_2E_z + a_5E_r)e^Q,$$
$$\sigma_r = C(1+2E_r)(a_6E_\theta + a_5E_z + a_3E_r)e^Q. \tag{4.105}$$

According to the equilibrium equation in cylindrical coordinates,

$$\frac{d\sigma_r}{dr} + \frac{\sigma_r - \sigma_\theta}{r} = 0 \tag{4.106}$$

with boundary condition

$$\sigma_r\big|_{r=r_i} = -P_i, \qquad \sigma_r\big|_{r=r_o} = -P_0 = 0 \tag{4.107}$$

where r_i, r_0 and P_i, P_0 are the radius and the pressure in the inner surface and outer surface, respectively. The internal pressure and the axial force acting on the sample can be calculated from Equations 4.100–106 by numerical integration:

$$p_i = \int_{r_o}^{r_i} c[(1+2E_r)(a_6E_\theta + a_5E_z + a_3E_r) - (1+2E_\theta)(a_1E_\theta + a_4E_z + a_6E_r)]\exp(Q)\frac{dr}{r} \tag{4.108}$$

and

$$F = -\pi r_i^2 p_i + 2\pi \int_{r_i}^{r_o} c[(1+2E_z)(a_4E_\theta + a_2E_z + a_5E_r)]\exp(Q)r\,dr. \tag{4.109}$$

Stress Analysis

The material parameters in Equation 4.100 of the individual layer in oesophagus can be determined from Equation 4.108 by non-linear least square method to minimise the differences between the experimental and the predicated internal pressure, respectively. In this paper, the Levenberg–Marquardt method is used to evaluate the constitutive constant from inflation experimental data and Equation 4.108.

With the material constants and the strain calculated from Equations 4.101–104 of different layers, the stress distribution in each layer can be easily obtained from Equation 4.105. Figure 4.20 shows a setup for obtaining data from the rat oesophagus. By means of this kind of analysis it has been confirmed that the submucosa layer is the stiffest. Therefore, the wall of oesophagus should be modelled as at least a two-layer system, each with its own constitutive equation. It is also clear that the stress distribution is non-linear within each layer and discontinuous at the interface between the layers (figure 4.21).

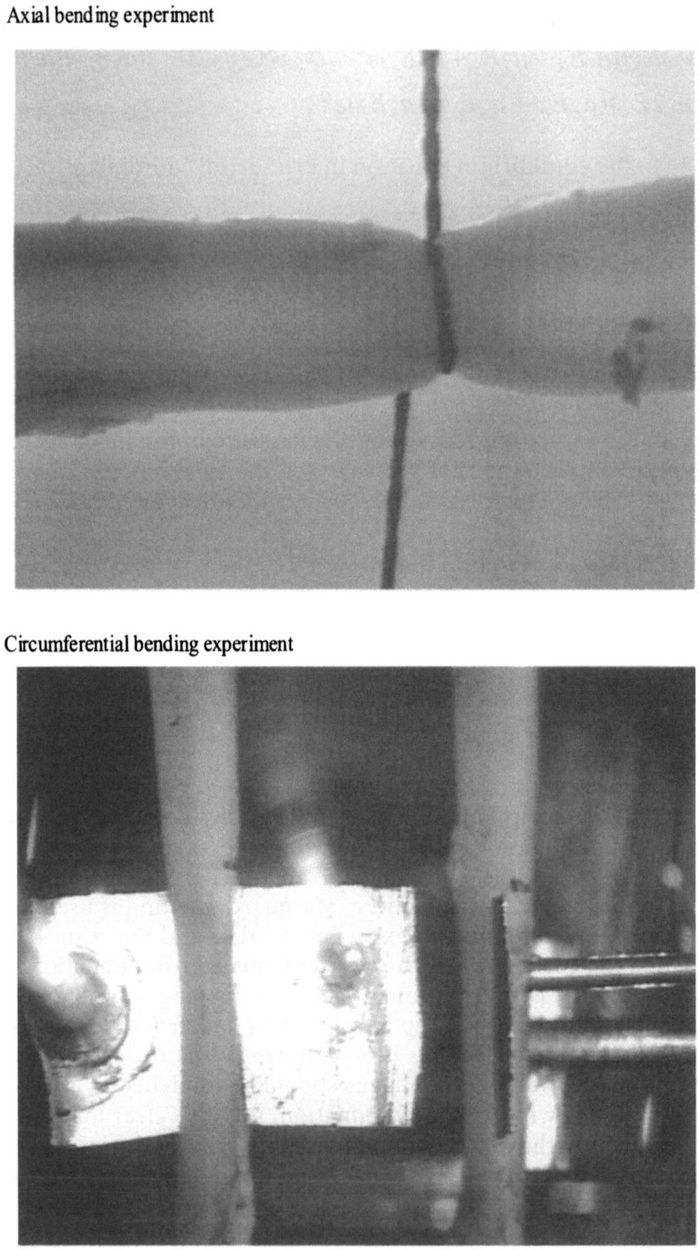

Figure 4.20. Two types of bending experiments on the isolated guinea-pig oesophagus.

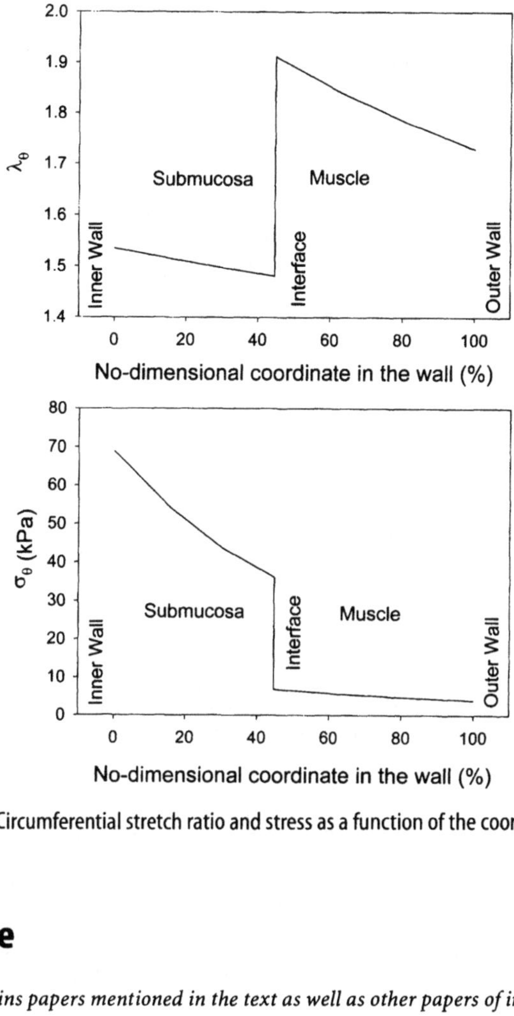

Figure 4.21. Circumferential stretch ratio and stress as a function of the coordinate in the wall.

4.4 Literature

The literature list contains papers mentioned in the text as well as other papers of interest.

Andersen, NT, Hobolth, A, Gregersen, H, Væth, M. 2000. Implicitly defined non-linear models of elastic properties of organs. *Research Report 2000-4*, August. Department of Biostatistics, University of Aarhus, Denmark.

Assentoft, JE, Gregersen, H, O'Brien ,WD. 2000. Determination of biomechanical properties in the guinea pig esophagus by means of high-frequency ultrasound and impedance planimetry. *Dig Dis Sci*, **45**, 1260–66.

Assentoft, JE, Gregersen, H, O'Brien Jr., WD. 2001. Propagation speed of sound assessment in the layers of the guinea pig esophagus by means of acoustic microscopy. *Ultrasonics*, **39**, 263–8.

Assentoft, JE, Jørgensen, CS, Gregersen, H, Christensen, LL, Djurhuus, JC, O'Brien, WD. 1996. Scanning laser acoustic microscopy as a method for characterizing the acoustic properties of the individual layers of biological tissue. *IEEE Eng Med Biol Mag*, **15**, 42–5.

Axenborg, J. 1979. An electrical impedance method for measurement of aortic cross-sectional area. Part 1: Principles and experimental evaluation. Thesis, Gothenburg University.

Bell, AM, Pemberton, JH, Hanson, RB, Zinsmeister, AR. 1991. Variations in muscle tone of the human rectum: recordings with an electromechanical barostat. *Am J Physiol*, **260**, G17–25.

Briggs, GAD, Wang, J, Gundle, R. 1993. Quantitative acoustic microscopy of individual living human cells. *J Microsc*, **172**, 3–12.

Briggs, GAD. 1992. A little elementary acoustics. In: *Acoustic Microscopy*, pp. 78–101. Oxford: Clarendon Press.

Cannon, WB. 1911. *The Mechanical Factors of Digestion*. Oxford.

Chen, CT, Millero, FJ. 1977. Speed of sound in seawater at high pressures. *J Acoust Soc Am*, **62**, 1129–35.

Chuong, CJ, Fung, YC. 1983. Three-dimensional stress distribution in arteries. *J Biomech Eng*, **105**, 268–74.

Chuong, CJ, Fung, YC. 1986. On residual stress in arteries. *J Biomech Eng*, **108**, 189–92.

Colstrup, H, Mortensen, SO, Kristensen, JK. 1983. A new method for the investigation of the closure function of the resting female urethra. *J Urol*, **130**, 507–11.

Colstrup, H, Mortensen, SO, Kristensen, JK. 1983. A probe for measurements of related values of cross-sectional area and pressure in the resting female urethra. *Urol Res*, **11**, 139–43.

Colstrup, H. 1984. Rigidity of the resting female urethra. Part I. Static measurements. *J Urol*, **132**, 78–81.

Colstrup, H. 1984. Rigidity of the resting female urethra. Part II. Dynamic measurements. *J Urol*, **132**, 82–6.

Colstrup, H. 1985. Voluntary contractions in the female urethra. *J Urol*, **134**, 902–6.

Dall, FH, Jørgensen, CS, Djurhuus, JC, Gregersen, H. 1991. Biomechanical wall properties of the porcine rectum. A study using impedance planimetry. *Dig Dis*, **9**, 347–53.

Dall, FH, Jørgensen, CS, Houe, D, Gregersen, H, Djurhuus, JC. 1993. Biomechanical wall properties of the human rectum. A study with impedance planimetry. *Gut*, **34**, 1581–6.

Deng, SX, Tomioka, J, Debes, JC, Fung, YC. 1994. New experiments on shear modulus of elasticity of arteries. *Am J Physiol*, **266**, H1–10.

Djurhuus, JC, Constantinou, CE. 1979. Assessment of pyeloureteral function using a flow velocity and cross-sectional diameter probe. *Invest Urol*, **17**, 103–7.

Duch, BU, Petersen, JAK, Gregersen, H. 1998. Luminal cross-sectional area and distensibility in the porcine bile duct. *Neurogastroenterol Motil*, **10**, 203–9.

Duch, BU, Petersen, JAK, Vinter-Jensen, L, Gregersen, H. 1996. Elastic properties in the circumferential direction in isolated rat small intestine. *Acta Physiol Scand*, **157**, 157–63.

Dunn, F, Edmonds, PD, Fry, WJ. 1969. Absorption and dispersion of ultrasound in biological media In: *Biological Engineering*, ed. Schwan, HP. New York: McGraw Hill.

Dunn, OJ, Clark, VA. 1987. *Applied Statistics: Analysis of Variance and Regression*. New York: John Wiley.

Elbrønd, H, Tøttrup, A, Forman, A. 1991. Mechanical properties of isolated smooth muscle from rabbit sphincter of oddi and duodenum. *Scand J Gastroenterol*, **26**, 289–94.

Fass, J, Silny, J, Braun, J, Heindrichs, U, Dreuw, B, Schumpelick, V, Rau, G. 1989. Ein neues verfahren zur quantitativen bestimmung oesophagealer motilitätsmuster mittels einer vielfachen impedanzmessung. In: *Chirurgiches forum 1989 f. experim. u. kliniche forschung*, ed. Hamelmann et al. Berlin: Springer-Verlag.

Fisher, MA, Hendrix, TR, Hunt, JN, Murrills, AJ. 1978. Relation between volume swallowed and velocity of the bolus ejected from the pharynx into the esophagus. *Gastroenterology*, **74**, 1238–40.

Flügge, W. 1960. *Stresses in Shells*. Berlin: Springer-Verlag.

Frøbert, O, Bagger, JP, Gregersen, H, Storkholm, JH. 1995. Methodologies for the measurement of mechanical properties of the aorta and the coronary arteries. *Surg Res Comm*, **17**, 181–99.

Frøbert, O, Gregersen, H, Bagger, JP. 1996. Mechanics of porcine coronary arteries ex vivo employing impedance planimetry: a new intravascular technique. *Ann Biomed Eng*, **24**, 148–55.

Frøbert, O, Schiønning, J, Gregersen, H, Baandrup, U, Petersen, JAK, Bagger, JP. 1997. Impaired human coronary artery distensibility by atherosclerotic lesions. A mechanical and histological investigation. *Int J Exp Pathol*, **78**, 421–8.

Frøbert, O, Storkholm, JH, Gregersen, H, Bagger, JP. 1996. In vivo assessment of luminal cross-sectional areas and circumferential tension–strain relations of the porcine aorta. *Scand J Thor Cardiovasc Surg*, **30**, 11–19.

Frøbert. O, Mikkelsen. EO, Gregersen. H, Nyborg. NCB, Bagger. JP. 1996. Porcine coronary artery pharmacodynamics in vitro evaluated by a new intravascular technique. Relation to axial stretch. *J Pharmacol Toxicol Methods*, **36**, 13–19.

Fung, YC, Liu, SQ, Zhou, JB. 1993. Remodeling of the constitutive equation while a blood vessel remodels itself under stress. *J Biomech Eng*, **115**, 1670–76.

Fung, YC, Liu, SQ. 1995. Determination of the mechanical properties of the different layers of blood vessel in vivo. *Proceedings of the National Academy of Sciences of the USA*, **92**, 2169–73.

Fung, YC. 1984. *Biodynamics. Circulation*. New York: Springer-Verlag.

Fung, YC. 1990. *A First Course in Continuum Mechanics*. Third Edition. New Jersey: Prentice Hall.

Fung, YC. 1990. *Biomechanics: Motion, Flow, Stress and Growth*. New York: Springer.

Fung, YC. 1993. *Biomechanics. Mechanical Properties of Living Tissues*. New York: Springer-Verlag.

Goss, SA, O'Brien Jr., W.D. 1979. Direct ultrasonic velocity measurements of mammalian collagen threads. *J Acoust Soc Am*, **65**, 507–11.

Gregersen, H, Andersen, MB. 1991. Impedance measurement system for quantification of cross-sectional area in the gastrointestinal tract. *Med Biol Eng Comput*, **29**, 108–10.

Gregersen, H, Barlow, J, Thompson, DG. 1999. Development of a computer-controlled tensiometer for real-time measurements of tension in tubular organs. *Neurogastroenterol Motil*, **11**, 109–18.

Gregersen, H, Christensen, LL. 1996. Pressure-cross-sectional area relations and elasticity in the rabbit esophagus. *Digestion*, **57**, 174–9.
Gregersen, H, Djurhuus, JC. 1991. Impedance planimetry. A new approach to biomechanical wall properties in the intestine. *Dig Dis*, **9**, 332–40.
Gregersen, H, Giversen, IM, Rasmussen, LM, Tøttrup, A. 1992. Biomechanical wall properties and collagen content in the partially obstructed opossum esophagus. *Gastroenterology*, **103**, 1547–51.
Gregersen, H, Giversen, IM, Rasmussen, LM, Tøttrup, A. 1992. Biomechanical wall properties and collagen content in the partially obstructed opossum esophagus. *Gastroenterology*, **103**, 1547–51.
Gregersen, H, Jensen, LS, Djurhuus, JC. 1988. Changes in oesophageal wall biomechanics after portal vein banding and variceal sclerotherapy measured by a new technique. An experimental study in rabbits. *Gut*, **29**, 1699–1704.
Gregersen, H, Jørgensen, CS, Dall, FH, Jensen, SL. 1992. Characteristics of spontaneous and evoked motility in the isolated perfused porcine duodenum. *J Applied Physiol*, **73**, 9–19.
Gregersen, H, Jørgensen, CS, Dall, FH, Jensen, SL. 1992. Spontaneous and evoked motility in the isolated perfused porcine duodenum. *J Applied Physiol*, **73**, 9–19
Gregersen, H, Jørgensen, CS, Dall, FH. 1992. Biomechanical wall properties in the isolated perfused porcine duodenum. An experimental study using impedance planimetry. *J Gastrointest Motil*, **4**, 125–35.
Gregersen, H, Kassab, GS, Pallencoea, E, Lee, C, Chien, S, Skalak, R, Fung YC. 1997. Morphometry and strain distribution in guinea pig duodenum with reference to the zero-stress state. *Am J Physiol*, **273**, G865–74.
Gregersen, H, Kassab, GS. 1996. Biomechanics of the gastrointestinal tract. *Neurogastroenterol Motil*, **8**, 277–97
Gregersen, H, Knudsen, L, Eika, B, Frøkiæ,r J, Djurhuus, JC. 1996. Regional differences exist in elastic wall properties in the ureter. *Scand J Urol Nephrol*, **30**, 343–8.
Gregersen, H, Knudsen, L, Eika, B, Salling Nerstrøm, L, Rasmussen, L, Jensen, LS. 1991. A time-dependent study of passive esophageal wall properties and collagen content in rabbits with esophageal varices. *Dig Dis Sci*, **36**, 1050–56.
Gregersen, H, Kraglund, K, Djurhuus, JC. 1990. Variations in duodenal cross-sectional area during the interdigestive migrating motility complex. *Am J Physiol*, **259**(*Gastrointest Liver Physiol*, **22**), G26–31.
Gregersen, H, Lee, C, Chien, S, Skalak, R, Fung, YC. 1999. Strain distribution in the layered wall of the esophagus. *J Biomed Eng*, **121**, 442–8.
Gregersen, H, Lee, TC, Chien, S, Skalak, R, Fung, YC. 1999. Strain distribution in the layered wall of the esophagus. *J Biomech Eng*, **121**, 442–8.
Gregersen, H, Lundby, L, Overgaard, J. 2002. Early and late effects of irradiation on the morphometry and residual strain of the mouse rectum. *Dig Dis Sci*, **47**, 1472–9.
Gregersen, H, Orvar, K, Christensen, J. 1992. Biomechanical properties of duodenal wall and duodenal tone during phase I and phase II of the MMC. *Am J Physiol*, **263** (*Gastrointest Liver Physiol*, **26**), G795–801.
Gregersen, H, Orvar, K, Christensen, J. 1992. Biomechanical wall properties and tone during phase I and phase II of the MMC. *Am J Physiol*, **263**, G795–801
Gregersen, H, Stødkilde-Jørgensen, H, Djurhuus, JC, Mortensen, SO. 1988. The four-electrode impedance technique: a method for investigation of compliance in luminal organs. *Clin Phys Physiol Meas*, **9** (Suppl. A), 61–4.
Gregersen, H, Vinter Jensen, L, Juhl, CO, Dajani, EZ. 1996. Impedance planimetric characterization of the distal oesophagus in the goettingen minipig. *J Biomech*, **29**, 63–8.
Harada, T. 1985. Experimental study on measurements of compliance and cross-sectional area in the urethra through field gradient principle. *Jpn J Urol*, **76**, 360–70.
Harris, JH, Therkelsen, EE, Zinner, NR. 1971. Electrical measurement of ureteral flow. In: *Urodynamics*, ed. Boyarsky, S, Tanagho, EA, Gottschalk, CW, Zimskind, PD, pp. 465–72. London: Academic Press.
Hausken, T, Ødegaard, S, Matre, K, Berstad, A. 1992. Antroduodenal motility and movements of luminal contents studied by duplex sonography. *Gastroenterology*, **102**, 1583–90.
Jørgensen, CS, Ahrensberg, JM, Gregersen, H, Flyvberg, A. 2001. Tension-strain relations and morphometry of the rat small intestine in experimental diabetes. *Dig Dis Sci*, **46**, 960–7.
Jørgensen, CS, Dall, FH, Jensen, SL, Gregersen, H. 1995. A new combined ultrasound-impedance planimetry measuring system for the quantification of organ wall biomechanics in vivo. *J Biomech*, **28**, 863–7.
Jørgensen, CS, Dall, FH, Jensen, SL, Storkholm, J, Gregersen, H. 1991. Viscoelastic properties of the isolated perfused porcine duodenum. *Dig Dis*, **9**, 401–7.
Jørgensen, CS, Gregersen, H. 1998. Scanning acoustic microscopy in biomedical research. In: *Aspects of Biomedical Engineering in Denmark*, ed. The Danish Society for Biomedical Engineering Society, pp. 61–9. Copenhagen.
Juhl, CO, Vinter-Jensen, L, Djurhuus, JC, Gregersen, H, Dajani, EZ. 1994. Biomechanical properties of the

oesophagus damaged by endoscopic sclerotherapy An impedance planimetric study in minipigs. *Scand J Gastroenterol*, **29**, 867–73.

Kessler, LW, Fields, SI, Dunn, F. 1974. Acoustic microscope of mammalian kidney. *J Clin Ultrasound*, **2**, 317–20.

Kessler, LW, Korpel, A, Palermo, PR. 1972. Simultaneous acoustic and optical microscopy of biological specimens. *Nature*, **239**, 111–12.

Knudsen, L, Gregersen, H, Eika, B, Frøkier, J. 1994. Biomechanical wall properties and collagen content in the ureter. An experimental study in pigs. *Neurourol Urodyn*, **13**, 597–608.

Korpel, A, Kessler, LW, Palermo, PR. 1971. Acoustic microscope operating 100 MHz. *Nature*, **232**, 110–11.

Krogh, K, Ryhammer, AM, Lundby, L, Gregersen, H, Laurberg, TS. 2001. Comparison of methods used for measurement of rectal compliance. *Dis Colon Rectum*, **44**, 199–206.

Lose, G, Colstrup, H, Saksager, K, Kristensen, JK. 1986. A new probe for measurement of related values of cross-sectional area and pressure in a biological tube. *Med Biol Eng Comput*, **24**, 488–92.

Lose, G, Colstrup, H, Saksager, K, Kristensen, JK. 1988. New method for static and dynamic measurement of related values of cross-sectional area and pressure in the female urethra. *Neurourol Urodyn*, **6**, 465–76.

Lose, G, Colstrup, H. 1990. Mechanical properties of the urethra in healthy and stress-incontinent females: dynamic measurements in the resting urethra. *J Urol*, **144**, 1258–62.

Lose, G, Colstrup, H. 1991. Pathophysiologic subdivision of genuine stress incontinence. *Neurourol Urodyn*, **10**, 247–56.

Lose, G, Colstrup, H. 1991. Urethral pressure and power generation during coughing and voluntary contraction of the pelvic floor in healthy females. *Br J Urol*, **67**, 573–9.

Lose, G, Schroeder, T. 1990. Pressure/cross-sectional area probe in the assessment of urethral closure function. Reproducibility of measurement. *Urol Res*, **18**, 143–7.

Lose, G. 1990. Impact on changes in posture and bladder filling on the mechanical properties of the urethra in healthy and stress-incontinent females. *Neurourol Urodyn*, **9**, 459–69.

Lose, G. 1991. Urethral pressure and power generation during coughing and voluntary contraction of the pelvic floor in females with genuine stress incontinence. *Br J Urol*, **67**, 580–5.

Lundby, L, Dall, FH, Gregersen, H, Overgård, J, Laurberg, S. 1999. Distensibility of the mouse rectum. Application of impedance planimetry for studying age-related changes. Int J *Colorectal Dis*, **1**, 34–41.

Meiss, RA. 1971. Some mechanical properties of cat intestinal muscle. *Am J Physiol*, **220**, 2000–7.

Mortensen, SO, Djurhuus, JC, Rask-Andersen, H. 1983. A system for measurements of micturition urethral cross-sectional areas and pressures. *Med Biol Eng Comput*, **21**, 482–8.

Mortensen, SO. 1989. Cross-sectional areas in the normal male urethra during voiding. Thesis. Mortensens Forlag, Charlottenlund, Denmark.

Orvar, KB, Gregersen, H, Christensen, J. 1993. Biomechanical characteristics of the human esophagus. *Dig Dis Sci*, **38**, 197–205.

Pedersen, JAK, Djurhuus, C, Koff, J, Vinter-Jensen, L, Gregersen, H. 1998. Endoscopic sclerotherapy in the porcine esophagus changes the luminal cross-sectional area and wall distensibility in a dose- and time-dependent way. *Dig Dis Sci*, **43**, 521–8.

Petersen, JAK, Duch, BU, Gregersen, H. 1999. Luminal cross-sectional area and wall distensibility in the isolated porcine oesophagus in situ. *Int J Surg Invest*, **1**, 23–8.

Pope II, CE. 1967. A dynamic test of sphincter strength: Its application to the lower esophageal sphincter. *Gastroenterology*, **52**, 779–86.

Prescott, J. 1946. *Applied Elasticity*. New York: Dover Publications.

Price, JM, Patitucci, P, Fung, YC. 1977. Mechanical properties of taenia coli smooth muscle in spontaneous contraction. *Am J Physiol*, **233**, C47–55.

Rao, SSC, Gregersen, H, Hayek, B, Summers, RV, Christensen, J. 1996. Unexplained chest pain: The hypersensitive, hyperreactive and poorly compliant esophagus. *Ann Internal Med*, **124**, 950–8.

Rask-Andersen, H, Djurhuus, JC. 1976. Development of a probe for endoureteral investigation of peristalsis by flow velocity and cross sectional area measurement. *Acta Chir Scand*, **472**, 59–65.

Rasmussen, OØ, Colstrup, H, Lose, G, Christiansen, J. 1990. A technique for the dynamic assessment of anal sphincter function. *Int J Colorect Dis*, **5**, 135–41.

Rouillon, J-M, Azpiroz, F, Malagelada, J-R. 1991. Reflex changes in intestinal tone: relationship to perception. *Am J Physiol*, **261**, G280–6.

Rouillon, J-M, Azpiroz, F, Malagelada, J-R. 1991. Sensorial and intestino-intestinal reflex pathways in the human jejunum. *Gastroenterology*, **101**, 1606–12.

Silny, J. 1991. Intraluminal multiple electric impedance procedure for measurement of gastrointestinal motility. *J Gastrointest Motil*, **9**, 151–62.

Storkholm, JH, Frøbert, O, Gregersen, H. 1997. Static elastic wall properties in the abdominal porcine aorta in vitro and in vivo. *Eur J Endovasc Surg*, **13**, 31–6.

Storkholm, JH, Jørgensen, CS, Dall, FH, Jensen, SL, Gregersen, H. 1994. Differences exist in passive elastic wall properties between segments of isolated guinea pig ileum and duodenum in vitro. *J Gastrointest Motil*, **6**, 21–7.

Storkholm, JH, Villadsen, GE, Krogh, K, Jørgensen, CS, Gregersen, H. 1997. Dimensions and mechanical properties in segments of porcine aorta determined by means of combined impedance planimetry-high-frequency ultrasound. *Med Biol Eng Comput*, **35**, 21–6.

Storkholm, JH, Villadsen, GV, Jensen, SL, Gregersen, H. 1995. Passive elastic wall properties in the isolated guinea pig small intestine. *Dig Dis Sci*, **40**, 976–82.

Storkholm, JH, Villadsen, GV, Jensen, SL, Gregersen, H. 1998. Mechanical properties and collagen content differ between isolated guinea pig duodenum, jejunum and distal ileum. *Dig Dis Sci*, **43**, 2034–41.

Tanko, L, Mikkelsen, EO, Frøbert, O, Bagger, JP, Gregersen, H. 1998. A new method for combined isometric and isobaric pharmacodynamics studies on porcine coronary arteries. *Clin Exp Pharmacol Physiol*, **25**, 919–27.

Tanko, LB, Simonsen, U, Frøbert, O, Gregersen, H, Bagger, JP, Mikkelsen, EO. 2000. Vascular reactivity to nifedipine and Ca^{++} in vitro. The role of preactivation, wall tension and geometry. *Eur J Pharmacol*, **387**, 305–14.

Tanko, LB, Simonsen, U, Matrougui, K, Gregersen, H, Frøbert, O, Bagger, JP, Mikkelsen, EO. 2001. Axial stretch modifies contractility of porcine coronary arteries by a protein kinase C-dependent mechanism. *Pharmacol Toxicol*, **88**, 89–97.

Tervola, KMU, Foster, SG, O'Brien Jr., WD. 1985. Attenuation coefficient measurement technique at 100 MHz with the scanning laser acoustic microscope. *IEEE Trans Sonics and Ultrasonics*, **32**, 259–65.

Tervola, KMU, Gummer, MA, Erdman Jr., JW, O'Brien Jr., WD. 1985. Ultrasound attenuation and velocities in rat liver as a function of fat concentration: A study at 100 MHz using a scanning laser acoustic microscope. *J Acoust Soc Am*, **77**, 307–13.

Tervola, KMU, O'Brien Jr., WD. 1985. Spatial frequency domain technique: An approach for analyzing the scanning laser acoustic microscope interferogram images. *IEEE Trans Sonics and Ultrasonics*, **4**, 544–54.

Thind, P, Colstrup, H, Lose, G, Kristensen, JK. 1991. Method for evaluation of the urethral closure mechanism in women during standardised changes of cross-sectional area. *Clin Phys Physiol Meas*, **12**, 163–70.

Thind, P, Lose, G, Colstrup, H, Andersson, K-E. 1992. The effect of alpha-adrenoceptor stimulation and blockade on the static urethral sphincter function in healthy females. *Scand J Urol Nephrol*, **26**, 219–25.

Thind, P, Lose, G, Colstrup, H. 1991. How to measure urethral elastance in a simple way. *Urol Res*, **19**, 241–4.

Thind, P, Lose, G, Colstrup, H. 1992. Resistance to forced opening of the resting urethra in healthy females assessed by experimental simulation of urine ingression. *Neurourol Urodyn*, **11**, 201–8.

Thind, P, Lose, G. 1992. Urethral stress relaxation phenomenon in healthy and stress-incontinent women. *Br J Urol*, **69**, 75–8.

Thind, P. 1992. Stress relaxation phenomenon in the healthy female urethra. *Br J Urol*, **69**, 71–4.

Tøttrup, A, Forman, A, Uldbjerg, N, Funch-Jensen, P, Andersson, KE. 1990. Mechanical properties of isolated human esophageal smooth muscle. *Am J Physiol*, **258**, G338–43.

Valori, RM, Collins, SM, Daniel, EE, Reddy, SN, Shannon, S, Jury, J. 1986. Comparison of methodologies for the measurement of antroduodenal motor activity in the dog. *Gastroenterology*, **91**, 546–53.

Valori, RM, Kumar, D, Wingate, DL. 1986. Effects of different types of stress and of "prokinetic" drugs on the control of the fasting motor complex in humans. *Gastroenterology*, **90**, 1890–900.

Villadsen, GE, Pedersen, JAK, Vinter-Jensen, L, Juhl, CO, Gregersen, H. 1995. Impedance planimetric characterization of the normal and diseased oesophagus. *Surg Res Comm*, **17**, 225–42.

Villadsen, GE, Storkholm, JH, Hendel, L, Vilstrup, H, Gregersen, H. 1997. Impedance planimetric characterization of esophagus in systemic sclerosis patients with severe involvement of esophagus. *Dig Dis Sci*, **4**, 2317–26.

Villadsen, GE, Storkholm, JH, Zachariae, H, Hendel, L, Bendtsen, F, Gregersen, H. 2001. Esophageal pressure–cross-sectional area distributions and secondary peristalsis in relation to subclasification of systemic sclerosis. *Neurogastroenterol Motil*, **13**, 199–210.

Vinter-Jensen, L, Duch, BU, Pedersen, JAK, Ryslev, A, Gregersen, H. 1996. Systemic treatment with epidermal growth factor in the rat. Biomechanical properties of the growing small intestine. *Reg Peptides*, **61**, 135–42.

Vinter-Jensen, L, Juhl, CO, Eika, B, Gregersen, H, Dajani, EZ. 1995. Epidermal growth factor attenuates the sclerotherapy-induced biomechanical properties of the oesophagus: An experimental study in the Goettingen minipig. *Scand J Gastroenterol*, **30**, 614–19.,

Vinter-Jensen, L, Juhl, CO, Gregersen, H. 1994. Regional differences in biomechanics of the oesophagus. An impedance planimetry study in pigs. *J Gastrointest Motil*, **6**, 233–8.

Vinter-Jensen, L, Juhl, CO, Gregersen, H. 1994. Regional differences in passive elastic wall properties of the esophagus: An impedance planimetric study in pigs. *Neurogastroenterol and Motility*, **6**, 233–8.

Yen, RT, Fung, YC, Bingham, N. 1980. Elasticity of small pulmonary arteries in the cat. *J Biomech Eng*, **102**, 170–77

Yu, Q, Zhou, J, Fung, YC. 1993. Neutral axis location in bending and Young's modulus of different layers of arterial wall. *Am J Physiol*, **265**, H52–60.

5. Gastrointestinal Smooth Muscle Mechanical Behaviour and Neural Circuits

The motor function of the gastrointestinal tract remains an empirically realised activity, incomprehensible and unpredictable to a high degree in comparison to other areas of systemic physiology. A useful scheme of the working organisation of the whole system could provide a theory to be used both to interpret the variety of normal or usual behaviours and to predict responses to exceptional conditions. Our ignorance about gastrointestinal motor physiology allows only the most superficial interpretations and predictions in health and disease. The empiricism surrounding gastrointestinal motility derives from the complexity of the biological and the physical processes, both the muscle and the mechanics. The peristaltic transport of material in the gastrointestinal tract is a neuromuscular function affected by a number of factors. The nervous system controls the twitch topology, frequency and time sequence. The intrinsic muscle mechanics, with its length–tension–velocity characteristics, complements the nervous function.

5.1 Models for the Study of Gastrointestinal Smooth Muscle Function, Motility and Mechanics

An overwhelming number of models ranging from isolated smooth muscle strips to in vivo preparations exist. Selection of an appropriate model is essential and should be based primarily on the hypothesis in question but will inevitably also depend on review of the literature, economics, laboratory facilities, etc. Some useful models for investigation of motility, sensory responses and biomechanics are described in the following sections.

From other parts of this book we know that selection of a proper model must depend on the mechanical environment for the mechanoreceptors. Previous studies have shown that the receptors are located both in the submucosa and in the muscle layers. From the zero-stress state studies (Chapter 6) we know that a gradient in strain exists from the mucosa to the serosa of the wall under physiological conditions; hence the mechanoreceptors from the different layers are exposed to different strains. This suggests that the receptors from different locations in the wall may have different zero-settings and respond differently to the same stimuli due to the variation in the magnitudes of tensile stresses and strain during distension and in the magnitudes of compressive stresses and strain during muscle contraction. A difference in the responses of mucosal and muscle receptors to stimuli have indeed been observed (Grundy and Scratcherd, 1990). This speaks in favour of studying the receptor kinematics by using experimental models on the intact organ (in vivo or in vitro). Obviously, the receptors also react to changes in the chemical environment; i.e., during ischaemia.

5.1.1 In Vivo Models

In vivo models have the advantage that anaesthetics often can be avoided. In animals and humans the response to mechanical, thermal, chemical and electrical stimulation can be evaluated. In animals the response can be measured as the activity in afferent nerves whereas in humans, sensation in terms of symptom and pain development can be reported, or evoked potentials can be measured from the skull. Animals with chronically implanted serosal electromyographic electrodes and tension-sensitive strain gauges are widely used in gastrointestinal motility research. In addition, many studies have been done creating intestinal loops and anastomosis. After the operation the animal is allowed to recover postoperatively for days or weeks dependent on the animal. Another widely used model is to create a chronic fistula. For the stomach and small intestine it can be done ad modem Thomas (1941). Hereby, easy access to the stomach and gut is obtained. It is also possible to create fistulas on the neck to the oesophagus or through the abdominal wall to the large bowel.

Whereas the in vivo models have been useful for studying the contractile patterns of the gastrointestinal tract, they have been less used for the study of elastic properties. The reasons for this are manifold. It is difficult to obtain complete relaxation of the smooth muscle in vivo and methodological developments were limited. Techniques like impedance planimetry and ultrasound have now been developed to overcome some of the limitations.

5.1.2 In Vitro Models

Isolated Perfused Gastrointestinal Segments

In recent years an extensive literature has been presented on isolated perfused organ models. This is due to many advances in surgical techniques and better understanding of the physiological and biochemical principles involved in organ preservation but also to improvements in measuring apparatus. In gastroenterology, isolated perfused preparations, e.g., the gastro-oesophageal sphincter (Thomas and Earlam, 1973), the stomach (Saffouri et al., 1980), the antrum (Holst et al., 1983), the pancreatico-duodenal block (Holst et al., 1981), the pancreas (Holst et al., 1980), the small bowel (Kavin et al., 1967; Holzer and Lembeck, 1979), the left colon (Ballantyne et al., 1989) and the liver (Hems et al., 1966) from various species have been used when it is desirable to avoid extrinsic neural effects and endocrine effects other than those under study or introduced by the investigator. Therefore, gastrointestinal motility can be studied in segments with preserved structural integrity in response to defined luminal, neural or blood-borne stimuli, while the analytical work is facilitated because of the lack of dilution of agents into the systemic circulation and the use of non-degrading synthetic perfusion media. Furthermore, pharmacological substances can be administered in exact concentrations to the arterial line of the segment. The ratio between contractions and artefacts seems favourable as compared to in vivo preparations because artefacts arising from body movements, movements of adjacent parts of the intestine and respiration are avoided. For biomechanical studies it is a great advantage that the transmural pressure is easy to determine and the outer dimensions easily measurable. In addition, secretion of bioactive compounds can be studied and correlated to motility (Holzer, 1989). Disadvantages with isolated perfused models are the cost of the perfusion media, which has to contain high molecular substances such as dextrans (Gregersen et al., 1992) and fluorocarbons (Faithfull, 1987), and the time-consuming anaesthetic and operative procedures. Furthermore, in contrast to smooth-muscle strips, the investigation time in isolated perfused organs is often limited to 1.5 to 5 hours (Gregersen et al., 1992), mainly because of a progressive oedema.

Two different isolation techniques have been used. One technique is to leave the segment in question in the animal, dissect extrinsic nerves and cannulate the major vessels for perfusion. Using the other technique the gastrointestinal segment is isolated and transferred to an organ bath for perfusion. For the purpose of illustrating the isolation and perfusion of the latter technique, the isolated perfused porcine duodenum is briefly described here. Preferentially, the animals should be fasted for at least 12 hrs. Surgical anaesthesia can be induced by 2.5% halothane without premedication whereafter it is replaced with intravenous infusion of α-chloralose. This is a widely used anaesthetic drug in animal experimental studies in gastroenterology, and it is often the drug of choice because of the minimal depressant action on autonomic functions (Balis and Monroe, 1964; Svendsen et al., 1990). The operative procedure is easiest through a long midline incision. After dissecting through the abdominal muscles and peritoneum, the spleen and the left adrenal gland can be excised. Hereafter, peritoneum along the aorta can be dissected and the lumbar arteries ligated and cut. Duodenum and the major vessels are at this point dissected from adjacent organs such as the stomach and transverse colon. It is a time-consuming procedure to dissect the pancreas from the portal vein due to numerous connecting veins. After further dissection to separate the caval vein, the duodenum en bloc along with a 6–8 cm segment of the aorta, comprising the coeliac trunk as well as the superior mesenteric artery, can be separated. Cutting duodenum at the level of

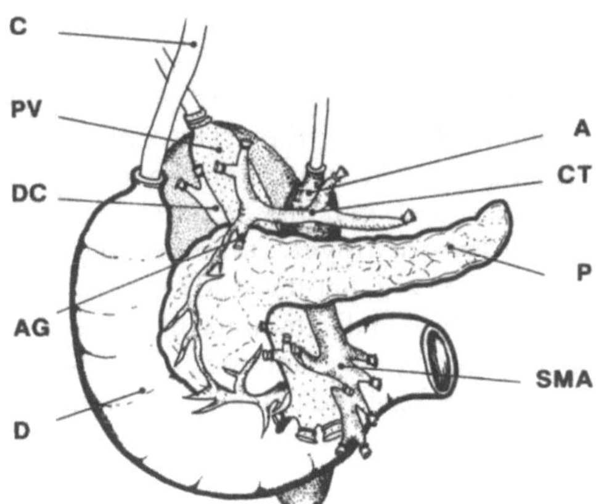

Figure 5.1. Schematic drawing of the isolated perfused porcine duodenum. A = aorta, AG = gastroduodenal artery, C = catheter for study of motility and balloon distension, CT = coeliac trunk, D = duodenum, DC = choledochal duct, P = pancreas, PV = portal vein, and SMA = superior mesenteric artery. With permission, Gregersen et al, *J Applied Physiol*, 1992; **73**:9–19.

pylorus and 3 cm distal to the Ligament of Treitz will provide a 20–25 cm-long duodenal segment. Hereafter, the portal vein must be isolated in porta hepatis, dissected and cannulated. As the last operative procedures, the aorta is ligated and cut distal to the superior mesenteric artery and the other end is dissected proximal to the coeliac trunc and cannulated. Instantly, the segment is transferred to the organ bath and perfused. The perfusion can be either single-pass or recirculating and different composites of perfusate can be used. Stable pH and temperature, high arterial oxygen saturation and high colloid–osmotic pressure are essential factors.

Isolated Non-perfused Gastrointestinal Segments

From a surgical point of view this is a simple model when compared to perfused segments. The organ bath contains a physiological solution with constant temperature and pH and it is bubbled with 5% O_2 and 95% CO_2. During anaesthesia or instantly after euthanising the animal, a midline incision is done and the gastrointestinal segment of choice is separated from adjacent organs. After mounting in the organ bath and a resting period, motility can be studied by the use of manometry, electromyography, radiographic techniques, etc. Segments usually exert spontaneous motility. Motility can also be evoked pharmacologically, by luminal perfusion and by intraluminal balloon distension. The solution in the organ bath must be continuously changed in order to avoid accumulation of pharmacological substances and metabolites. Small animals have to be used since oxygen does not penetrate deeply into tissues.

Isolated Smooth Muscle Tissue

Measurement in vitro is often done by the use of strain-gauge transducers in the investigation of different preparations of tissue strips (Meiss, 1971; Costa and Furness, 1976). The technique can be used both for studies of biomechanical properties and for studies of regulatory mechanisms. Different models have been described. The most widely used is for the study of smooth muscle strips sized approximately 1–2 mm wide and 1 cm long. Both ends are tied with silk thread for connection to the recording and stretching apparatus. The strips can be mounted vertically or horizontally in small organ baths containing as little as 5 ml of solution such as Tyrodes or Krebs solution (Elbrønd et al., 1991). It is bubbled with a mixture of 95% O_2 and 5% CO_2 and can be continuously changed. Temperature and pH must be constant at 37 °C and 7.40, respectively. The strip can be stimulated by stretch, pharmacologically and by transmural field stimulation. By careful dissection it may be possible to separate the longitudinal and circular smooth muscle layer of the gut. Compared to isolated perfused segments, the advantages of this technique are that the study time is much longer and that the cost is much lower due to the small amounts of solution and pharmacological substances required. Furthermore, both isotonic and isometric studies can be done. On the other hand, it is more artificial and the integrity of the gut wall is not preserved.

5.2 Smooth Muscle

5.2.1 The Structure of Smooth Muscle

Three layers of muscle generate the forces that produce flows in the luminal contents of the gut, two layers in the tunica muscularis and one in the mucosa. In all three layers, the visceral muscle cells, five microns wide and 100–200 microns long, are bound together more or less in parallel by a matrix of collagen fibres. The orientation of the cells in the muscle tissue determines the vector of the force generated by contractions of those cells and establishes the terminology of these muscle layers. Thus, the cells of the outer layer of muscle in the tunica muscularis are oriented along the axis of the cylinder, so that layer is called the longitudinal muscle layer. The cells of the inner layer in the tunica muscularis follow the circumference of the cylinder to form the circular muscle layer. The muscle layer in the mucosa is called the muscularis mucosae or the mucosal muscle. Its cells possess various orientations in the different organs.

A collagen framework, or stroma, confines the cells in visceral muscle, binding them into bundles that converge, diverge, and reconverge to make up the gross mass of the muscle. Strong mechanical bonds tie the muscle cells to the fibres of the collagen stroma and to one another so that the whole structure forms a single mechanical unit.

The muscle cells lie side by side with the thick central part of one cell adjacent to the thin extremities of others. The nuclei of the mononuclear cells thus exhibit no pattern to the microscopist, appearing to be randomly distributed within a field of cytoplasm lacking organelles or membranes that can be visualised with the usual tissue stains. This monotonous appearance is the origin of the term "smooth" muscle. Adjacent cells contact one another through both mechanical and functional linkages. The latter constitute gap junctions that allow the passage of electrical currents and small molecules between conjoined cells. Fibres of the contractile proteins, actin and myosin, function within the cells through a sliding filament mechanism much as they do in skeletal muscle. The globular head of the myosin molecule interacts with the actin molecule in a cross-bridge formation. Fibres of the structural protein, desmin, organise the cell interior and transmit the forces created by the interactions of the contractile filaments to the cell membrane. These fibres converge at dense bands on the cell membrane and dense bodies within the cytoplasm. Both these kinds of structure serve as points for the fixation of the several varieties of filaments to one another and to the cell membrane. The muscle cell also contains mitochondria and a smooth endoplasmic reticulum. The muscle cell membrane contains characteristic invaginations, the caveolae, thought to function in the electrochemical coupling necessary for the initiation of contraction by the electrical events of the membrane in smooth muscle.

5.2.2 Origin of Rhythmicity

The regions of the gut that exhibit rhythmic contractions also display an electrical phenomenon called the electrical slow waves (previously called the pacesetter potentials, the basic electrical rhythm, and the electrical control activity). As recorded by gross electrodes applied to the serosal surface of the stomach or intestine, these electrical signals possess the general appearance of the cardiac action potential. When recorded from a series of gross electrodes placed at close intervals along the intestine or distal stomach, they occur in a spatio-temporal sequence such that they appear to propagate along the organ. Similar signals can be recorded from the colon and the sphincter of Oddi, but only if the electrodes are buried deep in the muscular wall. Those parts of the gastrointestinal tract that do not generate rhythmic contractions lack such signals. The frequency of these electrical slow waves at any location corresponds to the fundamental frequency of the rhythmic contractions that occur at that site, and the direction and velocity of their migration conform to those characteristics of the peristaltic contractions of that region. For this reason, they are considered to be pacemaker signals establishing the rhythmicity of contractions in those regions (see Figure 2.5, Chapter 2).

For a long time, it was believed that these electrical signals arose in the smooth muscle itself, in the outer longitudinal muscle layer in the distal stomach and small intestine, and in the circular muscle layer in the colon. Now we know that they originate in close relationship to the interstitial cells of Cajal. However, the generation of slow waves requires the interaction of interstitial cells with smooth muscle cells. The exact nature of that interaction remains unknown. Interstitial cells resemble muscle cells in structure and can be viewed as specialised muscle cells.

When the muscle contracts in the neighbourhood of a gross recording electrode, the electrical record often shows a distinctive signal, a burst of one or more rapid electrical transients called the spike burst, superimposed upon the plateau of the electrical slow wave. This signal cannot be considered to be a true action potential, however, for small rhythmic contractions in the intestine can occur without an accompanying spike burst on the corresponding slow waves, and even strong contractions regularly lack associated spike bursts in some places, especially in the distal stomach and colon.

5.2.3 Origin of Slow Waves

Intracellular microelectrode recordings from smooth muscle cells in rhythmically active parts of the gut show an electrical slow wave as a signal with a relatively rapid depolarisation, a plateau, and a slower repolarisation. Spike bursts appear as a group of very rapid transients on the plateau of the slow wave. The depolarisation partly represents an inward current from the entry of calcium into the cell through the voltage-dependent and calcium-dependent L- and T-type calcium channels of the muscle. The magnitude of the depolarisation is determined by the modifications in the activity of those channels brought about by the changes in the membrane voltage and in the intracellular calcium concentration, and by the activation of calcium-dependent potassium channels that open to create an outward current. The plateau reflects a balance in the inward and outward currents. Repolarisation comes with the further rise in intracellular calcium concentration produced by calcium-induced calcium release from intracellular stores. This fully inactivates the calcium channels and activates the potassium channel. Smooth muscle cells also possess voltage-dependent potassium channels and leak chloride channels. The interactions among these five channels can fully account for the generation of both slow waves and spikes.

The gap junctions that link smooth muscle cells together ensure that the ion movements across membranes occur simultaneously within all muscle cells over a relatively large area, thus providing for synchronicity of slow waves and contractions; i.e., the muscle constitutes a functional syncytium. It is not clear, however, what factors establish the frequency of slow waves. Also, there is no theory to explain the mechanism that determines the way the electrical signals progress through an organ.

5.3 Synapses and Neural Circuits

Nerves affect one another and the non-neural tissues they control through synapses. Such contacts between nerves may be axo-dendritic, axo-somatic, or axo-axonal.

Both the presynaptic and post-synaptic structures exhibit morphological, physiological and biochemical specialisation at the synapse, related to the sequence of events that occurs in neurochemical transmission. These involve the synthesis, storage, release, diffusion, binding, and degradation or utilisation of the neurotransmitter substances.

All synaptic transmission in the mammalian gut makes use of neurochemical mechanisms. The vesicles within the presynaptic axonal varicosities store the neurotransmitters that are synthesised there by special biochemical systems. The arrival of the nerve impulse at the axon activates the influx of calcium ions through voltage-dependent ion channels in the membrane. The consequent increase in the calcium concentration of the axonal cytoplasm stimulates calcium-reactive centres on the vesicles, rupturing them to release the stored neurotransmitter, which then diffuses into the synaptic space. Some of the neurotransmitter encounters chemospecific sites on the postsynaptic membrane, receptors, to which it binds reversibly to initiate electrical transients in the postsynaptic membrane. Chemospecific enzymes simultaneously remove the released transmitter from the synaptic cleft.

The brief sketch given above describes transmission by the two best-studied neurotransmitters of the myenteric plexus, acetylcholine and noradrenaline. Other substances tentatively identified in the myenteric plexus as neurotransmitters include a variety of amino acids and small peptides, all larger molecules than acetylcholine and noradrenaline, and the gas, nitric oxide. Although the role of all such substances in neurotransmission must involve, in principle, the same basic steps (synthesis, storage, release, diffusion, binding, and degradation), these other neurotransmitters probably differ importantly from acetylcholine and noradrenaline in many aspects of the process. Thus, for example, it seems impossible for nerve terminals to store the unstable and highly diffusible gas nitric oxide. It must be synthesised on demand and diffuse instantly into the synaptic space. Most of these steps in neurotransmission remain to be described in detail for these tentative neurotransmitters.

Neuromuscular synapses in visceral muscle are diffuse rather than the focal formations that exist at such contacts in somatic muscle. An axon spreads widely throughout the smooth muscle mass, contacting thousands of cells in the syncytium. In this way its transmitter affects a large body of the muscle when it escapes from its varicosities. Furthermore, axons from different nerve cells, presumably of different kinds, share the same domain in the muscle mass. The result is an organisational structure of the neuromuscular unit in smooth muscle that bears essentially no resemblance to the motor unit organisation of somatic muscle.

5.4 Sensation

Although it was previously a common belief that the internal organs were devoid of sensory nerves, sensation arising directly from the viscera is widely accepted today. Most sensory nerves carrying information about the conscious sensation travel together with the sympathetic nerves, whereas the afferent sensory nerve supply that travels with the parasympathetic nerves predominantly carry non-painful information about tone, secretion and motor control that does not reach consciousness. Visceral pain is the most common form of pain produced by disease and one of the most frequent reasons why patients seek medical attention. Much of what we know about the mechanisms of pain derives from experimental studies of somatic and not visceral nociception. Although visceral pain is an important part of the normal sensory repertoire of all human beings and a prominent symptom of many clinical conditions, not much clinical research has been done in this field. The characteristics of the mechanoreceptors in the gut have been widely addressed in animal studies (Sengupta and Gebhart, 1995). Most fibres are non-myelinated (C-fibres) or small myelinated (Aδ-fibres). They are predominantly localised in the muscular layers, but the mucosa, serosa and mesentery are also supplied. The majority of the afferents cover *both* physiological *and* nociceptive ranges of stimulation, which suggests that the quality of sensation depends on the intensity of discharge. From animal studies we know that there may be three distinct classes of nociceptive sensory receptors that innervate internal organs, including the small intestine (Cervero, 1992). The first class of receptors has a high threshold to natural stimuli (mostly mechanical). The encoding function (the relation between stimulus intensity and nerve activity of these receptors) is evoked entirely by stimuli within the noxious range. The second class of receptors has a low threshold to natural stimuli. Finally, intensity-encoding receptors have a stimulus–response function that spans the range of stimulation intensity from innocuous to noxious. Furthermore, previous animal studies have also indicated that rapidly adapting mechanoreceptors may be located in the mucosal layer whereas slowly adapting mechanoreceptors are located in the intramuscular layer (Wood, 1987). Unfortunately, little evidence has been provided in human studies.

The mechano-sensitive fibres can also be divided into phasic and tonic populations, but conflicting results regarding the localisation of fibres responding to different distension speeds have been found in animal (Sengupta and Gebhart, 1995) and human studies (Lembo et al., 1994). Some fibres adapt to maintained distension, whereas others are non-adapting with the same response during, e.g., 30 seconds of distension (Su and Gebhart, 1998). Many mechano-sensitive fibres respond to both distension of the organ and contraction and are considered in-series with the muscle cells (Thewisen et al., 2000). However, most studies were performed in the period 1950–70 and suffer from serious methodological limitations (of which the biomechanical considerations are discussed elsewhere in this book). Thus, the localisation of the receptors and their characteristics in man is still a matter of debate.

It is well known that distension of the gastrointestinal tract elicits reflex-mediated inhibition and stimulation of motility via intrinsic or extrinsic neural circuits and induces visceral perception such as pain. Several studies have attempted to investigate the gastrointestinal motility and sensory properties by means of balloon distension (Rao et al., 1996). Unfortunately, the primary mechanism for symptoms evoked by gastrointestinal distension remains unclear. Previous studies demonstrated that mechanoreceptors located in the intestinal wall play an important role in the stimulus–response function. It is, however, a common mistake to believe that mechanoreceptors are sensitive to variations in pressure or volume. Large variations in the peristaltic reflex and perception when expressed as a function of pressure have been found in different studies and species, suggesting that pressure is not the direct stimulus. Instead, the receptors may be stimulated by mechanical forces and deformation acting in the intestinal wall due to changes in the transmural pressure. Thus, the mechanical distension stimulus and biomechanical tissue properties must be taken into account in studies of the sensorimotor function in the intestine.

As pain and other sensations related to the gut are common symptoms in gastroenterology, characterisation and knowledge of the mechanisms behind the sensory experience is of major importance in the diagnosis and assessment of organ

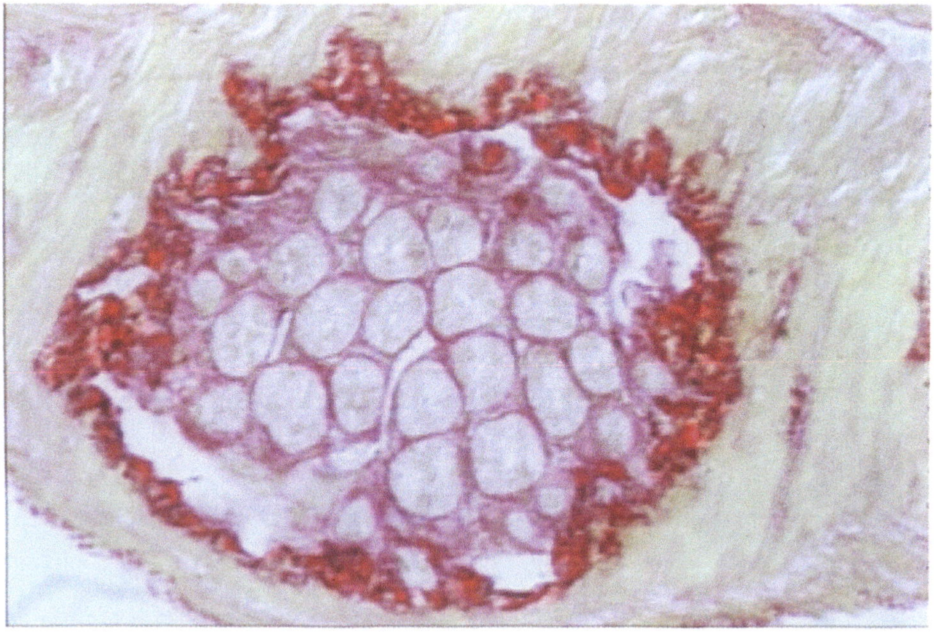

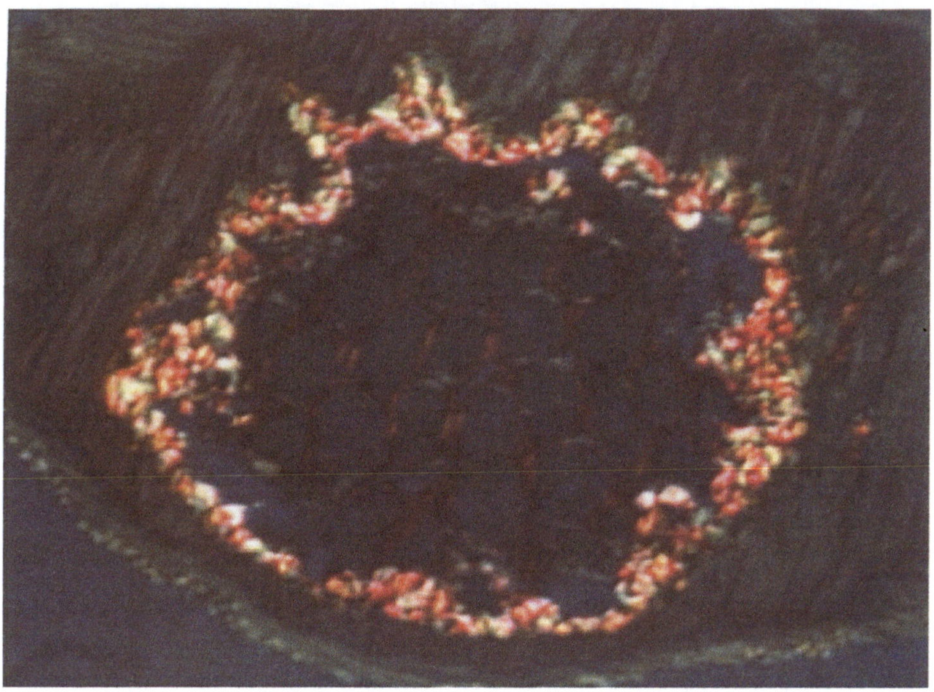

Two images of the same section obtained from the small intestine and stained with picrosirius red that stains collagen red and muscle yellow in brightfield microscopy (top) and collagen red and green using polarised microscopy (bottom). The cutting is made along the longitudinal axis and penetrates from the circular muscle layer into the submucosa and mucosa in the centre of the images. The submucosal collagen is easy to identify. Some collagen can also be identified in the subserosa.

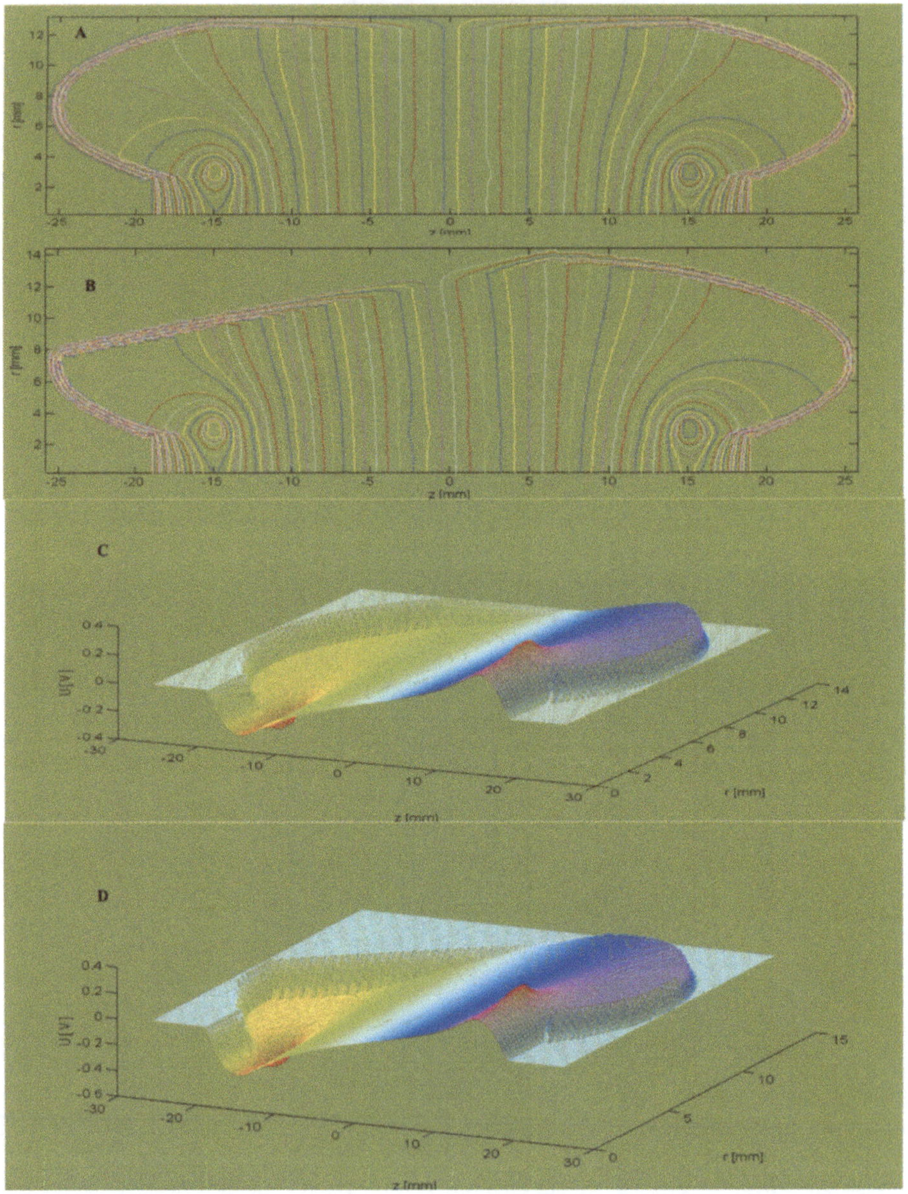

The potential distributions for default model with angle=0° (A, C) and angle=14° (B, D) .

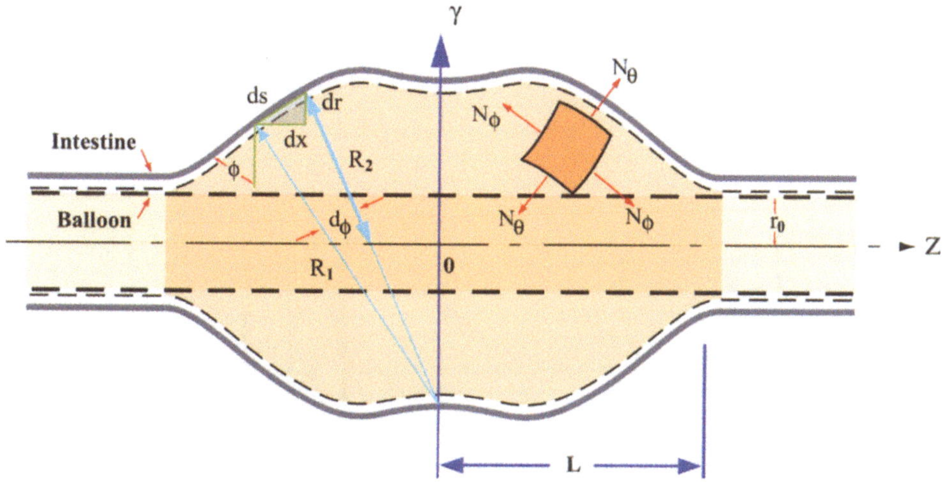

Schematic drawing of the distended segment of an intestine. N_ϕ and N_θ denote the tension in longitudinal and circumferential directions. To analyse the tension in the intestine as the balloon is inflated, we have to describe the geometry and define the notions of the stresses. Two systems of coordinates are convenient. (1) the 3-dimensional cylindrical polar coordinates (r, θ, z). (2) the two-dimensional latitude–longitude coordinates (ϕ, θ) on the surface of the intestine, similar to the one we use for our globe in geography. The latitudinal angle ϕ is defined as the angle between the normal vector to the intestinal surface and the axis of symmetry z. The longitudinal angle θ is the angle of rotation of the meridian cross-section about the z axis. Now there exists a normal vector at any point P on the surface with coordinate ϕ and θ. At a point neighbouring to P, with coordinates ($\phi+d\phi$ and θ), there is another normal vector. These two normal vectors will intersect at a distance r_1 from the surface. This r_1 is a principal radius of curvature of the surface in the longitudinal cross-section. On the other hand, the normals at (ϕ, θ) and ($\phi+d\theta$ and θ) will intersect at a distance r_2 from the surface. r_2 is also a principal radius of curvature of the surface. The radius of curvature of the curve of intersection of any other normal plane at P lies between r_1 and r_2. So the curvature of the intestinal surface is defined by r_1 and r_2.

The stresses of interest in an axisymmetric tube subjected to axisymmetric load are those lying in the wall, acting on normal cross-sections of the wall. In this figure, a rectangular element with sides df and dq is shown, similar to a sphere with a small latitudinal and longitudinal difference. The stresses s_{ff} and s_{qq} can be much larger than the pressure in the balloon (p). When the shear stress between the balloon and intestine is ignored, and the bending stresses are ignored because the intestinal wall thickness is small compared with the radii of curvature, then s_{ff} and s_{qq} are uniformly distributed throughout the intestinal wall. Multiplying the uniform stresses s_{ff} and s_{qq} by the wall thickness, h, we obtain $N_f = s_{ff}\,h$ and $N_q = s_{qq}\,h$, which are called membrane tensions in cross-sections whose normal vectors lie in the direction of f and q axes, respectively. The units of N_f, N_q are force per unit length, N/m, like those of surface tension, while the stresses have units N/m^2. N_f and N_q are by some called stress resultants or membrane stress resultants to recognise the fact that they are the integrals of stresses throughout the thickness of the membrane wall.

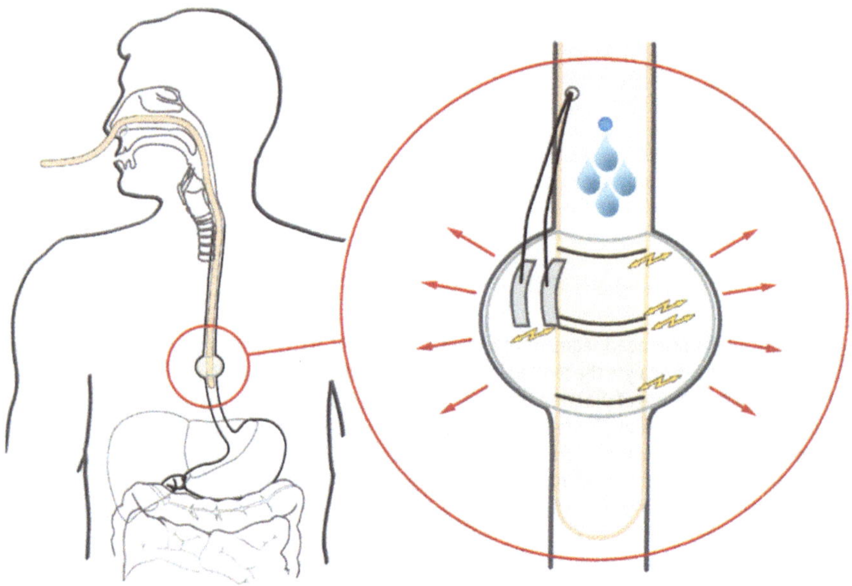

Schematic drawing of the multimodal probe. The probe contains a fluid-filled balloon that can be distended in a controlled way. The temperature in the fluid inside the balloon can also be controlled in the range from 5–60 °C. Electrical stimulation is done through flexible rectangular-shaped electrodes glued to the surface of the balloon. Chemical stimulation is done by infusion of the chemical substance (such as capsaicin or acid) through the side hole above the balloon. Modified from Am. J. Physiol 2002; 283: G95–G103.

dysfunction. Studies with focus on the development and manifestations of the sensory manifestations are highly needed. In the clinic, the study of pain is confounded by individual factors. Moreover, in most visceral diseases, either a large organ or several organs are typically involved together with systemic reactions such as fever and malaise. This leads to a diffuse clinical presentation of the symptoms. Experimental pain models are tools for characterisation and understanding of pain mechanisms with minor interference from such individual reactions to pain. The sensory functions of the small intestine in man have been disputed for several years. Several non-painful sensations such as fullness and satiety can be evoked by food or distension with balloons, but direct activation of the afferent nerves by electrical stimulation can also result in non-painful as well as painful sensations (Accarino et al., 1992).

Pain is a symptom in many diseases of the gut and in man painful sensations can be evoked in the small intestine by distension with balloons, electrical stimulation and chemical stimuli with capsaicin. The sensation can be assessed with different scales where intensity and pain quality by means of descriptors are most commonly used. Measurement of physiological parameters that are directly influenced by the pain stimulus, such as autonomic reflexes, the nociceptive withdrawal reflex and cerebral evoked potentials, are other possibilities of measuring pain intensity more objectively. It can be questioned, however, whether these methods give more reliable data than subjective assessment of pain intensity.

5.5 The Importance of Mechanical Data

In the gastrointestinal tract, mechanical data are important for an understanding of passive viscoelastic properties, peristaltic reflexes, perception, mechanoreceptor responses, tone, bolus transport mechanics, growth, and remodelling in experimental and clinical studies. These fields of investigation are in many ways related.

A large number of diseases in the gastrointestinal tract are associated with biomechanical dysfunction. This is no wonder considering the importance of the mechanical function of the gastrointestinal tract. Some of these diseases can be considered primary mechanical diseases whereas in other diseases, secondary mechanical remodelling may occur. As an example, partial obstruction in the gastrointestinal tract can have many causes, either inherent in the gastrointestinal tract or due to external factors. The partial obstruction creates a mass transport problem. The partial obstruction may in many cases easily be dealt with but the secondary changes, caused by the dilatation created by the contractile forces, may be irreversible, leaving the patient in a poor situation. In the following, attention is paid to understanding the interaction between mechanical stimuli and sensorimotor responses, to the oesophagus, due to the high prevalence of ageing-related diseases and symptoms, and to irritable bowel syndrome.

5.5.1 Mechanoreceptors, Peristaltic Reflex Activity and Sensory Function Testing

Distension of the gastrointestinal tract elicits reflex-mediated inhibition and stimulation of motility via intrinsic or extrinsic neural circuits. Many studies have focused on the stimulus–response function and the mechanisms involved in distension; for instance, in the oesophagus, the effect of distension has been studied in animal models, and in humans with normal oesophagus and with oesophageal chest pain (Paterson, 1991; Paterson et al., 1991; Sengupta et al., 1992; 1990; 1989; De Caestecker et al., 1992; Williams et al., 1993). Much of the basic work in the gastrointestinal tract was done in the late 1950s (Iggo, 1957; Paintal, 1957; Bülbring et al., 1958). The sensing elements are afferent neurons with the mechanoreceptors located in the gastrointestinal wall. For example, those receptors involved in the peristaltic reflex pathway are located in the muscularis externa, the myenteric plexus and mucosa (Iggo, 1957; Hukuhara and Fukuda, 1965; Yokoyama and Ozaki, 1980; Hukuhara et al., 1960). The variability in location may be important for the sensitivity to different mechanical stimuli. The mechanoreceptors have also been classified on the basis of their pressure or volume thresholds to balloon distension which in turn relates to whether the afferents project to the vagal nerve or to the spinal cord (Sengupta et al., 1992). Furthermore, multimodal receptors reacting to both chemical and mechanical stimuli exist in the mucosa.

Mechanoreceptors are not directly sensitive to variations in pressure or volume. A large variation in the pressure and volume thresholds needed to trigger the peristaltic reflex and perception have been found in various studies and species (Gregersen et al., 1992; 1992; Williams et al., 1993; Holzer, 1989; Akervall et al., 1989), suggesting that pressure is not the direct stimulus. Instead, the receptors are stimulated by forces and deformations acting in the intestinal wall due to changes in transmural pressure (Ginzel, 1959). The same mechanism accounts for the baroreceptors in the cardiovascular system. Despite the name (baro = pressure), the baroreceptor nerve activity is associated with wall deformation, because restricting wall distension abolishes baroreceptor-mediated reflexes (Hauss et al., 1949). In a similar fashion, it was demonstrated that stretch is a potent stimulus for gastrointestinal mechanoreceptors rather than increased transmural pressure (Ginzel, 1959; Kosterlitz and Robinson, 1959). The fact that the receptors are also stimulated by visceral tension and deformation developed actively during contractions in the small intestine (Iggo, 1957) further suggests that they are sensitive to deformation. However, it is important to recognise that different mechanoreceptors respond to different kinds of mechanical stimuli; e.g., mucosal receptors respond to distortion and pinching but are relatively insensitive to contraction and distension (Yuan et al., 1991).

Since pressure is not the direct stimulus, a large variation arises when expressing the receptor response as a function of pressure. Such an approach makes it difficult to draw conclusions. Gregersen et al. (1992) studied the ascending peristaltic reflex response as a function of circumferential wall tension in the isolated perfused porcine duodenum. Although this approach was important in that it quantitated the stimulus in terms of wall tension rather than pressure or volume, more detailed analysis is needed in future studies. Despite the vast number of studies in this area, the essential stimulus has not yet been defined. Further analysis must consider the fact that mechanical forces imposed on the tissue induce stresses and strains in various directions. Thus, distension will, in addition to circumferential tensile stress and strain, cause longitudinal tensile stress and strain and radial compressive stress and strain. For the study of mechanoreceptor kinematics, it is therefore important to express the stimulus in terms of magnitude and direction of stress, strain, and strain rate. Unfortunately, it is impossible to measure stresses and strains exactly at the receptor sites. Theoretical solutions based on experimentally determined mechanical properties in different layers of the gastrointestinal wall may therefore be necessary. A numerical solution to the peristaltic reflex of the small bowel is provided by Miftakhov and

Wingate (1994). By determining the nature of these mechanical stimuli we may learn more about the mechanoreceptors and how they adapt to the physical environment. This will allow us to establish the relation between the activity in the neurons, electrical responses in neurons, and the mechanical behaviour of the gut. On the other hand, if the appropriate mechanical stimulus is not known, the data are merely qualitative. Models that preserve the structural integrity of the tissue to the in vivo condition are most useful in this respect.

5.5.2 Disordered Gastrointestinal Motility

We now realise that disorders of motor function may represent primary abnormalities in the operations of the system of nerves and muscles in the gut rather than the effect of an abnormal psyche on a normal gut. We cannot yet confidently attribute any motor disorder to a specific abnormality of a particular motor process or mechanism. We do not yet understand the complex mechanisms involved in the motor function of the gastrointestinal tract. Its movements represent the end result of the interactions of non-linear chemical, electrical, and mechanical processes linked together at many levels. One cannot predict the end result of a change in one component process in such a complex matrix. In most cases, only the general nature of the abnormal motor process can yet be perceived.

This section presents the principal gastrointestinal motor disorders in terms of what is known about the pathogenesis. They fall into the generalised disorders related to systemic diseases, organ-specific disorders, and functional disorders. Little is known about the mechanics of these disorders. Likely, better diagnostic tools could reduce the number of the so-called functional disorders and help in subtyping patients, resulting in better therapeutic options.

Generalised Diseases Affecting Motility

Of the generalised diseases affecting motility, diabetes is likely the most common. Diabetes can seriously alter motility in many gastrointestinal organs when it is poorly controlled for a long period. The oesophageal motor abnormalities, which resemble those of esophageal spasm, usually cause only minor symptoms. Delayed gastric emptying is very common and troublesome, producing severe nausea with vomiting and leading to difficulty in the control of the diabetes. Diabetes can make the stomach and small intestine fail to generate the fasting motility pattern, resulting in small intestinal bacterial overgrowth that produces steatorrhoea. Colonic motility is generally depressed. These changes tend to reverse with improved control of the diabetes, leading most authorities to suspect a reversible neuropathy in the enteric innervation. Histological examination of the intrinsic nerves and muscles of the gut have failed to reveal morphological abnormalities to account for these alterations in motor function.

Organ-specific Disorders

Dysfunction of Pharyngo-oesophageal Striated Muscle

Oropharyngeal dysphagia, a sensation of difficulty in swallowing in the throat accompanied by choking and the nasopharyngeal reflux of liquids, signifies abnormal transit through the pharynx. Radiographic studies readily demonstrate the abnormal motility of the pharynx, if it is responsible, and the abnormal transit into the trachea and nasopharynx. Radiographic and manometric studies can also reveal abnormal function of the striated muscle in the proximal oesophagus, not surprisingly in view of the identity of the mechanism of peristalsis in the two organs and the equivalence of their innervation. When disordered function of the pharyngo-oesophageal striated muscle occurs with no evidence of dysfunction of any other somatic muscle, it indicates that the symptoms originate in the innervation specific to this musculature. Such dysphagia practically always represents damage to the brainstem nuclei that supply the somatic motor nerves. Such damage most commonly results from vascular insufficiency, a brainstem stroke, occasionally from a tumour or trauma. A selective defect in somatic muscle function producing oropharyngeal dysphagia can also occur in myasthenia gravis and in amyotrophic lateral sclerosis.

Dysfunction of Oesophageal Smooth Muscle

Oesophageal dysphagia, the sense of difficulty in swallowing felt behind the sternum without associated choking, signifies a hindrance to transit through the oesophagus. Of course, mechanical obstructions, such as tumours and strictures, can produce such dysphagia, but motor disorders of the smooth muscle do, too. The two disorders specific to oesophageal smooth muscle are achalasia and oesophageal spasm (described under ageing-related motor disorders).

In achalasia, two abnormalities produce the obstruction. The lower oesophageal sphincter fails to relax with swallowing, and the smooth muscle part of the oesophagus either fails to contract with swallowing or responds with a contraction that is simultaneous throughout the smooth muscle segment. In the special form of achalasia that results from infection with the parasite, Trypanosoma cruzi, found only in the tropical parts of the Americas, histopathology shows a paucity of ganglion cells in the oesophageal myenteric plexus. The same finding has been described in the idiopathic disease seen throughout the world. Hypersensitivity to cholinergic agonists exists, an abnormality attributed to denervation super-sensitivity. From these considerations, it seems reasonable to infer that achalasia represent a degree of denervation of the smooth muscle. The principal inhibitory innervation of the oesophagus acts through the release of nitric oxide, and so it seems likely that these inhibitory nerves are deficient or defective in achalasia. It may be confined to the oesophagus simply because of the relative paucity of ganglion cells in the normal organ. It seems possible that an incomplete loss of nerve cells brings the number to a critically low value in the oesophagus first simply because there are so few there to begin with. The cause of the ganglion cell loss in idiopathic achalasia remains unknown.

Reflux oesophagitis is dealt with in the section on ageing-related changes.

Postoperative Gastric Motor Dysfunction

Gastric operations condemned many patients to lives made difficult by the ensuing severe alterations in gastric motor function. The changes in gastric emptying caused by vagotomy led to a variety of modifications or additions to the operation. In some cases, a resection of the distal stomach (often done also in order to remove an ulcer or to remove the source of the hormone, gastrin) or pyloroplasty, the widening of the pylorus, accompanied the vagotomy. These were not very successful. Patients were still observed to empty liquids rapidly, leading to the "dumping syndrome", and to retain solids, leading to bezoar formation. Some complained of early satiety, the feeling of fullness early in a meal.

Modifications of the vagotomy itself were somewhat more successful. Selective division of the vagal branches to the stomach, rather than the vagal trunks, seemed to produce fewer difficulties, and the cutting of the vagal branches directed specifically to the antrum produced even less. Antral vagotomy has the least effect on emptying, confirming that the antrum has a minor place in governing emptying.

Motor Dysfunction of the Biliary Tract

Gallstone disease represents the consequence of more than one functional abnormality. One of them, without doubt, is abnormal motility of the gall bladder. The gall bladder exhibits a pattern of periodic emptying in the fasted state that clears it of microcalculi, debris, and bacteria. Any interference with its motility is expected to foster stone formation.

Biliary dyskinesia is a diagnosis to explain pain whose features suggest its origin in the biliary tract when no abnormality can be demonstrated. The concept lacked objective proof until manometric study of the sphincter of Oddi was made possible. Two abnormalities in the operation of this sphincter have been found to explain some cases of such pain: an abnormally forceful closure at rest and a paradoxical response to cholecystokinin (contraction rather than relaxation). The paradoxical response of the sphincter to cholecystokinin has been attributed to partial loss of the inhibitory innervation of the sphincter.

Motor Dysfunction of the Small and Large Intestine

Multiple jejunal diverticulosis, a disorder seen most commonly in old people, causes diarrhoea and, sometimes, vitamin B_{12} deficiency because of stasis with the overgrowth of coliform bacteria in the diverticula. Most cases are idiopathic, not associated with any conceivable cause except advanced age. These diverticula may well be neuropathic in origin. Other cases, however, represent the patchy degeneration of the visceral muscle in the organ and its replacement with collagen. This process is especially associated with scleroderma and related connective tissue diseases.

Diverticulosis of the colon is much more common than jejunal diverticulosis. It may be asymptomatic for long periods, producing sudden painless rectal bleeding, or it may give rise to a pain syndrome produced by infection in one or more of the diverticula. The sigmoid colon shows a predilection to develop the diverticula with fewer appearing in the descending colon. Occasionally, they occur throughout the colon. They form as pockets of the mucosa that herniate through the layers of the tunica muscularis at points of weakness, the places where the vessels penetrate the muscle. These sites lie always between the mesenteric tenia and the two antimesenteric teniae, so that the diverticula form rows along the lateral surfaces of the organ. Furthermore, they are spaced quite uniformly along the axis of the colon, roughly halfway between two haustral indentations. Hypertrophy of the circular muscle layer of the haustral septa in the region of the colon where diverticula occur, in fact, precedes the development of the pockets. Thus, diverticula form as a result of the hypertrophy of the septal circular muscle. Normal contractions of the colon in these regions generate abnormally high intraluminal pressures that force the mucosal pockets through the muscle layers at the points of least resistance, alongside the penetrating vessels. The primary abnormality, hypertrophy of the muscle, remains unexplained.

Hirschsprung disease is a congenital disorder characterised by the lack of relaxation of the internal anal sphincter and tonic contraction of the circular muscle layer for a variable length of the rectum. This produces a stenosis, with dilatation of the colon above the stenosis. The problem arises from a maturation arrest in the innervation of the narrowed regions.

Functional Disorders

Besides irritable bowel syndrome and non-cardiac chest pain (described elsewhere in this chapter) and several other unexplained gastrointestinal disorders (not the subject of this book), the functional order of interest is intestinal pseudo-obstruction.

In some patients presenting the symptoms and signs of intestinal obstruction, including visceral dilation on X-ray examination, surgeons may fail to find a responsible obstructing lesion at operation. This situation identifies intestinal pseudo-obstruction. Many different causes of this clinical circumstance define many different kinds of intestinal pseudo-obstruction. They can be classified generally as neurogenic and myogenic, denoting, respectively, a dysfunction that arises in the innervation of the muscle and a dysfunction that stems from a defective musculature. The commonest cases of intestinal pseudo-obstruction encountered in practice are iatrogenic, representing the side-effect of a neurotropic drug. Visceral myopathies and neuropathies can produce intestinal pseudo-obstruction. Some cases are familial and hereditary, while others are acquired. Some of the visceral neuropathies and myopathies are manifestations of more generalised and well-recognised disorders, while others seem to be idiotypic. Thus, "intestinal pseudo-obstruction" is not an etiological diagnosis but simply a reference to a clinical state that can reflect a variety of causes.

5.5.3 Ageing-related Changes in the Gastrointestinal Tract

More then 20% of our population is expected to exceed 85 years of age by the year 2030 (Katz, 1993). Hence, the gastroenterologist in the twenty-first century will be increasingly confronted with digestive diseases in the elderly population. Gastrointestinal disease is the second most common indication for hospital admission of elderly patients, a patient group that accounts for four times as many hospitalisations as younger patients (Almy, 1989).

During ageing, the digestive system exhibits alterations, and the incidence of age-specific diseases increases. In the oesophagus, decreased pressure and delayed relaxation of the upper oesophageal sphincter and reduction in the amplitude of oesophageal contraction may occur. Gastro-oesophageal reflux disease and symptoms such as dysphagia and heartburn are related to ageing. In the stomach, there is an increasing incidence of atrophic gastritis, a decline in gastric acid output, and prolonged gastric emptying of liquid. Gastroduodenal ulcers are more common in elderly people. In the small intestine, absorption of some vitamins and ions may be impaired and the prevalence of malabsorption is increased. In the large intestine, atrophy of the muscularis propria and an increase in the amount of fibrosis may be found, and the incidence of constipation is increased.

The oesophagus is an organ dedicated almost entirely to mechanical functions. It normally transports swallowed food to the stomach quickly and completely. At the same time, it prevents the contents of the stomach from flowing into the oesophagus, despite the existence of the normal pressure gradient between the abdomen and the thoracic cavity. When the organ fails to achieve either of these goals, symptoms ensue. Dysphagia usually represents failed transport and pyrosis commonly reflects gastro-oesophageal reflux. In neither case does the severity of the symptom accurately reflect the degree of abnormality. Significant, but easily detectable, degrees of dysfunction can persist for many years, producing such minor symptoms that the patients tolerate them, seeking no relief until some severe, and preventable, complications ensue. The start of these motor disorders generally increases in incidence with age and the late stages complicate the lives of many old people.

Two oesophageal motor disorders clearly increase in incidence with age, oesophageal spasm and reflux oesophagitis. They can be discussed separately although they seem to be related in some respect; that is, they often occur together.

Oesophageal Spasm

Oesophageal spasm is a disorder in which the transport function of the organ fails. The peristaltic contraction that normally sweeps the length of the organ once with each swallow becomes synchronous throughout the organ. The progressive nature is lost so that it becomes simultaneous at all levels. The problem underlying oesophageal spasm appears to represent a process confined to the smooth muscle because the short segment at the top of the organ that is composed of striated muscle never develops this problem.

The synchronicity of swallow-induced contraction all along the organ is not the sole abnormality in oesophageal spasm. The contractions are very often increased greatly in force, they may be segmental, and they may become repetitive, occurring in a sequence of up to ten or even more after a swallow. These variations in the behaviour of oesophageal smooth muscle in oesophageal spasm suggest that this rubric may involve a heterogeneous set of physiological problems. They are united under this term only because they all seem to involve most prominently the oesophageal smooth muscle and because the nature of the abnormal motor function in all of them remains enigmatic.

Reflux Oesophagitis

Reflux oesophagitis represents the failure of the organ to maintain the systems that protect it from the noxious actions of gastric contents. Adequate protection requires the normal operation of two motor mechanisms. *First*, the gating function of the gastro-oesophageal sphincter serves to minimise the volume of retrograde flow (reflux) that occurs under the influence of hydrostatic forces. *Second*, the entry of gastric contents into the oesophagus normally triggers the nerve-mediated reflex response called secondary peristalsis, a single peristaltic contraction that sweeps the length of the organ to promptly push the refluxed material back to the stomach. The failure of these two protective motor mechanisms in gastro-oesophageal reflux leads to acute and chronic inflammation because of the erosive action of the acidic and enzyme-laden gastric fluid on the oesophageal epithelium. The complications of such chronic inflammation include ulceration, fibrosis, abnormal epithelialisation, and cancer.

These two clinical entities are united, in terms of pathogenesis, by considering the two features that they have in common; their frequent concurrence and their clear increase in incidence with age. The pathogenic mechanism that it proposes is a change in the biomechanical properties of the oesophageal wall as a function of age. The biomechanical properties prominently arise from the connective tissue stroma of the tissue, and that stroma consists especially of collagen. Since age-related changes in the physical properties of collagen are well known, such changes occurring in the oesophagus may partially account for these two clinical disorders.

Generalised changes may occur in the oesophagus as a function of the changes in collagen that take place throughout the body in the ageing process. These changes alter the biomechanical characteristics of the organ. Even small modifications in these properties could change the operation of the oesophagus to such a degree that either or both the propulsive function and the barrier function of the organ could be compromised. Before these mechanisms can be unravelled, we need to get a better understanding of age-related changes in the structural components in the oesophageal wall and in the mechanical properties and their interaction.

5.5.4 A Biomechanical View on the Disordered Colorectal Sensation in Irritable Bowel Syndrome

Objective criteria for the diagnosis of irritable bowel syndrome (IBS) have never been determined. The fact that the diagnosis at present rests entirely on the complaints of the patients means that the accuracy of the diagnosis depends on the perceptions of the patients as well as those of their physicians. In the search for a physiological abnormality, we seek a tangible diagnostic criterion as well as a pathogenic mechanism. Lacking an objective abnormality, however, we must continue to be subjective. Many recent studies define study groups in terms of sets of subjective features that were drawn up in discussions among authorities with a broad experience and a special interest in this difficult matter. The two most widely used are the Manning criteria and the Rome criteria (Christensen, 1994).

Hypersensitivity in Irritable Bowel Syndrome

Many people now accept hypersensitivity to balloon distension in the distal colon and rectum as a common feature of IBS. A critical look at the evidence for rectal hypersensitivity in IBS seems prudent before making such commitment to a pathogenic mechanism in such a heterogeneous, poorly defined syndrome. And when the evidence involves a matter like visceral sensation, particular care to examine the evidence and the concepts underlying their interpretation is warranted.

Many investigators use balloon distension in the gut to seek evidence for irritability in IBS. Most studies have involved colonic or rectal distension. The evidence in this matter began with the investigation of the balloon rectometrogram. Varma and Smith (1986) proposed this procedure as a tool to investigate the pressure–volume relationship in the rectum, though the reproducibility of the measurement was poor. Kendall et al. (1990) similarly examined the variability of the rectal pressure–volume relationship during distension by latex balloon in healthy adult volunteers and in patients with IBS. A wide variation in tolerable volumes and pressures within groups was found. There was a large variation within subjects as well. The tolerance for balloon distension of the rectum increased progressively with repeated studies in the same individual. A number of other studies have been published (among others Ritchie, 1973; Whitehead et al., 1980; 1990; Lembo et al., 1994; Bernstein et al., 1996) but most of these studies ignored the variability in individual subjects detected by Varma and Smith and by Kendall et al. The conclusions from the studies seldom agree. Differences in their study groups (definition criteria, gender, etc.) and variations in the techniques (location and mode of distension, and the parameters evaluated) they used may, in part, account for this. Clearly, there are serious problems with this body of evidence. Beyond those criticisms, however, the investigators themselves seem unaware of several more basic difficulties that they face in the use of such a technique for such a purpose. These impediments stem both from the obscurities of colorectal sensory physiology and from the unfamiliarity of gastrointestinal biologists with the application of the principles of biomechanics to distension of the gastrointestinal tract with a balloon.

Anatomical, Physiological and Biomechanical Considerations

Most investigators assume that colorectal balloon distension excites mechanoreceptors that accounts for the pain produced by the distension. Classical teaching in physiology gives responsibility for pain perception to the sympathetic nervous system. Mechanoreceptive reflexes in the gut are usually considered to be parasympathetic, insofar as they involve extrinsic pathways at least, and the activation of mechanoreceptors is generally felt to be imperceptible. Although polymodality in intestinal mucosal chemoreceptors has been postulated (Grundy, 1993), that property remains to be established with respect to the responses to painful and mechanical stimuli in the colorectal region. Furthermore, there is no convincing evidence to support the idea that parasympathetic pathways transmit painful sensations arising from the gut. Thus, according to the usual ideas about these matters, the pain produced by distension in the gastrointestinal tract represents the excitation of pain receptors, not mechanoreceptors, and they transmit their signals to the brain over sympathetic pathways. It is often also assumed that the receptors lie only in the colorectal wall. However, the rectum is confined in the lesser pelvis along with the genitourinary organs. Even a small distension of the rectum must compress or displace these nearby structures to some degree. Furthermore, sensory receptors could lie in still other locations and be activated in other ways. The squamo-columnar epithelial junction in the anal canal contains a uniquely concentrated accumulation of all the morphological types of cutaneous receptors (Rogers, 1992). When the anal canal is closed, as it is almost all of the time, these receptors probably remain unstimulated. However, the distension of the rectum forces open the anal canal, everting it as seen from the interior. This deformation could well stimulate many of these different kinds of epithelial receptors as well as mechanoreceptors in the musculature of both the smooth and striated muscle of the two anal sphincters. There are muscle spindles in the external anal sphincter and in the puborectalis muscle, and the presacral space contains Pacinian corpuscles and other kinds of sensory endings (Li et al., 1990). Most such extracolonic receptors probably excite signals that traverse somatic sensory pathways. All of them could be excited by rectal balloon distension.

Hypersensitivity, by definition, represents the abnormal processing of normal nerve signals. Signal processing takes place at many levels in the sympathetic sensory pathway from the gut, including the sensory receptor itself and the synapses in the prevertebral ganglia, the spinal cord, and the brain. The postulated abnormal processing of signals could occur at any of these levels. The sympathetic nerve supply to the gut follows the perivascular nerves. The watershed between the inferior mesenteric artery and the branches of the iliac arteries in the blood supply of the rectum lies at about the colorectal junction. Thus, the course of sympathetic sensory signals from the rectum must lie along the iliac perivascular pathway, not the splanchnic pathway, excluding the prevertebral ganglia from consideration and so reducing the number of possible levels where abnormal processing could occur.

Finally, no one knows what stimuli can activate nerve fibres mediating painful sensations in the gut. While physiologists often speak about pain receptors, no characteristic structure has been recognised in the gut to which this specific modality of the sensation belongs, nor can the exact stimulus itself be identified. It is clear, however, that a variety of states can give rise to painful sensations in the gut, including excessive distension, inflammation, and ischaemia. Since ischaemia can elicit pain in the gut, then pain produced by colorectal distension in these experiments with IBS may represent ischaemia produced by capillary occlusion.

Many investigators monitor changes in balloon volume and pressure during colorectal distension. However, the gastrointestinal mechanoreceptors of most interest are those that respond to circumferential deformation in the wall of the distended organ. It is probably often assumed that balloon volume and pressure relate directly to circumferential strain. However, the relationship is complex and non-linear. Such simple assumptions can lead to considerable error. The thickness of the wall of the rectum, which changes with dilatation, must be taken into account, for example. The geometry of the distending balloon, the isotropy of the material of the wall, the maintenance of a circular outline in the cylinder, and the viscoelastic properties of the tissue all must affect the relationship in experiments of this nature. None of this, apparently, has been considered in the previous experiments.

The problem of IBS is too serious for us to allow ourselves to be seduced into premature conclusions because of inadequate experimentation. All reasonable proposals for a pathogenic mechanism that will explain the symptoms in IBS deserve serious study, but they must represent the proper application of the facts of anatomy, physiology, and biomechanics. The same requirement can be made, in fact, for theories about all disorders of gastrointestinal neuromuscular function, for the gut is a self-excitatory neuromuscular mechanical device, one whose controlling systems and mode of operation far exceed in complexity any machine made by man.

Several new methods that may become very useful in diagnostics of neuromuscular disorders are presented later in this chapter

5.6 Hill's Three-element Model

The three-dimensional distribution of stress and strain in the gastrointestinal tract, and the effect of stress and strain on the biology of mucosa, submucosa, and longitudinal and circumferential muscles have not been studied in detail. For this purpose, the overall scheme must briefly be described. Since the active contraction of the muscles is the most important feature of the peristaltic movement of the gastrointestinal tract, the method of handling the muscles must be described first.

The "three-element model" of Nobel Laureate A.V. Hill from 1939 considers the tissue as composed of a "contractile element" connected with a "series elastic element" to describe the active contraction of the muscle, and a "parallel element" to describe the connective tissue (such as collagen, elastin and other matrix molecules) (Figure 5.2). This model has been applied extensively to the mechanics of the heart, lung, blood vessels, and kidney (Hill, 1970). No competing model had better success. Applying to the gastrointestinal tract, we consider the mucosa, submucosa and the quiescent muscles as "parallel elements", whereas the contractile and series elastic elements belong to the active muscle. The contractile element is identical to the sliding actin–myosin molecules. At rest there is no tension in this element but active tension is generated when the number of cross-bridges increases. From isometric experiments (see later) it is known that an optimum sarcomere length exists where the active tension has its maximum. The number of cross-bridges explains this

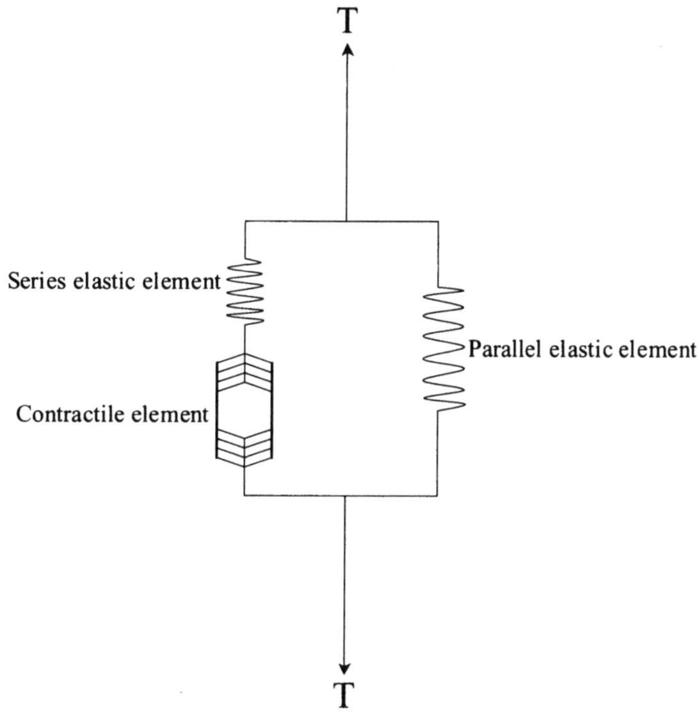

Figure 5.2. Illustration of Hill's model with the three elements. The parallel element represents the connective tissues and the quiescent muscle cells. The contractile element represents the active contraction, shortening and tension-generation machinery. The series element represents the elasticity of actin and myosin fibres and cross-bridges.

feature. When the muscle is too long, the actin and myosin molecules are pulled too far apart and the number of cross-bridges possibly decrease. If the muscle is too short, the actin filaments interfere with each other. If we assume that the contractile element does not contribute tension in the resting state, then the stress–strain history of the resting muscle determines the constitutive equation of the parallel element. Hence, the difference between the mechanical properties of the whole muscle and those of the parallel element characterise the combination of the contractile and series elastic elements. Due to the coupling in series between these two elements, the division of the strain between these elements is not unique. On the basis of further assumptions that have to be made, Fung (1993) provides the equations for a number of experimental conditions and also points out where the limitations are with Hill's model.

The effect of the tissue properties during loading and the zero-stress state on physiology may not be obvious until the full Hill's model is analysed. But it is clear that the zero-stress state (see Chapter 6) and the mechanical properties of collagen and other matrix components are a fundamental feature of the parallel element. The function of the contractile element depends on the parallel element. For example, the length of the muscle cell depends on the strain of the parallel element; hence, whether the muscle can achieve the optimum length of the length–tension relationship for its contraction or not depends on the zero-stress state of the parallel element. Work on

Hill's three-element model may help us to understand the interaction between the parallel and contractile elements. Further work is also needed to determine how the residual strains affect responses to physical and chemical stimulation in pathophysiological states. Since the gastrointestinal tract is essentially a muscular tube, alteration of the activity of the smooth muscle cells by agents acting on the muscles directly or on their innervation will change the zero-stress state of this organ. Hence, the effects of these agents on the gastrointestinal tract need investigations in order to model accurately the normal bolus transport in health and disease.

5.7 What Determines the Shape of the Mechanical Curves

Using Hill's model we can now explain what determines the shape of the distensibility curves. Smooth muscle cells contribute to the distensibility passively through the synthesis of extracellular collagen and actively through contraction. Though it is easy to record contractile activity in vitro and in vivo, it is, from a bioengineering point of view, easiest to analyse and interpret the passive behaviour, e.g., during administering of smooth muscle relaxants. As stated in the previous section about Hill's model, this is also a prerequisite for understanding the active smooth muscle behaviour. We shall first consider the passive state that can be studied in vitro and, to some extent, in vivo.

It is well known that the passive elastic behaviour of biological tissues is non-linear. This mechanical feature protects the tissue against over-distension and damage at high luminal pressure loads while distending easily and facilitating flow in the physiological pressure range. In arteries, it has been demonstrated that collagen bears circumferential loads at high stress levels (Dobrin, 1978; Roach and Burton, 1957). Due to the richness of collagen in gastrointestinal tissue (Gabella, 1987), collagen is likely the major determinant of the passive curve shapes. There exists only little data concerning the relation between elastic properties and morphology in the gastrointestinal tract. Previously, Fackler et al. (1981) and Orberg et al. (1983) showed the biaxial orientation of collagen fibres in submucosa and demonstrated how the fibres uncrimp to low levels of stress rather than being directly stretched. Other studies demonstrated correlation between the total collagen content and the biomechanical properties in obstruction models (Gregersen et al., 1992). It is also known that collagen deficiency changes the passive zero-stress state (Chapter 7).

Parameters such as wall tension, stress, and strain at equilibrium are likely to depend mainly on the elastic properties of tissue or the pseudo-elastic properties (Fung, 1993) since the stress–strain curve shows a high degree of non-linearity and different paths of the loading and unloading curves. The parameters are likely to reflect properties of the parallel elastic component according to the Hill model. The parallel elastic component seems to be anatomically located in the connective tissue since the contractile element of smooth muscle by itself does not contribute significantly to the passive tension (Conklin and Goyal, 1989; Murphy, 1976). Using this model the parallel elastic component determines the shape of the passive curve.

The distensibility in vivo depends not only on the passive properties but also on the physiological state of the muscle. Smooth muscle cells are oriented both circumferentially and longitudinally and their contractile state may change the geometry of the lumen and the wall stiffness considerably. For example, it is well known that longitudinal contraction increases the diameter of the intestine (Melville et al., 1975), which may affect its distensibility. Furthermore, the circular layer of the gastrointestinal tract contracts at the site of distension (Furness and Costa, 1987; Hukuhara et al., 1936). Such contractions are nerve-mediated via cholinergic pathways rather than being a direct smooth muscle response to distension (Gregersen et al., 1992). These contractions arise proximal to the distension site and propagate anally. As discussed earlier, this implies that the measured distensibility depends not only on the degree of phasic contractility but also on the dynamic properties of the instruments used for the measurements.

Although it is more difficult to study the passive properties in vivo due to contractility and difficulties in quantitating the transmural pressure, phasic contractions can, to a variable degree, be eliminated by administration of drugs such as butylscopolamin. Lowering the contractile activity to a level within the frequency range of the measuring equipment facilitates differentiation between passive and active properties. The passive tension can be evaluated in between the contractions where inertial forces are negligible under the assumptions of Laplace's law. However, active properties in the form of specific tone may still be present. This may cause a shift at the lower end of the tension–strain curves that may be difficult to differentiate from the passive properties. Animal studies are needed to evaluate the possible contribution of tone, to validate the baseline pressure determination and to elucidate the contribution of tethering to the transmural pressure in the organ wall during distension.

5.8 Active Tissue Properties

Novices tend to think that all aspects of gastrointestinal contractions represent the effects of nerves acting on an intrinsically passive musculature. However, many aspects of gastrointestinal motility arise from the special properties of the visceral muscle itself. These myogenic factors in motility bear especially upon two matters, the rhythmicity of phasic contractions – those lasting a matter of seconds with a rapid onset and offset – and the capacity of the muscle to generate tone – a sustained contraction. The former has been studied extensively whereas tone has largely escaped our attention, mainly due to methodological difficulties.

5.8.1 Phasic Contractility

Contractions in the gastrointestinal tract are commonly divided into phasic and tonic contractions. The former is defined as having definite beginnings and endings and with a duration of a few seconds while tonic contractions last longer with or without elevation of baseline pressure (Weisbrodt, 1987). Phasic contractions can be recorded from all parts of the gastrointestinal tract with the exception of the gastric fundus. The amplitude and duration of phasic contractions vary from one location to others and between species; e.g., contractions of the oesophagus and antrum often have amplitudes of more than 150 mmHg while those of the small intestine are of shorter duration and have amplitudes in the range of 20–50 mmHg. Phasic contractions may propagate in the aboral direction (antegrade peristalsis), in the oral direction (retrograde peristalsis) or be of the segmenting stationary type.

Spatial and Temporal Characteristics of Phasic Activity

Motility can be characterised according to the spatial and temporal properties. The latter refers to the chronological timing of contractions with features such as number, duration, and frequency of slow waves/spikes/contraction, and of clusters of contractions. Thus, temporal control refers to the control of contractions in time at a given site. Spatial properties refer to characteristics such as location, motion, and direction, distance and rate of spread of contractions/clusters of contractions. Thus, spatial control means the temporal relationship between the occurrence of contractions at adjacent sites. Spatial and temporal characteristics of contractions are, to a high degree, responsible for mixing and propulsion of intraluminal contents, e.g., propulsion is more effective when sequential contractions are recorded at adjacent sites than when they occur only localised. Other features of contractions such as amplitude and area under curve, whether the contractions are phasic/tonic, occluding/non-occluding or concentric/eccentric, can be described for both temporal and spatial patterns. Both properties are critically dependent on the electromyographic properties – especially frequency and propagation velocity of slow waves – , on neural control mechanisms and on chemical control mechanisms (Christensen et al., 1971; Sarna, 1985).

Various contractile and motor patterns of the gastrointestinal tract have been described and several terms are used. The contractile patterns describe specific patterns of activity like the MMC (phase III) in the fasted state. The contractile patterns can be divided into two groups: a) individual and b) grouped motor events. In contrast, motor pattern describes a complex intestinal motor activity like the fed motor pattern, which is composed of different contractile patterns.

It is beyond the scope of this book to give a detailed description of the contractile patterns in the different parts of the gastrointestinal tract. These patterns vary considerable from part to part. For example, regarding the stomach, the process of gastric emptying is a complex mechanism in which the gastric reservoir, the antral pump, the pyloric sphincter and the antroduodenal coordination play an important role; the interaction of these parameters has been described in several papers and reviews. The repertoire of the phasic activity of the small intestine is much larger than that of the stomach. Early fluoroscopic studies demonstrated that at least two motor patterns existed, segmentation and peristalsis. In recent years, recordings of mechanical or electrical activity using multiple closely spaced transducers or electrodes have shown that the small intestine is able to produce many patterns of activity (Ehrlein et al., 1987). The mechanical activity of the small intestine can be generally divided into patterns of individual motor events and patterns of grouped motor events. The term "stationary" designates a contractile event which occurs only in one of a longitudinal array of closely spaced sensors. The adjective "propagating" or "propagated" describes the spreading of an individual motor event along the longitudinal axis of the gut. The apparent aboral (antegrade) or oral (retrograde) movement of a grouped motor event is called migration. When results of electrical and mechanical recordings are compared, one must keep in mind that a burst of spike potentials is equivalent to a single mechanical event. According to this relationship, the appearance of the contractile patterns is much more clear-cut by recordings of mechanical signals compared with electrical signals. The motility patterns are also species-dependent; e.g., in the canine small intestine, three types of individual motor events can be differentiated: 1) propagating contractions, 2) stationary individual contractions, and 3) propagating power (giant) contractions.

How the Gut Induces Different Motor Patterns

The enteric nervous system plays a major role in the neural control of intestinal motor patterns. There is the concept that basic neural circuits exist within the enteric nervous system. These basic circuits represent the pathways for the peristaltic reflex. They are connected with each other by neuronal synaptic gates that might be open or closed. It is further postulated that inhibitory neurons of the enteric plexus suppress the ability of the slow waves to produce spike potentials (Wood, 1987). When this neural inhibition is effective, the motor activities of the gut cease. Contractions only occur when the neural inhibition is turned off and excitation is turned on. The excitation of an intestinal segment might activate the basic neural circuits and might open the synaptic gates between adjacent circuits. The sequential activation of the basic circuits results in a peristaltic wave. However, the smooth muscle cells only respond to the activated circuits if slow waves of the smooth muscle cells pass along the gut. Therefore, the slow waves determine the frequency and the propagation velocity of intestinal contractions. A concept on the generation of the different motor patterns of the gut has been presented; it is based on the different spatial and temporal excitation of the gut. Stationary segmenting contractions might be produced by a brief excitation of a short intestinal segment. If a slow wave passes over this short segment, spike potentials are induced and a local contraction occurs. Contraction waves might be produced by brief excitations of long intestinal segments. The slow wave, which moves over the excited segment, will induce spikes and will produce a contraction wave because several basic circuits are sequentially activated. Postprandially, this kind of excitation occurs irregularly in space and time; consequently, the contraction waves start at variable loci and spread over variable distances of the gut. Clusters of contractions might be produced by a prolonged excitation of a short intestinal segment. When consecutive slow waves move over the excited segment, synaptic gates are opened and short contraction waves are produced. Due to the prolonged excitation at the same locus, several consecutive slow waves induce repetitive short contraction waves, the characteristic feature of the cluster. When the excitation remains at the same locus, a stationary cluster occurs. When the segment of excitation migrates aborally, an aborally migrating cluster is produced.

Bolus Transport Mechanics

Many factors influence flow and bolus transport in distensible organs such as the gastrointestinal tract. The driving force is the pressure generated by the contractile peristaltic forces and, to a lesser extent, the hydrostatic force of gravity. Brasseur and co-workers (Ren et al., 1993) have demonstrated that bolus volume, viscosity and abdominal compression influence bolus pressure in the oesophagus. Biomechanical models predict that factors such as the shape and size of the luminal cross-section, tissue viscoelastic properties, such as the elastic modulus and shear modulus of the wall, are also important determinants of flow. Recent studies in pigs and guinea pigs have revealed that differences in luminal cross-sectional area and passive elastic properties exist between proximal and distal regions of the oesophagus and small intestine (Gregersen et al., 1996; Storkholm et al., 1995; Orloff Juhl et al., 1994). These findings are associated with differences in flow properties and motility between these regions (Schulze-Delrieu, 1991). The gastrointestinal tract is stretched in the vicinity of a bolus (Siegle and Ehrlein, 1989); i.e., the contents propelled in front of a peristaltic contraction bulge out the intestinal wall (Siegle and Ehrlein, 1989; Bayliss and

Starling, 1901). Thus, the intestinal reflexes initiated by localised stretch are of physiological significance for flow. Brasseur (1993) demonstrated that the bolus pressure in the human oesophagus ranges up to 3–5 kPa. Hence, the bolus pressure corresponds to the most elastic part of the distensibility curves and serves to reduce resistance to flow while the wall is much stiffer at higher pressure, which may prevent overstretching.

Distensibility studies may help us understand the complex fluid mechanical behaviour of the gastrointestinal tract. Thus far, oesophageal flow properties have been studied in detail by Brasseur and co-workers (Ren et al., 1993; Brasseur, 1993; Li et al., 1992), while gastric emptying, antroduodenal mechanics, and intestinal flow properties were described primarily by Christensen's group (Melville et al., 1975; Macagno et al., 1975; Macagno and Christensen, 1980), Ehrlein's group (Ehrlein et al., 1987; Siegle and Ehrlein, 1989), Meyer and co-workers (Meyer et al., 1986), Schulze-Delrieu and co-workers (Schulze-Delrieu, 1991; 1992; Schulze-Delrieu et al., 1991) as well as by other investigators (Keane et al., 1980; Kerlin et al., 1982). A promising method based on impedance measurements for evaluation of bolus shape in the oesophagus was introduced in gastroenterology by Fisher et al. (1978) and was later refined by Fass and co-workers (1989). A detailed review of the gastrointestinal flow properties is given by Weems (1987). Mathematical modelling of peristaltic transport in distensible tubes has also been attempted (Stavitsky et al., 1981; Srivastava et al., 1983; Fung and Yih, 1968; Li and Brasseur, 1993; and several papers by Miftakhov). The mathematical modelling of peristaltic transport is based on the premise that the interaction between gastrointestinal elasticity and bolus transport is best understood by examining the fluid dynamics equations (equations describing bolus transport) and equations describing gastrointestinal deformation (the constitutive equations). The interaction between these equations is governed by the boundary conditions.

Peristaltic Reflexes

Peristalsis is the coordinated movement of alimentary muscle resulting in the propulsion of intraluminal contents. The intestinal neural pathways controlling peristalsis are contained within the enteric nervous system, enabling propulsive motility to be studied in vitro (Trendelenburg, 1917; Costa and Furness, 1976). This evidence is obtained from studies on isolated segments and from studies where either the myenteric plexus was sectioned or neuronal transmission was blocked by the use of substances such as tetrodotoxin. It is, however, important to emphasise that extrinsic reflexes may have great influence on motility in vivo. Trendelenburg was the first to describe an in vitro method for studying peristalsis, using guinea pig ileum (Trendelenburg, 1917). By raising the intraluminal pressure, the segment of intestine emptied its contents by way of a contraction in the longitudinal layer, which was followed by a propagated contraction of the circular muscle. Thus, Trendelenburg distinguished between a preparatory phase and an emptying phase. By use of different modifications of the Trendelenburg method (Yokoyama and Ozaki, 1980; Hukuhara and Fukuda, 1968; Bülbring et al., 1958), it was demonstrated that the two phases had little in common except being elicited by the same stimulus (Kosterlitz and Lees, 1964). This stimulus was radial stretching (most likely Ginzel meant circumferential stretching!) rather than increased transmural pressure (Ginzel, 1959). The phases are also differently regulated (Yokoyama and North, 1983) and they are influenced by a large number of pharmacological and bioactive substances (Holzer and Lembeck, 1979; Barthó et al., 1982; Bennett et al., 1976; Kromer and Pretzlaff, 1979; Beleslin et

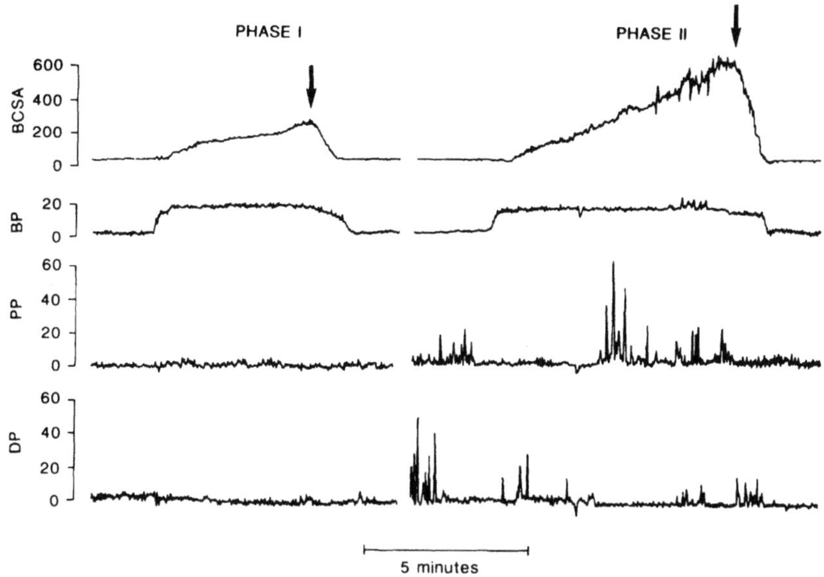

Figure 5.3. Illustration of the ascending excitatory peristaltic reflex and variation in tone between phase I and II in the human duodenum. The balloon cross-sectional area (BCSA), balloon pressure (BP), and pressure 2 cm proximal (PP) and 2 cm distal (DP) to the balloon were measured during balloon distension with a pressure of 20 cmH_2O. Balloon distension induced contractile activity proximal to the balloon. The low BCSA in early phase I indicates high smooth muscle tone. The arrows indicate when the balloon pressure was lowered. With permission. *Am J Physiol*, 1992;**263**:G795–801.

al., 1987; Holzer, 1989), thereby indicating the complex regulation of small-intestinal motility. Studies of the release of various substances such as dynorphines and tachykinins during distension have suggested a physiological role of some of these substances (Donnerer et al., 1984).

Although previously indicated by others (Nothnagel, 1882; Mall, 1896), Bayliss and Starling (1899) were the first to characterise the polarised reflexes by formulating the law of the intestine. Thus, an excitatory peristaltic reflex that initiates contractions proximal to the site of distension was distinguished from an inhibitory peristaltic reflex that initiates relaxation distal to the site of distension. Figure 5.3 shows an example of the ascending excitatory peristaltic reflex in the human duodenum during balloon distension. Such reflexes were initiated either by distension of the intestinal wall by use of a balloon or by intraluminal installation of liquid and by mechanochemical stimulation of the mucosa resulting in a net transport of luminal contents in the aboral direction (Bayliss and Starling, 1899; 1901; Costa and Furness, 1976; Hukuhara and Fukuda, 1965; Bülbring et al., 1958; Leek, 1977). The fact that the contents propelled in front of a peristaltic contraction bulge out the intestinal wall (Bayliss and Starling, 1901; Siegle and Ehrlein, 1989) supports the theory that the intestinal reflexes initiated by localised stretch have physiological significance. This is further supported from evidence that the distension-sensitive receptors are excited also by visceral tension developed actively during contractions in the small intestine (Iggo, 1957). Thus, propulsion occurs as a result of the sequential activation of the reflexes in a cascade fashion along the intestine.

5.8.2 Gastrointestinal Smooth Muscle Tone

This subject is reviewed in detail in Gregersen and Christensen (2000), from which the sections below is a summary. In the biomedical world, tone especially designates the stable resistance to stretch that characterises tissues containing muscle. Such resistance arises both from the fixed physical properties of the tissue (passive resistance) and from the energy-consuming activation of the contractile apparatus of the muscle (active resistance). Somatic muscle contracts in response to stretch, and the absence of such contraction – loss of tone – signifies neurological disease. For physicians and physiologists, tone refers only to the active resistance to stretch; that which arises from muscle contraction. For this reason, such terms as "active tone" and "passive tone" are confusing and unnecessary because other terms exist to make the distinction between active and passive mechanical properties. Per definition, tone is active.

The notion of muscle tone is an essential one in gastrointestinal physiology. Tone in gut muscle, however, has received little attention compared to striated muscle physiology until recent decades when its significance in the motility of the gastrointestinal tract became evident. By establishing the fundamental diameter of the conduit in gastrointestinal viscera, the capacity to generate tone is a crucial property of gastrointestinal muscle in sphincters and in other places in the gut, certainly having profound effects on gastrointestinal flow. Inadequate methods used in vivo can easily lead to erroneous conclusions. All investigations of tone must take into account the facts of biomechanics as well as those of anatomy and physiology. Such biomechanical complexities as anisotropy, non-linear behaviour, finite deformation, strain softening and viscoelasticity have only begun to enter the thinking of those who study tone in the gut. Biomechanics, a new discipline in itself, has much to teach us about the motility of the gastrointestinal tract, including the old but revitalised concept of tone in the muscle of the gut.

Myogenic Tone and Neurogenic Tone

Tone in somatic muscle is neurogenic. It represents a constant low level of discharge of its excitatory innervation. In contrast, in the gut, some tone represents a property of the muscle itself rather that the constant discharge of excitatory nerves. Thus, sphincter muscle exhibits myogenic tone, as evidenced by the fact that some part of the resistance to stretch is unaffected by tetrodotoxin and by drugs that block synaptic transmission (Anuras et al., 1974; Conklin and Christensen, 1975). Not all tone in the gut is myogenic. Hormones and excitatory nerves also modify the force of contraction in sphincters and elsewhere. There is some evidence that sphincter muscle is more sensitive to excitatory drugs than non-sphincter muscle (Christensen,1970). The same property that is responsible for myogenic tone may enhance its sensitivity to drugs acting through autonomic receptor mechanisms.

Tone is commonly contrasted to phasic contraction on the basis of temporal characteristics. The distinct differences between these two forms of gross behaviour preclude the need for precise quantitative criteria to distinguish them. The time between the onset and the offset of rhythmic contractions is measurable in terms of seconds with little variability, and such events recur at a reasonably constant frequency (Weisbrodt, 1987). In contrast, the duration of the contractions involved in tone can be measured in minutes or hours, and tonic contractions show no obvious pattern or regularity in occurrence.

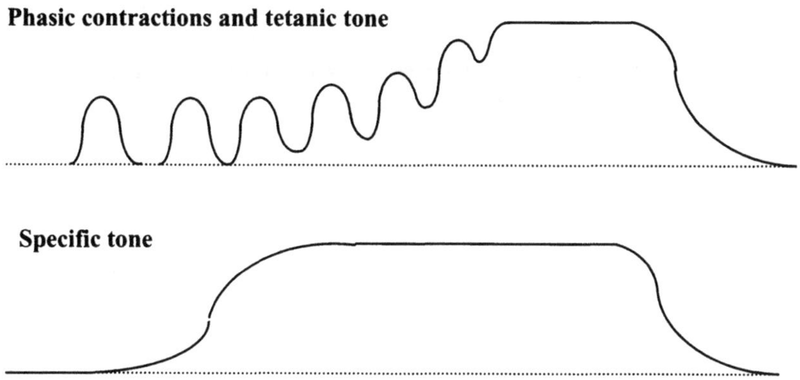

Figure 5.4. Schematic illustration of phasic contractions fusing into tetanic tone and specific tone. This classification was first outlined by Mandrek & Golenhofen. Specific tone is independent of phasic activity and electrical spike activity.

Tone arises in two different ways. From studies on isolated intestinal muscle strips, Mandrek and Golenhofen postulated a dual origin for tone. They presented evidence for two different contractile mechanisms producing, respectively, tonic and phasic (or rhythmic) contractions (Mandrek and Golenhofen, 1990) and they pointed out the existence of two types of tone. Tetanic tone represents fused phasic (or rhythmic) contractions at one locus. Tetanic tone represents rhythmic contractions of such frequency and force that they do not allow time for the repeating onsets and offsets of the rhythmic contractions to appear. The second type, specific tone, is characterised by a sustained contraction at one locus. Whereas tetanic tone seems likely to be at least partly neurogenic, specific tone appears to be fundamentally myogenic. Mandrek and Golenhofen pointed out that these two types of tone could be distinguished experimentally. Muscle cells that are involved in phasic contractions generate the slow changes in membrane potential called slow waves. Phasic contractions occur when the contractile machinery is activated by a Ca^{++} influx that is usually associated with spike potentials. Thus, the tetanic nature of a tonic contraction may be indicated by the presence of slow waves and spikes in the electromyogram. The absence of both such signals indicates the presence of specific or myogenic tone.

Mandrek and Golenhofen suggested that specific tone and tetanic tone could be distinguished on pharmacological grounds. The calcium channel antagonist nifedipine abolishes phasic contractions whereas tonic contractions usually remain unaltered or respond less. Nifedipine can completely suppress tetanic tone since it is controlled by spike discharges producing phasic contractions that fuse together. Sodium nitroprusside, by activating guanylate synthase, preferentially suppresses specific tone (Mandrek and Golenhofen, 1990).

Evidence for Tone in Vivo

The demonstration of tone in gastrointestinal muscle strips in vitro is relatively straightforward, but the measurement of its force in vivo is not. The conventional manometric method to record gastrointestinal motor function cannot detect tone in the intact gut except in sphincteric regions. Even there, the pressures registered cannot be taken as linear reflections of the degree of tone. Still, many investigators accept the manometrically recorded sphincter pressure or the volume of a non-sphincteric region at a constant pressure as faithful representations of tone. Of course, they are not. Such changes can have several causes. A critical look at the evidence for tone, especially in non-sphincteric regions, seems prudent before drawing clear-cut conclusions about the nature of tone in the various parts of the gut.

Tone in Gastrointestinal Sphincters

It is well known that gastrointestinal sphincters are specialised regions characterised by sustained muscle tone. Tonic contraction in the lower oesophageal sphincter normally generates a manometrically detected intraluminal pressure of about 30 mmHg. This seems to be specific tone since this sphincter exhibits no rhythmicity. Also, the tone is not associated with slow wave or spike activity. In the manometric definition of this sphincter, of course, the pressure represents the resistance to the flow of water through the side hole of the perfused tube produced by the approximation of the wall of the oesophagus to the catheter. It is difficult to accept the quantification of tone in this way as anything more that a distorted reflection both because such manometrically recorded pressure is not linearly related to the force produced by the contraction and because many passive factors enter into the measurement. These include the tethering of the oesophagus to external structures, the geometry of the sphincter, and the physical properties and geometry of the mucosa. A sleeve or cuff seems to give more consistent results in measuring tone in this sphincter (Gustavsson and Tucker, 1988; Siuri and Mittal, 1991). It simply integrates or averages the passive factors rather than eliminating them, and it does not make the relationship between tone and manometric pressure linear.

The demonstration of myogenic tone in the lower oesophageal sphincter immediately adjacent to muscle devoid of such tone in the oesophageal body makes the comparison of the two kinds of muscle obvious. There appears to be a greater proportion of collagen in the sphincter (Seelig and Goyal, 1978), but it has not been shown that this correlates with a greater passive resistance to stretch. The muscle itself in the lower oesophageal sphincter appears to differ from that of the adjacent oesophageal body. The most conspicuous difference is that the mitochondria occupy a greater fraction of space in the sphincter muscle and they tend to be more centrally located in the cells (Christensen and Roberts, 1983). Also, the myogenic tone of the sphincter is oxygen-dependent whereas the contraction of the adjacent oesophageal body is not (Christensen, 1982). The sphincteric muscle possesses more cytochrome C activity (Robison et al., 1984). All this points to a distinction in energy production in the genesis of myogenic tone. Recent studies suggest differences in the calcium-binding proteins and contractile proteins (Szymanski et al., 1998). Thus, the literature tends to confirm the idea that specific tone is myogenic, reflecting the activity of a different kind of muscle from that generating rhythmic contractions. The persisting belief that some distinction in enteric nerve function characterises tonic regions of the gut may be accounted for by the greater sensitivity of the sphincter to autonomic drugs acting through membrane receptors. The same special features that make the muscle generate tonic contraction also seem to make it more responsive to excitatory neurohormones. Finally, the existence of myogenic tone in the oblique muscle layer of the stomach, as distinct from the other two layers, was demonstrated long ago (Christensen and Torres, 1975), and it seems to be indistinguishable from that of the lower oesophageal sphincter. Indeed, the two structures seem to be anatomically continuous, the sphincter appearing, in part at least, to be a part of the so-called gastric sling fibres that constitute the edge of the oblique muscle layer. It seems probable, then, that a particular set of the characteristics of gut muscle establishes it as tonic muscle. This same set of features may well be found expressed, to a greater or lesser degree, in tonic muscle in the pylorus, at the ileo-caecal junction, and elsewhere throughout the gastrointestinal tract (Christensen et al., 1973; Conklin and Christensen, 1975).

Tone also characterises the pyloric sphincter and the ileo-caecal sphincter (Anuras et al., 1974; Conklin and Christensen, 1975), though little has been done to try to measure it in vivo in either location. Tone also characterises the internal anal sphincter (Hill et al., 1960). Here, the basic properties of the muscle closely resemble those of the lower oesophageal sphincter.

Tone in Non-sphincteric Regions of the Gastrointestinal Tract

Smooth muscle outside sphincteric regions is also capable of developing tonic contractions. Tone in such regions is more difficult to detect and to measure than tone in sphincters. In manometry, tonic contractions can occur without an apparent elevation of baseline pressure at the same time that phasic contractions, lasting a few seconds, may be seen. This happens because the intraluminal pressure generated by the tonic contraction of the muscle is stable, or reasonably so, and because the pressure dissipates into regions adjacent to the site of the recording when there is nothing to confine it. This occurrence is now widely termed the "common cavity" phenomenon. For this reason, tone, in most of the gastrointestinal tract, cannot even be detected, let alone measured, by means of ordinary perfusion manometry. Its detection requires a distending balloon that confines the generated pressures so that they can be recorded.

Extensive work has been done on tone in the gastric fundus where gastric relaxation mediated by non-adrenergic, non-cholinergic inhibitory nerves plays a major role in establishing the reservoir function of the proximal stomach (Azpiroz and Malagelada, 1985; 1990; De Ponti et al., 1987). This relaxation occurs from a weak tonic contraction mainly of the oblique muscle layer in the fundus that is maintained in the absence of slow waves and spike activity. This appears to be *specific tone*, an idea that is confirmed by the fact that this muscle in vitro exhibits myogenic tone (Christensen and Torres, 1975). Gastric tone is measured in vivo almost exclusively by the evaluation of volume variations in an intragastric balloon held at a constant low intraluminal pressure, in the so-called barostat technique. The limitations of this method are great (see below and elsewhere in this book).

There is also good evidence for tone in the small intestine. The fact that the small intestine becomes dilated in chronic intestinal pseudo-obstruction and during the administration of smooth muscle relaxants such as glucagon indicates that some level of basal tone exists there under normal conditions (Chernish et al., 1972; Diamant and Picazo, 1983). Furthermore, animal studies using serosal strain gauges and videofluoroscopy indicate that tonic contractions of the smooth muscle in the small intestine occur during the fasted state (Ehrlein et al., 1987; Reinke, 1964). The studies mentioned above are consistent with small intestinal tone but the data were qualitative by nature. A breakthrough in scientific studies of small intestinal tone came about in 1990 and 1992 where impedance planimetry was used in the human duodenum during fasting to show specific tone during phase I of the migrating motility complex (Gregersen et al., 1990; 1992). The tone in phase I was evident as a stable reduced luminal cross-sectional area that increased with the onset of phase II. This is consistent with a change in the active contractile state of the muscle between the two phases. The finding of tone in phase I is conspicuous because phase I is electromyographically devoid of spikes and contractions, or essentially so, whereas phase II is characterised electromyographically by constant bursts of spike activity and contractions. Thus, an inverse relationship exists between tone and spike activity (phasic contractions) in phase I and in phase II; tone is absent in phase II when spikes occur but present in phase I when they are absent. This is specific tone (Figure 5.3) These results favour the Mandrek–Golenhofen concept of dualism, i.e., that tone and phasic contractions represent different modes of activation of the contractile apparatus. Recent pressure–volume experiments with long balloons in the duodenum contradicted this conclusion, indicating that tone is higher in phase II (Rouillon et al., 1991). As reviewed below, however, this technique suffers from the possibility of an artefact that gives rise to fallacious results.

Long-lasting tonic contractions are frequently recorded in the colon and rectum. A few studies have also claimed the existence of tone in the oesophageal body (Mayrand and Diamant, 1993; Gonzalez et al., 1997).

A Mechanical Definition of Tone

Tone must be treated as a mechanical phenomenon. Consequently, physiologists must deal with the mechanical concepts of stress and strain. This demands consideration of the tissue mechanics and geometry of the various sites and organs organ under scrutiny.

Stress is the force per unit area, whereas strain is a dimensionless measure of deformation. In a body not containing muscle tissue, an applied force tends to deform that body and the non-linear stress–strain relationship can be derived. Muscle has the capacity to develop active force; that is, to contract, often in response to applied stress. This creates two kinds of stress, one kind arising from the passive mechanical properties of the tissue and another kind coming from the activation of the contractile apparatus of the muscle. Thus, there are both active and passive forms of stress. Tone is an active form of stress like phasic contractions. Looking at the issue of tone in uniaxial terms, tone (the active form of stress) in smooth muscle causes a sustained reduction in the length of the specimen (strain) under isotonic conditions, whereas isometric conditions, in which strain is prevented, allow the development of maximal active stress in the specimen. Gut smooth muscle mostly operates in vivo under essentially isotonic conditions, being allowed to shorten with little resistance. Thus, the amount of active stress that develops in the muscle is minimised. In manometry and in studies with balloons, the isotonic circumstance is violated to some extent. The presence of an object in the lumen that resists compression, by limiting the shortening of the muscle and so restricting the isotonic state, magnifies the active stress that develops. That is why the recorded pressure in the lower oesophageal sphincter increases with the diameter of the recording catheter.

Because the gastrointestinal tract is a reasonably uniform cylinder, essentially a deformable tube, it makes geometric sense to define tone as a sustained active stress tending to produce strain that leads to a reduction in luminal cross-sectional area and circumference, or surface area (Gregersen and Christensen, 2000). This stress represents the activation of the smooth muscle. In the irregularly globular parts of the gastrointestinal tract – the gastric fundus, the gallbladder and the caecum – it makes better sense to view the strain as one tending to reduce the surface area rather than volume because of the high degree of radial asymmetry in those regions. Even in spheres, volume and surface area are not linearly related. Their relationship is quite problematical in an irregular globular organ like the stomach.

Most studies of tone in vivo define tone in terms of changes in "strain" (volume at a constant predefined pressure, or a change in the slope of the compliance curve, $\Delta V/\Delta P$) without knowledge of the nature of the stresses, either active or passive (Gonzalez et al., 1997; Azpiroz and Malagelada, 1990; Steadman et al., 1992; Premen et al., 1984). Naturally, such a definition leads to much confusion because the relationship between stress and strain is complicated. The definition and measurement of tone in terms of stress rather than as strain or deformation gives rise to the need for the knowledge of forces in both the active and the passive components of the stress.

The passive properties of a tissue can contribute a large fraction of the resistance to stretch that characterises living muscle, and so be falsely identified as active properties that contribute to tone. That is, some phenomena that superficially look like tone may not actually be tone, the product of the contractile activity of muscle. For example, when the oesophagus is excised from the body it shortens to approximately half its length in vivo. This is not an expression of tone but merely a huge passive longitudinal stress in vivo, perhaps the largest found anywhere in the gut. The passive properties of gut muscle reflect the parallel elastic component described by the well known Hill model (Hill, 1970).

Why Tone Cannot be Measured by Pressure–Volume Measurements

Significant problems exist in the use of pressure–volume measurements in balloons to record tone. Such problems also apply to the recording of distensibility. Many investigators are unaware of the impediments they face both from the obscurities of gastrointestinal physiology and from the unfamiliarity of the mechanical principles that apply to the distension of the balloon within a deformable cylinder like the gastrointestinal tract. The problems relate to the probe design, the method of measurement, the geometry, balloon properties, and the assumptions made in the analysis. Actually, no form of pressure–volume measurement in the gut can accurately measure tone.

The whole surface area of the balloon ideally must be in contact with the tissue. This is never completely attained in volume measurement in the intestine and stomach, but efforts should be made to approach the ideal since the failure to achieve such contact can lead to major error. It is also important to ensure that the applied pressure is wholly transmitted to the tissue. This requires that the balloon is sufficiently large so that it does not contribute its own elastic properties to the results of the measurements. The pressure drop must be confined to the wall itself (as long as the surrounding structures do not contribute by tethering or by their own elastic properties. The hysteresis of the balloon and infusion system and the friction between the tissue and balloon wall may or may not be negligible. The balloon must be short in order to provide a localised distension. Studies making use of long balloons (such as Rouillon et al., 1991; Bell et al., 1991) provide data that are difficult to interpret since the simultaneous excitation and inhibition of phasic activity are likely to occur at the proximal and distal ends of the balloon. While short balloons may encounter fewer contractions, long balloons may sum up phasic contractions to produce a stable pressure or volume that is easily confused with tone. Volume measurements in a long balloon may consequently show no change even though the shape changes are considerable.

Any balloon tends to elongate rather than to distend the organ radially because the tubular viscus resists radial expansion more than it does expansion in the axial direction. The degree of elongation obviously depends on the balloon material. It is still seen that very compliant latex balloons are used in pressure–volume measurements. However, more often now, polyethylene or polyurethane bags are used, but some degree of elongation, no matter how small, inevitably occurs with any balloon. Also, pressure–volume techniques that distend balloons to study tone in tubular organs assume a circular cross-section in addition to a constant axial length in the balloon. Studies in animals, however, clearly show that the small intestine is elliptical in cross-section at low inflation pressures (Gregersen et al., 1997). The erroneous assumptions about elongation and cross-sectional configuration can have important results. Wall tension and compliance are overestimated with elongation. Even a small error in calculating volume in the low-pressure range will affect the results of any stress–strain analysis since the basal radius, r_0, is determined in this range in in vivo studies (where the true zero-stress state cannot be determined). As for the false assumption of a circular cross-section, the volume-to-surface area relationship is complex in an elliptical cylinder, and stress is not uniformly distributed in the circumference.

Since the surface area of a globular viscus cannot be measured directly in vivo, investigators have used methods to measure volume is such organs with the assumption that their balloon systems can follow changes in gastric tone by measuring variations in volume (representing length). This is fundamentally wrong since the stomach is not spherical and the tension induced in the stomach wall is not uniformly distributed because of the geometric asymmetry. Furthermore, phasic contractions moving through the distal stomach must affect the measured volume if part of the balloon extends into that region. Thus, such methods yield only an approximation of tone, and the errors may be quite large. Those who use long balloons in studying tone in the intestine have similar problems (Rouillon et al., 1991; Bell et al., 1991).

The distensibility of the gut in vivo depends upon both the passive properties of the muscle and its physiological state. Smooth muscle cells are oriented both circumferentially and longitudinally in all the gastrointestinal viscera and the contractile state of both layers independently must greatly affect both the wall stiffness and the geometry of the lumen. Longitudinal muscle contraction increases the diameter of the intestine (Melville et al., 1975) and tends to excite the circular layer to contract at the site of distension (Furness and Costa, 1987; Hukuhara et al., 1936). Such variables mean that distensibility as measured with balloons can be affected not only by the amount of phasic contractility and tone in the circular muscle layer, but also by an inaccessible variable, the relative activities of the two muscle layers. Phasic contractions in both layers can be eliminated, to a large degree, by administration of drugs which are assumed to have less or even little effect on myogenic tone. Still, it is difficult to study the tone and the passive properties of gastrointestinal muscle in vivo for the other reasons given above unrelated to the possible confusion that is created by simultaneous phasic contractions.

Many investigators give insufficient consideration to experimental protocol. They seem to be unaware that a specific loading pressure may be reached at different volumes, depending on the history of loading. The investigator must take into account the two forms of history-dependent mechanical behaviour (time-dependent viscoelasticity and load-dependent strain softening) in repeated distension protocols (Gregersen et al., 1998). It is important to carefully consider the reproducibility of distension protocols and design them specifically for the purpose intended.

5.9 Towards a New Standardised Way for Studying Tone, Secondary Peristalsis and Visceral Pain Mechanisms in Vivo

Visceral pain is one of the most frequent reasons why patients seek medical attention. It is well known that distension of the gastrointestinal tract elicits reflex-mediated inhibition and stimulation of motility via intrinsic or extrinsic neural circuits and induces visceral perception such as pain. Mechanoreceptors located in the intestinal wall play an important role in the sensory stimulus-response function. From animal studies it seems evident that some receptors have a high threshold to mechanical stimuli and an encoding function that is evoked by stimuli within the noxious range. Other receptors have a low threshold to mechanical stimuli and an encoding function that spans the range of stimulation intensity from innocuous to noxious. Furthermore, some evidence obtained in animal studies indicates that the mucosal nerve endings act as rapidly adapting mechanoreceptors, whereas the intramuscular endings act as slowly adapting mechanoreceptors. This suggests that rapid distension evokes the mucosal mechanoreceptor, whereas slow distension primarily evokes the muscular mechanoreceptors. Whether low and high threshold receptors exist in the human small intestine and how the receptors adapt to the velocity of mechanical stimulation still remain to be studied.

Mechanoreceptors are not directly sensitive to variation in pressure or volume. A large variation in the peristaltic reflex and perception has been found in various studies and species suggesting that pressure and volume are not the direct stimuli. Instead, the receptors may be activated by mechanical forces and deformations in the gastrointestinal wall secondary to changes in the transmural pressure. Thus, the mechanical distension stimulus and the biomechanical tissue properties must be taken into account in studies of the sensorimotor function in the gastrointestinal tract. Circumferential tension and strain are likely candidates as the direct receptor stimulus because in distensible biological tubes the tensile circumferential wall tension and strain are largest in that direction during distension. It is well known that stepwise balloon distension evokes non-painful and painful sensations in humans and nociception in animals. However, a ramp distension protocol may be more optimal than a stepwise distension protocol for determining which biomechanical parameter perception depends on and whether the sensory responses are dependent on the rate of distension.

These aspects of gastrointestinal function need to be challenged. In a bioengineering approach to analysis of gastrointestinal function and diseases, one should use the geometry and material properties of the gastrointestinal tract, apply the basic laws of physics and characterise the stimulus-response function. Hence, we need to know the stimulus and to measure the response with reliable tools.

The background for the new model with respect to promising clinical data obtained previously with impedance planimetry in patients with non-cardiac chest pain and systemic sclerosis and the new approach are described in detail below and some examples of the new approach are given. The new approach is based on controlled ramp balloon distension along with determination of tension and strain during active and passive conditions, secondary peristalsis and sensation or afferent nerve responses.

5.9.1 Clinical Studies with Pressure-controlled Stepwise Balloon Distension using Impedance Planimetry

Systemic Sclerosis

The first clinical data using impedance planimetry was obtained in patients with systemic sclerosis. This disease is a rare disorder with an incidence of 2.7 patients per million in the USA. The diagnosis of systemic sclerosis is based on the criteria set by the American Rheumatism Association (Anonymous, 1980). Systemic sclerosis is characterised by an excessive deposition of collagen and other connective tissue components in the skin and internal organs. The immune system plays an important role in initiating the fibrotic process and impaired regulation of collagen gene expression probably underlines the persistence of fibrosis. Excess extracellular matrix is deposited and organ dysfunction eventually results. Serious gastrointestinal involvement is frequent in systemic sclerosis. Smooth muscle cells appear morphologically normal but are decreased in number. In some cases, collagen replaces the muscle fibres without reducing the overall thickness of the involved muscle layer, in others the thickness of the fibrotic muscle is reduced. Therefore, oesophageal manometric studies often show amplitude reduction of the contraction.

Impedance planimetric studies in patients with severe systemic sclerosis showed dilatation and impaired secondary peristalsis in the distal part of the oesophagus. The upper part, on the other hand, the region that contains striated muscle, was not dilated and there was a tendency to hyperreactivity rather than to impairment of the distension-induced contraction (Villadsen et al., 1997). A second study on a more heterogeneous patient group showed the same kind of alterations as in the first study, though to a lesser degree. Allocating the systemic sclerosis patients into groups of subtypes with respect to the involvement of skin showed that type 3, the most severe form of systemic sclerosis characterised by diffuse cutaneous lesions, also had the most severe oesophageal changes (Villadsen et al., 2001). A major advantage of these studies was that the motility response was evaluated as a function of tension rather than as a function of pressure or volume. Distension-induced pain was not a significant symptom in the patient group.

Non-cardiac Chest Pain

Non-cardiac chest pain is characterised by recurrent and often incapacitating chest pain without demonstrable cardiac abnormalities. Up to 30% of patients with suspected angina may have this condition and the annual incidence is estimated to be 180,000 in the USA. Often, the oesophagus has been determined to be the source of the chest pain. pH-metry and manometry have been used to diagnose patients with gastro-oesophageal reflux and motility disorders. However, in many cases, the patients cannot be given a proper diagnosis.

On theoretical grounds, oesophageal balloon distension is a useful technique for studying patients with chest pain because sensation is induced, secondary peristalsis evoked and the distensibility tested. However, volume-controlled distension protocols did not prove themselves as a diagnostic tool because the overlap between normal controls and patients was too small. Impedance planimetric investigation based on a pressure-controlled distension protocol proved to be a much better approach. In the study by Rao et al. (1996), the patient group was carefully selected. Cardiologists referred the patients after extensive cardiac evaluations showing no cause for the chest pain (more than one episode weekly for at least three months). In addition, the patients went through oesophageal endoscopy, manometry and pH-metry and were only included if abnormalities were not found. Therefore, this rigorously defined patient group consisted of patients that could not be offered much more from a diagnostic and therapeutic point of view. Nevertheless, the impedance planimetric distension procedure showed abnormal responses in 85% of the patients. The major finding was that the patients responded with pain episodes during the distensions as those they experience during daily life. The threshold for induction of discomfort and pain was at a much lower mechanical load (in terms of the pressure applied) than in the control group. Some patients could be characterised as being hypersensitive whereas others had a hyperreactive oesophagus; i.e., increased secondary peristalsis was recorded when compared to that in normal controls. It is worthwhile to notice that the hyperreactivity was not observed in the manometric recordings during dry and wet swallowing. A provocative test like mechanical distension seems important. The fact that the patient population can be divided into subgroups renders possible a differentiated therapeutic approach. Clearly, the patients expressing hyperreactivity must be treated differently from the patients expressing hypersensitivity.

Though the studies mentioned above were very promising indeed, improvements in distension protocols can be made. These improvements must be based on theoretical considerations with respect to biomechanics. One may think that it is better to ramp the mechanical stimulus to avoid viscoelastic effects and to develop an in vivo test similar to the length–tension diagrams known from smooth muscle strips experiments in vitro.

First, let us select a way to measure the sensory response in human investigations, then the distension protocol is described.

5.9.2 Sensory Assessment

Sensory intensity is assessed continuously during the experiments using an electronic visual analogue scale (VAS). Sensory assessment on a VAS can be difficult, especially in visceral pain, which is diffuse and difficult to characterise. Therefore, it may be advantageous to train the patients in assessment of sensation to deep pressure at the muscles on the right forearm several times before the visceral stimulus is applied. Although still debated, most sensory afferents in the gut are probably polymodal. Whether or not afferents specific for high- and low-intensity stimuli exist is also a matter of debate, and in contrast to the skin, most visceral afferents probably encode both non-painful and painful sensations (Su and Gebhart, 1998). We therefore use a combined scale in the range of non-painful to painful sensations. This scale has proven to be robust in previous experiments with evoked visceral sensations in our laboratory. The intensity of the *non-painful* sensations are scored from 1–5, where the following descriptors are added to facilitate the ratings: 1 = vague perception of mild sensation; 2 = definite perception of mild sensation; 3 = vague perception of moderate sensation; 4 = definite perception of moderate perception and 5 = discomfort/pain threshold. The descriptors are chosen according to earlier studies using balloon distensions in the intestine (Serra et al., 1995). For the *painful* sensations, the patients use the scale from 5–10 anchored at 5 = discomfort/pain threshold to 10 = unbearable pain. To facilitate the scoring, the following anchor words are used: 6 = slight pain; 7 = moderate pain, 8 = medium pain intensity; 9 = intense pain and 10 = unbearable pain. These words were previously found to be equidistant from 1–5 on the pain scale

(Drewes et al., 1997). Hence, when the subject reports that the distension results in pain and /or severe discomfort (5 on the non-painful scale) they were asked to score the intensity from 5–10 on the VAS. The use of VAS was previously demonstrated to be useful to assess painful visceral stimuli to electrical current and distension in the stomach, small intestine and large intestine (Arendt-Nielsen et al., 1997; Drewes et al., 1997). The sensory response may also be evaluated on the basis of evoked brain potential in humans or by the recording of afferent nerve activity in animal studies.

5.9.3 Standardisation of the Mechanical Stimulus

In the first approach, a four-electrode impedance planimetric system is used. This allows determination of tension and strain. As will be shown later in this chapter, completely new and useful data can be obtained using this method. However, the method can be refined in several ways according to the other methods and directions given in this book. For example, the distension stimulus can be a strain ramp and the stress distribution can be determined everywhere in the distended area (see Chapter 4). Furthermore, more potent drugs for relaxation of smooth muscle may be available in the future.

Probe Design and Infusion System

In experiments done thus far, a four-electrode impedance measuring system (see Chapter 4) located inside a balloon on a 120 cm-long probe was used for measurements of CSA in the duodenum. The attached balloon was 40 mm long and was made of 50 μm-thick non-conducting polyurethane. The balloon was connected via an infusion channel to a pump that pumped fluid in and out of the balloon at a controlled flow rate.

The connecting tube between the pump and the probe was heated to 37 ^{0}C and contained 150 ml of fluid. A safety valve was placed on the tube so that the volunteer could deflate the balloon at any time. The fluid reservoir only contained 125 ml of fluid as a safety precaution.

Study Protocol

A questionnaire (see below) is administered to all participants to assess the characteristics of visceral perception before the study started. The balloon is positioned under fluoroscopic guidance into the third portion of the duodenum. Ramp-controlled distensions can then be initiated. The subjects must assess the sensation intensity on a VAS from zero to seven. At VAS = 7, the balloon is deflated using the same rate as during inflation (Figure 5.5). First, the tissue and the volunteer are preconditioned. The distensions can be performed at different infusion speeds to test whether the receptors and the mechanical properties are strain-rate dependent. After a period of rest, the distensions can be repeated during administration of the antimuscarinic drug butylscopolamine in order to relax the smooth muscle. The total butylscopolamine dose is guided by the degree of abolishment of contractions and by the development of classic anticholinergic side effects.

Computation of Active and Passive Tension

For simplicity in the first approach, the circumferential wall tension (T) is calculated according to Laplace's law for cylindrical structures as given in Chapter 3. The total tension (T_{total}) during distension (due to both active and passive tissue properties)

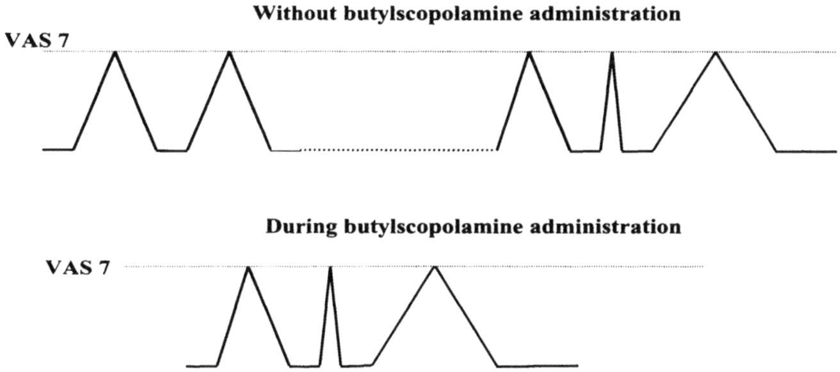

Figure 5.5. Example of ramp distension protocols. First, a number of distensions are done to precondition the tissue and to make the subject get used to the stimulation. Then, ramp tests can be done at different inflation rates. The test may be repeated during drug administration, e.g., with a relaxant drug.

can be determined from the distension test without the administration of butylscopolamine. The passive tension ($T_{passive}$) that results from passive components such as the extracellular collagen is obtained from the test with butylscopolamine. Assuming an additive relationship, the active tension (T_{active}) contributed by smooth muscle activity can then be computed using the equation

$$T_{total} = T_{active} + T_{passive}.$$

The circumferential strain (ε) is the fractional change in radius computed as

$$\varepsilon = \frac{r - r_0}{r_0}$$

where r is the radius at a given distension and r_0 is the reference radius at a wall tension of 2 kPa mm under the assumption that the geometry was circular. At the reference tension, it is easy to determine r_0 for the different subjects. The three tensions can be plotted as a function of strain to give information about the muscle function and tissue stiffness.

The volume, pressure, strain and tension are determined at the pain threshold (VAS = 5). As a statistical measure showing the relative variability of a trait, the coefficient of variation (CV) can be computed for determination of which parameter the mechanoreceptors depend on at the pain threshold. Furthermore, for some analyses, the normalised volume can be computed as the volume divided by the maximum volume where the pump was reversed (VAS = 7). The CSA, pressure, circumferential strain, circumferential tension and perception score can then be calculated for each interval of the normalised volume.

5.9.4 Sensory and Biomechanical Responses to Ramp-controlled Distension of the Human Duodenum

The CSA and pressure both with and without butylscopolamine administration increase as a function of volume at distension rates from 10 to 50 ml min^{-1}. Increasing

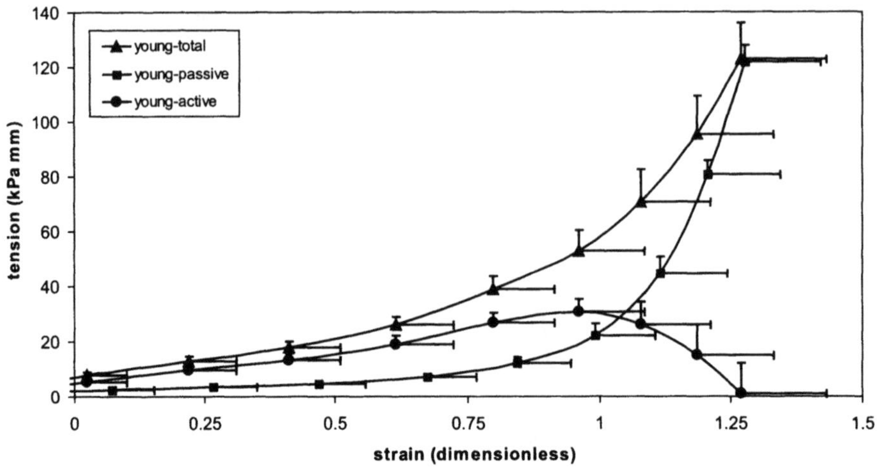

Figure 5.6. Tension–strain curves (total, passive and active tension) at a ramp distension rate of 25 ml/min in the duodenum of young healthy volunteers. Mean and SEM values are shown.

the distension rate resulted in fewer contractions when the distension is performed without butylscopolamine administration. Contractile activity is abolished by butylscopolamine administration in sufficient doses.

The mean total tension, passive tension and strain at the pain threshold (VAS = 5) were 110, 111 kPa mm and 1.1 respectively. When determining the coefficients of variation of VAS as functions of volume, pressure, tension and strain, it was found both before and during butylscopolamine infusion that the variation was lowest for the strain parameter. This indicates that the sensation receptors are strain-dependent. Some degree of strain-rate dependency was also demonstrated because the circumferential wall tension was lower at high distension rates compared to that at lower rates.

The length–tension diagrams were expressed in terms of tension–strain curves (Figure 5.6) with distension rates of 25 ml $\min^{-1}$. The total and passive tension increased exponentially as a function of strain whereas active tension increased to a maximum at a strain of 1.0. Active tension decreased at higher loads. The pain threshold appeared after the maximum on the tension–strain curve was reached. Figure 5.7 essentially shows the same phenomenon as Figure 5.6 but the VAS data are also plotted as functions of strain. The data representing the active curves are computed. From the figure it is clear that the sensation receptors react both to the active tension (i.e., compression) and to circumferential stretch.

In some experiments, the volume is kept constant when VAS 5 is reached during butylscopolamine infusion (Figure 5.8). In this case, it is observed that the perception intensity continues to increase after the infusion is stopped; i.e., volume and cross-sectional area do not change. It is seen that the pressure decreases due to stress relaxation behaviour. As a result, the passive tension must also decrease. Such data, therefore, are indicative of temporal summation.

The main conclusions that can be drawn from the results mentioned above are that; the ramp-controlled balloon distension was applicable to intestinal studies in humans; the passive elastic behaviour is exponential; in the physiological range, the

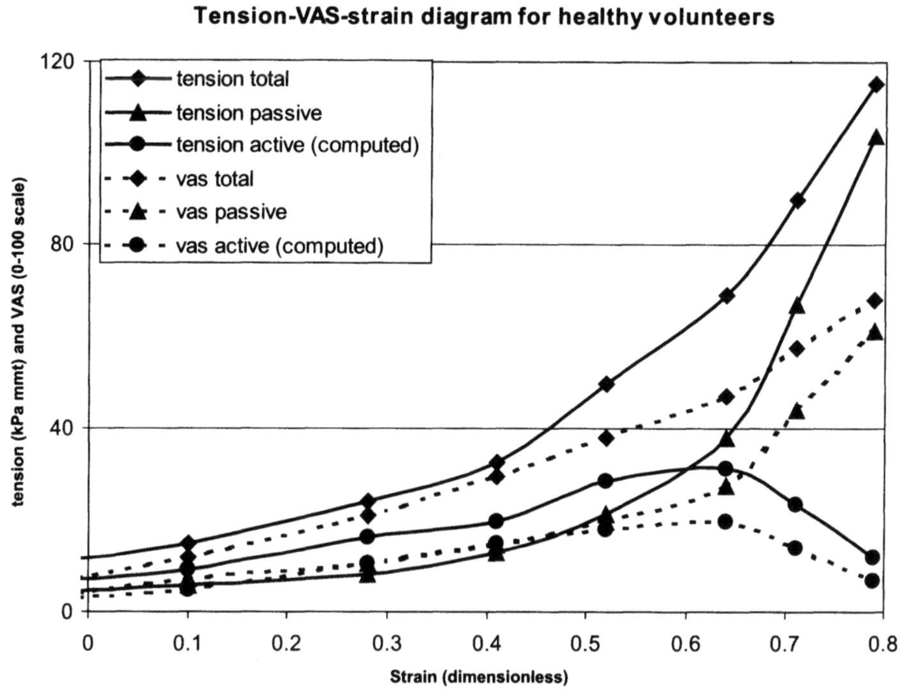

Figure 5.7. Total, passive and active tension–strain curves and corresponding VAS–strain curves (dotted lines) at a ramp distension rate of 25 ml/min in the duodenum of young healthy volunteers. Mean values are shown.

active tension curve has a major influence on the tissue behaviour; the biomechanical parameters showed some strain-rate dependency; and the mechano-sensitive receptors in the duodenal wall depend on circumferential wall strain rather than on volume, pressure and tension.

It is well known that the passive elastic behaviour of biological tissues is exponential. This mechanical feature protects the tissue against overdistension and damage at high luminal pressure loads while distending easily and facilitating flow in the physiological pressure range. In arteries, it has been demonstrated that collagen bears circumferential loads at high stresses. As gastrointestinal tissue is rich in collagen, it is likely that collagen is a major determinant of the curve shape. This study demonstrated that the passive elastic behaviour (tension–strain relation) of duodenum in vivo is exponential and hence can play a role in protecting tissue against high stress. At high loads the tissue elastic behaviour is contributed mainly by the passive tension curve, whereas at low stress levels, i.e., in the physiological range, the active tension curve dominates the tissue elastic behaviour, facilitating bolus transport. Thus, the distensibility in vivo depends not only on the passive properties but also on the physiological state of smooth muscle.

Several studies have been done in vitro for investigation of the active and passive mechanical properties of gastrointestinal smooth muscle, but no studies have been performed in vivo as it is more difficult to study the active and passive properties in vivo due to contractility and difficulties in quantitating the balloon dimensions. Butylscopolamine is often used to relax the smooth muscles, though it must be

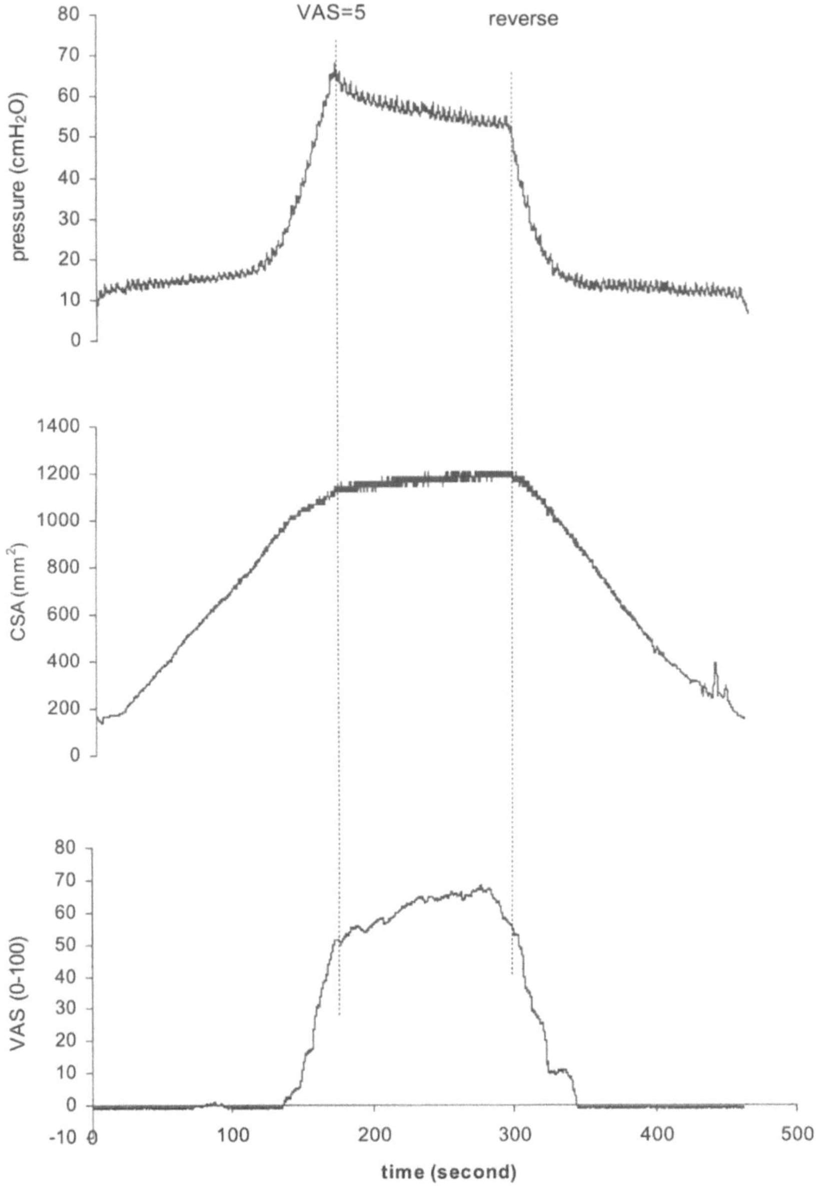

Figure 5.8. Balloon pressure (top), CSA (middle) and VAS (bottom) from a distension in a subject where the distension was stopped at VAS = 5 (shown as 50 in the figure). The sensation score continued to increase despite the stable CSA and the decrease in pressure (due to stress relaxation).

emphasised that in human studies it can be difficult to completely relax the muscle contractility. With these new distension protocols, active and passive length–tension components in the human intestine in vivo can be demonstrated. It was observed that the local active tension maximum always appeared before the gut wall was overstretched. The shape of the active component showed a maximum tension development followed by a decline with further stretch. This optimum tension is presumably reached at a level of optimum overlap between the sliding filaments. When the pain threshold was reached, the passive tension predominated. Thus, the gut wall can contribute its largest force to transport the bolus within the physiological range.

The data above shed some light on the discussion regarding the visceral sensory receptors in man. For many years gastroenterologists believed that the mechanoreceptors in the afferent pathways were pressure receptors. The current concept is that the mechanoreceptors are tension-sensitive receptors that lie in series or in parallel with the muscle cells. This concept is borrowed from striated muscle physiologists and should yet be regarded as a working hypothesis since no clear evidence supports it in gastrointestinal smooth muscle studies. It is basically a uniaxial model that does not account for more complex biomechanical properties such as the distribution of the deformation field and that different receptor populations exist. Furthermore, there seems to be no evidence that the in-series receptors respond to tension rather than to strain. In fact, our result demonstrated that mechanoreceptors with high thresholds evoked by high-intensity stimuli directly depended on wall strain or deformation rather than pressure and tension.

The intensity of pain kept increasing when loading was stopped and the same CSA withheld at the pain threshold. Because of stress-relaxation (Figure 5.8), pressure and tension decreases. Animal experiments in the cat small intestine have shown that the output from the peripheral receptors during constant load is either constant or decreases (so-called non-adapting and adapting receptor populations). Thus, the mechanoreceptors in the human duodenum are probably non-adapting at the peripheral level and, when the load is hold constant, central summation of the sustained input results in increasing pain. This explanation seems likely because the same phenomenon is not observed in recordings of afferent activity in the mesenteric nerves during balloon distension in the rat small intestine (Figure 5.9).

The stimulation of the intramuscular in-series receptors generally initiates vago-vagal reflexes. The responding vagal motor fibres discharge continuously for a long time, well beyond the duration of the stimulus. Vago–vagal reflexes may involve both excitation and inhibition of motor fibres. Some of these reflex responses reverse with the intensity of stimulation of the afferent input. That is, a stimulus that excites activity at a low intensity may inhibit activity at greater intensity.

5.9.5 The Effect of Ageing on the Tension–Strain Curves and Sensation

Ageing is a common reason for deterioration in organ function. During ageing the total number of neurons in the myenteric plexus decreases with an increase in the fibrous components of the myenteric ganglia. Animal studies suggest that senescent gastrointestinal muscle responds less to excitatory factors in vitro, and neural injury in older animals may result from apoptosis, defects of mitochondrial metabolism, and inadequate levels of or response to neurotrophines. Changes in the structure and functional integrity of the gastrointestinal tract depend on collagen content and the degree of cross linkage. Collagenous proteins seems to accumulate in old rats compared with middle-aged rats and this was accompanied by a decrease in the strength, which may again deteriorate the functional integrity of the gastrointestinal tract with age. Intestinal collagen at the fibril level is aligned at varying angles to the

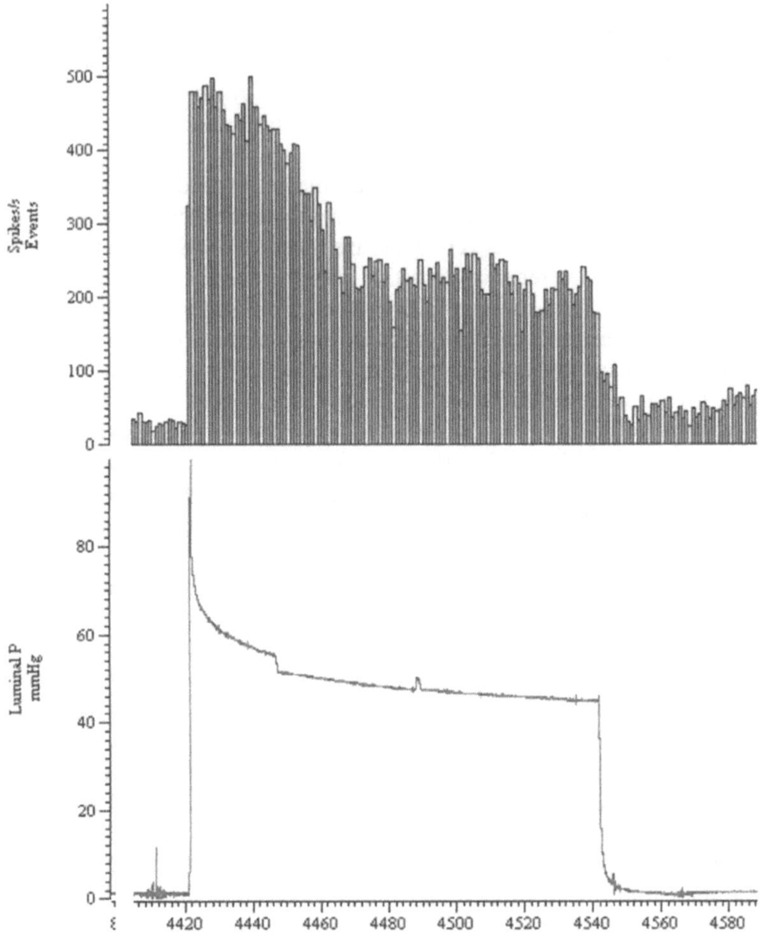

Figure 5.9. Recording of afferent nerve activity in the mesenteric nerves and pressure in a rat during intestinal balloon distension. A predetermined volume was injected into the balloon and kept constant; i.e., the circumferential strain was fairly constant. A decrease in pressure (due to stress relaxation) and afferent nerve activity was observed. The x-axis unit is seconds. The data were obtained by Wen Jiang and Xiao Lu.

longitudinal and circular directions of the intestine. Studies of the effect of age on fibre morphology showed that the length of the large fibre undulations increases during maturation but remains constant during ageing. The diameters of large and small fibres seem not to change with age, but more of the 4-micrometers fibres are loosely associated into larger fibres. Thus, the morphology of collagen and neuromuscular properties change during ageing. Consequently, sensory function and biomechanical properties in the gastrointestinal tract may also change.

Using the new approach it was found that the pain threshold in the younger group was significantly higher for most mechanical parameters than those in the older subjects. The CSA and strain of the younger group both without and with butylscopolamine administration were higher than that of the older group. In contrast, the

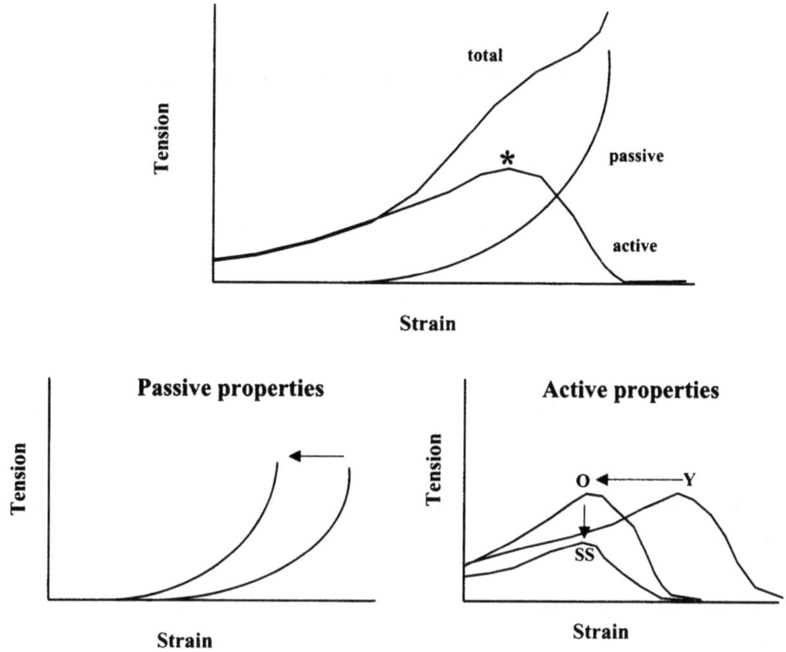

Figure 5.10. Schematic drawing of the alteration in tension–strain data in old volunteers and patients with systemic sclerosis when compared to young volunteers.

pressure and tension of these two groups both without and with butylscopolamine administration did not differ. In both groups the total and passive tension increased exponentially as a function of strain or normalised volume, whereas the active tension increased to a maximum at a strain of approximately 1.0 in the younger group and of 0.6 in the older group. This difference in strain was significant. The pain threshold appeared at a lower strain in the old group when compared to the young group. The passive tension curve was translated to the left for the older subjects when compared with the young subjects (Figure 5.10).

The data demonstrate differences in the sensorimotor and biomechanical properties of the duodenum between younger and older normal volunteers. The unloaded size of the duodenum did not differ between younger and older subjects. At higher loads, it was seen that the older group had lower CSAs than the younger volunteers, indicating that the duodenal wall is stiffest in older people. The increase in stiffness was clearly confirmed in the tension–strain graph (by comparing the passive curves in the tension–strain graph), since a translation to the left in the coordinate system and higher slope are consistent with a higher elastic modulus (Gregersen and Kassab, 1996). The increased stiffness in older subjects may be associated with previous findings of an accumulation of collagen in the intestine in old animal and human subjects since collagen has a high elastic modulus.

Another important finding relates to the position of the local maximum in the active tension–strain curve. The tension obtained at the local maximum did not differ between the two groups. This indicates that smooth muscle function is preserved in the older subjects since the muscle is capable of generating a high active force. However, the maximum active tension which is correlated with the smooth muscle contractile function appeared at lower strain (0.64) in the older subjects compared with the young individuals, where the strain at maximal active (muscle) tension was 0.96. The mechanism for the shift of the maximum towards a low strain in older individuals is unknown from the present data, but it can be explained by resetting of the mechano-sensitive receptors in the intestinal wall. Since the wall becomes stiffer, the strain sensitivity of the mechanoreceptors may reset as part of the structural and mechanical remodelling of the tissue. Whether it is a primary or secondary phenomenon is yet unknown. Since muscle function is preserved, the intestine apparently does not need to compensate for lack of muscle contractile force. However, the present study does not give information about the coordination of the contractility in the two groups. If the motility is more uncoordinated in older subjects, it may require a resetting of the mechanoreceptors to reach the same functional level as in young subjects. The data indicate that alterations appear in the receptor–neuromuscular pathway in the intestinal wall (intrinsic reflex). Hence, animal studies have shown that ageing resulted in a decreased number of neurons in the myenteric plexus, and the gastrointestinal muscle responds less to excitatory factors in vitro.

5.9.6 The Effect of Systemic Sclerosis on the Tension–Strain Curves and Sensation

As mentioned previously, systemic sclerosis (SS) is a rare disease characterised by an excessive deposition of collagen and other connective tissue components in the skin and internal organs. The gastrointestinal tract is often involved with structural changes and motility disturbances as a result. Thus, the alteration of the proportion between smooth muscle and collagen may change the biomechanical properties of the intestinal wall. The remodelling of the gut wall may be a reason for gastrointestinal dysfunction in SS patients.

Doing similar experiments as described above, it was found that total and passive tension in the systemic sclerosis patients and controls increased in an exponential-like way as a function of strain and that the curves for the systemic sclerosis patients were shifted to the left. The increased stiffness may be caused by the excess extracellular matrix deposited in the intestinal wall since collagen has a high elastic modulus.

The active tension increased and reached a maximum at a strain of approximately 0.4 in the patients and of 0.6 in the age-matched healthy volunteers. Furthermore, the maximum active tension and the corresponding strain were higher in the healthy volunteers when compared to the SS patients (Figure 5.10). The lower maximum tension in systemic sclerosis patients indicate that the muscle is not capable of generating the same force as in volunteers. This is consistent with atrophy of the muscle layers and their replacement with fibrous tissue. The mechanism for the shift of the maximum towards a lower strain in SS patients is unknown from the present data but resetting of the mechanosensitive receptors in the wall could be a feasible explanation. Since the wall become stiffer, the strain sensitivity of the mechanoreceptors may reset as part of the remodelling of the tissue. Whether it is a primary or secondary phenomenon is yet unknown. It also seems feasible that the intestine seeks to compensate for the lower maximum force generated by the muscle by initiating contractions at lower strains, hence peristalsis is likely activated by smaller boli in SS patients than in normal volunteers. The data indicate the alterations exist in the smooth muscle and in the receptor–neuromuscular pathway in the intestinal wall (intrinsic reflex). It seems likely that primarily the mechanoreceptors and the muscles are affected.

Another major finding is that the SS patients had higher perception scores than the normal subjects when they underwent the same level of mechanical stimulation. It suggests that the receptor located in the duodenal wall in SS patients is hypersensitive; i.e., resetting has occurred as indicated above, though some degree of central influence cannot be excluded. Thus, the resetting of the mechanoreceptors may result in allodynia; i.e., an increase of afferent activity from the periphery due to normal stimuli such as the passage of food. These may again increase central neuronal responsiveness and induce more permanent changes in the central nervous system such as have been demonstrated by repeated C-fibre input.

5.9.7 Preliminary Conclusions, Limitations and Perspectives

An impedance planimetric measurement system combined with a wide-range sensory scale covering non-painful and painful sensations was developed and validated for investigating the relation between the biomechanical and sensory functional properties in human small intestine. Preliminary data indicate that:

- The smooth muscle length–tension diagram known from in vitro studies can be reproduced in the human intestine in vivo.
- The receptors located in the gastrointestinal wall depend on wall circumferential strain rather than volume, pressure and tension.
- The high-threshold receptors are non-adapting.
- Ageing results in stiffening of the duodenal wall and resetting of the receptors in the gut, resulting in hypersensitivity in older age groups.
- The duodenum of patients with systemic sclerosis was stiffer and had less motor activity than the normal subjects. The intestinal wall in systemic sclerosis patients is hypersensitive, probably due to resetting at the receptor level.

Due to the fact that pain in the clinic is a multi-dimensional, highly individual perception with cognitive and emotional aspects, it is often difficult to quantify and validate in the patients. Experimental pain models are tools for better understanding of visceral pain mechanisms. Thus, the need for experimental models using well-defined and clinically relevant noxious visceral stimuli in man is obvious. However, these models have been troublesome to develop in comparison with pain models in the skin, mainly due to the localisation of the viscera which makes it difficult to apply the stimuli. Most human models used short-lasting mechanical or electrical gut stimuli. The electrical stimuli are non-specific, bypassing the receptor level, and the nature of the stimulus can be questioned in studies with the aim of mimicking clinical pain. Mechanical stimuli with balloon distension mimic diseases such as those seen during mechanical obstruction of the gut. Hence, such models may give valuable information about the pain system in the gut. The mechanical stimulus, however, needs to be well defined and reproducible and most previous methods suffer from technical limitations with respect to the force applied against the gut wall. Most previous balloon distension studies were based on either isovolumetric or isobaric methods. Although these methods can render valuable mechanical stimuli to the gut, the experimental parameters gained from these methods can not describe the significant findings in the circumferential direction of the gut wall. For instance, the circumferential tension and strain that may be more directly responsible for the mechanoreceptors in the gut wall can not be calculated. Impedance planimetry can provide us with a precise measurement of the CSA, by which both tension and strain can be calculated, and thus can cover the shortage of isometric and barostatic methods. Hence, the studies using impedance planimetry in combination with a sensory scale that spans the range from non-painful to painful sensations allowed us to establish a standardised experimental model to investigate the relationship between mechanical forces and visceral sensations.

There are limitations with this new approach, among other things because it is based on several assumptions. Some of these relate to the tension and strain measures (Chapter 3). A further limitation is that the wall thickness cannot be measured with the current method but imaging techniques can be employed to improve this aspect. It is also assumed that the relation between active, passive and total tension is additive. This assumption is valid for a uniaxial experiment but needs to be refined for the three-dimensional case. Finally, when comparing with uniaxial data obtained in vitro, it should be taken into account that smooth muscle is difficult to relax completely in vivo and that the maximum muscle contractile intensity is also difficult to obtain in vivo due to the limitations in the use of drugs in vivo and the phasic nature of gastrointestinal smooth muscle.

With these developments, if based on a bioengineering approach, this method could become a very powerful tool in basic research and diagnostics. Mechanical properties; smooth muscle properties; mechanoreceptor dynamics; intramural circuits; sensation, including temporal summation, being a human correlate to the early phase of the "wind-up" process may be a potent mechanism for the generation of visceral pain. Hence, the model may also be usable for testing basic central mechanisms and pharmacological substances targeted to modulate central temporal summation mechanisms.

5.10 Multimodal Stimulation Approach

It is clear from the previous sections in this chapter and the previous chapters that the receptors respond to several types of stimuli. To obtain a better understanding of visceral symptoms and pain, we used an experimental approach based on multimodal stimulation. In this sense, multimodal refers to mechanical, chemical, thermal (cold and warm) and electrical stimulation and the testing regime must make possible these modes in the same experiment. For this reason a new probe has been developed as sketched in Figure 5.11.

Preliminary data with the multimodal probe obtained in the human oesophagus demonstrates that the procedure can be done safely if proper security devices are built into the equipment and testing scheme. All modalities can induce pain separately and in combination. A stimulus such as acid infusion may change the sensitivity to mechanical stimulation, indicating the polymodal nature of at least some receptor populations. Furthermore, central phenomena such as temporal summation and viscero–somatic relations can be studied in the same experiment. This mimics the mechanisms known to be activated in clinical diseases of the gastrointestinal tract. Pharmacological modulation of the output variables is a potent option that should be tested in the future.

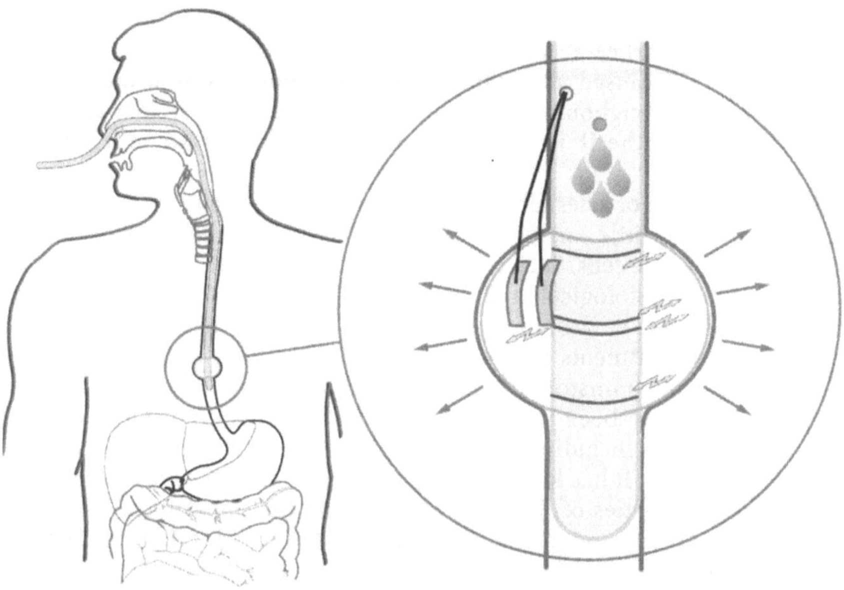

Figure 5.11. Schematic drawing of the multimodal probe. The probe contains a fluid-filled balloon that can be distended in a controlled way. The temperature in the fluid inside the balloon can also be controlled in the range from 5–60 °C. Electrical stimulation is done through flexible rectangular-shaped electrodes glued to the surface of the balloon. Chemical stimulation is done by infusion of the chemical substance (such as capsaicin or acid) through the side hole above the balloon. Modified from Am. J. Physiol 2002; 283: G95–G103. (see also Plate IV.)

5.11 Determination of the Effects of Mechanical Stretch on Smooth Muscle Cell Cultures

The cells of many tissues and organs respond both physiologically and biochemically to physical forces. While research has been conducted on the response of arterially derived smooth muscle cells to mechanical stress, little research has been conducted on intestinally derived smooth muscle cells. Such research is needed in order to have a better understanding of the physiological and biochemical mechanical stress responses that occur in the small intestine. Cell cultures can be exposed to physical forces in several ways. They can be seeded on stretchable membranes exposed to uniaxial stretch, equibiaxial stretch, and laminar shear stress in a static or dynamic fashion, and the biochemical and physical properties of cultured smooth muscle cells can be derived. This will lead to a better understanding of the response mechanisms occurring in the small intestine and could ultimately lead to improved treatments for intestinal disease and improved use of intestinally derived graft material. Unfortunately, the gastrointestinal literature is very sparse with regard to this kind of data. The cells of many other tissues and organs have demonstrated responses, both physiological and biochemical, to physical forces. For example, lung epithelial cells exposed to mechanical stretch respond by increasing surfactant secretion (Wirtz and Dobbs, 1990). External load has also been shown to play a critical role in determining cardiac muscle mass. Placing cardiac cells under mechanical stretch results in the rapid induction of a range of genes (Sadoshima and Izumo, 1993). Barbee et al. (1994) found that mechanical stretch on vascular smooth muscle cells caused alterations in cell morphology and in the organization of actin filaments throughout the cellular matrix. Mechanical stretch also increases the rate of DNA synthesis in cultured epithelial cells and smooth muscle cells (Karim et al., 1992).

While the response of arterially derived smooth muscle cells to mechanical stress has been examined, little analogous research has been conducted on intestinally derived smooth muscle cells. Such research is needed in order to have a better understanding of the physiological and biochemical mechanisms that occur in the mechanically dynamic environment of the small intestine. This could subsequently lead to improved treatments and procedures for intestinal injuries and intestinal disorders, such as inflammatory diseases. In addition, tissues derived from the small intestine have recently been considered as an alternative grafting material for a variety of applications including arterial grafting, bladder augmentation, and repair of the abdominal wall. It has long been suggested that graft patency is influenced by the mechanical properties of the graft material and that mechanical stimuli play a major role in the growth and maintenance of many tissue types (Liu and Fung, 1998). A better understanding of the remodelling process in intestinally derived tissues could lead to improved graft patency.

In the following, several methods are described for replicating and examining in vitro the micromechanical environment that small intestinal tissues experience. Three methods are described for inducing mechanical stress on a cultured monolayer of intestinal smooth muscle cells (SMCs): uniaxial stretch, equibiaxial stress, and laminar fluid flow shear stress. Within the application of each of these methods, biochemical properties such as collagenase and fibronectinase production can be determined and monitored by protein analysis methods. Also, the distribution of cellular components such as actin and collagen can be visualised and quantified by fluorescent microscopy. Cell viability, morphology, and density can be monitored by image analysis. Furthermore, two methods are described for testing the mechanical properties of individual cells within the culture. The first method measures viscoelastic deformation of the cell membrane of single cells, induced by micropipette manipulation. The second measures the force required to separate single cells from the overall monolayer matrix by micropipette aspiration.

For the cell monolayer to remodel in order to accommodate tensile stress or shear flow, it is expected that an initial release of collagenase by the cells will be observed. This is indeed what we have found in pilot studies. This subsequently reduces collagen linking and allows the cell mobility in order to alter morphology and orientation in response to the stress. Once the cells have achieved optimal morphology, collagenase production should resume. Such a process would also manifest itself in reduced viscoelastic properties of the cell surface and a reduced separation force initially, with a restoration of these properties over time. Cell viability may also be initially reduced in response to stress but may ultimately recover. This collagen/collagenase response can be examined under a variety of tensile stress and laminar shear stress conditions.

5.11.1 Cell Culture

Intestinal SMC lines are commercially available and can also be obtained directly from animal explant sources using established protocols (Graham et al., 1984;

Chamley-Campbell et al., 1979). Cells can be aseptically obtained from the centre of the circular layer of the muscularis propria of jejunal tissue obtained from pigs.

Several methods are available to confirm whether SMC cells have been properly obtained. SMCs exhibit a cobblestone morphology and a characteristic "Hill and Valley" growth formation at confluency (Chamley-Campbell et al., 1979). For further verification, tests for the confirmation of the production of angiotensin-converting enzyme, binding of fluorescent-labelled acetylated LDL, and the expression of Factor VIII-related antigen can be undertaken (Rosen et al., 1985).

Cells should ideally be maintained at 37 °C in a humidified atmosphere (5% CO_2: 95% air). Cells will be grown in standard culture flasks (T-flasks) containing Dulbecco's Modified Eagles Medium with 10% foetal calf serum and 25 mM Tricine (pH 7.4). At confluency, cells will be released from the flask surface by trypsinisation (0.1%) and subcultured at initial densities on 10^5 cells/ml. Cell counts can be determined by image analysis. Cell viability can be measured during the experiments using the trypan blue exclusion method.

5.11.2 Methods for Exposing Cells to Mechanical Stress

Uniaxial Stretching Apparatus

Uniaxial stretching of a cultured monolayer of intestinal SMCs can be done with a procedure previously used by Lyall et al. (1994) for mechanical stretching of vascular SMCs. Cells are grown on silicone sheets until confluency is reached. Figure 5.12 (top) shows intestinal SMCs grown to confluency on a collagen-pre-treated silicone sheet. The silicon sheet/cell monolayer is mounted on the stretching device (similar to a small vice), which has been previously autoclaved. The silicon sheets is then stretched by manually turning the handle of the device. For cyclic stretching, the device is attached to a servomotor to allow continuous repeated stretching over a period of time. The bottom image of figure 5.12 shows intestinal SMCs exposed to cyclic stretching (12 cycles per minute up to stretch ratio 1.2) for 48 hours. The cells now align themselves perpendicular to the direction of stretch.

Equiaxial Stretching Apparatus

Uniaxial stretching devices, as described above, can result in a anisotropic state of strain which will be perceived differently by each cell depending on their orientation relative to the strain and their position on the membrane. In order to determine the effect of equibiaxial stress, stretching experiments will be repeated using an equiaxial stretching device. This device, described by Hung and Williams (1994) and also by Schaffer et al. (1994), is designed to induce a uniform and equibiaxial strain on a cylindrical elastomeric sheet upon which cells are grown in a monolayer. The device also manages to keep the substrate in the same plane throughout the stretching cycle. A circular elastomeric membrane is placed on a stationary well and stretching is achieved by using a ringed indenter to push an annular portion of a the elastomeric membrane into a circular groove in the bottom of the well. Using this device, Hung and Williams (1994) achieved strains of 0.04–0.4%. Using a mean strain of 0.15%, a standard deviation of 0.01% was measured throughout the cultured portion of the membrane surface. Multi-well versions of this device have also been described (Schaffer et al., 1994), allowing for multiple culture samples.

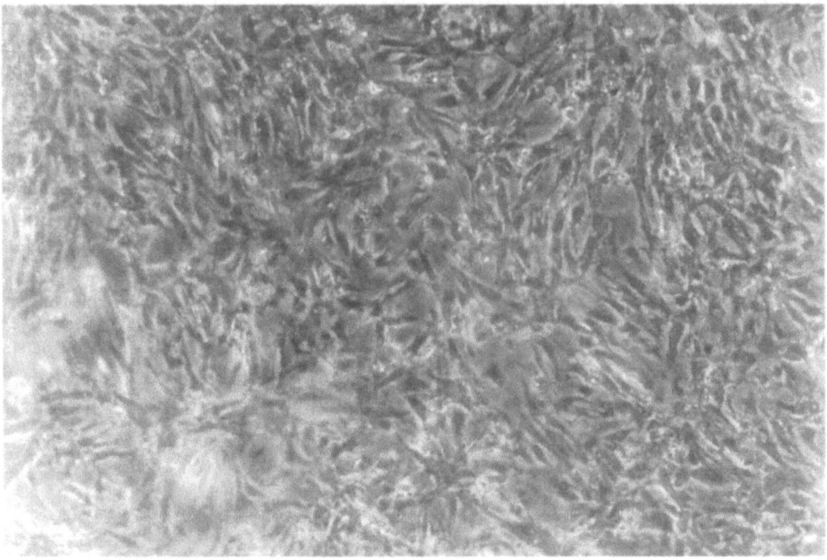

Control (Smooth muscle cells grown to confluency)

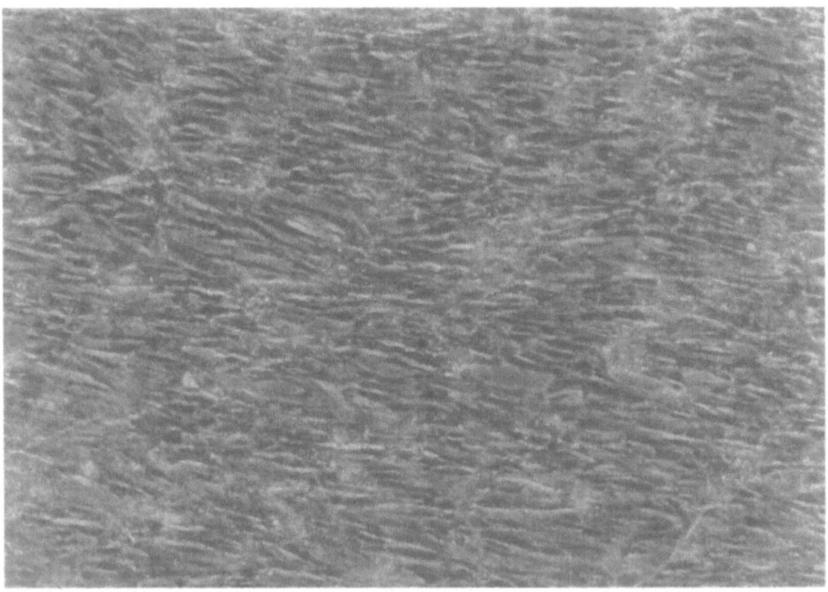

Smooth muscle cells after 48 hours cyclic stretching

Figure 5.12. Intestinal smooth muscle cell grown to confluency on a membrane. The top image is before stretching whereas the bottom image is obtained after 48 hours of cyclic stretch. Images kindly provided by Professor Xiao Lu, Aalborg Hospital, Denmark.

Stretching Procedures and Cell Stretching Surface

The methodology for applying mechanical stretch varies greatly in the literature. Lyall et al. (1994) conducted research using 0–20% single stretch on arterial SMCs. A cyclic stretch might more accurately mimic conditions experienced in vivo, and has be used by Dartsch et al. (1986) for arterial SMCs at rates of 50 stretches/min for a maximum stretch of 10%.

A range of commercially available silicon sheets of various thicknesses is available. Often, several types will have to be tested to determine the optimum material. These sheets must be able to withstand autoclaving, not impede smooth muscle growth, and be sufficiently translucent to allow the observation of the cell culture by microscope. The sheets can be cleaned by washing in phosphate-buffered saline and then rinsed thoroughly in distilled water. They are then wrapped in tinfoil and autoclaved. The autoclaved sheets will then be aseptically mounted onto the previously autoclaved stretching apparatus.

In order to facilitate the attachment of cells to the silicon sheet surface, the silicon sheets can be pre-treated with fibronectin (or collagen) prior to culturing. A volume of 10 ml of 25 μg/ml fibronectin solution, dissolved in serum-free medium, or 10 ml of a 0.01% collagen solution in 0.1 mol/L acetic acid, is applied to the silicon sheet (Lyall et al., 1994). The sheets will then be incubated at room temperature for 20 minutes, rinsed with serum-free medium, and allowed to dry at room temperature in a laminar-flow hood.

Once the silicon sheets are prepared, cells are plated on to the silicon sheets at a cell density of 10^5 cells/ml. The silicon sheet/cell monolayer can then be exposed to mechanical stretch once the culture has reached confluency. At all times throughout the procedure, the culture must be maintained in a sterile environment.

Laminar Shear Stress Apparatus

Sato et al. (1996) used a parallel-plate flow chamber. This has a flow section that is 0.5 mm high, 50 mm wide and 60 mm long. Flow is created by a peristaltic pump in circuit with the flow chamber. A reservoir is placed upstream of the chamber to act as a depulsator for the chamber. The flow chamber generates an applied shear stress of 2 Pa, representative of arterial shear stress levels. Through the use of pinch valves, flow direction can be controlled within the chamber. Cells grown as a monolayer on silicon sheets can be placed in the chamber, exposed to liquid flows, and then removed and analysed. The chamber can be made from clear plastic to allow for on-line observation of cell response to shear stress. To the best of my knowledge, data using laminar shear apparatus have not been published about gastrointestinal SMCs but preliminary studies in our labs have shown that sheared intestinal SMCs change the gene expression of collagenase in a time-dependent fashion.

5.11.3 Image Analysis

Throughout the stretching experiment, various parameters of the cell culture can be measured non-invasively by image analysis microscopy. These include:

- Strain measurements – by measuring to displacement of various markers on the membrane surface and relating this displacement to the amount of stress applied (Barbee et al., 1994).

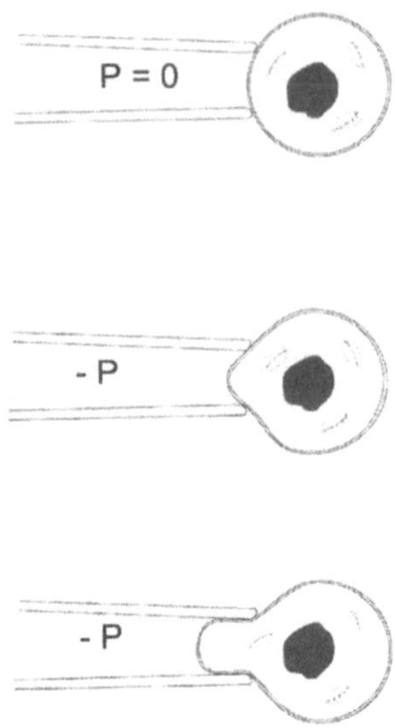

Figure 5.13. Schematic drawing of a pipette experiment where a cell is partly sucked into the pipette. From the geometric data and the pressure, it is possible to derive the viscoelastic surface properties.

- Angle of cell orientation – by measuring the deviation of the longest diagonal of the cell from the direction of strain or flow.
- Shape index – a non-dimensional parameter (circular 1.0 / highly elongated 0.0).
- Cell viability – measured by trypan blue dye exclusion.
- Cell density – by measuring cells/cm^2.

5.11.4 Viscoelastic Deformation Induced by Micropipette

A novel way of measuring the viscoelastic surface properties of cells is to apply a negative pressure directly to the surface of the cell membrane by using a micropipette (Sato et al., 1996). The micropipette tip is brought in direct contact with the cell membrane and a negative pressure is applied to the cell by drawing water from a reservoir in circuit with the micropipette tip (Figure 5.13). Using this method, a negative pressure of 10–40 mm H_2O has been used in previous experiments. Cells are tested while in suspension and care must be given to the method by which cells are detached from the growth surface prior to testing. Cells can be detached mechanically or by treatment with trypsin, EDTA or other reagents. Cell deformation can be monitored by image analysis and measured as a function of time and applied pressure. Specifically, the length of the aspirated portion of the cell can be related to time by:

$$L(t) = L_s[1 - k_2/(k_1 + k_2)\exp(-t/\tau)]h(t), \tag{5.1}$$

$$L_s = 2a\Delta P/\pi k_1, \tag{5.2}$$

where k_1 and k_2 are elastic constants, τ is the time constant, a is the radius of the micropipette, and h(t) is the unit step function, k_1 is determined from Equation 5.2 by measuring the asymptotic aspirated length L. The coefficient of viscosity, μ, is expressed by:

$$\mu = \tau k_1 k_2 / (k_1 + k_2) . \tag{5.3}$$

Fitting Equation 5.2 to experimental data by least squares will yield the values of k_2, k_1, τ, and μ. From this, calculated curves of normalised aspirated length (L) as a function of time can be plotted.

Separation Force Measurement by Micropipette Aspiration

The separation force required to remove a cell from the cell matrix can be measured by a procedure described by Shyy et al. (1993) and Tozeren et al. (1992). Micropipettes with an internal radius (R_p) of 1.6–3.3 um are manipulated using a hydraulic micromanipulator mounted on the stage of an inverted microscope. The tip of the pipette is positioned in the cell chamber and the wide end connected to a pressure-regulation system. Cells adhering to the surface are randomly chosen and held at the tip of the pipette by the application of an initial aspiration pressure. When the force due to aspiration is insufficient to balance the adhesive force, the cell will slip away from the micropipette. The aspiration pressure is increased stepwise until a critical separation pressure, P_c, is attained. The critical separation force can be calculated as $F_c = \pi R_p^2 P_c$.

5.12 Literature

The literature list contains papers mentioned in the text as well as other papers of interest.

Accarino, AM, Azpiroz, F, Malagelada, J-R. 1992. Symptomatic responses to stimulation of sensory pathways in the jejunum. Am J Physiol, **263**, G673–7.

Akervall, S, Fasth, S, Nordgren, S. 1989. Rectal reservoir and sensory function studies by graded isobaric distension in normal man. Gut, **30**, 496–502.

Almy, TP. 1989. The gastroenterologist and the graying of America. Am J Gastroenterol, **84**, 464–8

Anonymous. 1980. Preliminary criteria for the classification of systemic sclerosis (scleroderma). Subcommittee for scleroderma criteria of the American Rheumatism Association Diagnostic and Therapeutic Criteria Committee. Arthritis Rheum, **23**, 581–90.

Anuras, S, Cooke, AR, Christensen, J. 1974. An inhibitory innervation at the gastroduodenal junction. J Clin Invest, **54**, 529–35.

Arendt-Nielsen L, Drewes AM, Hansen JB, Tage-Jensen U. 1997. Plasticity of gut pain in man: an experimental investigation using short and long duration transmucosal electrical stimulation. Pain, **69**: 255–62.

Azpiroz F, Malagelada J-R. 1985. Intestinal control of gastric tone. Am J Physiol, **249**: G501–9.

Azpiroz F, Malagelada J-R. 1985. Physiological variations in canine gastric tone measured by an electronic barostat. Am J Physiol, **248**: G229–37.

Azpiroz F, Malagelada J-R. 1990. Isobaric intestinal distension in humans: sensorial relay and reflex gastric relaxation. Am J Physiol, **258**: G202–7.

Azpiroz F, Malagelada J-R. 1990. Perception and reflex relaxation of the stomach in response to gut distension. Gastroenterology, **98**: 1193–8.

Balis GU, Monroe RR. 1964. The pharmacology of chloralose. Psychopharmacologia, **6**: 1–30.

Ballantyne GH, Longo WE, Savoca PE, Adrian TE, Vukasin AP, Bilchik AJ, Sussman J, Modlin IM. 1989. Deoxycholate-stimulated release of peptide YY from the isolated perfused rabbit left colon. Am J Physiol, **257**:G715-G724.

Barbee KA, Macarak EJ, Thibault LE. 1994. Strain measurements in cultured vascular smooth muscle cells subjected to mechanical deformation. Ann Biomed Eng, **22**: 14–22.

Barthó L, Holzer P, Donnerer J, Lembeck F. 1982. Effects of substance P, cholecystokinin octapeptide, bombesin, and neurotensin on the peristaltic reflex of the guinea-pig ileum in the absence and in the presence of atropine. Naunyn-Schmiedeberg's Arch Pharmacol, **321**: 321–8.

Bartho L, Holzer P, Donnerer J, Lembeck F. 1982. Evidence for the involvement of substance P in the atropine-resistant peristalsis of the guinea-pig ileum. Neurosci Lett, **32**: 69–74.

Bayliss VM, Starling EH. 1901. The movements and innervation of the small intestine. J Physiol, **26**: 125–38.

Bayliss WM, Starling EH. 1899. The movements and innervation of the small intestine. J Physiol, **24**: 99–143.

Beleslin DB, Terzic B, Samardzic R. 1987. The effect of leucine-enkephalin on the peristaltic reflex of the isolated guinea-pig ileum. Acta Physiol Hung, **69**: 105–14.

Bell AM, Pemberton JH, Hanson RB, Zinsmeister AR. 1991. Variations in muscle tone of the human rectum: recordings with an electromechanical barostat. Am J Physiol, **260**: G17–25.

Bennett A, Eley KG, Stockley HL. 1976. Inhibition of peristalsis in guinea-pig isolated ileum and colon by drugs that block prostaglandin synthesis. Br J Pharmacol, **57**: 335–40.

Bernstein CN, Niaze N, Robert M et al. 1996. Rectal afferent function in patients with inflammatory and functional intestinal disorders. Pain, **66**: 151–61.

Brasseur JG. 1993. Mechanical studies of the esophageal function. Dysphagia, **8**: 384–6.

Bülbring E, Lin RCY, Schofield G. 1958. An investigation of the peristaltic reflex in relation to anatomical observations. Q J Exp Physiol, **43**: 26–37.

Cervero F. 1992. Sensory innervation of the viscera: Peripheral basis of visceral pain. Physiol Rev, **74**: 95–131.

Chamley-Campbell J, Campbell GR, Ross R. 1979. The smooth muscle cell in culture. Physiol Rev, **59**: 1–61.

Chernish SM, Miller RE, Rosenak BD, Scholz NE. 1972. Hypotonic duodenography with the use of glucagon. Gastroenterology, **63**: 392–8.

Christensen J, Conklin JL. 1973. Studies on the origin of the distinctive mechanics of smooth muscle at the esophagogastric junction. In: Proc Fourth International Symposium on Gastrointestinal Motility. pp. 63–71. Banff, Alberta, Canada. September.

Christensen J, Freeman BW, Miller JK. 1973. Some physiological characteristics of the esophagogastric junction in the opossum. Gastroenterology, **64**: 1119–25.

Christensen J, Glover JR, Macagno EO, Singerman RB, Weisbrodt NW. 1971. Statistics of contractions at a point in the human duodenum. Am J Physiol, **221**: 1818–23.

Christensen J, Roberts RL. 1983. Differences between esophageal body and lower esophageal sphincter in mitochondria of smooth muscle in opossum. Gastroenterology, **85**: 650–6.

Christensen J, Torres EI. 1975. Three layers of the opossum stomach. Response to nerve stimulation. Gastroenterology, **69**: 641–8.

Christensen J. 1970. Pharmacological identification of the lower esophageal sphincter. J Clin Invest, **49**: 681–91.

Christensen J. 1982. Oxygen dependence of contractions in esophageal and gastric pyloric and ileocecal muscle of opossum. Proc Soc Exp Biol Med, **170**: 194–202.

Christensen, J. 1994. Defining the irritable bowel syndrome. Perspect Biol Med, **38**: 21–35.

Christensen, J. 1994. The motility of the colon. In: Physiology of the Gastrointestinal Tract, 3rd edition, ed. Johnson LR, Alpers DH, Christensen, J, Jacobson, ED, Walsh, JH, pp. 991–1024. New York: Raven Press.

Conklin J, Goyal RK. 1989. Gastrointestinal smooth muscle. In: Gastrointestinal Disease, 4th edition, ed. Sleisenger MH, Fordtran JS, chapter 3. Philadelphia, PA: W.B Saunders Company.

Conklin JL, Christensen J. 1975. Local specialization at ileocecal junction of the cat and opossum. Am J Physiol, **228**: 1075–81.

Conklin JL, Christensen J. 1994. Motor functions of the pharynx and esophagus. In: Physiology of the Gastrointestinal Tract, 3rd edition, ed. Johnson LR, Alpers DH, Christensen, J, Jacobson, ED, Walsh, JH, pp. 903–28. New York: Raven Press.

Cook IJ. 1998. Colon and anorectum: constipation, urgency, pain syndromes. Clinical physiology. In: Functional Disorders of the Gut, ed. Phillips SF, Wingate DL, pp. 269–97. London: Churchill Livingstone.

Costa M, Furness JB. 1976. The peristaltic reflex: an analysis of the nerve pathways and their pharmacology. Naunyn-Schmiedeberg's Arch Pharmacol, **294**: 47–60.

Dartsch PC, Hammerle H, Betz E. 1986. Orientation of cultured arterial smooth muscle cells growing on cyclically stretched substrates, Acta Anat, **125**: 108–13.

de Caestecker JS, Pryde A, Heading RC. 1992. Site and mechanism of pain perception with oesophageal balloon distension and intravenous edrophonium in patients with oesophageal chest pain. Gut, **33**: 580–6.

De Ponti F, Azpiroz F, Malagelada J-R. 1987. Reflex gastric relaxation in response to distension of the duodenum. Am J Physiol, **252**: G595–601.

Diamant B, Picazo J. 1983. Spasmolytic action and clinical use of glucagon. In: Lefebre (ed.) Handbook of Experimental Pharmacology. pp. 614–43. Berlin: Springer-Verlag.

Dobrin PB. 1978. Mechanical properties of arteries. Physiol Rev, **58**: 397–460.
Donnerer J, Barthó L, Holzer P, Lembech F. 1984. Intestinal peristalsis associated with release of immunoreactive substance P. Neuroscience, **11**: 913–18.
Donnerer J, Holzer P, Lembeck F. 1984. Release of dynorphin, somatostatin and substance P from the vascularly perfused small intestine of the guinea-pig during peristalsis. Br J Pharmacol, **83**: 919–25.
Drewes AM, Arendt-Nielsen L, Jensen JH, Hansen JB, Krarup HB, Tage-Jensen U. 1997. Experimental pain in the stomach: a model with clinical significance. Gut, **41**: 753–7.
Ehrlein H-J, Scheemann M, Siegle M-L. 1987. Motor patterns of small intestine determined by closely spaced extraluminal transducers and videofluoroscopy. Am J Physiol, **253**: G259–67.
Elbrønd H, Tøttrup A, Forman A. 1991. Mechanical properties of isolated smooth muscle from rabbit sphincter of oddi and duodenum. Scand J Gastroenterol, **26**: 289–94.
Fackler K, Klein L, Hiltner A. 1981. Polarizing light microscopy of intestine and its relationship to mechanical behaviour. J Microscopy, **124**: 305–11.
Faithfull NS. 1987. Fluorocarbons. Current status and future applications. Anaesthesia, **42**: 234–42.
Fass J, Silny J, Braun J, Heindrichs U, Dreuw B, Schumpelick V, Rau G. 1989. Ein neues verfahren zur quantitativen bestimmung oesophagealer motilitätsmuster mittels einer vielfachen impedanzmessung. In: Chirurgiches forum 1989 f. experim. u. kliniche forschung, ed. Hamelmann et al. Berlin: Springer-Verlag.
Fisher MA, Hendrix TR, Hunt JN, Murrils AJ. 1978. Relation between volume swallowed and velocity of the bolus ejected from the pharynx into the esophagus. Gastroenterology, **74**: 1238–40.
Fung YC, Yih CS. 1968. Peristaltic transport. J Applied Mechanics, **35**: 669–75.
Fung YC. 1981. Biomechanics. Mechanical Properties of Living Tissues. New York: Springer-Verllag.
Furness JB, Costa M. 1987. The Enteric Nervous System. pp. 207–38. New York: Churchill Livingstone.
Furst DC, Clements PJ, Saab M, Sterz MG. 1984. Clinical and serological comparison of 17 chronic, progressive systemic sclerosis (PSS) and 17 CREST syndrome patients matched for sex, age, and disease duration. Ann Rheum Dis, **43**: 794–801.
Gabella G. 1976. Structure of the Autonomic Nervous System. London: Chapman and Hall.
Gabella G. 1987. Structure of muscles and nerves in the gastrointestinal tract. In: Physiology of the Gastrointestinal Tract. ed. Johnson LR, Christensen J, Jackson MJ, Jacobson ED, Walsh JH, pp. 335–82. New York: Raven Press.
Ginzel KH. 1959. Investigations concerning the initiation of the peristaltic reflex in the guinea-pig ileum. J Physiol, **148**: 75–6.
Gonzalez M, Mearin F, Vasconez C, Armengol JR, Malagelada JR. 1997. Oesophageal tone in patients with achalasia. Gut, **41**: 291–6.
Graham MF, Diegelmann RF, Elson CO, Bitar KN, Ehrlich HP. 1984. Isolation and culture of human intestinal smooth muscle cells, Proceedings of the Society for the Experimental Biology and Medicine, **176**: 503–7.
Gregersen H, Barlow J, Thompson DG. 1999. Development of a computer-controlled tensiometer for real-time measurements of tension in tubular organs. Neurogastroenterol Motil, **11**: 109–18.
Gregersen H, Christensen J. 2000. Gastrointestinal tone. Neurogastroenterol Motil, **12**: 501–8.
Gregersen H, Emery J, McCulloch AD. 1998. History-dependent mechanical behavior of the guinea-pig small intestine. Ann Biomed Eng, **26**: 1–9.
Gregersen H, Giversen IM, Rasmussen LM, Tøttrup A. 1992. Biomechanical wall properties and collagen content in the partially obstructed opossum esophagus. Gastroenterology, **103**: 1547–51.
Gregersen H, Jørgensen CS, Dall FH. 1992. Biomechanical wall properties in the isolated perfused porcine duodenum. An experimental study using impedance planimetry. J Gastrointest Motil, **4**: 125–35.
Gregersen H, Jørgensen CS, Dall FH. 1992. Spontaneous and evoked motility in the isolated perfused porcine duodenum. J Appl Physiol, **73**: 9–19.
Gregersen H, Kassab G, Pallencaoe E, Lee C, Chien S, Skalak R, Fung YC. 1997. Morphometry and strain distribution in guinea pig duodenum with reference to the zero-stress state. Am J Physiol, **273**: G865–74.
Gregersen H, Kassab GS. 1996. Biomechanics of the gastrointestinal tract. Neurogastroenterol Motil, **8**: 277–97.
Gregersen H, Kraglund K, Djurhuus JC. 1990. Variations in duodenal cross-sectional area during the interdigestive motility complex. Am J Physiol, **259**: G26–31.
Gregersen H, Lee C, Chien S, Skalak R, Fung YC. 1999. Strain distribution in the layered wall of the esophagus. J Biomed Eng, **121**: 442–8.
Gregersen H, Orvar K, Christensen J. 1992. Biomechanical wall properties and tone during phase I and phase II of the MMC. Am J Physiol, **263**: G795–801.
Gregersen H, Vinter-Jensen L, Juhl CO, Dajani EZ. 1996. Impedance planimetric characterization of the distal oesophagus in the Goettingen minipig. J Biomech, **29**: 63–8.

Grundy D. Scratcherd T. 1990. Sensory afferents from the gastrointestinal tract. Handbook of Physiology, **1**: 593–620.
Grundy, D. 1993. Mechanoreceptors in the gastrointestinal tract. J Smooth Muscle Res, **29**: 37–46.
Gustavsson S, Tucker R. 1988. Manometry. In: An Illustrated Guide to Gastrointestinal Motility. ed. Kumar D, Gustavsson S, pp. 67–76. New York: John Wiley & Sons.
Hauss WH, Kreuziger H, Asteroth H. 1949. Uber die reizung der pressorezeptoren im sinus caroticus beim hund. Z Kreislaufforsch, **38**: 28–33.
Hems R, Ross BD, Berry MN, Krebs HA. 1966. Gluconeogenesis in the perfused rat liver. J Biochem, **101**: 284–9.
Hill AV. 1970. First and Last Experiments in Muscle Mechanics. Cambridge University Press.
Hill JR, Kelley ML, Schlegel JF, Code CF. 1960. Pressure profile of the rectum and anus of healthy persons. Dis Colon Rectum, **3**: 203–9.
Holst JJ, Jensen SL, Knuhtsen S, Nielsen OV, Rehfeld JF. 1983. Effect of vagus, gastric inhibitory polypeptide, and HCl on gastrin and somatostatin release from perfused pig antrum. Am J Physiol, **244**: G515–22.
Holst JJ, Jensen SL, Nielsen OV, Schwartz TW. 1980. Oxygen supply, oxygen consumption, and endocrine and exocrine secretions of the isolated, perfused, porcine pancreas. Acta Physiol Scand, **109**: 7–13.
Holst JJ, Lauritsen K, Jensen SL, Nielsen OV, Schaffalitsky de Muckadell OB. 1981. Secretin release from the isolated vascularly perfused pig duodenum. J Physiol, **318**: 327–37.
Holst JJ, Schaffalitzky de Muckadell OB, Fahrenkrug J, Jensen SL, Nielsen OV, Schwartz TW. 1981. Nervous control of pancreatic endocrine secretion in pigs. Acta Physiol Scand, **111**: 15–22.
Holzer P & Lembeck F. 1979. Effect of neuropeptides on the efficiency of the peristaltic reflex. Naunyn-Schmiedeberg's Arch Pharmacol, **307**: 257–64.
Holzer P. 1989. Ascending enteric reflex: Multiple neurotransmitter systems and interaction. Am J Physiol, **256**: G540–5.
Hukuhara T, Fukuda H. 1965. The motility of the isolated guinea pig small intestine. Jpn J Physiol, **15**: 125–39.
Hukuhara T, Fukuda H. 1968. The electrical activity of guinea-pig small intestine with special reference to the slow wave. Jpn J Physiol, **18**: 71–86.
Hukuhara T, Masuda K, Kinose S. 1936. Über das Gesetz Des Darmes. Pflügers Arch, **237**: 619.
Hukuhara T, Nakayama S, Nanba R. 1960. Locality of receptors concerned with the intestino-intestinal extrinsic and intestinal muscular intrinsic reflexes. Jpn J Physiol, **10**: 420–26.
Hung CT, Williams JL. 1994. A method for introducing equi-biaxial and uniform strains in elastomeric membranes used as cell substrates. J Biomech, **27**: 227–32.
Iggo A. 1957. Gastro-intestinal tension receptors with un-myelinated afferent fibres in the vagus of the cat. Q J Exp Physiol, **42**: 130–43.
Jensen SL, Fahrenkrug J, Holst JJ, Nielsen OV, Schaffalitzky de Muckadell OB. 1978. Secretory effects of VIP on isolated perfused porcine pancreas. Am J Physiol, **235**: E387–91.
Jørgensen CJ, Gregersen H, Dall FH. 1992. A new method for measurement of stress–strain relations of the small intestine in vivo. J Gastrointest Motil, **4**: 225.
Jørgensen CS, Dall FH, Jensen SL, Storkholm J, Gregersen H. 1991. Elastic properties of the isolated perfused porcine duodenum. Dig Dis, **9**: 399–405.
Karim OMA, Pienta K, Seki N, Mostwin JL. 1992. Stretch-mediated visceral smooth muscle growth in vitro. Am J Physiol, **262**: R895–900.
KatzS. 1993. Gastrointestinaldiseaseoftheelderly: Introductiontotheseries. PracticalGastroenterology, **17**:9.
Kavin H, Levin NW, Stanley MM. 1967. Isolated perfused rat small bowel-technic, studies of viability, glucose absorption. J Applied Physiol, **22**: 604–11.
Keane FB, DiMagno EP, Dozois RR, Go VLW. 1980. Relationships among canine interdigestive exocrine pancreatic and biliary flow, duodenal motor activity, plasma pancratic polypeptide, and motilin. Gastroenterology, **78**: 310–16.
Kendall GPN, Thompson DG, Day SJ et al. 1990. Inter-and intraindividual variation in pressure volume relations of the rectum in normal subjects and patients with the irritable bowel syndrome. Gut, **31**: 1062–8.
Kerlin P, Zinsmeister A, Phillips S. 1982. Relationship of motility to flow of contents in the human small intestine. Gastroenterology, **82**: 701–6.
Kosterlitz HW, Lees GM. 1964. Pharmacological analysis of intrinsic intestinal reflexes. Pharmacol Rev, **16**: 301–39.
Kosterlitz HW, Robinson JA. 1959. Reflex contractions of the longitudinal muscle coat of the isolated guinea-pig ileum. J Physiol, **146**: 369–79.
Kromer W, Pretzlaff W. 1979. In vitro evidence for the participation of intestinal opioids in the control of peristalsis in the guinea pig small intestine. Naunyn-Schmiedeberg's Arch Pharmacol, **309**: 153–7.
Leek BF. 1977. Abdominal and pelvic visceral receptors. Br Med Bull, **33**: 163–8.
Lembo T, Munakata J, Mertz H, Niazi N, Kodler A, Nikas V, Mayer EA. 1994. Evidence for the hypersensitivity of lumbar splancnic afferents in irritable bowel syndrome. Gastroenterology, **107**: 1686–96.

Li L, Li Z, Hou HS, Wang HZ, Wang LY. 1990. Sensory nerve endings in puborectalis and anal region: normal findings in the newborn and changes in anorectal anomalies. J Pediat Surg, **25**: 658–64.
Li M, Brasseur JG, Kern MK, et al. 1992. Viscosity measurements of barium sulfate mixtures for use in motility studies of the pharynx and esophagus. Dysphagia, **7**: 17–30.
Li M, Brasseur JG. 1993. Non-steady peristaltic transport in finite-length tubes. J Fluid Mech, **248**: 129–51.
Liu SQ, Fung YC. 1998. Changes in the organization of the smooth muscle cells in rat vein grafts. Ann Biomed Eng, **26**: 86–95.
Lyall F, Deehan MR, Greer IA, Boswell F, Brown WC, McInnes GT. 1994. Mechanical stretch increases proto-oncogene expression and phosphoinositide turnover in vascular smooth muscle cells. J Hypertension, **12**: 1139–45.
Macagno E, Melville J, Christensen J. 1975. A model for longitudinal motility of the small intestine. Biorheology, **12**: 369–76.
Macagno EO, Christensen J. 1980. Fluid mechanics of the duodenum. Ann Rev Fluid Mech, **12**: 139–58.
Macagno EO, Christensen J. 1981. Fluid mechanics of gastrointestinal flow. In: Physiology of the Gastrointestinal Tract. ed. Johnson LR et al. New York: Raven Press.
Mall F. 1896. A study of the intestinal contraction. Johns Hopkins Hosp Rep, **1**: 37–75.
Mandrek K, Golenhofen K. 1990. Phasi--rhythmical and tonic components in gastrointestinal motility. In: Frontiers in Smooth Muscle Research. pp. 463–81. Alan R. Liss, Inc.
Mandrek K, Golenhofen K. 1990. Tonic duodenal contractions - a process for regulation of duodenal transit. In: Frontiers in Smooth Muscle Research. pp. 459–62. Alan R. Liss, Inc.
Mayrand S, Diamant NE. 1993. Measurement of human esophageal tone in vivo. Gastroenterology, **105**: 1411–20.
Meiss RA. 1971. Some mechanical properties of cat intestinal muscle. Am J Physiol, **220**: 2000–7.
Melville J, Macagno E, Christensen J. 1975. Longitudinal contractions in the duodenum: their fluid-mechanical function. Am J Physiol, **228**: 1887–92.
Meyer JH, Gu Y, Elashoff J, Reedy T, Dressman J, Amidon G. 1986. Effects of viscosity and fluid outflow on postcibal gastric emptying of solids. Am J Physiol, **250**: G161–4.
Meyer JH. 1987. Motility of the stomach and gastroduodenal junction. In: Physiology of the Gastrointestinal Tract, 2nd edition, volume 1. ed. Johnson, LR, pp. 613–30. New York: Raven Press.
Miftakhov R, Wingate D. 1994. Numerical simulation of the peristaltic reflex of the small bowel. Biorheology, **31**: 309–25.
Miftakhov RN, Abdusheva GR, Christensen, J. 1999. Numerical simulation of motility patterns of the small bowel. part i-formulation of a mathematical model. J Theoretical Biology, **197**: 89–112.
Miftakhov RN, Abdusheva GR, Wingate DL. 1996. Model predictions of myoelectrical activity of the small bowel. Biol Cybern, **74**: 167–79.
Miftakhov RN, Abdusheva GR. 1996. Effects of selective K+ channel agonists and antagonists on myoelectrical activity of a locus of the small bowel. Biol Cybern, **75**: 331–8.
Miftakhov RN, Abdusheva GR. 1996. Numerical simulation of excitation-contraction coupling in a locus of the small bowel. Biol Cybern, **74**: 455–67.
Miftakhov RN, Wingate DL. 1993. Mathematical modeling of the enteric nervous network. 2: facilitation and inhibition of cholinergic transmission. J Biomed Eng, **16**: 311–19.
Miftakhov RN, Wingate DL. 1994. Mathematical modeling of the enteric nervous network. 1: cholinergic neuron. J Med Eng Physics, **16**: 67–738.
Miftakhov RN, Wingate DL. 1994. Mathematical modeling of the enteric nervous network. 3: adrenergic neuron. J Med Eng Physics, **16**: 450–7.
Miftakhov RN, Wingate DL. 1994. Numerical simulation of the peristaltic reflex of the small bowel. J Biorheology, **31**: 309–25.
Miftakhov RN, Wingate DL. 1995. Mathematical modeling of the enteric nervous network. 4: analysis of the adrenergic transmission. J Med Eng Phys, **17**: 3–10.
Miftakhov RN, Wingate DL. 1995. Mathematical modeling of the enteric nervous network. 5: excitation propagation in a planar neural network. J Med Eng Phys, **17**: 11–19.
Miftakhov RN, Wingate DL. 1996. Electrical activity of the sensory afferent pathway in the enteric nervous system. Biol Cybern, **75**: 471–83.
Miftakhov RN. 1994. Mathematical modeling of the peristaltic reflex: a numerical experiment. J Math Sci, **71**: 2775–89.
Murphy RA. 1976. Contractile system function in mammalian smooth muscle. Blood Vessels, **13**: 1–23.
Nothnagel H. 1882. Zur chemischen reizung der glatten muskeln, zugleich als beitrag zur physiologie des darms. Arch Pathol Anat Physiol Klin Med, **88**: 1–11.
Orberg J, Baer E, Hiltner J. 1983. Organization of collagen fibers in the intestine. Connect Tissue Res, **11**: 285–97.

Orloff Juhl C, Vinter-Jensen L, Djurhuus JC, Gregersen H, Zapher Dajani E. 1994. Biomechanical properties of the oesophagus damaged by endoscopic sclerotherapy. An impedance planimetric study in minipigs. Scand J Gastroenterol, **29**: 867–73.

Paintal AS. 1957. Responses from mucosal mechanoreceptors in the small intestine of the cat. J Physiol, **139**: 353–68.

Paterson WG, Selucky M, Hynna-Liepert TT. 1991. Effect of intraesophageal location and muscarinic blockade on balloon distension-induced chest pain. Dig Dis Sci, **36**: 282–8.

Paterson WG. 1991. Neuromuscular mechanisms of esophageal responses at and proximal to a distending balloon. Am J Physiol, **260**: G148–55.

Premen AJ, Dobbins DE, Soika CY, Dabney JM. 1984. Relationship between substance P, intestinal wall compliance and vascular resistance in the canine ileum. Reg Peptides, **9**: 119–27.

Premen AJ, Soika CY, Dabney JM, Dobbins DE. 1984. Effects of gastrointestinal hormones on ileal vascular and visceral smooth muscle. Am J Physiol, **246**: G1–7.

Rao SSC, Gregersen H, Hayek B, Summers RV, Christensen J. 1996. Unexplained chest pain: The hypersensitive, hyperreactive and poorly compliant esophagus. Ann Intern Med, **124**: 950–58.

Reinke DA. 1964. Patterns and parameters of dog gastrointestinal contractile activity monitored by extraluminal force transducers. Thesis. University of Michigan.

Ren J, Massey BT, Dodds WJ, et al. 1993. Determinants of the bolus pressure during esophageal peristaltic bolus transport. Am J Physiol, **264**: G407–13.

Ritchie J. 1973. Pain from distention of the pelvic colon by inflating a balloon in the irritable colon syndrome. Gut, **14**: 125–32.

Roach MR, Burton AC. 1957. The reason for the shape of the distensibility curves of arteries. Can J Biochem Physiol, **35**: 681–90.

Robison BA, Percy WH, Christensen J. 1984. Differences in cytochrome C oxidase capacity in smooth muscle of opossum esophagus and lower esophageal sphincter. Gastroenterology, **87**: 1009–13.

Rogers J. 1992. Testing for and the role of anal and rectal sensation. Balliere's Clin Gastroenterol, **6**: 179–91.

Rosen EM, Mueller SN, Noveral J, Levine EM. 1985. Regulation of angiotensin I-converting enzyme activity in serially cultivated bovine endothelial cells. J Cell Physiol, **122**: 30–38.

Rouillon J-M, Azpiroz F, Malagelada J-R. 1991. Reflex changes in intestinal tone: relationship to perception. Am J Physiol, **261**: G280–86.

Rouillon J-M, Azpiroz F, Malagelada JR. 1991. Sensorial and intestinointestinal reflex pathways in the human jejunum. Gastroenterology, **101**: 1606–12.

Sadoshima J, Izumo S. 1993. Mechanical stratch rapidly activates multiple signal transduction pathways in cardiac mycocytes: potential involvement of an autocrine/paracrine mechanism. The EMBO Journal, **12**: 1681–92.

Saffouri B, Weir GC, Bitar KN, Makhlouf GM. 1980. Gastrin and somatostatin secretion by perfused rat stomach. functional linkage of antral peptides. Am J Physiol, **238**: G495–501.

Sarna SK. 1985. Cyclic motor activity: Migrating motor complex. Gastroenterology, **89**: 894–913.

Sato M, Oshima N, Nerem RM. 1996. Viscoelastic properties of cultured porcine aortic endothelial cells exposed to shear stress. J Biomech, **29**: 461–7.

Schaffer JL, Rizen M, L'Italien GJ, Benbrahim A, Megerman J, Gerstenfeld LC, Gray ML. 1994. Device for the application of a dynamic biaxially uniform and isotropic strain to a flexible cell culture membrane. J Orthop Res, **12**: 709–19.

Schulze-Delrieu K, Brown BP, Custer-hagen T. 1991. Contraction and accomodation of guinea pig duodenum in vitro. Am J Physiol, **261**: G364–72.

Schulze-Delrieu K. 1991. Intrinsic differences in the filling response of the guinea pig duodenum and ileum. J Lab Clin Med, **117**: 44–50.

Schulze-Delrieu K. 1992. Clearance patterns of the isolated guinea pig duodenum. Gastroenterology, **102**: 849–56.

Seelig LL, Goyal RK. 1978. Morphological evaluation of opossum lower esophageal sphincter. Gastroenterology, **75**: 51–8.

Sengupta JN, Gebhart GF. 1995. Mechanosensitive afferent fibers in the gastrointestinal and lower urinary tracts. In: Visceral Pain. ed. Gebhart GF, pp. 75–98. Seattle: IASP Press.

Sengupta JN, Kauvar D, Goyal RK. 1989. Characteristics of vagal esophageal tension-sensitive afferent fibers in the opossum. J Neurophysiol, **61**: 1001–10.

Sengupta JN, Saha JK, Goyal RK. 1990. Stimulus-response function studies of esophageal mechanosensitive nociceptors in sympathetic afferents of opossum. J Neurophysiol, **64**: 796–812.

Sengupta JN, Saha JK, Goyal RK. 1992. Differential sensitivity to bradykinin of esophageal distension-sensitive mechanoreceptors in vagal and sympathetic afferents of the opossum. J Neurophysiol, **68**: 1053–67.

Serra J, Azpiro F, Malagelada J-R. 1995. Perception and reflex responses to intestinal distention in humans are modified by simultaneous or previous stimulation. Gastroenterology, **109**: 1742–9.

Shyy YJ et al. 1993. Human monocyte colony-stimulating factor stimulates the gene expression of monocyte chemotactic protein-1 and increases the adhesion of monocytes to endothelial monolayers. J Clin Invest, **92**: 1745–51.

Siegle M-L, Ehrlein H-J. 1989. Effects of various agents on ileal postprandial motor patterns and transit of chyme in dogs. Am J Physiol, **257**: G698–703.

Siuri B, Mittal RK. 1991. Reverse-perfused sleeve: an improved device for measurement of sphincteric function of the crural diaphragm. Gastroenterology, **101**: 962–9.

Srivastava LM, Srivastava VP, Sinha SN. 1983. Peristaltic transport of a physiological fluid. Part I. Flow in non-uniform geometry. Biorheology, **20**: 153–66.

Stavitsky D, Macagno EO, Christensen J. 1981. Finite-element analysis of flow induced by contractions like those of the intestine. J Biomech, **14**: 183–93.

Steadman CJ, Phillips SF, Camilleri M, Talley NJ, Haddad A, Hanson R. 1992. Control of muscle tone in the human colon. Gut, **33**: 541–6.

Storkholm JH, Villadsen GE, Jensen SL, Gregersen H. 1995. Passive elastic wall properties in the isolated guinea pig small intestine. Dig Dis Sci, **40**: 976–82.

Su X, and Gebhart GF. 1998. Mechanosensitive pelvic nerve afferent fibers innervating the colon of the rat are polymodal in character. J Neurophysiol, **80**: 2632–44.

Svendsen P, Ainsworth M, Carter A. 1990. Acid-base status and cardiovascular function in pigs anaesthetized with α-chloralose. Scand J Lab Anim Sci, **17**: 89–95.

Szymanski PT, Chacko TK, Rovner AS, Goyal RK. 1998. Differences in contractile protein content and isoforms in phasic and tonic smooth muscle. Am J Physiol, **275**: C684–92.

Thewisen M, Ruhl A, Enck P. 2000. On the adequate stimulus for rectal mechanoception and perception: A study in cats and humans. *Neurogastroenterol Motil*, **12**: 43–52.

Thomas JE. 1941. An improved cannula for gastric and intestinal fistulas. *Proc Soc Exp Biol Med*, **46**: 260–61.

Thomas PA, Earlam RJ. 1973. The effect of the gastro-intestinal polypeptide hormones on the electrical activity and pressure of the isolated perfused canine gastro-oesophageal sphincter. *Proceedings from the Fourth International Symposium on Gastrointestinal Motility*. pp. 243–50. Vancouver: Mitchell Press.

Tøttrup A, Forman A, Uldbjerg N. 1990. Mechanical properties of isolated human esophageal smooth muscle. *Am J Physiol*, G338–43.

Tözeren A, Sung KLP, Sung LA, Dustin ML, Chan PY, Springer TA, Chien S. 1992. Micromanipulation of adhesion of a jurkat cell to a planar bilayer membrane containing lymphocyte function-associated antigen 3 molecules, *J Cell Biol*, **116**: 997–1006

Trendelenburg P. 1917. Physiologische und pharmacologische versuche über die dünndarmperistaltik. *Naunyn Schmiedeberg Arch Exp Pathol Pharmacol*, **81**: 55–129.

Varma JS, Smith AN. 1986. Reproducibility of the proctometrogram. *Gut*, **27**: 288–92.

Villadsen GE, Storkholm JH, Hendel L, Vilstrup H, Gregersen H. 1997. Impedance planimetric characterization of esophagus in systemic sclerosis patients with severe involvement of esophagus. *Dig Dis Sci*, **42**: 2317–26.

Villadsen GE, Storkholm JH, Zachariae H, Hendel L, Bendtsen F, Gregersen H. 2001. Esophageal pressure-cross-sectional area distributions and secondary peristalsis in relation to subclasification of systemic sclerosis. Gastroenterology. *Neurogastroenterol Motil*, **13**: 199–210.

Weems WA. 1987. Intestinal fluid flow: Its production and control. In: *Physiology of the Gastrointestinal Tract*. ed. Johnson LR, Christensen J, Jackson MJ, Jacobson ED, Walsh JH, pp. 571–93. New York: Raven Press.

Weisbrodt NW. 1987. Motility of the small intestine. In: *Physiology of the Gastrointestinal Tract*, 2nd edition. ed. Johnson LR, Christensen J, Jackson MJ, Jacobson ED, Walsh JH, pp. 631–64. New York: Raven Press.

Whitehead WE, Engle BT, Schuster MM. 1980. Irritable bowel syndrome. Physiological and psychological differences between diarrhea-predominant and constipation-predominant patients. *Dig Dis Sci*, **25**: 404–13.

Whitehead WE, Holtkotter B, Enck P et al. 1990. Tolerance for rectosigmoid distention in irritable bowel syndrome. *Gastroenterology*, **98**: 1187–92.

Williams D, Thompson D, Heggie L, Bancewicz J. 1993. Responses of the human oesophagus to experimental intraluminal distension. *Am J Physiol*, **265**: G196–203.

Wirtz HRW, Dobbs LG. 1990. Calcium mobilization and exocytosis after one mechanical stretch of lung epithelial cells. *Science*, **250**: 1266–9.

Wood JD. 1987. Neurophysiological theory of intestinal motility. *Jpn J Smooth Muscle Res*, **23**: 143–86.

Wood JD. 1987. Physiology of the enteric nervous system. In: *Physiology of the Gastrointestinal Tract* 2nd edition, volume 1. ed. Johnson LR, Christensen J, Jackson MJ, Jacobson ED, Walsh JH, pp. 67–110. New York: Raven Press.

Yamada H. 1970. *Strength of Biological Materials.* Baltimore: The Williams & Wilkins Company.
Yokoyama S, North A. 1983. Electrical activity of longitudinal and circular muscle during peristalsis. *Am J Physiol*, **244**: G83–8.
Yokoyama S, Ozaki T. 1980. Effects of gut distension on Auerbach's plexus and intestinal muscle. *Jpn J Physiol*, **30**: 143–60.
Yuan SY, Furness JB, Bornstein JC, Smith TK. 1991. Mucosal distorsion by compression elicits polarized reflexes and enhances responses of the circular muscle to distension in the small intestine. *J Autonom Nerv Syst*, **35**: 219–26.

6. The Zero-stress State of the Gastrointestinal Tract. The Concept of Residual Stress and Strain

6.1 Introduction

The function of the gastrointestinal tract is to propel food by peristaltic motion, which is a result of the interaction of the tissue forces in the wall and the hydrodynamic forces in the food bolus. To understand the tissue forces in the gastrointestinal tract, it is necessary to know the stress–strain relationships of the tissues. The stress–strain relationships must be measured with reference to the zero-stress state (the condition where neither external nor internal forces deform the tissue). The basic equations for computing stress and strain are given in Chapter 3. The zero-stress state of the tissue constitutes the standard state for describing tissue morphology because the tissue is not deformed by internal and external forces. The residual stress and strain cannot be assessed if the zero-stress state is not known, hence the determination of the zero-stress state of gastrointestinal tissue is the first step in the determination of the mechanical properties.

The zero-stress state of an organ is the state in which the organ is stress-free, meaning that all external and internal forces are removed. Until 1983 it was believed that biological organs were free of stress when all external loads were removed (now known as the no-load state). In 1983, Vaishnav and Vossoughi, and Fung independently, reported that the no-load state of a blood vessel was not the zero-stress state. This was demonstrated by an experiment where tissue rings that were cut radially opened up into sectors. The difference in stress and strain between the no-load (Figure 6.1, top) and zero-stress state (Figure 6.1, bottom) was called residual stress and residual strain, respectively, and it was found that the opening angle of the sector was a convenient way to quantitate the residual stress (Figure 6.1). Figure 6.1 shows the no-load and zero-stress states in schematics whereas Figure 6.2 shows photographs of small intestinal specimens in no-load and zero-stress states. Since 1983, a large number of studies have been published on the zero-stress state of different organ systems. The majority of these studies focused on the zero-stress state of systemic and pulmonary arteries, the heart and veins in normal animals and in various disease models such as experimental hypertension and diabetes and in venous grafts. Studies were also presented on the airways and ureter. Furthermore, the zero-stress state was used as the basic state to examine the remodelling of arterial tissue under stress by Fung and co-workers and theoretical analysis of structural components were made by several research groups. The first

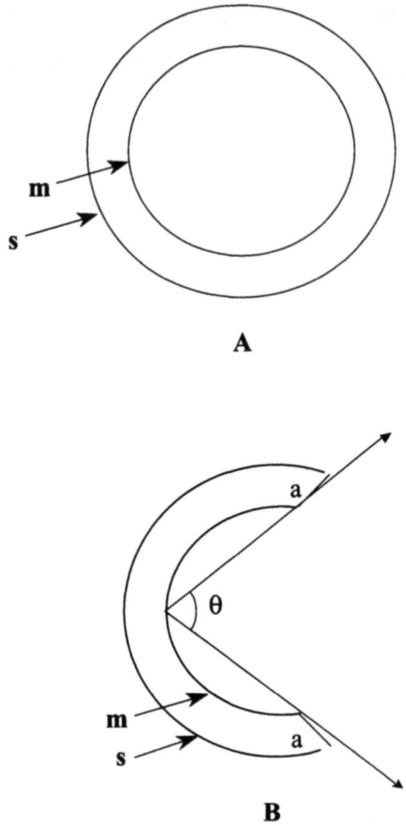

Figure 6.1. Schematic drawing of cross-section of a gastrointestinal specimen in no-load state (**A**) and zero-stress state (**B**). The zero-stress state is characterised by an opening angle θ defined as the angle subtended by two radii drawn from the mid-point of the inner wall to the inner tips of two ends of the sector in the zero-stress state. m and s denote mucosal and serosal surfaces. a is the line of cutting.

data on residual strains and stresses in the gastrointestinal tract were published in 1996 by Gregersen and Kassab. We realise now that the zero-stress configuration of the gastrointestinal tract is very different from that of the no-load condition and that the zero-stress state is sensitive to growth and remodelling. These studies opened up a whole new field of investigation in relation to gastrointestinal physiology and pathophysiology. Now data are quickly emerging. Data have been published on the zero-stress state of the oesophagus, the small and large intestines and the biliary tract (Chapters 8 and 9 summarise the majority of data obtained thus far). Table 6.1 is a list of all gastrointestinal studies. The stomach and sphincteric regions have not yet been studied.

The purpose of this chapter is to provide a basic understanding of the geometric characteristics of the zero-stress state, residual stress and strain and to discuss the implications of residual strain. Since most published data were obtained under passive muscle conditions, Hill's three-element model is briefly described in Chapter 5 in order to provide the reader with a reference frame for relating passive properties of the zero-stress state and the stress-strain relation to function.

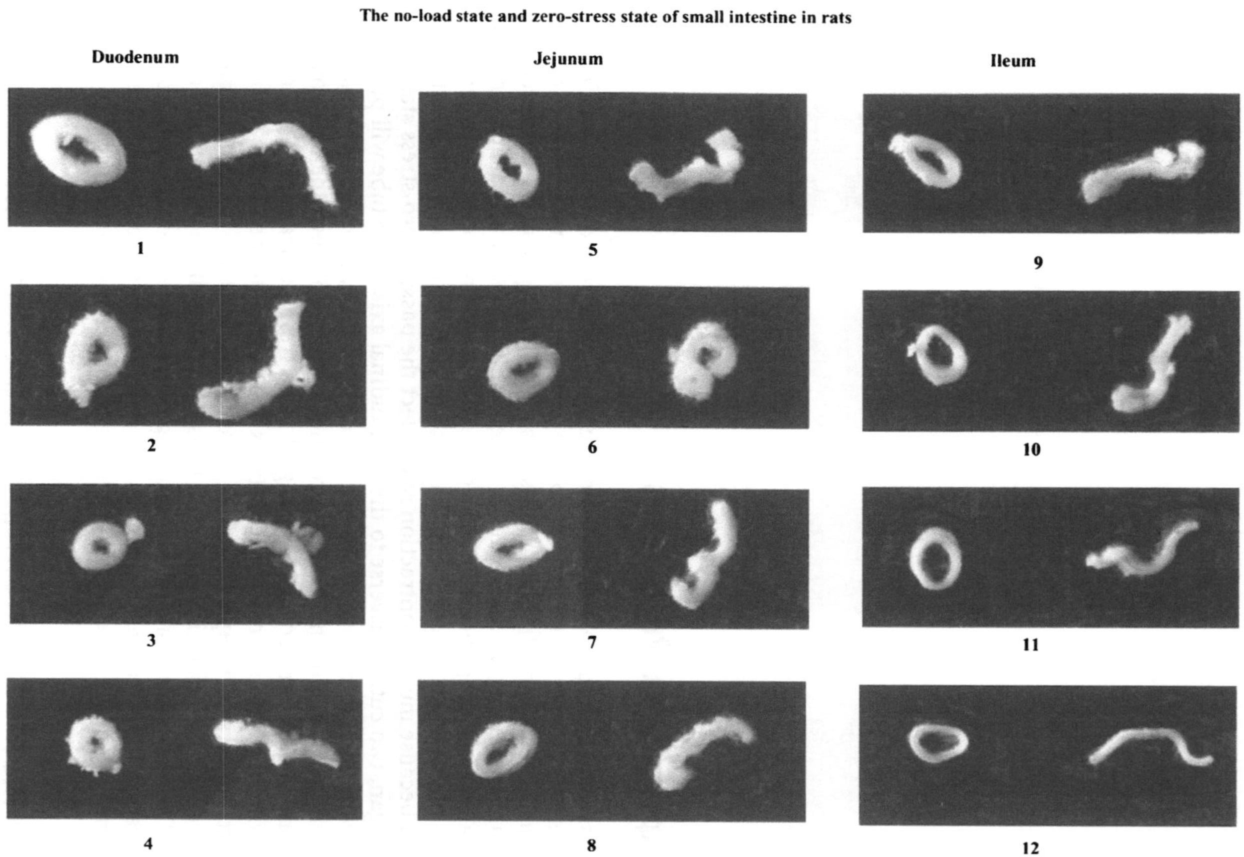

Figure 6.2. Photographs showing small intestinal specimens in the no-load state (closed ring) and in the zero-stress state (zero-bending moment state). The tissue was obtained from rats. Radial cuts in the tissue ring caused it to spring open into a sector which can be characterised by an opening angle (from Zhao et al., with permission).

Table 6.1. List of published papers on zero-stress state in the gastrointestinal tract

Organ	Species	Model	Publication Data
Whole gastro-intestinal tract (reviews)	several	circumferential and longitudinal, uni- and two-layer	Gregersen and Kassab, 1996 Gregersen, 2000 Gregersen et al., 2000
Oesophagus	guinea pig	circumferential uni- and two-layer	Gregersen et al., 1997
		circumferential uni-layer, ultrasound microscopy	Assentoft et al., 2001 Gregersen et al., 1997
	guinea pig		Lu and Gregersen, 2001
		circumferential two-layer	Gregersen et al., 200?
		longitudinal one-layer and circumferential two-layer	
	rabbit		
	mouse	circumferential one-layer, osteogenesis imperfecta mice	
Duodenum	guinea pig	circumferential uni-layer	Gregersen et al., 1997
	rat	circumferential uni-layer	Gao and Gregersen, 2000
Small intestine	rat	circumferential uni-layer	Gao et al., 2000
Large intestine	rat	circumferential uni-layer	Dou et al., 2002
Biliary tract	pig	circumferential uni-layer	Duch et al., 2002

6.2 Experimental Procedures

Zero-stress state experiments must be done under standardised conditions in vitro. After excision of one or more segments of the gastrointestinal tract, they are cleaned for luminal contents and adjacent tissue is dissected. The segment is transferred to an organ bath containing Krebs solution. Often, this solution is calcium-free and contains a relaxant drug such as EGTA to chelate intracellular calcium stores under otherwise physiological experimental conditions. Relaxation of muscle activity is important because muscle contraction may affect the passive zero-stress state. For a tubular organ, two cuts transverse to the longitudinal axis of the tube will produce a ring that will open into a sector when cut radially. Often, 30 minutes or more are needed before a steady opening angle is reached. Figure 6.3 shows the time course of the opening angle after cutting oesophageal rings from guinea pigs (a viscoelastic analysis with derivation of material constants from data obtained in the gastrointestinal tract appears in Chapter 7). It is clearly seen that the opening angle first increases quickly after which a slower phase follows until a new equilibrium is reached. The zero-stress state of the oesophagus has been shown to be independent of temperature in the range from ambient to body temperature. However, temperature dependency needs to be tested for other gastrointestinal tissues as well.

6.3 Geometric Characterisation of the Circumferential Zero-stress State

In order to establish the zero-stress state, it is necessary to make sufficient cuts to eliminate the internal stresses and strains. The state of zero stress-resultant and

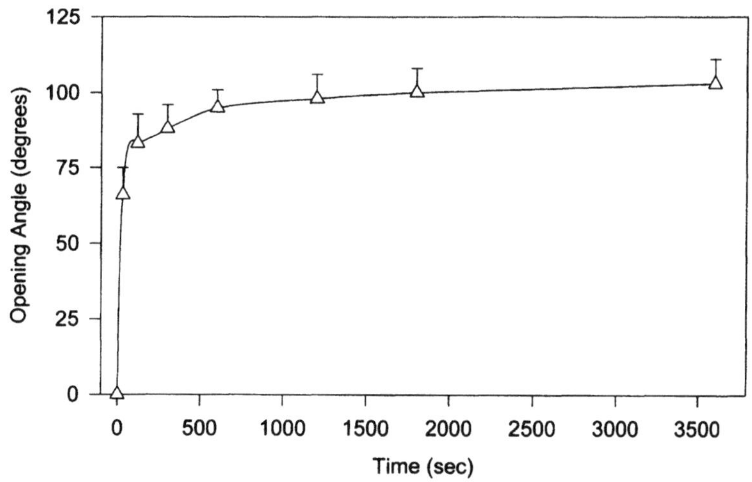

Figure 6.3. Time-dependence of the opening angle. The data are means and SE (n=9) and obtained from the guinea pig oesophagus. Steady state condition is often reached within 30 minutes of the radial cut where the viscoelastic phase has reached an asymptotic value. Modified from Gregersen et al. *J Biomech Eng*, 1999; 121: 442–448.

bending moment for a tubular organ is an open sector that can be conveniently described by an opening angle. Figure 6.1 (bottom) and Figure 6.4a show a schematic of a tissue sector in the zero-stress state. The subtended angle is called the opening angle, θ, and is defined as the subtended angle between the two radii from the midpoint of the arc of the inner wall of the intestinal ring to the inner tips of the sector.

The opening angle is a convenient measure but may not always be useful in practice. First, it requires that both the no-load and the zero-stress states be circular. However, neither configuration may be exactly circular. The opening angle is a measure of the angle between the two radii and does not account for the shape of the curve. It provides no information on the curvature of the wall. The angles between the tangents at the tip of the sector and the line joining the tips are designated as α and α', respectively (Figure 6.4a). The mean value $(\alpha + \alpha')/2$ is referred to as the mean tangential angle. If the sector is exactly circular, then $\theta = (\alpha + \alpha')/2$. The second difficulty with the opening angle is when its value exceeds 360° (i.e., the sector inverts inside out). Some organs such as the guinea pig duodenum often have θ greater than 360°. In fact, in the guinea pig duodenum, most rings, if sufficiently short, turn the mucosa to the outside even before the first radial cut is made, indicating large compressive forces in the mucosa in the no-load state. When the opening angle is 360° or greater, the tips of the sector come in contact after an inside-out deformation. In this case (or for θ smaller than 0°), it is necessary to make additional cuts (typically two or three) along the circumference to establish the zero-stress state. Figure 6.5 shows intestinal specimens after three radial cuts. When the opening angle is greater than 360°, it is more convenient to use the *tangent rotation angle* (as shown in Figure 6.4a) to describe the sector geometry in the zero-stress state. To define ϕ, one must fix a point on the outer rim of the cross-section and measure the length of the outer rim S as a curvilinear coordinate along the rim from the origin. Next, define a unit tangent vector T on the outer rim. As S increases, the tangent rotates. The tangent

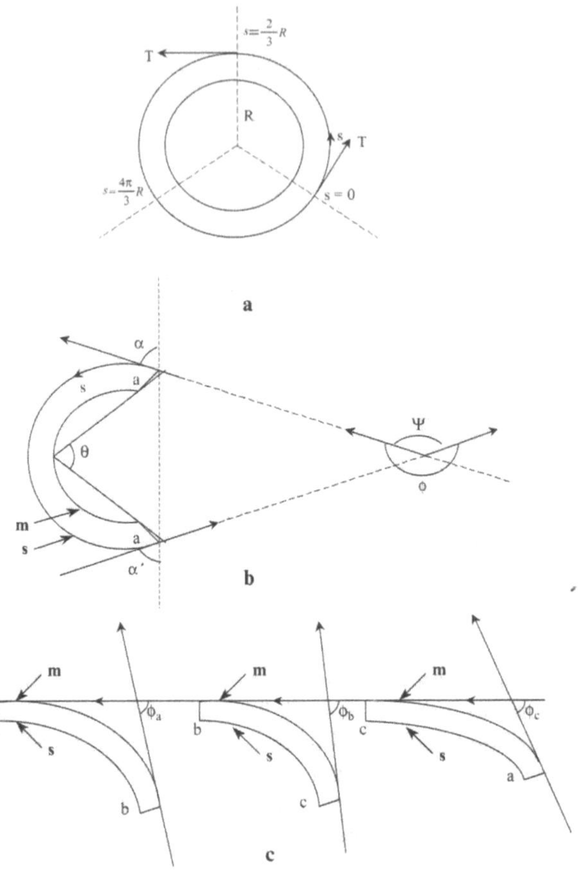

Figure 6.4. a) No-load state with lines of cut to produce the specimens in Figure 6.5c shown as dashed lines. T is a unit tangent vector T on the outer rim and s is a curvilinear coordinate of points on the curve. **b)** Definition of the zero-stress state. Traditionally, the zero-stress state is characterised by an opening angle defined as the angle subtended by two radii drawn from the mid-point of the inner wall to the inner tips of two ends of the sector in the zero-stress state. The opening angle is denoted by θ. The labels M and S denote the mucosal and serosal surfaces and a is the line of cut. The tangent rotation angle, which is described in the text, is denoted by ϕ. The angles between the tangents at the tip of the sector and the line joining the tips are designated as α and α', respectively. The mean value $(\alpha + \alpha')/2$ is referred to as the mean tangential angle. **c)** When a section becomes completely inside-out, the opening angle is 360°. The guinea pig duodenum often has α greater than 360° at zero-stress state. In fact, most rings turned the mucosa to the outside even before the first radial cut, indicating large compressive forces in the mucosa in the no-load state. When the opening angle is 360°, the cut ends came in contact again after an inside-out deformation. For α larger than 360°, it is simpler to make two to three cuts in each circumference to establish the zero-stress state. In the case illustrated, three pieces of a tissue ring produced by three radial cuts (at locations a, b, c) are shown. It is assumed that the tangent rotation angle turns the segments inside out, as indicated by the labels M (mucosa) and S (serosa). In this case it is easier to use the tangent rotation angle, denoted by ϕ, to describe the zero-stress state. When a section opens, the total angle of rotation from one tip to the other is smaller than 360°. The angle of rotation can be measured in segments. If the specimen is cut radially in three places to produce three non-overlapping pieces as shown in the figure and the tangents of the segments rotate by angles ϕ_a, ϕ_b, ϕ_c, then the total angle of rotation of the whole specimen is $\phi = \phi_a + \phi_b + \phi_c$. This resultant value of ϕ is independent of where the cuts are made in the original ring and does not require the original or final shapes to be circular. Modified from Gregersen et al., 1997 and Gregersen et al., 2000.

Figure 6.5. Duodenal specimens after two radial cuts along the circumference. They turn the inside out and, added together, have an opening angle bigger than 360°.

vector T is a unit vector field associated with any space curve. If *s* is a curvilinear coordinate of points on the curve, with significance of length measured along the curve from an origin on the curve, then the angle between T(*s*) and T(*o*) is the angle of the tangent between *s* and *o*. The derivative of rotation dT(*s*) / d*s* is the curvature of the curve, which is a quantity of major importance in the mechanics of the curve representing a shell. For a complete circle the tangent rotates 360° when S returns to the origin. When a section opens, the total angle of rotation from one tip to the other is less than 360°. The angle of rotation can be measured in segments. If the specimen is cut radially in three places to produce three non-overlapping pieces as shown in Figure 6.4b and the tangents of the segments rotate by angles ϕ_a, ϕ_b, ϕ_c, then the total angle of rotation of the whole specimen is $\phi = \phi_a + \phi_b + \phi_c$. The value of ϕ is independent of where the cuts are made in the ring and does not require the no-load or zero-stress shape to be circular. The geometry of the duodenum at the zero-stress state can also be characterised by the angle between the two tangents of the tip, ψ, as shown in Figure 6.4a. ψ refers to the angle between the tip tangents. The angle

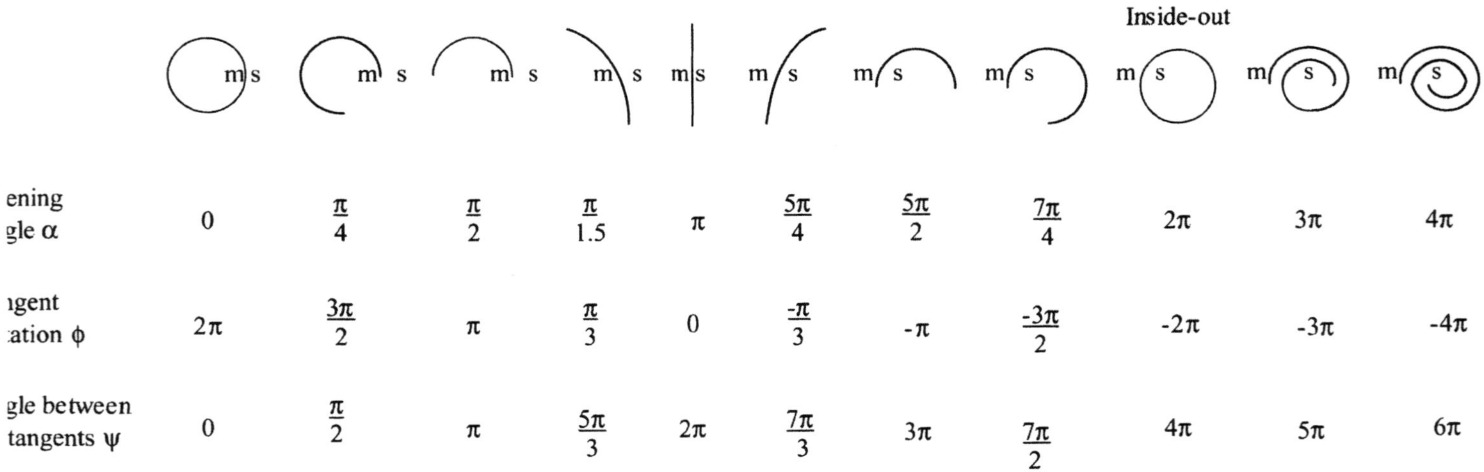

Figure 6.6. Relationship between the opening angle q, the tangent rotation f, and the angle between the tip tangents y. Mucosa and serosa are labelled m and s, respectively (With permission, Gregersen et al., Am. J. Physiol 1997; 273: G866–G874).

between the tangents at the tips is the angle between T(c) and T(o), c being the circumference. It is the integral of the curvature from s = o to s = c. Writing ϕ for the angle of rotation of the tangent from s = o to s = c, and ψ for the angle between the tip tangents, we have:

$$\phi = 2\pi - \psi, \quad \text{or} \quad \psi = 2\pi - \phi. \tag{6.1}$$

Clearly, the sum of ϕ and ψ is equal to 360°. All three measures, α, ϕ, ψ, are useful in the characterisation of the zero-stress state of tubular organs. Figure 6.6 demonstrates the relationship between the three measures.

6.4 Residual Strain Measures

So far, we have only considered angle measurements. Strain measurements, however, are fundamental in mechanical studies. As discussed in detail in Chapter 3, strain describes the mechanical deformation of a material. In a distensible biological tube like the gastrointestinal tract, the strain of principal interest is the circumferential strain. The loaded in vivo strain of the mucosa and serosa must be computed with reference to the mucosal and serosal circumferential lengths at zero stress. In the oesophagus, it may also be possible to measure the length of the interface between the mucosa–submucosa and the muscle layers. For a continuum subjected to finite deformation, strain can be defined in several different ways in relation to the deformation gradient. The Cauchy strain, ε, relative to the zero-stress state, is defined as:

$$\varepsilon_{\text{mucosa}} = \frac{C_{l-mucosa} - C_{z-mucosa}}{C_{z-mucosa}} \quad \text{and} \quad \varepsilon_{\text{serosa}} = \frac{C_{l-serosa} - C_{z-serosa}}{C_{z-serosa}} \tag{6.2}$$

where C is the length of distance between any two neighbouring material particles. Subscripts $_l$ and $_z$ refer to the loaded and zero-stress states, respectively. If the loaded state is replaced by the no-load state, the strains are called residual strains. The Cauchy strain is especially useful in the linearised theory of elasticity, which is valid when ε is small (less than 10%). For finite deformation, the strain defined by Green is more directly related to stress. The Green's strain in the circumferential direction is defined as:

$$E_{mucosa} = \frac{(\varepsilon_{mucosa}+1)^2-1}{2} \quad \text{and} \quad E_{serosa} = \frac{(\varepsilon_{serosa}+1)^2-}{2}. \tag{6.3}$$

Thus, from the circumferential lengths at the no-load, loaded and zero-stress states, we can compute the circumferential residual or loaded strain, respectively, at the mucosal and serosal surfaces either in the sense of Cauchy or Green. Negative strain implies that the tissue is in compression whereas positive strain implies extension.

6.5 Bilayer Models

Thus far, we have only considered monolayer models. The gastrointestinal tract, however, is not a homogeneous tube; rather it is a composite tube with differing properties in each layer. Data from individual layers adds another dimension to the hierarchy of the structure. Separation of layers is indeed possible in the oesophagus. A major difference between the structures of the oesophagus and blood vessels is that the submucosa and the muscle layers can be separated easily, unlike the intima, media, and adventitia. Hence, the oesophagus provides a unique model for studying remodelling of the zero-stress state. The layer dissection can be done with microsurgical technique without damaging the tissue. Figure 6.7 is a schematic of different states (distended, no-load, zero-stress and the separated state). The dissection can be done in the no-load or zero-stress state (the bilayer model is discussed further below). In guinea pigs, it was found that layer separation in the oesophagus caused the mucosa–submucosa layer to open up further whereas the muscle layer opening angle decreased (Figures 6.8 and 6.9). Hence, the previously observed zero-stress state of the non-separated oesophagus is, in fact, not the zero-stress state. When a whole oesophageal segment is cut radially, residual strains still exist in the various layers. The degree to which residual strains still exist depends on the difference in the opening angle of the different components. The separation experiments have shown that the inner layers increased whereas the outer layer decreased their opening angles. This additional bending implies that the inner layer is under less circumferential tension at physiologic pressures than would be indicated by a single layer model. This would further support the hypothesis that the organ wall remodels itself in order to equalise the strain distribution. Predictions of the stress distribution at physiologic pressures requires knowledge of the mechanical properties of each layer. These mechanical properties must be obtained with the knowledge of the zero-stress configuration of each layer.

A more abrasive way of obtaining data on the individual layers is to scrape off layers surgically; i.e., the mucosal layer or the muscle layers can be scraped off the submucosa. However, no data on the zero-stress state has yet appeared in the gastrointestinal literature with this technique.

6.6 Residual Strain Properties in Longitudinal and Radial Directions

The above description of deformation has mainly focused on the circumferential direction. However, due to the three-dimensional nature of biological organs, it is clear that residual stresses and strains may also exist in the longitudinal and radial directions. Residual strain in the radial direction may be analysed by putting markers on the transversely cut gastrointestinal segments in the no-load state. Analysing the displacements of the microdots after making the radial cut to obtain the zero-stress state will reveal the distribution of residual strains throughout the thickness of the wall. However, this has been found to be quite difficult due to the irregular surface. Such studies have been more successful in arteries (Han and Fung, 1996).

Figure 6.10 shows longitudinal strips of the rat small intestine. The longitudinal deformation can be described in a similar fashion to the circumferential deformation. In this special case, however, it may not be intuitive to define an opening angle. Instead, the opening angle per unit length may be defined. Strain data can be obtained by measuring mucosal and serosal length in the no-load and zero-stress states or by sprinkling ink or stainless steel beads onto the surface of the specimens.

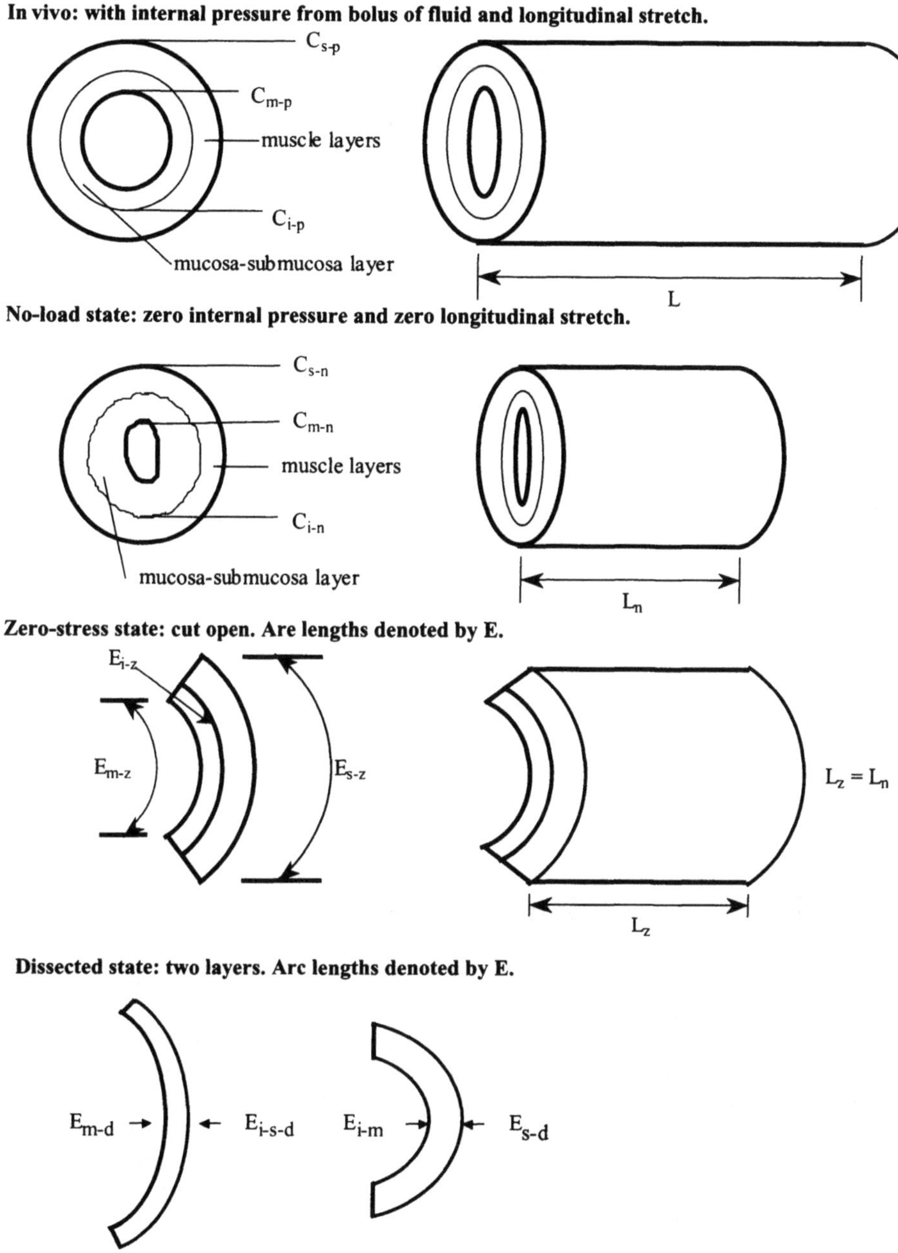

Figure 6.7. Illustration of the bilayer model. Schematic drawing of the distended state, no-load state, zero-stress state of the intact (non-separated) state and the oesophagus separated into layers (dissected state). Circumference, arc length and axial length denoted by C, E and L, respectively. Subscripts $_{m,\ i,\ s,\ p,\ n,\ z,}$ and $_{d}$ are the mucosal surface, interface between muscle and mucosa–submucosa, serosal surface, in vivo (pressurised), no-load, zero-stress and dissected state, respectively. In this case the dissected state produces layers with different opening angles. Hence, the dissected state is a better approximation of the true zero-stress state than is the zero-stress state of the non-separated oesophagus. Circumferential strains and residual strains can be computed from the circumferences and edge lengths. The figure is modified from Lu and Gregersen, 2001.

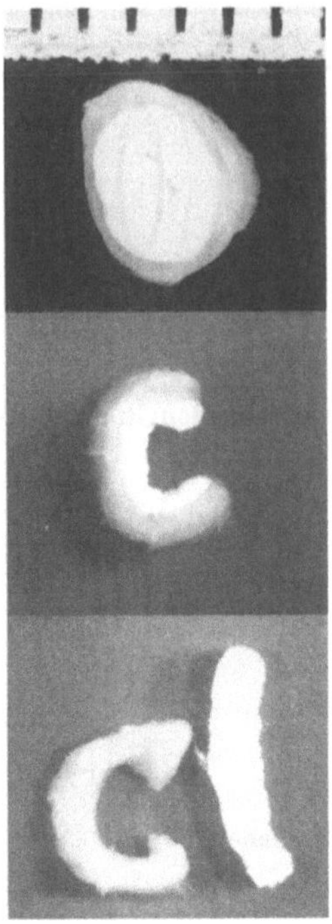

Figure 6.8. Oesophageal specimens in the no-load, zero-stress state of the non-separated oesophagus and the zero-stress state of the separated layers. The data were obtained in the guinea pig where the mucosa–submucosa has a higher opening angle than the muscle layers.

6.7 Implications of Residual Strain

There are several important implications of the zero-stress state, some of which are, as yet, hypothetical.

It is clear from the above discussion that using any state other than the zero-stress state will lead to error in the computed stresses and strains. Analysis of strain must refer to the zero-stress state in order to compare specimens obtained from different locations in the gastrointestinal tract or from different species, and in remodelling induced by diseases, growth or degeneration. Furthermore, since the zero-stress state is not influenced by external or internal forces, it provides a standard state for describing tissue morphology. Residual stresses in tissues seem to be caused by non-uniform growth and remodelling due to incompatibility of the growth strain field. The residual stresses in biological tissues have similarities with the residual stresses manufactured into gun barrels to strengthen them.

All gastrointestinal studies done thus far show that the rings open into sectors when cut radially. In fact, duodenal tissue from the guinea pig has thus far presented the largest residual stresses observed in soft tissues. The sectors imply that the mucosa is under compression in the no-load state (and at physiological conditions in the low-pressure range) whereas the muscle layrs are in tension. An example of this is given in Figure 6.11. The residual tension in the muscle may provide a more optimum

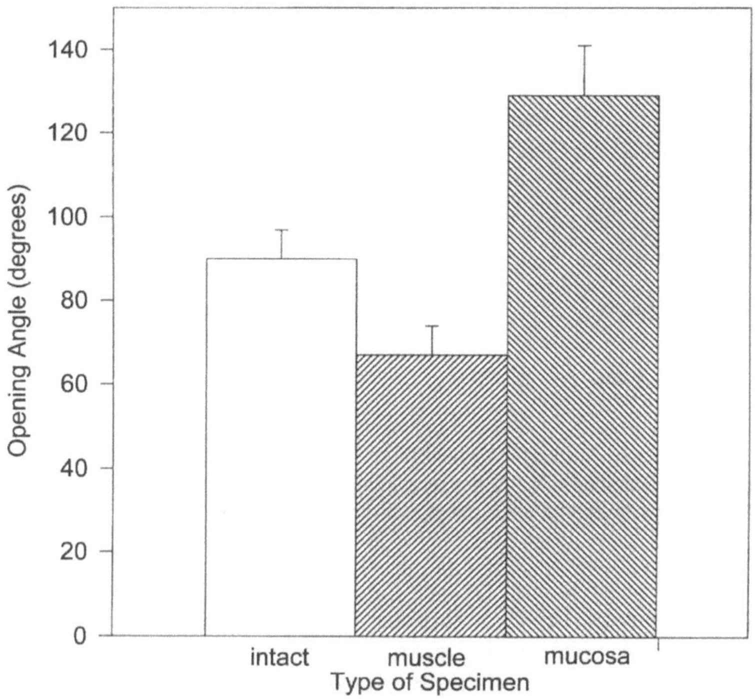

Figure 6.9. Mean and standard error of the opening angle in guinea pigs after anterior cut in the non-separated (intact) oesophagus and after separation into the muscle and mucosa–submucosa layers. Permission obtained from Gregersen et al., J Biomech Eng (ASME) 1999; 121: 442–448.

length for muscle contraction. The effect of the zero-stress state on physiology may not be obvious until the full Hill's model is analysed, but it is clear that it is a fundamental feature of the parallel element. The function of the contractile element depends on the parallel element: for example, the length of the muscle cell depends on the strain of the parallel element; hence, whether the muscle can achieve the optimum length of the length–tension relationship for its contraction or not depends on the zero-stress state of the parallel element in principle (see also Chapter 5 about Hill's model). Residual stress reduces the stress concentration at the inner wall of the gastrointestinal tract at no-load and homeostatic states (Gregersen et al., 1997; 1999; Lu and Gregersen, 2001; Gao et al., 2000). Thus, the compressed mucosa may be better protected against injury from the flow of luminal contents than an uncompressed mucosa. These protection mechanisms could be important when unphysiologically high pressures are reached, e.g., in mechanical obstruction. Since peristaltic bolus transport bulges out the intestinal wall in the vicinity of the bolus, the residual strain would likely influence the resistance to bolus flow under normal conditions. Since residual strains exist, the mucosa will inevitably be under compressive stress and strain in the no-load condition; i.e., where the transmural pressure is zero. The gastrointestinal muscles are usually passive since phasic contractions occur infrequently. Therefore, are there any other functions of the mucosa compression than those mentioned above? With respect to the oesophagus, the compression of the

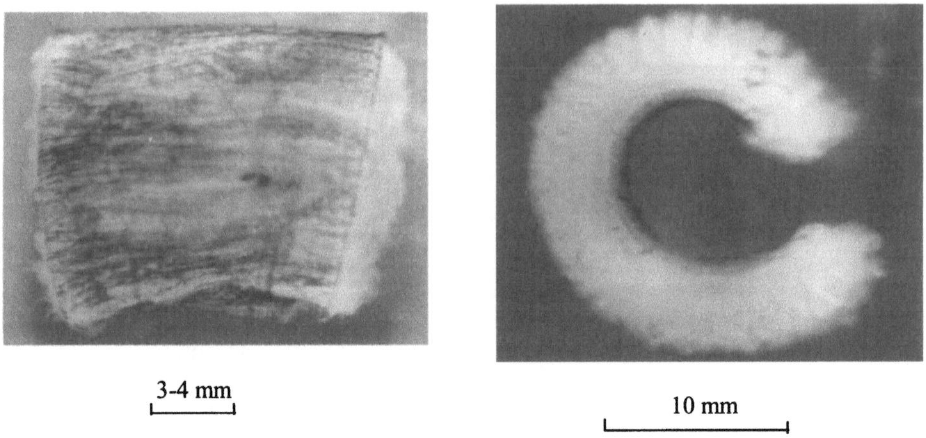

Figure 6.10. Photographs showing the longitudinal no-load and zero-stress states from the normal rat intestine. Left: Short segment of intestine before cutting several longitudinal strips. Right: A longitudinal strip at zero-stress state. The strip bends outwards due to the compression in the mucosa layer (Dou et al., unpublished data).

mucosa seems very pertinent. Reflux disease is associated with mucosal inflammation and symptoms such as heartburn. A high pressure in the lower oesophageal sphincter and secondary peristalsis is known to prevent reflux. Residual stress with buckling of the mucosa serves as a mechanism to close the oesophageal lumen and thereby prevents acid reflux from the stomach (see the end of this chapter for additional information about tissue buckling). In the small intestine, another function of mucosa compression may be important. It is well known that the mucosa in the small intestine is one of the tissues with the fastest turnover rate. The fast growth of the mucosal surface could easily cause mucosal compression and hence explain the presence of large residual strains. Likewise, the large compressive stresses may affect mucosal growth. Studies on the intestine and other organs have shown that mechanical stresses are important factors in regulating gene expression and growth (Fung, 1993; Gregersen et al., unpublished data). Well-known examples are cardiac hypertrophy caused by hypertension and muscle atrophy in space flight. Theoretically, absorption of luminal contents may also be affected by the residual compression. It is well known that a height gradient in the mucosal villi exists along the small intestine, with the highest villi found in the proximal duodenum (Gabella, 1987), and that small-intestinal absorption depends on the luminal pressure. Thus, there may exist a correlation between the residual strain gradient and the gradient in the height of villi. These interesting aspects of small intestinal function will the focus of future studies.

Other important issues pertain to the function of the mechanoreceptors (nerve endings of the sensory afferent nerves) in the gastrointestinal wall. These afferent nerve terminals initiate the sensory signals that are transmitted from the gastrointestinal tract via vagal and spinal pathways to the brain and bring about the perception of visceral events such as nausea, satiety and pain. Various receptor classes such as low- and high-threshold mechanoreceptors exist. The nerve terminals are also found in the primary afferent neurons that are components of the enteric nervous system and mediate local reflexes such as secondary peristalsis. Receptors are located both in the submucosa and in the muscle layers. The zero-stress state studies have shown that

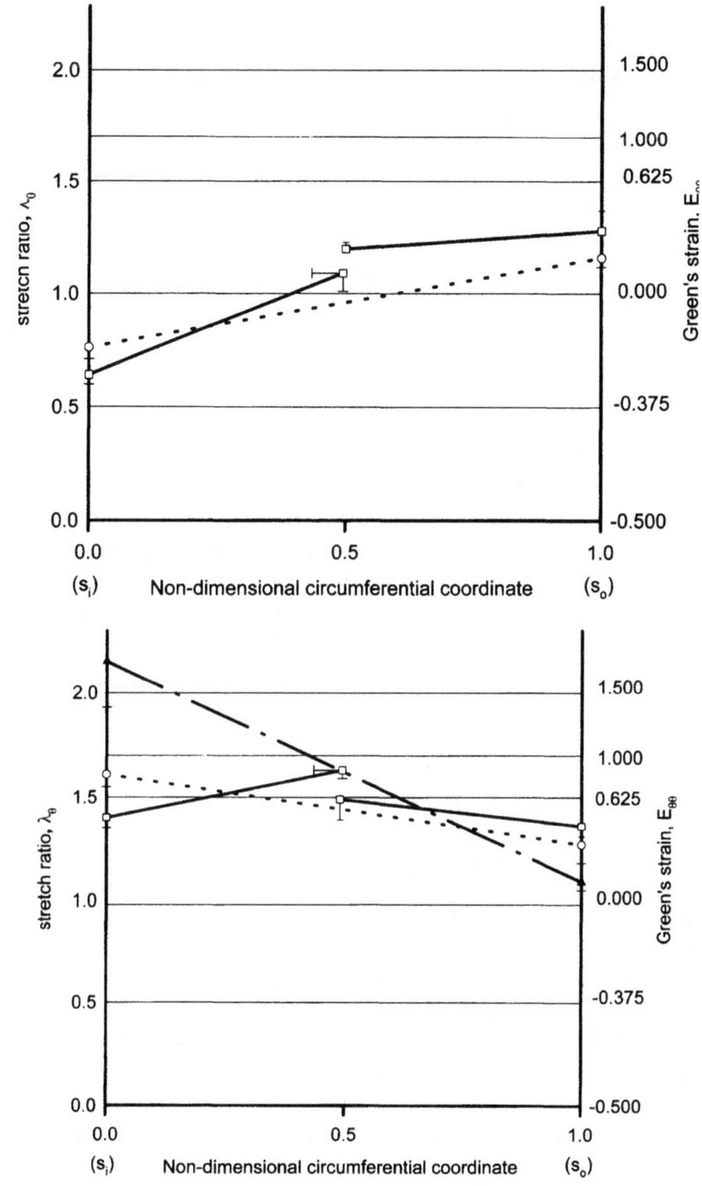

Figure 6.11. Relationship between circumferential stretch ratio and Green's strain in the oesophageal wall calculated according to the Bernoulli–Kirckhoff theory of the oesophageal wall and the non-dimensionalised radial coordinate with 0 (or S_i) indicating the inner (luminal) wall, and 1 (or S_0) indicating the outer wall. Stretch ratios and strains were also calculated for the interface between the mucosa–submucosa and muscle layers. Upper panel: Residual strain at the no-load state. Bottom panel: Strain at a homeostatic state with an internal pressure of 0.5 kPa. The solid lines and squared symbols represent strains computed for the bilayer model (submucosa and muscle). The dotted lines and circle symbols represent strains computed when the wall is considered homogeneous. The dot-dash lines and triangle symbols represent strain distribution when the residual strains are neglected. Residual strain decreases the strain at the submucosal surface and makes the strain distribution more uniform at the no-load and homeostatic states. The discontinuity at the interface between layers in the bilayer model likely represents shearing forces between the layers (with permission, Gregersen et al. J Biomech Eng (ASME) 1999; 121: 442–448).

a strain gradient exists from the mucosa to the serosa of the wall under physiological conditions. This suggests that the populations of receptors from different locations in the wall may have different zero-settings and respond differently to the same stimuli due to the variation in the magnitudes of tensile stresses and strains during distension and in the magnitudes of compressive stresses and strains during muscle contraction. A difference in the responses of mucosal and muscle receptors to stimuli have, in fact, been observed (Grundy, 1993). Furthermore, the present results emphasise the importance of studying the receptor kinematics by using experimental models on the intact organ approximating the in vivo conditions. Hence, data on receptor kinematics obtained from tissue strips or otherwise cut-open segments will likely give erroneous results.

The zero-stress state is sensitive to remodelling by disease, growth or degeneration. The change of the opening angle of the zero-stress state is a result of non-uniform tissue remodelling of the organ wall. Fung hypothesised that the opening angle increases when the inner layers have a higher growth rate than the outer layers or when the outer layers atrophy more than the inner layers, and vice versa. This has largely been found to be true for the gastrointestinal tract (see Chapter 8). Rodriguez et al. (1994) demonstrated by a theoretical analysis that volumetric tissue growth could alter residual stress. It was predicted that concentric hypertrophy, which increases the wall thickness-to-radius ratio, increased the opening angle and residual stress, and, conversely, eccentric hypertrophy decreased the opening angle. The opening angle increases when the inner layers have a higher growth rate than the outer layers or when the outer layers atrophy more than the inner layers, and vice versa. In other words, the opening angle of the zero-stress state does not change if homogeneous growth occurs in the tissue. Consistent with theoretical prediction (Rodriguez et al., 1994) and experimental studies on the aorta and left ventricular remodelling (Omens, 1988), the opening angle correlates significantly with the wall thickness-to-radius ratio in several studies conducted on the zero-stress state of the gastrointestinal tract. This implies that the gastrointestinal wall remodels itself in such a way as to maintain optimum circumferential stress, though the zero-stress state is a condition that is never experienced in vivo. However, the lack of correlation in some studies can possibly be due to the complexity of the layered gastrointestinal tissue. In addition to the mechanical factors, such as intestinal flow (shear) and pressure (circumferential tension), residual stress or strain also affects the histological changes during remodelling. Residual stress may affect the remodelling by varying the stress distribution in the organ. The data on residual strains in the gastrointestinal tract and the non-uniform remodelling may be used to improve surgery (e.g., procedures may be designed to pre-stress the tissue to make it more resistant to injury). An additional application may be the effect of drugs on tissue growth and remodelling.

Some of the most important future work is to complete the Hill's three-element model so that we can get an understanding of the interaction between the parallel element and contractile element. Further work is also needed to determine how the residual strains affect responses to physical and chemical stimulations in pathophysiological states, and to determine whether species variations exist. Since the duodenum is essentially a muscular tube, alteration of the activity of the smooth muscle cells by agents acting on the muscles directly or on their innervation will change the zero-stress state of this organ. Hence, the effects of these agents on the duodenum and other parts of the small intestine need investigation in order to model

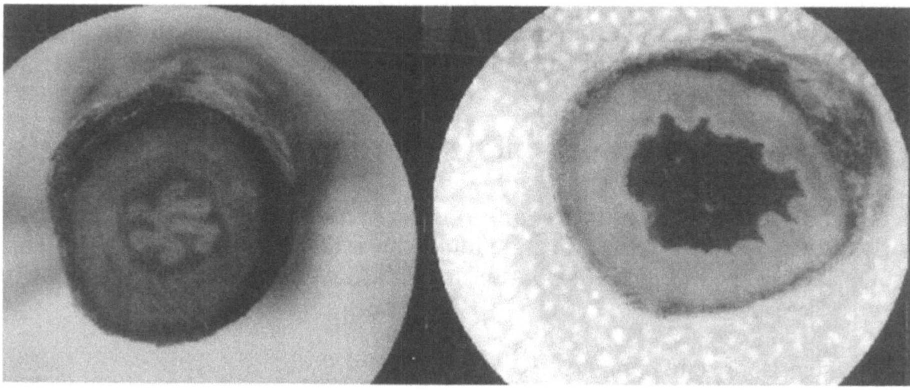

Figure 6.12. Photographs of snap-frozen sections of the oesophagus with zero luminal pressure (left) and 0.3 kPa luminal pressure applied (right). The outer surface and the fluid in the lumen are coloured with Evans blue. Large and small mucosal buckles are observed. The photographs were obtained by Jeff Cassin.

accurately the normal bolus transport and pathological situations of blockage or disease.

6.8 Additional Notes on Tissue Buckling

A string can be strong and tough in tension, but when it is relaxed or uniaxially compressed it has little resistance to deformation. This kind of large deformation is called buckling in engineering literature. The design of aeroplanes, spaceships, submarines, columns, tall buildings and tall bridges involve designs against buckling. Buckling or large deformation is often seen in tissues and organs when they are subjected to uniaxial compression. When the strain is referred to the zero-stress state, the tissue is in tension when $\lambda > 1$. It is in compression when $\lambda < 1$, and tissue buckling occurs when $\lambda < 1$. The assumption is that for very soft tissues in the neighbourhood of a state of zero-stress or no-load, the critical buckling stress tends to zero. Oesophagus, intestine, and blood vessels probably belong to this category. Buckling is a global or macroscopic phenomenon. Hence, the edge lengths E should be measured with a scale commensurate with the wall thickness of the tube, and not by fractal considerations. The gastrointestinal tract uses buckling to advantage in its normal function. Without food in the lumen, a small tension in the circumferential muscle is sufficient to cause large buckling of the mucosa and submucosa. Figure 6.12 shows the buckling pattern in the rabbit oesophagus. The mucosa buckles into a multi-lobed figure. The number of lobes is a characteristic number theoretically computable according to the first principles of mechanics provided that the constitutive equation (i.e., the stress–strain relationship of the material) is known. For the oesophagus, the condition of buckling are: 1) the tissues are incompressible; the volume of the tissue remains constant in the process of buckling, and 2) the compressive stresses and strains are small. Hence, the total length of the inner wall of the lumen remains almost constant. These conditions can be easily verified from analysis of photographs of specimens. Then the equations describing the balance of forces and conservation mass and the minimisation of strain energy yield the details of the buckling pattern.

Functionally, the closure of empty gastrointestinal lumen by buckling is important to the prevention of reflux. The required minimal tension, or tone, in the circumferential muscle to achieve this function can be computed.

6.9 Estimates of the Strain Distribution in the Loaded State

Because the sections of the duodenum considered herein are curved and the wall is fairly thick (compared to the mucosal radius), the strain distribution in the wall may be non-linear. The extent of the non-linearity may be estimated under the following idealised assumptions:

1. Plane sections remain plane.
2. All cross-sections are circular.
3. Mass is conserved and the length of the specimens in the axial direction does not change.

For the purposes of this analysis, the above assumptions are assumed to hold in the unloaded, loaded and stress-free states, and also in the transformation between these states. In this section we make an idealised geometry of the gastrointestinal segments. For purposes of this example, it is assumed that the tangent rotation angle ψ_2 lies between 360° and 720°. Then the angle ϕ is

$$\phi = \psi_2 - 360 = \psi_a + \psi_b + \psi_c \tag{6.4}$$

where ψ_2 is the tangent rotation angle. The equations given below are valid even if ψ_2 is greater than 720°.

The third assumption above leads to the conclusion that the cross-sectional area of the specimens is the same in all states. This leads to the relations:

$$r_s^2 - r_m^2 = R_s^2 - R_m^2 = \frac{\phi}{2\pi}\,(\rho_m^2 - \rho_s^2) \tag{6.5}$$

where r, R, and ρ refer to the radii in the unloaded, loaded and zero-stress states, respectively, and subscripts $_m$ and $_s$ refer to the mucosal and serosal surfaces, respectively. The angle ϕ in Equation 6.5 is expressed in radians.

The conservation of mass equation (6.5) may be written for any radius as

$$r^2 - r_m^2 = R^2 - R_m^2 = \frac{\phi}{2\pi}\,(\rho_m^2 - \rho^2) \tag{6.6}$$

Equation 6.6 allows the radii r, R, and ρ to be located within each cross-section.

The mucosal and serosal strains, ε_m and ε_s, respectively, in the loaded state may be expressed in the present case in the forms

$$\varepsilon_m = \frac{2\pi R_m - \phi\rho_m}{\phi\rho_m},\ \varepsilon_s = \frac{2\pi R_s - \phi\rho_s}{\phi\rho_s} \tag{6.7}$$

Similarly, the circumferential strains at any radius R in the loaded state are given by

$$\varepsilon = \frac{2\pi R - \phi\rho}{\phi\rho} \tag{6.8}$$

where ρ and R are the corresponding points determined from Equation 6.6.

As a numerical example of the use of the above equations, we assume the following data (which are approximately the mean data from a previous study in duodenum):

Zero-stress state edge lengths: $C_m = 19$ mm, $C_s = 13$ mm
Tangent rotation angle: $\psi_2 = 547°$
Loaded state circumferences: $C_m = 16$ mm, $C_s = 25$ mm.

The above data are all that is needed for the subsequent computations. It is assumed that the above data apply to the idealised circular model as well as the actual experiment. The above data were adjusted so that Equation 6.5 (conservation of mass) is satisfied to within 0.16%. From the above data it follows that the radii are:

Loaded state: $R_m = 2.54$ mm, $R_s = 3.98$ mm
Stress-free state: $\rho_m = 5.82$ mm, $\rho_s = 3.98$ mm

The angle ϕ is:

$\phi = 547 - 360 = 187° = 3.26$ radians.

Using the above values in Equation 6.7 gives strain values:

$\varepsilon_m = -0.1579, \varepsilon_s = 0.9231.$

The difference $(R_s - R_m)$ in the model is the wall thickness in the loaded state. The above figures give h = 1.4323 mm, which is similar to the value found by measurement. In the model, the wall thickness-to-mucosal circumference ratio is 0.0895, which is also similar to what was found by measurement.

From the above numerical values and the model, corresponding radii R and ρ can be computed by using Equation 6.6. Then Equation 6.8 can be used to compute the strain distribution as a function of radius R in the loaded state. Table 6.2 shows the results of these computations in the model example.

Table 6.2. Strains in the loaded state, based on the model

	R (mm)	ρ (mm)	e
Mucosa	2.55	5.82	−0.16
	2.8	5.59	−0.04
	3.2	5.16	0.19
	3.6	4.63	0.50
Serosa	3.98	3.98	0.92

6.10 Estimates of Mucosal and Serosal Strains in the No-load State

In some species and at certain experimental conditions it may not be possible to measure circumferences in the no-load state because the specimens turn inside out. However, available data are sometimes sufficient to estimate the mucosal and serosal strains in the no-load state under the assumption that the cross-sections are linearly elastic. This assumption is most probably not valid, and therefore the following estimates should be regarded as a first approximation until more complete experimental data are available.

Equations 6.7 and 6.8 hold for the unloaded state also, if R is simply replaced by r. If r_m were known, the unloaded state strains could be computed from the above equations for the idealised model. Since it was not possible to measure r_m (or the circumferences) in the unloaded state in the experiments reported here, it is of interest to develop an approximation of the unloaded state strains in the model.

The equilibrium equations in the unloaded state lead to the following equation in terms of circumferential stress σ:

$$\int_{r_m}^{r_s} \sigma dr = 0 \quad . \tag{6.9}$$

The integral in Equation 6.9 is, of course, the circumferential tension in the wall, which is zero by Laplace's law when the transmural pressure is zero. If the material of the duodenal wall is linearly elastic, then $\sigma = E\varepsilon$, where E is Young's modulus and Equation 6.9 leads to

$$\int_{r_m}^{r_s} \mathrm{E} dr = 0 \quad . \tag{6.10}$$

In principle, one can establish the dimensions in the unloaded state by assuming values of r_m, and checking Equation 6.10 until an r_m is found that satisfies Equation 6.10. After r_m has been determined, Equations. 6.6 and 6.8 can be used to find the strain distribution in the unloaded state.

The model results are uniform around the circumference in the circular model. In the actual experimental data, the curvatures are not uniform and hence it is likely that the strains are also not uniform around the circumference. However, the model values may be good approximations of the average over the circumference. In the case of the curvature itself, the model represents the mean curvature exactly since it reproduces the tangent rotation angle measured experimentally.

6.11. Literature

The literature list contains papers mentioned in the text as well as other papers of interest.

Assentoft JE, Gregersen H, O'Brien WD. 2000. Determination of biomechanical properties in the guinea pig esophagus by means of high-frequency ultrasound and impedance planimetry. *Dig Dis Sci*, **45**: 1260–66.

Assentoft JE, Gregersen H, O'Brien WD. 2001. Propagation speed of sound assessment in the layers of the guinea-pig esophagus in vitro by means of acoustic microscopy. *Ultrasonics*, **39**: 263–8.

Chuong CJ, Fung YC. 1986. On residual stresses in arteries. *J Biomech Eng*, **108**: 189–99.

Dou Y, Zhao J, Gregersen H. 2002. Morphology and stress–strain distribution along small intestine in the rat. *J Biomech Eng*, in press.

Fung YC, Liu SQ. 1992. Strain distribution in small blood vessels with zero-stress state taken into consideration. *Am J Physiol*, **262**: H544–52.

Fung YC. 1983. What principle governs the stress distribution in living organs? In: *Biomechanics in China, Japan and USA*, ed. Fung YC, Fukada E, Junjian W, pp. 1–13. Beijing, China: Science.

Fung YC. 1990. *Biomechanics: Motion, Flow, Stress, and Growth*. New York: Springer-Verlag.

Fung YC. 1993. *Biomechanics: Mechanical Properties of Living Tissue*. New York: Springer-Verlag.

Gabella G. 1987. Structure of muscles and nerves in the gastrointestinal tract. In: *Physiology of the*

Gastrointestinal Tract, ed. Johnson LR, Christensen J, Jackson MJ, Jacobson ED, Walsh JH, pp. 335–82. New York: Raven Press.

Gao C, Gregersen H. 2000. Biomechanical and morphological properties in rat large intestine. *J Biomech*, **33**: 1089–97.

Gao C, Zhao J, Gregersen H. 2000. Histomorphometry and strain distribution in pig duodenum with reference to the zero-stress state. *Dig Dis Sci*, **45**: 1500–8.

Gregersen H, Kassab GS, Fung YC. 2000. The zero-stress state of the gastrointestinal tract: Biomechanical and functional implications. *Dig Dis Sci*, **45**: 2271–81.

Gregersen H, Kassab GS, Pallencoea E, Lee C, Chien S, Skalak R, Fung YC. 1997. Morphometry and strain distribution in guinea pig duodenum with reference to the zero-stress state. *Am J Physiol*, **273**: G865–74.

Gregersen H, Kassab GS. 1996. Biomechanics of the gastrointestinal tract. *Neurogastroenterol Motil*, **8**: 277–97.

Gregersen H, Lee C, Chien S, Skalak R, Fung YC. 1999. Strain distribution in the layered wall of the esophagus. *J Biomed Eng*, **121**: 442–8.

Gregersen H, Weis S, McCulloch AD. 2001. Esophageal morphometry and residual strain in a mouse model of osteogenesis imperfecta. *Neurogastroenterol Motil*, **13**: 457–64.

Gregersen H. 2000. Residual strain in the gastrointestinal tract: a new concept. *Neurogastroenterol Motil*, **12**: 411–14.

Grundy, D. 1993. Mechanoreceptors in the gastrointestinal tract. *J Smooth Muscle Res*, **29**: 37–46.

Han HC, Fung YC. 1991. Residual strains in porcine and canine trachea. *J Biomech*, **24**: 307–15.

Han HC, Fung YC. 1996. Direct measurement of transverse residual strains in aorta. *Am J Physiol*, **270**: H750–9.

Han HC, Fung YC. 1991. Species difference of the zero-stress state of aorta: pig vs. rat. *J Biomech Eng*, **113**: 446–51.

Lu X, Gregersen H. 2001. Regional distribution of axial strain and circumferential residual strain in the layered rabbit oesophagus. *J Biomech*, **34**: 225–33.

Omens JH. 1988. Left ventricular strain in the no-load state due to the existence of residual stress. PhD thesis, Department of Bioengineering, University of California, San Diego.

Rachev A. 1997. Theoretical study of the effect of stress-dependent remodeling on arterial geometry under hypertensive conditions. *J Biomechanics*, **30**: 819–27.

Rodriquez EK, Hoger A, McCulloch AD. 1994. Stress-dependent finite growth in soft elastic tissues. *J Biomech*, **27**: 455–67.

Vaishnav RN, Vossoughi J. 1983. Estimation of residual strains in aortic segments. In: *Biomedical Engineering II. Recent Developments.*, ed. Hall CW, pp. 330–3. New York: Pergamon Press.

Vaishnav RN, Vossoughi J. 1987. Residual stress and strain – in aortic segments. *J Biomech*, **20**: 235–9.

Vossoughi J, Weizsacker HE, Vaishnav RM. 1985. Variation of aortic geometry in various animal species. *Biomedizinische Technik*, **30**: 48–54.

7. Mechanical Properties in Normal Gastrointestinal Tissue

Amazingly few data exist on the mechanical properties of gastrointestinal tissue, despite the fact that the motor function, and fluid transport, as encountered in secretion and absorption, are mechanical functions. Numerous studies attempted to describe pressure–volume relations, compliance, etc. Such data will not be covered in this book since they may be biased by sources of uncontrollable errors and do not have much bearing on our understanding of gastrointestinal function. The active muscle properties and data obtained in humans are presented in Chapter 5. Data from animal studies are also given in Chapters 3, 4 and 5. Most published data on the passive biomechanical properties consider the oesophagus and the small intestine from various species. Most attention in this chapter will be given to the zero-stress state and passive circumferential stress–strain relations. Until now no valid data exist on the zero-stress state and stress–strain properties of sphincteric regions and the stomach.

7.1 Preconditioning Behaviour. Strain Softening

In mechanical testing of living tissues in vitro, preconditioning is necessary to obtain repeatable results. This means that after the test specimen is prepared and mounted on the testing machine, the loading and unloading processes are repeated for a number of cycles until the stress–strain relationship becomes reproducible. The number of cycles depends on the tissue and the method of preparation. In gastrointestinal studies, the researchers often do not pay special attention to preconditioning; hence, the repeatability and reproducibility of their experiments are questionable. The basic rationale for preconditioning is that tissues are disturbed in the process of preparation by cutting, changes in stress, strain, temperature, blood flow, chemical environment, ischaemia, hypoxia, contracture of muscle, etc., and need to be restored to a stable condition by preconditioning (Deng et al., 1994). Figure 7.1. shows the preconditioning behaviour during the first two cycles of longitudinal stretch in the submucosal and muscle layers in the guinea pig oesophagus after separation of the layers. In these experiments, the two layers are stretched 100% of the length obtained in vitro at zero pressure, i.e., $\lambda = 2$, during ramp loading and unloading. A large hysteresis is observed during the first cycle. Cycle 2 has a much smaller hysteresis and the loading curve differs considerably from the loading curve of the first cycle. The largest difference is between the two first cycles and it can be seen that hysteresis is more prominent in the muscle tissue than in the submucosal layer. In gastrointestinal

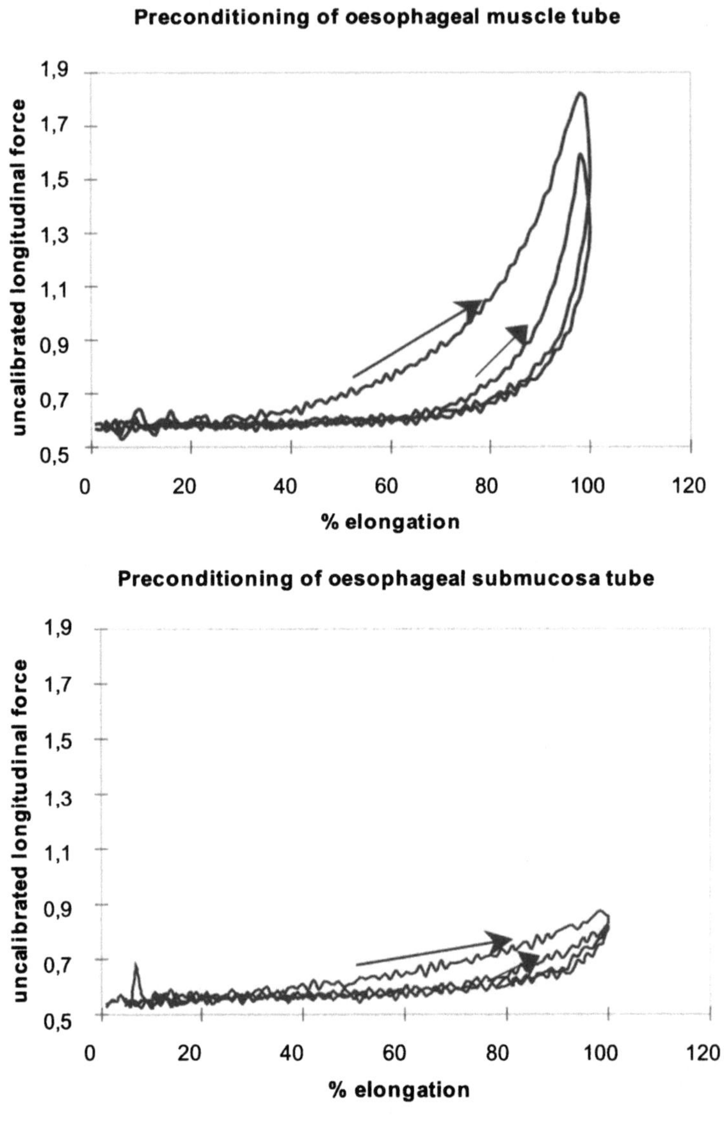

Figure 7.1. Longitudinal stretching of the separated muscle and submucosal tubes of the oesophagus. The protocol utilised a constant elongation rate. The force data are yet uncalibrated. Two cycles of stretching and shortening are shown in each diagram. The loading curve for the first cycle is indicated by the long arrow while the loading for the second cycle is indicated by the short arrow. A large hysteresis is found during the first cycle in the muscle tube. The curves are preconditioned after 3–4 cycles; i.e., the elongation–force relations become stable (not shown).

tissues (oesophagus and small intestine), the curves are not repeatable before 3–10 cycles have been carried out. The preconditioning behaviour cannot be explained by conventional viscoelastic theory. A strain softening theory has been developed to explain this behaviour (see below).

History-dependent Mechanical Behaviour of Small Intestine

In mechanical testing of living tissues in vitro, preconditioning is necessary to obtain repeatable results. After the test specimen is prepared and mounted on the testing machine, the loading cycles are repeated until the mechanical response is repeatable. Preconditioning is invariably associated with a loss of stiffness as the loading is repeated. The number of cycles depends on the tissue and the method of preparation. Preconditioning behaviour in soft tissues has traditionally been recognised as a viscoelastic property, like creep and relaxation (Fung, 1993). However, a different strain-history dependence has been demonstrated in certain rubber materials. Mullins (1947) found that the stress–strain relation in rubber depended on the maximum previous load experienced by the specimen, and this property is now known as strain softening or the Mullins effect. Certain elastomers become permanently softened after the load reaches a new maximum for the first time, which led to the suggestion that strain softening may explain preconditioning behaviour (Johnson and Beatty, 1993). Thus, some of the history dependence in soft tissues that has previously been attributed to viscoelasticity may actually be the result of strain softening. Emery et al. (1997) provided evidence that the preconditioning behaviour of the resting myocardium is due to strain softening and relatively independent of the time-history of the load. It is, therefore, of interest in gastrointestinal tissue to distinguish between passive properties of the tissue that depend on the time-history of load (viscoelastic effects) versus those that depend on the maximum previous load (strain softening effects). The results mentioned below are extracted from Gregersen et al. (1998).

Ten repeated pressure–volume relations were measured at each peak pressure level in isolated, passive guinea pig jejuni in vitro during balloon inflation and deflation cycles. With inflation to a new, higher peak pressure (ranging from 3 to 15 mmHg), the pressure–volume relation became less stiff, particularly in the low pressure range, without a significant change in unloaded jejunal volume. The normalised volume change as a function of the integrated volume–time history and maximum volume was computed.

Analysis of covariance followed by multiple linear regression analysis showed that approximately 90% of the history dependence could be attributed to the maximum volume. Most softening (loss of stiffness) happened in the low pressure range of the curve (0–3 mmHg). In the strain softening analysis, a simple theory proposed by Johnson and Beatty (1993) and recently used in cardiac mechanical studies was adopted. The effect of maximum previous pressure or volume is included by rescaling the volume by a "volume amplification factor" that is a function of maximum previous volume. The volume amplification factor was estimated as the ratio of the maximum volume for a given cycle to the volume at the same pressure for the cycle to highest maximum pressure experienced in protocol A. It was assumed that all curves were approximated by the softened curve and a constant volume scaling. The volume amplification factor varied as a linear function of the maximum previous volume.

The consistent changes in the pressure–volume relation with the history of maximum previous pressure support the hypothesis that gastrointestinal stiffness can decrease substantially as the maximum previous load increases. This altered stiffness was most evident at the low pressure range. The present data, therefore, have physiological relevance since the small intestine is at rest most of the time and the passive distension pressure during bolus transport is low compared to the contraction pressure. These results suggest that the preconditioning behaviour of the jejunum is due largely to strain softening and is only weakly dependent on the

the time-history of loading. Since strain softening effects were clearly greater than viscoelastic effects, the history-dependent changes in jejunal stiffness are more likely to involve alterations to elastic rather than viscous structures in the tissue. These effects must be taken into account when performing balloon distension studies in the gastrointestinal tract for studying physiological and pathophysiological problems in which loading conditions are altered, e.g., mechanoreceptor studies in normal intestine and in acute and chronic obstruction, in order to have an accurate description of the biomechanics.

Since strain softening in rubber materials is associated with "structural changes", one may think of it as damaging to the tissue. However, strain softening may be a beneficial tissue restructuring response as an adaptive mechanism to increased load to avoid injury. This function seems especially plausible in the gastrointestinal tract, which is subjected to dimensional changes. Thus, peristaltic bolus transport stretches the intestinal wall in the vicinity of the bolus at low pressure and strain softening may therefore have physiological relevance during normal bolus transport as suggested by the low pressure range for strain softening in this study. However, the time course of bolus mechanics and altered stiffness have to be studied in further detail. Strain softening may also serve as a protective mechanism to reduce stresses during repeatedly high stretching. The data indicates the importance of strain softening with large increases in volume. High pressures and dilatation are common in certain disease states associated with acute and chronic mechanical obstruction such as strictures due to systemic sclerosis and sclerotherapy-induced damage. The dilatation is likely caused by the repeated stretch during passage of foods due to the increased resistance to flow and the repetitive high-pressure contractions proximal to the site of obstruction. High-pressure waves generated by the smooth muscle layers subject the non-contracting regions to passive overstretching and may reduce its stiffness.

7.2 Zero-stress State and Stress–Strain Properties in the Oesophagus

Oesophageal data obtained in normal mice, rats, rabbits, guinea pigs and pigs have been published. At the no-load state, a tight buckling pattern is seen at the inner wall, indicating that the inner wall of the oesophagus is compressed (see Figures 6.8 and 6.12 in Chapter 6). Upon reducing the no-load state to the zero-stress state by radially cutting the ring, the ring becomes a sector. The additional buckling in the zero-stress state indicates that compression in the inner wall and tension in the outer wall are not completely relieved. The opening angle in the normal oesophagus is in the range of 60–100° with small variation along the oesophagus. The opening angle does not depend on the temperature, location of cut along the circumference, or presence of calcium (Gregersen et al., 1999). The opening angle shows a time-dependent response reaching a steady state value after 30 minutes. A residual strain gradient exists in the wall. The residual strain at the mucosal surface is negative whereas the residual strain at the serosal surface is tensile (Gregersen et al., 1999 and Figure 6.11). The data in the figure imply that the residual stress is a mechanism to provide more uniform stress and strain distributions during loading.

Evidence has been provided that the pre-stressing is not only found in the circumferential direction but also exists in the longitudinal direction. As in most other tissues, the circumferential and longitudinal stress–strain relation follows exponential courses. The oesophagus is stiffer in the longitudinal direction than in the circumferential direction. With respect to this, recall that the oesophagus is pre-stretched in vivo in the longitudinal direction (Gregersen et al., 1999). There seems to be axial variation in the longitudinal stiffness of the oesophagus with the abdominal part being stiffer than the rest (Lu and Gregersen, 2001).

7.2.1 Physiological Remodelling in the Oesophagus During Maturation and Ageing

More then 20% of our population is expected to exceed 85 years of age by the year 2030 (Katz, 1993), hence, the gastroenterologist in the twenty-first century will be increasingly confronted with digestive diseases in the elderly population. Gastrointestinal disease is the second most common indication for hospital admission of elderly patients, a patient group that accounts for four times as many hospitalisations as younger patients do (Almy, 1989). In the oesophagus, gastro-oesophageal reflux disease seems to be especially associated with age, therefore, it is important to determine the changes in morphology and biomechanical properties during maturation and ageing.

Although the origin of residual stress in soft tissues remains uncertain, experimental and theoretical evidence suggest that hypertrophy and geometric remodelling can alter it. In the rat aorta, changes in residual strain have been observed with age (Badreck-Amoudi et al., 1996) and with hypertension induced by abdominal aortic banding (Fung and Liu, 1989). Omens et al. (1998) demonstrated pronounced geometric remodelling with age in the rat heart. In a theoretical analysis, Rodriguez et al. (1994) demonstrated that volumetric tissue growth could alter residual stress. Specifically, it was predicted that concentric hypertrophy, which increases the wall thickness-to-radius ratio, increases the opening angle and residual stress, and, conversely, eccentric hypertrophy should decrease the opening angle. It has been shown in the rat abdominal aorta that changes in the opening angle with age correlated with changes in wall thickness-to-radius ratio (Badreck-Amoudi et al., 1996).

The geometry of the distended state (with internal pressure applied), the no-load state (oesophageal rings with no external forces or internal pressure applied), and the zero-stress state (cut-open configuration with internal stresses released by a radial cut) were studied in rats ranging from 1 to 32 weeks of age. The newborn rats gained weight rapidly during the first 15–20 weeks, after which the weight increase levelled off at approximately 450 grams at weeks 16–32. The oesophageal dimensions increased manyfold with age, e.g., the wall cross-sectional area increased eightfold (Figure 7.2). The opening angle was approximately 140° at one week and two weeks of age. It gradually decreased to approximately 80° after 16 weeks (Figure 7.2). Accompanying the geometric shape changes, there was a corresponding significant decrease in opening angle suggesting that residual strain and stress also decrease with growth. Consistent with the theoretical prediction (Rodriguez et al., 1994) and the experimental studies in the aorta (Badreck-Amoudi et al., 1996), the opening angle correlated with the wall thickness-to-radius ratio. A simple kinematic analysis of uniform transmural growth (Rodriguez et al., 1994) predicts that the opening angle is proportional to ln (h/r) because the opening angle is determined by the relative change in the thickness-to-radius ratio. At a given circumferential stress, the corresponding strain values became larger up to four weeks of age after which no further change was observed; i.e., the oesophagus became more compliant during the first four weeks of life. At a given longitudinal stress, the corresponding strain values

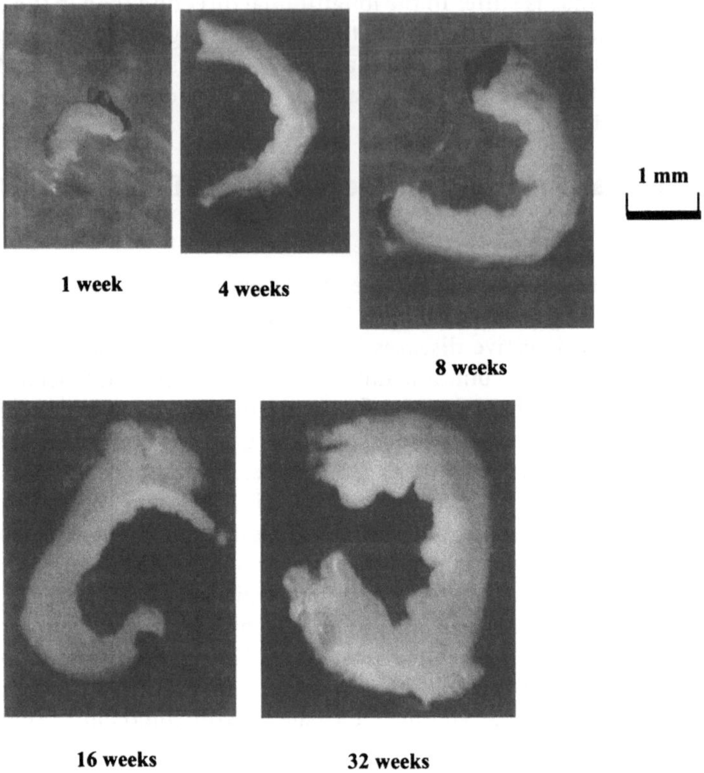

Figure 7.2. Images of oesophageal specimens in zero-stress state for rats aged 1–32 weeks.

became larger up to 16 weeks of age after which no further change was observed; i.e., the oesophagus became more compliant longitudinally during the first 16 weeks of life. The oesophagus was stiffer in the longitudinal direction compared to the circumferential direction.

Mechanical factors have been proposed to regulate growth of the adult heart in response to increases in loading conditions (Omens et al., 1998). An increase in wall stress has been postulated as a key signal in stimulating growth both in the embryonic and the adult heart, although tissue strain has also been proposed to be a stimulus to growth. Thus, increases in diastolic loads during physiologic growth are balanced by the appropriate geometric changes, which may be regulated by a combination of factors. Based on Laplace's law, the data mentioned above suggest that mean circumferential wall stress increases with age in the rat, since wall thickness–radius ratio decreases substantially, hence stress may play a central role as a regulatory factor. Although the present data do not confirm this hypothesis directly, a stress-dependent growth law can likely be derived that would help to uncover the actual stimulus to hypertrophic tissue growth.

7.2.2 Bilayer Models of the Oesophagus

The gastrointestinal tract is not a homogeneous tube. It is a composite tube with different properties in each layer (as can be seen in Figure 7.1). Data from individual layers adds another dimension to the hierarchy of the structure. Fortunately, the oesophagus can, apparently as the only organ in the body, readily be dissected into layers with microsurgical technique without damaging the tissue. Two papers have been published on the mechanical properties in the layered oesophagus (Gregersen et al., 1999; Lu et al., 2001). Both studies clearly show a difference in the opening angle and residual strains between the layers. Furthermore, the residual strain curves are discontinuous at the interface between the submucosal and muscle layers (see Figure 6.11). This is a clear indication of shear forces acting in this interface. Considering the longitudinal direction, when the mucosa–submucosa is attached to the muscle layers, it appears straight. When separated from the muscle layers, the mucosa–submucosa layer coils up. Hence, we must consider the three-dimensional distribution of residual strains in the gastrointestinal tissue.

7.3 Morphology, Residual Strains and Stress–Strain Properties along the Small Intestine

Despite the fact that the same kind of layers composes the whole small intestine, significant differences exist in its morphology. A number of studies have been published on this issue. Gabella was probably the one who paid most attention to it. The data provided, however, are often not very useful in bioengineering terms, simply due to the lack of quantitative data with respect to measures useful in a stress–strain analysis. Furthermore, the zero-stress state of the gastrointestinal tract was not known before 1996. As stated elsewhere in this book, biomechanical analysis of gastrointestinal physiology requires a full set of data describing the geometry (Gregersen and Kassab, 1996). Knowing the zero-stress configuration is essential in any mechanical analysis since it serves as the reference state for analysing stress and strain under physiological or pathophysiological conditions Gregersen et al. (1997) were first to describe the zero-stress configuration and the circumferential strain distribution at a low pressure level in the guinea pig duodenum. The serosal circumference, the duodenal wall thickness, and the ratio of the wall thickness-to-mucosal circumference decrease in the distal direction. This study taught us that very large residual strains exist in the guinea pig small intestine, in fact the largest found in any tissue yet. In the zero-stress state of the guinea pig duodenum, the total rotation of the tangent from one tip to the other is -500° to -850°, with the lowest absolute value being in the proximal duodenum (i.e., the duodenum usually turns itself inside-out). The duodenal residual strains obtained in pigs (Gao et al., 2000) and rats (see below) are not as large as in the guinea pig. In general, the opening angle in these species is less than 360°. New studies now have extended this kind of investigation to cover the whole small intestine and to also determine the distribution of stress–strain relations along the length of the small intestine.

The most detailed study was done by Dou et al. (2002) of which the following is a short summery. The rat small intestine varies along its length with respect to measures such as the intestinal mass per unit length (highest ratio in the duodenum and lowest in the ileum). In the no-load state, the wall thickness was highest (about 20%) and the inner circumferential length and luminal area lowest in the duodenum compared to more distal parts of the small intestine. The ratio of the wall thickness to the mucosal circumferential length (a measure related to circumferential stress) is therefore highest in the duodenum. In the zero-stress state, the duodenal rings are more likely to turn inside out. This results in relatively longer inner mucosal circumferential length in the duodenum compared to the more distal sites. With respect to the layers of the wall, submucosal thickness in the duodenum was much larger than in the jejunum and the ileum (approximately 0.15 mm vs. 0.03 mm and 0.04 mm). Both the circumferential and longitudinal muscle layer thicknesses were thinner in the jejunum than in the duodenum and ileum.

The opening angle (mean value in the duodenum approximately 175°) and the absolute values of the mucosal (mean value in the duodenum -0.22) and serosal (mean value in the duodenum 0.41) residual strains are highest in the duodenum and decreased in the distal direction (Figure 7.3). Hence, the serosal residual strains are tensile and the mucosal residual strains are compressive. The difference between mucosal and serosal residual strain normalised with respect to the wall thickness showed a similar decreasing pattern from duodenum to ileum. From the duodenum data we learn that the neutral axis (where residual strain is 0) is located closest to the mucosa. This is even more pronounced in the distal parts of the small intestine where it is located about 20% away from the mucosal surface. The implications of these data are clear. The zero-stress state must be accounted for in biomechanical studies of the small intestine. Furthermore, as it has been demonstrated in arteries, the residual stress reduces the stress concentration at the inner wall, thereby offering a better protection against injury due to contractile activity and for the duodenum against the flow of luminal contents ejected at intervals from the stomach. By contrast, the less-compressed ileal mucosa fulfils the protection in a region under relatively lower and stable pressure. The morphometric variation along the axis of the small intestine can largely be attributed to the variation in thickness of the mucosal and submucosal layers. The data suggests that a large proportion of the residual strain can be attributed to the mucosal–submucosal layers in compression. It still remains to be investigated whether mucosal growth is determined by the compression induced by residual stresses or whether the residual stress largely is a result of mucosal growth. It is of interest to note that Rodriguez et al. (1994) predicted how volumetric tissue growth alters residual stress. Specifically, it was predicted that concentric hypertrophy (increase in wall thickness-to-radius ratio) increases residual stress and strain. The prediction is in accordance with the data in this study. Furthermore, based on the location of the neutral axis in the wall and the variation in the residual strain normalised with respect to the wall thickness, it seems likely that the axial variation in residual stress also can result from a relative change in material properties between the layers.

The circumferential stress–strain curves for the duodenum, jejunum, and ileum were in order from the left to the right; i.e., the duodenum was the stiffest and the ileum the softest (Figure 7.3). In the longitudinal direction, the curves for the duodenum and jejunum were placed to the left whereas that for the ileum was shifted to the right. This indicated that the ileum was softer longitudinally than the proximal small intestine was. Furthermore, all segments were stiffer in the longitudinal direction than in the circumferential direction and all curves show an exponential behaviour. The exponential behaviour is consistent with mechanical studies on other biological tissues (Dobrin, 1978; Fung, 1993; Han and Fung, 1991). Our data also showed that the duodenal wall is stiffer than the jejunal and the ileal wall in both directions; i.e., the elastic properties differ among the small intestinal segments. This adds to the data obtained in guinea pigs (Storkholm et al., 1995; 1998), though a direct comparison between the studies can not be made due to differences in the analysis (i.e., in the selection of reference state, deformation theory and curve-fitting

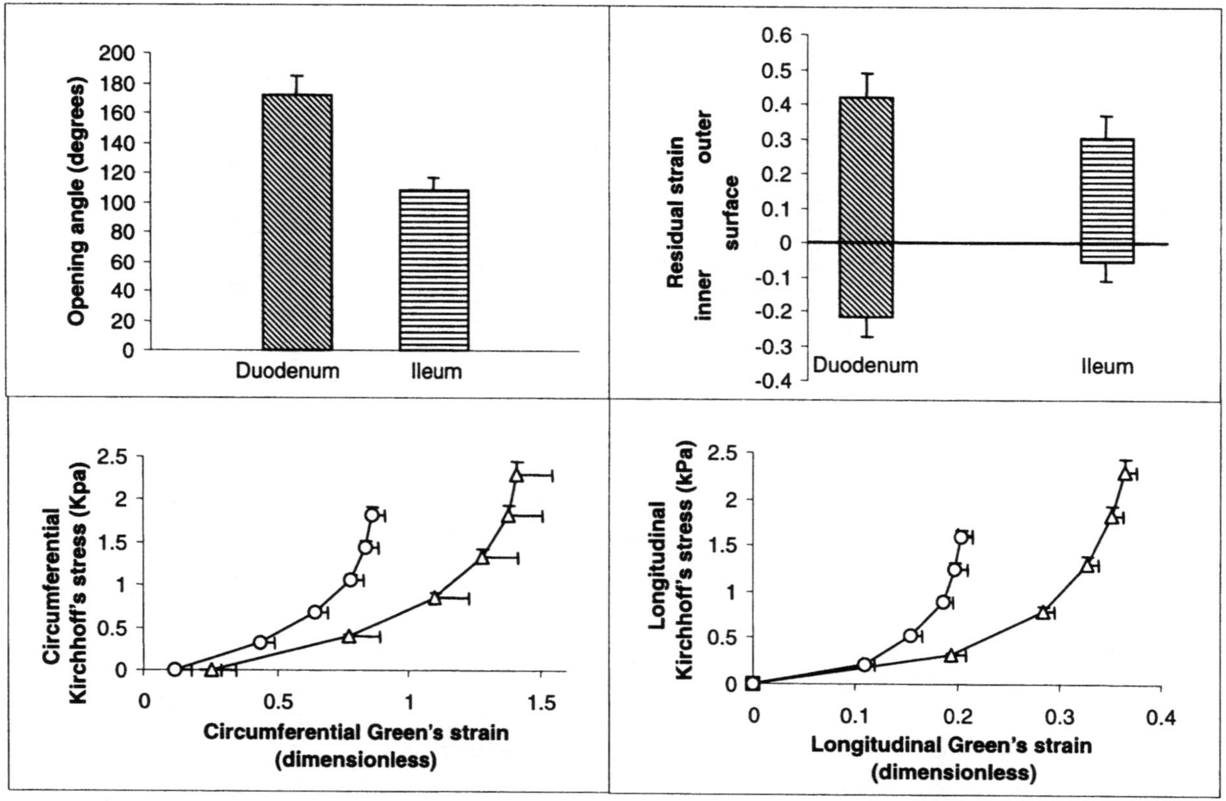

Figure 7.3. Biomechanical parameters in the duodenum and ileum; i.e., opening angles, residual strain and circumferential and longitudinal stress–strain relations. Values are means ± SE. The circumferential stress–strain relations in the duodenum were located to the left of that in the ileum both in the circumferential and longitudinal direction, demonstrating that the duodenal wall was stiffest in both directions. The intestinal wall was stiffer in the longitudinal direction than in the circumferential direction.

methods). The differences in stiffness are likely associated with the specialised functions of the proximal and distal segments of the small intestine; e.g., the duodenum has a large influence on gastric emptying and has been proposed to act as a capacitative resistor whereas the distal ileum acts as a reservoir (Schulze-Delrieu, 1991). The transit time is slower distally (Bueno et al., 1975), which may partly be explained by the difference in the viscosity of the chyme. However, the passive elastic properties may also contribute to differences in intestinal flow patterns: a passing bolus will be slowed to a lesser degree in the duodenum where the wall stiffness is high whereas the compliant ileal wall bulges leading to pooling of luminal contents and decreased flow.

7.3.1 Notes on the Determination of Longitudinal Residual Strain and Stress–Strain Relationship in Biaxial Experiments in Rat Small Intestine

The deformation of the intestine during contraction and relaxation is three-dimensional. To obtain a more thorough understanding of the intestinal transport function, we need to know the stresses and strains in multiple directions. In previous studies it was observed that the intestine bends outwards in the longitudinal direction when a radial cut is made. Therefore, to obtain a more detailed description of the stress-free state of the intestinal wall, longitudinal residual strain measurements are needed. It is clear that the opening angle or tangent rotation angle can not be used for determination of the longitudinal residual stress. The length of the strip cut parallel to the longitudinal axis of the intestine has an impact. The longitudinal residual stress can be characterised by a bending angle (unit: degrees per unit length and positive when bending outwards).

The bending angle (φ) is defined as the angle the tissue bends when a 1–2 mm-wide longitudinal strip was cut. If $\varphi > 360°$, the strip was cut short into two or three shorter segments each having a smaller 'φ' (φ_1, φ_2, φ_3). In that case, $\varphi = \varphi_1 + \varphi_2 + \varphi_3$. The bending angle per unit length is:

$$\varphi(°mm^{-1}) = \frac{\varphi}{(L_{n-m} + L_{n-s})/2} \tag{7.1}$$

where the subscripts $_m$, $_n$, and $_s$ refer to the no-load state, mucosal surface and the serosal surface. Residual strain, on the other hand, can be computed with the conventional method from the change in dimensions between the zero-stress state and the no-load state.

In a longitudinal stress–strain experiment it is often of interest to test the longitudinal properties at different pressure loads (in a biaxial way). This can be done by ligating the distal end of the segment, 2–3 cm in length, and connecting it to a thread that passes a pulley block and connects to hanging weights. The proximal end of the intestinal segment can be connected via a tube to a fluid-filled container level for applying pressures to the intestinal lumen.

The stress and strain in the segment under study can be computed under the assumptions that the intestinal wall was homogenous and the intestinal shape was circular and cylindrical. The calculations are also based on knowing the no-load state dimensions, the outer diameter and length of the specimen at varying pressures, and assuming incompressibility of the intestinal wall.

The longitudinal Kirchhoff's stress at the loaded condition can be computed according to the following equations (Fung, 1993):

Longitudinal Kirchhoff's stress due to longitudinal force (kPa):

$$S_1 = \frac{9.81W}{A_n \lambda_l} \tag{7.2}$$

where W is the weight and A is the wall cross-sectional area at no-load state. Longitudinal Kirchhoff's stress due to distension (kPa):

$$S_2 = \frac{\Delta \mathrm{P} r_{i-p}^2}{h\lambda_l^2 \left(r_{o-p} + r_{i-p}\right)}. \tag{7.3}$$

The total longitudinal Kirchhoff's stress (kPa):

$$S_l = S_1 + S_2. \tag{7.4}$$

The Green strain can be computed according to the formulas given in Chapter 3. The no-load state can be used as reference for the longitudinal strain because the mid-wall length did not change from the no-load state to the zero-stress state in the longitudinal direction.

The average bending angle is, in the whole small intestine, approximately 55°/mm. The longitudinal residual strain is tensile with values between 0.20 and 0.25 in the whole small intestine. It is compressive at the mucosal side with values between −0.25 and −0.20. A difference exists between the duodenum and ileum in longitudinal residual strain at the mucosal side. The mucosal longitudinal residual strain varies axially with larger absolute values observed in the duodenum than in the ileum, which may be related to the thicker mucosal layer and faster passage of food in the duodenum than in the ileum (Christensen, 1987). By comparison, the residual strains in the longitudinal direction measured in this study are smaller than those in the circumferential direction (Dou et al., 2002), especially at the mucosal side. This could be due to the fact that the largest tensile stress and strain are distributed in the circumferential direction when the intestine is distended during the passage of food. Hence, it is reasonable that larger residual strain exists in the circumferential direction. The longitudinal (total) stress–strain curves fit well with a mono-exponential function. The α constant increases as a function of pressure, indicating that longitudinal stiffness increases with the luminal pressure applied.

7.3.2 Physiological Remodelling of Small Intestinal Structure and Biomechanical Properties During Physiological Growth

As for the oesophagus, relatively little is know about the biomechanical remodelling in the small intestine during maturation and ageing. Much more information is available on morphometric alterations (several papers by Gabella). In rats studied during the first half-year of their life, the duodenal mass and dimensions increase manyfold with age; e.g., the wall cross-sectional area increases fourfold. The thickness of the duodenal wall, mucosal thickness, submucosal thickness, villus height and crypt depth increase rapidly in the first eight weeks with no further increase thereafter. The thickness of the duodenal muscle layer increases continuously until 24 weeks. The

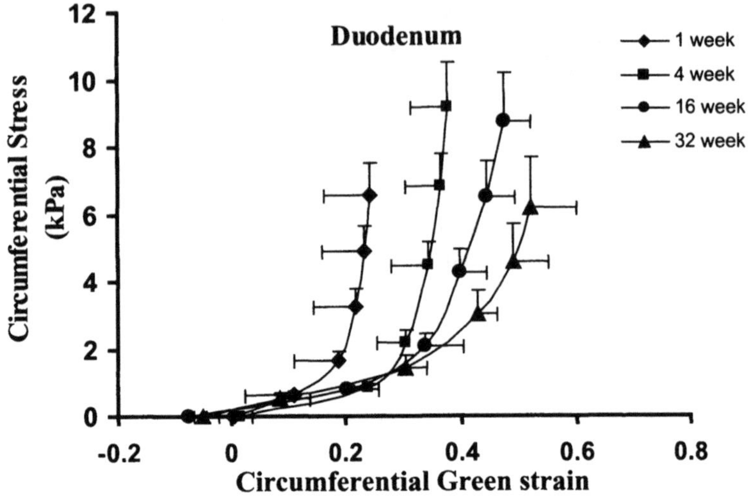

Figure 7.4. Stress–strain curves for rat duodenum from birth to 32 weeks old. Values are means ± SE.

opening angle of the duodenum obtained at zero-stress state is approximately 220° and 290° during the first and second week after birth and decreases to 170° at the other ages. The wall thickness-to-radius ratio decreases quickly during the first four weeks. The residual strains at the outer duodenal surface vary with age whereas the residual strains at the inner surface of the duodenum do not change. At a given circumferential stress, the corresponding strain values increase during the first eight weeks of age after which no further change is noticed; hence, the duodenum becomes more compliant during early life (Figure 7.4).

The zero-stress state is very sensitive to remodelling by disease, growth or degeneration. The change of the opening angle of the zero-stress state is a result of non-uniform tissue remodelling of the organ wall. The opening angle increases when the inner layers have a higher growth rate than the outer layers or when the outer layers atrophy more than the inner layers and vice versa (Rodriguez et al., 1994). In other words, the opening angle of the zero-stress state does not change if homogeneous growth occurs in the tissue. Most studies conducted on the zero-stress state of the gastrointestinal tract found a correlation between the opening angle and the thickness-to-radius ratio (Gregersen et al., 1997; 1999; Gao et al., 2000). In this study, this ratio was large in the first two weeks and decreased to a stable level later in the study span. The changes showed a pattern similar to changes in the opening angle and thus seem consistent with the theoretical prediction. The rather stable opening angle after eight weeks of life implies that the small intestinal physiological growth was homogeneous in the circumferential direction of the intestinal wall.

7.3.3 Time-dependent Viscoelastic Properties along the Rat Small Intestine

Making a radial cut in a gastrointestinal tissue ring will make it open into a sector. When the opening angle has reached a final, static angle, the tissue is considered to have attained a "zero-stress state". This section pertains to the description of the time-dependent course of the opening angle. The gradual increase of the opening

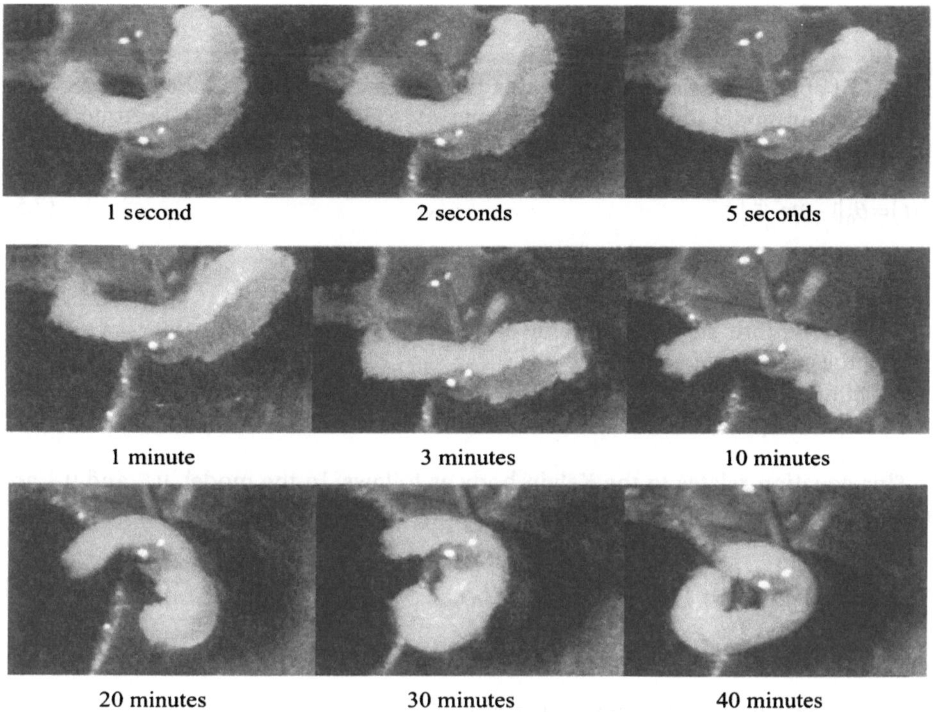

Figure 7.5. Images of intestinal tissue ring after making a radial cut.

angle over time until reaching a static position is defined as creep, and reflects the viscoelastic properties of the organ wall. Figure 7.5 shows an intestinal sample at several time points after making the radial cut. A Kelvin body (also called a Standard Linear Solid model, Figure 7.6) can model this viscoelastic behaviour.

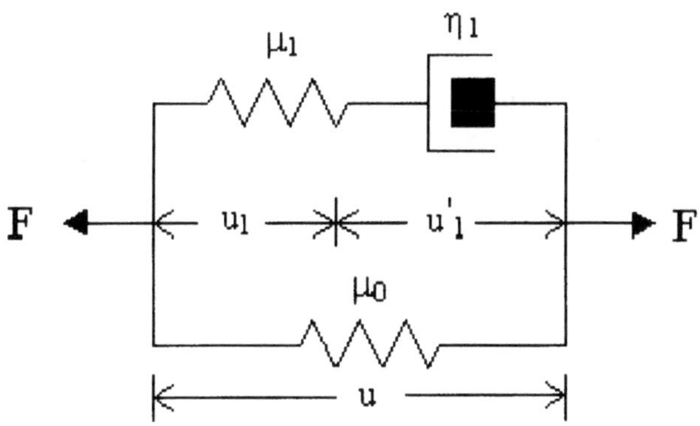

Figure 7.6. The Kelvin body, a mechanical model for a standard linear solid viscoelastic material.

Viscoelastic Analysis

Measured opening angles in the rat small intestine data were used to model the change of opening angle as a function of time, primarily by fitting the observed data to the Kelvin model for a standard linear solid (Figure 7.6) and were fitted to the exponential function:

$$\theta(t)=\theta_0\left[1-\eta e^{-\lambda t}\right] \tag{7.5}$$

where:

θ_0 = asymptotic steady state value of the opening angle,
λ = the creep rate ($s-^1$),
t = time after the initial cut (s),
η = the creep fraction (non-dimensional).

This equation relates to the Kelvin body as follows. In the model, μ_1 and μ_2 are spring constants and η_1 is a coefficient of viscosity. By letting

$$\tau_\varepsilon=\frac{\eta_1}{\mu_1},\ \tau_\sigma=\frac{\eta_1}{\mu_0}\left(1+\frac{\mu_0}{\mu 1}\right),\ E_R=\mu_0, \tag{7.6}$$

where

τ_ε = relaxation time for constant strain,
τ_σ = relaxation time for constant stress,
E_R = relaxed elastic modulus.

The equation describing this model can be obtained:

$$C(t)=\frac{1}{E_R}\left[1-\left(1-\frac{\tau_\varepsilon}{\tau_\sigma}\right)e^{-\frac{t}{\tau_\sigma}}\right], \tag{7.7}$$

letting

$$\alpha=\frac{1}{E_R};\eta=1-\frac{\tau_\varepsilon}{\tau_\sigma};\lambda=\frac{1}{\tau_\sigma} \tag{7.8}$$

The empirical constants η and λ can then be used to plot predicted creep curves and for comparison with other tissues.

For most rings the opening angle will in 45 minutes reach 90% of its maximum value. Subsequently, the opening angle data can be normalised by the angle at 45 minutes before fitting the data to the exponential function. In order to take this normalisation into account, the equation modelling a standard linear solid was modified as follows. Taking Equation 7.5 and letting θ (45 min) = $\alpha\,\theta_0$, we have:

$$\frac{\theta(t)}{\theta(45\text{min})}=\frac{1}{\alpha}\left[1-\eta e^{-\lambda t}\right] \tag{7.9}$$

The empirical constants α, η and λ can then be calculated. The change of opening angle over time for the small intestine segments indicates a slow creep phase whose

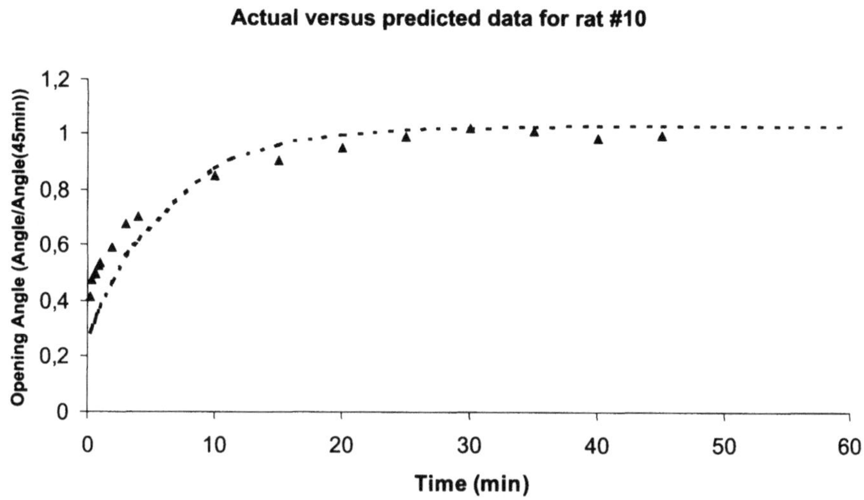

Figure 7.7. Comparison of predicted values using calculated coefficients to averaged opening angle vs. time data.

behaviour has been shown to fit well to the standard linear spring-dashpot model. For the ileum it was found that the constants were approximately 1.0, 0.72 and 1.6×10^{-3} s^{-1}. α and λ correspond to previous data obtained in the aorta and coronary arteries whereas η is somewhat higher in the ileum (Han and Fung, 1995; Frøbert et al., 1998). In Figure 7.7, the points represent experimental data of θ (t) / θ (45 min), and the calculated values of θ (t) / θ_0 using the empirical constants α, η and λ, and demonstrate how the predicted creep curves agree well with data obtained from this study.

7.4 Large Intestine

Numerous physiological and pharmacological studies have been done on taenia coli strips using the uniaxial setup. Only few data exist with respect to determination of the zero-stress state and stress–strain relations using biaxial techniques.

Gao and Gregersen (2000) found in rats that the opening angle varied between 40° and 125°, with the highest values in the ascending colon. As found in other parts of the gastrointestinal tract, the mucosa was in compression and the muscle layers in tension in the no-load state. Furthermore, the stress–strain relations, referenced to the zero-stress state, were exponential by nature and the colonic wall was stiffest in the longitudinal direction when compared to the circumferential direction. For both directions, the stiffest part was the transverse colon.

With respect to rectal properties during maturation and ageing, Gregersen et al. (2002) found in mice that the opening angle varied between 50° and 200° and that 15-week-old mice had 25–50% smaller opening angles than older mice (one year) had. The increase in opening angle with age could be related to findings of a larger rectum size in older subjects; i.e., to a higher stress level.

7.5 Biliary Tract

The biliary tract tissues also express residual stresses and residual strains. Hence, considering the zero-stress state is important.

In vitro ramp distensions in middle-sized pigs have been used for determination of circumferential and longitudinal stress–strain relations (Duch et al., 2002). The stress–strain relationship of the common bile duct is exponential for both the circumferential and longitudinal directions. An exponential stress–strain relationship ad modem Fung (1967) of the form

$$S = \beta[\exp(\alpha\varepsilon) - 1] \tag{7.10}$$

can be used to fit the experimental stress–strain data for the circumferential and longitudinal directions, respectively. Since strain is referred to the zero-stress state, stress is 0 when strain is 0, as satisfied by Equation 7.10. A least square fit is used to determine the values of α and β for the circumferential and longitudinal directions. α was found to be close to 9 in both directions whereas β was a factor 4 higher in the circumferential direction. The tangent modulus, E, is the slope of the stress–strain relationship (in the linear range of stress and strain, the tangent modulus is equivalent to Young's modulus) and can be computed analytically from Equation 7.10 as

$$E = \frac{dS}{d\varepsilon} = \alpha[\sigma + \beta]. \tag{7.11}$$

The tangent modulus varies exponentially with strain and linearly with stress. Examining the modulus as a function of stress may be advantageous, since it has a much narrower domain and shows a simple linear relationship. For a non-linear stress–strain relationship, it is meaningless to specify the modulus unless a strain or stress level is prescribed.

7.6 Concluding Remarks

There are substantial residual stresses and strains in the gastrointestinal tract, an organ that experiences low luminal pressures in resting conditions. In some cases, several radial cuts are needed to allow measurement of the residual strain in a segment. The configuration of the zero-stress state varies along the length of the gastrointestinal tract. The residual strain affects the strain distribution under resting conditions and influences the stress–strain distribution during bolus transport. It could also potentially affect mucosal growth and absorption. Data on the zero-stress state in the duodenum have physiological and clinical relevance since the biomechanical and morphological properties affect flow. These properties change in various diseases of the gastrointestinal tract, e.g., in mechanical obstruction of the small intestine, diabetes, and following partial resection of the small intestine.

It is clear that what is most needed are the constitutive equations of the gastrointestinal tract. Once the constitutive equation is known, the distribution of physical stresses and strains in the duodenum in vivo, and the function of bolus transport and fluid movement in the gastrointestinal tract can be analysed by methods of continuum mechanics. The analytical results will relate the stress and function with the geometric parameters (e.g., the circumferential length, wall thickness, tangent rotation angle, etc.). The task of determining the constitutive equations is not as daunting as it seems, because the mathematical form of most soft tissues is known (Fung, 1990; 1993). If the hypothesis that the constitutive equations of the gastrointestinal tissues are similar to those of other soft tissues can be confirmed, then the only task is to determine the material constants of the gastrointestinal tissues. Mechanical testing instruments are already available for accomplishing this task (see Chapters 4 and 5).

7.7 Literature

The literature list contains papers mentioned in the text as well as other papers of interest.

Almy TP. 1989. The gastroenterologist and the graying of America. *Am J Gastroenterol*, **84**: 464–8.

Badreck-Amoudi A, Patel CK, Kane TPC, Greenwald SE. 1996. The effect of age on residual strain in the rat aorta. *J Biomech Eng*, **118**: 440–4.

Bueno L, Fioramonti J, Ruckebusch Y. 1975. Rate of flow of digesta and electrical activity of the small intestine in dogs and sheep. *J Physiol*, **249**: 69–85.

Christensen J. 1987. Motility of the small intestine. In: In: *Physiology of the Gastrointestinal Tract*, 2nd edition, volume 1, ed. Johnson LR, Christensen J, Jackson MJ, Jacobson ED, Walsh JH, New York: Raven Press.

Christensen J. 1987. The oesophagus. In: *Physiology of the Gastrointestinal Tract*, 2nd edition, volume 1, ed. Johnson LR, Christensen J, Jackson MJ, Jacobson ED, Walsh JH, pp. 595–612. New York: Raven Press.

Deng SX, Tomioka J, Debes JC, Fung YC. 1994. New experiments on shear modulus of elasticity of arteries. *Am J Physiol*, **266**: H1–10.

Dobrin PB. 1978. Mechanical properties of arteries. *Physiol Rev*, **58**: 397–460.

Dou Y, Zhao J, Gregersen H. 2002. Morphology and stress-strain properties along the small intestine in the rat. *J Biomech Eng*, accepted.

Duch BU, Andersen H, Smith J, Kassab GS, Gregersen H. 2002. Structural and mechanical remodeling of the common bile duct after obstruction. *Neurogastroenterol Motility* **14**: 111–122.

Emery J L, Omens JH, McCulloch AD. 1997. Strain softening in rat left ventricular myocardium. *J Biomech Eng*, **119**: 6–12.

Frøbert O, Gregersen H, Bjerre J, Bagger JP, Kassab GS. 1998. Relation between zero-stress state and branching order of porcine left coronary arterial tree. *Am J Physiol*, **44**: H2283–90.

Fung YC, Liu SQ. 1989. Changes of residual strains in arteries due to hypertrophy caused by aortic constriction. *Circulation Research*, **65**: 1340–49.

Fung YC. 1967. Elasticity of soft tissues in simple elongation. *Am J Physiol*, **28**: 1532–44.

Fung YC. 1990. *Biomechanics. Motion, Flow, Stress, and Growth*. New York: Springer-Verlag.

Fung YC. 1993. *Biomechanics. Mechanical Properties of Living Tissues*. New York: Springer-Verlag.

Gabella G, Yamey A. 1977. Synthesis of collagen by smooth muscle in the hypertrophic intestine. *Quarterly J Exp Physiol*, **62**: 257–64.

Gabella G. 1981. On the musculature of the gastrointestinal tract of the guinea pig. *Anat Embryol*, **163**: 135–56.

Gabella G. 1987. Structure of muscles and nerves in the gastrointestinal tract. In: *Physiology of the Gastrointestinal Tract*, 2nd edition, ed. Johnson LR, Christensen J, Jackson MJ, Jacobson ED, Walsh JH. pp. 335–82.New York: Raven Press.

Gabella G. 1994. Anatomy. In: *Physiology of the Gastrointestinal Tract*, ed. Johnson LR, Christensen J, Jackson MJ, Jacobson ED, Walsh JH. New York: Raven Press.

Gao C, Gregersen H. 2000. Biomechanical and morphological properties in rat large intestine. *J Biomech*, **33**: 1089–97.

Gao C, Zhao J, Gregersen H. 1999. Histomorphometry and strain distribution in pig duodenum with reference to the zero-stress state. *Dig Dis Sci*, **45**: 1500–8.

Gregersen H, Emery J, McCulloch AD. 1998. History-dependent mechanical behavior of the guinea pig small intestine. *Annals Biomed Eng*, **26**: 1–9.

Gregersen H, Giversen IM, Rasmussen LM, Tøttrup A. 1992. Biomechanical wall properties and collagen content in the partially obstructed opossum esophagus. *Gastroenterology*, **103**: 1547–51.

Gregersen H, Kassab G, Pallencaoe E, Lee C, Chien S, Skalak R, Fung YC. 1997. Morphometry and strain distribution in guinea pig duodenum with reference to the zero-stress state. *Am J Physiol*, **273**: G865–74.

Gregersen H, Kassab GS, Fung YC. 2000. The zero-stress state of the gastrointestinal tract: biomechanical and functional implications. *Dig Dis Sci*, **45**: 2271–81.

Gregersen H, Kassab GS. 1996. Biomechanics of the gastrointestinal tract. *Neurogastroenterol Motility*, **8**: 277–97.

Gregersen H, Lee TC, Chien S, Skalak R, Fung YC. 1999. Strain distribution in the layered wall of the esophagus. *J Biomech Eng*, **121**: 442–8.

Gregersen H, Lundby L, Overgaard J. 2002. Early and late effects of irradiation on the morphometry and residual strain in the mouse rectum. *Dig Dis Sci*, **47**: 1472–9.

Han HC, Fung YC. 1991. Residual strains in porcine and canine trachea. *J Biomech*, **24**: 307–15.

Han HC, Fung YC. 1991. Species dependence of the zero-stress state of aorta: pig versus rat. *J Biomech.Eng*, **113**: 446–51.

Han HC, Fung YC. 1995. Longitudinal strain of canine and porcine aortas. *J Biomech*, **28**: 637–41.

Han HC. Fung YC. 1996. Direct measurement of transverse residual strains in aorta. *American Journal of Physiology*, **270**: H750–59.

Johnson M, Beatty M. 1993. The Mullins effect in uniaxial extension and its influence on the transverse vibration of a rubber string. *Continuum Mech Thermodyn*, **5**: 83–115.

Katz S. 1993. Gastrointestinal disease of the elderly: Introduction to the series. *Practical Gastroenterology*, **17**: 9.

Lu X, Gregersen H. 2001. Regional distribution of axial strain and circumferential residual strain in the layered rabbit esophagus. *J Biomech*, **34**: 225–33.

Mullins L. 1947. Effect of stretching on the properties of rubber. *J Rubber Res*, **16**: 275–89.

Omens JH, Fung YC. 1990. Residual strain in rat left ventricle. *Circulation Research*, **66**: 37–45.

Omens JH, Vaplon SM, Fazeli B, McCulloch AD. 1998. Left ventricular geometric remodeling and residual stress in the rat heart. *J Biomech Eng*, **120**: 715–19.

Rodriguez EK, Hoger A, McCulloch AD. 1994. Stress-dependent finite growth in soft elastic tissues. *J Biomech*, **27**: 455–67.

Rodriguez EK, Omens JH, Waldman LK, McCulloch AD. 1993. Effect of residual stress on transmural sarcomere length distributions in rat left ventricle. *American Journal of Physiology*, **264**: H1048–56.

Schulze-Delrieu K. 1991. Intrinsic differences in the filling response of the guinea pig duodenum and ileum. *J Lab Clin Med*, **117**: 44–50.

Storkholm JH, Jørgensen CS, Dall FH, Jensen SL, Gregersen H. 1994. Differences exist in passive elastic wall properties between segments of isolated guinea pig distal ileum and duodenum in vitro. *Neurogastroenterol Mot*, **6**: 21–7.

Storkholm JH, Villadsen GE, Jensen SL, Gregersen H. 1995. Passive elastic wall properties in the isolated guinea pig small intestine. *Dig Dis Sci*, **40**: 976–82.

Storkholm JH, Villadsen GE, Jensen SL, Gregersen H. 1998. Mechanical properties and collagen content differ between isolated guinea pig duodenum, jejunum, and distal ileum. *Dig Dis Sci*, **43**: 2034–41.

Storkholm JH. 1996. Aspects of biomechanics, morphology, and motility in the normal and the obstructed Guinea-pig small intestine. PhD thesis, Denmark, University of Aarhus.

Tøttrup A, Forman A, Uldbjerg N, Funch-Jensen P, Andersson KE. 1990. Mechanical properties of isolated human esophageal smooth muscle. *Am J Physiol*, **258**: G338–43.

Zabinski MP, Biancani P. 1977. Longitudinal force and stress of rat's esophagus: age-related changes. *Am J Physiol*, **1**: E580–3.

8. Growth and Remodelling in the Gastrointestinal Tract

8.1 Mechanical Forces, Growth, Remodelling, and Morphology

Mechanical forces are a major determinant of tissue behaviour and are likely the single most important factor influencing growth and remodeling of tissue. An understanding of this relationship is important in deciphering how the gastrointestinal tract responds and adapts to changes in the physical environment. The passive elastic properties of the intact organ must be studied since these properties reflect the state of the structural components determining a specific function or malfunction, e.g., the wall stress is modulated by changes in the wall structure due to obstructive diseases (Gregersen et al., 1992; Schulze-Delrieu et al., 1988; Ravinder et al., 1990). Therefore, it is desirable to study the biomechanical and morphological properties of the tissue simultaneously. The literature is relatively sparse in this regard. Gabella (1987; 1984) considered the structural apparatus for force transmission in smooth muscles, and Orberg et al. (1983) described the biaxial orientation of the collagen fibrils and the pattern of uncrimping in response to stress. Gregersen et al. (1992) found, in a partially obstructed model of the opossum oesophagus, a linear relation between the amount of collagen and the wall tension. It is likely that collagen synthesis was stimulated because of dilation and increased wall tension proximal to the site of obstruction in the oesophagus. Collagen may also be a major determinant of the biomechanical properties of the small intestine since large amounts are found in its wall. An increase in collagen content and wall thickening and a correlation between the elastic modulus and the collagen content proximal to the site of partial obstruction exists in the small intestine at high stress levels (see later in this chapter). Furthermore, the growth effect of epidermal growth factor on the biomechanical properties of the rat small intestine in vitro has been studied (Vinter-Jensen et al., 1996). However, despite these and several other studies, further studies are needed to clarify the relation between the biomechanical properties of the gut and its morphology in response to altered conditions and disease. Such studies should also utilise molecular techniques. The advent of molecular techniques provide an opportunity to study the effect of mechanical stimuli on the gene expression of various structural proteins such as collagens, fibronectin and stress proteins on the transcriptional level. This approach is now widely used in cardiovascular research and will undoubtedly provide us with important new insights into the mechanisms of structural remodeling of tissue in health and disease.

8.1.1 Clinical Perspectives

The clinical aspects of gastrointestinal biomechanics are mainly related to bolus transport and remodeling. Since distensibility is important for the physiologic function of the gut, it also plays an important role in diseases that affect wall structure and motility. There are several diseases that cause tissue remodeling that are closely related to changes in the mechanical properties. In obstructive diseases, proximal dilatation and wall thickening is a common phenomenon (Gregersen et al., 1992; Orloff Juhl et al., 1994; Tung et al., 1991). Hirschsprung's disease, ulcerative colitis, diabetes, systemic sclerosis and sclerotherapy-induced damage are some examples of diseases and treatments associated with remodelled morphology and biomechanical properties (Arhan et al., 1978; Villadsen et al., 1995; Denis et al., 1979; Nowak et al., 1990; Jensen et al., 1987). Mechanical tests able to detect the remodelling are also relevant in studying dysmotility, non-cardiac chest pain, and perception (Paterson et al., 1991; De Caestecker et al., 1992; Williams et al., 1993; Paterson, 1991). Mechanical tests, therefore, have potentially diagnostic and prognostic value. High-frequency ultrasound systems for imaging the wall and its layers may also be useful in clinical studies (Kimmey et al., 1989; Martin et al., 1989).

Mechanical data, in terms of stress–strain relations, have been published on the oesophagus, common bile duct, the small intestine and the large intestine. The stomach and sphincteric regions have not yet been studied. The sections below summarises data published recently on opening angles, residual strains and stress–strain relations in the process of remodelling.

8.2 Physiological Remodelling

Chapter 7 deals with the physiological remodelling observed in maturation and ageing. This section deals with the remodelling occurring during fasting and refeeding. Fasting is a common phenomenon in nature. It is part of the biological cycle of some animal species and appears in momentary food shortage caused by disasters. Clinically, it can result from anorexia, self-imposed food restriction, and serious conditions in which food ingestion is hampered or patients mainly receive parenteral nutrition.

8.2.1 Small Intestinal Remodelling During Fasting and Refeeding

During fasting the organism is subject to a negative nitrogen balance, metabolic changes, and weight loss (Wilson and Lamberts, 1979), therefore, it seems obvious that the gastrointestinal tract will also undergo structural and functional adaptation in response to fasting. The small intestine plays an important role in the food assimilation process. The structural and functional integrity of the intestine is largely dependent upon the presence of nutrients in the lumen (Bragg et al., 1991). Intestinal motility during fasting has only been studied rarely (Aldewachi et al., 1975). A number of studies have focused on morphological and metabolic changes when food is withheld, e.g., the intestinal and mucosal mass decreased more than the body weight; cell kinetics, metabolism and the amount of brush-border enzymes were reduced; and nutrient transport and absorption were altered (Kotler et al., 1980; Yamauchi et al., 1996). Fasting causes intestinal atrophy and functional adaptive changes such as reduction in villus height, crypt depth, epithelial turnover and migration, and in brush-border enzyme activities with concomitant nutrient absorptive changes (Clarke, 1975; Goodlad et al., 1988; Altmann, 1972; Yamauchi et al., 1996). Refeeding reverses the fasting-induced changes. The biomechanical remodelling process must be studied to learn more about the pronounced morphological changes during fasting and refeeding. Dou et al. (2002) recently studied the remodelling of the stress-free state and the stress–strain relation of the rat intestine during fasting and refeeding.

Rats were fasted for up to one week. Other groups of rats were fasted for one week and were allowed to survive up to one week after refeeding was initiated. The fasted animals lost weights in a linear fashion during the one-week fasting period. After one week of fasting the weight was only 75% of control animals, and the weight loss was 22% in relation to the pre-fasting weight. The re-fed animals regained the lost weight after seven days of refeeding. The duodenum seems to experience the greatest loss during fasting, i.e., the duodenal weight-to-length ratio gradually decreased by 38% when compared to fed control animals. The lost masses were regained after refeeding for two days in the duodenum and jejunum, and four days in the ileum. Refeeding not only rapidly normalised the morphometric measures but some of the measures increased to values larger than in the fed controls. The "over-recovery" observed as well as the overshoot in intestinal enzyme specific activity and augmented absorption capacities are likely associated with the rapid weight gain after the fast ended as observed in rats (Bjorntorp and Mu, 1982), and clinically in an obese patient undergoing weight-reduction therapy. It is worthwhile noticing that the normalisation by refeeding is time- and location-dependent. Duodenum recovers faster from the fasting–refeeding-induced changes than the jejunum and ileum. This is likely associated with the specialised functions of the various parts of the small intestine.

In the no-load state, it was found that all morphometric parameters (such as wall thickness and circumferences) invariably differed among the segments of the small intestine and significantly decreased during fasting. Histology confirmed this; i.e., the wall thickness and mucosal layer thickness decreased considerably during fasting (Figure 8.1). The insignificant changes in the submucosal and muscle layers may be attributed to the slow turnover rate of collagen and the short duration of this study. Most measures such as the outer circumferential length, wall thickness and wall area remodelled quickly back to normal; i.e., in the duodenum these values were restored after two days of refeeding while in the other segments the wall thickness was normalised at four days (jejunum) or seven days (ileum). The outer circumferential length and the wall area increased after refeeding and remained higher to the end of seven days of observation.

The opening angle increased in all segments after only two days of fasting and reached the highest level after fasting for four days (Figure 8.1). Then it remained constant in the duodenum and ileum, but decreased in the jejunum at the end of seven days of fasting. The opening angle was normalised after two days of refeeding in the duodenum and jejunum and seven days of refeeding in the ileum. The residual strain at the outer surface increased after fasting in the jejunum and ileum, but not in the duodenum. The values were normalised after two days of refeeding. The residual strain difference in the jejunum and ileum increased after fasting and maintained higher values to the end of seven days of refeeding. It seems logical that the residual strain and opening angle increase when the wall becomes softer because residual stress protects the wall against damage induced by forces. Since the mucosa reduces its thickness during fasting, the compressive forces at the inner wall will be unchanged or will vanish. Residual strain at the inner wall did not change. Despite the

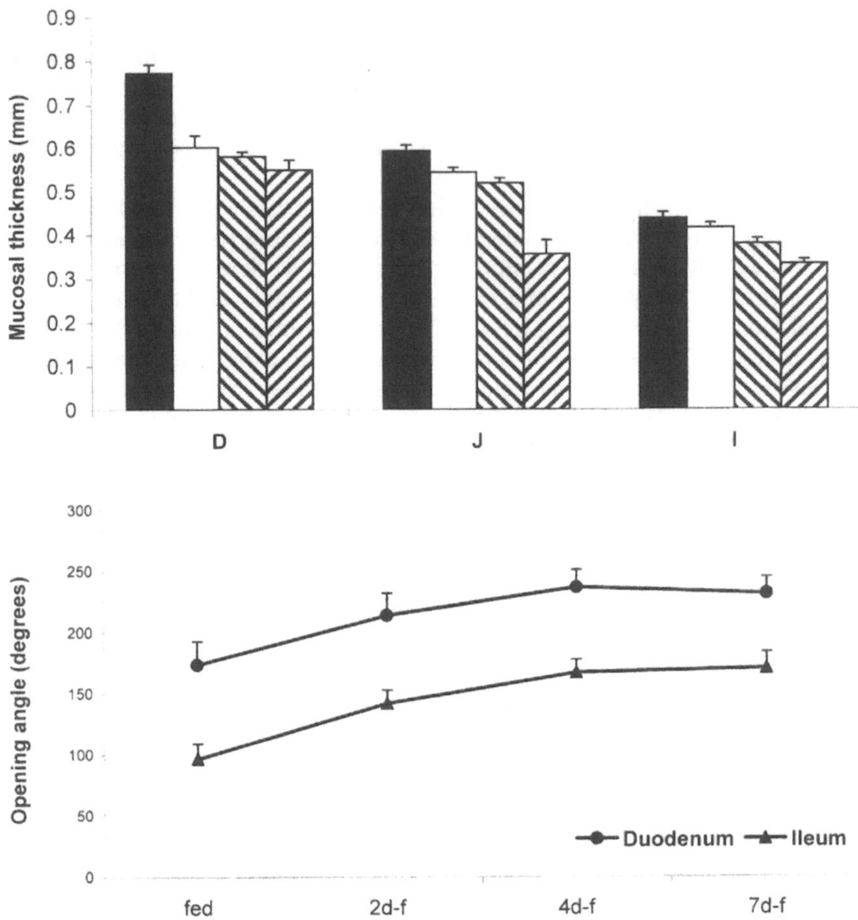

Figure 8.1. Mucosal thickness (top) and opening angle (bottom) data during fasting. The mucosal thickness decreased significantly after fasting in the whole small intestine. The mucosal remodelling was fastest in the duodenum but relatively more pronounced in the jejunum. The opening angle increased after fasting in the three segments (jejunal data not shown). One would expect the opening angle to decrease when the inner layer becomes smaller. This is not the case in this data material, indicating the complex mechanical behaviour of the small intestine. Values are means ± SE (Modified from Dou et al., 2002).

reduction in mucosal thickness, the opening angle and residual strains increased during fasting, hence, structural remodeling likely occurred in the submucosal and muscle layers as well, resulting in a larger tensile strain in the outer layers and a larger opening angle. With mechanoreceptor kinematics in mind it is of interest to note that the neutral axis (the radial location in the wall where residual strain is zero) moves outwards and that the residual strain difference per unit wall thickness increases during fasting. Hence, during distension and contraction the strain field changes in the vicinity of the deformation-sensitive mechanoreceptors located in the submucosa and muscle layers, resulting in altered motility. This could be a contributing factor to the motility changes reported previously.

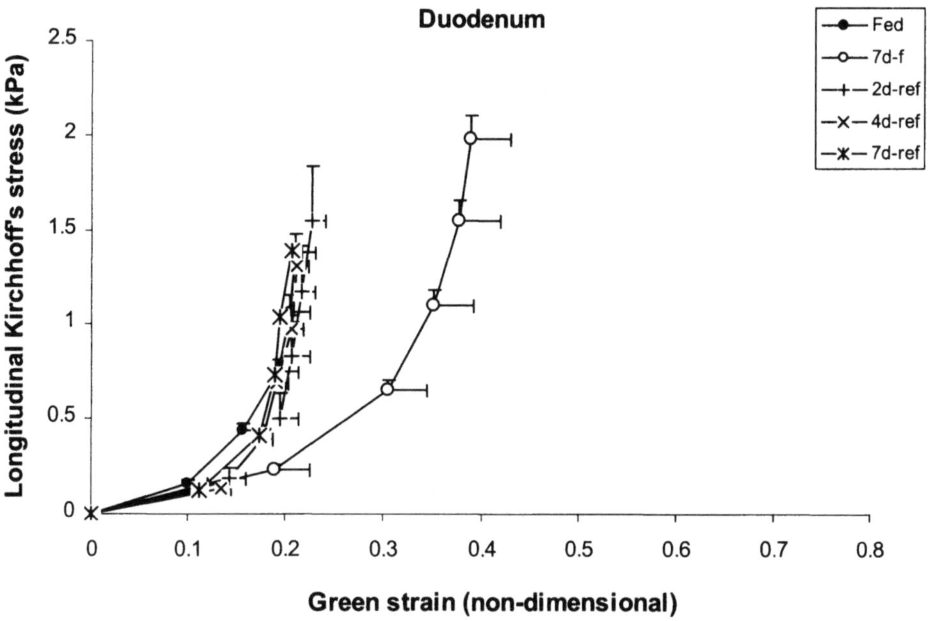

Figure 8.2. Longitudinal stress–strain data obtained during one week of fasting followed by up to one week of refeeding. The data are obtained in the duodenum. Fasting shifted the longitudinal stress–strain curves to the right, indicating that fasting decreases the wall stiffness. Refeeding reversed the fasting-induced remodelling within two days. Values are means ± SE (modified from Dou et al., 2002).

Fasting shifted the stress–strain curves to the right in all segments (the data for the duodenum are shown in Figure 8.2). This indicated that the intestinal wall became less stiff during fasting in both circumferential and longitudinal directions. However, the change was most pronounced in the longitudinal direction and in the jejunum and ileum. This remodeling seems logical since luminal loading during fasting is greatly reduced due to the lack of ingested food; i.e., the imposed stresses are smaller. The remodeling speed is amazingly fast considering that morphological changes likely precede the passive mechanical change. The change in stress–strain properties during fasting can only partly be explained by the thinning of the wall; i.e., the main thinning occurred in the mucosa that does not by itself support tensile stress due to villous structure. Refeeding rapidly reversed the change (Figure 8.2). The data do not reveal whether the mechanism for the changes is due to intestinal immobilisation and remobilization during fasting and refeeding or due to the lack of nutrients. The former mechanism would correspond to the skeletal muscle atrophy observed under weightless conditions and to the hypertrophy of the heart caused by a rise in blood pressure. The remodeling speed after refeeding is amazingly fast for the morphological and the mechanical parameters. Some humoral growth factors may contribute to the recovery, such as IGF-I, GH and IGFBP-1 since studies have demonstrated that they normalise as early as three hours after refeeding.

It seems reasonable that the residual strain and opening angle increase when the wall becomes thin and soft after fasting because the increased residual stress better protects the vulnerable wall against damage induced by forces. In a theoretical analysis, Rodriguez et al. (1994) predicted that concentric hypertrophy, which increases the wall thickness-to-radius ratio, increases the opening angle and residual stress, and, conversely, eccentric hypertrophy decreases the opening angle. Rodriguez' prediction, as mentioned before, does not agree with the experimental data since the opening angle was not associated with the wall thickness-to-inner circumferential length ratio. The reason could be related to the high complexity of the layers in the intestinal wall. The mucosal layer occupies a large proportion of the small intestinal wall. Its structure is unique with villi and crypts. The villi, only attached at the base, form the inner border of mucosa. Consequently, the mucosal layer does not support tensile stress, which may account for the discrepancy between our data and the prediction from the theoretical analysis. Since the mucosa reduces its thickness during fasting, the compressive forces at the inner wall will be unchanged or vanish. In this study, residual strain at the inner wall did not change during fasting. Despite the reduction in mucosal thickness, the opening angle and serosal residual strain increased during fasting. Hence, structural remodeling affecting the biomechanical properties during fasting occurred in the submucosal and muscle layers as well, resulting in a larger tensile strain in the outer layers and a larger opening angle. The normalisation process initiated by refeeding was the same but in a reversal direction.

8.3 Remodelling due to Interference with the Structural Components of the Wall

The effect of tissue remodeling on the zero-stress state has been studied in some detail. Physiological remodelling with age was studied in rats. A pronounced change during the first 32 weeks after birth was found with the opening angle decreasing over time (Lu and Gregersen, yet unpublished results). In order to understand the structure–mechanics relationship of gastrointestinal tissue, collagen-deficit mutant osteogenesis imperfecta mice have been studied (Gregersen et al., 2001). Despite the scarcity of data, it is generally believed that the collagen matrix in the submucosa layer is an important structural determinant of mechanical properties of resting gastrointestinal tissue due to the high collagen content in the submucosa (Orberg et al., 1982). The role of the collagen matrix in oesophageal mechanics is important because oesophageal disease can lead to alterations in both collagen structure and oesophageal mechanics.

8.3.1 Oesophageal Residual Strain Properties in Mutant Osteogenesis Imperfecta Mice

The collagen matrix in the submucosa layer is likely the structural determinant of mechanical properties in resting gastrointestinal tissue due to the high collagen content in the submucosa (Orberg et al., 1982). However, collagen fibrils are also abundant in the muscle layers where they form intramuscular septa (Gabella, 1987). While active contraction of muscles is the most important feature of the peristaltic movement of the intestine, the resting properties of the tissues are also significant. Hill's three-element model of muscle is still widely used (see Chapter 5). A model (such as the osteogenesis imperfecta murine (OIM) model mentioned in this section (Gregersen et al., 2002) is important in this regard because it remodels the passive element; i.e., the collagen. Collagen appears to mediate residual stresses whether directly or indirectly, and affects stresses under physiological loads. The role of the collagen matrix in oesophageal mechanics is important because oesophageal disease can lead to alterations in both collagen structure and oesophageal mechanics. The combined results from mechanical and structural studies using these mutant mice can be used to formulate a more precise model of the role of type I collagen fibres in oesophageal mechanics.

Human osteogenesis imperfecta is a rare heritable disease resulting from one of many known nucleotide substitutions or deletions on the genes encoding the pro-α1(I) and pro-α2(I) chains of type I collagen. Several mouse models of osteogenesis imperfecta have been used to investigate the effects of collagen deficiency on structure and mechanics in bone and tendon. The mutation in OIM mice is on the gene encoding the pro-α2 chain of type I collagen. This prevents association with the pro-α1(I) chains (Chipman et al., 1993) and causes the mutant procollagen chains to be structurally unstable and subsequently rapidly degraded. The homozygous OIM mouse produces only pro-α1(I) chains, and these form α1(I)-α1(I)-α1(I) homotrimers in some patients with osteogenesis imperfecta and in the OIM model (McBride et al., 1997). The OIM mouse exhibits many symptoms similar to human osteogenesis imperfecta, namely defective skeletal development, smaller stature, and skeletal fragility. In a recent study of tissue structure–function relationships in the OIM model, tail tendon had ~60% less collagen and ~50% lower tensile stiffness than control mice (Misof et al., 1997).

In Gregersen et al. (2002), the OIM model was used to examine the effects of type I collagen deficiency on the stress-free state of the oesophagus. It was hypothesised that the alteration in collagen in OIM mice alters the oesophageal mechanics. It was found that the mucosa circumferential length at no-load state was the same in OIM mice and controls compared with all the other dimensions, which were larger in the controls. The histological data showed that all four layers were thinner in OIM mice compared to controls. The layer thickness ratio between OIM and controls was lower for the submucosa (0.64) than for the other layers. The muscle thickness-to-wall thickness ratio was significantly higher and the wall thickness-to-inner hydraulic radius ratio lower in OIM mice than in controls. The average opening angle of the OIM oesophagus was approximately 120° compared to approximately 90° in the control mice. Hence, a decreased amount of collagen in the oesophagus of OIM mice may be associated with increased residual stress. In the OIM mice, the residual strains at the serosal and mucosal surfaces and at the interface between mucosa and submucosa and muscle layers were higher than in the controls; i.e., the neutral axis is displaced in the mucosal direction in OIM mice. The residual strain gradient per unit thickness was significantly higher in the mucosa–submucosa layer than in the muscle layer in both OIM and controls. Furthermore, the gradient in the submucosa was larger in OIM mice than in controls.

The data summarised above show that the most significant morphological and biomechanical alteration was found in the submucosa of OIM mice. As a general rule, the oesophageal wall will remodel itself in such a way as to maintain optimum circumferential stresses. The higher residual strain in the submucosa of OIM mice suggests that submucosal stresses would be higher relative to stresses in the muscle layers during luminal loading if residual strains did not exist. Thus, the higher opening angle is needed to maintain a uniform stress distribution. The larger residual strain in OIM may serve to effect strain gradients associated with higher strains during loading, if the OIM tissue is softer.

Similarity of the inner circumferences in the no load states was found (compared with all the other dimensions, which were larger in the normals suggesting anisotropic remodelling). It appears that even though the opening angles were larger in the OIM mice, the residual strain at the lumen was much less. This suggests that there may have been redistribution of residual stress between the layers of the wall. This could result from a relative change in material properties between the layers. However, for the tensile residual stresses in the muscle layers to be in equilibrium with the compressive residual stresses in the submucosal layer (as required since external forces are zero), this would suggest that submucosa stiffness relative to muscle stiffness must be increased in OIM. This seems unlikely given that collagen is a more predominant component of the submucosal tissue. An alternative explanation could be obtained if one accepts that after the cut, some residual stress remains in one or both layers (as has been shown for the guinea pig oesophagus (Gregersen et al., 1999)). Fung and others have suggested that muscle growth may be the main determinant of alterations in residual stress and strain. Although muscle and submucosal thicknesses were significantly lower in OIM than wild-type mice, muscle-to-wall thickness ratio was slightly higher. A relative thickening of the muscle layer would produce a greater residual tensile stress on the outside of the esophagus (Rodriguez et al., 1994), thus increasing the opening angle and shifting all residual strain upwards. Hence, the shift in the neutral axis may reflect the effects of muscle remodeling.

8.3.2 Collagenase- and Elastase-induced Remodelling in the Rat Oesophagus

Another way to interfere with the structural components in the tissue is to digest them enzymatically (Oxlund and Andreassen, 1980; Zeller and Skalak, 1998). The tissue can be treated with collagenase and elastase to digest collagen and elastin. Soybean trypsin inhibitor needs to be added to the solution with elastase. How will this affect the mechanical properties? In the gastrointestinal tract this has not yet been studied in detail. Preliminary data on the layers of the oesophagus do not show much effect on the opening angle in the intact oesophagus in rats of these two enzymes (added separately to the dishes the tissue rings were immersed in for 60 minutes). The passive components in the inner and outer layers, therefore, seem to counterbalance each other. The largest effect seems to occur in the mucosa–submucosa layer (after separation from the muscle) after treatment with elastase; i.e., the opening angle decreases to half the value of untreated specimens (Figure 8.3). Why elastase-induced changes are more prominent than those by collagenase on the mucosa–submucosal layer, known for its high content of collagen, is not yet clear.

8.4 Remodelling Caused by Experimental Diseases

Gastrointestinal disorders are common, both functional diseases and diseases where the cause can be determined. Some of these disorders can be imitated in animal experimental models, a valuable tool for learning more about the causes and mechanisms. One of the common diseases that affects the gastrointestinal tract is diabetes mellitus. As many as 75% of patients visiting diabetes clinics report significant gastrointestinal symptoms. The entire gastrointestinal tract can be affected by

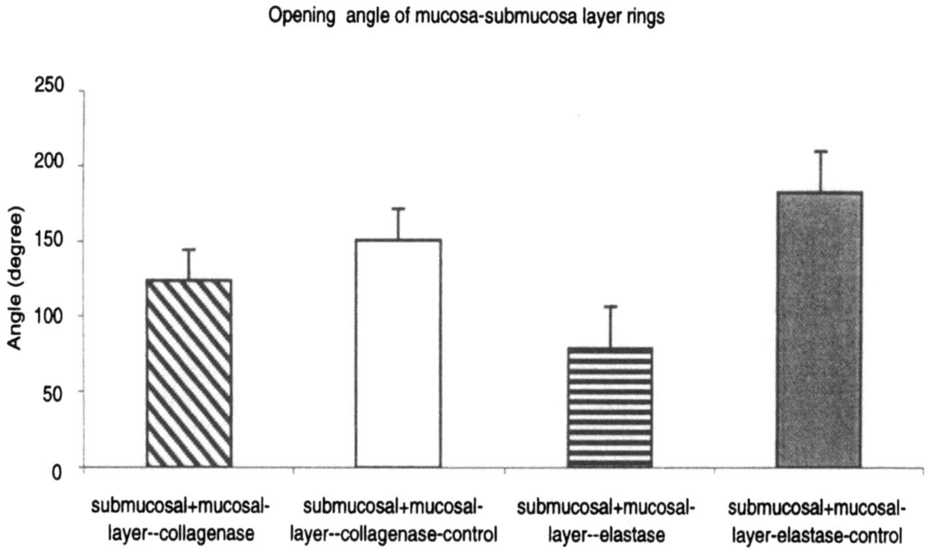

Figure 8.3. The effect of collagenase and elastase on the opening angle. The data were obtained by Yanhua Fan.

diabetes from the oesophagus to the anorectal region. Common complaints may include dysphasia, early satiety, reflux, constipation, abdominal pain, nausea, vomiting, and diarrhoea. Both acute and chronic hyperglycaemia can lead to specific gastrointestinal complications. As with other complications of diabetes, the duration of the disease and poor glycaemic control seem to be associated with more severe gastrointestinal problems. Diabetes is a disease that can be created in the laboratory and used for studying the remodelling of the mechanical properties.

8.4.1 Streptozotozin-induced Diabetes

Significant remodelling of the intestine occurs in diabetes. Histologically, many studies have demonstrated prominent proliferation of different layers during diabetes, especially mucosa in the small intestine and oesophagus (Zoubi et al., 1995; Mayhew et al., 1989). Few data about the relation between passive biomechanical properties and diabetic intestine have been reported. Jørgensen and co-workers demonstrated that diabetes altered the tension–strain relation of the intestine in a way consistent with stiffening of the intestinal wall (Jørgensen et al., 2001). There are, however, many ways to study the structural and mechanical remodelling of the intestine. With increased wall stiffness, it is likely that the opening angle becomes smaller.

Preliminary data in rats made diabetic with a single dose of streptozotozin and studied for up to five weeks show, as known from previous studies, a pronounced remodelling of the morphometric properties. The opening angle and residual strain decreased in the duodenum and increased in the jejunum and ileum for the diabetic rats compared to those in normal rats. After treatment with gliclazide (a well-known drug used in diabetic patients), these biomechanical parameters remodelled back towards the normal state.

Residual stress can be altered by many factors. If growth or resorption were uniform in the intestinal wall, then the structural change will cause no change in stress and strain in the intestine. However, if one part of the intestinal wall outgrows the rest, then it will compress the remainder, cause internal stress and non-uniform remodelling. The changes in the opening angle of the intestine during diabetes shown in this study suggest that the morphological and biomechanical remodelling of the different layers are an important determinant of growth and remodelling of the zero-stress state. Fung's hypothesis of non-uniform remodelling states that if the inner wall grows more than the outer wall, the opening angle will increase; whereas if the outer wall grows more than the inner wall, the opening angle will decrease (Fung, 1991). In this experiment, since the changes in the jejunum and ileum during the diabetes primarily are in the mucosa layers, the inner wall grows more than the outer wall and hence the opening angle increased. Correspondingly, the outer residual strain became more tensile whereas the inner residual strain became more compressive in these two segments. However, in the duodenal segment, the opening angle and absolute value of residual strain were decreased during diabetes. All layers of the duodenum increased after experimental diabetes; at this condition the biomechanical properties of the other layers may determine the remodelling of the zero-stress state.

8.4.2 Surgical Interventions

Remodelling Caused by Small Bowel Resection

The short-bowel syndrome caused by massive intestinal resection is a difficult clinical problem. An increased level of understanding of the physiology of the intestinal adaptation to resection could improve clinical treatments. Senn was probably the first in 1888 to demonstrate the ability of the intestinal remnant to undergo structural and functional adaptation in response to intestinal resection. During the last decades, extensive animal experimentation has demonstrated a huge adaptation process (remodelling) in the remnant intestine after massive small bowel resection. The remodeling is characterised by increased villus height and crypt depth as a result of accelerated mucosal cell proliferation, smooth muscle cell hypertrophy, and dilatation and elongation of the remnant bowel, leading to a greater absorptive surface area and prolonged transit time for maximising mucosal exposure to intra-luminal nutrition (Ziegler et al., 1998; Dowling, 1982; Williamson, 1982). The degree of remodeling depends on the degree and location of the resection. Although the morphological features of gut remodeling after resection are well characterised, the biomechanical properties in terms of stress and strain have to our knowledge not been studied in this particular case. Since intestinal adaptation occurs after resection, it can be used as a model for studying morphometric and biomechanical remodeling in the small intestine and to get a better understanding of the pathophysiology of the short-bowel syndrome.

Preliminary data were obtained in rats exposed to resection of the jejunum and ileum approximating to 67% of the whole small intestine. Control animals underwent no operation or an ileal transection with subsequent end-to-end anastomosis. The animals were followed for up to four weeks after the operation. Changes in biomechanical properties were studied in terms of residual strain, opening angle and stress–strain relations referenced to the zero-stress state. The resected animals gained less weight than the controls. The intestinal length and diameter increased more in the resected groups than the control groups, resulting in a larger absorptive area. The changes were time-dependent and were most prominent in the ileum. Resection induced profound gross morphometric changes and histological alterations characterised by proliferative increases in the tissue layers. The opening angle along with residual strain at the mucosal and serosal surface increased in the remnant small intestine. All changes increased as a function of postoperative time and were most prominent in the remnant ileum. The increased residual stress and residual strain may thus optimise the mechanical function of the remnant intestine under altered loading condition as found in the short-bowel.

The stress–strain relation does not remodel significantly after bowel resection. This indicates that the stiffness of the remnant intestine does not change for at least four weeks after the resection, although there are marked hypertrophic alterations in the layers of the intestinal wall. A fast remodeling of the zero-stress state may have contributed to the preserved stiffness during the study time. The lack of remodeling of the stress–strain relation is consistent with the hypothesis that soft tissue seeks to maintain a uniform stress level under various conditions. The remodeling after resection seems to be guided by the need for a greater absorptive surface area rather than for a change in the stress–strain properties. The observed change in opening angle and residual strain can be considered as both a result of the remodeling (especially in the mucosal layer), and a compensatory mechanism to maintain a uniform stress level.

Obstruction Models

Partial obstruction has been a popular model in gastroenterology. Numerous studies have been done in the oesophagus of various species, in particular the American opossum. The partial obstruction is created by placing a loose band around the distal part of the oesophagus during surgery. Schulze-Delrieu and co-workers contributed to the knowledge base with descriptions of the structural remodelling of the tissue. Unfortunately, not much was done on the mechanics of the partially obstructed oesophagus, though it would have been an excellent model for time-dependent studies of the remodelling process. Another study showed an association between the degree of dilation and the collagen content (Gregersen et al., 1992), a result that points in the direction of strain-dependent remodelling behaviour. This subject obviously needs more study.

Contrary to the oesophagus studies, the small intestinal studies of partial obstruction were done with the specific purpose of evaluating the passive elastic tissue properties in the circumferential direction. Unfortunately, the studies were conducted before anybody was aware of the large residual strains in the small intestine. The stress–strain curves obtained, however, may be valid because recent studies have shown that the mid-wall strain curves do not change much if the no-load state is used as a reference rather than the zero-stress state, but this assumption needs to be validated in tissue affected by obstruction.

Partial Obstruction in the Small Intestine

Small bowel obstruction is a frequent clinical problem associated with high rates of morbidity and mortality. It has been estimated that 5% of all general surgical admissions and 20% of emergency surgical admissions stem from small intestinal obstruction. Longstanding non-occluding obstruction of the small intestine results in tissue remodeling (morphological changes) with dilatation, wall thickening, increased collagen content, enlargement of enteric neurones, and hypertrophy of (especially) the tunica muscularis proximal to the site of obstruction (Herczel, 1886; Gabella, 1975; 1984). These changes are secondary to the accumulation of large amounts of luminal contents proximal to the obstruction resulting in increased loading forces and deformation of the intestine. Dysmotility in the proximal direction is a feature of obstruction with altered amplitudes, frequency and duration of contractions (Benson and Wingate, 1993; Schulze-Delrieu et al., 1995). Altered biomechanical properties are believed to play a major role in the pathophysiological process. Mechanical properties are a major determinant of tissue behaviour and are likely the single most important factor determining growth and remodelling of tissue. An understanding of this relationship is important in the interpretation of how gastrointestinal tissue responds and adapts to environmental changes.

Partial obstruction was created surgically in two groups of guinea pigs that lived for two and four weeks, respectively (Storkholm et al., 1998; Storkholm, 1996). When surgical anaesthesia was achieved, a small midline laparotomy was performed. A small loop of jejunum 50 cm distal to the pylorus was brought forward, and the mesenterium was incised close to the intestine to create a small window. Care was taken not to damage adjacent vessels or nerves. A Gore-Tex® band 3.5 mm wide was passed through the mesenterical window and closed antimesenterically with a small titanium clip. The band was closed at a fixed circumference of 14 mm and fitted loosely around the intestine without any apparent compression of the tissue. In the control group, the mesenteric incision was made and a suture in the mesenterium marked the location. Controls were shamoperated and lived for four weeks. No band was used in the control group. Data were obtained by using a pressure-recording system and a combined impedance planimetry–high-frequency ultrasound system was designed to measure the variables determining the mechanical wall properties in vitro of excised guinea pig intestinal segments along with measurements of total wall collagen and histology.

The spontaneous motility in the obstructed segments consisted of irregular phasic contractions with high amplitudes (approximately 125 cmH_2O) proximal to the site of obstruction (compared to approximately 50 cmH_2O in the control animals). The frequency, amplitude and area under curve of bolus-induced motility proximal to the site of obstruction were higher in banded animals than in controls. The most pronounced increase was observed in the area under the curve as a consequence of which both the frequency and amplitude increased. Atropine, tetrodotoxin and hexamethonium instantly abolished spontaneous as well as bolus-induced motility; i.e., the pharmacological data showed that spontaneous and secondary motility was neurogenic and mediated via muscarinic receptors. The dysmotility observed after obstruction can not be explained merely by altered muscle properties. The propagation time was significantly increased proximal to the obstruction site, which points towards neurogenic involvement. This is consistent with findings of disruption of the micro-anatomy of nerve plexuses in the small intestine during chronic obstruction (Ehrlam, 1971; Gabella, 1975). Due to thickening of the muscle layers, it was expected that the motility index would increase proximal to the obstruction site. Since a more pronounced motor response was induced in the obstructed animals than in controls, it is evident that the peristaltic reflex pathway is more sensitive to distension after obstruction.

The intestinal wall underwent massive hypertrophy proximal to the site of partial obstruction and dilated over the course of time; hence, the luminal cross-sectional area and the wall thickness of the proximal segments increased in a time-dependent manner proximal to the obstruction site. These morphometric data are largely consistent with previous data by Gabella (1984) and Schulze-Delrieu et al. (1995). More than a century ago, Herczel (1986) showed that the small intestine undergoes a remarkable hypertrophy when the flow of chyme is impaired.

The circumferential stress–strain curves from proximal segments in the four- and two-week intervention groups were shifted to the left when compared to those of the distal segments of the same groups and the controls, indicating increased wall stiffness in the segments proximal to the obstruction. The collagen content increased in a time-dependent manner proximally to the site of obstruction. This is in accordance with other investigators (Gabella and Yamey, 1977; Gabella, 1975; 1987). Organ function and morphology are to a large extent determined by the amount, distribution, organisation, and types of collagen present in the extracellular matrix (Bornstein, 1980). A derangement in one or more of these factors of the collagenous matrix may provoke a significant structural and functional disturbance in the organ. The main physiological function of collagen in the tissues is to resist distending forces – function in a protecting and stabilising manner – or to transmit forces – to act as a link between components of the contractile system (Fackler et al., 1981; Orberg et al., 1983; Gabella, 1987). A statistically significant association was found for the proximal location between the collagen content and the elastic modulus at a stress level of 70 kPa, indicating that collagen determines the mechanical properties in the intestine at high stress levels.

The proximal segments of the animals with partial obstruction showed marked hypertrophy of the smooth muscle cells, especially in the tunica muscularis. The ratio between the muscle layer thickness in the groups with obstruction and the controls was largest in the circular smooth muscle cell layer where the average ratio was approximately 2.1 and 5.1 in the two- and the four-week groups, respectively. The histological findings coincide with previous findings of gross hypertrophy especially of the circular muscle layer (Gabella and Yamey, 1977; Gabella, 1975; 1987). A possible explanation may be that the major force acts perpendicularly to the major axis of the smooth muscle cells of the longitudinal layer, whereas the circular layer increases along the axis of its cells; that is, the circular layer is strained more in the principal circumferential direction and therefore undergoes hypertrophy more readily (Gabella, 1975). This is in accordance with other investigators who have shown that the main deposit produced by the smooth muscle cells of the tunica muscularis, which is laid down in the intercellular space during stricture-induced hypertrophy, is mainly types IV and V collagen. This intercellular deposit has been characterised by means of electron microscopy as an amorphous meshwork that acts as an anchorage for smooth muscle cells, which lack proper tendons).

Obstruction of the Bile Duct

The most extensive work using partial obstruction models were, from a mechanical point of view, done in the biliary tract by Duch et al. (2001).

Biliary obstruction, a common clinical problem, may arise from cholelithiasis, neoplasm in the wall, or the compression of the lumen by structures outside the common bile duct. Gallstones, however, account for most obstructions. Twenty per cent of all men and 35% of all women have gallstones by the age of 75 years. 10–15% of the stones migrate to the common bile duct and give rise to such obstructive symptoms as acute pain in the right upper abdomen or epigastrium, nausea and vomiting. Persistent obstruction leads to prestenotic dilatation of the common bile duct and increase of its wall thickness (Steiner et al., 1989; Carlson et al., 1977). Several clinical and experimental studies describe the changes in the obstructed common bile duct (Lygidakis, 1984; Steiner et al., 1989; Carlson et al., 1977; Staritz et al., 1986). Although some studies provide measurements of pressure in the obstructed common bile duct (Carlsson, 1977; Staritz et al., 1986) and relate it to the degree of dilatation (Carlson, 1977), no thorough biomechanical study has been conducted in the obstructed common bile duct. Such an assessment requires a biomechanical analysis of the relation between stresses (force per area) and strains (non-dimensional deformation) in order to provide data on the tissue stiffness (elastic modulus).

The common bile duct was ligated in the pig and studied after 3 hours, 12 hours, 2 days, 8 days and 32 days (n=5 in each group). During the operation a pair of curved forceps were tunnelled under the common bile duct as close to the sphincter of Oddi as possible. The common bile duct was ligated approximately 1 mm from the duodenal wall by two ligatures to ensure complete obstruction. A scintigraphy was performed to verify that the obstruction was complete. The vascular supply to the duct was carefully maintained during the ligation procedures. At each scheduled time, it was mechanically tested in vitro with a computer-controlled volume infusion system to study the pressure–volume relationship of the common bile duct segment. A video camera provided simultaneous measurements of the outer dimensions of the common bile duct at the various pressures. Distension procedures and computation of Kirchhoff stress, Green strain and the tangent modulus were done according to the equations provided in Chapter 7.

The pressure in the common bile duct increased from 0.7 kPa in the control animals to approximately 3.0 kPa after 32 days of obstruction. The diameter and wall thickness of the common bile duct increased about threefold in the 32-day group as compared to the sham group. The circumferential stress–strain relationship differed between groups; it was shifted to the right, indicating softening, in the 3-hour, 12-hour, and 2-day groups and to the left, indicating stiffening, in the 8-day and 32-day groups, as compared to the sham group. The longitudinal stress–strain curves were all shifted to the left of the circumferential stress–strain curves, indicating a higher stiffness in the longitudinal direction. The increased stiffness following obstruction was also verified by the significant increase in the modulus after 8 and 32 days of obstruction. The collagen area increased during obstruction but no correlation between the size of the collagen area and the biomechanical parameters was found. A practical implication of the present study serves as a warning to surgeons. A reduction in the wall stiffness in the first several days of obstruction along with an increased duct diameter and a decreased wall thickness suggest that operative procedures such as suturing, anastomosis and procedures related to ERCP must be performed with care to avoid damage to the common bile duct.

8.4.3 Physical Intervention

Morphometry and Residual Strain in the Normal and Irradiated Mouse Rectum

Radiation injury of the rectum is a well-known complication of therapeutic radiation delivered to the pelvis (Widmark et al., 1994; Yeoh et al., 1996). Acute rectal reactions are usually transient and have no permanent impairment of the patient's quality of life. Late radiation damage may lead to obstruction, chronic ulceration, perforation and fistula formation, all symptoms causing severe morbidity (Pedersen et al., 1994; Eifel et al., 1995). A common feature of late radiation injury to the rectum is extensive submucosal fibrosis, ulcer formation, necrosis and vascular injury (Lundby, 1998; Hauer Jensen, 1990; Armour et al., 1996). Different methods and techniques have been developed in order to investigate late intestinal damage but most observations are based purely on histological assessment. However, the pathophysiological and morphological knowledge of rectal wall properties in relation to late radiation damage still suffers from the lack of appropriate methods.

This section summarises data on morphometry of the zero-stress and no-load states as well as residual strains and opening angles of the rectum in mice with acute and chronic radiation injury and in age-matched sham-irradiated control groups (Gregersen et al., 2002).

Two groups of mice were irradiated at the age of 15 weeks with a single dose of 20 Gy and the two other groups were sham-irradiated. Twenty Gy was selected on the basis of previous studies demonstrating a significant effect on rectal morphology and a high animal survival rate (Lundby, 1998). The two irradiated groups lived for five days (acute study) and for approximately eight months (chronic study). The two sham-irradiated groups served as age-matched controls for the acute and chronic irradiated mice.

The serosal and mucosal circumferential lengths and the wall thickness varied between groups with the longest serosal circumferential length and highest wall thickness in the chronic irradiated group and the smallest mucosal circumferential length in the acute irradiated group. Young mice had 25–50% smaller opening angles than the older mice and the irradiated groups had 25–75% smaller opening angles than the normal groups. The residual strain analysis showed mucosal compression and serosal tension in the no-load state. Both at the serosal and mucosal surfaces, variation was found between groups. Furthermore, proximal to the irradiated zone, the chronic irradiated group had significantly higher values compared to the normal group, indicating a higher stress level proximal to the irradiated zone.

The study demonstrates the presence of large residual strains in the normal and irradiated rectum. Previous studies showed that irradiation with high doses affected the normal growth of the rectum and resulted in reduction of the luminal cross-sectional area (Lundby et al., 1999), that the defecation pattern changes dose-dependently (Terry et al., 1983; Terry and Denekamp, 1984), and that rectal compliance (change in volume divided by the change in pressure) in mice decreases with increasing dose (Martin et al., 1991). By using the same method in rats, Tamou and Trott (1995) were not able to reproduce the results of Martin et al. They concluded that irradiation did not lead to a decrease in colorectal distensibility unless deep ulceration with scarring occurred. However, all studies so far suffer greatly from the fact that compliance is not a valid parameter for the determination of the elastic wall properties of tubular organs and that the morphometric and biomechanical measures must refer to the zero-stress state. In the present study, significant differences in morphometric and mechanical parameters were found between the irradiated groups and their controls. The most pronounced changes were the higher wall thickness in the chronic irradiated mice, the smaller opening angles in the irradiated mice compared to their controls, and the higher wall thickness-to-mucosal circumferential length ratio and lower residual strain at the mucosal surface proximal to the irradiated zone.

Further studies on the normal and irradiated rectum must provide detailed analysis of the association between stress and strain properties and morphological parameters referenced to the zero-stress state with the time course taken into account. Such an approach may lead to a better understanding of rectal physiology and the development of radiation injury.

8.5 Remodelling due to Growth Factors

8.5.1 Epidermal Growth Factor

Epidermal growth factor (EGF) is a polypeptide chain that belongs to an expanding group of growth factor ligands (Prigent and Lemoine, 1992). The EGF family and their related receptors are widely distributed in mammals and play an important role in development, regeneration and neoplastic growth. Mouse and human EGF were first discovered in 1962 and 1975 due to the ability of mouse EGF to accelerate eyelid opening in newborn mice (Cohen, 1962) and human EGF-induced inhibition of gastric acid secretion in humans (Gregory, 1975). Since a recombinant of human EGF is now available, extensive studies have been carried out to improve the knowledge of the effects of EGF on pharmacological, physiological, molecular-biological and cellular mechanisms. The gastrointestinal tract is one of the most sensitive organs to EGF treatment.

EGF is synthesized in the salivary glands from where it is secreted in an exocrine manner (Vinter-Jensen, 1999). Salivary EGF comes into contact with oesophageal and gastric mucosa and is a factor of importance for the functional integrity, mucosal protection, and repair (Nexø et al., 1984). Vinter-Jensen and coworkers and other groups studied the effect of long-term systemic treatment with EGF on a broad spectrum of tissues; e.g., long-term systemic EGF treatment increased the thickness and volume of the esophageal epithelia (Juhl et al., 1994; 1995). Vinter-Jensen and coworkers (1995) demonstrated that EGF treatment can attenuate the sclerotherapy-induced biomechanical properties of the oesophagus in minipigs using the impedance planimetric technique.

On the small intestine, it is well known that EGF has a trophic effect on the mucosa. It increases mucosal thickness (Jansen et al., 1997) and increases mucosal surface area (Vinter-Jensen et al., 1995). It has also been found that EGF changes electrical and enzyme activity and regulates glutamine transport in the small intestine (Opleta-Madsen et al., 1991; Ghishan et al., 1992; Salloum et al., 1993). Due to the effects on intestinal mucosa, the therapeutic potential of the systemic treatment of EGF has been demonstrated in animal experience; e.g., there is some evidence from animal trials that EGF can accelerate adaptive regrowth of the small intestine (Dunn et al., 1997; Liu et al., 1996; Rao and Porreca, 1996). Some reports of EGF treatment in human therapeutics also exist, e.g., EGF was administrated to accelerate healing in a child with necrotising enteritis. The intestinal biopsies revealed increased crypt mucosal proliferative activity and villus height in an eight-month-old girl treated for six days.

Despite all of the above studies, very few studies have addressed the effect of systemic treatment of EGF on the biomechanical function, in terms of stress–strain properties, of the gastrointestinal tract. Since changes in passive biomechanics reflect changes in the structure of a tissue that determine a specific motility function, acquiring biomechanical information is important. Vinter-Jensen et al. (1996) used impedance planimetry to examine the tension–strain properties of the intestinal wall after EGF treatment, and they found reduced circumferential wall stiffness during EGF treatment. However, this study suffered from the fact that the circumferential strain was referenced to the no-load state and data were only obtained in the circumferential direction.

EGF-induced Remodelling in the Oesophagus

Rats were treated with systemic EGF for up to 14 days. An osmotic mini-pump was placed subcutaneously in the neck and infused human recombinant EGF (150 ug/kg/ per day). Saline was infused in the control animals. The oesophagi were excised. One end was closed and the other end was cannulated so that a distension test could be performed. Kirchhoff's stress and Green's strain were computed for the circumferential direction according to the formulas given in Chapter 3.

The oesophageal wall thickness, inner and outer circumferences and wall cross-sectional area gradually increased during EGF treatment. The mucosa and submucosa thickness increased more than the muscle layer thickness. The opening angle and absolute value of residual strains increased during the EGF treatment, indicating that more residual forces reside in the tissue during EGF treatment. The opening angle increased from approximately 110° in the control animals to approximately 150° after 14 days of treatment. This is consistent with the fact that the growth in the inner layers was more pronounced than in the muscle layers according to Fung's hypothesis (1991). This was confirmed by doing a multiple linear regression analysis of the opening angle as a function of layer thicknesses (normalised). The stress-strain curve gradually shifted to the left during the first week of EGF treatment, indicating that the oesophageal wall stiffness increased. The stress–strain curve shifted slightly back towards the normal state after 14 days of EGF treatment. It is likely that the non-uniform growth during EGF treatment contributed to the remodelling of the biomechanical properties. The growth of the oesophagus was pronounced during the first week, while the speed of growth began to decrease after one week of EGF treatment. The pronounced growth of the oesophagus in the first week is likely due to the direct stimulation of exogenous EGF. Then the endogenous EGF production may decrease due to a negative feedback mechanism. It is yet unclear if this assumption is correct but similar results were obtained in another EGF experiment on the ileum by Vinter-Jensen and co-workers (1996). Furthermore, it was recently found that the mucosa–submucosa layer is stiffer than the muscle layer and the stress value in the mucosa–submucosa layer is higher than that in the muscle layer under the same loading conditions (Liao Donghua, personal communication). Therefore, the oesophageal wall became stiffer after EGF treatment because the mucosa and submucosa layers gained much more growth than the muscle layer.

EGF-induced Remodelling in the Small Intestine

Rats were treated with systemic EGF for up to 14 days using a subcutaneous mini-pump placed in the neck. A piece of the ileum was excised and tested biaxially in the test machine described in Chapter 4; i.e., both pressurisation and stretch were imposed. Kirchhoff's stress and Green's strain were computed for the circumferential and longitudinal directions according to the formulas given in Chapter 3. Data were also obtained on the zero-stress state morphometry. Biaxial material constants were obtained assuming a thin homogenous wall, a circular shape when loaded, and that the stress in the radial direction could be neglected. The functions given in section 3.4.2 were used. In brief, four constants are determined. The a_1 constant is the elastic modulus in the circumferential direction, affecting the curve for $S_{\theta\theta}$ and $E_{\theta\theta}$. The higher a_1, the stiffer the intestine wall. Constant a_2 is the elastic modulus in the longitudinal direction, affecting the curve for $S_{\phi\phi}$ and $E_{\phi\phi}$. Constant a_4 is the cross modulus. Constant C fixes the scale on the stress axis. The larger the value of a_1, a_2 and a_4, the smaller is C (Fung et al., 1979).

The wall thickness and wall area increased after four days of treatment and remained stable. The mucosa layer showed the largest increase in thickness. The opening angle increased after seven days of EGF treatment and continued to increase up to 14 days. The mucosal residual strain was compressive in all groups and decreased from four days of EGF treatment. The serosal residual strain was tensile with the relative higher values after 14 days of EGF treatment. The fitted circumferential stress–strain data are illustrated in Figure 8.4. In the circumferential direction,

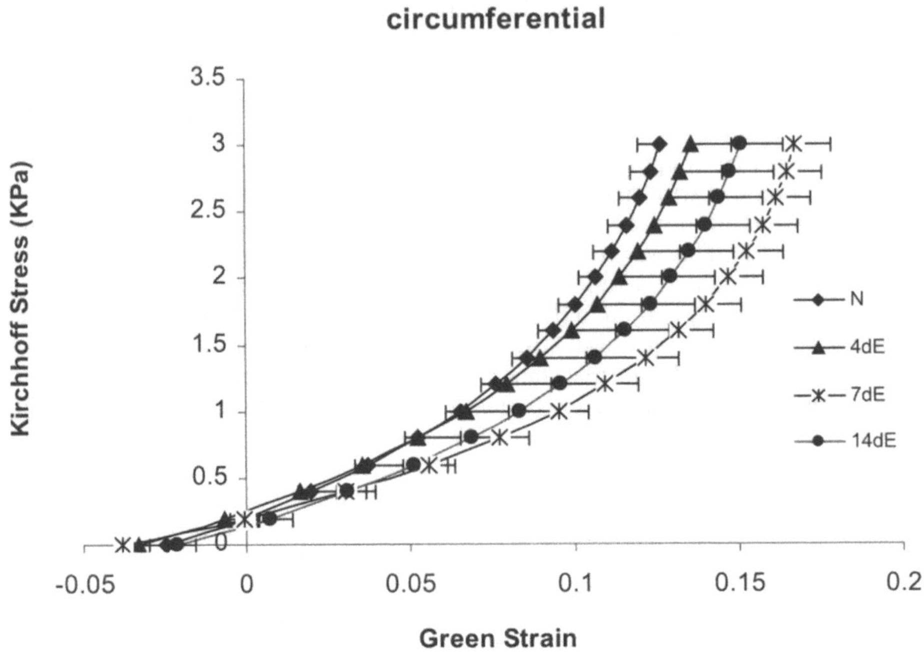

Figure 8.4. Fitted circumferential stress–strain curves obtained in the ileum during EGF administration. Translation of the stress–strain curve to the right in the coordinate system implies tissue softening (decreased stiffness) and visa versa. (Yang et al. Unpublished data).

the stress–strain curves representing four days and seven days of EGF treatment were shifted to the right, while the 14-day EGF treatment curve shifted slightly back to the control side. In the longitudinal direction, the stress–strain curves representing four days and seven days of EGF treatments shifted to the left, but the 14-day EGF treatment shifted to the control side. The derived material constants from the strain energy function a_1 and a_2 had the smallest variation coefficient and therefore were more stable than the other parameters. Significantly smaller a_1 values were observed after seven days and 14 days of EGF treatment and larger a_2 values can be seen in two days and seven days of EGF treatment. Hence, the EGF-induced remodelling is time-dependent and the mechanical properties are anisotropic.

It is not to be expected that the circumferential stiffness decreases initially after EGF treatment. Another intestinal study with a protocol similar to the oesophagus study mentioned above was conducted to test whether this effect represents the remodelling in the whole small intestine. It was found that it is only in the distal ileum that the softening happens since both the duodenum and jejunum exhibited increased stiffness in the first week of EGF treatment. The stress–strain curve shifted slightly back towards the normal state after 14 days of EGF treatment. In the longitudinal direction, the stress–strain curve was gradually shifted to the left during the first week of EGF treatment and after 14 days the curve tends to normalise again in the duodenal and jejunal segments. This study also showed, like the oesophagus study, that the remodelling of the opening angle is largely caused by the changes in the mucosa.

8.6 Tissue Remodelling

Tissue remodeling allows biological tissues to adapt to changes in the environment and to diseases. Living tissues maintain, restore, or improve tissue functions in response to functional stimuli through resorption and growth of cells and extracellular materials. The architecture of cancellous bone is probably the most classical example reflecting the relationship between structure and stress. As early as the nineteenth century, it was recognised that cancellous bone is moulded according to the stress trajectories of one particular loading. Its structure corresponds to the static pressure lines, which enables bone to resist external forces with a minimum material consumption (Fung, 1998). Soft tissues also remodel in response to stress and strain. For example, when the heart or the blood vessels are overloaded their muscle cells increase in size and the wall thickens. When one lung of a rabbit was excised, the remaining lung expanded to fill the thoracic cavity and it grew until it weighed approximately the weight of both lungs (Cowan and Crystal, 1995). Many diseases in the gastrointestinal tract, such as obstructive lesions, are also associated with profound tissue growth and remodeling, including prestenotic dilation, thickening of the wall, increased collagen synthesis, dysmotility, and altered wall stiffness (Gregersen et al., 1992). By proposing a "stress growth law", Fung suggests that non-uniform distribution of stress across a vessel wall in vivo is responsible for a remodelling process, which tends to reduce this non-uniformity (Fung, 1998). However, as it appears from the data referred to in this chapter, some rules may be established while others may only serve as explanations in some cases. For example, Rodriguez' hypothesis seems valid in some studies whereas negative associations were found in other studies. This can perhaps be explained by whether the assumptions for the hypothesis are valid for the gastrointestinal tract or not and by the fact that morphometric and biomechanical remodelling may occur at different time points. Fung's hypothesis seems to provide a more simple explanation and more valid way to provide some explanation of the relation between residual stress and tissue growth. Due to the high complexity of the gut structure and function, its growth and remodeling are affected by many factors and hence the stress–growth law may be complex. For example, it is unique for the intestine to have a layer, the mucosa, that contributes its mechanical properties in compression but does not support tensile stress during stretch due to the villous structure. The current data serve to inform us of this great complexity and that the properties are time-, direction- and location-dependent.

8.7 Literature

The literature list contains papers mentioned in the text as well as other papers of interest.

Aldewachi HS, Wright NA, Appleton DR, Watson AJ. 1975. The effect of starvation and refeeding on cell population kinetics in the rat small bowel mucosa. *J Anat*, **119**: 105–21

Altmann GG. 1972. Influence of starvation and refeeding on mucosal size and epithelial renewal in the rat small intestine. *Am J Anat*, **133**: 391–400.

Arhan P, Devroede G, Danis K, et al. 1978. Viscoelastic properties of the rectal wall in Hirschsprung's disease. *J Clin Invest*, **62**: 82–7.

Armour EP, White JR, DeWitt CC, Corry PM, Martinez A. 1996. Effects of continuous low-dose-rate brachytherapy on the rectum of the rat. *Radiat Res*, **145**: 474–80.

Benson MJ, Wingate DL. 1993. Ileus and mechanical obstruction. In: *An Illustrated Guide to Gastrointestinal Motility*, 2nd edition, ed. Kumar D, Wingate D, pp. 547–66. New York: Churchill Livingstone.

Bjorntorp P, Yang Mu. 1982. Refeeding after fasting in the rat: effects on body composition and food efficiency. *Am.J Clin Nutr*, **36**: 444–9.

Bornstein P. 1980. Structurally distinct collagen types. *Ann Rev Biochem*, **49**: 957–1003.

Bragg LE, Thompson JS, Rikkers LF. 1991. Influence of nutrient delivery on gut structure and function. *Nutrition*, **7**: 237–43.

Carlson E, Zukoski CF, Campbell J, Chvapil M. 1977. Morphologic, biophysical, and biochemical consequences of ligation of the common biliary duct in the dog. *Am J Pathol*, **86**: 301–20.

Chipman SD, Sweet HO, McBride Jr DJ, et al. 1993. Defective pro alpha 2(I) collagen synthesis in a recessive mutation in mice: a model of human osteogenesis imperfecta. *Proc Natl Acad Sci USA*, **90**: 1701–5.

Clarke RM. 1975. The time-course of changes in mucosal architecture and epithelial cell production and cell shedding in the small intestine of the rat fed after fasting. *J Anat*, **120**: 321–7.

Cohen S. 1962. Isolation of a mouse submaxillary gland protein accelerating incisor eruption and eyelid opening in the new-born animal. *J Biol Chem*, **237**: 1555–62.

Conklin JL, Goyal RK. 1989. Gastrointestinal smooth muscle. In: *Gastrointestinal Disease*, 4th edition, ed. Sleisenger MH, Fordtran JS. Philadelphia: W.B. Saunders Company.

Cowan MJ, Crystal RG. 1995. Lung growth after unilateral pneumonectomy: quantitation of collagen synthesis and content: *Am Rev Respir Dis*, **111**: 267–77.

de Caestecker JS, Pryde A, Heading RC. 1992. Site and mechanism of pain perception with oesophageal balloon distension and intravenous edrophonium in patients with oesophageal chest pain. *Gut*, **33**: 580–6.

Denis P, Colin R, Galmiche JP, Geffroy Y, Hecketsweiler P, Lefrancois R, Pasquis P. 1979. Elastic properties of the rectal wall in normal adults and in patients with ulcerative colitis. *Gastroenterology*, **77**: 45–8.

Dou Y, Gregersen S, Zhao J, Zhuang FY, Gregersen H. 2002. Effect of refeeding after starvation on biomechanical properties in rat small intestine. *J Med Eng Physics*, **23**: 557–66.

Dou Y, Gregersen S, Zhao J, Zhuang FY, Gregersen H. 2002. Morphometric and biomechanical intestinal remodeling induced by fasting in rats. *Dig Dis Sci*, **47**: 1158–68.

Dowling RH. 1982. Small bowel adaptation and its regulation. *Scand J Gastroenterol Suppl*, **74**: 53–74.

Dowling RH, Booth CC. 1967. Structural and functional changes following small intestinal resection in the rat. *Clin Sci*, **32**: 139–49.

Duch BU, Andersen HL, Smith J, Kassab GS, Gregersen H. 2002. Structural and mechanical remodeling of the common bile duct after obstruction. *Neurogastroenterol Motil*, **14**: 111–122.

Duch BU. 2000. Biomechanical and morphometric properties of the porcine common bile duct. PhD thesis, Denmark, University of Aarhus.

Dunn JC, Parungo CP, Fonkalsrud EW, McFadden DW, Ashley SW. 1997. Epidermal growth factor selectively enhances functional enterocyte adaptation after massive small bowel resection. *J Surg Res*, **67**: 90–93.

Ehrlam RJ. 1971. Ganglion cell changes in experimental stenosis of the gut. *Gut*, **12**: 393–8.

Eifel PJ, Levenback C, Wharton JT, Oswald MJ. 1995. Time course and incidence of late complications in patients treated with radiation therapy for figo stage Ib carcinoma of the uterine cervix. *Int J Radiat Oncol Biol Phys*, **32**: 1289–300.

Fackler K, Klein L, Hiltner AJ. 1981. Polarizing light microscopy of intestine and its relationship to mechanical behaviour. *J Microsc*, **124**: 305–11.

Fung YC, Fronek K, Patitucci P. 1979. Pseudoelasticity of arteries and the choice of mathematical expression. *Am J Physiol*, **237**: 620–31.

Fung YC. 1991. What are the residual stresses doing in our blood vessels? *Ann Biomed Eng*, **19**: 237–49.

Fung YC. 1993. *Biomechanics: Mechanical Properties of Living Tissues*, 2nd edition. New York: Springer-Verlag.

Fung YC. 1998. *Biomechanics: Motion, Flow, Stress, and Growth*. New York: Springer-Verlag.

Gabella G. 1975. Hypertrophy of intestinal smooth muscle. *Cell Tiss Res*, **163**: 199–214.

Gabella G. 1976. *Structure of the Autonomic Nervous System*. London: Chapman and Hall.

Gabella G. 1979. Innervation of the gastrointestinal tract. *Int Rev Cytol*, **59**: 129–93.

Gabella G. 1981. On the musculature of the gastrointestinal tract of the guinea pig. *Anat Embryol*, **163**: 135–56.

Gabella G. 1984. Hypertrophic smooth muscle. V. Collagen and other extracellular materials. Vascularisation. *Cell Tiss Res*, **235**: 275–83.

Gabella G. 1984. Structural apparatus for force transmission in smooth muscles. *Physiol Rev*, **64**: 455–77.

Gabella G. 1987. Structure of muscles and nerves in the gastrointestinal tract. In: *Physiology of the Gastrointestinal Tract*, 2nd edition, ed. Johnson LR, pp. 335–82. New York: Raven Press.

Gabella G. 1987. The cross-ply arrangement of collagen fibres in the submucosa of the mammalian small intestine. *Cell Tiss Res*, **248**: 491–7.

Gabella G, Yamey A. 1977. Synthesis of collagen by smooth muscle in the hypertrophic small intestine. *Q J Exp Physiol*, **62**: 257–64.

Ghishan FK, Kikuchi K, Riedel B. 1992. Epidermal growth factor up-regulates intestinal Na^+/H^+ exchange activity. *Proc Soc Exp Biol Med*, **201**: 289–95.
Goodlad RA, Plumb JA, Wright NA. 1988. Epithelial cell proliferation and intestinal absorptive function during starvation and refeeding in the rat. *Clin Sci*, **74**: 301–6.
Gregersen H, Giversen IM, Rasmussen LM, Tøttrup A. 1992. Biomechanical wall properties and collagen content in the partially obstructed opossum esophagus. *Gastroenterology*, **103**: 1547–51.
Gregersen H, Kassab GS, Fung YC. 2000. The zero-stress state of the gastrointestinal tract: biomechanical and functional implications. *Dig Dis Sci*, **45**: 2271–81
Gregersen H, Kassab GS. 1996. Biomechanics of the gastrointestinal tract. *Neurogastroenterol Motil*, **8**: 277–97.
Gregersen H, Lee TC, Chien S, Skalak R, Fung YC. 1999. Strain distribution in the layered wall of the esophagus. *J Biomech Eng*, **121**: 442–8.
Gregersen H, Lundby, L, Overgaard J. 2002. Early and late effects of irradiation on the morphometry and residual strain of the mouse rectum. *Dig Dis Sci*, **47**: 1472–9.
Gregersen H, Weis S, McCulloch AD. 2001. Esophageal morphometry and residual strain in a mouse model of osteogenesis imperfecta. *Neurogastroenterol Motil*, **13**: 457–64.
Gregory H. 1975. Isolation and structure of urogastrone and its relationship to epidermal growth factor. *Nature*, **257**: 325–7.
Gregory H. 1995. Isolation and structure of urogastrone and its relationship to epidermal growth factor. *Nature*, **257**: 325–7.
Hauer-Jensen M, Poulakos L, Osborne JW. 1990. Intestinal complications following accelerated fractionated X-irradiation. An experimental study in the rat. *Acta Oncol*, **29**: 229–34.
Herczel E. 1886. Experimentelle und histologische Untersuchungen über kompensatorische Muskelhypertrofie bei Darmstenosen. *Zeitschrift für Klinisches Medizin*, **11**: 321–41.
Jansen C, Ihse I, Axelson. 1997. Epidermal growth factor induces increased mucosal thickness of the small intestine in the mouse. *Eur Surg Res*, **29**: 447–54.
Jensen LS, Laurberg S, Andreassen TT. 1989. Endoscopic sclerotherapy of oesophageal varices in rabbits. The long-term effect on oesophageal collagen content and mechanical strength. *Scand J Gastroenterol*, **24**: 613–6.
Jensen LS, Laurberg S, Juhl CO, Andreassen TT. 1987. Esophageal collagen content and mechanical strength after endoscopic sclerotherapy of esophageal varices. *Scand J Gastroenterol*, **22**: 743–9.
Jørgensen CS, Ahrensberg JM, Gregersen H, Flyvberg A. 2001. Tension–strain relation and morphometry of rat small intestine in experimental diabetes. *Dig Dis Sci*, **46**: 960–7.
Juhl CO, Vinter-Jensen L, Djurhuus JC, Gregersen H, Dajani EZ. 1994. Biomechanical properties of the esophagus damaged by endoscopic sclerotherapy. An impedance planimetric study in mini-pigs. *Scand J Gastroenterol*, **29**: 867–73.
Juhl CO, Vinter-Jensen L, Poulsen SS, Ørntoft T, Dajani EZ. 1995. Chronic treatment with epidermal growth factor causes esophageal epithelial hyperplasia in pigs and rats. *Dig Dis Sci*, **41**: 2717–23.
Kimmey MB, Martin RW, Haggitt RC, Wang KY, Franklin DW, Silverstein FE. 1989. Histological correlates of gastrointestinal ultrasound images. *Gastroenterology*, **96**: 433–41.
Kotler DP, Levine GM, Shiau YF. 1980. Effects of nutrients, endogenous secretions, and fasting on in vitro glucose uptake. *Am J Physiol*, **238**: G219–27.
Levin RJ. 1970. The intestinal absorption of some essential and non-essential amino acids in fed and fasting rats. *Life Sci*, **9**: 61–8.
Liu CD, Rongione AJ, Shin MS, Ashley SW, McFaden DW. 1996. Epidermal growth factor improves intestinal adaptation during somatostatin administration in vivo. *J Surg Res*, **63**: 163–8.
Lundby L, Dall FH, Gregersen H, Overgaard J, Laurberg S. 1999. Distensibility of the mouse rectum: application of impedance planimetry for studying age-related changes. *Int J Colorectal Dis*, **1**: 34–41.
Lundby L. 1998. Biomechanical and morphometric evaluation of late radiation-induced changes in the mouse rectum. PhD thesis.
Lygidakis NJ. 1984. Incidence and significance of ultrastructural changes in the common bile duct in lithiasis with dilatation of the duct. *Acta Chir Scand*, **150**: 665–8.
Martin SG, Stratford MRL, Watfa RR, Miller GG, Murray JC. 1992. Collagen metabolism in the murine colon following X irradiation. *Radiat Res*, **130**: 38–47.
Martin SG, Vojnovic B, Murray JC. 1991. Determination of x-ray-induced damage to the murine colon using tissue compliance measurements. *Int J Radiat Biol*, **59**: 503–15.
Martin RW, Silverstein FE, Kimmey MB. 1989. A 20-MHz ultrasound system for imaging the intestinal wall. *Ultrasound Med Biol*, **15**: 273-80.
Mayhew TM, Carson FL, Sharma AK. 1989. Small intestinal morphology in experimental diabetic rats: a stereological study on the effects of an aldose reductase inhibitor (ponalrestat) given with or without conventional insulin therapy. *Diabetologia*, **32**: 649–54.

McBride Jr DJ, Choe V, Shapiro JR, Brodsky. 1997. Altered collagen structure in mouse tail tendon lacking the alpha 2(I) chain. *J Mol Biol*, **270**: 275–84.
Misof K, Landis WJ, Klaushofer K, Fratzl P. 1997. Collagen from the osteogenesis imperfecta mouse model (OIM) shows reduced resistance against tensile stress. *J Clin Invest*, **100**: 40–5.
Nexø E, Olsen PS, Poulsen K. 1984. Exocrine and endocrine secretion of renin and epidermal growth factor from the mouse submandibular glands. *Regul Pept*, **8**: 327–34.
Nowak TV, Harrington B, Weisbruch JP, Kalbfleish JH. 1990. Structural and functional characteristics of muscle from diabetic rodent small intestine. *Am J Physiol*, **258**: G690–8.
Opleta-Madsen K, Hardin J, Gall DG. 1991. Epidermal growth factor upregulates intestinal electrolyte and nutrient transport. *Am J Physiol*, **260**: 807–14.
Orberg J, Baer E, Hiltner A. 1983. Organization of collagen fibers in the intestine. *Conn Tiss Res*, **11**: 285–97.
Orberg JW, Klein L, Hiltner A. 1982. Scanning electron microscopy of collagen fibers in intestine. *Conn Tissue Res*, **9**: 187–93.
Orloff Juhl C, Vinter-Jensen L, Djurhuus JC, Gregersen H, Zapher Dajani E. 1994. Biomechanical properties of the oesophagus damaged by endoscopic sclerotherapy. An impedance planimetric study in mini-pigs. *Scand J Gastroenterol*, **29**: 867–73.
Oxlund H, Andreassen TT. 1980. The roles of hyaluronic acid, collagen and elastin in the mechanical properties of connective tissues. *J Anat*, **131**: 611–20.
Paterson WG, Selucky M, Hynna-Liepert TT. 1991. Effect of intraesophageal location and muscarinic blockade on balloon distension-induced chest pain. *Dig Dis Sci*, **36**: 282–8.
Paterson WG. 1991. Neuromuscular mechanisms of esophageal responses at and proximal to a distending balloon. *Am J Physiol*, **260**: G148–55.
Pedersen D, Bentzen SM, Overgaard J. 1994. Early and late radiotherapeutic morbidity in 442 consecutive patients with locally advanced carcinoma of the uterine cervix. *Int J Radiat Oncol Biol Phys*, **29**: 941–52.
Prigent SA, Lemoine NR. 1992. The type 1 (EGFR-related) family of growth factor receptors and their ligands. *Prog Growth Factor Res*, **4**: 1–24.
Rao R, Porreca F. 1996. Epidermal growth factor protects mouse ideal mucosa from Triton X-100-induced injury. *Eur J Pharmacol*, **15**: 209–12.
Ravinder KM, Ren J, McCallum RW, Shaffer HA, Sluss J. 1990. Modulation of feline esophageal contractions by bolus volume and outflow obstruction. *Am J Physiol*, **100**: 853–64.
Rodriguez EK, Hoger A, McCulloch AD. 1994. Stress-dependent finite growth in soft elastic tissues. *J Biomech*, **27**: 455–67.
Salloum RM, Stevens BR, Schultz GS, Souba WW. 1993. Regulation of small intestinal glutamine transport by epidermal growth factor. *Surgery*, **113**: 552–9.
Schulze-Delrieu K, Brown B, Herman B, Brown CK, Lawrence D, Shirazi S et al. 1995. Preservation of peristaltic reflex in hypertrophied ileum of guinea pig. *Am J Physiol*, **269**: G49–59.
Schulze-Delrieu K, Shirazi SS, Sprunger K, Noel S, Tung H. 1988. Functional denervation of the esophagus from experimental obstruction. *Gastroenterology*, **94**: A414.
Senn N. 1888. An experimental contribution to intestinal surgery with special reference to the treatment of intestinal obstruction. II:enterectomy. *Ann Surg*, **7**: 99–115.
Staritz M, Poralla T, Klose K, et al. 1986. Is the bile duct diameter a reliable parameter to diagnose extrahepatic cholestasis? Relationship between bile duct diameter and bile duct pressure. *Digestion*, **35**: 120–4.
Steiner D, Henning R, Lierse W. 1989. Bioconstruction of the extrahepatic biliary duct system in mini-pigs. Comparing normal condition with experimental obstruction. *Acta Anat Basel*, **136**: 159–64.
Steiner M, Bourges HR, Freedman LS, Gray SJ. 1968. Effect of starvation on the tissue composition of the small intestine in the rat: *Am J Physiol*, **215**: 75–7.
Storkholm JH. 1996. Aspects of biomechanics, morphology, and motility in the normal and the obstructed guinea pig small intestine. PhD thesis, Denmark, University of Aarhus.
Storkholm, JH, Villadsen GV, Jensen SL, Gregersen H. 1998. Peristaltic reflex activity and stress in the partially obstructed guinea pig small intestine. *Proc 16th Int Symp Gastrointestinal Motility*. Lorne, Australia.
Tamou S, Trott KR. 1995. The effects of local X-irradiation on the distensibility of the rectum in rats. *Br J Radiol*, **68**: 64–9.
Terry NHA, Denekamp J, Maughan RL. 1983. RBE values for colo-rectal injury after caesium 137 gamma-ray and neutron irradiation. I. Single doses. *Br J Radiol*, **56**: 257–65.
Terry NHA, Denekamp J. 1984. RBE values and repair characteristics for colo-rectal injury after caesium 137 gamma-ray and neutron irradiation. II. Fractionation up to ten doses. *Br J Radiol*, **57**: 617–92.
Tung H-N, Schulze-Delrieu K, Shirazi S, Noel S, Xia Q, Cue K. 1991. Hypertrophic smooth muscle in the partially obstructed opossum esophagus. The model: histological and ultrastructural observations. *Gastroenterology*,**100**: 853–64.

Villadsen GE, Petersen JAK, Vinter-Jensen L, Juhl CO, Gregersen H. 1995. Impedance planimetric characterisation of the normal and diseased oesophagus. *Surg Res Comm*, **17**: 225–42.

Vinter-Jensen L, Duch BU, Pedersen JAK, Ryslev A, Gregersen H. 1996. Systemic treatment with epidermal growth factor in the rat. Biomechanical properties of the growing small intestine. *Regul Pept*, **61**: 135–42.

Vinter-Jensen L, Juhl CO, Eika B, Gregersen H, Dajani EZ. 1995. Epidermal growth factor attenuates the sclerotherapy-induced biomechanical properties of the esophagus. An experimental study in mini-pig. *Scand J Gastroenterol*, **30**: 614–19.

Vinter-Jensen L. 1999. Pharmacological effects of epidermal growth factor (EGF) with focus on the urinary and gastrointestinal tracts. *APMIS*, **107** (Suppl 93): 1–42.

Widmark A, Fransson P, Tavelin B. 1994. Self-assessment questionnaire for evaluating urinary and intestinal late side effects after pelvic radiotherapy in patients with prostate cancer compared with an age-matched control population. *Cancer*, **74**: 2520–32.

Williams D, Thompson DG, Heggie L, Bancewitz J. 1993. Responses of the human esophaus to experimental intraluminal distension. *Am J Physiol*, **265**: G196–203.

Williamson RC. 1982. Intestinal adaptation: factors that influence morphology. *Scand J Gastroenterol* (Suppl.), **74**: 21–9.

Wilson JH, Lamberts SW. 1979. Nitrogen balance in obese patients receiving a very low calorie liquid formula diet. *Am J Clin Nutr*, **32**: 1612–16.

Yamauchi K, Kamisoyama H, Isshiki Y. 1996. Effects of fasting and refeeding on structures of the intestinal villi and epithelial cells in White Leghorn hens. *Br Poult Sci*, **37**: 909–21.

Yeoh E, Sun WM, Russo A, Ibanez L, Horowitz M. 1996. A retrospective study of the effects of pelvic irradiation for gynecological cancer on anorectal function. *Int J Radiat Oncol Biol Phys*, **35**: 1003–10.

Zeller PJ, Skalak TC. 1998. Contribution of individual structural components in determining the zero-stress state in small arteries. *J Vasc Res*, **35**: 8–17.

Ziegler TR, Mantell MP, Chow JC, Rombeau JL, Smith RJ. 1998. Intestinal adaptation after extensive small bowel resection: differential changes in growth and insulin-like growth factor system messenger ribonucleic acids in jejunum and ileum. *Endocrinology*, **139**: 3119–26.

Zoubi SA, Williams MD, Mayhew TM, et al. 1995. Number and ultrastructure of epithelial cells in crypts and villi along the streptozotocin-diabetic small intestine: a quantitative study on the effects of insulin and aldose reductase inhibition. *Virchows Arch*, **427** (2): 187–93

9. Perspectives

Gastrointestinal mechanics and motility continues to be probably the least rationally developed subject in medical physiology. The significance of this deficit lies in the prominence of disordered gastrointestinal motor function in clinical medicine. Such disorders constitute the root problem in most patients with gastrointestinal complaints. Nearly all oesophageal disease, for example, represents abnormal mechanics in that organ. Diverticulosis of the colon, an increasing problem in the ageing population, arises from motor dysfunction. Motor abnormality contributes importantly to gallstone disease. The many forms of the pseudo-obstruction syndrome arise from poorly understood defects in gastrointestinal motility. Motor disorders undoubtedly complicate and account for some of the most troublesome symptoms in the inflammatory bowel diseases. Irritable bowel syndrome, the commonest gastroenterological diagnosis made in general office practice, represents undefined gastrointestinal neuromuscular dysfunction where we need new tools developed on a rational bioengineering approach. Motor disorders often complicate long-standing diabetes, several connective tissue diseases, and many neurological disorders. The ubiquity and persistence of such clinical problems, together with our inability to deal with them rationally, leaves a vast population of chronically ill people who must deal daily with defective gastrointestinal functions with little hope of relief.

Rational management in medicine requires true knowledge of pathophysiology and that can come only from the application of real science to the matter. Now, at last, we are on the verge of having the ability to make such an application because of the progress foreseen in the mechanical determination of the motor performance of the various gastrointestinal organs. The accurate description of a behaviour must precede its analysis, here as in all dynamic systems, and we now have adequate descriptions of many, but not all, mechanical processes in the gut. Those who describe these behaviours also attempt to analyse and to explain them, but the explanations offered usually fail to pass the ultimate test of the truth of any theory – its ability to predict behaviour under prescribed circumstances. We still lack an adequate theoretical base. This failure comes mostly from the two problems I propose to begin to address in this research effort: i) our inability to perceive the operation of any complex system of non-linear processes without resort to the use of bioengineering and mathematical modelling, and ii) our nearly total ignorance of the biomechanical properties of the wall of the gastrointestinal tract.

Our knowledge of intestinal adaptation dates far back. Over the years, many important observations have been made about this process including changes in intestinal structure, functions and motor activities, cell morphology, metabolism and kinetics, and growth-regulating factors. However, many aspects are still not clear and most important of all, the mechanisms for these changes remain vague. It seems easier to understand from common sense that mechanics plays a crucial role in bone injury, repairing and remodeling processes. The importance of biomechanics in the remodeling of soft tissues, such as the heart, lungs and blood vessels has been shown experimentally. The gastrointestinal tract, as the hollow muscular tube it is, is subjected to changes in wall stresses and strains during its function. Hence, biomechanics must take an important role in the remodeling mechanism in response to environmental changes. The description of biomechanical properties in the remodeling is an important step to explore the mechanical pathway for intestinal adaptation. Among several other issues, further studies are needed to address problems such as: 1) the mechanoreceptors. The identification of mechanoreceptors will be beneficial to our better understanding of mechanisms for intestinal adaptation, as well as for injury, illness and repairing in the intestine. 2) The determination of constitutive equations of the gastrointestinal tract. Once the constitutive equation is known, the distribution of physical stresses and strains in the gastrointestinal tract in vivo, and the function of bolus transport and fluid movement can be analysed by methods of continuum mechanics. The analytical results will relate the stress and function with the geometric parameters. The task of determining the constitutive equations is not as difficult as it seems. If the hypothesis that the constitutive equations of the gastrointestinal tissues are similar to those of other soft tissues can be confirmed, then the only task is to determine the material constants of the intestinal tissues. Mechanical testing instruments have already been developed for accomplishing this task. It is the hope that this book can stimulate further research in the interesting subject of determining the mechanical behaviour of the gastrointestinal tract.

Index

J

K

L

M

MIX
Papier aus verantwortungsvollen Quellen
Paper from responsible sources
FSC® C105338

If you have any concerns about our products,
you can contact us on
ProductSafety@springernature.com

In case Publisher is established outside the EU,
the EU authorized representative is:
Springer Nature Customer Service Center GmbH
Europaplatz 3, 69115 Heidelberg, Germany

Printed by Libri Plureos GmbH
in Hamburg, Germany